Microbe Mediated Remediation of Environmental Contaminants

Woodhead Publishing Series in Food Science, Technology and Nutrition

Microbe Mediated Remediation of Environmental Contaminants

Edited by

Ajay Kumar

Vipin Kumar Singh

Pardeep Singh

Virendra Kumar Mishra

ELSEVIER

WP WOODHEAD PUBLISHING
An imprint of Elsevier

Woodhead Publishing is an imprint of Elsevier
The Officers' Mess Business Centre, Royston Road, Duxford, CB22 4QH, United Kingdom
50 Hampshire Street, 5th Floor, Cambridge, MA 02139, United States
The Boulevard, Langford Lane, Kidlington, OX5 1GB, United Kingdom

British Library Cataloguing-in-Publication Data
A catalogue record for this book is available from the British Library

Library of Congress Cataloging-in-Publication Data
A catalog record for this book is available from the Library of Congress

ISBN: 978-0-12-821199-1 (print)

ISBN: 978-0-12-823207-1 (online)

For information on all Woodhead Publishing publications
visit our website at https://www.elsevier.com/books-and-journals

Publisher: Charlotte Cockle
Acquisitions Editor: Nancy Maragioglio
Editorial Project Manager: Alice Grant
Production Project Manager: Surya Narayanan Jayachandran
Cover Designer: Mark Rogers

Typeset by MPS Limited, Chennai, India

Contents

List of Contributors

Mohd Aamir Department of Botany, M.M.V, Banaras Hindu University, Varanasi, India

Bashir Adelodun Department of Agricultural and Biosystems Engineering, University of Ilorin, Ilorin, Nigeria; Department of Agricultural Civil Engineering, Kyungpook National University, Daegu, Republic of Korea

Khyati Adlakha Department of Biotechnology, School of Bioengineering and Biosciences, Lovely Professional University, Phagwara, India

Waseem Ahmad Government Higher Secondary School Nehalpora, Srinagar, India

Fidelis O. Ajibade Department of Civil and Environmental Engineering, Federal University of Technology, Akure, Nigeria; Key Laboratory of Environmental Biotechnology, Research Centre for Eco-Environmental Sciences, Chinese Academy of Sciences, Beijing, P.R. China; University of Chinese Academy of Sciences, Beijing, P.R. China

Temitope F. Ajibade Department of Civil and Environmental Engineering, Federal University of Technology, Akure, Nigeria; University of Chinese Academy of Sciences, Beijing, P.R. China; Institute of Urban Environment, Chinese Academy of Sciences, Xiamen, P.R. China

Ana R.A. Alves Department of Geosciences, GeoBioTec, University of Aveiro, Aveiro, Portugal

Vandana Anand Division of Microbial Technology, CSIR-NBRI, Lucknow, India; Academy of Scientific and Innovative Research, AcSIR, Ghaziabad, India

Waquar Akhter Ansari Department of Botany, M.M.V, Banaras Hindu University, Varanasi, India

Neelam Atri Department of Botany, M.M.V, Banaras Hindu University, Varanasi, India

Megha Barot SRISTI Life Science, Grambharti, Gandhinagar, India

Monalisha Behera Department of Environmental Science, School of Earth Sciences, Central University of Rajasthan, Ajmer, India

Rahul Bhadouria Department of Botany, University of Delhi, New Delhi, India

Ram Naresh Bharagava Department of Microbiology (DM), Laboratory for Bioremediation and Metagenomics Research (LBMR), Babasaheb Bhimrao Ambedkar University (A Central University), Lucknow, India

Priyam Bharti Laboratory of Microbiology, Department of Microbiology, School of Life and Allied Sciences, Baba Farid Institute of Technology, Dehradun, India

Muhammad Bilal School of Life Science and Food Engineering, Huaiyin Institute of Technology, Huaian, P.R. China

Vidisha Bist Division of Microbial Technology, CSIR-NBRI, Lucknow, India; Academy of Scientific and Innovative Research, AcSIR, Ghaziabad, India

Anand Kumar Chaudhari Department of Botany, Centre of Advanced Study in Botany, Institute of Science, Banaras Hindu University, Varanasi, India

Roni Chaudhary Department of Biotechnology, School of Bioengineering and Biosciences, Lovely Professional University, Phagwara, India

Plaban Das Laboratory of Microbiology, Department of Microbiology, School of Life and Allied Sciences, Baba Farid Institute of Technology, Dehradun, India

Débora da Silva Vilar Graduate Program on Process Engineering, Tiradentes University (UNIT), Aracaju, Brazil

Shivika Datta Department of Zoology, Doaba College, Jalandhar, India

Vishal Kumar Deb Laboratory of Microbiology, Department of Microbiology, School of Life and Allied Sciences, Baba Farid Institute of Technology, Dehradun, India

María del Rosario Espinoza-Mellado Microscopy Central, National School of Biological Sciences, National Polytechnic Institute, México City, México

Vaishali Dhaka Department of Biotechnology, Lovely Professional University, Jalandhar, India

Daljeet Singh Dhanjal Department of Biotechnology, Lovely Professional University, Jalandhar, India

Himanshu Dwivedi Department of Botany, Maharaj Singh College, Saharanpur, India

Katlin Ivon Barrios Eguiluz Graduate Program on Process Engineering, Tiradentes University (UNIT), Aracaju, Brazil; Institute of Technology and Research (ITP), Tiradentes University (UNIT), Aracaju, Brazil

Oluniyi O. Fadare University of Chinese Academy of Sciences, Beijing, P.R. China; State Key Laboratory of Environmental Chemistry and Ecotoxicology, Research Centre for Eco-Environmental Sciences, Chinese Academy of Sciences, Beijing, P.R. China

Luiz Fernando Romanholo Ferreira Graduate Program on Process Engineering, Tiradentes University (UNIT), Aracaju, Brazil; Institute of Technology and Research (ITP), Tiradentes University (UNIT), Aracaju, Brazil

Marcial García-Pineda Botanical Garden, Faculty of Higher Studies Iztacala, National Autonomous University of Mexico, México City, México

Ajay Giri Centre of Advance Study in Science, Department of Botany, Institute of Science, Banaras Hindu University, Varanasi, India

Deen Dayal Giri Department of Botany, Maharaj Singh College, Saharanpur, India

Akanksha Gupta Institute of Environment and Sustainable Development, Banaras Hindu University, Varanasi, India

Madhuri Gupta Sardar Vallabhbhai Patel University of Agriculture and Technology, Meerut, India

Sanjay Kumar Gupta Environmental Engineering, Department of Civil Engineering, Indian Institute of Technology Delhi, Delhi, India

Hafiz M.N. Iqbal Tecnologico de Monterrey, School of Engineering and Sciences, Campus Monterrey, Monterrey, NL, CP, Mexico

Pooja Jaiswal Department of Botany, M.M.V, Banaras Hindu University, Varanasi, India

Prashant Kumar Jaiswal Department of Environmental Science, School of Earth Sciences, Central University of Rajasthan, Ajmer, India

Mathew Jyothis School of Biosciences, Mahatma Gandhi University, Kottayam, India

Ratna Katiyar Botany Department, Lucknow University, Lucknow, India

Jasvinder Kaur Division of Microbial Technology, CSIR-NBRI, Lucknow, India

Paramdeep Kaur Department of Biotechnology, Panjab University, Chandigarh, India

Bhupendra Koul Department of Biotechnology, School of Bioengineering and Biosciences, Lovely Professional University, Phagwara, India

Ram Krishna Directorate of Onion and Garlic Research, Pune, India

Ajay Kumar Agriculture Research Organization, Ministry of Agriculture and Rural Development, Volcani Center, Rishon LeZion, Israel

Naveen Kumar School of Life Sciences, The University of Warwick, Coventry, United Kingdom

Nirmal Kumar PG Department of Environment Science and Technology, ISTAR, Anand, India

Vijay Kumar Department of Chemistry, Regional Ayurveda Research Institute for Drug Development, Gwalior, India

Nisha Kumari Department of Environmental Science, School of Earth Sciences, Central University of Rajasthan, Ajmer, India

Kayode H. Lasisi Department of Civil and Environmental Engineering, Federal University of Technology, Akure, Nigeria; University of Chinese Academy of Sciences, Beijing, P.R. China; Institute of Urban Environment, Chinese Academy of Sciences, Xiamen, P.R. China

María Fernanda López-Gómez Biological Variation and Evolution Laboratory, Zoology Department, National School of Biological Sciences, National Polytechnic Institute, México City, México

Edgar Oliver López-Villegas Microscopy Central, National School of Biological Sciences, National Polytechnic Institute, México City, México

Abdul Malik Department of Agricultural Microbiology, Faculty of Agricultural Science, Aligarh Muslim University, Aligarh, India

Sebastian Jose Midhun School of Biosciences, Mahatma Gandhi University, Kottayam, India

Manisha Mishra University Department of Botany, T. M. Bhagalpur University, Bhagalpur, India

Virendra Kumar Mishra Institute of Environment and Sustainable Development, Banaras Hindu University, Varanasi, India

Arpan Modi Agriculture Research Organization, Ministry of Agriculture and Rural Development, Volcani Center, Rishon LeZion, Israel

Deepanshu Monga Department of Biotechnology, Panjab University, Chandigarh, India

Sahadevan Neethu School of Biosciences, Mahatma Gandhi University, Kottayam, India

Kin Israel R. Notarte Faculty of Medicine and Surgery, University of Santo Tomas, Manila, Philippines

Luís A.B. Novo Scotland's Rural College, Edinburgh, United Kingdom

Nathaniel A. Nwogwu Department of Agricultural and Bioresources Engineering, Federal University of Technology, Owerri, Nigeria

Rui S. Oliveira Department of Life Sciences, Centre for Functional Ecology, University of Coimbra, Coimbra, Portugal

Sushree Pallavi Division of Microbial Technology, CSIR-NBRI, Lucknow, India

K.D. Pandey Department of Botany, Institute of Science, Banaras Hindu University, Varanasi, India

Shweta Pandey School of Agriculture, Graphic Era Hill University, Dehradun, India

Parul Parihar Department of Botany, Lovely Professional University, Jalandhar, India

Ahmad Rabbani Laboratory of Microbiology, Department of Microbiology, School of Life and Allied Sciences, Baba Farid Institute of Technology, Dehradun, India

E.K. Radhakrishnan School of Biosciences, Mahatma Gandhi University, Kottayam, India

Kumari Rashmi L.N. Mishra Institute Patna, Patna, India

Devendra Singh Rawat Laboratory of Biotechnology, Department of Biotechnology, School of Life and Allied Sciences, Baba Farid Institute of Technology, Dehradun, India

Angélica Rodríguez-Dorantes Plant Physiology Laboratory, Botany Department, National School of Biological Sciences, National Polytechnic Institute, México City, México

Aída Verónica Rodríguez-Tovar Mycology Laboratory, Microbiology Department, National School of Biological Sciences, National Polytechnic Institute, México City, México

Aditi Roy Division of Environmental Sciences, CSIR-NBRI, Lucknow, India

Giancarlo Richard Salazar-Banda Graduate Program on Process Engineering, Tiradentes University (UNIT), Aracaju, Brazil; Institute of Technology and Research (ITP), Tiradentes University (UNIT), Aracaju, Brazil

Gaurav Saxena Laboratory of Microbiology, Department of Microbiology, School of Life and Allied Sciences, Baba Farid Institute of Technology, Dehradun, India; Department of Environmental Microbiology (DEM), School for Environmental Sciences (SES), Babasaheb Bhimrao Ambedkar (Central) University, Lucknow, India

Avimanu Sharma Department of Environmental Science, School of Earth Sciences, Central University of Rajasthan, Ajmer, India

Reetika Shukla Institute of Environment and Sustainable Development, Banaras Hindu University, Varanasi, India

Eduardo F. Silva Department of Geosciences, GeoBioTec, University of Aveiro, Aveiro, Portugal

Anand Vikram Singh Centre for Advanced Study in Botany, Banaras Hindu University, Varanasi, India

Anubhuti Singh Institute of Environment and Sustainable Development, Banaras Hindu University, Varanasi, India

Baljinder Singh Department of Biotechnology, Panjab University, Chandigarh, India

Divya Singh Central Sericultural Research and Training Institute, Mysore, India

Gurudatta Singh Institute of Environment and Sustainable Development, Banaras Hindu University, Varanasi, India

Joginder Singh Department of Microbiology, School of Bioengineering and Biosciences, Lovely Professional University, Jalandhar, India; Department of Biotechnology, Lovely Professional University, Jalandhar, India

Jyoti Singh Department of Botany, Institute of Sciences, Banaras Hindu University, Varanasi, India

Kaushalendra Kumar Singh Department of Zoology, Pachhunga University College, Mizoram University, Aizawl, India

Manoj Kumar Singh Chemical Division, Intellectual Property Office, Delhi, India

Monika Singh Department of Botany, Centre of Advance Study in Science, Institute of Science, Banaras Hindu University, Varanasi, India; Centre for Advanced Study in Botany, Banaras Hindu University, Varanasi, India

Pallavi Singh Division of Microbial Technology, CSIR-NBRI, Lucknow, India

Prashant Kumar Singh Department of Biotechnology, Pachhunga University College, Mizoram University, Aizawl, India

Prem Pratap Singh Centre for Advanced Study in Botany, Banaras Hindu University, Varanasi, India

Priyanka Singh Institute of Environment and Sustainable Development, Banaras Hindu University, Varanasi, India

Rajesh Kumar Singh Banaras Hindu University, Varanasi, India

Rishikesh Singh Institute of Environment and Sustainable Development, Banaras Hindu University, Varanasi, India

Ritu Singh Department of Environmental Science, School of Earth Sciences, Central University of Rajasthan, Ajmer, India

Sandeep Kumar Singh Department of Botany, Centre of Advance Study in Science, Institute of Science, Banaras Hindu University, Varanasi, India; Centre for Advanced Study in Botany, Banaras Hindu University, Varanasi, India

Simranjeet Singh Department of Biotechnology, Lovely Professional University, Jalandhar, India; Punjab Biotechnology Incubators, Mohali, India; Regional Advanced Water Testing Laboratory, Mohali, India

Sushil Kumar Singh Department of Agricultural Biotechnology, Assam Agricultural University, Jorhat, India

Vipin Kumar Singh Department of Botany, Centre of Advanced Study in Science, Institute of Science, Banaras Hindu University, Varanasi, India; Centre for Advanced Study in Botany, Banaras Hindu University, Varanasi, India

Akhileshwar Kumar Srivastava Central Food Technological Research Institute, Mysuru, India

Sonal Srivastava Division of Microbial Technology, CSIR-NBRI, Lucknow, India; Academy of Scientific and Innovative Research, AcSIR, Ghaziabad, India

Subhi Srivastava Plant Molecular Biology Division, CSIR-NBRI, Lucknow, India

Suchi Srivastava Division of Microbial Technology, CSIR-NBRI, Lucknow, India; Academy of Scientific and Innovative Research, AcSIR, Ghaziabad, India

Ishaq D. Sulaymon University of Chinese Academy of Sciences, Beijing, P.R. China; Huairou Eco-Environmental Observatory, Chinese Academy of Sciences, Beijing, P.R. China

Pooja Taak Department of Biotechnology, School of Bioengineering and Biosciences, Lovely Professional University, Phagwara, India

Ramya Thangamani Department of Chemical Engineering, King Mongkut's University of Technology North Bangkok, Bangkok, Thailand

Nádia Hortense Torres Institute of Technology and Research (ITP), Tiradentes University (UNIT), Aracaju, Brazil

Adamu Y. Ugya Department of Environmental Management, Kaduna State University, Kaduna, Nigeria; College of New Energy and Environment, Jilin University, Changchun, P.R. China

Sunita Varjani Gujarat Pollution Control Board, Gandhinagar, India

Hariom Verma Centre for Advanced Study in Botany, Banaras Hindu University, Varanasi, India; Department of Botany, Banaras Hindu University, Varanasi, India

Lakshmanaperumal Vidhya Department of Chemical Engineering, Sethu Institute of Technology, Virudhunagar, India

Pratibha Vyas Department of Microbiology, College of Basic Sciences and Humanities, Punjab Agricultural University, Ludhiana, India

Aijie Wang Key Laboratory of Environmental Biotechnology, Research Centre for Eco-Environmental Sciences, Chinese Academy of Sciences, Beijing, P.R. China

Hong Cheng Wang Key Laboratory of Environmental Biotechnology, Research Centre for Eco-Environmental Sciences, Chinese Academy of Sciences, Beijing, P.R. China

Wang Wenjing State Key Laboratory of Cotton Biology, Henan Key Laboratory of Plant Stress Biology, School of Life Science, Henan University, Kaifeng, China

Deepanker Yadav Agriculture Research Organization, Ministry of Agriculture and Rural Development, Volcani Centre, Rishon LeZion, Israel

Surabhi Zainith Department of Environmental Microbiology (DEM), School for Environmental Sciences (SES), Babasaheb Bhimrao Ambedkar (Central) University, Lucknow, India

Mohammad Tarique Zeyad Department of Agricultural Microbiology, Faculty of Agricultural Science, Aligarh Muslim University, Aligarh, India

Chapter 1

Emerging contaminants in environment: occurrence, toxicity, and management strategies with emphasis on microbial remediation and advanced oxidation processes

Débora da Silva Vilar[1], Nádia Hortense Torres[2], Ram Naresh Bharagava[3], Muhammad Bilal[4], Hafiz M.N. Iqbal[5], Giancarlo Richard Salazar-Banda[1,2], Katlin Ivon Barrios Eguiluz[1,2] and Luiz Fernando Romanholo Ferreira[1,2]

[1]*Graduate Program on Process Engineering, Tiradentes University (UNIT), Aracaju, Brazil,* [2]*Institute of Technology and Research (ITP), Tiradentes University (UNIT), Aracaju, Brazil,* [3]*Department of Microbiology (DM), Laboratory for Bioremediation and Metagenomics Research (LBMR), Babasaheb Bhimrao Ambedkar University (A Central University), Lucknow, India,* [4]*School of Life Science and Food Engineering, Huaiyin Institute of Technology, Huaian, P.R. China,* [5]*Tecnologico de Monterrey, School of Engineering and Sciences, Campus Monterrey, Monterrey, NL, CP, Mexico*

1.1 Introduction

For some decades, studies have attracted the attention to the possible harm that emerging contaminants (ECs) can cause to water resources, motivated by the increasing number of generation of these products (Gogoi et al., 2018). This fact boosted the interest of the scientific community in search of solutions for identification, monitoring, and removal of these substances. ECs are an allusion to toxic products that are not removed or eliminated by traditional processes of water treatment for human consumption. Among these contaminants are the drugs of various compositions, such as nanomaterials, bactericides, insecticides, algicides, herbicides, cleaning and personal hygiene products, sunscreens, textile dyes, and endogenous hormones (Tran et al., 2018). In general, they are recognized as new classes of water contaminants, due to their proven or potential adverse effects on aquatic ecosystems and human health (Barroso et al., 2019).

According to Lamastra et al. (2016), approximately 700 different emerging pollutants and metabolites are disposed of irregularly in the environment. This fact is aggravated by the lack of information about the destination and transportation of these new chemicals, which include their toxic potential for the recipients, in addition to hindering the development of standards controlled by regulatory agencies that provide the management of contaminants in the environment. (Taheran et al., 2018). However, the constant increase in these in the environment still occurs due to the consumerist profile of modern society (Gogoi et al., 2018). However, because these are recent studies, there is a deficiency of data regarding the occurrence of many ECs in environmental matrices, since these "new products" are more toxic than their original molecules, which makes it more difficult for your detection or identification. This is possible due to the limited information available regarding their interactions with toxicological impacts on water bodies.

From this problem the challenge remains to quest new technologies to develop more efficient and sustainable products in order to reduce or eliminate the use and generation of hazardous substances for human health and the environment. In this case a green alternative is the biological process, since it allows to replace traditional processes through the use of enzymes produced by microorganism (Przystaś et al., 2018). In addition, they are able to degrade and/or mineralize various recalcitrant aromatic compounds. For this reason, these processes have gained extensive recognition as a

Microbe Mediated Remediation of Environmental Contaminants. DOI: https://doi.org/10.1016/B978-0-12-821199-1.00001-8

promising alternative used in chemical processes, with the differential that these can eliminate or decrease the number of toxic substances, in addition to meeting the needs of a specific application. The literature also reports that a new set of conventional treatments (adsorption, coagulation/flocculation, chemical oxidation, ozonization, etc.) has already been introduced in the treatment of these contaminants, which adds to the market products with a high level of performance and with sustainable characteristics (Wu et al., 2016; Rodriguez-Narvaez et al., 2017; Paździor et al., 2018). However, to be economically viable and produced on a large scale, it is essential that most of these processes are environmentally compatible, which is limited as they generate toxic by-products due to the use of various types of chemicals (Elisandra et al., 2018). From this the use of advanced oxidative processes (AOPs) emerges as an excellent technology in the degradation of organic products, especially those that resist biological treatment, in which these are oxidized via hydroxyl radicals (•OH) (Salazar et al., 2017).

In this scenario, in order to attend the principles of technological development processes and sustainable products, this project aims to review the main ECs found in the aquatic environment. It also presents an overview of the sources, its path to receptors, destination, and toxicological impacts on the biota, besides focusing on treatment mechanisms from biological remediation and oxidative processes. This new approach is an alternative method to reduce costs and increase process efficiency and, consequently, be more attractive to industries of interest as they use innovative processes in the treatment of these contaminants.

1.2 Occurrence of emerging contaminants in environment: causes, effects, and analysis

ECs, micropollutants, or contaminants of emerging concern are chemicals of growing concern, the toxicity of which and persistence of these in the environment alter the metabolism of a living being (Gogoi et al., 2018). These are generated by human (domestic effluent) and industrial activities (hospitals, livestock, and aquaculture), such as pharmaceuticals and personal care products (PPCPs), polycyclic aromatic hydrocarbons, solar filters, cosmetics, pesticides, biocides, hormones, nanomaterials, heavy metals, plasticizers, food additives, wood preservatives, surfactants, disinfectants, flame retardants, and other high-organic load by-products found in wastewater (Petrie et al., 2014; Richardson and Kimura, 2017). In addition, these ECs can be of natural or synthetic origin and are generally not detected in the environment, since most of these also do not have standard regulations, which promote lethal effects on human and aquatic life, even in small concentrations (range from L^{-1} to $\mu g/L$), due to its bioaccumulation (Noguera-Oviedo and Aga, 2016; Taheran et al., 2018).

In general, water treatment is necessary in the presence of ECs from the use of several conventional processes of the sewage treatment plant (STP), according to the applicable operating conditions, since they are excreted or disposed of in wastewater and underground, which also enter rivers and drinking water supplies (Fig. 1.1). In addition to wastewater, agricultural runoff can also be a source of entry of these ECs into the environment (Richardson and Kimura, 2017).

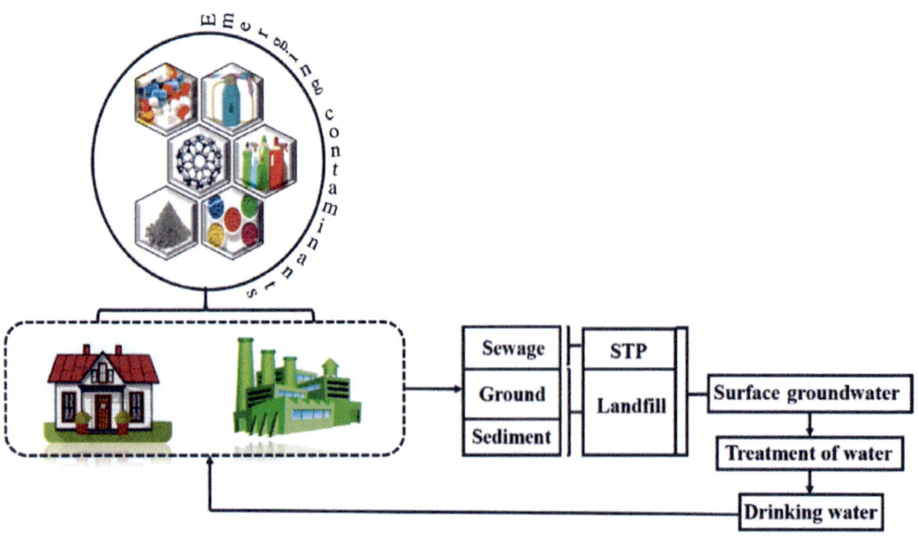

FIGURE 1.1 Origins of emerging pollutants and routes to the environment.

Effluent treatment can be divided into preliminary, primary, secondary, and tertiary. Preliminary treatment consists of the removal of coarse solids. The primary has the function of removing suspended solids. Secondary is the step responsible for reducing organic matter. Tertiary is responsible for the effective removal of the remaining compounds from the previous steps (Carmalin Sophia and Lima, 2018).

Treatment processes (physical, chemical, and biological) can also be combined, in which they occur simultaneously and synergistically, increasing the efficiency of the contaminant removal process (Rodriguez-Narvaez et al., 2017). However, the more unit processes are included, the desired effluent quality is achievable, but treatment costs are increased. Therefore the development of the appropriate technology for the treatment of ECs has attracted great interest, since the main objective is the efficiency in the removal and degradation of these, recently recognized as significant water pollutants. In this context, the main groups of priority ECs were studied, which are pharmaceuticals, pesticides/ biocides, textile dyes, and personal hygiene products.

1.2.1 Pharmaceuticals

Pharmaceuticals have received increasing attention as one of the main bioactive chemical contaminants in the environment, since their consumption in the world has increased significantly in recent years, especially those sold without prescription medical (Rodriguez-Narvaez et al., 2017). The presence of these products in groundwater promotes the dissemination of genes of various microorganisms as well as directly impacts human health, since the compounds of these contaminants are collected in wastewater effluents after their consumption (Sharma et al., 2019). Some of these compounds are more bioactive than their metabolic precursor.

It is estimated that about 3000 different chemicals are used in the manufacture of pharmaceutical products, which include antiinflammatory and analgesics (paracetamol, acetylsalicylic acid, ibuprofen, and diclofenac); antibiotics (tetracyclines, macrolides, beta-lactamase, penicillins, quinolones, sulfonamides, fluoroquinolones, chloramphenicol, and imidazole derivatives); antidiabetics; beta-blockers (atenolol, propranolol, and metoprolol); contraceptives; antiulcer drugs and antihistamines (ranitidine and famotidine); lipid regulators; synthetic hormones; and antidepressants (benzodiazepines) (Rivera-Utrilla and Sánchez-Polo, 2013; Rodriguez-Narvaez et al., 2017; Salomão et al., 2019).

Fig. 1.2 has shown over the past 20 years, the summary of the main pharmaceutical products consumed in the world, which are led by China, India, Indonesia, and Brazil. In addition, global drug use will reach 4.5 trillion doses by 2020, with 50% of the world's population consuming more than 1 dose/person/day (Patel et al., 2019).

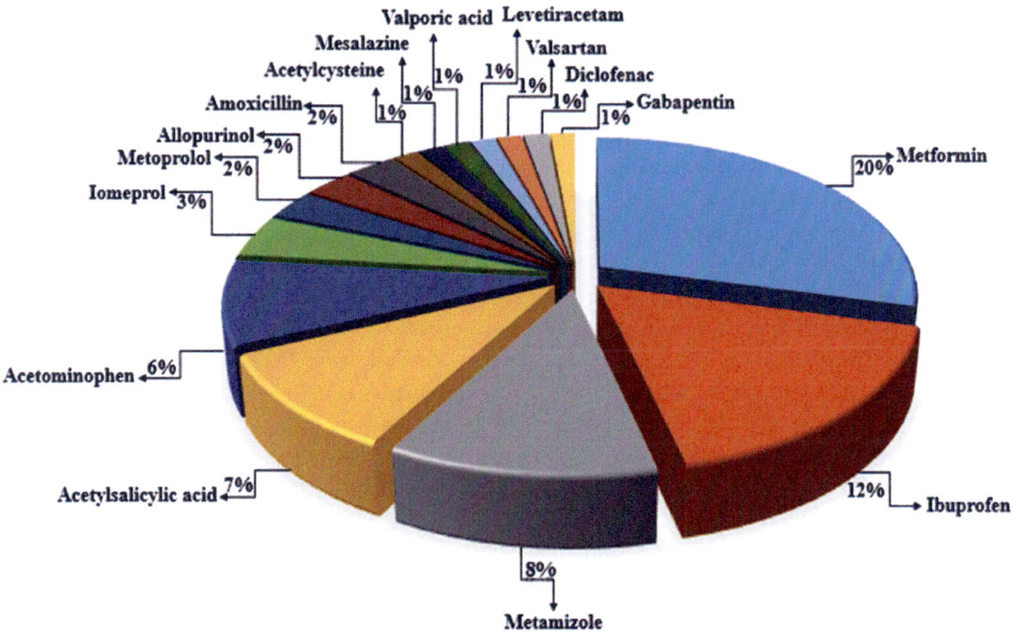

FIGURE 1.2 Consumption pattern of several pharmaceutical products in the world. *Adapted from Patel, M., Kumar, R., Kishor, K., et al., 2019. Pharmaceuticals of emerging concern in aquatic systems: chemistry, occurrence, effects, and removal methods, Chem. Rev. 119, 3510–3673, Based on IMS. 2013. Global Pharmaceutical Sales, MIDAS database.*

In general, these products have been present in wastewater for decades, however, only recently began to be quantified and recognized as toxic potentials for the ecosystem (Bilal et al., 2020). Thus it is evident the social concern towards public health, since there are few reports about the toxic effects caused by these contaminant products, which are ingested in the long term and which are present in drinking water associated with various xenobiotic compound mixtures. For these reasons the great challenge for the chemical industry and the like in the 21st century is to find viable alternatives to achieve the effective removal of pharmaceutical products, together with other priority pollutants present in wastewater, before their discharge.

1.2.2 Pesticides and biocides

The most representative synthetic chemicals for agricultural purposes are known as pesticide, which include herbicides, insecticides, and fungicides, as well as are classified according to their chemical composition and biological characteristics specific under target species (Morillo and Villaverde, 2017). Os agrotóxicos mais utilizados na agricultura são os carbamatos, organoclorados, organofosforados, piretróides e triazinas. With regard to chemicals used in urban environments, biocides that are antibacterial disinfectants, sprays, sterilizers, and condoms are also considered ECs, since they are incorporated into surface waters and underground water flow. In general, these agricultural pesticides are considered a vital element for modern agriculture that integrates technologies associated with crop development, which results in greater food production capacity. However, soil, water, and air dispersed are produced and dispersed in large quantities, in which they promote bioaccumulation in organic tissues, besides having high persistence that involves risks associated with adverse defects in human health, such as cancer, diseases neurodegenerative and congenital defects (Jallow et al., 2017).

Because they are considered potentially toxic, they still have properties to block the vital metabolic process for organisms. The considerable and abusive increase in the use of these agrochemicals is the result of the imposition of the policy that meets the innovations of the green revolution and promotes agility in the release of substances, as well as the use of transgenic crops, increased pests, exemption from improper taxes and storage, handling, and disposal (Lewis et al., 2016). On the other hand, the reduction of these contaminants is motivated by the fragility in supervision and the lack of policies that do not match the development of sustainable agriculture. Consequently, these factors are associated with the physicochemical properties of pesticides that promote leaching, sorption, volatilization, photodegradation, microbial or chemical degradation of soil, and wastewater (Cycoń et al., 2017). Thus, as the population is exposed to several pesticides (simultaneously or sequentially), the need to perform pesticide risk assessments goes beyond the approach to regulatory standards.

1.2.3 Textile dye

Textile dyes fall within the category of emerging pollutants since they are defined as any chemical that has not been included in monitoring programs and legislation relevant to environmental quality (Torres et al., 2019). These pollutants are usually organic compounds, of synthetic origin, soluble in water, the main characteristic of which is the absorption of visible light (Roy et al., 2018). In addition, they present distinct molecular weights and complex aromatic structures as well as are used in a wide range of applications: in the food, pharmaceutical, plastic, cosmetic, textile and leather industries, photographic printing, and in paper and petroleum products (Hernández et al., 2018).

Although they present fruitful characteristics for textile industrial use, when discarded without treatment in water bodies they promote pollution in the entire ecosystem and consequently cause the death of aquatic organisms as well as impair photosynthetic activity, resulting in decreased oxygen dissolved in water (Mondal et al., 2017). In addition, since they are considered stable and difficult to degrade, they can also generate acute toxic effects on humans, mainly due to their metabolites directly influencing carcinogenesis, teratogenesis, and mutagenesis (Roy et al., 2018; Tang et al., 2018). In general, about 10%–90% of the effluents correspond to all synthetic dyes that do not fix tissue fibers and therefore are easily released into watery environments, promoting an increase in chemical oxygen demand (COD) and biochemical oxygen demand (BOD) (Kumar and Saravanan, 2017).

The classification of dyes is based on the chemical structure of their chromogenic groups (nitrogen, heterocyclic, anthrachinons, indigoids, cyanin, xantenos, and triphenylmetanos) and the way of fixation with the substrate (acids, basic, reactive, direct, sulfurous, premetallized, bleaching, and nitrogen), which basically involves four types of molecular interactions: ionic, covalent, hydrogen, and van der Waals (Kumar and Saravanan, 2017). However, azo dyes are the largest class of synthetic dyes and are the most used in commercial applications, as they present satisfactory characteristics in terms of resistivity to light and moisture, in addition to promoting high fixation dyeing. However, it is necessary to use diversified techniques in the treatment of these textile dyes, since conventional methods are not considered satisfactory.

1.2.4 Personal care products

Personal hygiene chemicals (PPCPs) followed by drugs are the most commonly found household contaminants in urban environments, as well as in STPs, and include fragrances, toothpastes, shampoos, soaps, sunscreens, detergents, soaps, disinfectants, packaging plasticizers, preservatives, insect repellents, and antifungal agents (Yang et al., 2017). These contaminants are associated for health, beauty, and cleanliness purposes, and defined as any synthetic or natural compound, present in surface waters and receiving soil, with the potential to promote adverse health effects in humans and microorganisms, due to its inherent ability to induce physiological effects, even at low doses (ng/L−μg/L), the treatment of which does not have routine monitoring (Montes-Grajales et al., 2017).

The main entry of these chemicals occurs through domestic sewage due to inadequate disposal and/or inefficient treatment in urban water treatment plants (Noguera-Oviedo and Aga, 2016). PPCPs may be partially metabolized or biodegraded incompletely in water matrices. Table 1.1 summarizes the various common PPCPs detected in surface, groundwater, and drinking waters. In addition, PPCPs residues can also enter the environment through the hydrological cycle.

TABLE 1.1 Concentration of personal hygiene products (PPCPs) and their representative compounds.

Classification of PPCPs	Compound	Concentration (ng/L)	Reference
Bactericides, disinfectants	TCS	547	Tohidi and Cai (2017)
		75	Emnet et al. (2015)
		157	Carmona et al. (2014)
	TCC	540	Blair et al. (2015)
		75	Blair et al. (2013)
		12	Gottschall et al. (2012)
	Methyltriclosan	74	Zhao et al. (2014)
Insect repellents	DEET	78.4	Lin et al. (2016)
		298.4	Tran et al. (2014)
		600	Sui et al. (2010)
Sunscreen UV filters	4MBC	45.1	Emnet et al. (2015)
		169	Tsui et al. (2014)
	EHMC	150	Meador et al. (2016)
		462	Tsui et al. (2014)
	BP	10.3	Emnet et al. (2015)
	OC	8	Tsui et al. (2014)
		283	Pintado-Herrera et al. (2013)
Fragrances	PDL	178	Arbulu et al. (2011)
	HHCB	84	Cavalheiro et al. (2013)
		57	Matamoros and Salvadó (2012)
	AHTN	32	Cavalheiro et al. (2013)
		13	Matamoros and Salvadó (2012)
		550	Lee et al. (2010)

4MBC, 4-Methyl-benzilidine-camphor; *AHTN*, tonalide; *BP*, benzophenone; *EHMC*, 2-ethyl-hexyl-4-trimethoxycinnamate; *HHCB*, galaxolide; *OC*, octocrylene; *PDL*, exaltolide; *TCC*, triclocarban; *TCS*, triclosan; *DEET*, N-diethyl-m-toluamide.

The presence of these organic contaminants in treatment plants varies according to the pattern of consumption of the population, as well as the social and economic pattern of the region (Roberts et al., 2016). Moreover, most of these polluting agents cannot be completely degraded by the conventional and unitary treatment process, which makes it worrisome, since they are persistent, bioactive, bioaccumulative, and endocrine compounds. The fact that it has strict regulations and implemented to minimize the consumption of these PPCPs does not prevent their continued use due to their beneficial properties for health purposes for humans and animals. Therefore it is necessary to evaluate the performance of the different methods of treatment of PPCPs, as well as the deleterious effects caused by these contaminants.

1.3 Treatment technologies

Conventional water treatment technologies are not able to completely remove contaminants, since they differ widely from their chemical, biological, and physical properties, such as chemical reactivity, biodegradability, and hydrophobicity. Thus typical treatments for contaminant removal include processes through biological remediation and advanced oxidation processes (AOPs). Several authors, from the use of microorganisms, have studied the importance of bioremediation since they present potential for direct transformation of xenobiotic compounds during the remediation process, in some cases with significant mineralization rates. In addition, this treatment is considered an excellent alternative in the replacement of traditional chemical processes, since it allows the reduction of the production of undesirable by-products and requires fewer steps to achieve the final product. Although this treatment has many advantages, a single process does not remove all contaminants. However, AOPs become particularly interesting as simple, safe, economical, and environment-friendly technologies for the degradation of these pollutants into aqueous environment.

1.3.1 Biological processes

Bioremediation is one of the most viable options to minimize or remove xenobiotics from soil and water from the use of various microorganisms (fungi, bacteria, yeasts, algae, and plants) (Przystaś et al., 2018). The literature reports that studies focused on biotechnology indicate that fungi and bacteria are considered the main microorganisms efficient in the degradation of pollutants, since their efficiency is related to the high potential for action capable of recovering contaminated environments (Garcia-Rodríguez and Matamoros, 2014; Boshir et al., 2017; Przystaś et al., 2018). In addition, this process presents low-cost technology, high potential for degradation of recalcitrant compounds, has resistance mechanisms in adverse environmental conditions, and is environmentally compatible. The most effective metabolic process is microbial, since microorganisms are able to convert external polymers into smaller molecules, where they are metabolized or cometabolized. Thus this process allows to remove or decrease most impurities present in contaminants, besides offering advantages with regard to greater production savings, possibility of optimization, and reduction of the use of chemical additives (Kanaujiya et al., 2019).

Generally, these processes are associated with physical operations of concentration and separation of solids. The primary physicochemical process can also complete the secondary system consisting of the biological reactor in the presence of activated sludge, in order to increase the removal of contaminants (Rodriguez-Narvaez et al., 2017). In addition, this biological remediation is divided into two different redox conditions: anaerobic and anoxic aerobic. Aerobic applications include activated sludge, membrane bioreactor (MBR), and sequential batch reactor, while the anoxic methods use anaerobic sludge reactors and anaerobic film reactors, while the anoxic condition uses anaerobic reactors sludge in the presence of nitrates (Kanaujiya et al., 2019). However, depending on the selected condition, the degradation route may change due to the variation of the consortium of microorganisms present in the medium, as well as the type of contaminant used. Table 1.2 shows the summary of the main biological processes used for removal in water, along with the type of contaminant and the removal efficiency achieved.

Since then, bioremediation is promoted in greater potential and efficiency by specific enzyme reactions (Saha et al., 2016). Microorganisms still have the ability to easily adapt their metabolism to different carbon sources in order to promote a large amount of extracellular enzymes and consequently degrade and/or mineralize toxic compounds (Rouches et al., 2016). In general, this unique treatment is not sufficient to completely remove hazardous particles from wastewater, so physicochemical treatment methods and AOPs should be added and/or combined (Paździor et al., 2018). Other processes also used in biological remediation are microbial biomass adsorption, degradation with activated sludge, MBR, aerobic bioreactor, anaerobic bioreactor, microalgae bioreactor, fungal bioreactor with pure or mixed cultures, biofilm, rotating biological reactor, nitrification, enzymatic treatment, phytoremediation, and biosorption (Paździor et al., 2018). In fact, the main difficulty in the application of biological processes is related to the lack of defined

TABLE 1.2 Removal efficiency of emerging contaminants (ECs) using biological processes.

Type	System	EC	Removal efficiency (%)	Reference
Anaerobic digestion	Activated sludge	AS-C12	92	Abril et al. (2019)
		AS-C14	85	
		AS-C16	103	
		AS-C18	89	
		TCB	96	
		TCS	83	
		BP-1	82	
		BP-2	94	
		BP-3	79	
		BP-6	74	
		BP-8	98	
		4-OH-BP	97	
Adsorption	Activated sludge	Indigo dye	87.54	Haddad et al. (2018)
	Activated sludge	Tetracycline	30	Ocampo-Pérez et al. (2012)
	Microbial biomass	Acid Red 27	85	Ali and El-Mohamedy (2012)
Anaerobic digestion	Activated sludge	Paracetamol	71	Yang et al. (2016)
		Primidone	23	
		Fluoxetine	10	
		Clozapine	1	
		Verapamil	3	
		Amitriptyline	3	
		Carbamazepine	56	
		Naproxen	82	
		Diuron	27	
		Ibuprofen	40	
		Bisphenol	27	
		Diclofenac	43	
		Gemfibrozil	28	
		Triclosan	16	
		Triclocarban	7	
Anaerobic digestion	Biofilm reactor	Diclofenac	41	Casas and Bester (2015)
		Propranolol	94	
		Iopromide	58	
		Iohexol	57	
		Iomeprol	85	
Aerobic	Soil filtration	Estrogens	26	Carr et al. (2011)
		17ß-Estradiol	99	
		17ß-Ethinyl estradiol	27	
		Triclosan	90	
		Ibuprofen	18	
Anaerobic digestion	Enzyme degradation	Green Domalan BL	88.6	Hidalgo et al. (2011)

BP, Benzophenone; *TCS*, triclosan.

analytical methodologies, which are able to identify or quantify complex micropollutants, which confers an emerging development area (Rodriguez-Narvaez et al., 2017).

1.3.2 Advanced oxidation processes

Advanced oxidative processes (AOPs) in recent years have gained prominence among alternative treatment methods in the degradation of toxic and/or recalcitrant organic compounds present in wastewater, since they present environmental compatibility (when operating under mild conditions), versatility, high efficiency, low energy cost, and ease of automation (Chaplin, 2014; Sirés et al., 2014; Bassyouni et al., 2017). In addition, they are able to fully mineralize persistent and refractory organic pollutants into CO_2, water, and inorganic ions, or convert them into less toxic and easy-to-degrade for subsequent combined treatments, as they can be used before or after conventional treatment, depending on the characteristics of the effluent. The reason for this is the fact that these processes are based on the use of electric current and possess the ability to control and produce free radicals in situ (HO, O_2^-, $HO_2^{\cdot}$), without the addition of chemical reagents or large amounts of catalyst in the medium (Moreira et al., 2017).

These radicals, especially hydroxyl radicals ($OH^{\cdot}$), are characterized by nonselective and highly reactive with the ability to attack most organic molecules, since they have high constants of high reaction speed (range 10^6–10^{11} M/s), in addition to being considered a strong oxidizing agent in aqueous phase ($E^\circ = 1.9$–2.8 V/SHE) (Sopaj and Rodrigo, 2015; Moreira et al., 2017). However, the generation of these hydroxyl radicals can be caused by the action of strong oxidizing agents such as H_2O_2, O_3, irradiation (with ultraviolet light or ultrasound), and catalysts (such as Fe^{2+}). Tertiary AOPs may also be applied before and/or after biological processes (Wu et al., 2016).

Thus several analytical methods have been developed to evaluate the profiles and patterns of occurrence of ECs during the last decade, such as ozonization (O_3, O_3/H_2O_2, O_3/activated carbon, O_3/biological treatment), photooxidation (UV, UV/H_2O_2, UV/TiO_2, UV/H_2O_2/TiO_2, UV/TiO_2/activated carbon, photo-Fenton. Fig. 1.3 compares the different types of AOPs used to remove ECs present in water.

In general, AOPs can be adversely affected by several factors, such as complexity of the water matrix, type and concentration of contaminant and oxidants and catalysts, and reactor configuration (Dewil et al., 2017). Therefore, among these processes, it is important to select the best technology using the parameters analyzed in electrochemical oxidation reactions, since the most effective method promotes greater operational stability and has high oxidation power contaminants, as well as their remarkable characteristics are related to high yield and biocompatibility with contaminants.

1.4 Environmental toxicology of emerging contaminants

In order to analyze the results of degradation, especially in contaminants that have a complex composition, the ecotoxicological test is considered an efficient method, since biological responses show synergistic effects, additives, antagonistic nature, and potentiation that do not appear in physicochemical analyses or in isolated evaluations of contaminants (Figuerêdo et al., 2016). Nevertheless, toxicity tests do not replace traditional chemical analyses of contaminants

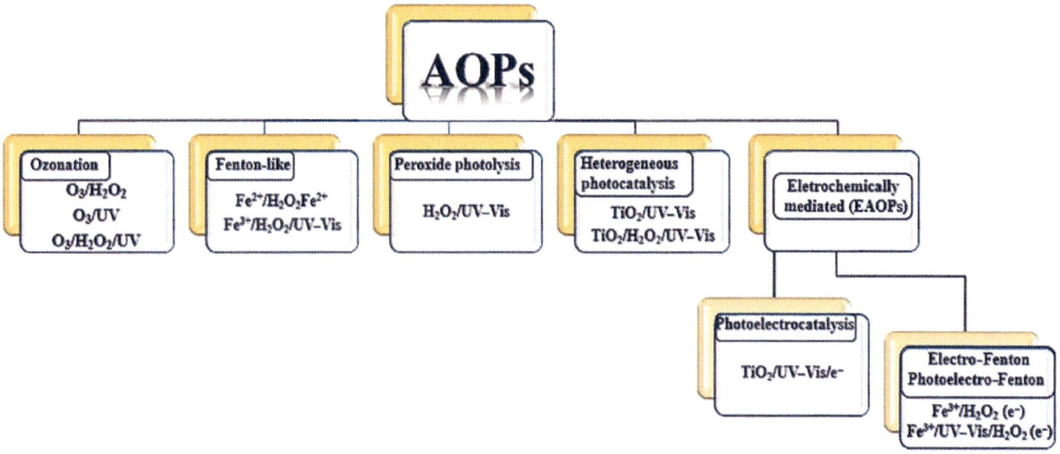

FIGURE 1.3 Types of advanced oxidation processes widely applied for water and wastewater treatments. *Adapted from Khan, J.A., Sayed, M., Khan, S., Shah, N.S., 2020. Advanced Oxidation Processes for the Treatment of Contaminants of Emerging Concern. Elsevier Inc. (Khan et al., 2020).*

(Vilar et al., 2018). Since toxicity tests assess the effect of toxic substances on biological systems, while chemical analyses identify and quantify the concentrations of these substances. Thus chemical analyses and toxicity tests complement each other, allowing an integrated assessment of the environmental quality of marine areas.

Bioassays are carried out in laboratory or in the field under controlled and specific experimental conditions, in which test organisms are exposed to different concentrations and/or doses of an environmental sample and the effects of toxic agents produced are estimated. In addition, they are characterized by being simple, fast, and reliable tests, as well as presenting low cost (Menz et al., 2013; Hassan et al., 2016). However, toxicity tests can be classified as acute and chronic, depending on the exposure time to test organisms (Seeland et al., 2012), and the result is expressed in medium lethal concentration, mean effective concentration (EC_{50}), or inhibition concentration (Clément et al., 2013; Medeiros et al., 2013). Moreover, the general toxicity approach of contaminants is classified according to the tolerable limits by standard organisms, and those with EC_{50} between 10 and 100 mg/L are harmful, from 1 to 10 mg/L are toxic, and less than 1 mg/L very toxic. The test organisms used in these bioassays must belong to different trophic levels and taxonomic groups, already standardized. In general, toxicity studies are rare; however, it is possible to verify that this reality follows a predominantly exponential trend in recent years (Table 1.3), with the use of different test organisms applied in treatment processes of the main contaminants present in water bodies and soils.

In this context, the study of toxicology is based on the protection of the biotic integrity of organisms (plants and animals) and on the control of measures to prevent or remedy their poisoning, since they are subjected to chemical stressors from various contaminants. Thus this approach is an alternative method to reduce costs and increase the efficiency of the treatment process and, consequently, make the final product more attractive to biotechnological and related industries.

TABLE 1.3 Application of bioassays with different test organisms in the evaluation of different ECs.

Type of ECs	Composite	Organism	Response	Reference
Pesticide	Chlorpyrifos	*Danio rerio*	EC_{50} de 29 mg/L	Rodrigues et al. (2019)
	Acetamiprid	*Hexagenia* spp.	LC_{50} de 780 μg/L	Bartlett et al. (2018)
	Fenvalerate	*D. rerio*	IC_{50} de 100 μg/L	Han et al. (2017)
	Diuron	*Raphidocelis subcapitata*	IC_{50} de 10.4 μg/L	Mansano et al. (2017)
	Atrazine	*Lactuca sativa*	EC_{50} de 10 mg/L	Pinto et al. (2017)
	Carbofuran	*R. subcapitata*	IC_{50} de 7426.5 μg/L	Mansano et al. (2017)
Pharmaco	Sertraline	*R. subcapitata*	IC_{50} de 150.7 μg/L	Minguez et al. (2018)
	Fluoxetine	*R. subcapitata*	IC_{50} de 202.9 μg/L	Yegin et al. (2017)
	Propranolol	*Daphnia magna*	EC_{50} de 5.53 mg/L	De Oliveira et al. (2016)
	Chlorpromazine	*D. magna*	EC_{50} de 1.80 mg/L	De Oliveira et al. (2016)
	Ciprofloxacin	*Lemna minor*	EC_{50} de 0.19 mg/L	Martins et al. (2012)
Textile dye	Sunset Yellow	*D. rerio*	LC_{50} de 38.93 mM	Jiang et al. (2020)
	Reactive Red 195	*Artemia salina*	EC_{50} de 200 mg/L	Malik et al. (2018)
	Acid Black 24	*D. magna*	LC_{50} de 58.22 mg/L	Croce et al. (2017)
	Reactive Blue	*R. subcapitata*	IC_{50} de 70%	Bedoui et al. (2015)
	Remazol Brilliant Blue R	*L. sativa*	EC_{50} de 28%	Silva et al. (2013)
PPCPs	Benzophenone	*Ceriodaphnia dubia*	EC_{25} de 0.041 mg/L	Tamura et al. (2017)
	DEET	*R. subcapitata*	EC_{10} de 16 mg/L	Tamura et al. (2017)
	TCS	*D. magna*	EC_{50} de 0.33 mg/L	Peng et al. (2013)

EC, Emerging contaminant; *TCS*, triclosan.

1.5 Future challenges and directions

Contamination of water by ECs is inevitable, since they are indispensable in modern society. However, these pollutants released inadequately and indiscriminately in wastewater cause adverse negative effects on human health and target organisms, combined with environmental insecurity due to their high toxic potential. Although there has been much progress in identifying, quantifying, understanding environmental destination, and developing remediation strategies for emerging environmental contaminants, many challenges remain, as there are still many contaminants to be identified in the environment. For these reasons, future research should be focused on the development of economic treatment strategies, as well as on prevalence and destination, since adequate pretreatment is required for different types of contaminants. The biological process was considered a promising technology, since conventional physical–chemical treatment technologies are inefficient in treating eminent micropollutants. Moreover, it allows the remediation/removal of ECs by different processes, since they present potential for direct transformation of xenobiotic compounds, in some cases with significant rates of mineralization of pollutant. However, although this treatment has many advantages, due to its limitations, it has not yet been used on an industrial scale and depending on the compound used under study, additional treatment is required, depending on the conditions of applicable operation. For this, it is essential to combine methods of treatments, mainly, however, advanced oxidation processes (AOPs) are particularly interesting, such as simple, safe, economical, and environmentally compatible technologies for the degradation of these pollutants in aqueous environment. It is also essential to deepen the current understanding of the toxicological implications of chronic exposure of these complex contaminants in target and nontarget organisms. Therefore toxicity is an not been elucidated area, which needs to direct studies aimed at integrated environmental quality assessment, as well as develop new protocols for ecotoxicity testing, in order to meet industrial sustainability and social.

Acknowledgments

The authors would like to thank the Institute of Technology and Research at Tiradentes University (ITP-UNIT), the Foundation of Support to Research and Technological Innovation of the State of Sergipe (FAPITEC/SE), the Coordination for the Improvement of Higher Education Personnel (CAPES) under the Finance Code 001, and the National Council for Scientific and Technological Development (CNPq) for financial support (process nos. 315405/2018-0; 421147/2016-4).

References

Abril, C., Santos, J.L., Martín, J., Aparicio, I., Alonso, E., 2019. Occurrence, fate and environmental risk of anionic surfactants, bisphenol A, perfluorinated compounds and personal care products in sludge stabilization treatments. Sci. Total Environ. 711, 135–148.

Ali, N.F., El-Mohamedy, R.S.R., 2012. Microbial decolourization of textile waste water. J. Saudi Chem. Soc. 16 (2), 117–123. Available from: https://doi.org/10.1016/j.jscs.2010.11.005.

Arbulu, M., Sampedro, M.C., Unceta, N., et al., 2011. A retention time locked gas chromatography–mass spectrometry method based on stir-bar sorptive extraction and thermal desorption for automated determination of synthetic musk fragrances in natural and wastewaters. J. Chromatogr. A 1218 (20), 3048–3055. Available from: https://doi.org/10.1016/j.chroma.2011.03.012.

Barroso, P.J., Santos, J.L., Martín, J., Aparicio, I., Alonso, E., 2019. Emerging contaminants in the atmosphere: analysis, occurrence and future challenges. Crit. Rev. Environ. Sci. Technol. 49 (2), 104–171.

Bartlett, A.J., Hedges, A.M., Intini, K.D., et al., 2018. Lethal and sublethal toxicity of neonicotinoid and butenolide insecticides to the may fly, *Hexagenia* spp. Environ. Pollut. 238, 63–75. Available from: https://doi.org/10.1016/j.envpol.2018.03.004.

Bassyouni, D.G., Hamad, H.A., El-Ashtoukhy, E.-S.Z., Amin, N.K., El-Latif, M.M.A., 2017. Comparative performance of anodic oxidation and electrocoagulation as clean processes for electrocatalytic degradation of diazo dye Acid Brown 14 in aqueous medium. J. Hazard. Mater. 335, 178–187. Available from: https://www.sciencedirect.com/science/article/pii/S0304389417302947?via%3Dihub.

Bedoui, A., Tigini, V., Ghedira, K., Varese, G.C., Chekir Ghedira, L., 2015. Evaluation of an eventual ecotoxicity induced by textile effluents using a battery of biotests. Environ. Sci. Pollut. Res. 22 (21), 16700–16708.

Bilal, M., Mehmood, S., Rasheed, T., Iqbal, H.M.N., 2020. Antibiotics traces in the aquatic environment: persistence and adverse environmental impact. Environ. Sci. Health 13, 68–74. Available from: https://doi.org/10.1016/j.coesh.2019.11.005.

Blair, B.D., Crago, J.P., Hedman, C.J., Klaper, R.D., 2013. Pharmaceuticals and personal care products found in the Great Lakes above concentrations of environmental concern. Chemosphere 93 (9), 2116–2123. Available from: https://doi.org/10.1016/j.chemosphere.2013.07.057.

Blair, B., Nikolaus, A., Hedman, C., Klaper, R., Grundl, T., 2015. Evaluating the degradation, sorption, and negative mass balances of pharmaceuticals and personal care products during wastewater treatment. Chemosphere 134, 395–401. Available from: https://doi.org/10.1016/j.chemosphere.2015.04.078.

Boshir, M., Zhou, J.L., Hao, H., et al., 2017. Progress in the biological and chemical treatment technologies for emerging contaminant removal from wastewater: a critical review. J. Hazard. Mater. 323, 274–298. Available from: https://doi.org/10.1016/j.jhazmat.2016.04.045.

Carmalin Sophia, A., Lima, E.C., 2018. Removal of emerging contaminants from the environment by adsorption. Ecotoxicol. Environ. Saf. 150, 1–17. Available from: https://doi.org/10.1016/j.ecoenv.2017.12.026.

Carmona, E., Andreu, V., Picó, Y., 2014. Occurrence of acidic pharmaceuticals and personal care products in Turia River Basin: from waste to drinking water. Sci. Total Environ. 484 (1), 53–63. Available from: https://doi.org/10.1016/j.scitotenv.2014.02.085.

Carr, D.L., Morse, A.N., Zak, J.C., Anderson, T.A., 2011. Biological degradation of common pharmaceuticals and personal care products in soils with high water content. Water Air Soil Pollut. 217, 127–134.

Casas, M.E., Bester, K., 2015. Can those organic micro-pollutants that are recalcitrant in activated sludge treatment be removed from wastewater by biofilm reactors (slow sand filters)? Sci. Total Environ. 506–507, 315–322.

Cavalheiro, J., Prieto, A., Monperrus, M., Etxebarria, N., Zuloaga, O., 2013. Determination of polycyclic and nitro musks in environmental water samples by means of microextraction by packed sorbents coupled to large volume injection-gas chromatography–mass spectrometry analysis. Anal. Chim. Acta 773, 68–75. Available from: https://doi.org/10.1016/j.aca.2013.02.036.

Chaplin, B.P., 2014. Critical review of electrochemical advanced oxidation processes for water treatment applications. Environ. Sci. Processes Impacts 16 (6), 1182–1203.

Clément, L., Hurel, C., Marmier, N., 2013. Toxicity of TiO_2 nanoparticles to cladocerans, algae, rotifers and plants – effects of size and crystalline structure. Chemosphere 90 (3), 1083–1090.

Croce, R., Cinà, F., Lombardo, A., Crispeyn, G., 2017. Aquatic toxicity of several textile dye formulations: acute and chronic assays with *Daphnia magna* and *Raphidocelis subcapitata*. Ecotoxicol. Environ. Saf. 144, 79–87.

Cycoń, M., Mrozik, A., Piotrowska-Seget, Z., 2017. Bioaugmentation as a strategy for the remediation of pesticide-polluted soil: a review. Chemosphere 172, 52–71.

Dewil, R., Mantzavinos, D., Poulios, I., Rodrigo, M.A., 2017. New perspectives for advanced oxidation processes. J. Environ. Manage. 195, 93–99. Available from: https://doi.org/10.1016/j.jenvman.2017.04.010.

Elisandra, G., Napoleão, D.C., Oliveira, C. De, et al., 2018. Degradation of textile dyes Remazol Yellow Gold and reactive Turquoise: optimization, toxicity and modeling by artificial neural networks. Spectrochim. Acta, A: Mol. Biomol. Spectrosc. 1–12.

Emnet, P., Gaw, S., Northcott, G., Storey, B., Graham, L., 2015. Personal care products and steroid hormones in the Antarctic coastal environment associated with two Antarctic research stations, McMurdo Station and Scott Base. Environ. Res. 136, 331–342. Available from: https://doi.org/10.1016/j.envres.2014.10.019.

De Figuerêdo, L.P., Nilin, J., Da Silva, A.Q., Loureiro, S., Costa-Lotufo, L.V., 2016. Development of a short-term chronic toxicity test with a tropical mysid. Mar. Pollut. Bull. 106 (1–2), 104–108. Available from: https://doi.org/10.1016/j.marpolbul.2016.03.020.

Garcia-Rodríguez, A., Matamoros, V., 2014. The ability of biologically based wastewater treatment systems to remove emerging organic contaminants – a review. Environ. Sci. Pollut. Res. 21, 11708–11728.

Gogoi, A., Mazumder, P., Kumar, V., et al., 2018. Occurrence and fate of emerging contaminants in water environment: a review. Groundwater Sustain. Dev. 6, 169–180. Available from: https://doi.org/10.1016/j.gsd.2017.12.009.

Gottschall, N., Topp, E., Metcalfe, C., et al., 2012. Pharmaceutical and personal care products in groundwater, subsurface drainage, soil, and wheat grain, following a high single application of municipal biosolids to a field. Chemosphere 87 (2), 194–203. Available from: https://doi.org/10.1016/j.chemosphere.2011.12.018.

Haddad, M., Abid, S., Hamdi, M., Bouallagui, H., 2018. Reduction of adsorbed dyes content in the discharged sludge coming from an industrial textile wastewater treatment plant using aerobic activated sludge process. J. Environ. Manag. 223, 936–946. Available from: https://doi.org/10.1016/j.jenvman.2018.07.009.

Han, J., Ji, C., Guo, Y., et al., 2017. Mechanisms underlying melatonin-mediated prevention of fenvalerate-induced behavioral and oxidative toxicity in zebrafish. J. Toxicol. Environ. Health, A 1–11. Available from: https://doi.org/10.1080/15287394.2017.1384167.

Hassan, S.H.A., Ginkel, S. W. Van, Hussein, M.A.M., Abskharon, R., Oh, S.E., 2016. Toxicity assessment using different bioassays and microbial biosensors. Environ. Int. 92–93, 106–118. Available from: https://doi.org/10.1016/j.envint.2016.03.003.

Hernández, V.A., Galleguillos, F., Thibaut, R., Müller, A., Hernández, V.A., 2018. Fungal dyes for textile applications: testing of industrial conditions for wool fabrics dyeing. J. Text. Inst. 1–6. Available from: https://doi.org/10.1080/00405000.2018.1460037.

Hidalgo, N., Mangiameli, G., Manzano, T., et al., 2011. Oxidation and removal of industrial textile dyes by a novel peroxidase extracted from post-harvest lentil (L.) stubble. Biotechnol. Bioprocess. Eng. 829, 821–829.

IMS. 2013. Global Pharmaceutical Sales, MIDAS database.

Jallow, M.F.A., Awadh, D.G., Albaho, M.S., Devi, V.Y., Thomas, B.M., 2017. Pesticide risk behaviors and factors influencing pesticide use among farmers in Kuwait. Sci. Total Environ. 574, 490–498. Available from: https://doi.org/10.1016/j.scitotenv.2016.09.085.

Jiang, L., Li, K., Yan, D., et al., 2020. Toxicity assessment of 4 azo dyes in zebrafish embryos. Int. J. Toxicol. 1, 1–9.

Kanaujiya, D.K., Paul, T., Sinharoy, A., Pakshirajan, K., 2019. Biological treatment processes for the removal of organic micropollutants from wastewater: a review. Curr. Pollut. Rep. 5 (3), 112–128.

Khan, J.A., Sayed, M., Khan, S., Shah, N.S., 2020. Advanced Oxidation Processes for the Treatment of Contaminants of Emerging Concern. Elsevier Inc.

Kumar, P.S., Saravanan, A., 2017. Sustainable wastewater treatments in textile sector. Sustain. Fibres Text. 323–346. Available from: https://doi.org/10.1016/B978-0-08-102041-8.00011-1.

Lamastra, L., Balderacchi, M., Trevisan, M., 2016. Inclusion of emerging organic contaminants in groundwater monitoring plans. MethodsX 3 (16), 459–476. Available from: https://doi.org/10.1016/j.mex.2016.05.008.

Lee, I., Lee, S., Oh, J., 2010. Occurrence and fate of synthetic musk compounds in water environment. Water Res. 44 (1), 214–222. Available from: https://doi.org/10.1016/j.watres.2009.08.049.

Lewis, K.A., Tzilivakis, J., Warner, D.J., Green, A., 2016. An international database for pesticide risk assessments and management. Hum. Ecol. Risk Assess. 7039, 1050–1064.

Lin, T., Yu, S., Chen, W., 2016. Occurrence, removal and risk assessment of pharmaceutical and personal care products (PPCPs) in an advanced drinking water treatment plant (ADWTP) around Taihu Lake in China. Chemosphere 152, 1–9. Available from: https://doi.org/10.1016/j.chemosphere.2016.02.109.

Malik, N.H., Zain, H., Ali, N., 2018. Organismic-level acute toxicology profiling of reactive azo dyes. Environ. Monit. Assess. 190, 612.

Mansano, A.S., Moreira, R.A., Dornfeld, H.C., Freitas, E.C., Vieira, E.M., 2017. Effects of diuron and carbofuran and their mixtures on the microalgae *Raphidocelis subcapitata*. Ecotoxicol. Environ. Saf. 142, 312–321. Available from: https://doi.org/10.1016/j.ecoenv.2017.04.024.

Martins, N., Pereira, R., Abrantes, N., et al., 2012. Ecotoxicological effects of ciprofloxacin on freshwater species: data integration and derivation of toxicity thresholds for risk assessment. Ecotoxicology 21, 1167–1176.

Matamoros, V., Salvadó, V., 2012. Evaluation of the seasonal performance of a water reclamation pond-constructed wetland system for removing emerging contaminants. Chemosphere 86, 111–117.

Meador, J.P., Yeh, A., Young, G., Gallagher, E.P., 2016. Contaminants of emerging concern in a large temperate estuary. Environ. Pollut. 213, 254–267. Available from: https://doi.org/10.1016/j.envpol.2016.01.088.

Medeiros, L.S., Souza, J.P., Winkaler, E.U., Carraschi, S.P., 2013. Acute toxicity and environmental risk of teflubenzuron to *Daphnia magna*, *Poecilia reticulata* and *Lemna minor* in the absence and presence of sediment. J. Environ. Sci. Health, B: Pestic. Food Contam. Agric. Wastes 48 (7), 600–606.

Menz, J., Schneider, M., Kümmerer, K., 2013. Toxicity testing with luminescent bacteria - Characterization of an automated method for the combined assessment of acute and chronic effects. Chemosphere 93 (6), 990–996. Available from: https://doi.org/10.1016/j.chemosphere.2013.05.067.

Minguez, L., Bureau, R., Halm-Lemeille, M.-P., 2018. Joint effects of nine antidepressants on *Raphidocelis subcapitata* and *Skeletonema marinoi*: a matter of amine functional groups. Aquat. Toxicol. 196, 117–123. Available from: http://linkinghub.elsevier.com/retrieve/pii/S0166445X18300237.

Mondal, P., Baksi, S., Bose, D., 2017. Study of environmental issues in textile industries and recent wastewater treatment technology. World Sci. N. 61 (2), 98–109.

Montes-Grajales, D., Fennix-Agudelo, M., Miranda-Castro, W., 2017. Occurrence of personal care products as emerging chemicals of concern in water resources: a review. Sci. Total Environ. 595, 601–614. Available from: https://doi.org/10.1016/j.scitotenv.2017.03.286.

Moreira, F.C., Boaventura, R.A.R., Brillas, E., Vilar, V.J.P., 2017. Electrochemical advanced oxidation processes: a review on their application to synthetic and real wastewaters. Appl. Catal. B: Environ. 202, 217–261.

Morillo, E., Villaverde, J., 2017. Advanced technologies for the remediation of pesticide-contaminated soils. Sci. Total Environ. 586, 576–597. Available from: https://doi.org/10.1016/j.scitotenv.2017.02.020.

Noguera-Oviedo, K., Aga, D.S., 2016. Lessons learned from more than two decades of research on emerging contaminants in the environment. J. Hazard. Mater. 316, 242–251. Available from: https://www.sciencedirect.com/science/article/pii/S0304389416303995.

Ocampo-Pérez, R., Rivera-Utrilla, J., Gómez-Pacheco, C., Sánchez-Polo, M., López-Peñalver, J.J., 2012. Kinetic study of tetracycline adsorption on sludge-derived adsorbents in aqueous phase. Chem. Eng. J. 213, 88–96.

De Oliveira, L.L.D., Antunes, S.C., Gonçalves, F., Rocha, O., Nunes, B., 2016. Acute and chronic ecotoxicological effects of four pharmaceuticals drugs on cladoceran *Daphnia magna*. Drug. Chem. Toxicol. 39 (1), 13–21.

Patel, M., Kumar, R., Kishor, K., et al., 2019. Pharmaceuticals of emerging concern in aquatic systems: chemistry, occurrence, effects, and removal methods. Chem. Rev. 119, 3510–3673.

Paździor, K., Bilińska, L., Ledakowicz, S., 2018. A review of the existing and emerging technologies in the combination of AOPs and biological processes in industrial textile wastewater treatment. Chem. Eng. J. 376, 120–597. Available from: https://doi.org/10.1016/j.cej.2018.12.057.

Peng, Y., Luo, Y., Nie, X.P., 2013. Toxic effects of Triclosan on the detoxification system and breeding of *Daphnia magna*. Ecotoxicology 22 (9), 1384–1394.

Petrie, B., Barden, R., Kasprzyk-Hordern, B., 2014. A review on emerging contaminants in wastewaters and the environment: current knowledge, understudied areas and recommendations for future monitoring. Water Res. Available from: https://doi.org/10.1016/j.watres.2014.08.053.

Pintado-Herrera, M.G., González-Mazo, E., Lara-Martín, P.A., 2013. Environmentally friendly analysis of emerging contaminants by pressurized hot water extraction – stir bar sorptive extraction – derivatization and gas chromatography – mass spectrometry. Anal. Bioanal. Chem. 405, 401–411.

Pinto, C.F., Antonelli, R., Araújo, K.S., De, Fornazari, A.L.D.E.T., 2017. Experimental-design-guided approach for the removal of atrazine by sono-electrochemical-UV-chlorine techniques. Environ. Technol. (UK) 3330, 1–11.

Przystaś, W., Zabłocka-Godlewska, E., Grabińska-Sota, E., 2018. Efficiency of decolorization of different dyes using fungal biomass immobilized on different solid supports. Braz. J. Microbiol. 49 (2), 285–295.

Richardson, S.D., Kimura, S.Y., 2017. Emerging environmental contaminants: challenges facing our next generation and potential engineering solutions. Environ. Technol. Innov. 8, 40–56. Available from: https://doi.org/10.1016/j.eti.2017.04.002.

Rivera-Utrilla, J., Sánchez-Polo, M., Ferro-García, M.Á., Prados-Joya, G., Ocampo-Pérez, R., 2013. Pharmaceuticals as emerging contaminants and their removal from water. A review. Chemosphere 93, 1268–1287.

Roberts, J., Kumar, A., Du, J., et al., 2016. Pharmaceuticals and personal care products (PPCPs) in Australia's largest inland sewage treatment plant, and its contribution to a major Australian river during high and low flow. Sci. Total Environ. 541, 1625–1637.

Rodrigues, E.T., Varela, A.T., Pardal, M.A., Oliveira, P.J., 2019. Cell-based assays seem not to accurately predict fish short-term toxicity of pesticides. Environ. Pollut. 252, 476–482. Available from: https://doi.org/10.1016/j.envpol.2019.05.033.

Rodriguez-Narvaez, O.M., Peralta-Hernandez, J.M., Goonetilleke, A., Bandala, E.R., 2017. Treatment technologies for emerging contaminants in water: a review. Chem. Eng. J. 323, 361–380. Available from: https://doi.org/10.1016/j.cej.2017.04.106.

Rouches, E., Herpoël-Gimbert, I., Steyer, J.P., Carrere, H., 2016. Improvement of anaerobic degradation by white-rot fungi pretreatment of lignocellulosic biomass: a review. Renew. Sustain. Energy Rev. 59, 179–198. Available from: https://doi.org/10.1016/j.rser.2015.12.317.

Roy, D.C., Biswas, S.K., Saha, A.K., Sikdar, B., 2018. Biodegradation of Crystal Violet dye by bacteria isolated from textile industry effluents. PeerJ 1–15.

Saha, B.C., Qureshi, N., Kennedy, G.J., Cotta, M.A., 2016. Biological pretreatment of corn stover with white-rot fungus for improved enzymatic hydrolysis. Int. Biodeterior. Biodegrad. 109, 29–35. Available from: https://doi.org/10.1016/j.ibiod.2015.12.020.

Salazar, R., Ureta-Zañartu, M.S., González-Vargas, C., Nascimento, C., Martinez-Huitle, C.A., 2017. Electrochemical degradation of industrial textile dye disperse yellow 3: role of electrocatalytic material and experimental conditions on the catalytic production of oxidants and oxidation pathway. Chemosphere 198, 21–29.

Salomão, G.R., Américo-Pinheiro, J.H.P., Isique, D., et al., 2019. Diclofenac removal in water supply by adsorption on composite low-cost material. Environ. Technol. 1–47. Available from: https://doi.org/10.1080/09593330.2019.1692078.

Seeland, A., Oetken, M., Kiss, A., Fries, E., Oehlmann, J., 2012. Acute and chronic toxicity of benzotriazoles to aquatic organisms. Environ. Sci. Pollut. Res. 19 (5), 1781–1790.

Sharma, B.M., Bečanová, J., Scheringer, M., et al., 2019. Health and ecological risk assessment of emerging contaminants (pharmaceuticals, personal care products, and artificial sweeteners) in surface and groundwater (drinking water) in the Ganges River Basin, India. Sci. Total Environ. 646, 1459–1467.

Silva, M.C., Torres, J.A., Vasconcelos De SÁ, L.R., Chagas, P.M.B., 2013. The use of soybean peroxidase in the decolourization of Remazol Brilliant Blue R and toxicological evaluation of its degradation products. J. Mol. Catal. B: Enzym. 89, 122–129. Available from: https://doi.org/10.1016/j.molcatb.2013.01.004.

Sirés, I., Brillas, E., Oturan, M.A., Rodrigo, M.A., Panizza, M., 2014. Electrochemical advanced oxidation processes: today and tomorrow. A review. Environ. Sci. Pollut. Res. 21, 8336–8367.

Sopaj, F., Rodrigo, M.A., 2015. Influence of the anode materials on the electrochemical oxidation efficiency. Application to oxidative degradation of the pharmaceutical amoxicillin. Chem. Eng. J. 262, 286–294.

Sui, Q., Huang, J., Deng, S., Yu, G., Fan, Q., 2010. Occurrence and removal of pharmaceuticals, caffeine and DEET in wastewater treatment plants of Beijing, China. Water Res. 44 (2), 417–426. Available from: https://doi.org/10.1016/j.watres.2009.07.010.

Taheran, M., Naghdi, M., Brar, S.K., Verma, M., Surampalli, R.Y., 2018. Emerging contaminants: here today, there tomorrow!. Environ. Nanotechnol. Monit. Manage. 10, 122–126. Available from: https://doi.org/10.1016/j.enmm.2018.05.010.

Tamura, I., Yasuda, Y., Kagota, K., Yoneda, S., Nakada, N., 2017. Contribution of pharmaceuticals and personal care products (PPCPs) to whole toxicity of water samples collected in effluent-dominated urban streams. Ecotoxicol. Environ. Saf. 144, 338–350. Available from: https://doi.org/10.1016/j.ecoenv.2017.06.032.

Tang, A.Y.L., Lo, K.Y., Kan, C., 2018. Textile dyes and human health: a systematic and citation network analysis review. Color. Technol. 134, 245–257.

Tohidi, F., Cai, Z., 2017. Fate and mass balance of triclosan and its degradation products: comparison of three different types of wastewater treatments and aerobic/anaerobic sludge digestion. J. Hazard. Mater. 323, 329–340. Available from: https://doi.org/10.1016/j.jhazmat.2016.04.034.

Torres, N.H., Souza, B.S., Ferreira, L.F.R., et al., 2019. Real textile effluents treatment using coagulation/flocculation followed by electrochemical oxidation process and ecotoxicological assessment. Chemosphere 236, 124–309.

Tran, N.H., Li, J., Hu, J., Ong, S.L., 2014. Occurrence and suitability of pharmaceuticals and personal care products as molecular markers for raw wastewater contamination in surface water and groundwater. Environ. Sci. Pollut. Res. 21, 4727–4740.

Tran, N.H., Reinhard, M., Gin, K.Y.H., 2018. Occurrence and fate of emerging contaminants in municipal wastewater treatment plants from different geographical regions-a review. Water Res. 15.

Tsui, M.M.P., Leung, H.W., Lam, P.K.S., Murphy, M.B., 2014. Seasonal occurrence, removal efficiencies and preliminary risk assessment of multiple classes of organic UV filters in wastewater treatment plants. Water Res. 53, 58–67. Available from: https://doi.org/10.1016/j.watres.2014.01.014.

Vilar, D.S., Carvalho, G.O., Pupo, M.M.S., et al., 2018. Vinasse degradation using *Pleurotus sajor-caju* in a combined biological – electrochemical oxidation treatment. Sep. Purif. Technol. 192, 287–296.

Wu, L., Liang, D., Xu, Y., et al., 2016. A robust and cost-effective integrated process for nitrogen and bio-refractory organics removal from landfill leachate via short-cut nitrification, anaerobic ammonium. Bioresour. Technol. 212, 296–301. Available from: https://www.sciencedirect.com/science/article/pii/S0960852416305284.

Yang, S., Mcdonald, J., Hai, F.I., et al., 2016. Occurrence of trace organic contaminants in wastewater sludge and their removals by anaerobic digestion. Bioresour. Technol. 210, 153–159. Available from: https://doi.org/10.1016/j.biortech.2015.12.080.

Yang, Y., Sik, Y., Kim, K., Kwon, E.E., Fai, Y., 2017. Occurrences and removal of pharmaceuticals and personal care products (PPCPs) in drinking water and water / sewage treatment plants: a review. Sci. Total Environ. 596–597, 303–320. Available from: https://doi.org/10.1016/j.scitotenv.2017.04.102.

Yegin, Y., Yegin, C., Oh, J.K., Orr, A., Zhang, M., 2017. Ecotoxic effects of paclitaxel-loaded nanotherapeutics on freshwater algae, *Raphidocelis subcapitata* and *Chlamydomonas reinhardtii*. Environ. Sci.: Nano 4 (5), 1077–1085. Available from: http://xlink.rsc.org/?DOI = C6EN00332J.

Zhao, X., Chen, Z., Wang, X., 2014. PPCPs removal by aerobic granular sludge membrane bioreactor. Appl. Microbiol. Biotechnol. 98, 9843–9848.

Chapter 2

Role of bacteria and mycorrhizal fungi in phytomining: status and future perspectives

Ana R.A. Alves[1], Rui S. Oliveira[2], Eduardo F. Silva[1] and Luís A.B. Novo[3]

[1]*Department of Geosciences, GeoBioTec, University of Aveiro, Aveiro, Portugal,* [2]*Department of Life Sciences, Centre for Functional Ecology, University of Coimbra, Coimbra, Portugal,* [3]*Scotland's Rural College, Edinburgh, United Kingdom*

2.1 Introduction

Population growth and the expectancy of improved living standards lead to the rising demand of chemical elements, since they are fundamentally linked to industrial development (Whiting et al., 2002). Mine sites are the primary source for recovering valuable metals suited for applications, such as clean energy technologies and medicinal industry. However, conventional methods of extraction take a negative toll on the environment, producing large amounts of mine waste, causing water and soil contamination, destruction of habitats, and loss of biodiversity (Thijs et al., 2017). Commercial mining also requires large capital investments and the existence of easily accessible ores that possess a high concentration of target metals. Ore bodies of this nature only occur in small localized areas and are declining due to its disarrayed extraction (Sheoran et al., 2009). The issues associated with conventional mining and the need of a global shift toward sustainable development have prompted the advance of phytomining during the last three decades. Phytomining is an alternative technology based on the use of certain plants to recover valuable metals from mineralized or polluted soils (Brooks et al., 1998). This innovative, solar-driven, environment–friendly, and low-cost approach offers the possibility of exploiting resources from low-grade ores, which are subeconomic for conventional mining. Moreover, phytomining is virtually harmless to the environment, promotes its remediation, and is aesthetically pleasing, thus having high probability of public acceptance (Nedelkoska and Doran, 2000; Novo et al., 2017). Plants used for phytomining can accumulate astonishing amounts of metals, such as nickel (Ni), cobalt (Co), thallium (Tl), and rhenium (Re), in their shoots without showing significant signs of toxicity (Novo et al., 2018a,b; van der Ent et al., 2013). These plants, called hyperaccumulators, constitute an example of the curious and outstanding adaptations of flora to metalliferous environments, which are adverse to normal plants (Whiting et al., 2002). Species can be classified as hyperaccumulators if their shoot concentrations surpass the thresholds defined for each element (van der Ent et al., 2013). Nickel hyperaccumulators, for instance, are defined as plants that accumulate Ni in their dry leaf tissues to concentrations surpassing 1000 mg/kg, that is, 0.1% of their leaf dry mass (Brooks et al., 1977). The basic mechanisms governing hyperaccumulation are a high rate of metal uptake from the soil, an efficient root-to-shoot translocation, and a great ability to detoxify and sequester metals in leaves (Rascio and Navari-Izzo, 2011).

The phytomining process consists in the cultivation of hyperaccumulators (preferably native or naturalized noninvasive species) in selected metal-rich substrates, followed by the harvest of the aboveground parts and transformation of the biomass to obtain valuable end products (Kidd et al., 2018; Novo et al., 2017). There are two main approaches to be considered for phytomining: (1) phytomining on degraded mine soils and quarries, as part of a simultaneous rehabilitation strategy (along with phytoremediation), where the revenue obtained through phytomining finances its costs and (2) phytomining in natural metal–rich soils unsuitable for conventional agriculture and grazing (e.g., serpentine soils) as part of an integrated agricultural chain known as agromining (van der Ent et al., 2015). For this phytotechnology to be economically profitable, high plant biomass yield and elevated metal concentrations in aerial plant parts are required, since these aspects govern the quantity of metal to be harvested from each plant (harvestable mount) (Novo et al.,

Microbe Mediated Remediation of Environmental Contaminants. DOI: https://doi.org/10.1016/B978-0-12-821199-1.00002-X

2015). Nickel phytomining is particularly promising due to the high market price of this element, the great variety of Ni hyperaccumulator species (which comprise nearly 90% of the reported hyperaccumulators), and the abundance of Ni-rich ultramafic soils worldwide (van der Ent et al., 2015). Furthermore, there are numerous studies on Ni phytomining showing its great economic potential (Nkrumah et al., 2016, 2019; Novo et al., 2017). Nonetheless, there are limitations to the application of this phytotechnology due to slow growth of hyperaccumulators, low biomass production, shallow root systems, and different factors influencing metal uptake such as the solubility and availability of elements in soil (Ghosh and Singh, 2005). Thus it is imperative to study agronomical practices for optimization of phytomining systems, which may lead to the improvement of metal extraction, and thereby to greater economic profits.

Microbial-assisted phytomining is a relatively recent concept that consists in the incorporation of microbial inoculants into phytomining systems, for overcoming limitations in this process. The rhizosphere of all plants, including hyperaccumulators, is inhabited by a high diversity of microorganisms, some of which are shown to have the ability for assisting plant growth and regulating metal uptake (Thijs et al., 2017). Microorganisms play a very important part in nutrient uptake and transport, attenuate stress, and protect the plant from pathogens (Lugtenberg and Kamilova, 2009; Rashid et al., 2016). Serpentine soil microbes, such as bacteria and fungi, are considered important drivers of ecosystem diversity, often featuring critical roles in soil nutrient cycling and stimulating plant growth. It should be noted that serpentinic environments are poor in nutrients essential to plants [such as nitrogen (N), phosphorus (P), and potassium], and organic matter, and rich in metals (such as Ni, Co, and chromium) (Alves et al., 2019; Benizri and Kidd, 2018; Kidd et al., 2018). Microorganisms can also contribute for stress alleviation in plants, metal tolerance, mobilization, and uptake (Thijs et al., 2017).

The characterization and study of the rhizosphere microbiome of Ni hyperaccumulators is crucial for assessing endemic soil biological- and ecological-friendly phytomining techniques (Sheoran et al., 2016; Thijs et al., 2017). Plant–microbe interactions play an important role in plant biomass yield and metal accumulation, and, therefore, the application of inoculants into phytomining systems may overcome several limitations in this process. In addition, the use of inoculants can minimize the application of other agronomic optimizing agents, such as inorganic fertilizers, organic compounds, phytohormones, and chelating substances, and consequently reduce the corresponding costs (Álvarez-López et al., 2016; Thijs et al., 2017). This chapter overviews the status of microbial-assisted phytomining, focusing on the roles of plant growth–promoting rhizobacteria (PGPR) and mycorrhizal fungi, and their potential to enhance Ni phytomining.

2.2 Rhizosphere and its microorganisms

The rhizosphere is defined as the volume of soil adjacent to living plant roots that is physically and chemically influenced by root activity and growth (Benizri and Kidd, 2018; Nguyen, 2003). It constitutes a very dynamic microsystem in which roots, microorganisms, and soil interact (Thijs et al., 2017). There is a high microbial diversity and activity in the rhizosphere that contrasts with the microbial communities inhabiting nonrhizosphere or bulk soil (Baudoin et al., 2003). This is a result of the overall plant–rhizosphere microbes interaction and of several biotic and abiotic factors, such as root exudation, inorganic nutrients, competition for iron (Fe) and carbon (C), soil pH, and temperature (Baudoin et al., 2003; Benizri and Kidd, 2018; Khan, 2005). Root exudates are especially important for rhizosphere microorganisms because they represent a source of C (and possibly N) and energy, driving microbial activity, diversity, and growth (Bowen and Rovira, 1999). Thereby, compared to bulk soil the rhizosphere is characterized by higher concentrations of nutrients and labile organic C, representing a unique "hot spot" in terms of microbial ecology (Hinsinger et al., 2006). It was estimated that the rhizosphere contains up to 10^{11} microbial cells per gram soil, from thousands of different species, comprising bacteria, fungi, and archaea (Mendes et al., 2013). The important flux of C in the rhizosphere results of large amounts of this element that is transferred to the root and lost through rhizodeposition in the soil (Nguyen, 2003). There are five types of rhizodeposits: secretions, diffusates, lysates, gases, and mucilage (Nguyen, 2003). Rhizodeposits serve as a nutrient source and are key determinants of rhizosphere microbial community structure, since the type and composition of secretions are responsible for altering soil microbial diversity (Chaparro et al., 2012). Plants provide vitamins, hormones, toxins, and antagonistic compounds that have a selective effect on microorganisms, favoring species beneficial to plants, that promote health and productivity (Benizri and Kidd, 2018; Warembourg, 1997). Thus, while plants support microbial growth, bacteria and fungi promote plant development by improving nutrient uptake, attenuating abiotic stress, and protecting the plant from pathogens (Lugtenberg and Kamilova, 2009; Rashid et al., 2016). Fig. 2.1 displays the main benefits of plant growth–promoting microorganisms.

Rhizosphere microorganism drives the cycle of important elements (such as C, N, and P) through their enzyme activity, affecting soil structure and functions (Benizri and Kidd, 2018). They are often responsible for increasing the

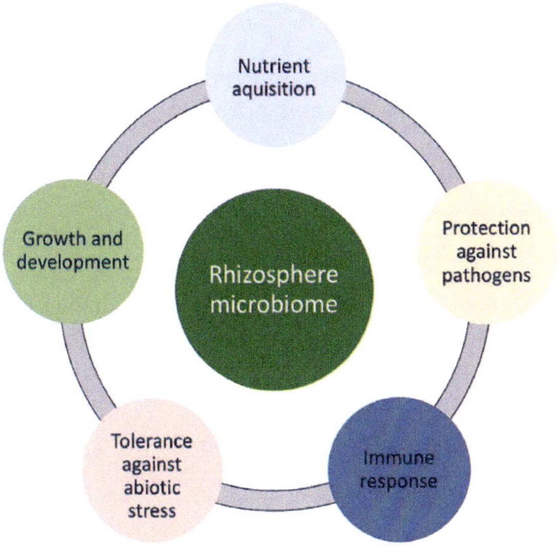

FIGURE 2.1 Main functions of plant beneficial rhizosphere microorganisms.

availability of important plant macronutrients and the extent of the root system, leading, in both cases, to the improvement of nutrient uptake by plants. Rhizosphere beneficial microbiome may also influence plant–pathogen interactions and plant growth, as previously mentioned, by several mechanisms, including the production of hormones and antibiotics (Benizri and Kidd, 2018).

2.2.1 Rhizosphere microbiome of hyperaccumulator plants

Despite the occurrence of high concentrations of metals in soils, the rhizosphere of hyperaccumulator plants is inhabited by a large amount of plant beneficial microorganisms (Abou-Shanab et al., 2003a,b). These microorganisms are responsible for assisting plants in tolerating high metal toxicity (Manoj et al., 2020). Natural metal-rich sites, like ultramafic soils, constitute an abiotic factor that modulates microbial activity in the rhizosphere (Kidd et al., 2018). Besides, it is known that Ni hyperaccumulator species select metal-tolerant microbial strains that are linked to metal hyperaccumulation, with some unique microbial groups detected in large quantities at these metal-rich sites (Abou-Shanab et al., 2003a,b; Mengoni et al., 2001; Thijs et al., 2017).

The beneficial effects of rhizosphere microbiome on plant growth, functions, physiology, and efficiency in metal accumulation have made the study of these microorganisms very attractive during the last couple of decades (Ma et al., 2011; Weyens et al., 2009). However, there is still little information about the diversity and functions of microbial communities in metal-rich sites (Kidd et al., 2018). Microorganisms can have a large effect on phytoremediation and phytomining and, therefore, have the potential to be used as inoculants to improve the efficiency of these phytotechnologies. Reports show that PGPR and fungal symbionts contribute for plant growth and metal tolerance in ultramafic environments and enhance Ni bioavailability and accumulation by hyperaccumulator plants (Idris et al., 2004; Kidd et al., 2018).

2.3 Plant growth–promoting rhizobacteria

PGPR are plant-associated, free-living, soilborne bacteria that are responsible for enhancing plant growth (Manoj et al., 2020). In order to classify as PGPR, bacteria must be root colonizers and capable of surviving, multiplying, adapting, and competing with other microorganisms until expressing plant growth promotion (Kloepper, 1994). PGPR comprise a great number of bacterial genera that can be categorized according to their functional properties as: (1) biofertilizers (increase availability of nutrients to the plant); (2) phytostimulators (promote plant growth through the production of hormones); (3) rhizoremediators (break down organic contaminants and modulate metal solubilization); and (4) biopesticides (control diseases and plant pathogens by synthetizing antibiotics and antifungal metabolites) (Somers et al., 2004). PGPR may also have the capacity to increase metal availability and accumulation by producing compounds such as siderophores, organic acids, and biosurfactants (Sessitsch et al., 2013). The compounds can acidify the rhizosphere, decrease soil pH, generate oxidation–reduction reactions, and alter the mobility of metals trough chelation, precipitation, and immobilization (Sessitsch et al., 2013). Microbial siderophores are considered the major metal chelating agents in metal solubilization (Ahemad, 2014).

PGPR can influence plant growth through direct and indirect mechanisms (Table 2.1). The former consists in the promotion of plant growth by facilitating nutrient uptake or via synthesis of phytohormones (Novo et al., 2018a,b). Bacteria strains can stimulate root growth, cell division, and elongation and boost metabolic stabilization (Manoj et al., 2020). Indirect mechanisms are responsible for reducing or suppressing plant pathogens by producing antagonistic compounds, improving induction of systemic resistance, and suppressing metal toxicity in the case of plants that inhabit metal-rich soils (Saraf et al., 2014). Indirect mechanisms are not involved directly in plant growth but play a role in reducing plant stress caused by biotic (e.g., pathogenic attacks) and abiotic (e.g., heavy metal toxicity) factors (Manoj et al., 2020). PGPR direct mechanisms consist in N fixation, phosphate solubilization, Fe sequestration, and in the production of phytohormones (such as auxins, cytokinins, and gibberellins) and modulation. Indirect mechanisms comprise biocontrol via the production of compounds known as allelochemicals (such as antibiotics, lytic enzymes, and siderophores) and in eliciting induced systemic resistance (ISR) (Glick, 2012; Novo et al., 2018a,b).

2.3.1 Nitrogen fixation, phosphate solubilization, and iron sequestration

PGPR can increase availability of nutrients that are important to plants such as N, phosphate, and Fe. Nitrogen has an important role in plant growth and development; however, plants are unable to utilize atmospheric N_2 directly (Novo et al., 2018a,b). Nitrogen-fixing bacteria are responsible for continuously converting atmospheric N_2 into plant-available forms such as ammonia or nitrate. This process is carried out by a complex enzyme system called nitrogenase (Hoffman et al., 2014). Phosphorus also constitutes a very important macronutrient for plants and although it is relatively abundant, it is generally in an insoluble form, not available to plants (Glick, 2012). Bacteria play an important role in P solubilization, supplying bioavailable P to plants (Novo et al., 2018a,b). This process occurs through the secretion of enzymes (such as phosphatases and C–P lyases) and low-molecular-weight organic acids (such as acetic, citric, lactic, and gluconic) (Goswami et al., 2016). Some PGPR are able to produce low-molecular-weight compounds known as siderophores (Novo et al., 2018a,b). Siderophores are proteins that exhibit a very high affinity for Fe^{3+} (Rajkumar et al., 2010). They are responsible for converting Fe^{3+}, which is unavailable for plants, into Fe^{2+} that is available for plant uptake and required for several metabolic activities (Manoj et al., 2020). Siderophores are also able to bind to a wide range of metals (besides Fe), thus enhancing their availability for plant uptake, and potentially improving metal hyperaccumulation (Sheoran et al., 2016).

2.3.2 Phytohormone production and modulation

PGPR can provide plants with a variety of phytohormones, including, auxins, cytokinins, and gibberellins, and produce enzymes that reduce ethylene levels (Goswami et al., 2016; Manoj et al., 2020). These substances enhance plant growth, even at low concentrations, by modulating several physiological and metabolic functions of the plant under metal stress (Manoj et al., 2020). Phytohormones synthetized by PGPR may prevent detrimental effects of metals in plants inhabiting metal-rich sites and consequently increase metal accumulation (Goswami et al., 2016; Ma et al., 2016b).

Auxins possess a valuable role in root system development and PGPR synthetize several forms of auxins, with the most important being indole-3-acetic acid (IAA) (Manoj et al., 2020). This auxin is responsible for stimulating root growth by cell division and tissue expansion and stimulates seed and tuber germination, among other beneficial effects (Goswami et al., 2016; Novo et al., 2018a,b). Cytokinins are the major bacterial phytohormones after auxins and can improve plant growth by, for instance, eliciting cell division and differentiation and enhancing leaf expansion (Bhattacharyya and Jha, 2012). Gibberellins are able to promote shoot growth by cell elongation, division, and differentiation (Wang et al., 2015). Lastly,

TABLE 2.1 Plant growth–promoting rhizobacteria mechanisms.

Mechanism type	Function	Synthesized compound
Direct	Nitrogen fixation	Nitrogenase
	Phosphate solubilization	Low-molecular-weight organic acids, phosphatases, C–P lyases
	Iron sequestration	Siderophores
	Phytohormones production and modulation	Auxins, cytokinins, gibberellins, ethylene
Indirect	Biocontrol	Antibiotics, lytic enzymes, siderophores

PGPR are also responsible for producing an enzyme called 1-aminocyclopropane-1-carboxylate (ACC) deaminase that regulates ethylene levels in plants (Ma et al., 2016a). Ethylene is a phytohormone that plays an important role in plant growth and development and its synthesis depends on ACC synthase. However, under stress conditions, such as metal toxicity, plants end up producing extreme amounts of ethylene, which can negatively affect, for instance, plant metabolisms and performance (Manoj et al., 2020). Therefore ACC deaminase is crucial to normalize ethylene levels and protect plants (Ma et al., 2016a).

2.3.3 Antibiotics, enzymes, siderophore production, and induced systemic resistance

There are a great number of antibiotic compounds produced by PGPR strains that stimulate plant defense mechanisms against a variety of pathogens (Glick, 2012). This trait is effective in preventing the proliferation of other bacteria and fungi (Novo et al., 2018a,b). PGPR are also known to produce hydrogen cyanide, which, despite its reduced individual biocontrol activity, enhances the effect of bacterial antibiotics when acting synergistically (Beneduzi et al., 2012; Glick, 2012). Other important biocontrol agents produced by PGPR are lytic enzymes (such as proteases and cellulases) (Manoj et al., 2020; Novo et al., 2018a,b). These enzymes avoid the proliferation of pathogenic fungi by hydrolyzing components in their cell wall, reducing their populations (Glick, 2012). Besides biocontrol, lytic enzymes are involved in nutrient recycling through organic matter decomposition, leading to increased nutrient availability in the rhizosphere (Karthik et al., 2017). Siderophores produced by PGPR play a role in Fe sequestration, as aforementioned, and can also contribute to biocontrol activity (Glick, 2012). Bacterial strains use siderophores for preventing pathogens (generally fungi) from acquiring sufficient amounts of Fe, thus hindering their proliferation (Rajkumar et al., 2010).

PGPR are able to induce plant resistance against a large array of pathogens, such as bacteria, fungi, viruses, and pests through ISR (Lugtenberg and Kamilova, 2009). ISR is activated by different signaling pathways that include a great variety of molecules (Novo et al., 2018a,b). This mechanism is responsible for a faster and more efficient reaction against pathogens (Beneduzi et al., 2012). Lastly, some biocontrol strains simply outcompete pathogens by efficiently colonizing the plant rhizosphere and consuming resources; this way, the use of most available nutrients makes it difficult for pathogens to grow (Whipps, 2001).

2.4 Mycorrhizal fungi

Mycorrhiza is described as the mutualistic symbiotic association between soilborne fungi and the roots of plants (Sieverding et al., 1991). Mycorrhizal fungi are widespread and form associations with the great majority of plants. A characterizing trait of these fungi is that their mycelium overgrows the soil surrounding plant roots (Turnau et al., 2005). The extensive hyphal net around the roots is responsible for significantly increasing the volume of soil that can be explored for nutrients (Thijs et al., 2017). For efficient nutrient uptake, most plants establish a symbiotic relation with mycorrhizal fungi that play a crucial role in supplying important nutrients, augmenting drought tolerance, and increasing plant yields and health (Bonfante and Anca, 2009). Promotion of plant growth can be accomplished by enhancing nutrient availability and, via the production of phytohormones, mycorrhizal fungi can also be responsible for protecting plants and enhancing their resistance against biotic and abiotic stress (Waqas et al., 2012). Mycorrhizal fungi have been reported to enhance metal solubilization by producing several organic compounds such as citric and oxalic acids (Coninx et al., 2017). Several types of mycorrhizas have been described, and they can be mainly classified as endomycorrhizas or ectomycorrhizas (Bonfante and Anca, 2009). The main type of endomycorrhizas is formed by arbuscular mycorrhizal fungi (AMF), which penetrate root cells and establish intracellular symbioses (Turnau et al., 2005). In ectomycorrhizas the mycelium forms a somewhat compact fungal mantle on the surface of the roots of trees and shrubs (Turnau et al., 2005).

AMF can be found in natural and agricultural ecosystems, including heavy metal–rich sites (Sheoran et al., 2016). AMF are incredibly widespread among plant taxa, occurring in 80% of plant species, being by far the most studied type of fungi in hyperaccumulator plants (Göhre and Paszkowski, 2006; Schussler et al., 2001). In this association the plant is responsible for supplying C to the fungi, while AMF enhance the plant ability to uptake important nutrients, leading to increased plant yields (Miransari, 2011). These fungi comprise an important component that significantly improves the ecosystem efficiency (Miransari, 2011), since they promote soil structure, enhance resistance to pathogens, and alleviate stress (such as metal toxicity) (Miransari, 2011; Turnau et al., 2005). There is no consensus about the mechanisms used by fungi to avoid metal phytotoxicity, meaning that there is the possibility that it might happen indirectly, through the facilitation of nutrient uptake in adverse metal-rich environments, stimulating plant development (Kidd et al., 2018). It is also posited that it may happen directly, by regulation of specific metal tolerance mechanisms

(Kidd et al., 2018). There is also no unequivocal way of action of AMF regarding the effect in metal uptake and distribution (Kidd et al., 2018). Reports show that several AMF species can increase the uptake of metals by plants and their translocation from root to shoot (Cao et al., 2008). Yet, some species are known to protect the plant by decreasing metal uptake (Rozpądek et al., 2017).

Ectomycorrhizal fungi may occur in metal-rich environments and are considered drivers of ecosystem diversity just like AMF. They play an important role in stimulating plant yields in these adverse environments (Kidd et al., 2018). Ectomycorrhizal fungi can improve plants' physiological status and modulate heavy metal tolerance (Luo et al., 2014). In some cases, they are responsible for inducing immobilization of certain metals on the surface of the hyphae (component that forms the mycelium) by complexation of the elements with molecular compounds released in the rhizosphere or by using chemical binding groups in the cell wall (Luo et al., 2014).

2.5 Microbe-assisted phytomining

As previously stated, some plant rhizosphere beneficial microorganisms, such as PGPR and mycorrhizal fungi, inhabit metal-rich areas and contribute for the growth, health, and yield of hyperaccumulator plants (Benizri and Kidd, 2018). They may play a role in heavy metal tolerance, availability, uptake, and distribution within the plant, making them the focus of many studies in recent times (Benizri and Kidd, 2018). Reports show an improvement in hyperaccumulator plants' growth and metal accumulation after soil, seed, or plant inoculations (Kidd et al., 2018). The potential benefits of microbes associated with hyperaccumulator plants led to the creation of the concept of microbial-assisted phytomining/phytoextraction (Thijs et al., 2016). Microbial-assisted phytomining consists in an agronomical strategy that uses microbes beneficial to hyperaccumulators, to improve phytomining yields, resulting in a revenue increase. This approach aims to enhance metal recovery by inoculating hyperaccumulator plants with plant-associated microorganisms that are able to increase the harvestable amount by enhancing the plants' biomass yield and/or the bioavailability, uptake, and translocation of the target metal (Benizri and Kidd, 2018). What's more, microbial-assisted phytomining could replace or complement other agronomical optimization approaches, such as the application of fertilizers or chelating agents, reducing the operational costs.

2.5.1 Microbe-assisted nickel phytomining

For the effective application of plant beneficial microbiota to enhance Ni phytomining, it is fundamental to have a thorough understanding of the variety, structure, and functions of microbes inhabiting the rhizosphere of hyperaccumulators from ultramafic soils (Thijs et al., 2017). Several studies were conducted over the years with a number of PGPR (Table 2.2) and mycorrhizal fungi, under metal-rich substrates and with Ni hyperaccumulator species.

Reports show that bacterial strains collected from ultramafic soils, associated to Ni hyperaccumulators, tolerate much higher concentrations of Ni in comparison to strains from other soils (Schlegel et al., 1991; Turgay et al., 2012). Ma et al. (2009a) studied the effects of several strains of PGPR on plant growth and Ni uptake by the species *Brassica juncea* and *Brassica oxyrrhina*. The results were positive for IAA production and P solubilization, and almost all strains exhibited siderophore production. Some PGPR also showed ACC deaminase activity. The inoculation of the bacterial strains increased the biomass yield of both plant species, with strains *Pseudomonas* sp. SRI2, *Psychrobacter* sp. SRS8, and *Bacillus* sp. SN9 showing the best results. SN9 was also responsible for enhancing Ni concentrations in the root and shoot of both plant species. It was concluded that the inoculation with the bacterial strain SN9 increases Ni accumulation directly and indirectly (by promoting biomass production) in the species *B. juncea* and *B. oxyrrhina*. Another study from 2016 demonstrates that the PGPR *Bacillus subtilis* strain SJ-101 has the ability to promote Ni accumulation, tolerance and growth of the plant *B. juncea* (Zaidi et al., 2006). It was determined that this strain produced IAA and solubilized P.

Regarding Ni hyperaccumulator species, a study from 2003 reported that *Microbacterium arabinogalactanolyticum* has an important role in enhancing Ni availability and therefore its accumulation by *Alyssum murale* (Abou-Shanab et al., 2003a,b). Inoculation resulted in an increase of 32.4% in shoot Ni concentration, in comparison with the uninoculated specimens. Another study, with the same plant species, but with the inoculation of the another bacterial strain, *Microbacterium oxydans* AY509223, also reported a rise in Ni accumulation (Aboushanab et al., 2006). The bacteria were tested on plants in low, medium, and high Ni-contaminated soils. Results showed that *M. oxydans* AY509223 enhances foliar Ni in these soils from 82.9, 261.3, and 2829.3 to 129.7, 430.7, and 3914.3 mg/kg, respectively, when compared to the control plants. The authors concluded that bacteria are important for hyperaccumulation and the use of inoculants can be developed for improving commercial Ni phytomining.

TABLE 2.2 Summary of some studies with plant growth–promoting rhizobacteria.

PGPR strains	Plant species	Mechanism	Main effects	References
Microbacterium arabinogalactanolyticum	*Alyssum murale*	Siderophore production	↑ Ni availability ↑ Ni accumulation	Abou-Shanab et al. (2003a,b)
Bacillus subtilis SJ-101	*Brassica juncea*	IAA production P solubilization	↑ Ni accumulation ↑ Metal tolerance ↑ Biomass	Zaidi et al. (2006)
Microbacterium oxydans AY509223	*A. murale*	P solubilization Siderophore production	↑ Ni shoot	Aboushanab et al. (2006)
Pseudomonas sp. Ps29C *Bacillus megaterium* Bm4C	*B. juncea*	IAA production P solubilization Siderophore production ACC deaminase activity	↑ Metal tolerance ↑ Biomass	Rajkumar and Freitas (2008)
Pseudomonas sp. SRI2 *Psychrobacter* sp. SRS8 *Bacillus* sp. SN9	*B. juncea* *Brassica oxyrrhina*	IAA production P solubilization Siderophore production ACC deaminase activity	↑ Ni root and shoot ↑Bbiomass	Ma et al. (2009a)
Psychrobacter sp. SRA1/SRA2 *Bacillus cereus* sp. SRA10/SRP4 *Bacillus weihenstephanensis* SRP12	*B. juncea* *B. oxyrrhina*	IAA production P solubilization Siderophore production ACC deaminase activity	↑ Ni root and shoot ↑ Biomass ↑ Availability	Ma et al. (2009b)
Mixed: *Bacillus pumilus–Micrococcus* sp.	*Noccaea caerulescens*	Ni resistance	↓ Solubility ↑ Biomass ↑ Ni yield	Aboudrar et al. (2013)
Microbacterium sp. *Arthrobacter* sp. *Streptomyces* sp.	*Alyssum pintodasilvae*	IAA production P solubilization Siderophore production	↑ Ni shoot ↑ Biomass ↑ Ni yield	Cabello-Conejo et al. (2014)
Variovorax paradoxus	Intercropped: *Bornmuellera tymphaea–Noccaea tymphaea B. tymphaea–A. murale*	IAA production P solubilization Siderophore production	↑ Ni accumulation ↑ Biomass	Durand et al. (2016)

ACC, 1-Aminocyclopropane-1-carboxylate; *IAA*, indole-3-acetic acid; *PGPR*, plant growth–promoting rhizobacteria.

Durand et al. (2016), isolated *Variovorax* strains, from an ultramafic rhizosphere of the mixed cover of *Bornmuellera tymphaea*–*Noccaea tymphaea* and *B. tymphaea*–*A. murale*. They reported that the total and shoot biomass were significantly superior in the inoculated plant association *B. tymphaea*–*N. tymphaea*. The PGPR significantly increased the accumulation of Ni in roots and shoots, when compared to uninoculated plants. This effect was attributed to the production of IAA and ACC deaminase. Concerning Ni accumulation in the shoots, there was a 39.9% and 79.6% increase in *B. tymphaea* –*A. murale* and *B. tymphaea*–*N. tymphaea*, respectively, when inoculated with bacterial strains. The combination *B. tymphaea*–*N. tymphaea* inoculated with *Variovorax* NB24 presented interesting results to be tested in the field.

In 2017 the strains *Arthrobacter* sp. LA44, *Arthrobacter* sp. SBA82, and *Variovorax paradoxus* AB30 were assessed in a field-scale phytomining study with the species *A. murale* (Pardo et al., 2017). The results showed that inoculated plants increased the soil cation-exchange capacity and organic C content, and SBA82 and LA44 did not alter the composition of soil bacterial communities. These strains were responsible for increasing plant coverage and the latter also enhanced significantly the aboveground biomass of *A. murale*. The harvestable amount of Ni per plant was significantly higher with the inoculation of LA44 and SBA82, that is, the use of these strains has the potential to improve phytomining systems, without compromising the structure of soil bacterial communities.

A pot study was conducted to evaluate the effects of the concurrent use of bacterial inoculation and soil fertilization or soil amendment with cow manure, on Ni phytoextraction by *Odontarrhena bracteata*, *Odontarrhena inflata*, and *Odontarrhena serpyllifolia* (Ghasemi et al., 2018). For this purpose, serpentine soil from the Melide ultramafic complex (NW Spain) was inoculated with five strains of rhizobacteria, previously isolated from *O. serpyllifolia*, and the plants were allowed to grow for 3 months, before harvesting and analysis. The results showed greater biomass yield, enhanced nutritive conditions, and higher accumulation of Ni in plant tissue. The authors noted that selected bacterial inoculants improved Ni phytoextraction as consequence of stimulated plant growth and/or increased Ni levels in aerial plant parts, although contingent to plant species, type of soil, and inoculant. In at least one soil treatment, *Pseudarthrobacter oxydans* strain SBA82 boosted the shoot dry mass yield of all plant species, whereas *Stenotrophomonas* sp. strain MA98 increased the growth of *O. bracteata and O. serpyllifolia*. Plants subjected to rhizobacterial inoculation displayed reduced activity of antioxidative enzymes, and lesser levels of oxidative stress metabolites, denoting an improved physiological response. The authors emphasized the potential of the joint application of rhizobacteria and manure to ameliorate the growth, development, and Ni uptake ability of *Odontarrhena* hyperaccumulators.

Lopez et al. (2019) studied the rhizosphere bacteria of several Ni hyperaccumulator species from the ultramafic soils of Albania. The investigated plant species were *Noccaea ochroleuca*, *Odontarrhena smolikana*, *Odontarrhena rigida*, and *Odontarrhena chalcidica*. The study showed that Proteobacteria, Actinobacteria, and Acidobacteria were the predominant bacterial phyla in the soil. This work was successful in characterizing the rhizosphere microbial community of hyperaccumulator species and in evaluating their potential influence in soil biogeochemical parameters. The authors suggested the future step to be the determination of traits that allow the bacteria to colonize ultramafic environments, providing significant advantages for sustainable hyperaccumulator production and the enhancement of the phytomining process.

Before the early 2000s hyperaccumulator plants were thought to be nonmycorrhizal, therefore, regarding the application of mycorrhizal fungi for enhancing hyperaccumulation yields, there is still little information in comparison to studies on bacterial inoculants (Benizri and Kidd, 2018). The first Ni hyperaccumulator plant in which the presence of AMF was reported was *Berkheya coddii* (Turnau and Mesjasz-Przybylowicz, 2003). In this study, several Ni hyperaccumulator species from ultramafic soils in South Africa were analyzed for AMF root colonization. It was found that the plants were all arbuscular mycorrhizal. In the species *B. coddii*, which exhibits great potential in the context of phytomining, well developed arbuscular mycorrhizas were observed. All plant species inoculated with AMF in greenhouse trials showed a higher biomass production and Ni content in relation to the uninoculated plants.

A study from 2011 consisted in the analyses of the effect of AMF in *B. coddii* growth and element uptake (Orłowska et al., 2011). The plants were grown in the laboratory on ultramafic soils, and the AMF used in this experiment were from several different origins. Results showed that inoculations affected significantly plant growth and survival, especially with the application of an indigenous strain. The fungi were responsible for increasing concentrations of K, Fe, P, and calcium (Ca) in the shoots and Zn, manganese (Mn), Ca, and P in roots. The levels of Mn, Co, and Ni decreased in plant shoots. However, due to the high biomass production with AMF inoculations, the Ni harvestable amount was much higher (20-fold greater) in mycorrhizal plants when compared to nonmycorrhizal controls. The authors proposed that AMF may help boosting Ni phytomining.

2.6 Future perspectives

The rhizosphere microbiome, particularly PGPR and mycorrhizal fungi, can play a pivotal role in enhancing plant growth, health, and metal uptake. Nonetheless, there is still a wealth of information to be uncovered about rhizosphere microbial diversity, functions, and the mechanisms affecting hyperaccumulator plants. Hence, future studies focusing on the influence of mycorrhizal fungi on hyperaccumulators and the scrutiny of the mechanisms responsible for enhanced Ni extraction would be particularly relevant, since the information about these subjects is relatively scarce. The number of bacterial strains examined to date is also quite limited. Hence, further phytomining efforts involving hyperaccumulators and PGPR would certainly provide valuable insights. Field trials are of particular importance for these deliver results under real-world conditions and facilitate a deeper understanding of the processes involved, without tampered variables. Several central aspects, such as the compatibility of the selected host species and inoculant, degree of plant species specificity, and the ability of microbes to perform and proliferate in the soil, need to be studied under natural field conditions for longer time periods, since they dictate the efficiency of inoculants (Kidd et al., 2018). Furthermore, inoculation methods, such as the frequency, volume, and timing of inoculation, and the application in soil, seed, or plant, need to be considered because these are aspects that are likely to influence the outcome of phytomining yields. It is also important to assess the effects of inoculants in the natural rhizosphere microbiome. Lastly, phytomining could also profit from the application of advanced gene editing techniques such as clustered regularly interspaced short palindromic repeats (CRISPR) to obtain more efficient or tailor-made PGPR (Basharat et al., 2018). Given that genes responsible for the synthesis of phytohormones, nitrogenase, and siderophores have been already identified, the prospect of designing PGPR with improved direct (phytohormones production, phosphate solubilization, N fixation, and Fe sequestration) and indirect (biocontrol) mechanisms may not be so far-fetched.

Acknowledgments

This study was financially supported by national funds through the Portuguese Foundation for Science and Technology (FCT) under grants nos. UIDB/04035/2020 and UID/BIA/04004/2020.

References

Aboudrar, W., Schwartz, C., Morel, J.L., Boularbah, A., 2013. Effect of nickel-resistant rhizosphere bacteria on the uptake of nickel by the hyperaccumulator *Noccaea caerulescens* under controlled conditions. J. Soils Sediment 13, 501–507. Available from: https://doi.org/10.1007/s11368-012-0614-x.

Aboushanab, R., Angle, J., Chaney, R., 2006. Bacterial inoculants affecting nickel uptake by *Alyssum murale* from low, moderate and high Ni soils. Soil Biol. Biochem. 38, 2882–2889. Available from: https://doi.org/10.1016/j.soilbio.2006.04.045.

Abou-Shanab, R.A., Angle, J.S., Delorme, T.A., Chaney, R.L., Van Berkum, P., Moawad, H., et al., 2003a. Rhizobacterial effects on nickel extraction from soil and uptake by *Alyssum murale*. N. Phytol. 158, 219–224. Available from: https://doi.org/10.1046/j.1469-8137.2003.00721.x.

Abou-Shanab, R.I., Delorme, T.A., Angle, J.S., Chaney, R.L., Ghanem, K., Moawad, H., et al., 2003b. Phenotypic characterization of microbes in the rhizosphere of *Alyssum murale*. Int. J. Phytorem. 5, 367–379. Available from: https://doi.org/10.1080/15226510309359043.

Ahemad, M., 2014. Remediation of metalliferous soils through the heavy metal resistant plant growth promoting bacteria: paradigms and prospects. Arabian J. Chem. 12, 1365–1377. Available from: https://doi.org/10.1016/j.arabjc.2014.11.020.

Álvarez-López, V., Prieto-Fernández, Á., Cabello-Conejo, M.I., Kidd, P.S., 2016. Organic amendments for improving biomass production and metal yield of Ni-hyperaccumulating plants. Sci. Total Environ. 548–549, 370–379. Available from: https://doi.org/10.1016/j.scitotenv.2015.12.147.

Alves, A.R.A., Silva, E.F., Novo, L.A.B., 2019. Morais ultramafic complex: a survey towards nickel phytomining. Resources 8, 144. Available from: https://doi.org/10.3390/resources8030144.

Basharat, Z., Novo, L., Yasmin, A., 2018. Genome editing weds CRISPR: what is in it for phytoremediation? Plants 7, 51. Available from: https://doi.org/10.3390/plants7030051.

Baudoin, E., Benizri, E., Guckert, A., 2003. Impact of artificial root exudates on the bacterial community structure in bulk soil and maize rhizosphere. Soil Biol. Biochem. 35, 1183–1192. Available from: https://doi.org/10.1016/S0038-0717(03)00179-2.

Beneduzi, A., Ambrosini, A., Passaglia, L.M.P., 2012. Plant growth-promoting rhizobacteria (PGPR): their potential as antagonists and biocontrol agents. Genet. Mol. Biol. 35, 1044–1051. Available from: https://doi.org/10.1590/S1415-47572012000600020.

Benizri, E., Kidd, P.S., 2018. The role of the rhizosphere and microbes associated with hyperaccumulator plants in metal accumulation. In: Van der Ent, A., Echevarria, G., Baker, A.J.M., Morel, J.L. (Eds.), Agromining: Farming for Metals, Mineral Resource Reviews. Springer International Publishing, Cham, pp. 157–188. Available from: http://doi.org/10.1007/978-3-319-61899-9-9.

Bhattacharyya, P.N., Jha, D.K., 2012. Plant growth-promoting rhizobacteria (PGPR): emergence in agriculture. World J. Microbiol. Biotechnol. 28, 1327–1350. Available from: https://doi.org/10.1007/s11274-011-0979-9.

Bonfante, P., Anca, I.-A., 2009. Plants, mycorrhizal fungi, and bacteria: a network of interactions. Annu. Rev. Microbiol. 63, 363–383. Available from: https://doi.org/10.1146/annurev.micro.091208.073504.

Bowen, G.D., Rovira, A.D., 1999. The rhizosphere and its management to improve plant growth. Adv. Agron. 63, 1–102. Available from: https://doi.org/10.1016/S0065-2113(08)60425-3.

Brooks, R., Lee, J., Reeves, R., Jaffre, T., 1977. Detection of nickeliferous rocks by analysis of herbarium specimens of indicator plants. J. Geochem. Explor. 7, 49–57. Available from: https://doi.org/10.1016/0375-6742(77)90074-7.

Brooks, R.R., Chambers, M.F., Nicks, L.J., Robinson, B.H., 1998. Phytomining. Trends Plant Sci. 3, 359–362.

Cabello-Conejo, M.I., Becerra-Castro, C., Prieto-Fernández, A., Monterroso, C., Saavedra-Ferro, A., Mench, M., et al., 2014. Rhizobacterial inoculants can improve nickel phytoextraction by the hyperaccumulator *Alyssum pintodasilvae*. Plant Soil 379, 35–50. Available from: https://doi.org/10.1007/s11104-014-2043-7.

Cao, L., Jiang, M., Zeng, Z., Du, A., Tan, H., Liu, Y., 2008. *Trichoderma atroviride* F6 improves phytoextraction efficiency of mustard (*Brassica juncea* (L.) Coss. var. *foliosa* Bailey) in Cd, Ni contaminated soils. Chemosphere 71, 1769–1773. Available from: https://doi.org/10.1016/j.chemosphere.2008.01.066.

Chaparro, J.M., Sheflin, A.M., Manter, D.K., Vivanco, J.M., 2012. Manipulating the soil microbiome to increase soil health and plant fertility. Biol. Fertil. Soils 48, 489–499. Available from: https://doi.org/10.1007/s00374-012-0691-4.

Coninx, L., Martinova, V., Rineau, F., 2017. Mycorrhiza-assisted phytoremediationt. Adv. Bot. Res. 83. Available from: https://doi.org/10.1016/bs.abr.2016.12.005.

Durand, A., Piutti, S., Rue, M., Morel, J.L., Echevarria, G., Benizri, E., 2016. Improving nickel phytoextraction by co-cropping hyperaccumulator plants inoculated by plant growth promoting rhizobacteria. Plant Soil 399, 179–192. Available from: https://doi.org/10.1007/s11104-015-2691-2.

Ghasemi, Z., Ghaderian, S.M., Rodríguez-Garrido, B., Prieto-Fernández, Á., Kidd, P.S., 2018. Plant species-specificity and effects of bioinoculants and fertilization on plant performance for nickel phytomining. Plant Soil 425, 265–285. Available from: https://doi.org/10.1007/s11104-017-3553-x.

Ghosh, M., Singh, L., 2005. A review on phytoremediation of heavy metals and utilization of its byproducts. Appl. Ecol. Environ. Res. 3, 1–18. Available from: https://doi.org/10.15666/aeer/0301-001018.

Glick, B.R., 2012. Plant growth-promoting bacteria: mechanisms and applications. Scientifica (Cairo) 2012, 1–15. Available from: https://doi.org/10.6064/2012/963401.

Göhre, V., Paszkowski, U., 2006. Contribution of the arbuscular mycorrhizal symbiosis to heavy metal phytoremediation. Planta 223, 1115–1122. Available from: https://doi.org/10.1007/s00425-006-0225-0.

Goswami, D., Thakker, J.N., Dhandhukia, P.C., 2016. Portraying mechanics of plant growth promoting rhizobacteria (PGPR): a review. Cogent Food Agric. 2, 1–19. Available from: https://doi.org/10.1080/23311932.2015.1127500.

Hinsinger, P., Plassard, C., Jaillard, B., 2006. Rhizosphere: a new frontier for soil biogeochemistry. J. Geochem. Explor. 88, 210–213. Available from: https://doi.org/10.1016/j.gexplo.2005.08.041.

Hoffman, B.M., Lukoyanov, D., Yang, Z.-Y., Dean, D.R., Seefeldt, L.C., 2014. Mechanism of nitrogen fixation by nitrogenase: the next stage. Chem. Rev. 114, 4041–4062. Available from: https://doi.org/10.1021/cr400641x.

Idris, R., Trifonova, R., Puschenreiter, M., Wenzel, W.W., Sessitsch, A., 2004. Bacterial communities associated with flowering plants of the Ni hyperaccumulator *Thlaspi goesingense*. Appl. Environ. Microbiol. 70, 2667–2677. Available from: https://doi.org/10.1128/AEM.70.5.2667-2677.2004.

Karthik, C., Barathi, S., Pugazhendhi, A., Ramkumar, V.S., Thi, N.B.D., Arulselvi, P.I., 2017. Evaluation of Cr(VI) reduction mechanism and removal by *Cellulosimicrobium funkei* strain AR8, a novel haloalkaliphilic bacterium. J. Hazard. Mater. 333, 42–53. Available from: https://doi.org/10.1016/j.jhazmat.2017.03.037.

Khan, A.G., 2005. Role of soil microbes in the rhizospheres of plants growing on trace metal contaminated soils in phytoremediation. J. Trace Elem. Med. Biol. 18, 355–364. Available from: https://doi.org/10.1016/j.jtemb.2005.02.006.

Kidd, P.S., Bani, A., Benizri, E., Gonnelli, C., Hazotte, C., Kisser, J., et al., 2018. Developing sustainable agromining systems in agricultural ultramafic soils for nickel recovery. Front. Environ. Sci. 6, 44. Available from: https://doi.org/10.3389/fenvs.2018.00044.

Kloepper, J.W., 1994. Plant growth-promoting rhizobacteria (other systems). In: Okon, Y. (Ed.), Azospirillum/Plant Associations. CRC Press, FL, pp. 137–166.

Lopez, S., Goux, X., Echevarria, G., Calusinska, M., Morel, J.L., Benizri, E., 2019. Community diversity and potential functions of rhizosphere-associated bacteria of nickel hyperaccumulators found in Albania. Sci. Total Environ. 654, 237–249. Available from: https://doi.org/10.1016/j.scitotenv.2018.11.056.

Lugtenberg, B., Kamilova, F., 2009. Plant-growth-promoting rhizobacteria. Annu. Rev. Microbiol. 63, 541–556. Available from: https://doi.org/10.1146/annurev.micro.62.081307.162918.

Luo, Z.-B., Wu, C., Zhang, C., Li, H., Lipka, U., Polle, A., 2014. The role of ectomycorrhizas in heavy metal stress tolerance of host plants. Environ. Exp. Bot. 108, 47–62. Available from: https://doi.org/10.1016/j.envexpbot.2013.10.018.

Ma, Y., Oliveira, R.S., Freitas, H., Zhang, C., 2016a. Biochemical and molecular mechanisms of plant-microbe-metal interactions: relevance for phytoremediation. Front. Plant Sci. 7. Available from: https://doi.org/10.3389/fpls.2016.00918.

Ma, Y., Prasad, M.N.V., Rajkumar, M., Freitas, H., 2011. Plant growth promoting rhizobacteria and endophytes accelerate phytoremediation of metalliferous soils. Biotechnol. Adv. 29, 248–258. Available from: https://doi.org/10.1016/j.biotechadv.2010.12.001.

Ma, Y., Rajkumar, M., Freitas, H., 2009a. Improvement of plant growth and nickel uptake by nickel resistant-plant-growth promoting bacteria. J. Hazard. Mater. 166, 1154–1161. Available from: https://doi.org/10.1016/j.jhazmat.2008.12.018.

Ma, Y., Rajkumar, M., Freitas, H., 2009b. Isolation and characterization of Ni mobilizing PGPB from serpentine soils and their potential in promoting plant growth and Ni accumulation by *Brassica* spp. Chemosphere 75, 719–725. Available from: https://doi.org/10.1016/j.chemosphere.2009.01.056.

Ma, Y., Rajkumar, M., Zhang, C., Freitas, H., 2016b. Inoculation of *Brassica oxyrrhina* with plant growth promoting bacteria for the improvement of heavy metal phytoremediation under drought conditions. J. Hazard. Mater. 320, 36–44. Available from: https://doi.org/10.1016/j.jhazmat.2016.08.009.

Manoj, S.R., Karthik, C., Kadirvelu, K., Arulselvi, P.I., Shanmugasundaram, T., Bruno, B., et al., 2020. Understanding the molecular mechanisms for the enhanced phytoremediation of heavy metals through plant growth promoting rhizobacteria: a review. J. Environ. Manage. 254, 109779. Available from: https://doi.org/10.1016/j.jenvman.2019.109779.

Mendes, R., Garbeva, P., Raaijmakers, J.M., 2013. The rhizosphere microbiome: significance of plant beneficial, plant pathogenic, and human pathogenic microorganisms. FEMS Microbiol. Rev. 37, 634–663. Available from: https://doi.org/10.1111/1574-6976.12028.

Mengoni, A., Barzanti, R., Gonnelli, C., Gabbrielli, R., Bazzicalupo, M., 2001. Characterization of nickel-resistant bacteria isolated from serpentine soil. Environ. Microbiol. 3, 691–698. Available from: https://doi.org/10.1046/j.1462-2920.2001.00243.x.

Miransari, M., 2011. Hyperaccumulators, arbuscular mycorrhizal fungi and stress of heavy metals. Biotechnol. Adv. 29, 645–653. Available from: https://doi.org/10.1016/j.biotechadv.2011.04.006.

Nedelkoska, T.V., Doran, P.M., 2000. Characteristics of heavy metal uptake by plant species with potential for phytoremediation and phytomining. Miner. Eng. 13, 549–561. Available from: https://doi.org/10.1016/S0892-6875(00)00035-2.

Nguyen, C., 2003. Rhizodeposition of organic C by plants: mechanisms and controls. Agron 23375–39623, 375–396. Available from: https://doi.org/10.1007/978-90-481-2666-8_9.

Nkrumah, P.N., Baker, A.J.M., Chaney, R.L., Erskine, P.D., Echevarria, G., Morel, J.L., et al., 2016. Current status and challenges in developing nickel phytomining: an agronomic perspective. Plant Soil . Available from: https://doi.org/10.1007/s11104-016-2859-4.

Nkrumah, P.N., Tisserand, R., Chaney, R.L., Baker, A.J.M., Morel, J.L., Goudon, R., et al., 2019. The first tropical 'metal farm': some perspectives from field and pot experiments. J. Geochem. Explor. 198, 114–122. Available from: https://doi.org/10.1016/j.gexplo.2018.12.003.

Novo, L.A.B., Castro, P.M.L., Alvarenga, P., da Silva, E.F., 2017. Phytomining of rare and valuable metals. In: Ansari, A.A., Gill, S.S., Gill, R., Lanza, G., Newman, L. (Eds.), Phytoremediation. Springer International Publishing, Cham, pp. 469–486. Available from: http://doi.org/10.1007/978-3-319-52381-1-18.

Novo, L.A.B., Castro, P.M.L., Alvarenga, P., da Silva, E.F., 2018a. From phyto to agromining: past, present, and future scope. Recent Advances in Environmental Science From the Euro-Mediterranean and Surrounding Regions. Springer International Publishing, pp. 99–101. Available from: http://doi.org/10.1007/978-3-319-70548-4-34.

Novo, L.A.B., Castro, P.M.L., Alvarenga, P., da Silva, E.F., 2018b. Plant growth–promoting rhizobacteria-assisted phytoremediation of mine soils. Bio-Geotechnologies for Mine Site Rehabilitation. Elsevier, pp. 281–295. Available from: http://doi.org/10.1016/b978-0-12-812986-9.00016-6.

Novo, L.A.B., Mahler, C.F., González, L., 2015. Plants to harvest rhenium: scientific and economic viability. Environ. Chem. Lett. 13, 439–445. Available from: https://doi.org/10.1007/s10311-015-0517-3.

Orłowska, E., Przybyłowicz, W., Orlowski, D., Turnau, K., Mesjasz-Przybyłowicz, J., 2011. The effect of mycorrhiza on the growth and elemental composition of Ni-hyperaccumulating plant *Berkheya coddii* Roessler. Environ. Pollut. 159, 3730–3738. Available from: https://doi.org/10.1016/j.envpol.2011.07.008.

Pardo, T., Benizri, E., Saad, R., Rodríguez-Garrido, B., Echevarria, G., Prieto-Fernandez, A., et al., 2017. Assessment of the use of bacterial inoculants for improving the agromining potential of Ni-hyperaccumulating plant species at field-scale. J. Geol. Geophys. 51–52. Available from: https://doi.org/10.4172/2381-8719-C1-014.

Rajkumar, M., Ae, N., Prasad, M.N.V., Freitas, H., 2010. Potential of siderophore-producing bacteria for improving heavy metal phytoextraction. Trends Biotechnol. 28, 142–149. Available from: https://doi.org/10.1016/j.tibtech.2009.12.002.

Rajkumar, M., Freitas, H., 2008. Effects of inoculation of plant-growth promoting bacteria on Ni uptake by Indian mustard. Bioresour. Technol. 99, 3491–3498. Available from: https://doi.org/10.1016/j.biortech.2007.07.046.

Rascio, N., Navari-Izzo, F., 2011. Heavy metal hyperaccumulating plants: how and why do they do it? And what makes them so interesting? Plant Sci. 180, 169–181. Available from: https://doi.org/10.1016/j.plantsci.2010.08.016.

Rashid, M.I., Mujawar, L.H., Shahzad, T., Almeelbi, T., Ismail, I.M.I., Oves, M., 2016. Bacteria and fungi can contribute to nutrients bioavailability and aggregate formation in degraded soils. Microbiol. Res. 183, 26–41. Available from: https://doi.org/10.1016/j.micres.2015.11.007.

Rozpądek, P., Domka, A., Ważny, R., Nosek, M., Jędrzejczyk, R., Tokarz, K., et al., 2017. How does the endophytic fungus *Mucor* sp. improve *Arabidopsis arenosa* vegetation in the degraded environment of a mine dump? Environ. Exp. Bot. 147. Available from: https://doi.org/10.1016/j.envexpbot.2017.11.009.

Saraf, M., Pandya, U., Thakkar, A., 2014. Role of allelochemicals in plant growth promoting rhizobacteria for biocontrol of phytopathogens. Microbiol. Res. 169, 18–29. Available from: https://doi.org/10.1016/j.micres.2013.08.009.

Schlegel, H.G., Cosson, J.-P., Baker, A.J.M., 1991. Nickel-hyperaccumulating plants provide a niche for nickel-resistant bacteria. Bot. Acta 104, 18–25. Available from: https://doi.org/10.1111/j.1438-8677.1991.tb00189.x.

Schussler, A., Schwarzott, D., Walker, C., 2001. A new fungal phylum, the Glomeromycota: phylogeny and evolution. Mycol. Res. 105, 1413–1421. Available from: https://doi.org/10.1017/S0953756201005196.

Sessitsch, A., Kuffner, M., Kidd, P., Vangronsveld, J., Wenzel, W.W., Fallmann, K., et al., 2013. The role of plant-associated bacteria in the mobilization and phytoextraction of trace elements in contaminated soils. Soil Biol. Biochem. 60, 182–194. Available from: https://doi.org/10.1016/j.soilbio.2013.01.012.

Sheoran, V., Sheoran, A.S., Poonia, P., 2009. Phytomining: a review. Miner. Eng. 22, 1007–1019. Available from: https://doi.org/10.1016/j.mineng.2009.04.001.

Sheoran, V., Sheoran, A.S., Poonia, P., 2016. Factors affecting phytoextraction: a review. Pedosphere 26, 148–166. Available from: https://doi.org/10.1016/S1002-0160(15)60032-7.

Sieverding, E., Friedrichsen, J., Suden, W., 1991. Vesicular-arbuscular mycorrhiza management in tropical agrosystems. In: Schriftenreihe der GTZ.

Somers, E., Vanderleyden, J., Srinivasan, M., 2004. Rhizosphere bacterial signalling: a love parade beneath our feet. Crit. Rev. Microbiol. 30, 205–240. Available from: https://doi.org/10.1080/10408410490468786.

Thijs, S., Langill, T., Vangronsveld, J., 2017. The bacterial and fungal microbiota of hyperaccumulator plants: small organisms, large influence, in: Advances in Botanical Research. pp. 43–86. https://doi.org/10.1016/bs.abr.2016.12.003.

Thijs, S., Sillen, W., Rineau, F., Weyens, N., Vangronsveld, J., 2016. Towards an enhanced understanding of plant–microbiome interactions to improve phytoremediation: engineering the metaorganism. Front. Microbiol. 7, 1–15. Available from: https://doi.org/10.3389/fmicb.2016.00341.

Turgay, O.C., Görmez, A., Bilen, S., 2012. Isolation and characterization of metal resistant-tolerant rhizosphere bacteria from the serpentine soils in Turkey. Environ. Monit. Assess. 184, 515–526. Available from: https://doi.org/10.1007/s10661-011-1984-z.

Turnau, K., Jurkiewicz, A., Lingua, G., Barea, J.M., Gianinazzi–Pearson, V., 2005. Role of arbuscular mycorrhiza and associated microorganisms in phytoremediation of heavy metal-polluted sites, in: Trace Elements in the Environment. pp. 235–252. https://doi.org/10.1201/9781420032048.ch13.

Turnau, K., Mesjasz-Przybylowicz, J., 2003. Arbuscular mycorrhiza of *Berkheya coddii* and other Ni-hyperaccumulating members of Asteraceae from ultramafic soils in South Africa. Mycorrhiza 13, 185–190. Available from: https://doi.org/10.1007/s00572-002-0213-6.

van der Ent, A., Baker, A.J.M., Reeves, R.D., Chaney, R.L., Anderson, C.W.N., Meech, J.A., et al., 2015. Agromining: farming for metals in the future? Environ. Sci. Technol. 49, 4773–4780. Available from: https://doi.org/10.1021/es506031u.

van der Ent, A., Baker, A.J.M., Reeves, R.D., Pollard, A.J., Schat, H., 2013. Hyperaccumulators of metal and metalloid trace elements: facts and fiction. Plant Soil 362, 319–334. Available from: https://doi.org/10.1007/s11104-012-1287-3.

Wang, G.-L., Que, F., Xu, Z.-S., Wang, F., Xiong, A.-S., 2015. Exogenous gibberellin altered morphology, anatomic and transcriptional regulatory networks of hormones in carrot root and shoot. BMC Plant Biol. 15, 290. Available from: https://doi.org/10.1186/s12870-015-0679-y.

Waqas, M., Khan, A.L., Kamran, M., Hamayun, M., Kang, S.M., Kim, Y.H., et al., 2012. Endophytic fungi produce gibberellins and indoleacetic acid and promotes host-plant growth during stress. Molecules 17, 10754–10773. Available from: https://doi.org/10.3390/molecules170910754.

Warembourg, F.R., 1997. The 'rhizosphere effect': a plant strategy for plants to exploit and colonize nutrient-limited habitats. Bocconea 7, 187–194.

Weyens, N., van der Lelie, D., Taghavi, S., Newman, L., Vangronsveld, J., 2009. Exploiting plant–microbe partnerships to improve biomass production and remediation. Trends Biotechnol. 27, 591–598. Available from: https://doi.org/10.1016/j.tibtech.2009.07.006.

Whipps, J.M., 2001. Microbial interactions and biocontrol in the rhizosphere. J. Exp. Bot. 52, 487–511.

Whiting, S.N., Reeves, R.D., Baker, A.J.M., 2002. Mining, metallophytes and land reclamation. Min. Environ. Manage. 10, 11–16.

Zaidi, S., Usmani, S., Singh, B.R., Musarrat, J., 2006. Significance of *Bacillus subtilis* strain SJ-101 as a bioinoculant for concurrent plant growth promotion and nickel accumulation in *Brassica juncea*. Chemosphere 64, 991–997. Available from: https://doi.org/10.1016/j.chemosphere.2005.12.057.

Chapter 3

Biotization and in vitro plant cell cultures: plant endophyte strategy in response to heavy metals knowledge in assisted phytoremediation

María del Rosario Espinoza-Mellado[1], Edgar Oliver López-Villegas[1], María Fernanda López-Gómez[2], Aída Verónica Rodríguez-Tovar[3], Marcial García-Pineda[4] and Angélica Rodríguez-Dorantes[5]

[1]*Microscopy Central, National School of Biological Sciences, National Polytechnic Institute, México City, México,* [2]*Biological Variation and Evolution Laboratory, Zoology Department, National School of Biological Sciences, National Polytechnic Institute, México City, México,* [3]*Mycology Laboratory, Microbiology Department, National School of Biological Sciences, National Polytechnic Institute, México City, México,* [4]*Botanical Garden, Faculty of Higher Studies Iztacala, National Autonomous University of Mexico, México City, México,* [5]*Plant Physiology Laboratory, Botany Department, National School of Biological Sciences, National Polytechnic Institute, México City, México*

3.1 Introduction

Callus culture is one of the most important plant science techniques for the development that involves plant regeneration and genetic manipulation. A functional association between phytobacteria and cultured plants has been successful, and coculture of tissue explants with beneficial microbes induces developmental and metabolic alterations giving tolerance to growth under abiotic and biotic stresses. This review focused on a particularly biotization process analyzing the effect to cadmium by the response on growth and development of the desert plant *Fouquieria splendens* callus cocultivated with endophyte bacteria.

3.2 Plant cell culture and cocultives

Tissue culture has been defined as a process where small pieces of living tissue named explants isolated from an organism grown aseptically on a nutrient medium under control conditions for indefinite periods; formerly, it is known as in vitro cultivation of plants. This process allows the proliferation of an undifferentiated cell mass called callus characterized by thin-walled parenchyma cells derived from parental plant tissue (Dodds and Robert, 1985). It involves a dramatic change related to the appearance and metabolism of plant cells (Aitchison et al., 1978); also, it is important to note that the concentrations of plant growth regulators present are very critical to guide growth and morphogenesis. Phytohormones such as auxins and cytokinins promote callus formation in tissue culture (Skoog and Armstrong, 1970; Letham, 1974; Akiyoshi et al., 1983). Auxins are associated with plant growth and morphogenesis, plant cells elongation, and division and differentiation (Dietz et al., 1990); meanwhile, cytokinins promote the plant cell enlargement in certain tissues (Rayle et al., 1982; Ross and Rayle, 1982). Skoog and Miller (1957) noted that the balance between auxins and cytokinins establish controls regarding the in vitro differentiation of callus tissue. Thus callus culture is considered as one of the most important plant science techniques for the development of clonal populations, plant regeneration, and genetic manipulation according to Reinert and Bajaj (1976). The cocultivation of tissue explants with beneficial microbes actually it is known induces developmental and metabolic alterations in the resulting plantlets; this complex interaction also gives tolerance to growth under abiotic and biotic stresses. The kind of microorganisms adequate to this purpose is the plant growth-promoting bacteria (PGPB) studied and applied by Kloepper et al. (1991) as

Microbe Mediated Remediation of Environmental Contaminants. DOI: https://doi.org/10.1016/B978-0-12-821199-1.00003-1

bioinoculants for the increase of plant productivity. Nowak (1998) also suggests that this relationship can increase the plant cell biomass and survival rate of tissue cultures. Some in vitro cocultive experiments reported successfully symbiotic relationships for *Daucus carota* (Varga et al., 1994), *Fragaria* sp. (Preininger et al., 1997; Vestberg et al., 2004), *Saccharum officinarum* (Oliveira et al., 2002), *Crocus sativus* L. (Parray et al., 2015), and *Swertia chirayita* (Sharma et al., 2015) showing significant effects in morphogenetic responses toward plant growth.

3.3 Biotization process in plant cell cultures

Nowak (1998) defines "biotization" as the in vitro coculture of tissue explants with beneficial microbes, a term that has been modified very recently by Herman (1996a,b). In vitro biotization in plant tissue culture not only made a screening of positive plant–microbe interaction under in vitro conditions to guarantee to be effective for the plant growth under field conditions; it also induces a resistance response against microbial inoculants (Benhamou et al., 1996a,b; Hunt and Ryals, 1996; Kloepper et al., 1997). It is known that artificial symbiotic associations could be included in this category (Varga et al., 1994; Preininger et al., 1997). Biotization prefers the employ of nonpathogenic organisms in plant tissue culture systems; thus they could act not only as inducers of stress resistance responses, because they also find and occupy microsites on host plants making them unavailable to pathogens. There are some reports regarding in vitro plants biotization that have employed microbial inoculants applied to plant micropropagation techniques (Hooker et al., 1994; Elmeskaoui et al., 1995; Nowak et al., 1995; Balla et al., 1997; Lazarovits and Nowak, 1997; Murphy et al., 1997; Wilhelm et al., 1997). Particularly bacterization techniques have been applied, and bacteria stimulate plantlets growth favoring developmental changes (Frommel et al., 1991; Liu et al., 1995; Nowak et al., 1995; Lazarovits and Nowak, 1997; Pillay and Nowak, 1997; Bensalim et al., 1998), where live bacterial cells are required. Nowak et al. (1995, 1997), Lazarovits and Nowak (1997), Pillay and Nowak (1997), and Bensalim et al. (1998) noted that both in vitro and ex vitro benefits of bacterization depended on plant species and growth conditions. DeFreitas and Germida (1990) reported the modification of morphology and stimulation in root tissue culture of wheat by various bacterial species and strains; Chanway and Nelson (1991) analyzed the *Glycine max* callus growth promotion on a vitamin-free medium by *Pseudomonas putida*. A *Brassica napus* and *Lactuca sativa* seedling bioassay demonstrated the root development stimulation by *Rhizobium leguminosarum* (Noel et al., 1996). Conn et al. (1997) successfully reported the assessment of growth-promoting ability from various bacterial isolates and mutants of nonfluorescent *Pseudomonas* species, and also this bioassay was employed to determine plant genotypic responses to particular strains or their combinations.

3.4 Plant growth-promoting bacteria functional attributes

PGPB are multifunctional organisms that are important for plants in many successful aspects as nutrition, growth, morphogenesis, and development, and all of the responses of them to biotic and abiotic stress also allow interactions with other organisms in agroecosystems (Ahmad et al., 2008; Babalola, 2010; Bhattacharyya and Jha, 2012; Berendsen et al., 2012; Gaiero et al., 2013; Philippot et al., 2013; Vacheron et al., 2013; Berg et al., 2014; du Jardin, 2015). It is known that the mechanisms involved in plants' growth promotion associate direct and indirect effects; direct effects are those when PGPB produce phytohormones, the most closer and important tool for sustainable agricultural (Dubeikovsky et al., 1993; Janzen et al., 1992; Srinivasan et al., 1996; Siddiqui, 2006; Erturk et al., 2010). Thus functional associations between bacteria and cultured plants under in vitro conditions have been very successful (Bertrand et al., 2001). It is important to consider that these microorganisms are a group of bacteria that actively colonize plants and stimulate their growth involving mechanisms such as symbiotic nitrogen fixation (Dobbelaere et al., 2003) and solubilization of phosphates (Cattelan et al., 1999). Enhance of growth and development of plants by some PGPB strains is possible because they could modify the concentration of phytohormones (Khalid et al., 1997, 2004). The employ of these kinds of bacteria for the acclimatization and hardening of in vitro raised plants under field conditions also have been reported (Thomas et al., 2010).

3.5 Analyzing the biotization of *Fouquieria splendens* callus cultures exposed to cadmium and their physiological response: strategy of phytoremediation knowledge

Plants possess at least two definitive cellular strategies to begin the process of regeneration, by the reactivation of undifferentiated cells and by the reprogramming of differentiated somatic cells. In both cases, regeneration is based on cellular plasticity, considered as the ability to respecify the cell program; knowing also that mode of plant regeneration is the de novo organogenesis, in which at first by the activity of apical meristems, shoots, and roots subsequently is developed; even this event proceeds either directly from parental tissues or indirectly via callus formation (Ikeuchi et al., 2016). Keeping in view the present knowledge, cocultivation of in vitro cells or tissues with PGPB strains is one of the appropriate approaches for mass propagation of plants (Parray et al., 2015). Considering that the use of in vitro systems as Golan-Goldhirsh et al. (2004) mention the advantages of plant tissue cultures, their reproducible growth during long periods and the sterile cultures allow the measurement of physiological activities that can be controlled by environmental conditions; the functional associations between cultured plants under in vitro conditions and bacteria have been successful as Bertrand et al. (2001) recommended.

Preininger et al. (1997) noted that there has been an increasing interest in understanding the symbiotic mechanisms in coculture between PGPB and plant tissue cultures not only to favoring a better propagation of plants; meanwhile, they apply as a strategy against heavy metals contributing to the basic knowledge of phytoremediation.

In this review a particular case is described here; the effect on growth and development of the desert plant *F. splendens* callus cocultivated with *Staphylococcus pasteuri* strain Fs2 (access number MG920267) exposed to cadmium. This endophyte bacterium was classified according to Khalid et al. (2004) by its in vitro production of indole acetic acid as a medium producer (13 μg/mL). *F. splendens* callus were obtained from leaves of plants, and in vitro cocultives of *F. splendens* callus and the endophyte bacteria were established according to Sharma et al. (2015) divided into two sequential phases: in the first phase the inoculum was streaked at the left and right sides of central division on Petri dishes containing specific culture medium and the second phase of the established cocultures initiated after the incubation of the endophyte strain when colonies appeared on the medium and pieces of the obtained *F. splendens* callus where deposited in the central division of the plates, with or without the endophyte bacteria. It is known that endophytes could interact with their hosts plant tissues effectively, because these are neither organ nor host specific (Elmeskaoui et al., 1995; Nowak et al., 1995 Balla et al., 1997; Lazarovits and Nowak, 1997). As a starting point, it should be emphasized that all available evidence indicates that endophytic PGPB employ plant growth promotion mechanisms and colonize the inside of plant tissues promptly (Frommel et al., 1991; Liu et al., 1995; Wilhelm et al., 1997). In vitro coculture of tissue explants with beneficial microbes may enhance protection and tolerance against a variety of abiotic stresses such as the presence of heavy metals. Progress of callus development under both not cocultivated and cocultivated with or without cadmium conditions (Fig. 3.1) was shown according to the rating system proposed by Souissi and Kremer (1994; 1998) based in visual observations of callus appearance (scale "0–4" without and with cell damage); in this bioassay, only the callus cultured in Murashige-Skoog "E" (MSE) medium (Murashige and Skoog, 1962) MS 1/4 salts medium + 1 mg/L NAA (naphthalene acetic acid) and + 1.5 mg/L KIN (kinetin); MSE + Cd 0.1 mM; MSE + Fs2; and MSE + Cd 0.1 mM + Fs2 were categorized in rating 0–2 with no visible callus damage, no discoloration, and an obvious not growth reduction. The rest of the callus where the increase in cadmium concentration under cocultivated or not cocultivated conditions showed a diminished growth and browning color categorized at the highest rating callus damage between 3 and 4, with characteristics such as tissue color change, cellular leakage, callus disintegration, and severe growth reduction (Table 3.1). In this study the nature of phytohormones assayed in *F. splendens* callus in the media tested (NAA and KIN) was determinant; Smolenskaya et al. (2007) and Meftahizade et al. (2000) particularly noted that employ of NAA in cell cultures ensures the growth and viability of them, and Sharma et al. (2015) commented that effects of higher concentration of cytokinins reduce the number of shoots generated per explant. In this bioassay, scanning electron microscopy analysis (Fig. 3.1) of selected *F. splendens* callus in no cocultivated conditions showed an evidence of an extracellular matrix (ECM) covering all the mass of cells, giving an early shoots development, as a differentiation of cells. The cell wall of higher plants is a dynamic cellular compartment with structural, protective, and growth regulation functions (Baluška et al., 2003). This active structure, referred to as the ECM (Roberts, 1994), is an integral part of the ECM–plasma membrane–cytoskeleton continuum and plays fundamental roles in intercellular interactions of multicellular organisms. Recently, there has been considerable interest in the ECM surface network; its chemical composition and the structural arrangement on the cell surface may play significant roles in morphological processes (Popielarska-Konieczna et al., 2010). This structure, which is referred to as ECM or ECM surface network (ECMSN), has been found during the early stages of somatic embryogenesis and organogenesis (Šamaj et al., 1999; Verdeil et al., 2001). The exact role of this structure is still not known, but its regulatory and coordinating functions during the early stages of morphogenesis have been suggested (Verdeil et al., 2001; Bobák et al., 2003; Konieczny et al., 2005; Popielarska-Konieczna et al., 2008a,b; Blehová et al., 2010; Pilarska et al., 2014). Popielarska-Konieczna et al. (2010) noted that cultivation of cell and tissue in vitro activates different cellular response mechanisms that could be related to mediating cell adaptation under new changing culture conditions. One of them generates the formation of the ECM surface layer or network on the outer cell wall, and also Bobák et al. (2003, 2004) suggest that these layers of complex noncellular material outside the outer cell wall of the cells on the callus surface are related to a stress

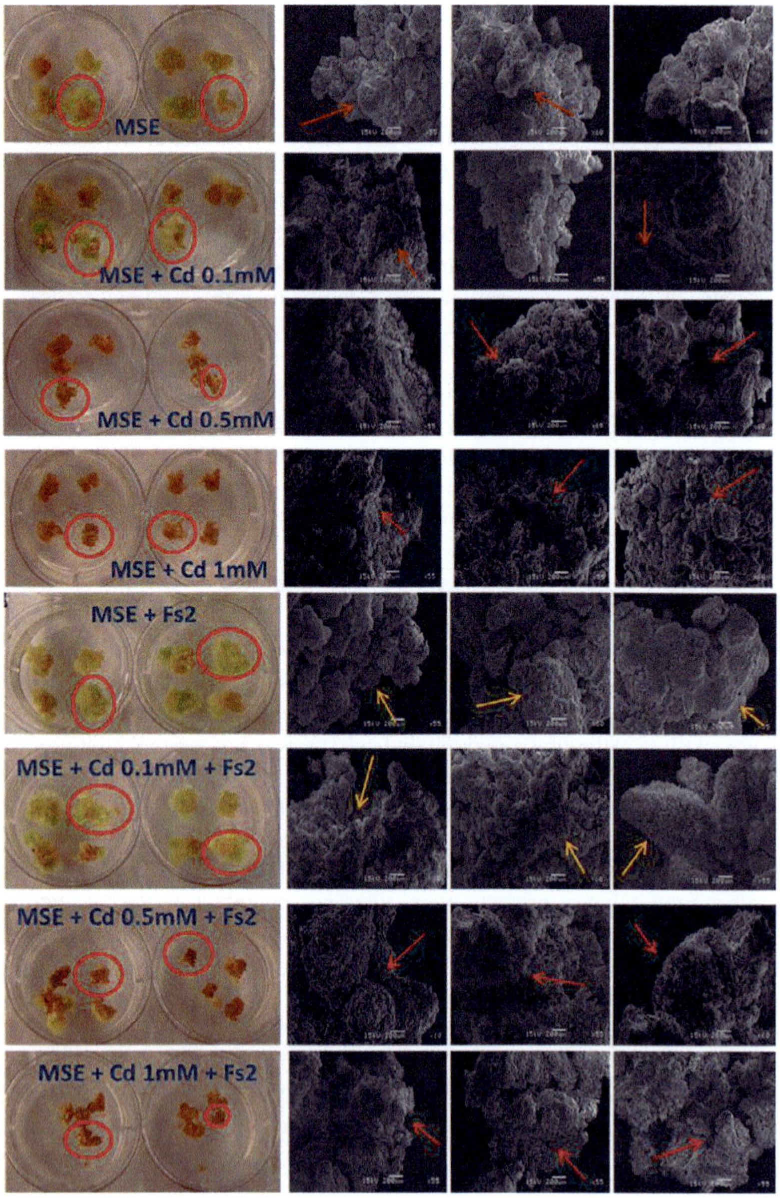

FIGURE 3.1 *Fouquieria splendens* visual callus development not cocultivated and cocultivated with *Staphylococcus pasteuri* strain Fs2. Scanning electronic micrographs of each experimental condition revealed the presence of heterogeneous material covering the callus surface (*orange arrows*); particular morphological changes in callus clusters showed an early appearance of shoots and roots (*yellow arrows*) and the highest damage with a constraint of callus exposed to cadmium 0.5 and 1 mM concentrations (*red arrows*).

TABLE 3.1 *Fouquieria splendens* callus damage after exposure to cadmium.

	Callus damage				
Experiment	"0"	"1"	"2"	"3"	"4"
MSE	16	3	5	0	0
MSE + Fs2	16	4	2	1	0
MSE + Cd 0.1 mM	20	14	5	4	5
MSE + Cd 0.1 mM + Fs2	21	1	2	0	0
MSE + Cd 0.5 mM	0	0	0	7	37
MSE + Cd 0.5 mM + Fs2	0	0	0	0	14
MSE + Cd 1.0 mM	0	0	0	33	3
MSE + Cd 1.0 mM + Fs2	0	0	0	10	13

response of explants, and it might protect the surface of callus clusters and favored the integration and cell recognition of them. In *F. splendens* callus, there was a discontinuous amorphous layer of complex noncellular material outside the outer cell wall of the cells on the callus surface. Popielarska et al. (2006) and Popielarska-Konieczna et al. (2008a,b, 2010) revealed in previous scanning electron microscopy studies on kiwifruit organogenesis the presence of a membranous layer covering the outer cell wall. The role of ECMSN cannot be generalized, but it is evident that it is formed on the surface of cultured in vitro tissues regardless of its morphogenetic competence. Therefore we agree with the suggestion that has been made by several authors (Dubois et al., 1992; Šamaj et al., 1995) that ECMSN formation could be a stress response of plant tissues that is triggered by specific culture conditions and that covering callus with extracellular material could provide the protection against external factors (Pilarska et al., 2014). Callus of *F. splendens* developed in MSE medium in cocultivated with the endophyte bacteria showed morphological changes where some cluster cells change acquiring a slight appearance of early shoots (MSE + Fs2) or roots (MSE + Cd 0.1 mM + Fs2).

Fig. 3.2 showed the cellular mass gain of callus that also is related to the effect of cadmium concentrations in medium. It is important to note that hormesis response (Bagur-González et al., 2011) was obtained in this bioassay, regarding the lowest concentration of cadmium employed, because callus fresh weight biomass and the response of this cellular mass increased; also the interaction cocultivated with the endophyte *S. pasteuri* strain Fs2 enhances the gain and development of *F. splendens* callus that did not tolerate and resist the other two cadmium concentrations tested: 0.5 and 1 mM, with a diminish in growth.

Cadmium is known to be a nonessential heavy metal and can cause oxidative stress in plants (Benavides et al., 2005; Szőllősi et al., 2009). Benavides et al. (2005) and Sgherri et al. (2001) established that the stage of ontogenesis when the plant is stressed by heavy metals seems to be an important point along with the stressing period. It is known that cadmium as a nonessential toxic heavy metal induces morphological and physiological responses in plants (Chaoui et al., 1997), and also the presence of this heavy metal generates oxidative stress with the formation of reactive oxygen species (ROS). Gallego et al. (1996, 1999), Mishra et al. (2006), Ranieri et al. (2005), and Smeets et al. (2005) reported also a decrease in the activity of enzymatic and nonenzymatic antioxidants. A biological antioxidant has been defined as "any substance that, when present at low concentrations compared to those of an oxidisable substrate, significantly delays or prevents oxidation of that substrate" (Halliwell and Gutteridge, 1995). This definition is clear and covers every member of the antioxidant defense team. However, unless an antioxidant prevents the generation of ROS, a redox reaction still occurs (Gutteridge, 1994; Halliwell, 1994). The difference is that the oxidizing species reacts with the

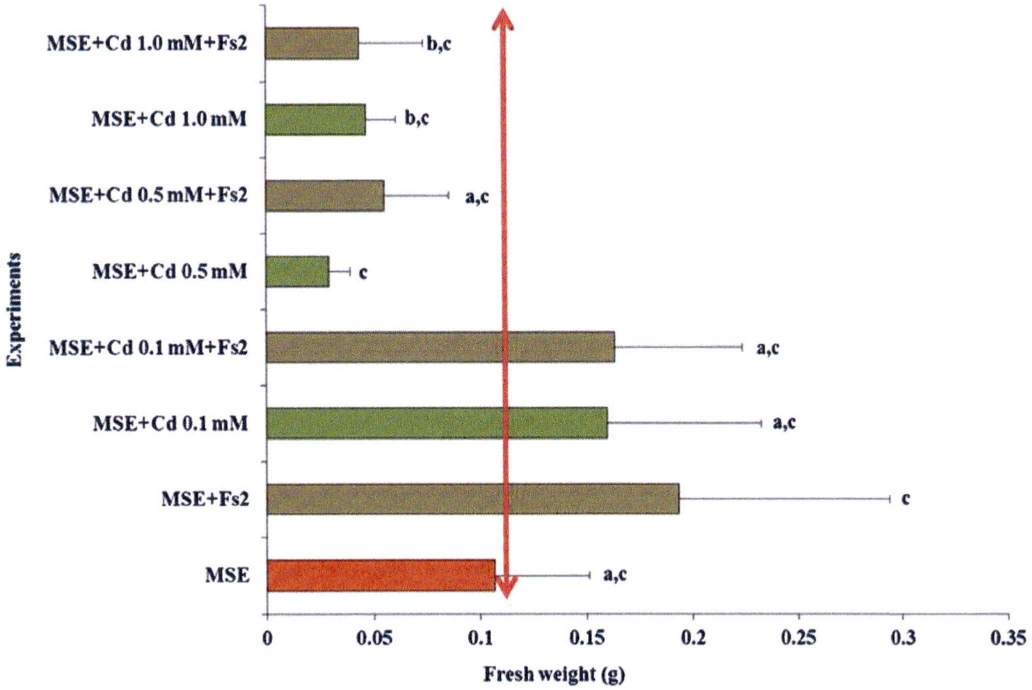

FIGURE 3.2 Fresh weight of *Fouquieria splendens* callus growth under not cocultivated and cocultivated with *Staphylococcus pasteuri* strain Fs2 [$n = 24$, the different lower case letters show the significant differences founded ($P < .001$)].

antioxidant instead of the "substrate." In this context, antioxidant power may be referred to analogously as reducing ability. This being the case, a method using reductants in a redox-linked colorimetric method employing an easily reduced oxidant in stoichiometric excess could offer a simple way of assessing this ability. The method described measures the ferric reducing ability of plasma (FRAP). At low pH, when a ferric-trotripyridyltriazine (Fe^{III}–TPTZ) complex is reduced to the ferrous (Fe^{II}) form, an intense blue color with an absorption maximum at 593 nm develops (Liu et al., 1982; Benzie, 1996). The reaction is nonspecific, and any half reaction that has a less-positive redox potential, under reaction conditions, than the Fe^{III}/Fe^{II}–TPTZ half reaction will drive Fe^{III}–TPTZ reduction. Test conditions favor reduction of the complex and, thereby, color development, provided that a reductant (antioxidant) is present (Benzie and Strain, 1996). To measure the total antioxidant power or capacity of homogenates, which includes the enzymatic and nonenzymatic antioxidants, a simple, quick, and semiquantitative FRAP method has been modified and applied for plant material (Varga et al., 2000; Schützendübel and Polle, 2002; Szőllősi and Varga, 2002). FRAP assay applied to the callus of *F. splendens* allows to measure a particularly indirect biochemical response regarding to the redox potential analyzed in heavy metal conditions tested in this bioassay; powerfully free radicals as indirect measure by FRAP show how the callus responds against not only to cadmium but also the presence of endophyte. It is clear that all the experimental conditions favored the response to external agents even the heavy metal or the bacteria, with the highest values presented in MSE + Fs2 condition, followed by MSE + 0.1 mM Cd + Fs2 and MSE + 0.5 mM Cd conditions (Fig. 3.3).

It is important to note, as Parray et al. (2015) mention, that the effects of PGPB are considered to be highly specific with respect to plant and bacterial genotypic combination, and also Yang et al. (1991) reported another possible reason for the induction of some callus changes is the release of certain compounds from the bacteria. These authors also mention that, besides the phytobacterial strains, there were several factors influencing the initiation and differentiation of callus; the quality and especially the age of the callus employed were of great importance, because some young callus lines are considered more resistant and contribute to the initiation and growth of callus with a great regeneration potential. Fig. 3.4 shows principal component analysis (PCA), comprising two principal components (PC1 78.78% and PC2 13.09%), accounted for 91.87% of the total variance. In this figure, there was a particular association of these experimental conditions: MSE, Fs2, Cd 0.1 mM and Cd 0.1 mM + Fs2; all of them do not present any callus damage. The other association clustered: Cd 0.5 mM, Cd 0.5 mM + Fs2, Cd 1.0 mM, and Cd 1.0 mM + Fs2, with the highest callus damage, and the PCA clearly shows the relationship between this coculture conditions with the heavy metal and also

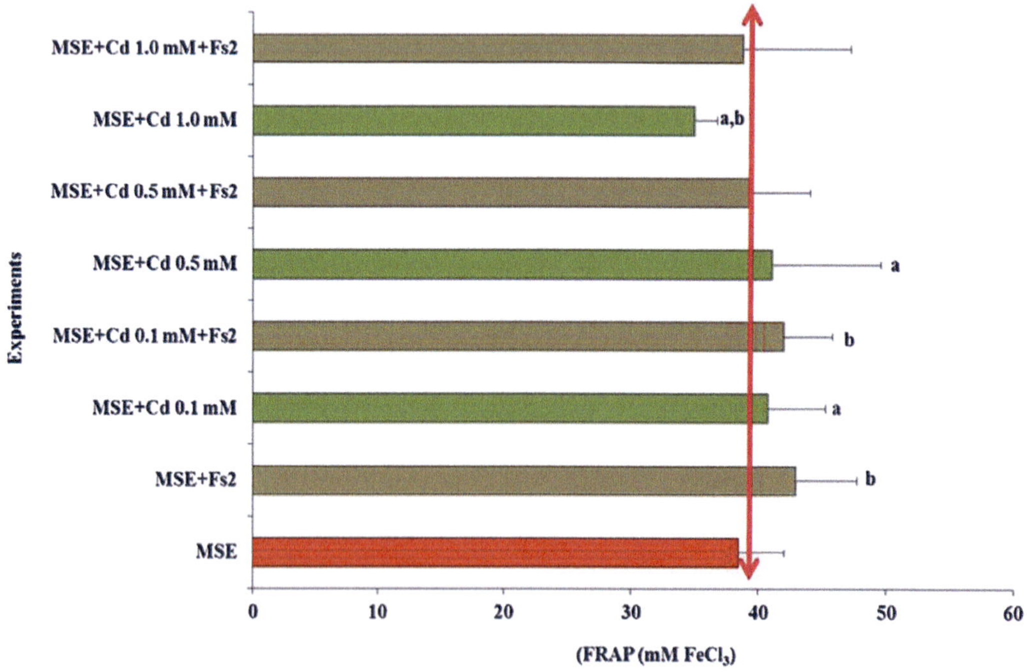

FIGURE 3.3 FRAP assay of *Fouquieria splendens* callus growth under not ocultivated and cocultivated with *Staphylococcus pasteuri* strain Fs2 [$n = 24$, the different lower case letters show the significant differences founded ($P < .001$)]. *FRAP*, Ferric reducing ability of plasma.

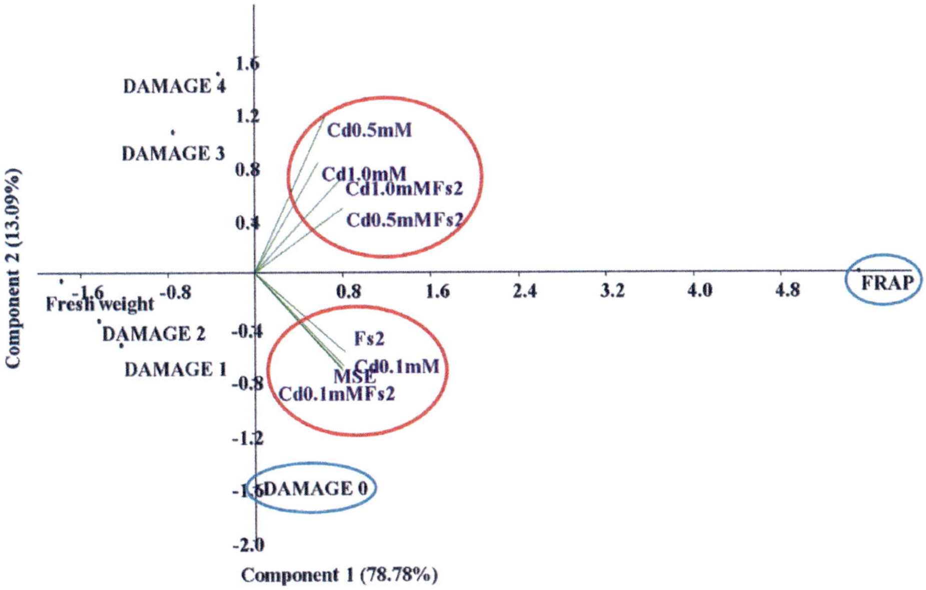

FIGURE 3.4 Principal component analysis of *Fouquieria splendens* callus growth under not cocultivated and cocultivated with *Staphylococcus pasteuri* strain Fs2 comparing the measured parameters.

the FRAP assay. Finally, the composition of the culture medium has a great influence on cocultures, giving an initiation, proliferation, and development of the callus tested.

3.6 Conclusion

In this work the established cocultive between *F. splendens* callus and the presence of the streaked endophyte *S. pasteuri* strain Fs2 on culture medium showed that it could be a complex network of responses between the plant growth regulators present in culture medium and the release of phytohormones produced by the phytobacteria that at least showed a notably effect on the growth and differentiation response of the callus tested. As phytoremediation strategy, the basic studies and knowledge obtained from in vitro strategies favored the employ of these systems to understand the mechanisms that could explain the plant cell response against inorganic contaminants as protecting relationship to respond cadmium presence.

Acknowledgments

The authors are grateful to the research projects SIP-20181504 and SIP-20202094 of the Secretaría de Investigación y Posgrado del Instituto Politécnico Nacional, Comisión de Operaciones y Fomento de Actividades Académicas (COFAA-IPN), EDI (Estímulo al Desempeño de Investigadores-I.P.N.), and Sistema Nacional de Investigadores (SNI-CONACyT), the fellowships for its support.

References

Ahmad, I., Pichtel, J., Hayat, S., 2008. Plant-bacteria interactions. Strategies and Techniques to Promote Plant Growth. WILEY-VCH Verlag GmbH and Co. KGaA, Weinheim.

Aitchison, P.A., MacLeod, A.J., Yeoman, M.M., 1978. Growth patterns in tissue (callus) cultures. In: Street, H.E. (Ed.), Plant Cell Culture. Blackwell, Science Publishers, Oxford, pp. 267–306.

Akiyoshi, D.E., Morris, R.O., Hinz, R., Mischke, B.S., Kosuge, T., Garfinkel, D.J., et al., 1983. Cytokinin/auxin balance in crown gall tumors is regulated by specific loci in the T-DNA. Proc. Natl. Acad. Sci. U.S.A. 80, 407–411.

Babalola, O.O., 2010. Beneficial bacteria of agricultural importance. Biotechnol. Lett. 32, 1559–1570.

Bagur-González, M.G., Estepa-Molina, C., Martín-Peinado, F., Morales-Ruano, S., 2011. Toxicity assessment using *Lactuca sativa* L. bioassay of the metal(loid)s As, Cu, Mn, Pb and Zn in soluble-in-water saturated soil extracts from an abandoned mining site. J. Soil Sediment. 11, 281–289.

Balla, I., Virtesy, J., Ktives-Pechy, K., 1997. Acclimation results of micropropagated black locust (*Robinia pseudoacacia* L.) improved by use of microorganisms. In: Cassells, A.L. (Ed.), Pathogen and Microbial Contamination Management in Micropropagation. Kluwer Acad. Publ, Dordrecht, pp. 351–354.

Baluška, F., Šamaj, J., Wojtaszek, P., Volkmann, D., Menzel, D., 2003. Cytoskeleton–plasma membrane–cell wall continuum in plants. Emerging links revisited. Plant Physiol. 133, 482–491.

Benavides, M.P., Gallego, S.M., Tomaro, M.L., 2005. Cadmium toxicity in plants. Braz. J. Plant Physiol. 17, 21–34.

Benhamou, N., 1996a. Elicitor-induced plant defence pathways. Trends Pharmacol. Sci. 1, 233–240.

Benhamou, N., Kloepper, J.W., Quadt-Hallman, A., Tuzun, S., 1996b. Induction of defence-related ultra-structural modifications in pea root tissues inoculated with endophytic bacteria. Plant Physiol. 112, 919–929.

Bensalim, S., Nowak, J., Asiedu, S.K., 1998. A plant growth promoting rhizobacterium and temperature effects on performance of 18 clones of potato. Am. J. Potato Res. 75, 145–152.

Benzie, I.F.F., Strain, J.J., 1996. The ferric reducing ability of plasma (FRAP) as a measure of "antioxidant power": the FRAP assay. Anal. Biochem. 239, 70–76.

Benzie, I.F.F., 1996. An automated, specific, spectrophotometric method for measuring ascorbic acid in plasma (EFTSA). Clin. Biochem. 29, 111–116.

Berendsen, R.L., Pieterse, C.M., Bakker, P.A., 2012. The rhizosphere microbiome and plant health. Trends Plant Sci. 17, 1360–1385.

Berg, G., Grube, M., Schloter, M., Smalla, K., 2014. Unraveling the plant microbiome: looking back and future perspectives. Front. Microbiol. 5, 1–7.

Bertrand, H., Nalin, R., Bally, R., Marel, J.C.C., 2001. Isolation and identification of the most efficient plant growth promoting bacteria associated with canola. Biol. Fertil. Soils 3, 152–156.

Bhattacharyya, P.N., Jha, D.K., 2012. Plant growth-promoting rhizobacteria (PGPR): emergence in agriculture. World J. Microbiol. Biotechnol. 28, 1327–1350.

Blehová, A., Bobák, M., Šamaj, J., Hlinková, E., 2010. Changes in the formation of an extracellular matrix surface network during early stages of indirect somatic embryogenesis in *Drosera spathulata*. Acta Bot. Hung. 52, 23–33.

Bobák, M., Šamaj, J., Hlinková, E., Hlavačka, A., Ovečka, M., 2003. Extracellular matrix in early stages of direct somatic embryogenesis in leaves of *Drosera spathulata*. Biol. Plant. 47, 161–166.

Bobák, M., Šamaj, J., Pretóvá, A., Blehová, A., Hlinkova, E., Ovečka, M., et al., 2004. The histological analysis of indirect somatic embryogenesis on *Drosera spathulata* Labill. Acta Physiol. Plant. 26, 353–361.

Cattelan, A.J., Hartel, P.G., Fuhrmann, J.J., 1999. Screening of plant growth promoting rhizobacteria to promote early soybean growth. Soil Sci. Soc. Am. J. 63, 1670–1680.

Chanway, C.P., Nelson, L.M., 1991. Tissue culture bioassay for plant growth promoting rhizobacteria. Soil Biol. Biochem. 23, 331–333.

Chaoui, A., Mazhoudi, S., Ghorbal, M.H., El Ferjani, E., 1997. Cadmium and zinc induction of lipid peroxidation and effects on antioxidant enzyme activities in bean (*Phaseolus vulgaris* L). Plant Sci. 127, 139–147.

Conn, K.L., Nowak, J., Lazarovits, G.A., 1997. Gnotobiotic bioassay for studying interactions between potato and plant growth-promoting rhizobacteria. Can. J. Microbiol. 43, 801–808.

DeFreitas, J.R., Germida, J.J., 1990. A root tissue culture system to study Winter wheat-rhizobacteria interactions. Appl. Microbiol. Biotechnol. 33, 589–595.

Dietz, A., Kutschera, U., Ray, P.M., 1990. Auxin enhancement of mRNAs in epidermis and internal tissues of the pea stem and its significance for control of elongation. J. Plant Physiol. 93, 432–438.

Dobbelaere, S., Vanderleyden, J., Okon, Y., 2003. Plant growth promoting effects of dizotrophs in the rhizosphere. Crit. Rev. Plant Sci. 22, 149–197.

Dodds, J.H., Robert, L.W., 1985. Experiment in Plant Tissue Culture, second ed. Cambridge University Press, Cambridge.

du Jardin, P., 2015. Plant biostimulants: definition, concept, main categories and regulation. Sci. Hortic. 196, 3–14.

Dubeikovsky, A.N., Mordukhova, E.A., Kochetkov, V.V., Polikarpova, F.Y., Boronin, A.M., 1993. Growth promotion of blackcurrant softwood cuttings by recombinant strain *Pseudomonas fluorescens* BSP53a synthesizing an increased amount of indole-3acetic acid. Soil Biol. Biochem. 25, 1277–1281.

Dubois, T., Dubois, J., Guedira, M., Diop, A., Vasseur, J., 1992. SEM characterization of an extracellular matrix around somatic proembryos in roots of *Cichorium*. Ann. Bot. 70, 119–124.

Elmeskaoui, A., Damont, J.P., Poulin, M.J., Piché, Y., Desjardins, Y., 1995. A tripartite euhure system for endomycorrhizal inoculation of micropropagated strawberry plantlets *in vitro*. Mycorrhiza 5, 313–319.

Erturk, Y., Ercisli, E., Haznedar, A., Cakmakc, R., 2010. Effects of plant growth promoting rhizobacteria (PGPR) on rooting and root growth of kiwifruit (*Actinidia deliciosa*) stem cuttings. Biol. Res. 43, 91–98.

Frommel, M.I., Nowak, J., Lazarovits, G., 1991. Growth enhancement and developmental modifications of *in vitro* grown potato (*Solanum tuberosum* spp. *tuberosum*) as affected by a nonfluorescent *Pseudomonas* sp. Plant Physiol. 96, 928–936.

Gaiero, J.R., McCall, C.A., Thompson, K.A., Dayu, N.J., Best, A.S., Dunfield, K.E., 2013. Inside the root microbiome: bacterial root endophytes and plant growth promotion. Am. J. Bot. 100, 1738–1750.

Gallego, S.M., Benavides, M.P., Tomaro, M.L., 1996. Effect of heavy metal ion excess on sunflower leaves: evidence for involvement of oxidative stress. Plant Sci. 121, 151–159.

Gallego, S.M., Benavides, M.P., Tomaro, M.L., 1999. Effect of cadmium ions on antioxidant defense system in sunflower cotyledons. Biol. Plant. 42, 49–55.

Golan-Goldhirsh, L.A., Barazani, O., Nepovim, A., Soudek, P., Smrcek, S., Dufkova, L., et al., 2004. Plant response to heavy metals and organic pollutants in cell culture and at whole plant. J. Soil Sediment. 4, 133–140.

Gutteridge, J.M.C., 1994. Biological origin of free radicals, and mechanisms of antioxidant protection. Chem. Biol. Interact. 91, 133–140.

Halliwell, B., 1994. Antioxidants and human disease: a general introduction. Nutr. Rev. 53, 253–265.

Halliwell, B., Gutteridge, J.M.C., 1995. The definition and measurement of antioxidants in biological systems methods. Free Radical Biol. Med. 18, 125–126.

Herman, E.B., 1996a. Beneficial effects of bacteria and fungi on plant tissue cultures. Agricell Rep. 27, 26–27.

Herman, E.B., 1996b. Microbial contamination of plant tissue cultures, Recent Advances in Plant Tissue Culture, vol. IV. Agriteeh Cons., Inc, Shrub Oak, NY.

Hooker, J.E., Gianinazzi, S., Vestberg, M., Barea, J.M., Atkinson, D., 1994. The application of arbuscular mycorrhizal fungi to micropropagation systems: an opportunity to reduce chemical inputs. Agric. Sci. Finl. 3, 227–232.

Hunt, M.D., Ryals, J.A., 1996. Systemic acquired resistance signal transduction. Crit. Rev. Plant Sci. 15, 583–606.

Ikeuchi, M., Ogawa, Y., Iwase, A., Sugimoto, K., 2016. Plant regeneration: cellular origins and molecular mechanisms. Development 143, 1442–1451.

Janzen, R.A., Rood, S.B., Dormaar, J.F., McGill, W.B., 1992. *Azospirillum brasilense* produces gibberellin in pure culture on chemically-defined medium and in co-culture on straw. Soil Biol. Biochem. 24, 1061–1064.

Khalid, A., Arshad, M., Zahir, Z.A., Khaliq, A., 1997. Potential of plant growth promoting rhizobacteria for enhancing wheat yield. J. Anim. Plant Sci. 7, 53–56.

Khalid, A., Arshad, M., Zahir, Z.A., 2004. Screening plant growth promotin rhizobacteria for improving growth and yield of wheat. J. Appl. Microbiol. 96, 473.

Kloepper, J.W., Rodriguez-Kabana, R., McInroy, J.A., Collins, D.J., 1991. Analysis of population and physiological characterization of microorganism in rhizosphere of plants with antagonistic properties to phytopathogenic nematodes. Plant Soil 136, 95–102.

Kloepper, J.W., Tuzun, S., Zehnder, G.W., Wei, G., 1997. Multiple disease protection by rhizobacteria that induce systemic resistance--historical perspective. Phytopathology 87, 136–137.

Konieczny, R., Bohdanowicz, J., Czaplicki, A.Z., Przywara, L., 2005. Extracellular matrix surface network during plant regeneration in wheat anther culture. Plant Cell Tissue Organ. Cult. 83, 201–208.

Lazarovits, G., Nowak, J., 1997. Rhizobacteria for improvement of plant growth and establishment. HortScience 32, 188–192.

Letham, D.S., 1974. Regulators of cell division in plant tissues. XX. The cytokinins of coconut milk. J. Plant Physiol. 32, 66–70.

Liu, T.Z., Chin, N., Kiser, M.D., Bigler, W.N., 1982. Specific spectrophotometry of ascorbic acid in serum or plasma by use of ascorbate oxidase. Clin. Chem. 28, 2225–2228.

Liu, Z., Pillay, V., Nowak, J., 1995. *In vitro* culture of watermelon and cantaloupe with and without beneficial bacterium. Acta Hortic. 402, 58–60.

Meftahizade, H., Lotfi, M., Moradkhani, H., 2000. Optimization of micropropagation and establishment of cell suspension culture in *Melissa officinalis* L. Afr. J. Biotechnol. 9, 4314–4321.

Mishra, S., Srivastava, S., Tripathi, R.D., Govindarajan, R., Kuriakose, S.V., Prasad, M.N.V., 2006. Phytochelatin synthesis and response of antioxidants during cadmium stress in *Bacopa monnieri* L. Plant Phys. Biochem. 44, 25–37.

Murashige, T., Skoog, F., 1962. A revised medium for rapid growth and bioassays with tobacco tissues cultures. Physiol. Plant. 15, 473–497.

Murphy, J., Mark, L., Periappuram, C., 1997. Microbial characterisation and preparation of inoculum for *in vitro* mycorrhization of strawberry in autotrophic culture. In: Cassells, A.L. (Ed.), Pathogen and Microbial Contamination Management in Micropropagation. Kluwer Acad. Publ., Dordrecht, pp. 345–350.

Noel, T.C., Sheng, C., Yost, C.K., Pharis, R.P., Hynes, M.F., 1996. *Rhizobium leguminosarum* as a plant growth promoting rhizobacterium: direct growth promotion of canola and lettuce. Can. J. Microbiol. 42, 279–283.

Nowak, J., 1998. Benefits of *in vitro* "biotization" of plant tissue cultures with microbial inoculants. In Vitro Cell Dev. Biol. Plant 34, 122–130.

Nowak, J., Asiedu, S.K., Lazarovits, G., 1995. Enhancement of *in vitro* growth and transplants, tress tolerance of potato and vegetable plants cocultured with a plant growth promoting rhizobacterium. In: Carre, E., Chagvardieff, P. (Eds.), Ecophysiology and Photosynthetic *In Vitro* Cultures. CEA, Aix-en-Provence, pp. 173–180.

Nowak, J., Asiedu, S.K., Bensalim, S., 1997. From laboratory to applications: challenges and progress with *in vitro* dual cultures of potato and beneficial bacteria. In: Cassells, A.L. (Ed.), Pathogen and Microbial Contamination Management in Micropropagation. Kluwer Acad. Publ., Dordrecht, pp. 321–329.

Oliveira, A.D., Urquiaga, S., Döbereiner, J., Baldani, J.I., 2002. The effect of inoculating endophytic N_2-fixing bacteria on micropropagated sugarcane plants. Plant Soil 242, 205–215.

Parray, J.A., Kamili, A.N., Reshi, Z.A., Qadri, R.A., Jan, S., 2015. Interaction of rhizobacterial strains for growth improvement of *Crocus sativus* L. under tissue culture conditions. Plant Cell Tissue Organ 121, 325–334.

Philippot, L., Raaijmakers, J.M., Lemanceau, P., Putten, W.H.V.D., 2013. Going back to the roots: the microbial ecology of the rhizosphere. Nat. Rev. Microbiol. 11, 789–799.

Pilarska, M., Popielarska-Konieczna, M., Ślesak, H., Kozieradzka-Kiszk, M., Góralski, G., Konieczny, R., et al., 2014. Extracellular matrix surface network is associated with nonmorphogenic calli of *Helianthus tuberosus* cv. Albik produced from various explants. Acta Soc. Bot. Pol. 83, 67–73.

Pillay, V.K., Nowak, J., 1997. Inoculum density, temperature and genotype effects on epiphytic and endophytic colonization and in vitro growth promotion of tomato (*Lycopersicon esculentum* L.) by a pseudomonad bacterium. Can. J. Microbiol. 43, 354–361.

Popielarska, M., Ślesak, H., Góralski, G., 2006. Histological and SEM studies on organogenesis in endosperm-derived callus of kiwifruit (*Actinidia deliciosa* cv. Hayward). Acta Biol. Cracov. Ser. Bot. 48, 97–104.

Popielarska-Konieczna, M., Kozieradzka-Kiszkurno, M., Świerczyńska, J., Góralski, G., Ślesak, H., Bohdanowicz, J., 2008a. Ultrastructure and histochemical analysis of extracellular matrix surface network in kiwifruit endosperm-derived callus culture. Plant Cell Rep. 27, 1137–1145.
Popielarska-Konieczna, M., Kozieradzka-Kiszkurno, M., Świerczyńska, J., Góralski, G., Ślesak, H., Bohdanowicz, J., 2008b. Are extracellular matrix surface network components involved in signaling and protective function? Plant Signal. Behav. 3, 707–709.
Popielarska-Konieczna, M., Bohdanowicz, J., Starnawska, E., 2010. Extracellular matrix of plant callus tissue visualized by ESEM and SEM. Protoplasma 247, 121–125.
Preininger, Ė., Zatyko, J., Szucs, P., Koranyi, P., Gyurjan, I., 1997. *In vitro* establishment of nitrogen-fixing strawberry (*Fragaria ananassa*) via artificial symbiosis with *Azomonas insignis*. In Vitro Cell Dev. Biol. Plant 33, 190–194.
Ranieri, A., Castagna, A., Scebba, F., Careri, M., Zagnoni, I., Predieri, G., et al., 2005. Oxidative stress and phytochelatin characterisation in bread wheat exposed to cadmium stress. Plant Phys. Biochem. 43, 45–54.
Rayle, D.L., Ross, C.W., Robinson, N., 1982. Estimation of osmitc parameters accompanying zeatin-induced growth of detached cucumber cotylendons. J. Plant Physiol. 70, 1634–1636.
Reinert, J.Y., Bajaj, P.S., 1976. Applied and fundamental aspects of plant cell, tissue and organ culture. J. Exp. Bot. 27, 144–147.
Roberts, K., 1994. The plant extracellular matrix: in a new expansive mood. Curr. Opin. Cell Biol. 6, 688–694.
Ross, C.W., Rayle, D.L., 1982. Evaluation of H^+ secretion relative to zeatin-induced growth of detached cucumber cotyledons. J. Plant Physiol. 70, 1470–1474.
Šamaj, J., Baluška, F., Bobák, M., Volkmann, D., 1999. Extracellular matrix surface network of embryogenic units of friable maize callus contains arabinogalactan-proteins recognized by monoclonal antibody JIM4. Plant Cell Rep. 18, 369–374.
Šamaj, J., Bobák, M., Blehová, A., Krištin, J., Auxtová-Šamajová, O., 1995. Developmental SEM observations on an extracellular matrix in embryogenic calli of *Drosera rotundifolia* and *Zea mays*. Protoplasma 186, 45–49.
Schűtzendűbel, A., Polle, A., 2002. Plant responses to abiotic stresses: heavy metal-induced oxidative stress and protection by mycorrhization. J. Exp. Bot. 53, 1351–1365.
Sgherri, C., Milone, A.M.T., Clijsters, H., Navari-Izzo, F., 2001. Antioxidative enzymes in two wheat cultivars, differently sensitive to drought and subjected to subsymptomatic copper doses. J. Plant Physiol. 158, 1439–1447.
Sharma, V., Kamal, B., Srivastava, N., Negi, Y., Dobriyal, A.N., Jadon, V.S., 2015. Enhancement of *in vitro* growth of *Swertia chirayita* Roxb. Ex Fleming co-cultured with plant growth promoting rhizobacteria. Plant Cell Tissue Organ. Cult. 121, 215–225.
Siddiqui, Z., 2006. Plant Growth Promoting Bacteria (PGPR). Springer, p. 318.
Skoog, F., Armstrong, D.J., 1970. Cytokinin. Annu. Rev. J. Plant Physiol. 21, 359–384.
Skoog, F., Miller, E.O., 1957. Chemical regulation of growth and organ formation in plant tissues cultured *in vitro*. J. Exp. Biol. 11, 118–131.
Smeets, K., Cuypers, A., Lambrechts, A., Semane, B., Hoet, P., Van Laere, A., et al., 2005. Induction of oxidative stress and antioxidative mechanisms in *Phaesolus vulgaris* after Cd application. Plant Phys. Biochem. 43, 437–444.
Smolenskaya, I.N., Reshetnyak, O.V., Smirnova, Y.N., Chernyak, N.D., Nosov, A.M., Nosov, A.V., 2007. Opposite effects of synthetic auxins, 2,4-dichlorophenoxyacetic acid and 1-naphthalene acetic acid on growth of true ginseng cell culture and synthesis of ginsenosides. Russ. J. Plant Physiol. 54, 215–223.
Souissi, T., Kremer, R.J., 1994. Leafy spurge (*Euphorbia esula*) cell cultures for screening deleterious rhizobacteria. Weed Sci. 42, 310–315.
Souissi, T., Kremer, R.J., 1998. A rapid microplate callus bioassay for assessment of rhizobacteria for biocontrol of leafy spurge (*Euphorbia esula* L.). Biocontrol Sci. Technol. 8, 83–92.
Srinivasan, M., Petersen, D.J., Holl, F.B., 1996. Influence of indolacetic-acid-producing *Bacillus isolates* on the nodulation of *Phaseolus vulgaris* by *Rhizobium etli* under gnotobiotic conditions. Can. J. Microbiol. 42, 1006–1014.
Szőllősi, R., Varga, I.S., 2002. Total antioxidant power in some species of Labiatae (adaptation of FRAP method). Acta Biol. Szeged. 46, 125–127.
Szőllősi, R., Varga, I.S., Erdei, L., 2009. Cadmium-induced oxidative stress and antioxidative mechanisms in germinating Indian mustard (*Brassica juncea* L.) seeds. Ecotoxicol. Environ. Saf. 72, 137–1342.
Thomas, J., Ajay, D., Kumar, R.R., Mandal, A.K.A., 2010. Influence of beneficial microorganisms during *in vivo* acclimatization of *in vitro*-derived tea (*Camellia sinensis*) plants. Plant Cell Tissue Organ 101, 365–370.
Vacheron, J., Desbrosse, G., Bouffaud, M.L., Touraine, B., Moënne-Loccoz, Y., Muller, D., et al., 2013. Plant growth-promoting rhizobacteria and root system functioning. Front. Plant Sci. 4, 1–19.
Varga, S.S., Koranyi, P.A., Preininger, E., Gyruran, I., 1994. Artificial associations between *Daucus* and nitrogen-fixing *Azotobacter* cells *in vitro*. Plant Physiol. 90, 789–790.
Varga, I.S., Szőllősi, R., Bagyánski, M., 2000. Estimation of total antioxidant power in medicinal plants (adaptation of FRAP). Curr. Top. Biophys. 24, 219–225.
Verdeil, J.L., Hocher, V., Huet, C., Grosdemange, F., Escoute, J., Ferriere, N., 2001. Ultrastructural changes in coconut calli associated with the acquisition of embryogenic competence. Ann. Bot. 88, 9–18.
Vestberg, M., Kukkonen, S., Saari, K., Parikka, P., Huttunen, J., Tainio, L., et al., 2004. Microbial inoculation for improving the growth and health of micropropagated strawberry. Appl. Soil Ecol. 27, 243–258.
Wilhelm, E., Arthofer, W., Schafleitner, R., 1997. *Bacillus subtilis* an endophyte of chestnut (*Castanea sativa*) as antagonist against chestnut blight (*Cryphonectaria parasitica*). In: Cassells, A.L. (Ed.), Pathogen and Microbial Contamination Management in Micropropagation. Kluwer Acad. Publ., Dordrecht, pp. 331–337.
Yang, Y.S., Wada, K., Goto, M., 1991. *In vitro* formation of nodular calli in soybean (*Glycine max* L.) induced by co-cultivated *Pseudomonas maltophilia*. Jpn. J. Breed. 41, 595–604.

Chapter 4

Environmental contaminants and their management using microorganisms

Manisha Mishra[1], Sandeep Kumar Singh[2], Monika Singh[2], Vipin Kumar Singh[2], Hariom Verma[2], Kumari Rashmi[3] and Ajay Kumar[4]

[1]*University Department of Botany, T.M. Bhagalpur University, Bhagalpur, India,* [2]*Centre for Advanced Study in Botany, Banaras Hindu University, Varanasi, India,* [3]*L.N.Mishra Institute Patna, Patna, India,* [4]*Agriculture Research Organization, Ministry of Agriculture and Rural Development, Volcani Center, Rishon LeZion, Israel*

4.1 Introduction

From the last few decades, environmental contaminants in the soil, water, or environment have been rising continuously that poses serious threat to the global climate change, agricultural productivity, human health hazards as well as socio-economic status of human beings. After industrialization, rapid growth in the establishment of new industries such as textile, leather, petrochemicals, metalloids has been found in both developing and developed countries, and this rapid growth adversely exploits the natural environment by their toxic exhaust and effluents. However, toxicity of the effluents mainly depends upon the nature of the industries and has both long- and short-term effect on the health and environment of human beings (Farhadian et al., 2008).

Currently, there is a range of environmental contaminants including both inorganic (heavy metals) and organic nature (poly aromatic hydrocarbons, industrial effluents, oil spills, petroleum products, pesticides, xenobiotics anthropogenic waste) have been present in each ecosystem that affect, texture and productivity of soil as well as health or environment of living organism (Zhu et al., 2012). However, some of the contaminants even get accumulated in the body and magnified with the rise in trophic levels and causes complete extinction of species. To mitigate the rising environmental contaminants, various physiochemical remediation technologies such as chemical oxidation (Yan and Lo, 2013), electro-kinetic remediation (Ma et al., 2010), and nanomaterial remediation (Shih et al., 2011) landfill have been used. The traditional methods include incarnation of contaminants at high temperature, stabilization using lime but the limitation of land resources, higher cost, toxic residual products is still a matter of concern for the environmentalist and researchers (Kümmerer et al., 2019).

In this regard "bioremediation" or use of biological means to mitigate these environmental contaminants is a chief, ecofriendly and sustainable approach. The term bioremediation commonly used for the remediation of environmental contamination by the use of biological means such as microbes, plants, microbial products, (Ye et al., 2017). The term bioremediation made up of words "bio" and "remediation"; and their literal meaning is same. In the current scenario of global climate change and raising environmental pollution, bioremediation of any contaminants is considered safe, economic, and eco-friendly in context of the human health as well as surrounding environments. Bioremediation is a broad term, and in this approach, various biological means such as microbes, microbial consortia, and microbial enzymes have been directly or indirectly utilized to mitigate or detoxify the harmful or complex environmental contaminants (Kertesz et al., 1994).

The process of bioremediation generally carried out in two ways: (1) in situ or (2) ex situ. In situ bioremediation involves the remediation at the site of origin whereas ex situ involves mitigation elsewhere from the site of origin. Generally, the term ex situ is employed for the treatment of soil and groundwater, which is taken at another site via excavation or pumping respectively. Bioremediation includes different types of remediation strategies such as phytoremediation, rhizoremediation, and mycoremediation. The term phytoremediation used for the plant assisted remediation; rhizoremediation includes rhizosphere microbes or exudates roots of plants for contamination management. The plant

Microbe Mediated Remediation of Environmental Contaminants. DOI: https://doi.org/10.1016/B978-0-12-821199-1.00004-3

roots secrete an array of chemical compounds including proteins, lipids, carbohydrates acids, and these compounds effectively help in accumulation and remediation of chemical pollutants. Similarly, the term mycoremediation used for the remediation of pollutants by using fungal species (Rajkumar et al., 2012; McCutcheon and Jørgensen, 2008; Chang et al., 2009; Vettori et al., 2012).

4.2 Bioremediation and environmental contaminants

In the current scenario, sustainable method for environmental management is the urgent needs and bioremediation using microbial strains or microbial products is the economic and sustainable approach (Singh et al., 2020a, 2020b; Gupta et al., 2020). In bioremediation, contaminants, which are consisted of inorganic or organic components, degraded or transformed from complex or toxic state to less toxic or even complete degradation of pollutants (Margesin and Schinner, 2001). In microbial degradation the microbial community or species either be indigenous to the contaminated sites or may be transferred from culture to the contamination origin site. However, for the effective and efficient degradation, growth of microbial mass at the site of contamination is the prime concern. In microbial-based remediation, transformation of complex organic contaminants to simpler one is a part of metabolic process and generally carried out through various microbial enzymes, exopolysaccharides (EPS), microbial surfactants, and microbial metabolites (Vidali, 2001). Each of the environmental contaminants has specific physicochemical properties and, due to this, has different migration potential, absorption, and transformation; mode of these contaminants varies in the soil, which also influences procedures and efficiency of remediation (Ye et al., 2017). In addition, sometimes presence of more than two contaminants showed complicated response against remediation as well as efficiency potential. The interaction of organic and inorganic pollutants might change the bioavailability, solubility, and speciation of pollutants in comparison to singly presence (Ye et al., 2017) and also reduce the efficiency of remediation through inhibiting the growth of microbes via increased competition for binding sites of adsorbents and lowering enzymes production. However, some authors also reported cocontaminated soil with heavy metals and organic contaminants, showed lower remediation potential due to higher concentration of heavy metals slow down the microbial metabolism, biomass, enzyme production of microbes that ultimately reduce the bioremediation potential of organic contaminants (Sun et al., 2011). However, in contrary to organic contaminants the uptake of heavy metals increased due to changes in membrane permeability and this may be caused due to the interaction between hydrophobic components of cell membrane and PAHs (polycyclic aromatic hydrocarbons) (Ye et al., 2017; Gong et al., 2009)

4.3 Beneficial microbes

Microbes are ubiquitous in nature and approximated that 1 g of bulk soil contains 10^5-10^7 bacterial species; however, the rhizosphere region of the plants or rhizosphere soil contains approximately 10^7-10^{10} microbial population (Kumar et al., 2015). There are various microbial species present that even grown in the very extreme conditions such as cold, harsh, temperature, alkalinity, salinity, or the stress conditions (Vidali, 2001; Prescott et al., 2002). During microbial-based remediation, chemical nature of contaminants and the growth of microbes at the contaminated sites are the prime responsible factor for bioremediation, which include direct contact between substrate (contaminants) and the microbial strains for the degradation or transformation

Microorganisms utilize contaminants for their vital physiological and biochemical processes, under certain conditions different microbial strains can coordinate with each other for the degradation of pollutants. Microbial strain used in the bioremediation may be pure or mixed consortia of the strains such as bacteria–bacterial, bacteria–fungi, cyanobacteria–bacteria. There are various reports present in which pure microbial consortia of bacteria, fungi, and cyanobacteria have been used for the remediation of various environmental contaminants (Singh et al., 2019). However, drawback of pure consortia utilization during contamination degradation is the lack of proper enzymes or secretes of particular contaminants or the substrate specificity, and due to this condition, pure cultures sometimes are unable to secrete specific enzymes, in that condition mixed consortia play an important role in the degradation. The mixed culture secretes range of enzymes that help in degradation more efficiently than pure cultures. There are various reports present that showed effective remediation potential of mixed cultures in comparison to pure cultures. In a study (Santoso et al., 2011) reported about carbamazepine stable compounds, which are not detoxified by pure strain while the mixed culture of basidiomycete significantly detoxified. Similarly, Popa et al. (2014) reported about the mixed culture of *Streptomyces* with the supplementation of glucose help in effective degradation of carbamazepine. Diclofenac appears as resistance against biological degeneration in activated sludge system. However, utilization of white rot fungi completely remove the diclofenac up to safety limits without any addition of supplement (Hata et al., 2010).

4.4 Microbes in the environmental contaminants

There are numerous reports present that showed excellent remediation potential of different microbial strain in the variety of environmental pollutants such as petroleum products, oil spills, heavy metals, anthropogenic pollutants, and degradation of pesticides and plastic. Some of the commonly used microbial strains used in environment pollutants degradation are *Pseudomonas*, *Rhodococcus*, *Mycobacterium*, methylotrophs, ligninolytic fungi, marine fungi cyanobacterial strains, etc. (El-Bestawy et al., 2007). The microbial bioremediation is achieved through several processes in which microbes interact with the contaminants and also with the niche and followed several mechanisms such as biostimulation, bioaugmentation, bioaccumulation, biosorption, and also through production of range of enzymes, biofilms, biosurfactants (Fig. 4.1, Table 4.1) (Riccardi et al., 2005; Vasudevan et al., 2001; Tyagi et al., 2011; Gupta et al., 2000; Singh et al., 2006, 2019a; Datta et al., 2020).

The biodegradation of any organic contaminant depends upon their chemical and physical constituents as well growth and accessibility of specific microbes at the site of contamination (Foght et al., 1996; Guthrie et al., 1999). Environment contains diverse nature of contaminants and each has specific properties and these properties either limit or enhance the remediation potential by the growth of microbes. As for example hydrophobic nature and long chain of hydrocarbons in petroleum products do not support optimum growth of most of the microbes, this factor limits the biodegradation of petroleum products. Efficiency and biodegradation of these contaminants can be enhanced by modifying environmental condition, growth support of microbes (Babaarslan et al., 2003; Yeom et al., 1997; Lee et al., 2001).

Microbial remediation of inorganic contaminants includes a range of compounds in which toxicity of heavy metals is one of the severe challenges that are generally outcome of various processes such as industries, mining, metallurgy (Goyal et al., 2003; Srivastava et al., 2014; Singh et al., 2018). The presence of heavy metals above the concentration either in the soil or aquatic ecosystem is a severe concern for the living organism due to higher toxicity and long persistence power (Kaewsarn and Yu, 2001). From the last few decades, various researchers continuously use microbial techniques or microbial consortia for the mitigation of heavy metal toxicity in sustainable ways (Srivastava et al., 2014). The microbes or the microbial techniques during heavy metal stress tolerance detoxify or transform higher oxidation

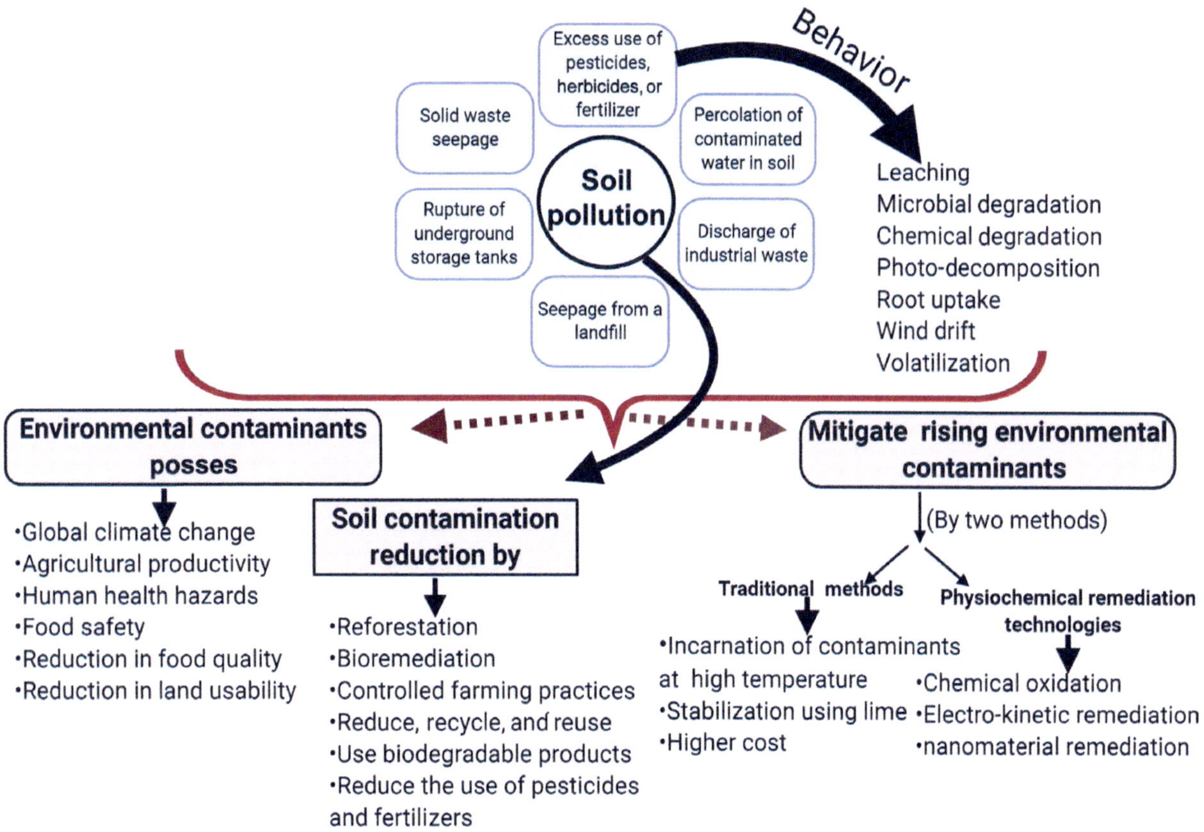

FIGURE 4.1 An overview of environmental contaminants and their mitigation strategies.

TABLE 4.1 Environmental contaminants and their mitigation using microorganism.

S. no.	Microbes	Degradation condition	References
1.	*Raoultella* sp.	Cadmium bioremediation in soil	Xu et al. (2019)
2.	*Providencia vermicola, Ochrobactrum* sp.	Radionuclide containing soil	Shukla et al. (2019)
3.	*Enterobacter cloacae*	Metal mines	Kang et al. (2015)
4.	*Bacillus polymyxa*	Crude oil polluted soil	Odokuma, and Inor (2002)
5.	*Pseudomonas* sp.	Atrazine-contaminated soil	Newcombe and Crowley (1999)
6.	*Bacillus* sp. and *Staphylococcus capitis*	Chromium containing wastewater	Zahoor and Rehman (2009)
7.	*Cupriavidus taiwanensis KKU2500-3* and *Pseudomonas aeruginosa*	Cadmium from paddy fields	Siripornadulsil and Siripornadulsil (2013)
8.	*Geobacter* sp. and sulfate-reducing bacteria	Uranium in ground water	Barlett et al. (2012)
9.	*Agrobacterium radiobacter*	Herbicide atrazine (Laboratory condition)	Struthers et al. (1998)
10.	*Bacillus subtilis*	Crude oil-polluted soils	Cubitto et al. (2004)
11.	*Coliform bacteria*	Waste Hydrocarbons	Khanafer et al. (2017)
12.	*Acinetobacter* sp., *Pseudomonas* sp., *Microbacterium* sp.	Aromatic hydrocarbons	Simarro et al. (2013)
13.	*B. subtilis, Corynebacterium*	Phenols	Singh et al. (2013)
14.	*Pseudomonas putida*	Monocyclic aromatic hydrocarbons	Safiyanu et al. (2015)
15.	*P. aeruginosa, Aeromonas* sp.	U, Cu, Ni, Cr	Sinha et al. (2011)
16.	Fungal strains	Hydrocarbons	Al-Jawhari (2014)
17.	*Penicillium chrysosporium*	Phenol	Wolski et al. (2012)
18.	*Penicillium chrysogenum*	Azo dye	Durruty et al. (2015)
19.	*Penicillium* sp.	Aliphatic hydrocarbon degradation	Al-Hawash et al. (2018)
20.	*Daedaleopsis* sp.	Petroleum products	Pourfakhraei et al. (2018)
21.	*Oscillatoria* sp., *Synechococcus* sp., *Nodularia* sp., *Nostoc* sp., and *Cyanothece* sp.	Industrial effluents	Dubey et al. (2011)
22.	*Chlorella* sp., *Phormidium* sp., and *Oscillatoria* sp.	Hexavalent chromium	Shukla et al. (2012)
23.	*Chlorella* sp.	Cadmium-polluted seawater	Matsunaga et al. (1999)
24.	*Anabaena flos-aquae*	Chromium in tannery effluent treatment	Kannan et al. (2012)
25.	*Phormidium valderianum* BDU	Phenol degradation	Shashirekha et al. (1997)
26.	*Synechocystis* sp.	Chlorpyrifos degradation	Singh et al. (2011)
27.	*Spirulina platensis*	Degradation of chlorpyrifos	Thengodkar and Sivakami (2010)
28.	*Oscillatoria annae*	Tannery effluent	Shankar et al. (2013)
29.	*Oscillatoria salina; Plectonema terebrans; Aphanocapsa* sp.	Bombay High crude oil	Raghukumar et al. (2001)
30.	*Oscillatoria limnetica*	Herbicide glyphosate	Salman and Abdul-Adel (2015)
31.	*Spirulina* sp.	Polyphosphonates from wastewater	Forlani et al. (2011)
32.	*S. platensis*	Bioremediation of cobalt and chromium	Murali et al. (2014)
33.	*Chroococcus* sp.	Cyanide bioremediation	Naghavi et al. (2012)
34.	*Arthrospira platensis*	Tannery effluent containing Cr	Balaji et al. (2015)
35.	*Fischerella* sp.	Pesticide methyl parathion	Tiwari et al. (2017)

states of the metals ion to lower oxidation states or in the stable form, which further metabolize or accumulate in the bacterial cell or may be used as microbial nutrients (Garbisu and Alkorta, 2003). During remediation, microbes released some chelating agents such as siderophores, EPS (extracellular polymeric substances), which alters the physicochemical properties of the surroundings and enhance the bioavailability of metal ions (Abou-Shanab et al. 2003; Abou-Shanab et al. 2006).

However, during mitigation of heavy metal stress, the attachment or contact between the heavy metal ions and the microbial cells, termed as biosorption, this leads to adsorption followed by ion exchange between microbial cells and metal ions and forms a complex which finally transform or precipitated into the less toxic form (Deng et al., 2007; Gadd, 2009; Wang and Chen, 2009; Han and Gu, 2010). There are various reports present that effectively showed bioremediation potential of heavy metals using bacterial strains (Naraian et al., 2012; El-Deeb, 2009; Srivastava et al., 2014). In a study, Wood and Wang (1985) reported the efficiency of *Pseudomonas putida* in tolerance of Cu metal stress, similarly Tan and Ting (2012) reported the use of *Trichoderma asperellum* in the removal of Cu(II) ion from aqueous solution. Samuelson et al. (2000) reported the use of *Staphylococcus carnosus* in the remediation of Cd and Ni from the soil or sediment (Khomarbaghi et al., 2019).

4.5 Fungi in bioremediation

The applications of fungal microbes in bioremediation have been well known as fungal bioremediation or mycoremediation. These fungal strains secrete enzymes or other metabolites that have a strong capability to decompose or transform complex organic and inorganic pollutants including heavy metals, metalloids, petrochemicals, chemical pesticides into less toxic or even complete degradation of the contaminants in elemental form (Singh et al., 2019b; Bhadouria, et al., 2020). Fungi possess biochemical and ecological capacity to degrade environmental or organic chemicals to reduce the risk associated with metals, metalloids, and radionuclides, either by chemical modification or through influencing chemical bioavailability. However, to date, bioremediation has tended to disregard the ecological demands and eco-physiological strengths of fungi. In most of the cases, fungal strain utilizes the degrading products as a source of energy and immobilizes or stores in the microsphere and translocates to the different parts through hyphae. The action mechanism of these fungal strains are very much similar to the bacteria strain in terms of enzymes, surfactant, extracellular polymeric substances (EPS), biofilms, siderophore, hydrogen cyanide (HCN) production, and these secretes help in mobilization, precipitation, solubilization, and transformation of organic contaminants from a higher state to low toxic or even complete degradation (Harms et al., 2011; Balabanova et al., 2018). Some of the most common fungal strains such as *Pleurotus* sp., *Phanerochaete chrysosporium*, *Trichoderma harzianum*, *Mucor* had been reported by various authors for the effective bioremediation potential (Singh et al., 2020a). These fungal strains showed differential mode of action against various organic such as persistent organic pollutants, PAHs, pharmaceuticals and personal care products, textile dyes, petroleum hydrocarbons (PHCs), pesticides, pulp, and paper industry effluents, leather tanning effluents (Deshmukh et al., 2016; Majeau et al., 2010; Divya et al., 2014). In study, some reports mentioned that fungal strains have also been seen to perform process of esterification, dioxygenation, hydroxylation, dehydrogenation during the transformation of different pesticides during degradation (Deng et al., 2015; Pinto et al., 2012). However, some fungal strains also showed capacity to degrade polychlorinated biphenyls by involvement of nonligninolytic enzymes (Tigini et al., 2009).

4.6 Cyanobacteria in bioremediation

Cyanobacteria a group of photosynthetic bacteria and currently used as an effective biological agent for the mitigation of diverse range of environmental contaminants (Kertesz et al., 1994; Gupta et al., 2018). Like bacterial and fungal species, cyanobacteria also produce several enzymes, metabolites extracellular polymeric substances (EPS), these secretes generally present on the outer surface of the cell and play effective role in protection from the desiccation and heavy metal stress present in the contaminated water. The mitigating rate of metal stress by EPS secreting cyanobacterial species is very high in comparison to the other microorganism due to minimal requirement of nutrients and high growth rate. There are various reports present that strengthen the use of cyanobacterial species in the bioremediation of environmental contaminants. *Cyanothece* sp. CCY 0110, a unicellular and marine cyanobacterium previously shown to be a highly efficient EPS producer and help in effective bio remediation. In this work, the effects induced by the metals in the cell growth, internal ultrastructure, and EPS production were evaluated (Mota et al., 2015). El-Bestawy et al. (2007) reported various cyanobacterial groups such as *Nostoc*, *Oscillatoria*, *Anabaena*, and *Microcystis* that mitigate the lindane residues by cracking them in simpler molecules. Similarly, strains such as *Synechococcus elongatus*, *Microcystis aeruginosa*, and *Anacystics nidulans* had been reported to potentially remove the organochlorine and organophosphorus insecticides (Vijayakumar, 2012). However, Zhou et al. (2012) reported the efficiency of cyanobacterial strain *Chlorella pyrenoidosa* and *Scenedesmus obliquus* in tolerance of zinc and copper stress. The higher concentration of zinc and copper reduces the growth, photosynthetic pigment of both the cyanobacteria. However, both the cyanobacteria strains showed the zinc and copper efficiency. In some cases, cyanobacterial consortia singly or in combination with another bacterial consortium have been used to remove various toxic substances. In a study consortium of *Oscillatoria*- and Gammaproteobacteria can degrade phenanthrene, dibenzothiophene, pristine, and *n*-octadecane (Abed and Köster, 2005). Similarly, *Phormidium valderianum* and *Oscillatoria boryana* had been utilized to remove phenol and melanoidin, respectively, from the effluents of distillery (Shashirekha et al., 1997; Kalavathi et al., 2001). In a study, Thengodkar and Sivakami (2010) studied different green algae and cyanobacterial strains and evaluate their degradation potential against fenamiphos pesticides and found that all the cyanobacterial strains were able to oxidized fenamiphos to its primary oxidation product, fenamiphos sulfoxide (FSO).

4.7 Conclusion

In the few recent decades, exploration of novel and sustainable approach for the mitigation of environmental pollutants is a critical need. Currently, the whole globe is suffering from the problem of global warming and climate change. The microbial-based bioremediation is the best alternative in the field of environmental contamination management. However, there is still a need for improvement and exploration of microbial genera that may have significant and effective role in the mitigation of complex and toxic contamination. There should also be a need to explore the limiting factors such as growth supplements of microbes, pH, temperature, conditions for the better adaptability of microbial species on the contaminated sites. The microbes synthesize diverse range of metabolites, enzymes, nanoparticles, so exploration of these microbial products will make easy and specific in the contamination degradation.

References

Abed, R.M., Köster, J., 2005. The direct role of aerobic heterotrophic bacteria associated with cyanobacteria in the degradation of oil compounds. Int. Biodeterior. Biodegrad. 55, 29–37.

Abou-Shanab, R.A., Angle, J.S., Delorme, T.A., Chaney, R.L., Van Berkum, P., Moawad, H., et al., 2003. Rhizobacterial effects on nickel extraction from soil and uptake by *Alyssum murale*. N. Phytol. 158, 219–224.

Abou-Shanab, R.A.I., Angle, J.S., Chaney, R.L., 2006. Bacterial inoculants affecting nickel uptake by *Alyssum murale* from low, moderate and high Ni soils. Soil Biol. Biochem. 38, 2882–2889.

Al-Hawash, A.B., Alkooranee, J.T., Abbood, H.A., Zhang, J., Sun, J., Zhang, X., et al., 2018. Isolation and characterization of two crude oil-degrading fungi strains from Rumaila oil field, Iraq. Biotechnol. Rep. 17, 104–109.

Al-Jawhari, I.F.H., 2014. Ability of some soil fungi in biodegradation of petroleum hydrocarbon. J. Appl. Environ. Microbiol. 2, 46–52.

Babaarslan, C., Tekeli, A., Mehmetoğlu, T., 2003. Biodegradation of BTEX compounds by a mixed culture obtained from petroleum formation water. Energy Sources 25, 733–742.

Balabanova, L., Slepchenko, L., Son, O., Tekutyeva, L., 2018. Biotechnology potential of marine fungi degrading plant and algae polymeric substrates. Front. Microbiol. 9, 1527.

Balaji, S., Kalaivani, T., Rajasekaran, C., Siva, R., Shalini, M., Das, R., et al., 2015. Bioremediation potential of *Arthrospira platensis* (Spirulina) against chromium (VI). CLEAN—Soil, Air, Water 43 (7), 1018–1024.

Barlett, M., Zhuang, K., Mahadevan, R., Lovley, D., 2012. Integrative analysis of *Geobacter* spp. and sulfate-reducing bacteria during uranium bioremediation. Biogeosciences. 9, 1033–1040.

Bhadouria, R., Das, S., Kumar, A., Singh, R., Singh, V.K., 2020. Mycoremediation of agrochemicals. Agrochem. Detect. Treat. Rem. Available from: https://doi.org/10.1016/B978-0-08-103017-2.00022-2.

Chang, J.S., Yoon, I.H., Kim, K.W., 2009. Heavy metal and arsenic accumulating fern species as potential ecological indicators in As-contaminated abandoned mines. Ecol. Indic. 9, 1275–1279.

Cubitto, M.A., Moran, A.C., Commendatore, M., Chiarello, M.N., Baldini, M.D., Sineriz, F., 2004. Effects of *Bacillus subtilis* O9 biosurfactant on the bioremediation of crude oil-polluted soils. Biodegradation 15, 281–287.

Datta, S., Singh, S., Kumar, V., Dhanjal, D.S., Sidhu, G.K., Amin, D.S., et al., 2020. Endophytic bacteria in xenobiotic degradation. In: Kumar, A., Singh, V.K. (Eds.), Microbial Endophytes: Prospects of Sustainable Agriculture. Elsevier, pp. 125–156.

Deng, J., Liao, B., Ye, M., Deng, D., Lan, C., Shu, W., 2007. The effects of heavy metal pollution on genetic diversity in zinc/cadmium hyperaccumulator *Sedum alfredii* populations. Plant Soil 297, 83–92.

Deng, S., Chen, Y., Wang, D., Shi, T., Wu, X., Ma, X., et al., 2015. Rapid biodegradation of organophosphorus pesticides by *Stenotrophomonas* sp. G1. J. Hazard. Mater. 297, 17–24.

Deshmukh, R., Khardenavis, A.A., Purohit, H.J., 2016. Diverse metabolic capacities of fungi for bioremediation. Indian J. Microbiol. 56, 247–264.

Divya, L.M., Prasanth, G.K., Sadasivan, C., 2014. Potential of the salt-tolerant laccase-producing strain *Trichoderma viride* Pers. NFCCI-2745 from an estuary in the bioremediation of phenol-polluted environments. J. Basic Microbiol. 54, 542–547.

Dubey, S.K., Dubey, J., Mehra, S., Tiwari, P., Bishwas, A.J., 2011. Potential use of cyanobacterial species in bioremediation of industrial effluents. Afr. J. Biotechnol. 10, 1125–1132.

Durruty, I., Fasce, D., González, J.F., Wolski, E.A., 2015. A kinetic study of textile dyeing wastewater degradation by *Penicillium chrysogenum*. Bioprocess Biosyst. Eng. 38, 1019–1031.

El-Bestawy, E.A., El-Salam, A.Z.A., Mansy, A.E.R.H., 2007. Potential use of environmental cyanobacterial species in bioremediation of lindane-contaminated effluents. Int. Biodeterior. Biodegrad. 59, 180–192.

El-Deeb, B., 2009. Natural combinations of genetic systems for degradation of phenol and resistance to heavy metals in phenol and cyanide assimilating bacteria. Malays. J. Microbiol. 5, 94–103.

Farhadian, M., Vachelard, C., Duchez, D., Larroche, C., 2008. In situ bioremediation of monoaromatic pollutants in groundwater: a review. Bioresour. Technol. 99, 5296–5308.

Foght, J.M., Westlake, D.W., Johnson, W.M., Ridgway, H.F., 1996. Environmental gasoline-utilizing isolates and clinical isolates of *Pseudomonas aeruginosa* are taxonomically indistinguishable by chemotaxonomic and molecular techniques. Microbiology 142, 2333–2340.

Forlani, G., Prearo, V., Wieczorek, D., Kafarski, P., Lipok, J., 2011. Phosphonate degradation by Spirulina strains: cyanobacterial biofilters for the removal of anticorrosive polyphosphonates from wastewater. Enzyme Microb. Technol. 48, 299–305.

Gadd, G.M., 2009. Biosorption: critical review of scientific rationale, environmental importance and significance for pollution treatment. J. Chem. Technol. Biotechnol. 84, 13–28.

Garbisu, C., Alkorta, I., 2003. Basic concepts on heavy metal soil bioremediation. Eur. J. Min. Process. Environ. Prot. 3, 58–66.

Gong, J.L., Wang, B., Zeng, G.M., Yang, C.P., Niu, C.G., Niu, Q.Y., et al., 2009. Removal of cationic dyes from aqueous solution using magnetic multi-wall carbon nanotube nanocomposite as adsorbent. J. Hazard. Mater. 164, 1517–1522.

Goyal, N., Jain, S.C., Banerjee, U.C., 2003. Comparative studies on the microbial adsorption of heavy metals. Adv. Environ. Res. 7, 311–319.

Gupta, R., Ahuja, P., Khan, S., Saxena, R.K., Mohapatra, H., 2000. Microbial biosorbents: meeting challenges of heavy metal pollution in aqueous solutions. Curr. Sci. 78, 967–973.

Gupta, A., Singh, M., Pandey, K.D., Kumar, A., 2018. Biodegradation of xenobiotics using cyanobacteria. J. Sci. Res. 62, 35–40.

Gupta, A., Singh, S.K., Singh, V.K., Singh, M.K., Modi, A., Zhimo, V.Y., et al., 2020. Endophytic microbe approaches in bioremediation of organic pollutants. In: Kumar, A., Singh, V.K. (Eds.), Microbial Endophytes. Woodhead Publishing, pp. 157–174.

Guthrie, E.A., Bortiatynski, J.M., van Heemst, J.D., Richman, J.E., Hardy, K.S., Kovach, E.M., et al., 1999. Determination of [13C] Pyrene Sequestration in Sediment Microcosms Using Flash Pyrolysis–GC–MS and 13C NMR. Environ. Sci. Technol. 33, 119–125.

Han, X., Gu, J.D., 2010. Sorption and transformation of toxic metals by microorganisms. In: Mitchell, R., Gu, J.D. (Eds.), Environmental Microbiology, second ed., Wiley-Blackwell, pp. 153–176.

Harms, H., Schlosser, D., Wick, L.Y., 2011. Untapped potential: exploiting fungi in bioremediation of hazardous chemicals. Nat. Rev. Microbiol. 9, 177.

Hata, T., Kawai, S., Okamura, H., Nishida, T., 2010. Removal of diclofenac and mefenamic acid by the white rot fungus *Phanerochaete sordida* YK-624 and identification of their metabolites after fungal transformation. Biodegradation 21, 681–689.

Kaewsarn, P., Yu, Q., 2001. Cadmium (II) removal from aqueous solutions by pre-treated biomass of marine alga *Padina* sp. Environ. Poll. 112, 209–213.

Kalavathi, D.F., Uma, L., Subramanian, G., 2001. Degradation and metabolization of the pigment—melanoidin in distillery effluent by the marine cyanobacterium *Oscillatoria boryana* BDU 92181. Enzyme Microb. Technol. 29, 246–251.

Kang, C.H., Oh, S.J., Shin, Y., Han, S.H., Nam, I.H., So, J.S., 2015. Bioremediation of lead by ureolytic bacteria isolated from soil at abandoned metal mines in South Korea. Ecol. Eng. 74, 402–407.

Kannan, V., Vijayasanthi, M., Chinnasamy, M., 2012. Bioremediation of chromium in tannery effluent by filamentous Cyanobacteria *Anabaena flos-aquae* West. Int. J. Environ. Sci. 2, 2360–2366.

Kertesz, M.A., Cook, A.M., Leisinger, T., 1994. Microbial metabolism of sulfurand phosphorus-containing xenobiotics. FEMS Microbiol. Rev. 15, 195–215.

Khanafer, M., Al-Awadhi, H., Radwan, S., 2017. Coliform bacteria for bioremediation of waste hydrocarbons. BioMed. Res. Int. 2017.

Khomarbaghi, Z., Shavandi, M., Amoozegar, M.A., Dastgheib, S.M.M., 2019. Bacterial community dynamics during bioremediation of alkane-and PAHs-contaminated soil of Siri island, Persian Gulf: a microcosm study. Int. J. Environ. Sci. Technol. 16, 7849–7860.

Kumar, A., Vandana, R.S., Yadav, A., Giri, D.D., Singh, P.K., Pandey, K.D., 2015. Rhizosphere and their role in plant–microbe interaction. Microbes in Soil and Their Agricultural Prospects. Nova Science Publisher, Inc, Hauppauge, pp. 83–97.

Kümmerer, K., Dionysiou, D.D., Olsson, O., Fatta-Kassinos, D., 2019. Reducing aquatic micropollutants—increasing the focus on input prevention and integrated emission management. Sci. Total Environ. 652, 836–850.

Lee, P.H., Ong, S.K., Golchin, J., Nelson, G.S., 2001. Use of solvents to enhance PAH biodegradation of coal tar. Water Res. 35, 3941–3949.

Ma, J.W., Wang, F.Y., Huang, Z.H., Wang, H., 2010. Simultaneous removal of 2, 4-dichlorophenol and Cd from soils by electrokinetic remediation combined with activated bamboo charcoal. J. Hazard. Mater. 176, 715–720.

Majeau, J.A., Brar, S.K., Tyagi, R.D., 2010. Laccases for removal of recalcitrant and emerging pollutants. Bioresour. Technol. 101, 2331–2350.

Margesin, R., Schinner, F., 2001. Biodegradation and bioremediation of hydrocarbons in extreme environments. Appl. Microbiol. Biotechnol. 56, 650–663.

Matsunaga, T., Takeyama, H., Nakao, T., Yamazawa, A., 1999. Screening of marine microalgae for bioremediation of cadmium-polluted seawater. J. Biotechnol. 70, 33–38.

McCutcheon, S.C., Jørgensen, S.E., 2008. Phytoremediation. In: Fath, S.E.J.D. (Ed.), Encyclopedia of Ecology. Academic Press, Oxford, pp. 2751–2766.

Mota, R., Pereira, S.B., Meazzini, M., Fernandes, R., Santos, A., Evans, C.A., 2015. Effects of heavy metals on Cyanothece sp. CCY 0110 growth, extracellular polymeric substances (EPS) production, ultrastructure and protein profiles. J. Proteomics. 120, 75–94.

Murali, O., Shaik, G., Mehar, S.K., 2014. Assessment of bioremediation of Cobalt and Chromium using cyanobacteria. Indian J. Fund. Appl. Life Sci. 4, 252–255.

Naghavi, N.S., Mazrouei, B., Afsharzadeh, S., 2012. Analysis of cyanide bioremediation using cyanobacterium; *Chroococcus* isolated from steel manufacturing industrial wastewater. Int. J. Biol. Chem. 6, 113–121.

Naraian, R., Ram, S., Kaistha, S.D., Srivastava, J., 2012. Occurrence of plasmid linked multiple drug resistance in bacterial isolates of tannery effluent. Cell Mol. Biol. 58, 134–141.

Newcombe, D.A., Crowley, D.E., 1999. Bioremediation of atrazine-contaminated soil by repeated applications of atrazine-degrading bacteria. Appl. Microbiol. Biotechnol. 51, 877–882.

Odokuma, L.O., Inor, M.N., 2002. Nitrogen fixing bacteria enhanced bioremediation of a crude oil polluted soil. Glob. J. Pure Appl. Sci. 8, 455–470.

Pinto, A.P., Serrano, C., Pires, T., Mestrinho, E., Dias, L., Teixeira, D.M., et al., 2012. Degradation of terbuthylazine, difenoconazole and pendimethalin pesticides by selected fungi cultures. Sci. Total. Environ. 435, 402–410.

Popa, C., Favier, L., Dinica, R., Semrany, S., Djelal, H., Amrane, A., et al., 2014. Potential of newly isolated wild *Streptomyces* strains as agents for the biodegradation of a recalcitrant pharmaceutical, carbamazepine. Environ. Technol. 35, 3082–3091.

Pourfakhraei, E., Badraghi, J., Mamashli, F., Nazari, M., Saboury, A.A., 2018. Biodegradation of asphaltene and petroleum compounds by a highly potent *Daedaleopsis* sp. J. Basic Microbiol. 58, 609–622.

Prescott, L.M., Harley, J.P., Klein, D.A. (Eds.), 2002. Microbiology. fifth ed. McGraw Hill, New York.

Raghukumar, C., Vipparty, V., David, J., Chandramohan, D., 2001. Degradation of crude oil by marine cyanobacteria. Appl. Microbiol. Biotechnol. 57, 433–436.

Rajkumar, M., Sandhya, S., Prasad, M.N.V., Freitas, H., 2012. Perspectives of plant-associated microbes in heavy metal phytoremediation. Biotechnol. Adv. 30, 1562–1574.

Riccardi, C., Papacchini, M., Mansi, A., Ciervo, A., Petrucca, A., La Rosa, G., et al., 2005. Characterization of bacterial population coming from a soil contaminated by polycyclic aromatic hydrocarbons (PAHs) able to degrade pyrene in slurry phase. Ann. Microbiol. 55, 85–90.

Safiyanu, I., Isah, A.A., Abubakar, U.S., Rita Singh, M., 2015. Review on comparative study on bioremediation for oil spills using microbes. Res. J. Pharm. Biol. Chem. Sci. 6, 783–790.

Salman, J.M., Abdul-Adel, E., 2015. Potential use of cyanophyta species *Oscillatoria limnetica* in bioremediation of organophosphorus herbicide glyphosate. Mesop. Environ. J. 1, 15–26.

Samuelson, P., Wernerus, H., Svedberg, M., Sthal, S., 2000. Staphylococcal surface display of metal binding polyhistidyl peptides. Appl. Environ. Microbiol. 66, 1243–1248.

Santoso, H., Gunawan, T., Jatmiko, R.H., Darmosarkoro, W., Minasny, B., 2011. Mapping and identifying basal stem rot disease in oil palms in North Sumatra with QuickBird imagery. Precis. Agric. 12, 233–248.

Shankar, A.M., Henciya, S., Malliga, P., 2013. Bioremediation of tannery effluent using fresh water cyanobacterium *Oscillatoria annae* with coir pith. Int. J. Environ. Sci. 3, 1881.

Shashirekha, S., Uma, L., Subramanian, G., 1997. Phenol degradation by the marine cyanobacterium *Phormidium valderianum* BDU 30501. J. Ind. Microbiol. Biotechnol. 19, 130–133.

Shih, Y.H., Chen, M.Y., Su, Y.F., 2011. Pentachlorophenol reduction by Pd/Fe bimetallic nanoparticles: effects of copper, nickel, and ferric cations. Appl. Catal. B: Environ. 105, 24–29.

Shukla, D., Vankar, P.S., Srivastava, S.K., 2012. Bioremediation of hexavalent chromium by a cyanobacterial mat. Appl. Water Sci. 2, 245–251.

Shukla, A., Parmar, P., Saraf, M., Patel, B., 2019. Isolation and screening of bacteria from radionuclide containing soil for bioremediation of contaminated sites. Environ. Sustain. 1–10.

Simarro, R., González, N., Bautista, L.F., Molina, M.C., 2013. Assessment of the efficiency of in situ bioremediation techniques in a creosote polluted soil: change in bacterial community. J. Hazard. Mater. 262, 158–167.

Singh, R., Paul, D., Jain, R.K., 2006. Biofilms: implications in bioremediation. Trends Microbiol. 14, 389–397.

Singh, D.P., Khattar, J.I.S., Nadda, J., Singh, Y., Garg, A., Kaur, N., et al., 2011. Chlorpyrifos degradation by the cyanobacterium *Synechocystis* sp. strain PUPCCC 64. Environ. Sci. Poll. Res. 18, 1351–1359.

Singh, A., Kumar, V., Srivastava, J.N., 2013. Assessment of bioremediation of oil and phenol contents in refinery waste water via bacterial consortium. J. Pet. Environ. Biotechnol. 4, 1–4.

Singh, V.K., Singh, A.L., Singh, R., Kumar, A., 2018. Iron oxidizing bacteria: insights on diversity, mechanism of iron oxidation and role in management of metal pollution. Environ. Sustain. 1 (3), 221–231.

Singh, P., Kumar, A., Borthakur, A. (Eds.), 2019. Abatement of Environmental Pollutants: Trends and Strategies. Elsevier.

Singh, S.K., Singh, P.P., Gupta, A., Singh, A.K., Keshri, J., 2019a. Tolerance of heavy metal toxicity using PGPR strains of *Pseudomonas* species. In: Singh, A.K., Kumar, A., Singh, P.K. (Eds.), PGPR Amelioration in Sustainable Agriculture. Woodhead Publishing, pp. 239–252.

Singh, V.K., Singh, M., Singh, S.K., Kumar, C., Kumar, A., 2019b. Sustainable agricultural practices using beneficial fungi under changing climate scenario. In: Choudhary, K.K., Kumar, A., Singh, A.K. (Eds.), Climate Change and Agricultural Ecosystems. Woodhead Publishing, pp. 25–42.

Singh, D., Singh, S.K., Singh, V.K., Gupta, A., Aamir, M., Kumar, A., 2020a. Plant growth–promoting bacteria and their role in environmental management. In: Singh, P., Kumar, A., Borthakur, A. (Eds.), Abatement of Environmental Pollutants. Elsevier, pp. 161–175.

Singh, P., Singh, V.K., Singh, R., Borthakur, A., Madhav, S., Ahamad, A., et al., 2020b. Bioremediation: a sustainable approach for management of environmental contaminants. In: Singh, P., Kumar, A., Borthakur, A. (Eds.), Abatement of Environmental Pollutants. Elsevier, pp. 1–23.

Sinha, S.N., Biswas, M., Paul, D., Rahaman, S., 2011. Biodegradation potential of bacterial isolates from tannery effluent with special reference to hexavalent chromium. Biotechnol. Bioinf. Bioeng. 1, 381–386.

Siripornadulsil, S., Siripornadulsil, W., 2013. Cadmium-tolerant bacteria reduce the uptake of cadmium in rice: potential for microbial bioremediation. Ecotoxicol. Environ. Saf. 94, 94–103.

Srivastava, J., Naraian, R., Kalra, S.J.S., Chandra, H., 2014. Advances in microbial bioremediation and the factors influencing the process. Int. J. Environ. Sci. Technol. 11, 1787–1800.

Struthers, J.K., Jayachandran, K., Moorman, T.B., 1998. Biodegradation of atrazine by *Agrobacterium radiobacter* J14a and use of this strain in bioremediation of contaminated soil. Appl. Environ. Microbiol. 64, 3368–3375.

Sun, Y., Zhou, Q., Xu, Y., Wang, L., Liang, X., 2011. Phytoremediation for co-contaminated soils of benzo[a]pyrene (B[a]P) and heavy metals using ornamental plant *Tagetes patula*. J Hazard. Mater. 186, 2075–2082.

Tan, W.S., Ting, A.S., 2012. Efficacy and reusability of alginate-immobilized live and heat inactivated *Trichoderma asperellum* cells for Cu (II) removal from aqueous solutions. Bioresour. Technol. 123, 290–295.

Thengodkar, R.R.M., Sivakami, S., 2010. Degradation of chlorpyrifos by an alkaline phosphatase from the cyanobacterium *Spirulina platensis*. Biodegradation 21, 637–644.

Tigini, V., Prigione, V., Di Toro, S., Fava, F., Varese, G.C., 2009. Isolation and characterisation of polychlorinated biphenyl (PCB) degrading fungi from a historically contaminated soil. Microb. Cell Fact. 8, 5.

Tiwari, B., Chakraborty, S., Srivastava, A.K., Mishra, A.K., 2017. Biodegradation and rapid removal of methyl parathion by the paddy field cyanobacterium *Fischerella* sp. Algal Res. 25, 285–296.

Tyagi, M., da Fonseca, M.M.R., de Carvalho, C.C., 2011. Bioaugmentation and biostimulation strategies to improve the effectiveness of bioremediation processes. Biodegradation 22, 231–241.

Vasudevan, P., Padmavathy, V., Tewari, N., Dhingra, S.C., 2001. Biosorption of heavy metal ions. J. Sci. Ind. Res. 60, 112–120.

Vettori, L., Missaglia, A., Felici, C., Russo, A., Tamantini, I., Carrozza, G.P.C., et al., 2012. "Bioremedation and phytoremedation: synergism in lead extraction from contamined soils" [J. Biotechnol. 150S (2010) S509–S510]. J. Biotechnol. 3 (160), 268.

Vidali, M., 2001. Bioremediation. An overview. Pure Appl. Chem. 73, 1163–1172.

Vijayakumar, S., 2012. Potential applications of cyanobacteria in industrial effluents—a review. J. Bioremed. Biodegrad. 3, 1–6.

Wang, J., Chen, C., 2009. Biosorbents for heavy metals removal and their future. Biotechnol. Adv. 27, 195–226.

Wolski, E.A., Barrera, V., Castellari, C., González, J.F., 2012. Biodegradation of phenol in static cultures by *Penicillium chrysogenum* ERK1: catalytic abilities and residual phytotoxicity. Rev. Argent. Microbiol. 44, 113–121.

Wood, J.M., Wang, H.K., 1985. Strategies for microbial resistance to heavy metals. Chemical Processes in Lakes. John Wiley and Sons, New York, pp. 81–98.

Xu, S., Xing, Y., Liu, S., Huang, Q., Chen, W., 2019. Role of novel bacterial *Raoultella* sp. strain X13 in plant growth promotion and cadmium bioremediation in soil. Appl. Microbiol. Biotechnol. 103, 3887–3897.

Yan, D.Y., Lo, I.M., 2013. Removal effectiveness and mechanisms of naphthalene and heavy metals from artificially contaminated soil by iron chelate-activated persulfate. Environ. Pollut. 178, 15–22.

Ye, S., Zeng, G., Wu, H., Zhang, C., Dai, J., Liang, J., et al., 2017. Biological technologies for the remediation of co-contaminated soil. Crit. Rev. Biotechnol. 37, 1062–1076.

Yeom, S.H., Kim, S.H., Yoo, Y.J., Yoo, I.S., 1997. Microbial adaptation in the degradation of phenol by Alcaligenes xylosoxidans Y234. Korean J. Chem. Eng. 14 (37–40).

Zahoor, A., Rehman, A., 2009. Isolation of Cr (VI) reducing bacteria from industrial effluents and their potential use in bioremediation of chromium containing wastewater. J. Environ. Sci. 21, 814–820.

Zhou, G.J., Peng, F.Q., Zhang, L.J., Ying, G.G., 2012. Biosorption of zinc and copper from aqueous solutions by two freshwater green microalgae *Chlorella pyrenoidosa* and *Scenedesmus obliquus*. Environ. Sci. Poll. Res. 19, 2918–2929.

Zhu, Z.Q., Yang, X.E., Wang, K., Huang, H.G., Zhang, X., Fang, H., et al., 2012. Bioremediation of Cd-DDT co-contaminated soil using the Cd-hyperaccumulator *Sedum alfredii* and DDT-degrading microbes. J. Hazard. Mater. 235, 144–151.

Chapter 5

Environmental factors affecting the bioremediation potential of microbes

Manisha Mishra[1], Sandeep Kumar Singh[2] and Ajay Kumar[3]

[1]*University Department of Botany, T. M. Bhagalpur University, Bhagalpur, India,* [2]*Centre for Advanced Study in Botany, Banaras Hindu University, Varanasi, India,* [3]*Agriculture Research Organization, Ministry of Agriculture and Rural Development, Volcani Center, Rishon LeZion, Israel*

5.1 Introduction

Pollutant or contaminant refers to any unwanted substance introduced into the environment. These contaminants lead to pollution that has deleterious effects both on the flora and fauna. There has been a rapid increase in the use of chemical substances such as petroleum oil, hydrocarbons including aliphatic, aromatic, polycyclic aromatic hydrocarbons (PAHs), BTEX (benzene, toluene, ethyl benzene, and xylenes), PCB (polychlorinated biphenyls), TCE (trichloroethylene and perchloroethylene), nitroaromatic compounds, and organophosphorus compounds. Human beings are responsible for 90% of this oil pollution (Zhu et al., 2001). Hydrocarbons are released into the environment majorly by anthropogenic causes and to a much lesser extent naturally. Some hydrocarbons also occur naturally due to diagenesis of bacteriohopanetetrol (a constituent of membrane) to form hopane and hopanoic acids (Stout et al., 2001). Organic chemicals undergo nonstop volatilization and condensation, so they are discernable even in rain, fog, snow, etc. (Dubus et al., 2000). The contaminants that are released into the soil may be present in association with the soil particle, dissolved in soil water, or may be suspended in soil atmosphere. This multiphasic distribution of contaminants makes the bioremediation process complex. Therefore successful bioremediation process involves an interplay of several disciplines including microbiology, engineering, ecology, geology, and chemistry (Boopathy, 2000).

Effluents coming out from industries such as petrochemical, textile, and carpet, generate enormous amount of waste, oily sludge including petroleum waste. These together constitute hazardous waste that needs to be removed by their management (Farhadian et al., 2008). Industries generate huge amount of sludge. Conventionally, industrial sludge is removed by creating landfills, drying at high temperature, drying of sludge by spreading on land surface, stabilization by adding lime, composting and burning, etc. Sludge management is a costly affair. Therefore industries discharge their effluents into agricultural lands, fallow lands, sometimes open dumps, sanitary landfills, etc. The pollutants seep into the soil and find way into the various trophic levels in the food chain. When it comes to metal pollutants, it has been reported that they lead to significantly reduced plant growth. Therefore supply of food is challenged and results in food insecurity (Chibuike and Obiora, 2014).

Microorganisms in their metabolic activities produce energy that forms the key to bioremediation process. The continued presence of desired microbe with appropriate catabolic and anabolic capacity is an important prerequisite for bioremediation process (Singh et al., 2020; Gupta et al., 2020). Greater numbers of oil degrading microbes have been isolated in hydrocarbon polluted soil. Immediately after exposure to hydrocarbons, microbial communities start colonizing and adapt. Only those microbes become abundant that have the capability to bring about higher biodegradation. Hydrocarbon degradation from oil spills requires a range of bacterial groups so that a much wider range of hydrocarbons can be degraded. The oil spill is a serious concern for the environment as well as health of human beings due high toxicity and low degradation capacity. Figs. 5.1 and 5.2 describes the detail effect of oil spill on the human beings as well as fauna.

The diversity of microbes is immense. Soils contaminated by polyaromatic hydrocarbons are first of all colonized by microorganisms that are tolerant to these compounds and these gradually predominate (Huertas et al., 1998). Although a few have been cultured and many more remain to be grown under culture conditions (Curtis et al., 2002). The degrading capabilities of the microbial communities also vary. For example, aromatic hydrocarbons are degraded

Microbe Mediated Remediation of Environmental Contaminants. DOI: https://doi.org/10.1016/B978-0-12-821199-1.00005-5

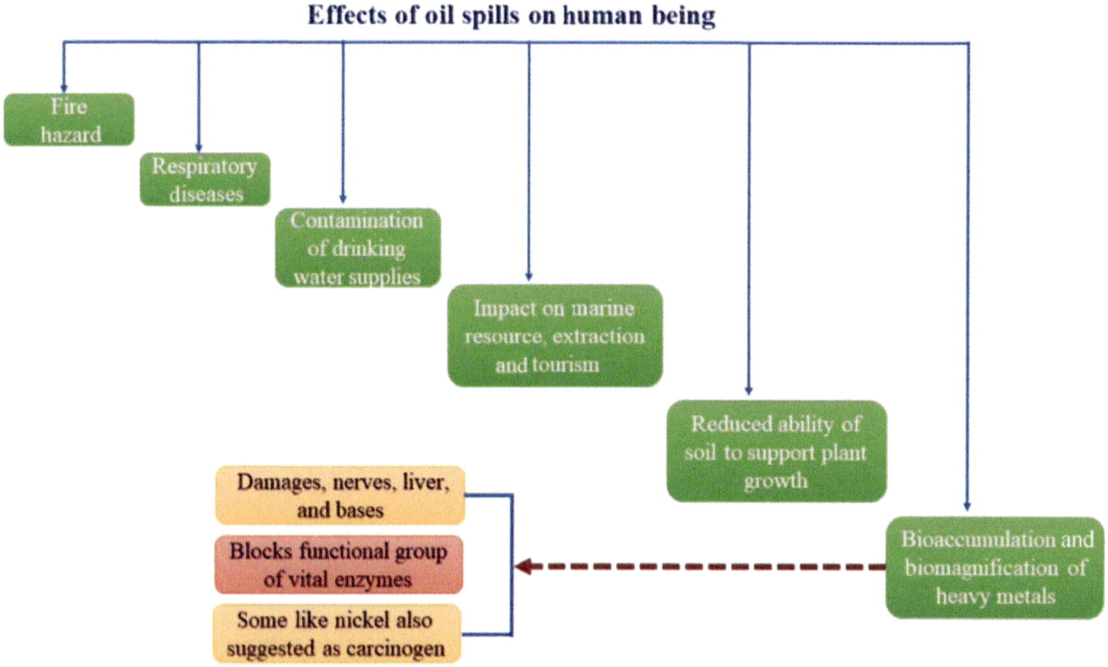

FIGURE 5.1 Effects of oil spills on human beings.

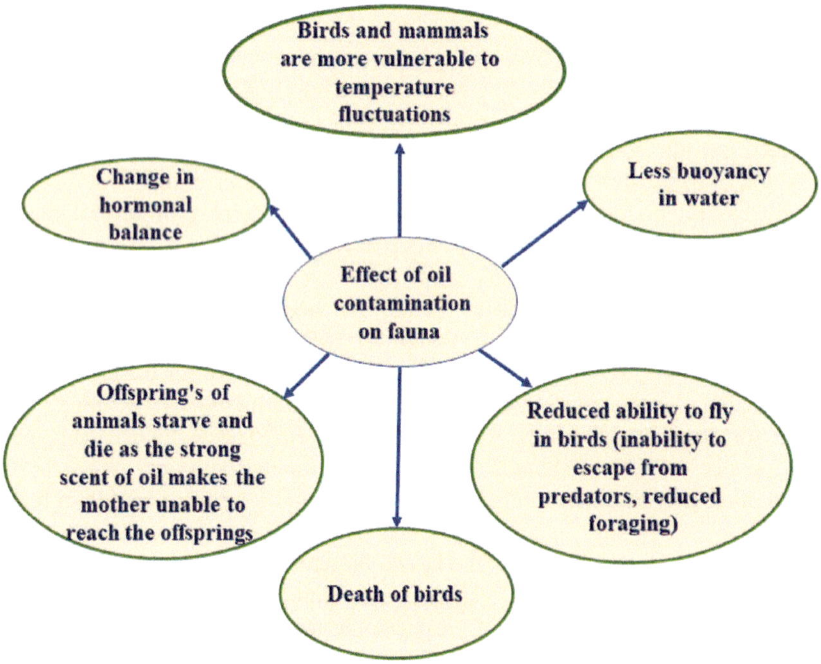

FIGURE 5.2 Oil contamination and its effects on fauna.

in an excellent way by mycobacteria, which can remove maximum up to five benzene rings (Megharaj et al., 2011). However, algae and microalgae cannot withstand aromatic hydrocarbons. Toluene specifically disrupts cell membrane and, therefore, toxic. When bioremediation occurs it is not only a single strain of microorganism but a plethora of microbial communities that may be antagonistic or synergistic (Megharaj et al., 2011).

The use of green plants, microorganisms, or their enzymes so that the polluted site is brought back to their original condition is called bioremediation (Glazer and Nikaido, 1995). The quest for microorganisms that can degrade pollutants has been persistent. Such novel microorganisms that can bring about degradation are now made more feasible due

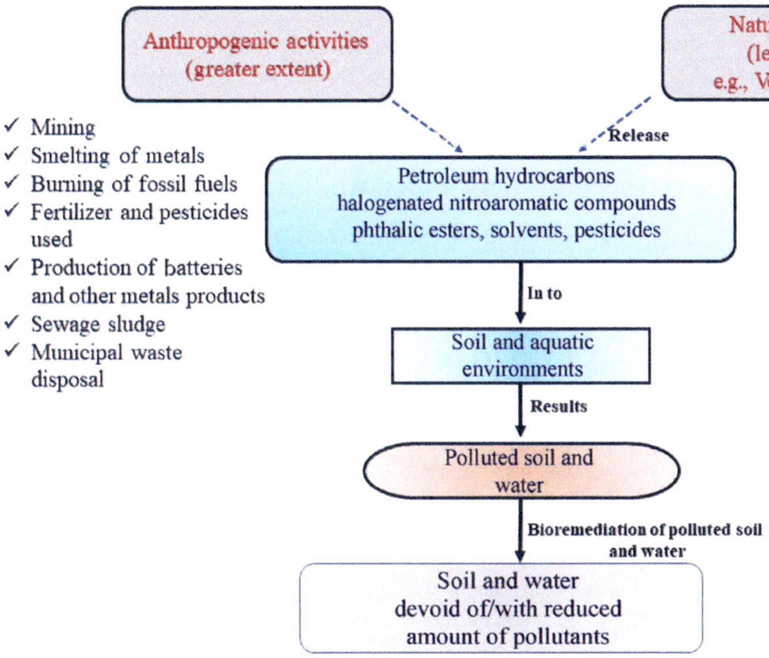

FIGURE 5.3 Overview of the bioremediation process including the source of contaminants.

to advances in metagenomics and whole genome sequencing (Golyshin et al., 2003). Although other living organisms can also bring about bioremediation, microbial biodegradation is the preferred one. Groundwater pollution adversely affects human health. Especially in times of population explosion, where density is quite high, it is difficult to offer lands for landfills. We also come across governments making various laws and amendments regarding policies for discarding wastes. So, reducing the toxic environmental wastes becomes important to achieve sustainable development (Kümmerer et al., 2019). As discussed earlier, creating landfills will not be a wise solution due to already reducing available landmasses, and the use of sewage treatment plants is economically not viable. Besides being expensive, treating excessive amounts of sewage is also difficult. Hence the need of the hour is the development of methods that are ecofriendly, sustainable, and cost-effective.

In bioremediation microbes growing naturally in the area get adapted to the noxious wastes. Enzymatic activities of microbes are responsible for breakdown of xenobiotics and in compliance with various organisms and processes may be called rhizomediation or phytoremediation (Singh et al., 2019a,b and Datta et al., 2020). If plants are used in the removal of pollutants, the process is called phytoremediation and if microbes do this at the interface of plant roots and appropriate microorganisms, it is called rhizomediation (Rajkumar et al., 2012). The conversion of organic wastes to substances that bring about nutrient enrichment by utilizing worms is called vermicomposting and vermifiltration. These approaches allow domestic waste water and industrial waste management in a sustainable way (Benitez et al., 2002) (Fig. 5.3).

5.2 Creating wholesome and safe environment by bioremediation

Since several decades, the world is experiencing indiscriminate land resource utilization. This is solely attributed to rapid industrialization and urbanization. Ground water, surface water, terrestrial lands, and atmosphere as a result are under continuous threat. Industrial effluents, pesticides, heavy metals, and petroleum hydrocarbons adversely affect soil and water contained in aquatic bodies. Bioremediation in the past few decades is considered as a sustainable and eco-friendly process for the restoration of pollutant lands. In this, pollutants at the contaminated site are used as carbon sources for the microorganisms by either cometabolism or complete mineralization. Technological advances also allow manipulating the genetic material or DNA of the microbe for enhancing the efficiency of the bioremediation process.

5.3 Bioremediation strategy for hydrocarbon contaminated water and soil

Coming into picture are the BTEX compounds requiring strict attention as they are categorized as priority pollutants requiring strict regulation by Environmental Protection Agency. According to De Graaff et al (2011), these compounds are both carcinogenic and neurotoxic. The structure and functioning of microbial communities is greatly affected by the release of BTEX compounds (Jin et al., 2013). In a system having multiple components the surplus components may affect the biodegradation of any components. So, hydrocarbon containing petroleum should be degraded by a consortium of microbes.

5.4 Treatment of heavy metal contaminated water by bioremediation

Heavy metals are metals with relatively high density and include elements with atomic numbers greater than 20. Some of the heavy metals are required by organisms such as Co, Cu, Fe, Mn, Mo, Ni, V, and Zn. Excessive amounts of these heavy metals may become deleterious to organisms, but minute quantities of these are required by microorganisms. Metals such as Pb, Cd, Hg, and As are harmful to both plants and animals. pH of the soil mainly contributes to heavy metal availability. Other factors contributing to heavy metal availability in soil includes density, soils relative surface area, type of charge in soil colloids, and degree of complexation with ligands. Heavy metals also show antagonistic and synergistic behavior (Chibuike and Obiora, 2014). Through the complex processes of biosorption, precipitation, surface complexation, and oxidoreduction, microbes can mobilize or immobilize or diminish the concentration of heavy metals (Teixeira et al., 2014; Singh et al., 2018). In soils contaminated by heavy metals, microbes evolve and that is how resistance develops (Luo et al., 2014). Even very minute levels of heavy metals either in suspended or dissolved state can be removed by microbes by various processes such as bioaccumulation, surface complexation, and biosorption. When it comes to heavy metal–contaminated soil, bioremediation emerges as a nondisruptive method and makes the remediated site fit for growth of plants (Chibuike and Obiora, 2014). Although heavy metals are preferentially removed by phytoextraction, the efficiency of bioremediation is enhanced when bacteria are used along with the plants.

5.5 Degradation of dye using bioremediation

Effluents from textile-processing industries and dyeing industries contains up to 50% of the originally used dye concentration. Along with the dye, there are various other constituents including dispersing agents, fixatives, heavy metals, and inorganic salts. Dyes in water lead to reduction in dissolved oxygen. Decreased light penetrance also diminishes photosynthetic capacity of submerged plants. Various processes such as adsorption, precipitation, flocculation, and oxidation followed by filtration and electrochemical methods are used to remove color from the waste water. However, these methods are not too pocket friendly and there are operational troubles involved too (Kapdan et al., 2000). Dyes that are discharged along with the waste water constitute 90% of the reactive dyes (Abadulla et al., 2000). Treatment of water containing dyes also demands plenty of expensive chemicals and extensive power consumption.

In light of the difficulties discussed previously, bioremediation of effluents containing dues is gaining momentum. In order to achieve enhanced biodegradation, manipulation of the bacterial genome is also being carried out in various labs worldwide. Dye degradation can be carried out under various aerobic and anaerobic conditions. Depending upon the microorganisms contained in the consortium and the chemical nature of the dye, the degradation of latter can be fine-tuned.

5.6 Biodegradation of pharmaceutical and personal care products

Past few years have witnessed the contamination of water sediments and biota by the unintentional release of pharmaceutical and personal care products (PPCPs). PPCPs find extensive use in human and veterinary medicine and so released continuously in the environment (Ebele et al., 2017). Biodegradation of PPCPs is possible through pure cultures of several organisms. Sulfamethoxazole, including various other sulfonamides, can be mineralized by *Achromobacter denitrificans* (Reis et al., 2018). Paracetamol can be mineralized by *Delftia tsuruhatensis* (Dang et al., 2013; Almeida et al., 2013). Few PPCPs are not compliant with certain microorganisms so do not get degraded by pure cultures. This may be attributed to the substrate specificity of the microorganisms involved. Maintenance of vital metabolic functioning of the microbes requires carbon and energy sources. Sometimes breakdown of carbamazepine that is not completely degraded by any microorganism because of its stability, the breakdown is possible by a mixed consortium consisting of a basidiomycete member which has not been identified as yet (Santoso et al., 2011) and *Streptomyces* MIUG (Popa et al., 2014) when glucose was supplemented.

One constituent which needs mention here is diclofenac, which finds extensive use in pain relief tablets. Diclofenac exhibits extreme resistance to degradation in activated sludge. However, it is possible to completely degrade diclofenac by using white rust fungus (Hata et al., 2010). It is essential to understand microbes must synthesize specific enzymes.

For instance, for the faster degradation of triclosan, the bacteria *Nitrosomonas europaea* must synthesize ammonia monooxygenase (Roh et al., 2009). It is also essential here to highlight that certain PPCPs, such as ciprofloxacin, tetracycline, and trimethoprim do not undergo biodegradation because they do not induce microbes to produce degradative enzymes. It has been, however, found that mixed cultures are more efficient in degrading PPCPs. The degradation of a single PPCP is slower as compared to mixed PPCP.

5.7 Removal of pesticides by bioremediation

Bioremediation is a technology, whereas biodegradation is a natural process. Both of them are somehow related as they involve metabolism of pesticides by microorganisms. As regards the pesticides, their extensive use harms the natural environment as well as it affects human life. (Bhanti and Taneja, 2007). Of special mention here are the organochlorine compounds. These compounds are nonbiodegradable, so they undergo bioaccumulation at various trophic levels. Too much use of these pesticides has led to the development of pesticide resistance and hazardous effect on nontarget organisms (Carson, 1962). A bacterial strain should be chosen that can degrade maximum amount of pollutant to the minimum level in a limited time frame. However, difficulty arises due to very limited knowledge about the population dynamics of degrading organisms in relation to other microorganisms that reside in the same habitat (Singh, 2008).

5.8 Activated sludge process

Activated sludge process finds great application in the biological treatment of waste water. During the degradation of PPCPs by this method a host of processes such as volatilization, surface binding or adsorption and microbial decomposition takes place. However, PPCP degradation is minimal using volatilization and adsorption (Li et al., 2015). For enhanced and optimal levels of degradation the microbial density has to be maintained. Bioaugmentation can be done by acclimatizing the microbes to contaminants (Wang et al., 2002). Whether pure or mixed cultures, the microbial communities can be made to bring about degradation by adaptation and bioaugmentation.

5.9 Removal of solid wastes by vermicomposting

The disposal of industrial solids also greatly contributes to surface and ground water pollution. Landfills and incineration were traditionally used for the removal of sludge (Hashemimajd et al., 2006). According to Lalander et al. (2013), vermicomposting is a sustainable for degrading solid and human wastes. Vermicomposting converts wastes into compost using worms and the resulting vermicompost greatly enhances plant nutrition (Gómez-Brandón et al., 2013). The generation of organic matter rich amendments is, therefore, possible by vermicomposting.

5.10 Factors affecting bioremediation

Monitoring physical and chemical properties of soil is time taking. Instead the extent of soil pollution may be measured by evaluating microbiological and biochemical properties of soil after contamination (Nannipieri et al., 1997).

According to Guthrie and Pfaender (1998), in order to achieve enhanced benefits of the bioremediation process, it becomes essential to consider the factors affecting the process. The complexity of the process magnifies due to noxious and hydrophobic nature of the hydrocarbons contained in the petrochemicals. The intricacies are further diversified due to prevailing environmental conditions and microbial community diversity. Various physical, chemical, and biological factors seem to limit the bioremediation process as evident from the literature. pH, temperature, nutrient status, dissolved oxygen content, presence of electron donors and acceptors, dissolved oxygen content, contaminant load, etc. all affect the efficiency of the bioremediation process (Mohan et al., 2006). Degradation rates of individual contaminants are also affected by substrate interactions (Wang et al., 2007). Bacterial metabolism is greatly affected by substrate interactions at different concentrations. It has been found that catabolic enzymes can improve decomposition rates of various components of the contaminant due to synergistic actions. The BTEX compounds beyond a certain concentration, due to their complex interactions, show inhibitory effect to microbial actions (Mathur and Majumder, 2010). The subsequent sections will highlight the major limiting factors affecting the degradation of petrochemical wastes.

Bioremediation process depends on a host of factors for optimal removal of the contaminant. These are discussed below.

5.10.1 Concentration of the contaminant

Concentration of the contaminant affects the microbial activity. If the concentration of the contaminant is low, bacterial degradation enzymes are prevented from induction. Toxicity effects are, however, manifested by very high concentrations of contaminant (Adams et al., 2015). The decomposition rates of catabolic enzymes are improved by the synergistic interactions between different components of a contaminant. At high substrate concentration, in batch culture, the growth rate of *Pseudomonas putida* was found to be decreased (Abuhamed et al., 2004). Due to complex microbiological interactions BTEX compounds showed inhibitory effect (Mathur and Majumder, 2010) to the biodegradation process.

5.10.2 Nutrient availability

Microbes require nutrients such as carbon, nitrogen, phosphorus, potassium, and calcium for their growth. In addition to this, the relative concentrations of the available nutrients also impacts degradation of the contaminant. A negative impact on biodegradation resulted due to excess of nitrogen, potassium, and phosphorus on the degradation of hydrocarbons (Van Hamme et al., 2003). The rate of biodegradation is also determined by whether organic pollutants are accessible to microorganisms, which is also referred to as bioavailability. When mass transfer is a limiting factor, higher biotransformation is not achieved in spite of high microbial conversion (Boopathy et al., 1998). It has been found that even after 50 years, the contaminating explosives in soil did not undergo biodegradation and rigorous mixing of the soil and dividing the coarse soil particles stimulated biodegradation (Mannig et al., 1995). Aging or weathering refers to reduction in bioavailability with the progress of time. Aging happens due to several physical and chemical processes such as:

1. Incorporation of contaminant into natural organic matter.
2. Reduced diffusion into very small pores and absorption into organic matter.
3. Formation of a not so solid layer around NAPL (nonaqueous phase liquids) (Boopathy, 2003).
4. It has been found that certain surfactants of food grade can overcome the bioavailability problem.

5.10.3 Surfactants: enhancers of bioavailability

In order to increase mass transfer of hydrophobic organic contaminants, surfactants are added to polluted soils (Laha et al., 2009; Rosen, 1989). Chemical surfactants which are derived from petroleum includes—Triton X 100, Tween 80, and SDS. However, there are food grade surfactants including T-MAZ 28, T-MAZ 10, and T-MAZ 60 (Shiau et al., 1995). The pericarp of *Sapindus mukorossi* and humic acid is used as a plant-based surfactant (Roy et al., 1997). Some microbes also produce surfactants and used in the bioremediation technology for the mitigation of environmental contaminants (Fig. 5.4).

5.10.4 Characteristics of the contaminated site

Bioremediation process is also greatly affected by the characteristics of the contaminated site. Parameters such as soil texture, permeability, pH, water holding capacity, soil temperature, nutrient, and oxygen content affect bioremediation process. Clayey soil is mixed with amendments or bulking materials such as straw and sawdust because of low permeability (Sihag et al., 2014).

5.10.4.1 pH

pH of the contaminated site is related to the contaminant and optimal pH is required for the bioremediation process. Generally a pH range of 6–8 (International Centre for Soil and Contaminated Sites, 2006) is considered optimal (Adams et al., 2015). It also needs to be emphasized that at the contaminated site, microorganisms are tolerant of the site conditions and may metabolize polyaromatic hydrocarbons at high pH (Sihag et al., 2014). Neutral pH is favorable for the mineralization of petroleum hydrocarbons. As per the available studies, biodegradation process may be affected by even the temperature range of 15°C–20°C (Mueller et al., 1989). It has been observed that increased temperatures enhance the solubility of organic hydrocarbons in the medium, making the petrochemical hydrocarbons readily available to a slight variation in pH. According to Stapleton et al. (1998), biodegradation potential of certain fungi and acidophilic microbes are higher in acidic environments.

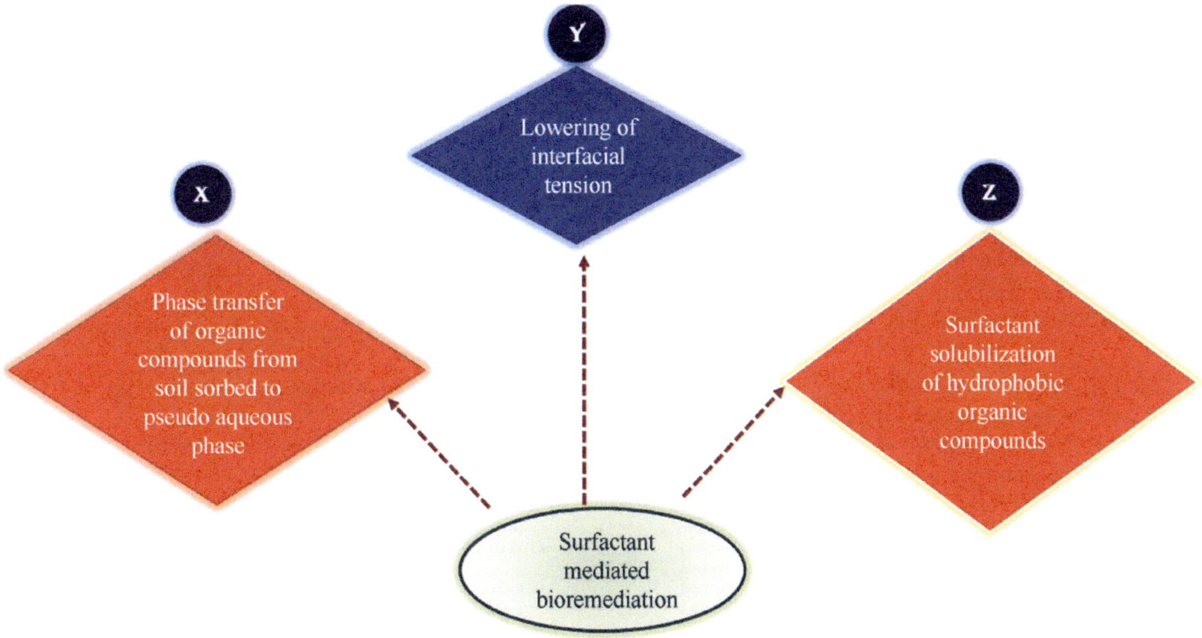

FIGURE 5.4 Mechanism of surfactant-mediated bioremediation.

5.10.4.2 *Temperature*

The degradation of the contaminant specially the hydrocarbon under both in situ and ex situ conditions is affected by temperature (Margesin and Schinner, 2001). It has been observed that higher temperature ranges such as 30°C–40°C enhance soil biodegradation and the same is true in aqueous or marine environments in microorganisms.

5.10.4.3 *Oxygen availability*

The availability of oxygen decides whether the bioremediation process is aerobic or anaerobic. The few initial steps of aerobic metabolism of PAH oxygen is required for the action of monooxygenase and dioxygenase enzymes in the oxidation of aromatic ring (Sihag et al., 2014). During anaerobic oxidation of aromatic compound, ferrous iron, nitrate, and sulfate in their substituted form are required as electron acceptors. However, the process leads to emission of sufficient levels of phosphorus and ferrous ion, which in turn again contaminated the ecosystem. The anaerobic degradation of petrochemical hydrocarbons release greenhouse gases such as methane and nitrogen dioxide along with increased pH (Bamforth and Singleton, 2005). Bioremediation in submerged marine sediments and aquifers occurs mostly by anaerobic process (Coates et al., 1996).

The extent of biodegradation depends on the following factors:

1. Biodegradability of contaminant
2. Whether biodegradability of contaminant is taking place on its own, that is, naturally
3. Suitability of environmental conditions for biodegradation to occur
4. In the event of failure of biodegradation the sink for the waste.

Surface soils have higher organic matter content. This is due to the fact that there is greater diversity of microorganisms and their cont. However, in contrast, the organic matter content of the subsurface soil and ground water sediments is lower due to less diversity and number of microorganisms (Adriaens and Hickey, 1993). Bacteria being able to use alternative electron acceptors become more prevalent than other microbial communities with increasing depth. Temperature, moisture content, and dissolved oxygen are other factors influencing the extent to spread of microbial populations (Fig. 5.5).

The effectiveness of bioremediation depends to a large extent on choosing appropriate microorganisms. It should also be emphasized that for the technology to be effective, the environmental factors should also be conducive. Conventionally, pollutant eradication involved creating landfills and incineration. However, this has now been replaced by bioremediation.

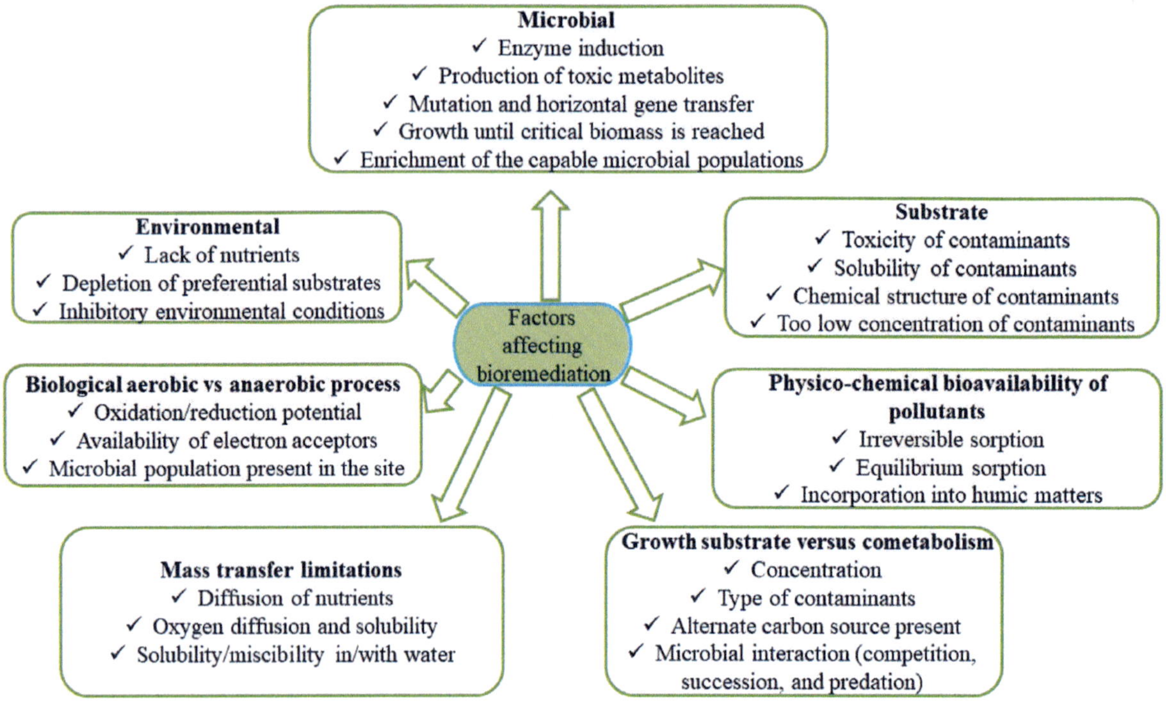

FIGURE 5.5 Factors affecting bioremediation process.

5.11 Understanding biodegradation kinetics of organic constituents

It becomes essential to understand the fact that several microorganisms utilize hydrocarbons as the source of carbon and energy. During the course of biodegradation, both aliphatic and aromatic hydrocarbons are brought about by more than 40 species of microorganisms. The environmental factors responsible for biodegradation should be optimized to achieve maximized decomposition. Hence kinetics studies of biodegradation of organic constituents have to be performed. Oil degrading ability of microorganisms is partly resident on the plasmids. For the breakdown of *n*-octane and other 5–12 carbon linear alkanes, the OCT plasmid of *Pseudomonas oleovorans* contains two operons, alk BFGHJKL and alk ST. These operons code for proteins that bring about degradation of said hydrocarbons and octane gets easily degraded as compared to branched isomers like isooctane (Juhasz et al., 2000).

5.12 Metabolic cooperation and networking—the systems biology approach

Instead of looking for a single isolate, microbial consortia are desired to bring about bioremediation in ecologically relevant conditions (Supaphol et al., 2006). It is necessary to go by the systems biology approach for understanding complex metabolic and regulatory interaction even in a single isolate (Pazos et al., 2003; Trigo et al., 2008). For studying the organization of biological systems and predicting their responses to internal and external environmental variations, network theory becomes vital to understand (Feist and Palsson, 2008). UM-BBD (University of Minnesota Biocatalysis/Biodegradation Database) contains information on about 900 compounds, 600 enzymes, 1000 reactions, and 350 microbial entities (Ellis et al., 2006). The reaction rules for particular functional groups predict the metabolic pathway. A database called "Metarouter" was developed by Pazos et al. (2005) based on information available at UM-BBD. The authors suggested that a correlation existed between common chemotypes contained in organochemical compounds and the metabolic capacity of microorganisms to degrade (Fig. 5.6). A predictive tool (https://www.pdg.cno.uam.es/BDPSERVER) was developed to provide insight into the contaminant type as to whether they are biodegradable or recalcitrant. The complete system required for biodegradation multiple biotic and abiotic factors of which the latter consists of a host of microbial communities (Abraham et al., 2002).

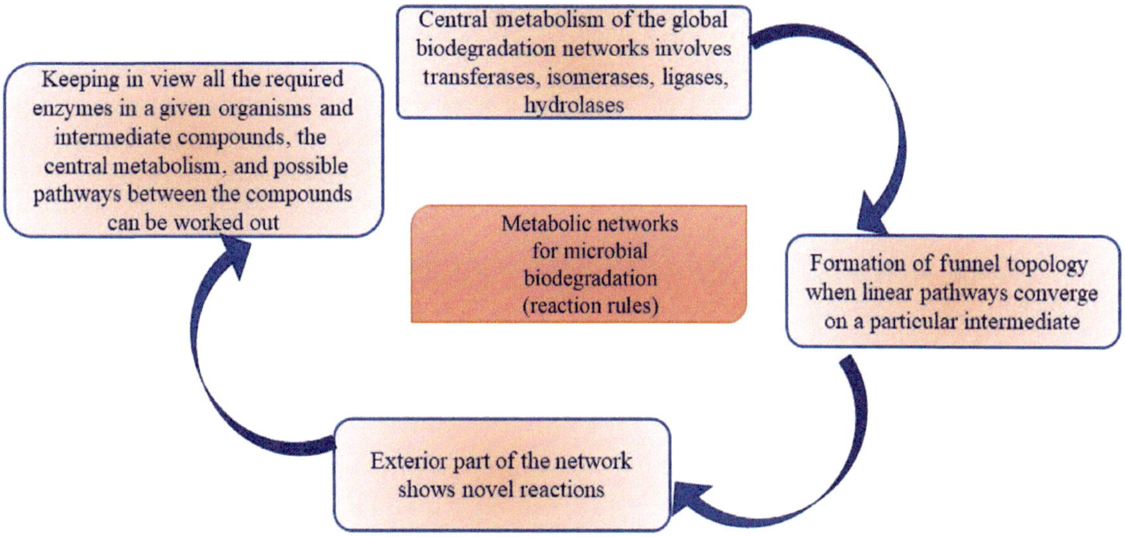

FIGURE 5.6 Reaction rules for metabolic networks for microbial biodegradation.

5.13 Stable isotope probing analysis

This helps in identifying microorganisms that have been cultured and those remaining to be cultured and also assists in associating function with microbial diversity (Cupples, 2011). [flowchart for recent advances in characterization of contaminated site, identification of new degradative activities, and monitoring bioremediation efficiency (Megharaj et al., 2011)].

5.14 Advantages of bioremediation

The advantages of bioremediation are as follows:

1. It can be done on site.
2. It is less expensive.
3. In sharp contrast to landfills that requires large proportion of lands, bioremediation site disruption is minimal.
4. Elimination of waste is permanent and long-term liability is reduced.
5. Public acceptance of the technology is greater with certain regulatory encouragement.
6. It can be also clubbed with certain physical or chemical methods.

5.15 Disadvantages of bioremediation

The disadvantages of bioremediation are as follows:

1. Heavy metals, radionuclides, and some other chlorinated compounds cannot be removed by bioremediation.
2. Microbial metabolism may produce certain toxic end products.
3. Before proceeding to the actual cleanup of the sites, treatability studies have to be performed as bioremediation is a scientifically intensive procedure.

5.16 Conclusion

Chemical pollutants have been in extensive use since the discovery of DDT's. Discovery of chemicals often is not followed by a knowledge of its toxicity and recalcitrance. Keeping this in view, policymakers should come up with laws to decide about the immediate necessity and practicality of the bioremediation process (Alcock et al., 2011). Therefore stringent regulatory measures are required right from their disposal to their effective removal by bioremediation process. Some contaminated sites may exhibit multiple contaminants. In such cases, restoration of the total functionality appears not too practical approach. In addition, different contaminants require different strategies for removal and each strategy has its own pros and cons. Hence, the adopted strategies (either a single strategy or a mix of strategies) would require to be selected with great care. Speaking about the microorganisms even within a single microbial guild, the biological response to pollutants vary (Ramakrishnan et al., 2011). In spite of all this, the systems biology approach demands knowledge about metabolic cooperation on biological communities. Community fingerprinting and environmental genomics can throw some light on structure—functional relationship of microbial communities at the site of contamination. Despite all the previously mentioned variables, Government being the law-making body can come up with regulatory considerations regarding the use of bioremediation technologies. However, for *sites* that are contaminated by multiple pollutants, in such cases restoring the total functionality appears not so practical approach.

References

Abadulla, E., Tzanov, T., Costa, S., Robra, K.H., Cavaco-Paulo, A., Gübitz, G.M., 2000. Decolorization and detoxification of textile dyes with a laccase from *Trametes hirsuta*. Appl. Environ. Microbiol. 66, 3357–3362.

Abraham, W.R., Nogales, B., Golyshin, P.N., Pieper, D.H., Timmis, K.N., 2002. Polychlorinated biphenyl-degrading microbial communities in soils and sediments. Curr. Opin. Microbiol. 5, 246–253.

Abuhamed, T., Bayraktar, E., Mehmetoğlu, T., Mehmetoğlu, Ü., 2004. Kinetics model for growth of *Pseudomonas putida* F1 during benzene, toluene and phenol biodegradation. Process Biochem. 39, 983–988.

Adams, G.O., Fufeyin, P.T., Okoro, S.E., Ehinomen, I., 2015. Bioremediation, biostimulation and bioaugmention: a review. Int. J. Environ. Biorem. Biodegrad. 3, 28–39.

Adriaens, P., Hickey, W.J., 1993. In: Stone, D.L. (Ed.), Biotechnology for the Treatment of Hazardous Waste. Lewis Publishers, Ann Arbor, MI, pp. 97–120.

Alcock, R.E., MacGillivray, B.H., Busby, J.S., 2011. Understanding the mismatch between the demands of risk assessment and practice of scientists—the case of Deca-BDE. Environ. Int. 37, 216–225.

Almeida, B., Kjeldal, H., Lolas, I., Knudsen, A.D., Carvalho, G., Nielsen, K.L., et al., 2013. Quantitative proteomic analysis of ibuprofen-degrading *Patulibacter* sp. strain I11. Biodegradation 24, 615–630.

Bamforth, S.M., Singleton, I., 2005. Bioremediation of polycyclic aromatic hydrocarbons: current knowledge and future directions. J. Chem. Technol. Biotechnol.: Int. Res. Process Environ. Clean. Technol. 80 (7), 723–736.

Benitez, E., Sainz, H., Melgar, R., Nogales, R., 2002. Vermicomposting of a lignocellulosic waste from olive oil industry: a pilot scale study. Waste Manage. Res. 20, 134–142.

Bhanti, M., Taneja, A., 2007. Contamination of vegetables of different seasons with organophosphorous pesticides and related health risk assessment in northern India. Chemosphere 69, 63–68.

Boopathy, R., 2000. Factors limiting bioremediation technologies. Bioresour. Technol. 74, 63–67.

Boopathy, R., 2003. Use of anaerobic soil slurry reactors for the removal of petroleum hydrocarbons in soil. Int. Biodeterior. Biodegrad. 52, 161–166.

Boopathy, R., Manning, J., Kulpa, C.F., 1998. Biotransformation of explosives by anaerobic consortia in liquid culture and in soil slurry. Int. Biodeterior. Biodegrad. 41, 67–74.

Carson, R., 1962. Silent Spring. Honghton Mifflin Co, Boston, MA, p. 368.

Chibuike, G.U., Obiora, S.C., 2014. Heavy metal polluted soils: effect on plants and bioremediation methods. Appl. Environ. Soil Sci. 2014.

Coates, J.D., Anderson, R.T., Lovley, D.R., 1996. Oxidation of polycyclic aromatic hydrocarbons under sulfate-reducing conditions. Appl. Environ. Microbiol. 62, 1099–1101.

Cupples, A.M., 2011. The use of nucleic acid based stable isotope probing to identify the microorganisms responsible for anaerobic benzene and toluene biodegradation. J. Microbiol. Methods 85, 83–91.

Curtis, T.P., Sloan, W.T., Scannell, J.W., 2002. Estimating prokaryotic diversity and its limits. Proc. Natl. Acad. Sci. U.S.A. 99, 10494–10499.

Dang, D., Wang, D., Zhang, C., Zhou, W., Zhou, Q., Wu, H., 2013. Comparison of oral paracetamol versus ibuprofen in premature infants with patent ductus arteriosus: a randomized controlled trial. PLoS One 8 (11).

Datta, S., Singh, S., Kumar, V., Dhanjal, D.S., Sidhu, G.K., Amin, D.S., et al., 2020. Endophytic bacteria in xenobiotic degradation. In: Kumar, A., Singh, V.K. (Eds.), Microbial Endophytes: Prospects of Sustainable Agriculture. Elsevier, pp. 125–156.

De Graaff, M., Bijmans, M.F., Abbas, B., Euverink, G.J., Muyzer, G., Janssen, A.J., 2011. Biological treatment of refinery spent caustics under halo-alkaline conditions. Bioresour. Technol. 102, 7257–7264.

Dubus, I.G., Hollis, J.M., Brown, C.D., 2000. Pesticides in rainfall in Europe. Environ. Pollut. 110, 331–344.

Ebele, A.J., Abdallah, M.A.E., Harrad, S., 2017. Pharmaceuticals and personal care products (PPCPs) in the freshwater aquatic environment. Emerging Contam. 3, 1–16.

Ellis, L.B., Roe, D., Wackett, L.P., 2006. The University of Minnesota biocatalysis/biodegradation database: the first decade. Nucl. Acids. Res 34 (Suppl_1), D517–D521.

Farhadian, M., Vachelard, C., Duchez, D., Larroche, C., 2008. In situ bioremediation of monoaromatic pollutants in groundwater: a review. Bioresour. Technol. 99, 5296–5308.

Feist, A.M., Palsson, B.Ø., 2008. The growing scope of applications of genome-scale metabolic reconstructions using *Escherichia coli*. Nat. Biotechnol. 26, 659–667.

Glazer, A.N., Nikaido, H., 1995. Microbial Biotechnology: Fundamentals of Applied Microbiology. Freeman, New York, p. 656.

Golyshin, P.N., Dos Santos, V.A.M., Kaiser, O., Ferrer, M., Sabirova, Y.S., Lünsdorf, H., et al., 2003. Genome sequence completed of *Alcanivorax borkumensis*, a hydrocarbon-degrading bacterium that plays a global role in oil removal from marine systems. J. Biotechnol. 106, 215–220.

Gómez-Brandón, M., Lores, M., Domínguez, J., 2013. Changes in chemical and microbiological properties of rabbit manure in a continuous-feeding vermicomposting system. Bioresour. Technol. 128, 310–316.

Gupta, A., Singh, S.K., Singh, V.K., Singh, M.K., Modi, A., Zhimo, V.Y., et al., 2020. Endophytic microbe approaches in bioremediation of organic pollutants. In: Kumar, A., Singh, V.K. (Eds.), Microbial Endophytes: Prospects of Sustainable Agriculture. Elsevier, pp. 157–174.

Guthrie, E.A., Pfaender, F.K., 1998. Reduced pyrene bioavailability in microbially active soils. Environ. Sci. Technol. 32, 501–508.

Hashemimajd, K., Kalbasi, M., Golchin, A., Knicker, H., Shariatmadari, H., Rezaei-Nejad, Y., 2006. Use of vermicomposts produced from various solid wastes as potting media. Eur. J. Hortic. Sci. 71, 21.

Hata, T., Kawai, S., Okamura, H., Nishida, T., 2010. Removal of diclofenac and mefenamic acid by the white rot fungus *Phanerochaete sordida* YK-624 and identification of their metabolites after fungal transformation. Biodegradation 21, 681–689.

Huertas, M.J., Duque, E., Marqués, S., Ramos, J.L., 1998. Survival in soil of different toluene-degrading *Pseudomonas* strains after solvent shock. Appl. Environ. Microbiol. 64, 38–42.

International Centre for Soil and Contaminated Sites, 2006. Manual for biological remediation techniques. 81pp.

Jin, H.M., Choi, E.J., Jeon, C.O., 2013. Isolation of a BTEX-degrading bacterium, *Janibacter* sp. SB2, from a sea-tidal flat and optimization of biodegradation conditions. Bioresour. Technol. 145, 57–64.

Juhasz, A.L., Stanley, G.A., Britz, M.L., 2000. Degradation of high molecular weight PAHs in contaminated soil by a bacterial consortium: effects on microtox and mutagenicity bioassays. Biorem. J. 4, 271–283.

Kapdan, I.K., Kargi, F., McMullan, G., Marchant, R., 2000. Decolorization of textile dyestuffs by a mixed bacterial consortium. Biotechnol. Lett. 22, 1179–1181.

Kümmerer, K., Dionysiou, D.D., Olsson, O., Fatta-Kassinos, D., 2019. Reducing aquatic micropollutants—increasing the focus on input prevention and integrated emission management. Sci. Total Environ. 652, 836–850.

Laha, S., Tansel, B., Ussawarujikulchai, A., 2009. Surfactant–soil interactions during surfactant-amended remediation of contaminated soils by hydrophobic organic compounds: a review. J. Environ. Manage. 90, 95–100.

Lalander, C.H., Hill, G.B., Vinneras, B., 2013. Hygienic quality of faeces treated in urine diverting vermicomposting toilets. Waste Manage. 33, 2204–2210.

Li, W., Shi, Y., Gao, L., Liu, J., Cai, Y., 2015. Occurrence, fate and risk assessment of parabens and their chlorinated derivatives in an advanced wastewater treatment plant. J. Hazard. Mater. 300, 29–38.

Luo, X., Zeng, X.C., He, Z., Lu, X., Yuan, J., Shi, J., et al., 2014. Isolation and characterization of a radiation-resistant bacterium from Taklamakan Desert showing potent ability to accumulate Lead (II) and considerable potential for bioremediation of radioactive wastes. Ecotoxicology 23, 1915–1921.

Mannig, J.F., Boopathy, R., Kulpa, C.F. 1995. A Laboratory Study in Support of the Pilot Demonstration of a Biological Soil Slurry Reactor, Argonne National Laboratory. SFIM-AECTS-CR-9408.

Margesin, R., Schinner, F., 2001. Biodegradation and bioremediation of hydrocarbons in extreme environments. Appl. Microbiol. Biotechnol. 56, 650–663.

Mathur, A.K., Majumder, C.B., 2010. Kinetics modelling of the biodegradation of benzene, toluene and phenol as single substrate and mixed substrate by using *Pseudomonas putida*. Chem. Biochem. Eng. Q. 24, 101–109.

Megharaj, M., Ramakrishnan, B., Venkateswarlu, K., Sethunathan, N., Naidu, R., 2011. Bioremediation approaches for organic pollutants: a critical perspective. Environ. Int. 37, 1362–1375.

Mohan, S.V., Kisa, T., Ohkuma, T., Kanaly, R.A., Shimizu, Y., 2006. Bioremediation technologies for treatment of PAH-contaminated soil and strategies to enhance process efficiency. Rev. Environ. Sci. Biotechnol. 5, 347–374.

Mueller, J.G., Chapman, P.J., Pritchard, P.H., 1989. Creosote-contaminated sites. Their potential for bioremediation. Environ. Sci. Technol. 23, 1197–1201.

Nannipieri, P., Badalucco, L., Landi, L., Pietramellara, G., 1997. Measurement in assessing the risk of chemicals to the soil ecosystem. Ecotoxicology: Responses, Biomarkers and Risk Assessment. OECD Workshop. SOS. Publ, Fair Haven, NY, pp. 507–534.

Pazos, F., Valencia, A., De Lorenzo, V., 2003. The organization of the microbial biodegradation network from a systems-biology perspective. EMBO Rep. 4, 994–999.

Pazos, F., Guijas, D., Valencia, A., De Lorenzo, V., 2005. MetaRouter: bioinformatics for bioremediation. Nucleic Acids Res. 33 (Suppl_1), D588–D592.

Popa, C., Favier, L., Dinica, R., Semrany, S., Djelal, H., Amrane, A., et al., 2014. Potential of newly isolated wild *Streptomyces* strains as agents for the biodegradation of a recalcitrant pharmaceutical, carbamazepine. Environ. Technol. 35, 3082–3091.

Rajkumar, M., Sandhya, S., Prasad, M.N.V., Freitas, H., 2012. Perspectives of plant-associated microbes in heavy metal phytoremediation. Biotechnol. Adv. 30, 1562–1574.

Ramakrishnan, B., Megharaj, M., Venkateswarlu, K., Sethunathan, N., Naidu, R., 2011. Mixtures of environmental pollutants: effects on microorganisms and their activities in soils, Reviews of Environmental Contamination and Toxicology, vol. 211. Springer, New York, pp. 63–120.

Reis, A.C., Čvančarová, M., Liu, Y., Lenz, M., Hettich, T., Kolvenbach, B.A., et al., 2018. Biodegradation of sulfamethoxazole by a bacterial consortium of *Achromobacter denitrificans* PR1 and *Leucobacter* sp. GP. Appl. Microbiol. Biotechnol. 102, 10299–10314.

Roh, H., Subramanya, N., Zhao, F., Yu, C.P., Sandt, J., Chu, K.H., 2009. Biodegradation potential of wastewater micropollutants by ammonia-oxidizing bacteria. Chemosphere 77, 1084e1089.

Rosen, M.J., 1989. Surfactants and Interfacial Phenomena. Wiley, New York.

Roy, D., Kommalapati, R.R., Mandava, S.S., Valsaraj, K.T., Constant, W.D., 1997. Soil washing potential of a natural surfactant. Environ. Sci. Technol. 31, 670–675.

Santoso, H., Gunawan, T., Jatmiko, R.H., Darmosarkoro, W., Minasny, B., 2011. Mapping and identifying basal stem rot disease in oil palms in North Sumatra with Quick Bird imagery. Precis. Agric. 12, 233e248.

Shiau, B.J., Sabatini, D.A., Harwell, J.H., 1995. Properties of food grade (edible) surfactants affecting subsurface remediation of chlorinated solvents. Environ. Sci. Technol. 29, 2929–2935.

Sihag, S., Pathak, H., Jaroli, D.P., 2014. Factors affecting the rate of biodegradation of polyaromatic hydrocarbons. Int. J. Pure Appl. Biosci. 2, 185–202.

Singh, D.K., 2008. Biodegradation and bioremediation of pesticide in soil: concept, method and recent developments. Ind. J. Microbiol. 48, 35–40.

Singh, V.K., Singh, A.L., Singh, R., Kumar, A., 2018. Iron oxidizing bacteria: insights on diversity, mechanism of iron oxidation and role in management of metal pollution. Environ. Sustainability 1 (3), 221–231.

Singh, M., Singh, D., Gupta, A., Pandey, K.D., Singh, P.K., Kumar, A., 2019a. Plant growth promoting rhizobacteria: application in biofertilizers and biocontrol of phytopathogens. In: Singh, A.K., Kumar, A., Singh, P.K. (Eds.), PGPR Amelioration in Sustainable Agriculture. Woodhead Publishing, pp. 41–66.

Singh, V.K., Singh, M., Singh, S.K., et al., 2019b. Sustainable agricultural practices using beneficial fungi under changing climate scenario. In: Choudhary, K.K., Kumar, A., Singh, A.K. (Eds.), Climate Change and Agricultural Ecosystems. Woodhead Publishing, pp. 25–42.

Singh, P., Singh, V.K., Singh, R., Borthakur, A., Madhav, S., Ahamad, A., et al., 2020. Bioremediation: a sustainable approach for management of environmental contaminants. In: Singh, P., Kumar, A., Borthakur, A. (Eds.), Abatement of Environmental Pollutants. Elsevier, pp. 1–23.

Stapleton, R.D., Savage, D.C., Sayler, G.S., Stacey, G., 1998. Biodegradation of aromatic hydrocarbons in an extremely acidic environment. Appl. Environ. Microbiol. 64, 4180–4184.

Stout, S.A., Magar, V.S., Uhler, R.M., Ickes, J., Abbott, J., Brenner, R., 2001. Characterization of naturally-occurring and anthropogenic PAHs in urban sediments-Wycoff/Eagle harbor superfund site. Environ. Forensics 2, 287–300.

Supaphol, S., Panichsakpatana, S., Trakulnaleamsai, S., Tungkananuruk, N., Roughjanajirapa, P., O'Donnell, A.G., 2006. The selection of mixed microbial inocula in environmental biotechnology: example using petroleum contaminated tropical soils. J. Microbiol. Methods 65, 432–441.

Teixeira, S., Vieira, M.N., Marques, J.E., Pereira, R., 2014. Bioremediation of an iron-rich mine effluent by *Lemna minor*. Int. J. Phytorem. 16, 1228–1240.

Trigo, A., Valencia, A., Cases, I., 2008. Systemic approaches to biodegradation. FEMS Microbiol. Rev. 33, 98–108.

Van Hamme, J.D., Singh, A., Ward, O.P., 2003. Recent advances in petroleum microbiology. Microbiol. Mol. Biol. Rev. 67, 503–549.

Wang, J.L., Quan, X.C., Wu, L.B., Qian, Y., Hegemann, W., 2002. Bioaugmentation as a tool to enhance the removal of refractory compound in coke plant wastewater. Process Biochem. 38, 777–781.

Wang, L., Barrington, S., Kim, J.W., 2007. Biodegradation of pentyl amine and aniline from petrochemical wastewater. J. Environ. Manage. 83, 191–197.

Zhu, X., Venosa, A.D., Suidan, M.T., Lee, K., 2001. Guidelines for the Bioremediation of Marine Shorelines and Freshwater Wetlands. US Environmental Protection Agency.

Chapter 6

Microbial enzymes and their exploitation in remediation of environmental contaminants

Sandeep Kumar Singh[1], Manoj Kumar Singh[2], Vipin Kumar Singh[1], Arpan Modi[3], Pooja Jaiswal[4], Kumari Rashmi[5] and Ajay Kumar[3]

[1]Department of Botany, Centre of Advanced Study in Science, Institute of Science, Banaras Hindu University, Varanasi, India, [2]Chemical Division, Intellectual Property Office, Delhi, India, [3]Agriculture Research Organization, Ministry of Agriculture and Rural Development, Volcani Center, Rishon LeZion, Israel, [4]Department of Botany, M.M.V, Banaras Hindu University, Varanasi, India, [5]L.N.Mishra Institute Patna, Patna, India

6.1 Introduction

From last few decades, human population continuously is increasing, and it has been estimated up to the year of 2050 will be reached to 7 billion, and this increasing population creates extra pressure on the agriculture and environment for their livelihood; however, their anthropogenic activities enhance disturbances in the environment via rapid industrialization, release of toxic effluents, and waste products arise through various resources (Raghunandan et al., 2014, 2018; Ahuti, 2015). The rapid industrialization results in the deposition of various contaminants, xenobiotics, accidental leakage and discharge of toxic chemicals, heavy metals, and chemical pesticides, in the soil, water, and air ecosystem that not only affects the growth and productivity of agricultural products but also severally affects the health of human beings (Mgbemene et al., 2016).

Traditionally, various physiochemical methods have been practiced for contaminant remediation in which filling of waste inside the ground is the most common practice, but it is very challenging due to limited space as well as searching new places at every time of dump. However, further high-temperature incineration and chemical decomposition are carried out for waste management. Although they can be very effective in reducing a wide range of contaminants, at the same time, they have several drawbacks due to complexity and cost-effectiveness. The associated drawbacks in the traditional methods, open, have focused efforts toward harnessing modern-day bioremediation process as a suitable alternative (Karigar and Rao, 2011). Bioremediation is a microbiological well-organized procedural activity, applied to break down or transform toxic contaminates to less toxic or nontoxic or even in elemental forms. Bioremediations is a biological agent-mediated process used for remediation in order to clean up contaminated sites. The employability of various organisms such as bacteria, fungi, algae, and plants is used for the efficient bioremediation of pollutants (Vidali, 2001; Strong and Burgess, 2008). The involvement of plants in remediation of environmental pollutants is termed as phytoremediation and considered as an emerging green technology that facilitates the removal or degradation of the toxic chemicals from soils, sediments, groundwater, surface water, and air. Genetically, engineered plants are also in use during remediation process. Leung (2004) reported details of arsenic management through genetically modified plant *Arabidopsis thaliana* that expresses involvement of two bacterial genes, one allows the plant to modify arsenate into arsenite and the second one binds the modified arsenite and stores it in the vacuoles.

Enzyme molecule has been used for the broad range of soil and water contaminants, due to effectivity, efficiency as well as restoring natural surrounding without secreting toxic residuals (Demnerová, et al., 2005). To date, enzymes have been applied in a broad range of practical applications mainly in analytical chemistry and medicine (Uhlig, 1998), but now attempts are underway to employ these biological catalysts for the protection of environment (Karam and Nicell, 1997). Enzymatic processes are expected to replace traditional industrial technologies to reduce the emission of toxic by-products and to eliminate hazards associated with industries, anthropogenic activities. Enzymes also have the

Microbe Mediated Remediation of Environmental Contaminants. DOI: https://doi.org/10.1016/B978-0-12-821199-1.00006-7

potential for bioremediation of polluted environments (Alexander, 1999). The process of bioremediation depends upon the nature of microorganisms, which enzymatically attack the pollutants and convert them to innocuous products. The efficiency of bioremediation also depends upon the optimum growth as well as activity of the microbes. The rationale for developing enzymatic cleanup is the ultimate cause of pollutant degradation during bioremediation procedures. Enzymatic treatment is expected to have many advantages such as the ease and simplicity of application, short treatment time, ability to target-specific pollutants, low sludge volume, and the absence of ecological hazards associated with their use (Nannipieri and Bollag, 1991). The application of enzymes has several advantages over microbial treatment such as enzymes do not require an acclimatization phase as is often required for microorganisms. Unlike microorganisms, enzymes may be effective at both low and high pollutant concentrations and can penetrate microporous sites in the soil matrix where microorganisms cannot be entered. Enzymes are resistant to many inhibitors that can affect microbial metabolism. Furthermore, following isolation from genetically engineered microorganisms, enzymes can be released safely into the environment. Some of the challenges in enzymatic treatment are of high cost of isolation from microbial cells, storage, and prolong stability under harsh environment.

Another disadvantage is the need for cofactors, without which many enzymes do not show functional activity. The most desired enzymatic action is complete mineralization of the toxic compounds. Unfortunately, as discussed in detail later, the majority of enzymes are not capable of mineralization when applied individually. Theoretically, mineralization may be possible if several different enzymes are applied simultaneously, but all the enzymes that are known to be involved in established mineralization pathways require expensive cofactors. With individual enzymes, one can expect only a certain rate of pollutant degradation or transformation. However, as long as the transformation products are less toxic than the parent compounds, or, in other words, as long as there is a detoxification effect, the enzymatic treatment may be considered successful, especially when the degradation products are susceptible to further transformation by microorganisms. The process of bioremediation is a very slow process. Only certain species of bacteria and fungi have proven their ability as potent pollutant degraders. Many strains are known to be effective as bioremediation agents but limit only in the laboratory conditions. The limitation of bacterial growth is under the influence of pH, temperature, oxygen, soil structure, moisture, and appropriate level of nutrients. Most bioremediation systems operate under aerobic conditions, but anaerobic environments may also permit microbial degradation of recalcitrant molecules. Both bacteria and fungi rely on the participation of different intracellular and extracellular enzymes, respectively, for the remediation of recalcitrant and lignin and organ pollutants (Vidali, 2001; Cadisch, and Giller, 1997).

6.2 Enzymes

Enzymes are also known as a biocatalyst that catalyzes various biological processes and enhances the rate of reaction. Besides in vivo, enzymes are also utilized in the in vitro analysis to enhance the rate of reaction of various chemical reactions in industries, particularly, pharmaceuticals, paper, leather food, and many more industries and also have proven utility in bioconversions and bioremediation (Nigam, 2013).These industries utilized enzymes at the time of processing of raw materials/substrates at the any stage of productions. The selection of enzymes during application is highly specific and selected on the basis of stability, specificity and having high rate of reaction (Jegannathan and Nielsen, 2013). Currently, more than 80% of the enzymes are produced through various industrial processes, and there is a need for deep insight, extensive study, and knowledge of substrates or resources to produce or synthesize optimum quantities of enzymes (Sarrouh et al., 2012).

In this regard, enzymes are produced through microbial resources, nowadays gaining high attention from all the leading industries of pharmaceutics, alcohol, food, leather, etc. due to low cost of productions and easily available resources. The enzymes are discovered from the microbial resource in the 20th century; however, in the present scenario, isolation, identification, and their applications of microbial enzymes have been updated regularly to achieve the maximum reach and explore novel resources and possible applications in broad fields (Hasan, et al., 2006).

Currently, more than 3000 enzymes have been listed, and most of them had been isolated from the mesophilic organism (Pandey et al., 1999), and each day novel microbes or microbial resources have been explored for enzymes production. Generally, for the industrial process, enzymes have been synthesized from the growing bacteria and fungi in submerged or solid-state fermentation, and these enzymes worked under certain circumstances such as a limited range of pH, temperature-specific substrates (Nigam, 2013).

6.3 Bioremediation and bioremediation types

The term bioremediation is used for the remediation of environmental contaminants through the biological means such as plants, microbes, and microbial products. In bioremediation, environmental contaminants are either completely degraded or transformed from complex or high toxicity to simpler or less toxic state via the use of plant products such as enzymes, hormones, and microbes or their microbial products such as consortia, enzymes, and surfactants. Microbial bioremediation includes two types of procedures, in situ or ex situ bioremediation (Marykensa, 2011). In situ remediation emphasizes "at the site" treatment or at the origin site and are generally applied for the petroleum contaminants (Gomes et al., 2013; Farhadian, et al., 2008).

During in situ bioremediation, some limiting factors are present that can check the extent of bioremediation via nutrient limitation, oxygen, and surrounding environment, in order to overcome this situation, and nutrients or nutrient sources such as nitrogen, carbon, phosphorous constantly supplemented to the sites of in situ remediation for the proper growth and metabolic activities of microorganism (Brown et al., 2017; Menendez-Vega et al., 2007). However, the survivability of the microorganism as well as slow rate of contaminants degradation is the concern of in situ bioremediation (Hao Chang et al., 2017).

However, ex situ describes the treatments of contaminates, other places from the site of origin or the contaminants once excavated from the original site of contamination (Megharaj et al., 2014). In this treatment, water, air, nutrients, and other reagents are mixed properly with soil in a bioreactor that supports the proper growth for microbes (Aislabie et al., 2006; Lin et al., 2010; Chagas-Spinelli et al., 2012; Van Deuren et al., 2002). Ex situ remediation technologies are highly used for the treatment of pesticides and toxins originated from various sources and considered as efficient and effective compared to other types of treatments (Coulon et al., 2010).

6.4 Mechanism of the bioremediation processes

Microorganisms are applied variably for the various processes such as regulating biogeochemical cycles perpetuating atmosphere growth and developments of plant as well as biocontrol agents against various pests or pathogens (Griggs et al., 2013; Singh et al., 2019a). Meanwhile, these also play pivotal role in cleaning environmental pollutants (Singh et al., 2020; Datta et al., 2020; Gupta et al., 2020). Microbial-mediated bioremediation is of great significance because it promises a low cost, simpler, and more environment-friendly method in comparison to commonly employed "muck, suck, and truck" nonbiological remedial methods. However, the promise of bioremediation is yet to be fully realized. The microbial processes that remediate pollutants under lab conditions may fail to perform adequately in the field, and reasons for such failures are, however, unclear, and as a result, unwilling to use bioremediation is an option for environmental cleanup. Moreover, mechanisms that control the growth and activity of microorganisms in the contaminated environments are not well understood, thereby limiting the implementation of bioremediation. Dynamic behavior, nutritional flexibility, and adaptation to extreme environmental conditions make microbes, the most suitable life forms for endurance. This feature of microbes is advantageous and beneficial to the mankind particularly when it comes to elimination of pollutants and other toxic compounds from the environment. An extensive list of the microbes that carry out the bioremediation processes is available and published by various authors such as Singh et al. (2018; 2019b,c) and Gupta et al. (2020). Microorganisms can carry out environmental restoration through a diverse array of processes, such as binding, oxidation–reduction, volatilization, and immobilization, or also through chemical transformation of the pollutants. Oxygen is the most common electron acceptor for microbial respiration as well as the agent of aerobic degradation against a wide variety of organic pollutants ranging from arenes such as benzene to xenobiotic (pesticides) has been studied in detail (Wackett and Hershberger, 2001). While a vast phylogenetic diversity is able to degrade aerobic pollutants (Wackett and Hershberger, 2001), but the different types of *Pseudomonas* species and their close associates are the highly investigated organisms and reported for their ability to degrade many different contaminants.

6.5 Enzymes used in bioremediation

It has been assumed that bioremediation especially microbial degradation of various contaminants is considerably a slow process in comparison to other physicochemical processes (Ghosh et al., 2017). But the utilization of microbial-synthesized enzymatic overcomes time-consuming situation and nowadays effectively applied for effective degradation over range of organic and inorganic contaminants. The application of microbial enzymes showed beneficial and effective performance in regards to detoxification or degradation of environmental contaminants (Fig. 6.1) (Thatoi et al., 2014). Enzymes are biological complex molecules that enhance the rate of reaction and act as biocatalyst and their application lower activation energy of the molecules and speed up rate of reaction. Microbial enzymes have been broadly utilized in the detoxification or degradation of various environmental contaminations (Kalogerakis et al., 2017; Ruggaber and Talley, 2006). Catalytic activities largely depend upon the enzymes secreted by microbial strains rather than their purity and growth of microorganism. The enzymes are highly specific for the substrate rather than direct microbial applications (Gianfreda and Bollag, 2002). Different types of enzymes, which have been utilized in the

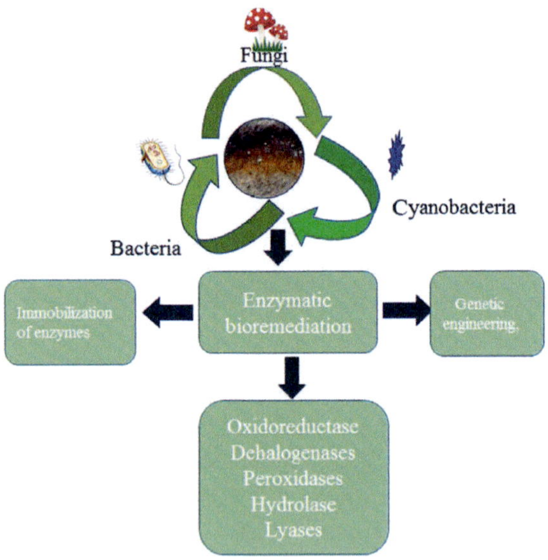

FIGURE 6.1 Overview of combinatory methodologies of enzymatic bioremediation.

bioremediation, are described in Table 6.1. As early in 1965, Kearney and Kaufman reported about soil bacterium identified as a species of *Pseudomonas* possessed an enzyme that could hydrolyze the herbicide isopropyl *N*-(3-chlorophenyl)carbamate and several other biologically active phenylcarbamates. Further, Munnecke (1976) reported enzymatic hydrolysis of parathion, and the enzyme was isolated from a mixed culture of soil bacteria that are dominating by *Pseudomonas* species.

Currently, different kinds of microbial-derived enzymes had been reported that act against a range of contaminates. Some of the most important enzymes are listed in the following sections.

6.5.1 Oxidoreductases

Oxidoreductases are the enzymes that degrade or detoxify the contaminants via oxidation coupling or oxidation of the compounds. In this enzymatic reaction process, energy is liberated that has been utilized by microorganism for their metabolic process (Sharma et al., 2018). Many microbial species such as bacteria and fungi are capable to synthesize oxidoreductase and utilized them in biodegradation of environmental contaminants. Many reports are available that state the role of oxidoreductase enzymes or their potentially role in oxidation or to remove hazardous compounds from the environment (Torres et al. 2003).

6.5.1.1 Laccases

Laccases constitute multicopper proteins that belong to the family blue oxidase enzymes that can oxidize a variety of aromatic compounds along with the concomitant reduction of oxygen to water molecules (Torres et al. 2003). These enzymes are reported very early to oxidize phenolic compounds and broad applications against various industrial sectors (Madhavi and Lele 2009; Giardina et al. 2010; Pointing, 2001). Laccase enzyme can also transform xenobiotics pesticides, polynitrated aromatic compounds (Ramos et al. 2005), dyes, polycyclic aromatic hydrocarbons (PAHs) into simpler one, which is a major source of soil and water contamination. Many microbial communities such as bacteria and fungi significantly produce laccase enzymes. Dodor et al. (2004) reported laccases produced by the strains *Trametes versicolor* and *Pleurotus ostreatus*, which metabolized polychlorinated biphenyls into constituents elements; and similarly, Mayer and Staples (2002) reported laccase-synthesizing strain *Pycnoporus sanguineus* and their effective use in degradation of bromophenol blue dye. In the same way, Soares et al. (2001) also reported laccase enzymes produced from *Flavodon flavus* and potential application against Azure B and Brilliant Blue R dyes at the low nitrogen concentration. Chandra and Chowdhary (2015) reported bacterial-synthesized laccase enzymes and their potential application against remediation of industrial wastes. In a study, Bollag and Myers (1992) reported laccase enzymes found from white-rot fungus, *Trametes hirsuta* and used to oxidize alkenes.

TABLE 6.1 Microbial enzymes and their role in contamination management.

Enzymes	Source	Contaminants	Mechanism	References
Peroxidases	*Phanerochaete chrysosporium*	(2,4,6-Trinitrotoluene) nitroaromatic compounds	2amDNT and 4amDNT is the initial, intermediate products of TNT biodegradation, which is further mineralized into simpler compounds.	Cameron et al. (2000)
Lignin peroxidase	*P. chrysosporium*	Pyrene	Higher decrease in contaminants.	Hammel et al. (1986)
Manganese peroxidase	*Phanerochaete laevis*	Fluorene	Quinone intermediates of PAH metabolism were degraded.	Bogan and Lamar (1996)
Chromium reductase	*Pseudomonas, Bacillus, Enterobacter, Deinococcus, Shewanella, Agrobacterium, Escherichia*	Chromium	Transform Cr (VI) into Cr (III) in the presence of oxygen via a two-step reaction; first Cr(VI) accepts one electron from one molecule of NADH to generate Cr(V) and then Cr(V) accepts two electrons to form Cr(III).	Thatoi et al. (2014)
Horseradish peroxidase	*Armoracia rusticana*	Phenol, chlorophenol	Horseradish peroxidase enzyme degrades the phenolic compounds by both meta- and ortho-pathways. Phenol is converted into aldehyde and pyruvate after degradation by meta pathway.	Flock et al. (1999); Bilal et al. (2017)
Laccase	*Trametes villosa*	2,4,6-Trinitrotoluene	100% decrease in contaminants in presence of catechol.	Wang et al. (2002)
Manganese peroxidase	*Nematoloma frowardii* and *Stropharia rugosoannulata*	Aminodinitrotoluenes	Higher concentration of L-cysteine led to decrease of 2-AmDNT and 4-AmDNT.	Scheibner and Hofrichter, 1998)
Phytase	*Aspergillus niger* NCIM 563	Organophosphate	It can release inorganic phosphorus by the degradation of organophosphate-containing compounds	Shah et al. (2017)
Arylsulfotransferase	*Clostridium innocuum*	Phenols	Arylsulfotransferase capable of sulfating a wide range of phenol-containing compounds, Higher decrease in contaminants.	Mozhaev et al. (2002)
Laccase, manganese peroxidase, lignin peroxidase	*Trametes versicolor*	Delor 106 PCB	Degradation of polycyclic aromatic hydrocarbons (anthracene and pyrene), a polychlorinated biphenyl mixture (Delor 106) and a number of synthetic dyes.	Novotny et al. (2004)
Laccase	*T. versicolor, Pleurotus ostreatus*	Polychlorinated biphenyls	PCBs metabolized into its constituent elements and laccase are able to dechlorinate chlorophenols by oligomerization of the substrate.	Dodor et al. (2004)
Carboxylesterases	*Pseudomonas aeruginosa* PA1	Malathion and Parathion	Carboxylesterase transforms malathion into malathion monocarboxylic acid and dicarboxylic acid.	Qiao et al. (2003); Singh et al. (2012)
Atrazine dechlorinase, triazine hydrolase	*Nocardioides* sp. C190, *Pseudomonas, Rhodococcus erythropolis*	Triazine herbicides	Metabolized atrazine into chloride, ammonia, and carbon dioxide with the help of enzymes.	Scott et al. (2010)
Laccase	*P. ostreatus*	2,6-dimethoxyphenol	Detoxify a variety of phenolic pollutants	Hublik and Schinner (2000)
Laccase	*T. versicolor*	Anthracene and benzo(a)pyrene	Degradation of 90%–100% of these contaminants in 24 h.	Collins et al. (1996)

2amDNT, 2-Amino-4,6-dinitrotoluene; *4amDNT*, 4-amino-2,6-dinitrotoluene.

6.5.1.2 *Tyrosinase*

Tyrosinase enzyme is known as a monophenol oxidase or catecholase and is ubiquitously distributed in the organisms (Claus and Decker, 2006). Tyrosinases catalyze two types of reactions that occur sequentially: the first one is the *o*-hydroxylation of phenols with molecular oxygen in order to produce catechols (cresolase activity) and the second is the subsequent oxidation of catechols with oxygen in order to form *o*-quinones (catecholase activity), *o*-quinones are unstable compounds and spontaneously polymerize in the nonenzymatic reaction to produce insoluble melanin-like products (Demarche et al., 2012).

6.5.1.3 *Peroxidases*

Peroxidases use hydrogen peroxide or other organic hydroperoxides as a cosubstrate to oxidize a variety of organic and inorganic substrates and frequently applied in lignin degradation. There are various peroxidase enzymes reported, in which most commonly utilized peroxidases are lignin peroxidase (LIP), manganese peroxidase (MnP), horseradish peroxidase (HRP), ascorbate peroxidase (APX), and chloroperoxidase (CPO) (Torres et al., 2003; Koua et al., 2009). Peroxidases share some common structural and catalytic features—they are glycosylated proteins having an iron protoporphyrin IX (heme) prosthetic group located at the active site (Mougin et al., 2009). In a study, Novotný et al. (2004) reported the abilities of *Phanerochaete chrysosporium*, *T. versicolor*, *Coriolopsis polyzona*, and *P. ostreatus* in degradation of PCBs that had been grown in a nitrogen-limited mineral medium.

6.5.1.4 *Lignin peroxidases*

It is a heme-containing monomeric protein compounds, secreted by various fungal and bacterial strains and broadly utilized in the degradation or treatment of wastewater contaminants (Xu et al., 2014 Tuomela and Hatakka, 2011).

6.5.1.5 *Manganese peroxidases*

It is also a heme-containing extracellular enzyme produced by some specific lignin degradation fungal strains and has the capability to oxidize and metabolize Mn2þ into Mn3þ through several intermediary reactions. Besides this, these enzymes have the capability to decolorize various dyes such as indigo, anthraquinone, azo, and triphenyl methane. In a study, Zhanga et al. (2016) reported MnPTra-48424 enzyme isolated from the white-rot fungi *Trametes* sp. 48424 had strong potential to decolorize various dyes as well as effective degradation of various polycyclic aromatic hydrocarbons.

6.5.2 Oxygenases

Generally, oxygenases are used for the degradation of aromatic compounds aerobically and catalyze the rate of reaction in terms of enhanced degradation rate of organic contaminants (Sharma et al., 2018). Oxygenases enhance the rate of reaction by adding oxygen molecules in the aromatic ring that results in the easy cleavage of complex contaminant molecules.

The enzyme on the basis of addition of oxygen in the aromatic rings is classified mainly in two types; first one is monooxygenase that catalyzes reaction rate by adding one O_2 molecule and plays effective role in oxidation (Arora et al., 2010). There are various microorganisms reported about monooxygenase production and their practical application in contaminants degradation. In a study, Roccatano (2015) studies P450 monooxygenase isolated from *Bacillus megaterium* and their potential application for the degradation of aromatic compounds as well as fatty acids. The second one is dioxygenase that catalyzes the rate of reaction by addition of two O_2 molecules. Toluene dioxygenase (TOD) is one of common examples of dioxygenase produced by *Pseudomonas putida* F1 that effectively degrades toluene by enhancing the rate of reaction (Wackett and Gibson, 1988). In another study, Gennaro et al. (1997) reported naphthalene dioxygenase isolated from the *P. putida* effectively degrade the naphthalene. There are several reports available that showed the potency of oxygenase enzymes in the bioremediation of various organic pollutants (Karigar and Rao, 2011; Singh and Ward, 2004; Sivaperumal et al., 2017).

6.5.3 Hydrolases

This is one of the most common enzymes used in the treatment of various contaminants, including chemical pesticides and effectively degrade the contaminants by dissolving interconnecting bonds between the compounds, which results in easy degradation or detoxification of contaminants (Karigar and Rao, 2011).The hydrolytic enzymes are used for the degradation of oil spills, insecticides; however, these enzymes also are used in the food and chemical industries.

6.5.3.1 *Lipases*

Lipases are ubiquitous enzymes that have considerable physiological and industrial significance as they catalyze hydrolysis of triacylglycerol into glycerol and free fatty acids (Suneetha and Khan, 2011). Microbial lipases have been biotechnological applied in the organic synthesis, food and flavor industries, detergent manufacturing, the removal of oil and grease from wastewaters, and also in biodiesel refining (Demarche et al., 2012).

6.5.3.2 *Cellulases*

This is one of the most common enzymes that are commonly used for the degradation of cellulose molecules into simple sugars. Several types of cellulases had been reported that effectively degrade various contaminants (Annamalai et al., 2016; Bhardwaj et al., 2017). The microbial synthesized cellulases can be extracellular, cell bound, or the on the cell envelope (Yang et al., 2016). In a study, Imran et al. (2016) reported cellulase produced by *Humicola* species and their potential application in the degradation of detergent and washing industrial contaminants.

6.5.4 Carboxylesterases

Generally, carboxylesterase enzymes carboxylesterases have been applied for the degradation of chemical pesticides at industrial level by hydrolyzing the carboxyl ester bond present by the addition of H_20 molecules (Singh, 2014). In a study, Scott et al. (2008) reported enzyme carboxylesterase E3 extracted from the strain *Lucilia cuprina* and their potential role in organophosphorus degradation.

6.6 Enzymes as decontaminating agents

Enzymes are biocatalysts that increase the rate of a reaction and accelerate the conversion of substrates into products over millions of times faster through lowering activation energy of the reaction. The active site (region) of the enzyme is directly involved in the catalytic process (Daniel and Danson, 2010). An enzyme may have one or more groups that are prerequisite for catalytic activity and associated with the active sites through either covalent or noncovalent bonds. Being catalytic, the enzymes are immensely effective and can detoxify the contaminants having weight million times higher than their own weight in the seconds or minutes. The biocatalytic nature and detoxification abilities of enzymes make them efficient decontamination agents (Rao et al., 2010). Enzymes have several beneficial characteristics and catalyze the rate with either narrow (chemo-, region-, and stereoselectivity) or broad specificity, and therefore they can be applied to a large range of the compounds in the mixture as well. They may produce extensive transformations of structural and toxicological properties of contaminants and even their complete conversion into innocuous inorganic end products. Moreover, the utilization of enzymes has many more advantages over traditional technologies in xenobiotics degradation. Indeed, enzymes are not inhibited by inhibitors of microbial metabolism and can be used under extreme condition, which may limit the microbial activity

6.7 Enzymes showing the ability to degrade xenobiotics

Xenobiotics is a broad term used for chemical compounds such as pesticides and polynuclear aromatic hydrocarbons present in the soil, and their presence is considered as significant environmental problem, which have negative impact on human health and agriculture. Xenobiotics may be used by microorganisms as sources of energy and nutrients for growth. Another important mode of microbial transformation is metabolism, in which microorganisms transform the xenobiotic molecules, however, unable to proliferate on the degradation products. According to literature data, oxidoreductases and hydrolases play a major role in both the metabolic and catabolic transformation of xenobiotics (Karam and Nicell, 1997; Johnson and Talbot, 1983; Gianfreda et al., 1999). The decontamination potential of some enzymes is illustrated next by the transformation pathways of four common contaminants (2,4-D, atrazine, naphthalene, and 4-chlorobiphenyl). Based on various literature data, the herbicide 2,4-D is transformed by a dioxygenase from *Alcaligenes eutrophus* to 2,4-dichlorophenol with the release of glyoxylate (Tiedje and Alexander, 1999; Fukumori and Hausinger, 1993). In turn, 2,4-dichlorophenol is hydroxylated through oxygenase, and aromatic ring of the resulting dichlorocatechol is cleaved by dioxygenase. In further, study of degradation pathway for atrazine involves dehalogenation and hydroxylation of herbicide by a hydrolase from *Pseudomonas* (de Souza et al., 1996). Biodegradation of naphthalene may proceed through the oxidation of the compound to dihydroxynaphthalene through dioxygenase from *Pseudomonas* sp. and then by a dehydrogenase (Eaton and Chapman, 1992; Barriault et al., 1999). Similarly, for 4-chlorobiphenyl, the dioxygenases, dehydrogenases, and hydrolases from *Pseudomonas* sp. (Eltis et al., 1993; Barriault et al., 1998; Hofer et al., 1993) can mediate the transformation of it.

6.8 Advantages of enzyme in bioremediation

The microbial resources used for bioremediation is one of the best options with some rate-limiting factors. However, in some cases the use of microbial consortia as a remediation process is challenging due to possible contamination, Furthermore, some physiochemical parameters such as chemical shock, extremes of pH, temperature variation, and higher concentrations of the pollutants may irreversibly damage or metabolically inactivate the microbial cells. Thus most of these adverse factors can either be eliminated or mitigated by the utilization of enzymes (preferably immobilized) instead of the microbes (Glaser et al., 1989). Enzymes are complex proteins that exhibit specificity and are characterized at an optimum temperature and pH for their actions. Similar to other catalysts, they accelerate the chemical reaction rate by lowering the activation energy for a particular reaction.

The application of microbial enzymes in bioremediation technology faces critical challenges due to sensitivity against pH variation, fragile nature heat sensitive nature, slight variation in any factors changed, and active conformational state of the enzymes, which results the in inactiveness of enzyme or loss of enzymatic activity (Nigam and Shukla, 2015). In natural conditions the limited amount of enzyme production is still a matter of concern, and the researchers are continuously working in order to enhance the quantity as well as quality of enzyme production that will be used for the mitigation of various environmental contaminants (Baweja et al., 2016).

6.8.1 Genetic engineering

Under normal conditions, the amount of enzyme production is very low, and researchers continuously tried to enhance the rate or amount of production. In this regard, genetic engineering has been applied in order to modify or enhance the rate of enzyme production through several processes such as isolation of genes, transfer of expressing gene in the specific host (Gupta and Shukla, 2015). DNA recombinant technology is also cost-effective method for the enhancement of enzyme production, which increases stability as well as amount of production (Alcalde et al., 2006). The genetically modified or recombinant DNA enzymes in comparison to normal secreted enzymes had been found more stable and highly efficient in degradation of various environmental contaminants under specific environmental condition (Dua et al., 2002). In table 3, we have discussed some commonly utilized recombinant enzymes used in bioremediation

6.8.2 Enzyme engineering

As the name suggests enzyme engineering, it involves change or modification in the sequence of amino acid via, insertion, deletion, or replacement for the improvement in the stability against various stresses such as temperature heat, enzymatic activity, tolerance (Rayu et al., 2012). Recombinant technology has also been involved in genetic engineering to bring modification in the sequence of amino acids (Singh et al., 2013). There are several reports present that describe the use of enzyme engineering in the mitigation of heavy metal stress (Dhanya, 2014). In a study, Ju and Parales (2006) reported modification in amino acid residue at the position of 293 of nitrobenzene 1,2-dioxygenase increased the rate by 2.5-fold. Similarly, Singh et al. (2008) reported that modification in the position of 258 in 2-nitrotoluene dioxygenase enhances the conversion of nitrotoluene to 3-methylcatechol and nitrite.

6.9 Immobilization of enzymes

The term immobilization refers for "to restrict the mobilization or movement" of any cell, enzymes, molecules either completely or partially on any solid surface (Yu-Qung et al., 2004), and nowadays, immobilization of enzymes has been applied to improve degradation potential of various environmental contaminants in sustainable ways (Margaritis and Merchant 1984; Bayat, et al.,2015). Microbial-based bioremediation needs to maintain the survivability, retention as well as higher biomass of microbes at the contaminated sites, and in this context, immobilization of enzymes provides better platform for remediation (Scott, 1987; Cassidy et al., 1996; Bayat et al., 2015). The immobilization of cell or enzymes system benefitted through providing higher cell biomass reduces the need of cell washout, higher flow rate;

allows higher volumetric productivity, stability, resistance from variation in temperature, pH; and also permits stable microclimatic conditions (Bayat et al., 2015).

Immobilization of whole cells can be classified into three physiological states consisting of dead, living, and growing states, and the selection of stages depends upon the choice of application (Rahman et al., 2006). The thermal and functional stability of immobilized microbial cells has more than the commonly used enzyme systems. Many different forms of cell immobilization have been reported such as adsorption, covalent binding, entrapment, and encapsulation, and among all, entrapping has been widely investigated (Jack and Zajic, 1977). Degradation of environmental contaminants depends upon affinity between substrate and hydrophobic immobilization materials, better interaction enhanced the rate of degradation (Bayat et al., 2015). In a study, Omar and Rehm (1988) reported immobilized strains of *Candida parapsilosis* and *Penicillium frequetans* that effectively degrade hydrocarbon chain of C12–C18, *n*-alkanes (up to 80%–85%) in comparison to free bacterial cells (only maximum 15%). Quek et al. (2006) have reported the immobilization and performance of *Rhodococcus* sp. on polyurethane foam (PUF) in the bioremediation of petroleum hydrocarbons.

6.10 Conclusion

The rapid industrialization releases various types of toxic and harmful inorganic and organic by-products, and their presence in the surrounding environment causes severe concern of climate as well as human health. The traditional physiochemical techniques used for contaminant mitigation are not only costly but management of their residual toxic products is still a matter of concern. However, the uses of microbes or microbial products in the remediation techniques are economic, effective and sustainable, and the use of microbial enzymes in contaminants management is of great importance due to their rapid, efficient, and sustainable. Besides this, microbial enzymes also are utilized in the various industrial processes of different sectors. However, there is a need for extensive study of these microbial enzymes and their purification process for efficient utilization.

References

Ahuti, S., 2015. Industrial growth and environmental degradation. Int. Educ. Res. J. 1, 5–7.

Aislabie, J., Saul, D.J., Foght, J.M., 2006. Bioremediation of hydrocarbon contaminated polar soils. Extremophiles 10 (3), 171e179.

Alcalde, M., Ferrer, M., Plou, F.J., Ballesteros, A., 2006. Environmental biocatalysis: from remediation with enzymes to novel green processes. Trends Biotechnol. 24, 281e287.

Alexander, M., 1999. Biodegradation and Bioremediation. Academic Press, San Diego, CA.

Annamalai, N., Rajeswari, M.V., Balasubramanian, T., 2016. Thermostable and alkaline cellulases from marine sources. New and Future Developments in Microbial Biotechnology and Bioengineering 91–98.

Arora, P.K., Srivastava, A., Singh, V.P., 2010. Application of monooxygenases in dehalogenation, desulphurization, denitrification and hydroxylation of aromatic compounds. J. Biorem. Biodegrad. 1, 112.

Barriault, D., Durand, J., Maaroufi, H., Eltis, L.D., Sylvestre, M., 1998. Degradation of polychlorinated biphenyl metabolites by naphthalene-catabolizing enzymes. Appl. Environ. Microbiol. 64 (12), 4637–4642.

Barriault, D., Vedadi, M., Powlowski, J., Sylvestre, M., 1999. *cis*-2,3-Dihydro-2,3-dihydroxybiphenyl dehydrogenase and *cis*-1,2-dihydro-1,2-dihydroxynaphathalene dehydrogenase catalyze dehydrogenation of the same range of substrates. Biochem. Biophys. Res. Commun. 260 (1), 181–187.

Baweja, M., Nain, L., Kawarabayasi, Y., Shukla, P., 2016. Current technological improvements in enzymes toward their biotechnological applications. Front. Microbiol. 7.

Bayat, Z., Hassanshahian, M., Cappello, S., 2015. Immobilization of microbes for bioremediation of crude oil polluted environments: a mini review. Open Microbiol. J. 9, 48.

Bhardwaj, V., Degrassi, G., Bhardwaj, R.K., 2017. Bioconversion of cellulosic materials by the action of microbial cellulases. Int. Res. J. Eng. Technol. 4 (8), 1–10.

Bilal, M., Iqbal, H.M.N., Hu, H., Wang, W., Zhang, X., 2017. Development of horseradish peroxidase-based cross-linked enzyme aggregates and their environmental exploitation for bioremediation purposes. J. Environ. Manage. 188, 137e143.

Bogan, B.W., Lamar, R.T., 1996. Polycyclic aromatic hydrocarbon-degrading capabilities of *Phanerochaete laevis* HHB-1625 and its extracellular ligninolytic enzymes. Appl. Environ. Microbiol. 62 (5), 1597–1603.

Bollag, J.M., Myers, C., 1992. Detoxification of aquatic and terrestrial sites through binding of pollutants to humic substances. Sci. Total. Environ. 117, 357–366.

Brown, L.D., Cologg, D.L., Gee, K.E., Ulrich, A.C., 2017. Bioremediation of oil spills on land. In: Oil Spill Science and Technology, second ed., pp. 699–729.

Cadisch, G., Giller, K.E., 1997. Driven by Natureplant Litter Quality and Decomposition (No. 631.4 I5).

Cameron, M.D., Timofeevski, S., Aust, S.D., 2000. Enzymology of *Phanerochaete chrysosporium* with respect to the degradation of recalcitrant compounds and xenobiotics. Appl. Microbiol. Biotechnol. 54 (6), 751–758.

Cassidy, M.B., Lee, H., Trevors, J.T., 1996. Immobilized microbial cells: a review. J. Ind. Microbiol. Biotechnol. 16, 79–101.

Chagas-Spinelli, A.C., Kato, M.T., de Lima, E.S., Gavazza, S., 2012. Bioremediation of a tropical clay soil contaminated with diesel oil. J. Environ. Manage. 113, 510e516.

Chandra, R., Chowdhary, P., 2015. Properties of bacterial laccases and their application in bioremediation of industrial wastes. Environ. Sci.: Process. Impacts 17 (2), 326–342.

Claus, H., Decker, H., 2006. Bacterial tyrosinases. Syst. Appl. Microbiol. 29, 3–14.

Collins, P.J., Kotterman, M.J.J., Field, J.A., Dobson, A.D., 1996. Oxidation of anthracene and benzoa[*a*]pyrene by laccases from *Trametes versicolor*. Appl. Environ. Microbiol. 6212, 4562–4567.

Coulon, F., Al Awadi, M., Cowie, W., Mardlin, D., Pollard, S., Cunningham, C., et al., 2010. When is a soil remediated? Comparison of biopiled and windrowed soils contaminated with bunker-fuel in a full-scale trial. Environ. Pollut. 158 (10), 3032e3040.

Daniel, R.M., Danson, M.J., 2010. A new understanding of how temperature affects the catalytic activity of enzymes. Trends Biochem. Sci. 35 (10), 58491.

Datta, S., Singh, S., Kumar, V., Dhanjal, D.S., Sidhu, G.K., Amin, D.S., et al., 2020. Endophytic bacteria in Xenobiotic degradation. In: Kumar, A., Singh, V.K. (Eds.), Microbial Endophytes: Prospects of Sustainable Agriculture. Elsevier, pp. 125–156.

Demarche, P., Junghanns, C., Nair, R.R., Agathos, S.N., 2012. Harnessing the power of enzymes for environmental stewardship. Biotechnol. Adv. 30, 933–953.

Demnerová, K., Mackova, M., Speváková, V., Beranova, K., Kochánková, L., Lovecká, P., et al., 2005. Two approaches to biological decontamination of groundwater and soil polluted by aromatics—characterization of microbial populations. Int. Microbiol. 8 (3), 205–211.

de Souza, M.L., Sadowsky, M.J., Wackett, L.P., 1996. Atrazine chlorohydrolase from *Pseudomonas* sp. strain ADP: gene sequence, enzyme purification, and protein characterization. J. Bacteriol. 178 (16), 4894–4900.

Dhanya, M.S., 2014. Advances in microbial biodegradation of chlorpyrifos. J. Environ. Res. Dev. 9 (1), 232.

Dodor, D.E., Hwang, H.M., Ekunwe, S.I., 2004. Oxidation of anthracene and benzo[*a*] pyrene by immobilized laccase from *Trametes versicolor*. Enzyme Microb. Technol. 35, 3210e3217.

Dua, M., Singh, A., Sethunathan, N., Johri, A.K., 2002. Biotechnology and bioremediation: successes and limitations. Appl. Microbiol. Biotechnol. 59, 143e152.

Eaton, R.W., Chapman, P.J., 1992. Bacterial metabolism of naphthalene: construction and use of recombinant bacteria to study ring cleavage of 1,2-dihydroxynaphthalene and subsequent reactions. J. Bacteriol. 174 (23), 7542–7554.

Eltis, L.D., Hofmann, B., Hecht, H.J., Lünsdorf, H., Timmis, K.N., 1993. Purification and crystallization of 2,3-dihydroxybiphenyl 1,2-dioxygenase. J. Biol. Chem. 268 (4), 2727–2732.

Farhadian, M., Vachelard, C., Duchez, D., Larroche, C., 2008. In situ bioremediation of monoaromatic pollutants in groundwater: a review. Bioresour. Technol. 99 (13), 5296–5308.

Flock, C., Bassi, A., Gijzen, M., 1999. Removal of aqueous phenol and 2-chlorophenol with purified soybean peroxidase and raw soybean hulls. J. Chem. Technol. Biotechnol. 74 (4), 303e309.

Fukumori, F., Hausinger, R.P., 1993. Purification and characterization of 2,4-dichlorophenoxyacetate/alpha-ketoglutarate dioxygenase. J. Biol. Chem. 268 (32), 24311–24317.

Gennaro, P.D., Galli, E., Albini, G., Pelizzoni, F., Sello, G., Bestetti, G., 1997. Production of substituted naphthalene dihydrodiols by engineered *Escherichia coli* containing the cloned naphthalene 1,2-dioxygenase gene from *Pseudomonas fluorescens* N3. Res. Microbiol. 148, 355e436.

Ghosh, A., Dastidar, M.G., Sreekrishnan, T.R., 2017. Bioremediation of chromium complex dyes and treatment of sludge generated during the process. Int. Biodeterior. Biodegrad. 119, 448e460.

Gianfreda, L., Bollag, J.-M., 2002. Isolated enzymes for the transformation and detoxification of organic pollutants. In: Burns, R.G., Dick, R. (Eds.), Enzymes in the Environment: Activity, Ecology and Applications. Marcel Dekker, New York, pp. 491–538.

Gianfreda, L., Xu, F., Bollag, J.M., 1999. Laccases: a useful group of oxidoreductive enzymes. Bioremed. J. 3 (1), 1–26.

Giardina, P., Faraco, V., Pezzella, C., Piscitelli, A., Vanhulle, S., Sannia, G., 2010. Laccases: a neverending story. Cell Mol. Life Sci. 67, 369–385.

Glaser, J.A., Lzm?r, R., Dietrich, D., Kirk, T.K., 1989. Treatment of wood preserving soil contaminants by white rot fungus. Presented at: Fifteenth Annual Research Symposium on Remedial Action, Treatment and Disposll of Hazardous Wastes, Cincinnati, OH, April 10–12.

Gomes, H.I., Ferreira, C.D., Ribeiro, A.B., 2013. Overview of in situ and ex situ remediation technologies for PCB-contaminated soils and sediments and obstacles for full-scale application. Sci. Total. Environ. 445e446, 237e260.

Griggs, D., Stafford-Smith, M., Gaffney, O., Rockström, J., Öhman, M.C., Shyamsundar, P., et al., 2013. Policy: Sustainable development goals for people and planet. Nature 495 (7441), 305.

Gupta, S.K., Shukla, P., 2015. Advanced technologies for improved expression of recombinant proteins in bacteria: perspectives and applications. Crit. Rev. Biotechnol. 36, 1089e1098.

Gupta, A., Singh, S.K., Singh, V.K., Singh, M.K., Modi, A., Zhimo, V.Y., et al., 2020. Endophytic microbe approaches in bioremediation of organic pollutants. In: Kumar, A., Singh, V.K. (Eds.), Microbial Endophytes: Prospects of Sustainable Agriculture. Elsevier, pp. 157–174.

Hammel, K.E., Kalyanaraman, B., Kirk, T.K., 1986. Oxidation of polycyclic aromatic hydrocarbons and dibenzo [p]-dioxins by *Phanerochaete chrysosporium* ligninase. J. Biol. Chem. 261 (36), 16948–16952.

Hao Chang, C., Yang, H.Y., Min Hung, J., Jen Lu, C., Min-Hsin, L., 2017. Simulation of combined anaerobic/aerobic bioremediation of tetrachloroethylene in groundwater by a column system. Int. Biodeter. Biodegrad. 117, 150e157.

Hasan, F., Shah, A.A., Hameed, A., 2006. Industrial applications of microbial lipases. Enzyme Microb. Technol. 39 (2), 235–251.

Hofer, B., Eltis, L.D., Dowling, D.N., Timmis, K.N., 1993. Genetic analysis of a Pseudomonas locus encoding a pathway for biphenyl/polychlorinated biphenyl degradation. Gene 130 (1), 47–55.

Hublik, G., Schinner, F., 2000. Characterization and immobilization of the laccase from *Pleurotus ostreatus* and its use for the continuous elimination of phenolic pollutants. Enzyme Microb. Technol. 27 (3-5), 330–336.

Imran, M., Anwar, Z., Irshad, M., Asad, M.J., Ashfaq, H., 2016. Cellulase production from species of fungi and bacteria from agricultural wastes and its utilization in industry: a review. Adv. Enzym. Res. 4 (02), 44.

Jack, T.R., Zajic, J.E., 1977. The immobilization of whole cells. Adv. Biochem. Eng. Biotechnol. 5, 125–145.

Jegannathan, K.R., Nielsen, P.H., 2013. Environmental assessment of enzyme use in industrial production–a literature review. J. Clean. Prod. 42, 228–240.

Johnson, L.M., Talbot, H.W., 1983. Detoxification of pesticides by microbial enzymes. Experientia 39 (11), 1236–1246.

Ju, K.S., Parales, R.E., 2006. Control of substrate specificity by active-site residues in nitrobenzene dioxygenase. Appl. Environ. Microbiol. 72 (3), 1817e1824.

Kalogerakis, N., Fava, F., Corvine, P.F., 2017. Bioremediation advances. N. Biotech. 38, 41e42.

Karam, J., Nicell, J.A., 1997. Potential applications of enzymes in waste treatment. J. Chem. Technol. Biotechnol.: Int. Res. Process, Environ. Clean. Technol. 69 (2), 141–153.

Karigar, C.S., Rao, S.S., 2011. Role of microbial enzymes in the bioremediation of pollutants: a review. Enzyme Res. 2011, 1–11.

Koua, D., Cerutti, L., Falquet, L., et al., 2009. PeroxiBase: a database with new tools for peroxidase family classification. Nucleic Acids Res. 37, D261eD266.

Leung, M., 2004. Bioremediation: techniques for cleaning up a mess. BioTeach J. 2, 18–22.

Lin, T.-C., Pan, P.T., Cheng, S.S., 2010. Ex situ bioremediation of oil contaminated soil. J. Hazard. Mater. 176, 27e34.

Madhavi, V., Lele, S.S., 2009. Laccase: properties and applications. BioResources 4, 1694–1717.

Margaritis, A., Merchant, F.J.A., 1984. Advances in ethanol production using immobilized cell systems. Crit. Rev. Biotechnol. 2, 339–393.

Marykensa, V., 2011. Bioremediation – an overview. J. Ind. Pollut. Contr. 27, 161e168.

Mayer, A.M., Staples, R.C., 2002. Laccase: new functions for an old enzyme. Phytochem. 60 (6), 551e565.

Megharaj, M., Venkateswarlu, K., Naidu, R., 2014. Bioremediation. In: Encyclopedia of Toxicology, third ed., Elsevier, pp. 485–489.

Menendez-Vega, D., Gallego, J.L.R., Pelaez, A.I., de Cordoba, G.F., Moreno, J., Munoz, D., et al., 2007. Engineered in situ bioremediation of soil and groundwater polluted with weathered hydrocarbons. Eur. J. Soil Biol. 43 (5), 310e321.

Mgbemene, C.A., Nnaji, C.C., Nwozor, C., 2016. Industrialization and its backlash: focus on climate change and its consequences. J. Environ. Sci. Technol. 9, 301–316. Available from: https://doi.org/10.3923/jest.2016.301.316.

Mougin, C., Boukcim, H., Jolivalt, C., 2009. Soil bioremediation strategies based on the use of fungal enzymes. In: Singh, A., Kuhad, R.C., Ward, O. P. (Eds.), Advances in Applied Bioremediation, Soil Biology, vol. 17. Springer, Heidelberg, pp. 123–149.

Mozhaev, V.V., Khmelnitsky, Y.L., Sanchez-Riera, F., Maurina-Brunker, J., Rosson, R.A., Grund, A.D., 2002. Arylsulfotransferase from *Clostridium innocuum*—a new enzyme catalyst for sulfation of phenol-containing compounds. Biotechnol. Bioeng. 78 (5), 567–575.

Munnecke, D.M., 1976. Enzymatic hydrolysis of organophosphate insecticides, a possible pesticide disposal method. Appl. Environ. Microbiol. 32 (1), 7–13.

Nannipieri, P., Bollag, J.M., 1991. Use of enzymes to detoxify pesticide-contaminated soils and waters. J. Environ. Qual. 20 (3), 510–517.

Nigam, P., 2013. Microbial enzymes with special characteristics for biotechnological applications. Biomolecules 3 (3), 597–611.

Nigam, V.K., Shukla, P., 2015. Enzyme based biosensors for detection of environmental pollutants—a review. J. Microbiol. Biotechnol. 25 (11), 1773e1781.

Novotny, C., Svobodova, K., Erbanova, P., Cajthamal, T., Kasinath, A., Lang, E., et al., 2004. Ligninolytic fungi in bioremediation: extracellular enzyme production and degradation rate. Soil Biol. Biochem. 36, 1545e1551.

Omar, S.H., Rehm, H.J., 1988. Degradation of *n*-alkanes by *Candida parapsilosis* and *Penicillium frequentans* immobilized on granular clay and aquifer sand. Appl. Microbiol. Biotechnol. 28 (1), 103–108.

Pandey, A., Selvakumar, P., Soccol, C.R., Nigam, P., 1999. Solid state fermentation for the production of industrial enzymes. Curr. Sci. 149–162.

Pointing, 2001. Feasibility of bioremediation by white-rot fungi. Appl. Microbiol. Biotechnol. 57 (1–2), 20–33.

Qiao, C.L., Huang, J., Li, X., Shen, B.C., Zhang, J.L., 2003. Bioremediation of organophosphate pollutants by a genetically-engineered enzyme. Bull. Environ. Contam. Toxicol. 70 (3), 0455e0461.

Quek, E., Ting, Y.P., Tan, H.M., 2006. Rhodococcus sp. F92 immobilized on polyurethane foam shows ability to degrade various petroleum products. Bioresour. Technol., 97 (1), 32–38.

Raghunandan, K., Kumar, A., Kumar, S., Permaul, K., Singh, S., 2018. Production of gellan gum, an exopolysaccharide, from biodiesel-derived waste glycerol by Sphingomonas spp. 3Biotech 8, 71. Available from: https://doi.org/10.1007/s13205-018-1096-3.

Raghunandan, K., McHunu, S., Kumar, A., Kumar, K.S., Govender, A., Permaul, K., et al., 2014. Biodegradation of glycerol using bacterial isolates from soil under aerobic conditions. J. Environ. Sci. Health A: Tox. Hazard. Subst. Environ. Eng. 49, 85–92. Available from: https://doi.org/10.1080/10934529.2013.824733.

Rahman, R.N.Z., Ghazali, F.M., Salleh, A., et al., 2006. Biodegradation of hydrocarbon contamination by immobilized bacterial cells. J. Microbiol. 4, 354–359.

Ramos, J.L., Gonzalez-Perez, M.M., Caballero, A., van Dillewijn, P., 2005. Bioremediation of polynitrated aromatic compounds: plants and microbes put up a fight. Curr. Opin. Biotechnol. 16, 275–281.

Rao, M., Scelza, R., Scotti, R., Gianfreda, L., 2010. Role of enzymes in the remediation of polluted environments. J. Soil. Sci. Plant. Nutr. 10 (3), 33353.

Rayu, S., Karpouzas, D.G., Singh, B.K., 2012. Emerging technologies in bioremediation: constraints and opportunities. Biodegradation 23 (6), 917e926.

Roccatano, D., 2015. Structure, dynamics, and function of the monooxygenase P450 BM-3: insights from computer simulations studies. J. Phys. Condens. Matter 27, 273102.

Ruggaber, T.P., Talley, J.W., 2006. Enhancing bioremediation with enzymatic processes: a review. Pract. Period. Hazard. Toxic Radioact. Waste Manage. 10 (2), 73e85.

Sarrouh, B., Santos, T.M., Miyoshi, A., Dias, R., Azevedo, V., 2012. Up-to-date insight on industrial enzymes applications and global market. J. Bioprocess. Biotech. S 4 (002).

Scheibner, K., Hofrichter, M., 1998. Conversion of aminonitrotoluenes by fungal manganese peroxidase. J. Basic. Microbiol. 38 (1), 51–59.

Scott, C.H., 1987. Immobilized cells: a review of recent literature. Enzyme Microb. Technol. 9, 66–79.

Scott, C., Pandey, G., Hartley, C.J., Jackson, C.J., Cheesman, M.J., Taylor, M.C., et al., 2008. The enzymatic basis for pesticide bioremediation. Indian J. Microbiol. 48 (1), 65.

Scott, C., Lewis, S.E., Milla, R., Taylor, M.C., Rodgers, A.J.W., Dumsday, G., et al., 2010. A free-enzyme catalyst for the bioremediation of environmental atrazine contamination. J. Environ. Manage. 91, 2075e2078.

Shah, P.C., Kumar, V.R., Dastager, S.G., Khire, J.M., 2017. Phytase production by *Aspergillus niger* NCIM 563 for a novel application to degrade organophosphorus pesticides. AMB Exp. 7, 66.

Sharma, B., Dangi, A.K., Shukla, P., 2018. Contemporary enzyme based technologies for bioremediation: a review. J. Environ. Manage. 210, 10–22.

Singh, B., 2014. Review on microbial carboxylesterase: general properties and role in organophosphate pesticides degradation. Biochem. Mol. Biol. 2, 1e6.

Singh, A., Ward, O.P., 2004. Biotechnology and bioremediation—an overview. Biodegradation and Bioremediation. Springer, Berlin, Heidelberg, pp. 1–17.

Singh, S., Seung, H.K., Ashok, M., Wilfred, C., 2008. Bioremediation: environmental clean-up through pathway engineering. Curr. Opin. Biotechnol. 19, 437e444.

Singh, B., Kaur, J., Singh, K., 2012. Biodegradation of malathion by *Brevibacillus* sp. strain KB2 and Bacillus cereus strain PU. World J. Microbiol. Biotechnol. 28 (3), 1133e1141.

Singh, R.K., Tiwari, M.K., Singh, R., Lee, J.K., 2013. From protein engineering to immobilization: promising strategies for the upgrade of industrial enzymes. Int. J. Mol. Sci. 14 (1), 1232e1277.

Singh, V.K., Singh, A.L., Singh, R., Kumar, A., 2018. Iron oxidizing bacteria: insights on diversity, mechanism of iron oxidation and role in management of metal pollution. Environ. Sustain. 1 (3), 221–231.

Singh, M., Singh, D., Gupta, A., Pandey, K.D., Singh, P.K., Kumar, A., 2019a. Plant growth promoting rhizobacteria: application in biofertilizers and biocontrol of phytopathogens. PGPR Amelioration in Sustainable Agriculture. Woodhead Publishing, pp. 41–66.

Singh, S.K., Singh, P.P., Gupta, A., Singh, A.K., Keshri, J., 2019b. Tolerance of Heavy Metal toxicity using PGPR strains of Pseudomonas Species. In: Singh, A.K., Kumar, A., Singh, P.K. (Eds.), PGPR Amelioration in Sustainable Agriculture. Woodhead Publishing, pp. 239–252.

Singh, V.K., Singh, M., Singh, S.K., et al., 2019c. Sustainable agricultural practices using beneficial fungi under changing climate scenario. In: Choudhary, K.K., Kumar, A., Singh, A.K. (Eds.), Climate Change and Agricultural Ecosystems. Woodhead Publishing, pp. 25–42.

Singh, P., Singh, V.K., Singh, R., Borthakur, A., Madhav, S., Ahamad, A., et al., 2020. Bioremediation: a sustainable approach for management of environmental contaminants. In: Singh, P., Kumar, A., Borthakur, A. (Eds.), Abatement of Environmental Pollutants. Elsevier, pp. 1–23.

Sivaperumal, P., Kamala, K., Rajaram, R., 2017. Bioremediation of industrial waste through enzyme producing marine microorganisms. Adv. Food Nutr. Res. 80, 165e179.

Soares, G.M., Costa-Ferreira, M., de Amorim, M.P., 2001. Decolorization of an anthraquinone-type dye using a laccase formulation. Bioresour. Technol. 79 (2), 171–177.

Strong, P.J., Burgess, J.E., 2008. Treatment methods for wine-related and distillery wastewaters: a review. Bioremed. J. 12 (2), 70–87.

Suneetha, V., Khan, Z.A., 2011. Actinomycetes: sources for soil enzymes. In: Shukla, G., Varma, A. (Eds.), Soil Enzymology. Springer, Heidelberg, pp. 259–269.

Thatoi, H., Das, S., Mishra, J., Rath, B.P., Das, N., 2014. Bacterial chromate reductase, a potential enzyme for bioremediation of hexavalent chromium: a review. J. Environ. Manage. 146, 383–399.

Torres, E., Bustos-Jaimes, I., Le Borgne, S., 2003. Potential use of oxidative enzymes for the bioremediation of organic pollutants. Appl. Catal. B: Environ. 46, 1–15.

Tuomela, M., Hatakka, A., 2011. Oxidative fungal enzymes for bioremediation. In: Comprehensive Biotechnology: Environmental Biotechnology and Safety, second ed., 6.

Uhlig, H. ed., 1998. Industrial Enzymes and Their Applications. John Wiley & Sons.

Van Deuren, J., Lloyd, T., Chhetry, S., Liou, R., Peck, J., 2002. Remediation technologies screening matrix and reference guide. Fed. Remediation Technol. Roundtable .

Vidali, M., 2001. Bioremediation. An overview. Pure Appl. Chem. 73 (7), 1163–1172.

Wackett, L.P., Gibson, D.T., 1988. Degradation of trichloroethylene by toluene dioxygenase in whole-cell studies with *Pseudomonas putida* F1. Appl. Environ. Microbiol. 54 (7), 1703–1708.

Wackett, L.P., Hershberger, C.D., 2001. Biocatalysis and Biodegradation: Microbial Transformation of Organic Compounds (No. QP517. B5 W33). ASM Press, Washington, DC.

Wang, C.J., Thiele, S., Bollag, J.M., 2002. Interaction of 2, 4, 6-trinitrotoluene (TNT) and 4-amino-2, 6-dinitrotoluene with humic monomers in the presence of oxidative enzymes. Arch Environ Con Tox. 42 (1), 1–8.

Xu, P., Liu, L., Zeng, G., Huang, D., Lai, C., Zhao, M., et al., 2014. Heavy metal-induced glutathione accumulation and its role in heavy metal detoxification in *Phanerochaete chrysosporium*. Appl. Microbiol. Biotechnol. 98 (14), 409e6418.

Yang, C., Xia, Y., Qu, H., Li, A.D., Liu, R., Wang, Y., et al., 2016. Discovery of new cellulases from the metagenome by a metagenomics-guided strategy. Biotechnol. Biofuels 9 (1), 138.

Yu-Qung, Z.T., Mei-Lin, S., Wei-De, Z., et al., 2004. Immobilization of Lasparaginase of the microparticles of the natural silk serum protein and its characters. Biomaterials 25, 3751–3759.

Zhanga, H., Zhanga, S., Hea, F., Qina, X., Zhanga, X., Yang, Y., 2016. Characterization of a manganese peroxidase from white-rot fungus *Trametes* sp. 48424 with strong ability of degrading different types of dyes and polycyclic aromatic hydrocarbons. J. Hazard. Mater. 320, 265e277.

Chapter 7

Plant growth-promoting bacteria: application in bioremediation of salinity and heavy metal–contaminated soils

Divya Singh[1], Sandeep Kumar Singh[2], Vipin Kumar Singh[2], Hariom Verma[3], Manisha Mishra[4], Kumari Rashmi[5] and Ajay Kumar[6]

[1]*Central Sericultural Research and Training Institute, Mysore, India,* [2]*Department of Botany, Centre of Advance Study in Science, Institute of Science, Banaras Hindu University, Varanasi, India,* [3]*Department of Botany, Banaras Hindu University, Varanasi, India,* [4]*University department of Botany, T.M. Bhagalpur University, Bhagalpur, India,* [5]*L.N. Mishra Institute Patna, Patna, India,* [6]*Agriculture Research Organization, Ministry of Agriculture and Rural Development, Volcani Center, Rishon LeZion, Israel*

7.1 Introduction

Agricultural yield and productivity are adversely affected by frequent exposure to different stress factors (biotic and abiotic). The term biotic stress includes phytopathogens, which can affect the growth and productivity of plants/crops by causing various types of diseases (Singh et al., 2019a); however, abiotic stress is a broad term and generally comprises accumulation of heavy metals, salinity, drought, variation in pH, and deposition of environmental contaminants in the soil, water, or natural ecosystem. The presence of any of these stresses severely affects natural productivity of plants, soil, and environment causes serious threat for health of human beings (Wani et al., 2016; Seherm and Coakley, 2003).

The physicochemical properties of soil are specific and act as limiting factors that also determined texture and productivity of soil. The presence of any stress such as salinity, drought, pH, disturbs their natural properties that ultimately affect the growth and productivity of crops. It is estimated that approximately 800 million ha of land have been used for agriculture, of which the large area of this productive land severely is affected by abiotic stresses, in which salinity and accumulation of heavy metals are the most prominent ones (Munns and Tester, 2008; Ilangumaran and Smith, 2017). According to the report approx. 15% of the arid or semiarid region and approx. 40% of irrigated land severely are under the threat of salinity throughout the world (Orhan, 2016). Currently, salinization of arable land continuously is expanding at the rate of 10% each year, and it is estimated to reach up to 50% of total arable land up to the year of 2050 (Jamil et al., 2012). The saturation limit (ECe) of saline soil at the root zone [in terms of electrical conductivity (EC)] is more than 4 dS/mL or approximately 40 mM NaCl at the 25°C, and the exchangeable Na is approximately 15% (Shrivastava and Kumar, 2015). The threshold limit of salinity varies among the plant species, for example, thresholds levels in maize are 2 dS/m and wheat 6 dS/m (Saeed et al., 2001). The lower value of ECe from the critical point adversely affect the growth and productivity of plant and soil (Munns, 2005; Jamil et al., 2012).

There are several factors responsible for the deposition of salts in the soil, in which long-term depositions of natural salts through the breakdown of rock salts and sea salt through surface runoff, wind, etc. are the some prime factors (Shrivastava and Kumar, 2015). The presence of saline conditions adversely affects the morphology and physiological pathways of plants (Jung & McCouch, 2013; Han and Lee, 2005). However, metallurgical industries that are rapid industrialization released various types of contaminants in the form of heavy metals to the environment, which results in harmful impact on the ecology as well as human health (Carlos et al., 2016; Singh et al., 2020a,b; Gupta et al., 2020). Metals are required in biological activities of living organism as stimulators, enzymatic cofactors, etc. Metals having specific weight >5 g/cm^3 are characterized as heavy metals and grouped as toxic metals such as As, Cd, Zn Cr, Cu, Hg, Ni, and Pb; and precious metals such as Ag, Au, Pd, Pt.; and radionuclides such as Ra, Th, and U (Nies, 1999; Bishop, 2002). The high persistence capacity and low degradation potential of heavy metals result in soil accumulation that affect texture of the soil, nutrient availability, growth and yield of plants/crops, and natural microbial community

Microbe Mediated Remediation of Environmental Contaminants. DOI: https://doi.org/10.1016/B978-0-12-821199-1.00007-9

as well as health of human beings (Etesami, 2018; Singh et al., 2018b, 2019b). However, toxicity of some heavy metals appears at the range of concentration between 1 and 10 mg/L, whereas Cd and Hg display toxicity at the range 0.001−0.1 mg/L (Rajkumar et al., 2012; Carlos et al., 2016). The exposure of mercury (Hg) can reduce the transpiration rate, chlorophyll synthesis, and photosynthesis rate causing loss of magnesium, manganese, potassium, and iron accumulation (Boening, 2000). High concentration of lead (Pb) can strongly inhibit seed germination and development stages of seedling, root elongation, chlorophyll formation, transpiration, and water as well as protein content. However, excess accumulation of nickel (Ni) affects the seed germination, root, and shoot growth, distorts plant parts, and decreases the production of biomass, leaf spotting, and abnormal flower shape. Palacios et al. (1998) reported that the presence of Ni (5, 15, and 30 mg/L) affected growth and reduced yield of dry matter of tomato plants. There are many reports present that show and describe how to cope up with the salinity stress as well as heavy metals contamination. To remediate heavy metal contamination, various physiochemical paths are comprised such as chemical oxidation, precipitation, reduction, filtration, ion exchange, and electrochemical treatment. However, these practices have an adverse effect on the environment and health concern of living organisms (Vijayaraghavan et al., 2006). In this aspect the use of beneficial microbes as pant or soil inoculants is considered as one of the best choices to withstand the salinity stress as well as to mitigate heavy metals problems (Datta et al., 2020; Singh et al., 2019b, 2019c).

7.2 Salinity stress and plants

The productivity of plants is severely affected by the presence of salt and their concentration. Salinity tolerance in the plants mainly depends upon concentration, duration of exposure as well as the surrounding environmental conditions that ultimately affect yields and productivity of the crop via affecting physiology, morphology, and reproductive structures (Munns, 2005). Generally, most of the plant species are sensitive to salinity and their severities are species dependent. Each species has specific tolerance limits for the protection they tend to exclude salts from roots and other physiological processes (Zhu, 2007).

The presence of high amount of salt around the roots, beyond threshold value, first impairs the osmotic balance that results in water deficiency and ultimately affects the growth and productivity of the plants, but their immediate impact appears on the root and shoots. In later stages, it excludes through root parts, which are present above the tolerance limits; however, certain extent gets deposited in the cytosol that disrupts cellular structures, osmotic balance as well as several physiological functions (Munns and Tester, 2008). For stress tolerance, integrated signaling pathways are crucial and plant system developed regulates various mechanisms such as synthesizing antioxidant enzymes superoxide dismutase, glutathione reductase, catalase, and peroxidases, for the control or prevents tissues from oxidative damages through quenching and detoxifying reactive oxygen species (ROS) (Gill and Tuteja, 2010).

7.3 Plant growth-promoting rhizobacteria

Microbes are ubiquitous in nature in which rhizosphere of the plant is hot spot for microbial communities such as bacteria, fungi, actinomycetes, and they interact with each other through various chemical signals released from the rhizosphere parts of the plants. Rhizosphere of the plant is the region of root surface, which is surrounded by thin layer of soil and secretes a significant amount of exudates that act as signaling molecules during plant−microbe interaction. Root exudates secrete array of chemical compounds, which mainly constituted of amino acids, carbohydrates flavonoids, lipids, organic acids, and phenols, and these act as chemoattractants and help in effective colonization of microbial communities in the rhizosphere of the plants (Bauer and Mathesius, 2004; Badri and Vivanco, 2009; Kumar et al., 2015). Some of the microbes enter inside the plant tissues through natural opening without causing any external sign of infection and resides as endophytes (Kumar et al., 2016b). The rhizosphere and endophytic microorganism have diverse roles in terms of growth promotion, disease management, and stress management (Kumar et al., 2015, 2017, 2018, Singh et al., 2017, 2018a). Some of the microbial genera that are commonly present in the rhizosphere or endophytes are *Pseudomonas, Bacillus, Acinetobacter, Klebsiella, Azotobacter, Burkholderia* while *Trichoderma, Acromonium*, etc. are the most commonly found fungal species (Kumar et al., 2016a, 2016b; Singh et al., 2018b). These microbial genera are directly or indirectly associated with the plants for the growth promotion, disease management as well as management of different kinds of biotic and abiotic stresses (Kumar et al., 2017, 2018; Singh et al., 2015, 2018b). Many researchers have reported the ability of plant growth-promoting rhizobacteria (PGPR) strains in reducing the effects of salinity and helping plants in tolerating the stress conditions Table 7.1.

TABLE 7.1 An overview of plant growth-promoting bacteria mediated salt tolerance.

PGPB	Experiments	Results	References
Pseudomonas fluorescens and *Pseudomonas migulae*	Two ACCD + bacterial endophytes were tested on tomato plants under salt stress	Plants pretreated with wild-type ACC deaminase–containing endophytic strains were healthier and grew to a much larger size	Ali et al. (2014)
Azospirillum	Pot experiments; nitrogenase activity (in situ) was assayed, acetylene reduction assay was used	Restricted Na^+ uptake and increased K^+ and Ca^{2+} uptake	Hamdia et al. (2004)
Pseudomonas syringae, P. fluorescens, Enterobacter aerogenes	Pot and jar experiments, pot experiments, characterization of strains	Improving the growth and yield under salinity stress condition, it also increases total chlorophyll content (a, b, and carotenoids)	Nadeem et al. (2007)
P. fluorescens	Serial dilution technique was used to isolate pseudomonas strains, DNA extraction and amplification of *Pseudomonas* sp., amplification of ACC deaminase, isolation, cloning and sequencing of 16S rDNA of *P. fluorescens* strain TDK1	Enhanced ACC deaminase activity	Saravanakumar and Samiyappan (2007)
Pseudomonas mendocina	Seeds were grown for 15 days in peat substrate under nursery conditions, without any fertilization treatment, CAT activity was determined	ACC deaminase activity and enhanced uptake of essential nutrients	Kohler et al. (2009)
Rhizobium, Pseudomonas	Pot experiments, determination of phosphate-solubilizing bacteria was done	Decreased electrolyte leakage and, increase in proline production, maintenance of relative water content of leaves, and selective uptake of K^+ ion	Bano and Fatima (2009)
Pseudomonas putida	Bacterial characterization, 16 S rDNA identification, extraction and estimation of the IAA, plant growth promotion by rs-198 in the fields	Increase the absorption of the Mg^{2+}, K^+, and Ca^{2+} and decrease the uptake of the Na_2	Yao et al. (2010)
Streptomyces sp.	Screening of actinobacterial isolates for NaCl tolerance, assessment of tomato growth promotion under NaCl stress	ACC deaminase activity and IAA production and phosphate solubilization	Palaniyandi et al. (2014)

7.4 Plant growth-promoting bacteria and salinity tolerance

Plant growth-promoting bacteria (PGPB) are directly or indirectly associated with the plants for managing their growth, yields, productivity via various mechanisms such as enriching the soil with nutrients such as nitrogen phosphorous, iron via mechanism nitrogen fixation, phosphorus solubilization, and siderophore. The production of HCN, NH_3 manages the growth of various phytopathogens, mitigation of various abiotic stresses such as salinity, drought, pH, etc. and also degradation of various environmental contaminants (Kumar et al., 2015, 2017; Singh et al., 2018b).

PGPB regulate various physiological activities such as water potential and transpiration rate in the plants under saline condition. Besides these functions, PGPB induce accumulation of osmolyte as well as various signaling molecules that regulate phytohormone synthesis to overcome the salinity stress. There are various reports available that show the potency of PGPR applications in the mitigation of salinity stress. Marulanda et al. (2010) studied the inoculation impact of *Bacillus megaterium* in maize plants and found enhancement in the hydraulic conductivity in root compared to control without inoculated plants after exposed to (2.59 dS/m) of salinity, Similarly, Nautiyal et al. (2013) reported enhanced salinity tolerance limits in the rice plant under hydroponic conditions after inoculation of *Bacillus amyloliquefaciens* strain.

Many PGPR have been investigated for their role in improving relation between plant water, growth rate, carbohydrate accumulation, and photosynthetic efficiency in plants under salt stress. In a study, Upadhyay et al. (2012) found enhanced level of sugar content in wheat after inoculation of *Bacillus aquimaris* strain under saline conditions; similarly, Rojas-Tapias et al. (2012) reported improved K^+ uptake and Na^+ exclusion in the maize plants after inoculation of *Azotobacter* strains under saline conditions.

In plant system, ROS are generated due to various stresses such as salinity, heavy metal, and drought. The level of ROS depends upon the severity of stress and rise with increase of stresses (Choudhary et al, 2007). The removal of free radicals generated during high salt conditions in plant cell is an important process for the normal physiology and morphology of plants. PGPB secrete various hormones that can ameliorate in growth promotion and mitigation of biotic and abiotic stresses that impact on plant system (Dimkpa et al., 2009). Under stress conditions, PGPR secrete various antioxidant enzymes that scavenge free radicals or ROS generated by increasing saline conditions (Radhakrishnan and Baek, 2017).

7.5 Heavy metals and plant growth-promoting bacteria

From the last few decades, heavy metal stress has become one of the major concerns for the environment. The accumulation of heavy metals in the soil adversely affects the texture and productivity of soil, physiological activities of plants that ultimately affect the agricultural production (Hassan et al., 2017; Singh et al., 2018b). Some of the heavy metals such as Mo, Mn, Cu, Zn, and Ni are required as an essential element for the various biological processes (Salla et al., 2011; Shahid et al., 2015). However, some metals such as arsenic (As), lead (Pb), cadmium (Cd), and mercury (Hg) are highly toxic, and the presence of which in slight high severely affects the crop production and also their presence in the water, soil, and environment adversely affects the food chain and health of human beings (Xiong et al., 2018; Pierart et al., 2015; Tiwari and Lata, 2018).

The mitigation of heavy metal stress is a prime concern for environmentalists and researchers. There are various known mechanisms that are used for the remediation of heavy metal stress, but cost-effectiveness and their toxic residue are still a major challenge for the environment management (Ali et al., 2014; Ullah et al., 2015).

However, biological remediation technique using beneficial microbes is a sustainable approach for heavy metal stress management (Singh et al., 2018b). The use of PGPB in heavy metal stress management is a novel approach and continuously has been practicing from last few decades. PGPR microbes possess various biological activities such as siderophore production, biosurfactant molecules, biofilm, EPS that helps in chelation, precipitation, acidification, oxidation–reduction, and ultimately in the remediation of heavy metal stress from the soil or environment (Singh et al., 2018b; Mishra et al., 2017; Zubair et al., 2016). In the recent past, various bacterial species have been used and applied as inoculants in the soil for heavy metal stress remediation. *Pseudomonas* and *Bacillus* are the most commonly used bacterial strain in the various fields of agriculture and stress management. There are various reports present that show the efficiency of microbial strains in remediation of heavy metal stress. Pandey et al. (2013) reported Pb- and As-resistant *Bacillus* species having potential role in remediation of the metals as well as growth-promoting potential in the rice field.

7.6 Conclusion

From the last few decades, rapid industrialization and increasing global population cause emergence huge amount of environmental contaminants that severely affect the environment as well as health concern of the entire living organism. Salinity and accumulation of heavy metals are the two most common stresses that limit the optimum crop production by influencing normal physiological and morphological parameters of the plants. Currently, beneficial microbes have been broadly utilized as soil or plant inoculants to mitigate the challenges of these stresses. These microbes besides stress management also play role in growth promotion in sustainable ways. Meanwhile, the limited knowledge, exploration of these microbes, including bacteria, fungi, cyanobacteria, and actinomycetes, limits their effective application in stress management.

References

Ali, S., Charles, T.C., Glick, B.R., 2014. Amelioration of high salinity stress damage by plant growth-promoting bacterial endophytes that contain ACC deaminase. Plant. Physiol. Biochem. 80, 160–167.

Badri, D.V., Vivanco, J.M., 2009. Regulation and function of root exudates. Plant Cell Environ. 32 (6), 666–681.

Bano, A., Fatima, M., 2009. Salt tolerance in *Zea mays* (L.) following inoculation with *Rhizobium* and *Pseudomonas*. Biol. Fertil. Soils 45, 405–413.

Bauer, W.D., Mathesius, U., 2004. Plant responses to bacterial quorum sensing signals. Curr. Opin. Plant. Biol. 7 (4), 429–433.

Bishop, P.L., 2002. Pollution Prevention: Fundamentals and Practice. Tsinghua University Press, Beijing, China.

Boening, D.W., 2000. Ecological effects, transport, and fate of mercury: a general review. Chemosphere 40 (12), 1335–1351.

Carlos, M.H.J., Stefani, P.V.Y., Janette, A.M., Melani, M.S.S., Gabriela, P.O., 2016. Assessing the effects of heavy metals in ACC deaminase and IAA production on plant growth-promoting bacteria. Microbiol. Res. 188, 53–61.

Choudhary, M., Jetley, U.K., Khan, M.A., Zutshi, S., Fatma, T., 2007. Effect of heavy metal stress on proline, malondialdehyde, and superoxide dismutase activity in the cyanobacterium *Spirulina platensis*-S5. Ecotoxicol. Environ. Saf. 66 (2), 204–209.

Datta, S., Singh, S., Kumar, V., Dhanjal, D.S., Sidhu, G.K., Amin, D.S., et al., 2020. Endophytic bacteria in Xenobiotic degradation. In: Kumar, A., Kumar Singh, V. (Eds.), Microbial Endophytes: Prospects of Sustainable Agriculture. Elsevier, pp. 125–156.

Dimkpa, C., Weinand, T., Asch, F., 2009. Plant–rhizobacteria interactions alleviate abiotic stress conditions. Plant Cell Environ. 32 (12), 1682–1694.

Etesami, H., 2018. Bacterial mediated alleviation of heavy metal stress and decreased accumulation of metals in plant tissues: mechanisms and future prospects. Ecotoxicol. Environ. Saf. 147, 175–191.

Gill, S.S., Tuteja, N., 2010. Reactive oxygen species and antioxidant machinery in abiotic stress tolerance in crop plants. Plant. Physiol. Biochem. 48 (12), 909–930.

Gupta, A., Singh, S.K., Singh, V.K., Singh, M.K., Modi, A., Zhimo, V.Y., et al., 2020. Endophytic microbe approaches in bioremediation of organic pollutants. In: Kumar, A., Kumar Singh, V. (Eds.), Microbial Endophytes: Prospects of Sustainable Agriculture. Elsevier, pp. 157–174.

Hamdia, M.A.E.S., Shaddad, M.A.K., Doaa, M.M., 2004. Mechanisms of salt tolerance and interactive effects of *Azospirillum brasilense* inoculation on maize cultivars grown under salt stress conditions. Plant. Growth Regul. 44 (2), 165–174.

Han, S.H., Lee, J.H., 2005. An overview of peak-to-average power ratio reduction techniques for multicarrier transmission. IEEE Wirel. Commun. 12 (2), 56–65.

Hassan, T.U., Bano, A., Naz, I., 2017. Alleviation of heavy metals toxicity by the application of plant growth promoting rhizobacteria and effects on wheat grown in saline sodic field. Int. J. Phytoremediat. 19 (6), 522–529.

Ilangumaran, G., Smith, D.L., 2017. Plant growth promoting rhizobacteria in amelioration of salinity stress: a systems biology perspective. Front. Plant. Sci. 8.

Jamil, M., Bashir, S., Anwar, S., Bibi, S., Bangash, A., Ullah, F., et al., 2012. Effect of salinity on physiological and biochemical characteristics of different varieties of rice. Pak. J. Bot. 44 (2012), 7–13.

Jung, J.K., McCouch, S., 2013. Getting to the roots of it: genetic and hormonal control of root architecture. Front. Plant. Sci. 4.

Kohler, J., Hernandez, J.A., Caravaca, F., Roldan, A., 2009. Induction of antioxidant enzymes is involved in the greater effectiveness of a PGPR versus AM fungi with respect to increasing the tolerance of lettuce to severe salt stress. Environ. Exp. Bot. 65, 245–252.

Kumar, A., Singh, V.K., Tripathi, V., Singh, P.P., Singh, A.K., 2018. Plant growth-promoting rhizobacteria (PGPR): perspective in agriculture under biotic and abiotic stress. In:. Crop Improvement Through Microbial Biotechnology. 333–342. Available from: https://doi.org/10.1016/B978-0-444-63987-5.00016-5.

Kumar, A., Vandana, Yadav, A., Giri, D.D., Singh, P.K., Pandey, K.D., 2015. Rhizosphere and their role in plant –microbe interaction. In: Chaudhary, K.K., Dhar, D.W. (Eds.), Microbes in Soil and Their Agricultural Prospects. Nova Science Publisher, Inc, pp. 83–97.

Kumar, A., Verma, H., Singh, V.K., Singh, P.P., Singh, S.K., Ansari, W.A., et al., 2017. Role of *Pseudomonas sp.* in sustainable agriculture and disease management. Agriculturally Important Microbes for Sustainable Agriculture. Springer, Singapore, pp. 195–215.

Kumar, A., Singh, R., Yadav, A., Giri, D.D., Singh, P.K., Pandey, K.D., 2016a. Isolation and characterization of bacterial endophytes of *Curcuma longa* L. 3 Biotech 6, 60.

Kumar, A., Vandana, Singh, M., Singh, P.P., Singh, S.K., Singh, P.K., et al., 2016b. Isolation of plant growth promoting rhizobacteria and their impact on growth and curcumin content in *Curcuma longa* L. Biocatal. Agric. Biotech. 8, 1–7.

Marulanda, A., Azcón, R., Chaumont, F., Ruiz-Lozano, J.M., Aroca, R., 2010. Regulation of plasma membrane aquaporins by inoculation with a *Bacillus megaterium* strain in maize (*Zea mays* L.) plants under unstressed and salt-stressed conditions. Planta 232 (2), 533–543.

Mishra, J., Singh, R., Arora, N.K., 2017. Alleviation of heavy metal stress in plants and remediation of soil by rhizosphere microorganisms. Front. Microbiol. 8, 1706.

Munns, R., Tester, M., 2008. Mechanisms of salinity tolerance. Annu. Rev. Plant. Biol. 59, 651–681.

Munns, R., 2005. Genes and salt tolerance: bringing them together. N. Phytol. 167 (3), 645–663.

Nadeem, S.M., Zahir, Z.A., Naveed, M., Arshad, M., 2007. Preliminary investigation on inducing salt tolerance in maize through inoculation with rhizobacteria containing ACC-deaminase activity. Can. J. Microbiol. 53, 1141–1149.

Nautiyal, C.S., Srivastava, S., Chauhan, P.S., Seem, K., Mishra, A., Sopory, S.K., 2013. Plant growth-promoting bacteria *Bacillus amyloliquefaciens* NBRISN13 modulates gene expression profile of leaf and rhizosphere community in rice during salt stress. Plant. Physiol. Biochem. 66, 1–9.

Nies, D.H., 1999. Microbial heavy-metal resistance. Appl. Microbiol. Biotechnol. 51, 730–750.

Orhan, F., 2016. Alleviation of salt stress by halotolerant and halophilic plant growth-promoting bacteria in wheat (*Triticum aestivum*). Braz. J. Microbiol. 47 (3), 621–627.

Palacios, G., Gomez, I., Carbonell-Barrachina, A., Pedreno, J.N., Mataix, J., 1998. Effect of nickel concentration on tomato plant nutrition and dry matter yield. J. Plant. Nutr. 21 (10), 2179–2191.

Palaniyandi, S.A., Damodharan, K., Yang, S.H., Suh, J.W., 2014. *Streptomyces* sp. strain PGPA39 alleviates salt stress and promotes growth of 'Micro Tom' tomato plants. J. Appl. Microbiol. 117, 766–773.

Pandey, S., Barai, P.K., Maiti, T.K., 2013. Influence of heavy metals on the activity of antioxidant enzymes in the metal resistant strains of *Ochrobactrum* and *Bacillus* sp. J. Environ. Biol. 34 (6), 1033.

Pierart, A., Shahid, M., Séjalon-Delmas, N., Dumat, C., 2015. Antimony bioavailability: knowledge and research perspectives for sustainable agricultures. J. Hazard. Mater. 289, 219–234.

Radhakrishnan, R., Baek, K.H., 2017. Physiological and biochemical perspectives of non-salt tolerant plants during bacterial interaction against soil salinity. Plant. Physiol. Biochem. 116, 116–126.

Rajkumar, M., Sandhya, S., Prasad, M.N.V., Freitas, H., 2012. Perspectives of plant-associated microbes in heavy metal phytoremediation. Biotechnol. Adv. 30, 1562–1574.

Rojas-Tapias, D., Moreno-Galván, A., Pardo-Díaz, S., Obando, M., Rivera, D., Bonilla, R., 2012. Effect of inoculation with plant growth-promoting bacteria (PGPB) on amelioration of saline stress in maize (*Zea mays*). Appl. Soil Ecol. 61, 264–272.

Saeed, M.M., Ashraf, M., Asghar, M.N., Bruen, M., Shafique, M.S., 2001. Root Zone Salinity Management Using Fractional Skimming Wells With Pressurized Irrigation, 46. International Water Management Institute (IWMI), Lahore.

Salla, V., Hardaway, C.J., Sneddon, J., 2011. Preliminary investigation of *Spartina alterniflora* for phytoextraction of selected heavy metals in soils from Southwest Louisiana. Microchem. J. 97, 207–212.

Saravanakumar, D., Samiyappan, R., 2007. ACC deaminase from *Pseudomonas fluorescens* mediated saline resistance in groundnut (*Arachis hypogea*) plants. J. Appl. Microbiol. 102, 1283–1292.

Seherm, H., Coakley, S.M., 2003. Plant pathogens in a changing world. Austr. Plant. Pathol. 32 (2), 157–165.

Shahid, M., Khalid, S., Abbas, G., Shahid, N., Nadeem, M., Sabir, M., et al., 2015. Heavy metal stress and crop productivity. Crop Production and Global Environmental Issues. Springer, Cham, pp. 1–25.

Shrivastava, P., Kumar, R., 2015. Soil salinity: a serious environmental issue and plant growth promoting bacteria as one of the tools for its alleviation. Saudi J. Biol. Sci. 22 (2), 123–131.

Singh, D., Singh, S.K., Singh, V.K., Gupta, A., Aamir, M., Kumar, A., 2020b. Plant growth–promoting bacteria and their role in environmental management. In: Singh, P., Kumar, A., Borthakur, A. (Eds.), Abatement of Environmental Pollutants. Elsevier, pp. 161–175.

Singh, M., Singh, D., Gupta, A., Pandey, K.D., Singh, P.K., Kumar, A., 2019a. Plant growth promoting rhizobacteria: application in biofertilizers and biocontrol of phytopathogens. In: Singh, A.K., Kumar, A., Singh, P.K. (Eds.), PGPR Amelioration in Sustainable Agriculture. Woodhead Publishing, pp. 41–66.

Singh, P., Singh, V.K., Singh, R., Borthakur, A., Madhav, S., Ahamad, A., et al., 2020a. Bioremediation: a sustainable approach for management of environmental contaminants. In: Singh, P., Kumar, A., Borthakur, A. (Eds.), Abatement of Environmental Pollutants. Elsevier, pp. 1–23.

Singh, R.P., Jha, P., Jha, P.N., 2015. The plant-growth-promoting bacterium *Klebsiella* sp. SBP-8 confers induced systemic tolerance in wheat (Triticumaestivum) under salt stress. J. Plant Physiol. 184, 57–67.

Singh, S.K., Singh, P.P., Gupta, A., Singh, A.K., Keshri, J., 2019b. Tolerance of heavy metal toxicity using PGPR strains of *Pseudomonas* species. In: Singh, A.K., Kumar, A., Singh, P.K. (Eds.), PGPR Amelioration in Sustainable Agriculture. Woodhead Publishing, pp. 239–252.

Singh, V.K., Singh, A.K., Kumar, A., 2017. Disease management of tomato through PGPB: current trends and future perspective. 3 Biotech. 7 (4), 255.

Singh, V.K., Singh, A.K., Singh, P.P., Kumar, A., 2018a. Interaction of plant growth promoting bacteria with tomato under abiotic stress: a review. Agric. Ecosyst. Environ. 267, 129–140.

Singh, V.K., Singh, A.L., Singh, R., Kumar, A., 2018b. Iron oxidizing bacteria: insights on diversity, mechanism of iron oxidation and role in management of metal pollution. Environ. Sustain. 1 (3), 221–231.

Singh, V.K., Singh, M., Singh, S.K., Kumar, C., Kumar, A., 2019c. Sustainable agricultural practices using beneficial fungi under changing climate scenario. In: Choudhary, K.K., Kumar, A., Singh, A.K. (Eds.), Climate Change and Agricultural Ecosystems. Woodhead Publishing, pp. 25–42.

Tiwari, S., Lata, C., 2018. Heavy metal stress, signaling, and tolerance due to plant-associated microbes: an overview. Front. Plant Sci. 9, 452.

Ullah, A., Heng, S., Munis, M.F.H., Fahad, S., Yang, X., 2015. Phytoremediation of heavy metals assisted by plant growth promoting (PGP) bacteria: a review. Environ. Exp. Bot. 117, 28–40.

Upadhyay, S.K., Singh, J.S., Saxena, A.K., Singh, D.P., 2012. Impact of PGPR inoculation on growth and antioxidant status of wheat under saline conditions. Plant. Biol. 14 (4), 605–611.

Vijayaraghavan, K., Palanivelu, K., Velan, M., 2006. Biosorption of copper(II) and cobalt(II) from aqueous solutions by crab shell particles. Bioresour. Technol. 97, 1411–1419.

Wani, S.H., Kumar, V., Shriram, V., Sah, S.K., 2016. Phytohormones and their metabolic engineering for abiotic stress tolerance in crop plants. Crop. J. 4 (3), 162–176.

Xiong, Q., Zhou, M., Liu, M., Jiang, S., Hou, H., 2018. The transformation behaviors of heavy metals and dewaterability of sewage sludge during the dual conditioning with Fe^{2+}-sodium persulfate oxidation and rice husk. Chemosphere 208, 93–100.

Yao, L., Wu, Z., Zheng, Y., Kaleem, I., Li, C., 2010. Growth promotion and protection against salt stress by *Pseudomonas putida* Rs-198 on cotton. Eur. J. Soil Biol. 46, 49–54.

Zhu, J.K., 2007. Plant Salt Stress. Elsevier.

Zubair, M., Shakir, M., Ali, Q., Rani, N., Fatima, N., Farooq, S., et al., 2016. Rhizobacteria and phytoremediation of heavy metals. Environ. Technol. Rev. 5 (1), 112–119.

Chapter 8

Role of plant–microbe systems in remediation of petrochemical-contaminated water and soil environment

Simranjeet Singh[1,2,3,†], Vijay Kumar[4,†], Shivika Datta[5], Daljeet Singh Dhanjal[1], Parul Parihar[6] and Joginder Singh[1]

[1]*Department of Biotechnology, Lovely Professional University, Jalandhar, India,* [2]*Punjab Biotechnology Incubators, Mohali, India,* [3]*Regional Advanced Water Testing laboratory, Mohali, India,* [4]*Department of Chemistry, Regional Ayurveda Research Institute for Drug Development, Gwalior, India,* [5]*Department of Zoology, Doaba College, Jalandhar, India,* [6]*Department of Botany, Lovely Professional University, Jalandhar, India*

8.1 Introduction

Soil and water are the natural resources important for life (Kirkby, 2016). Both these resources have the vital role in both agriculture and livelihood of human beings (Rockström et al., 2017). They both help in fulfilling the basic requirements such as food, fiber, and shelter for various organisms (Sarvade et al., 2019). Various dramatic events of oil spills in past have received significant attention and prompted for remediation response, as the large petrochemical-contaminated surface area and water have remained untreated (Jernelv, 2010). Accidental spillage of these petrochemicals during their exploration, extraction, storage, and transportation leads to both water and soil pollution (Das and Chandran, 2011). Discharging of these petrochemicals causes disturbances in natural equilibrium of the environment and living species. This polluted water and soil further enter our food chain as the same natural resources are used for culturing plant and water is used for drinking (Tchounwou et al., 2012; Frone and Frone, 2015). Due to intake of these pollutants, it causes health issues, which affects the environment as well as economy as depicted in Fig. 8.1 (Kelishadi, 2012). Petrochemical Pollutants are member of carcinogenic and neurotoxic pollutants. The major reason for the exponential increase of petroleum spillage cases is due to rapid urbanization and industrialization (Xu et al., 2018).

Soil and water being the valuable natural resources demand the protective measures as well as evaluation methods for effective remediation (Aqeel et al., 2014; Connor et al., 2017). As the conventional methods are outdated as well as ineffective in treating the petrochemical-contaminated soil and water (Pichtel, 2016; Sharma et al., 2018). These conventional approaches include the techniques such as burying, dispersion, evaporation, mechanical, and washing (Das and Chandran, 2011). Recently, plant–microbe-associated system has gained the significant attention for remediating the petrochemical-contaminated water and soil. In this, plants as well as microbes are used for detoxifying and removing the contaminants via different metabolic processes (Ojuederie and Babalola, 2017). This being cost-effective in comparison to other methods has made it a potential approach. The success of these approaches also depends on the conditions maintained favoring the rapid degradation of the petrochemicals at the contaminated site (Gkorezis et al., 2016). In this view, this chapter focuses on underlining the composition, properties, and harmful effects of petrochemical on water and soil. Further, it will also discuss about plant–microbe remediation approach for restoring the contaminated environment.

†. Equal contribution.

Microbe Mediated Remediation of Environmental Contaminants. DOI: https://doi.org/10.1016/B978-0-12-821199-1.00008-0

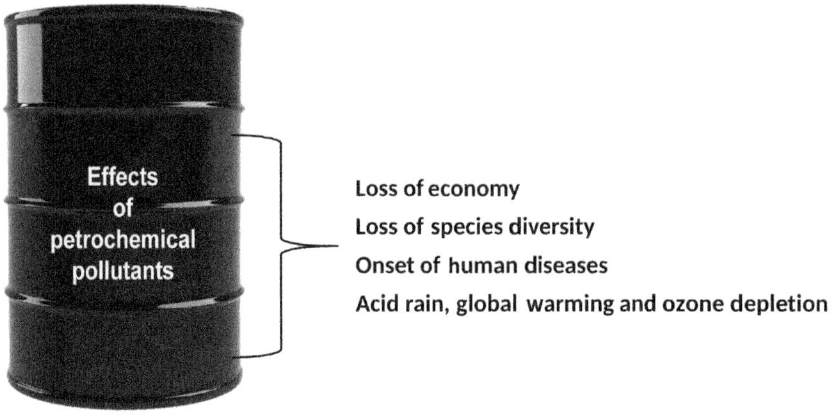

FIGURE 8.1 Diagrammatic representation of the effect of petrochemical pollutants.

8.2 Composition and properties of petrochemical-contaminated water and soil

The chemical petroleum is the amalgam of different hydrocarbons that can be classified into aromatic hydrocarbons, nonhydrocarbon compounds, and saturated hydrocarbons (Varjani, 2017). The saturated hydrocarbons are composed of C–C and C–H bond that can be easily dissociated. Moreover, the melting point of these saturated hydrocarbons is relatively low; therefore by natural volatilization these compounds get readily disappeared from soil (Mamy et al., 2015). On the other hand, aromatic hydrocarbons have complex structure, as they have benzene ring in it. Due to the presence of benzene ring, their boiling point increases and makes its removal difficult from soil (Adeniji et al., 2018). In addition, polycyclic aromatic hydrocarbons (PAHs) are the persistent organic contaminants found in diverse environment. Environmental Protection Agency, United States, has comprehended major 16 PAHs named acenaphthene, acenaphthylene, anthracene, benzene [ghi]perylene, benzo[*a*]anthracene, benzo[*a*]pyrene, benzo[*b*]fluoranthene, benzo[*k*]fluoranthene, diphenyl[*a*,*h*]anthracene, fluoranthene, fluorene, phenanthrene, indene, indeno(1,2,3-*cd*)pyrene, naphthalene, and pyrene (Yan et al., 2004; Abdel-Shafy and Mansour, 2016).

Out of three different types of hydrocarbons, nonhydrocarbons are composed with high number of carbon atoms. Even these nonhydrocarbons are insoluble in water and boiling as well as fusion point of these compounds is relatively high (Hsu et al., 2019). These properties of nonhydrocarbons make the remediation of these compounds difficult from soil, which makes it responsible for mutagenicity and environmental toxicity (Chibwe et al., 2015). Hence, this is the reason that petrochemical-contaminated soil chiefly contains nonhydrocarbon compounds and PAHs (Koshlaf and Ball, 2017). Table 8.1 comprehends the different types of petrochemical compounds and Fig. 8.2 highlights the composition of petrochemical compounds. The accidental spillage of these petrochemical in water and soil leads to water as well as soil pollution. These pollutants not only degrade the quality of water and soil but also make the environment incompatible for microbes and living organisms (Jernelv, 2010).

8.3 Harmful effects of petrochemical-contaminated water and soil

Extensive literature is available regarding the petrochemical-contaminated water and soil, which chiefly discusses about the effects as described in the following sections.

8.3.1 Effect of petrochemical-contaminated soil and water on their quality

The petrochemical compounds have low density as well as emulsifying ability and high viscosity, which readily allow it to be absorbed in soil and alter the porosity and permeability of soil (Klamerus-Iwan et al., 2015), Few components of these petrochemical compounds get mixed in the water, while some petrochemicals being insoluble in nature obstruct the life of marine organisms (Villegas et al., 2016). In addition, these petrochemicals contain carbon and nitrogen compound in it, as a resultant they also change the texture and composition of soil (Panno et al., 2005). These changes further affect C/N ratio, C/P ratio, conductivity, EH, pH, and salinity of the soil as well as water (Cheema et al., 2015). And, heavy metals (such as nickel and vanadium) and salts present in petroleum mixtures also affect the properties of the soil as well as water (Akinyemi et al., 2006).

TABLE 8.1 Different types of petrochemical compounds.

Name	Type	Boiling point	Structure	Examples
Asphaltene and resin	Nonhydrocarbons	Higher	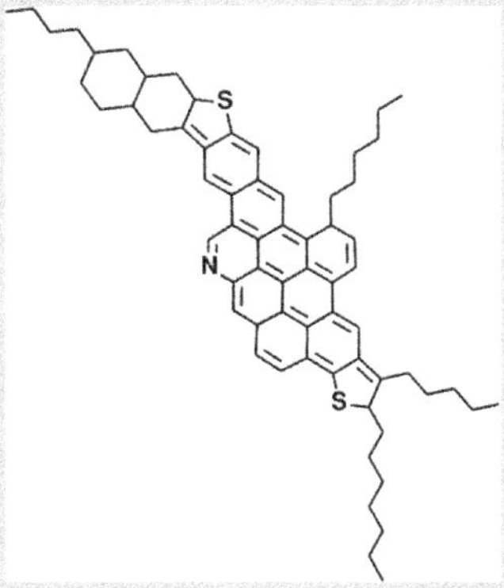	Naphthenic aromatic nucleus and polycyclic aromatic nucleus
N-containing compounds	Nonhydrocarbons	Higher	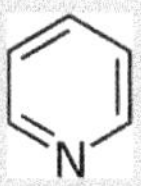	Pyridine, quinoline
Organometallic compounds	Nonhydrocarbons	Higher		Porphyrin-like compound
Oxy-compounds	Nonhydrocarbons	Higher	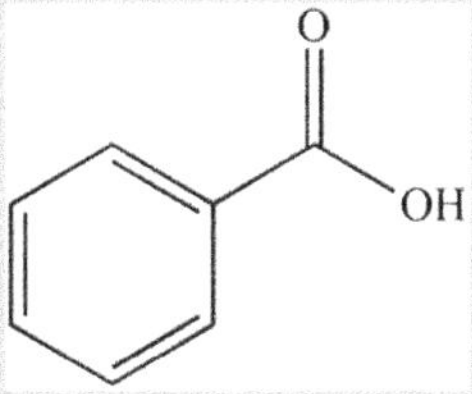	Aromatic, naphthenic, and saturated acids
Saturated	Hydrocarbons	Lower	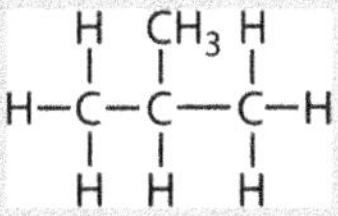	Cycloparaffin, isoparaffin, and *n*-alkanes
Sulfo-compounds	Nonhydrocarbons	Higher	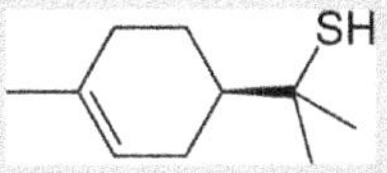	Di-sulfide, mercaptan, monosulfide, and thiophene derivatives
Unsaturated	Hydrocarbons	High	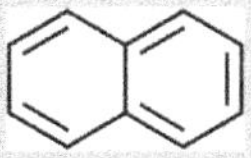	Aromatic and naphthenoaromatic hydrocarbons

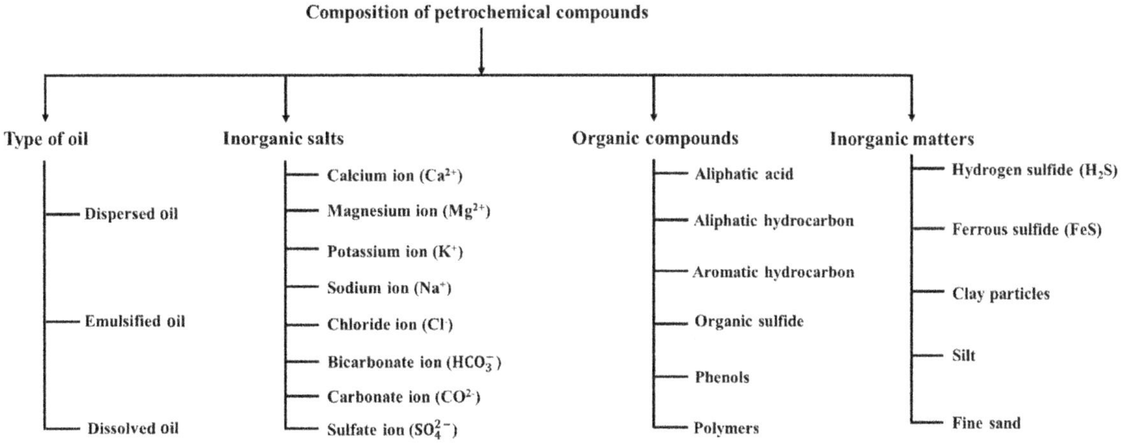

FIGURE 8.2 Outline of the composition of petrochemical compounds.

8.3.2 Effect of petrochemical-contaminated soil and water on microbes

As the microbes are omnipresent in the environment and present in abundance in clean soil and water (Ahmad et al., 2018). Generally, in ordinary condition the microbes that have the ability to resist the stress induced by petrochemical compounds are not found (Ortíz-Castro et al., 2009). But, in contaminated soil and water, these microbes adapt themselves and synthesize the enzyme for the degradation of petrochemical compounds and increase their number because of synergic effect (Thijs et al., 2016). Various studies have highlighted that petrochemical contaminants alter the microbial diversity and enzymes in both water and soil in accordance to their inhibitory action (Zhang et al., 2019).

8.3.3 Effect of petrochemical-contaminated soil and water on crops

The contaminated soil and water are known to obstruct the normal development of crops as it affects crops' fertility and germination and weakens their resistance against diseases and pest (Altieri and Nicholls, 2003). Furthermore, petrochemical compound present in both water and soil reacts with inorganic N and P, causing the limited nitrification as well as remediation of phosphoric acid. As a resultant, the effective N and P decrease in soil, which subsequently affects its absorption by crops (Mobin et al., 2018; Zhao and Shen, 2018).

Also, PAHs present are petrochemical compounds that show carcinogenic, teratogenic, mutagenic, and other lethal effects (Yu, 2002). The lethal contaminants enter in our body by the intake of crops grown from contaminated soil by using the contaminated water, which degrades the normal function of organs such as kidney and liver, and eventually impose as health threat on humans (Jan et al., 2015). Lastly, the oil pollutants enter atmosphere, as constituent of petrochemicals, have low boiling point, which promptly evaporates it. Later through runoff and infiltration it enters surface water and finally in soil and then via food chain in human bodies (Yang et al., 2009).

8.4 Plant–microbe remediation technology classification and application

Plant-assisted microbial remediation means to use the metabolism of both plant and microbes together to degrade the contaminants to nontoxic compounds to regain the quality of healthy soil and water (Ojuederie and Babalola, 2017). This method is more effective than the other methods comprehended in Table 8.2. This approach uses the targeted plants and microbes for assimilation, degradation, extraction, sequestration, and transformation of both organic and inorganic pollutants (Azubuike et al., 2016). Plant-assisted microbial remediation involves unique mechanism for detoxifying organic contaminants, which is also stated to be "green liver concept" (Reichenauer and Germida, 2008). For this, parameters such as environmental conditions, petrochemical-degrading strain, and physiochemical traits of contaminants play vital roles in overall success of plant-assisted microbial remediation (Xu et al., 2018). During this process, major attention is given on stimulating the growth of petrochemical-degrading microbes. The roots of plants provide the microenvironment as well as the carbon source for petrochemical-degrading microbes (Tahri et al., 2013). In reimbursement, microbes improve the plant growth by supplying nutrients and providing protection against degraded contaminants as well as plant pathogens. It has been well comprehended that plant-assisted microbial remediation helps in dissipation of organic contaminants from petrochemical-contaminated soils (Correa-García et al., 2018).

8.5 Plant–microbe systems toward remediation of petrochemical-contaminated soil

For survival of plants in petrochemical-contaminated environment, they need to (1) have the ability to tolerate the one or more petrochemical compounds, (2) grow rapidly, (3) possess the ability to synthesize as well as secrete the petrochemical-degrading enzyme, and (4) have high compatibility (Gkorezis et al., 2016). During this, plant shows positive influence because of the presence of microbes that produce plant hormones such as cytokinins, gibberellins, and indole-3-acetic acid and restrain the ethylene synthesis by the action of 1-aminocyclopropane-1-carboxylate deaminase (Olanrewaju et al., 2017; Tsukanova et al., 2017; Kudoyarova et al., 2019). Nowadays, an increasing numbers of evidences are available, which highlight the association of plant–microbe system for remediation of petrochemical-contaminated soil. Various studies using the plants species such as *Festuca arundinacea, Hordeum sativum, Hordeum vulgare, Lolium multiflorum, Lolium perenne, Lotus corniculatus, Lupinus arboreus, Medicago sativa, Oryza sativa, Phragmites australis, Pisum sativum, Populus* sp., *Secale cereale, Sorghum bicolor, Trifolium repens, Triticum aestivum, Zea mays* along with microbes for remediation of petrochemical-contaminated soil have been comprehended (Kuiper et al., 2002; Johnson et al., 2004; Huang et al., 2004; Anokhina et al., 2004; Muratova et al., 2005, 2009, 2010; Reed and Glick, 2005; Sheng and Gong, 2006; Child et al., 2007a,b; Li et al., 2008; Chouychai et al., 2009, 2012;

TABLE 8.2 List of the remediation methods for remediating the organic contaminants from soil and water.

Methods	Description	Contaminants									References
		BTEX	CS	CB	CC	Pes	PH	VOCs	HVOCs	NHVOCs	
Natural attenuation	Conversion, reduction, and immobilization of the contaminants to less toxic state via natural processes such as adsorption, dispersion, dilution, and microbial degradation and volatilization	√	√								Mulligan and Yong (2004)
Bioaugmentation	Supplementing of competent or genetically modified microbes that catabolize the specific contaminant in contaminated medium					√	√				Lebeau et al. (2008), Herrero and Stuckey (2015)
Biopiles	Treatment in which contaminated soil is piled and later stimulated by aerobic microbes via subsequent aeration, addition of nutrient and minerals to elevate the performance of aerobic microbes					√		√	√	√	Coulon et al. (2010)
Bioremediation	Breakdown of organic contaminant to CO_2 and H_2O or other less toxic compounds via the action of biological agent in normal conditions		√	√		√	√				Singh et al. (2017)
Biostimulation	Stimulation of degradation process facilitated by microbes by providing the ideal environmental conditions					√	√				Nikolopoulou and Kalogerakis (2009), Tyagi et al. (2011)
Bioventing	Supplementation of enhanced ventilation or oxygen to solid space in polluted soil through pump to elevate the multiplication of aerobic microbes				√			√			Hoeppel et al. (1991), Xiao and Zytner (2019), Ossai et al. (2020)

BTEX, Benzene, toluene, ethylbenzene, and xylene; *CB*, chlorinated biphenyls; *CC*, chlorinated compounds; *CS*, chlorinated solvents; *HVOCs*, halogenated volatile organic compounds; *NHVOCs*, nonhalogenated volatile organic compounds; *Pes*, pesticides; *PH*, petroleum hydrocarbons; *VOCs*, volatile organic compounds.

Germaine et al., 2009; Andria et al., 2009; Weyens et al., 2010a,b; Yousaf et al., 2010; Teng et al., 2011; Yu et al., 2011; Golubev et al., 2011; Hong et al., 2011; Afzal et al., 2012). Taking into account, the plant and their associated microbes for remediation of various petrochemical compounds have been summarized in Table 8.3.

TABLE 8.3 List of the selected plant–microbe system for remediation of petrochemical compounds.

Plant species	Microbes	Petrochemical compound	References
Festuca arundinacea	*Pseudomonas* strains, UW3 and UW4	Hydrocarbons	Gurska et al. (2009)
F. arundinacea	*Azospirillum brasilense* Cd, *Enterobacter cloacae* CAL2, *Pseudomonas putida*UW3, *P. putida*, *Flavobacterium* sp., *Pseudomonas aeruginosa*	Hydrocarbons	Huang et al. (2004)
Hordeum sativum	*P. putida* KT2440	Various PAHs	Child et al. (2007a,b)
H. sativum	*Mycobacterium* sp. strain KMS	Pyrene	Child et al. (2007b)
H. sativum	*Pseudomonas fluorescens*, *Pseudomonas aureofaciens*	Phenanthrene	Anokhina et al. (2004)
Hordeum vulgare	*Pseudomonas* strains, UW3 and UW4	Hydrocarbons	Gurska et al. (2009)
Lolium multiflorum	*Pantoea* sp. strain BTRH79	Hydrocarbons	Afzal et al. (2012)
L. multiflorum	*Acinetobacter* sp. strain	Hydrocarbons	Yu et al. (2011)
L. multiflorum	*Pantoea* sp. strain BTRH79, *Pseudomonas* sp. strainITRH76	Hydrocarbons	Yousaf et al. (2010)
L. multiflorum	*Rhodococcus* sp. strain ITRH43	Hydrocarbons	Andria et al. (2009)
L. multiflorum	*P. putida* PCL1444	Naphthalene and various PAHs	Kuiper et al. (2002)
Lolium perenne	*Pseudomonas* strains, UW3 and UW4	Hydrocarbons	Gurska et al. (2009)
Lotus corniculatus	*Pantoea* sp. strain BTRH79, *Pseudomonas* sp. strainITRH76	Diesel oil and hydrocarbons	Yousaf et al. (2010)
Lupinus arboreus	*Burkholderia cepacia* VM1468	Trichloroethylene	Weyens et al. (2010a)
Medicago sativa	*Rhizobium meliloti* (strain ACCC 17519)	Various PAHs	Teng et al. (2011)
M. sativa	*A. brasilense* SR80	Crude oil and hydrocarbons	Muratova et al. (2010)
Oryza sativa	*Acinetobacter* sp.	Various PAHs	Li et al. (2008)
Phragmites australis	*Pseudomonas asplenii* AC	Hydrocarbons	Reed and Glick (2005)
Pisum sativum	*P. putida*	Naphthalene	Germaine et al. (2009)
Populus sp.	*P. putida* W619-TCE	Trichloroethylene	Weyens et al. (2010a,b)

(Continued)

TABLE 8.3 (Continued)

Plant species	Microbes	Petrochemical compound	References
Secale cereale	*A. brasilense* SR80	Hydrocarbons	Muratova et al. (2010)
Sorghum bicolor	*Sinorhizobium meliloti*	Hydrocarbons	Golubev et al. (2011)
S. bicolor	*S. meliloti* P221	Phenanthrene	Muratova et al. (2009)
Trifolium repens	*Rhizobium leguminosarum*	Chrysene	Johnson et al. (2004)
Triticum aestivum	*Pseudomonas* sp. GF3	Phenanthrene	Sheng and Gong (2006)
T. aestivum	*Azospirillum lipoferum* sp.	Hydrocarbons	Muratova et al. (2005)
Zea mays	*Pseudomonas* sp. UG14Lr	Hydrocarbons	Chouychai et al. (2012)
Z. mays	Rhizobacterium *Gordonia* sp. S2RP-17	Hydrocarbons	Hong et al. (2011)
Z. mays	*P. putida* MUB1	Hydrocarbons	Chouychai et al. (2009)

PAHs, Polycyclic aromatic hydrocarbons.

8.6 Conclusion and future prospects

Overall it can be concluded that the major threat to the environment is the hydrocarbon contamination, which is the resultant of activities ongoing into petrochemical industries. Apart from these industries, accidental spilling of petroleum product (during transport, storage, as well as exploration) is another point of concern regarding contamination by petrochemicals. These chemical compounds have several carcinogenic effects, degrade the soil quality, and in turn disturb the biodiversity. Currently methods such as disposal or burial are being employed but these methods are very expensive when it comes to disposal of large amount of contaminants. In this context, plant-assisted bioremediation technology serves as the cheap and best method due to the biodegradation of these harmful petrochemicals into complete minerals (process called mineralization). We have also seen that these bioremediants degrade these contaminants either by bioaugmentation or by biostimulation. Although a huge amount of work has been done on these bioremediants but it has also been noticed that they all are genetically modified bacteria, but several indigenous species could also perform an adequate function in degrading these chemicals. Due to our limited knowledge regarding the remediation mechanism, we are not able to tap the complete potential of these wild species. So, this kind of work could be taken into consideration in future and some innovations could be done in bioremediation technology.

References

Abdel-Shafy, H.I., Mansour, M.S.M., 2016. A review on polycyclic aromatic hydrocarbons: Source, environmental impact, effect on human health and remediation. Egypt. J. Pet. 25, 107–123.

Adeniji, A.O., Okoh, O.O., Okoh, A.I., 2018. Analytical methods for polycyclic aromatic hydrocarbons and their global trend of distribution in water and sediment: a review. Recent Insights in Petroleum Science and Engineering. InTech.

Afzal, M., Yousaf, S., Reichenauer, T.G., Sessitsch, A., 2012. The inoculation method affects colonization and performance of bacterial inoculant strains in the phytoremediation of soil contaminated with diesel oil. Int. J. Phytoremediat. 14, 35–47. Available from: https://doi.org/10.1080/15226514.2011.552928.

Ahmad, M., Pataczek, L., Hilger, T.H., et al., 2018. Perspectives of microbial inoculation for sustainable development and environmental management. Front. Microbiol. 9. Available from: https://doi.org/10.3389/fmicb.2018.02992.

Akinyemi, O.D., Olowofela, J.A., Akinlade, O.O., Akande, O.O., 2006. Thermal conductivity of soils with heavy metals concentration from the Niger Delta region of Nigeria. J. Zhejiang Univ. Sci. B 7, 615–618. Available from: https://doi.org/10.1631/jzus.2006.b0615.

Altieri, M.A., Nicholls, C.I., 2003. Soil fertility management and insect pests: Harmonizing soil and plant health in agroecosystems. Soil. Tillage Res. 72, 203–211. Available from: https://doi.org/10.1016/S0167-1987(03)00089-8.

Andria, V., Reichenauer, T.G., Sessitsch, A., 2009. Expression of alkane monooxygenase (alkB) genes by plant-associated bacteria in the rhizosphere and endosphere of Italian ryegrass (*Lolium multiflorum* L.) grown in diesel contaminated soil. Environ. Pollut. 157, 3347–3350. Available from: https://doi.org/10.1016/j.envpol.2009.08.023.

Anokhina, T.O., Kochetkov, V.V., Zelenkova, N.F., et al., 2004. Biodegradation of phenanthrene by *Pseudomonas* bacteria bearing rhizospheric plasmids in model plant-microbial associations. Appl. Biochem. Microbiol. 40, 568–572. Available from: https://doi.org/10.1023/B:ABIM.0000046992.01220.35.

Aqeel, M., Jamil, M., Yusoff, I., 2014. Soil contamination, risk assessment and remediation. Environmental Risk Assessment of Soil Contamination. InTech.

Azubuike, C.C., Chikere, C.B., Okpokwasili, G.C., 2016. Bioremediation techniques–classification based on site of application: principles, advantages, limitations and prospects. World J. Microbiol. Biotechnol. 32.

Cheema, S., Lavania, M., Lal, B., 2015. Impact of petroleum hydrocarbon contamination on the indigenous soil microbial community. Ann. Microbiol. 65, 359–369. Available from: https://doi.org/10.1007/s13213-014-0868-1.

Chibwe, L., Geier, M.C., Nakamura, J., et al., 2015. Aerobic bioremediation of PAH contaminated soil results in increased genotoxicity and developmental toxicity. Environ. Sci. Technol. 49, 13889–13898. Available from: https://doi.org/10.1021/acs.est.5b00499.

Child, R., Miller, C.D., Liang, Y., et al., 2007a. Pyrene mineralization by *Mycobacterium* sp. strain KMS in a Barley rhizosphere. J. Environ. Qual. 36, 1260–1265. Available from: https://doi.org/10.2134/jeq2007.0008.

Child, R., Miller, C.D., Liang, Y., et al., 2007b. Polycyclic aromatic hydrocarbon-degrading Mycobacterium isolates: their association with plant roots. Appl. Microbiol. Biotechnol. 75, 655–663. Available from: https://doi.org/10.1007/s00253-007-0840-0.

Chouychai, W., Thongkukiatkul, A., Upatham, S., et al., 2009. Plant-enhanced phenanthrene and pyrene biodegradation in acidic soil. J. Environ. Biol. 30, 139–144.

Chouychai, W., Thongkukiatkul, A., Upatham, S., et al., 2012. Effect of corn plant on survival and phenanthrene degradation capacity of *Pseudomonas* Sp. UG14Lr in two soils. Int. J. Phytoremediat. 14, 585–595. Available from: https://doi.org/10.1080/15226514.2011.587478.

Connor, J.A., Paquette, S., McHugh, T., et al., 2017. Application of natural resource valuation concepts for development of sustainable remediation plans for groundwater. J. Environ. Manage. 204, 721–729. Available from: https://doi.org/10.1016/j.jenvman.2017.03.053.

Correa-García, S., Pande, P., Séguin, A., et al., 2018. Rhizoremediation of petroleum hydrocarbons: a model system for plant microbiome manipulation. Microb. Biotechnol. 11, 819–832. Available from: https://doi.org/10.1111/1751-7915.13303.

Coulon, F., Whelan, M.J., Paton, G.I., et al., 2010. Multimedia fate of petroleum hydrocarbons in the soil: oil matrix of constructed biopiles. Chemosphere 81, 1454–1462. Available from: https://doi.org/10.1016/j.chemosphere.2010.08.057.

Das, N., Chandran, P., 2011. Microbial degradation of petroleum hydrocarbon contaminants: an overview. Biotechnol. Res. Int. 2011, 1–13. Available from: https://doi.org/10.4061/2011/941810.

Frone, D.-F., Frone, S., 2015. The importance of water security for sustainable development in the Romanian agri-food sector. Agric. Agric Sci. Procedia 6, 674–681. Available from: https://doi.org/10.1016/j.aaspro.2015.08.120.

Germaine, K.J., Keogh, E., Ryan, D., Dowling, D.N., 2009. Bacterial endophyte-mediated naphthalene phytoprotection and phytoremediation. FEMS Microbiol. Lett. 296, 226–234. Available from: https://doi.org/10.1111/j.1574-6968.2009.01637.x.

Gkorezis, P., Daghio, M., Franzetti, A., et al., 2016. The interaction between plants and bacteria in the remediation of petroleum hydrocarbons: An environmental perspective. Front. Microbiol. 7.

Golubev, S.N., Muratova, A.Y., Wittenmayer, L., et al., 2011. Rhizosphere indole-3-acetic acid as a mediator in the *Sorghum bicolor*-phenanthrene-*Sinorhizobium meliloti* interactions. Plant. Physiol. Biochem. 49, 600–608. Available from: https://doi.org/10.1016/j.plaphy.2011.03.007.

Gurska, J., Wang, W., Gerhardt, K.E., et al., 2009. Three year field test of a plant growth promoting rhizobacteria enhanced phytoremediation system at a land farm for treatment of hydrocarbon waste. Environ. Sci. Technol. 43, 4472–4479. Available from: https://doi.org/10.1021/es801540h.

Herrero, M., Stuckey, D.C., 2015. Bioaugmentation and its application in wastewater treatment: a review. Chemosphere 140, 119–128.

Hoeppel, R.E., Hinchee, R.E., Arthur, M.F., 1991. Bioventing soils contaminated with petroleum hydrocarbons. J. Ind. Microbiol. 8, 141–146.

Hong, S.H., Ryu, H.W., Kim, J., Cho, K.S., 2011. Rhizoremediation of diesel-contaminated soil using the plant growth-promoting rhizobacterium *Gordonia* sp. S2RP-17. Biodegradation 22, 593–601. Available from: https://doi.org/10.1007/s10532-010-9432-2.

Hsu, C.S., Robinson, P.R., Hsu, C.S., Robinson, P.R., 2019. Chemical composition. Petroleum Science and Technology. Springer International Publishing, pp. 39–56.

Huang, X.D., El-Alawi, Y., Penrose, D.M., et al., 2004. A multi-process phytoremediation system for removal of polycyclic aromatic hydrocarbons from contaminated soils. Environ. Pollut. 130, 465–476. Available from: https://doi.org/10.1016/j.envpol.2003.09.031.

Jan, A., Azam, M., Siddiqui, K., et al., 2015. Heavy metals and human health: mechanistic insight into toxicity and counter defense system of antioxidants. Int. J. Mol. Sci. 16, 29592–29630. Available from: https://doi.org/10.3390/ijms161226183.

Jernelv, A., 2010. The threats from oil spills: now, then, and in the future. Ambio 39, 353–366. Available from: https://doi.org/10.1007/s13280-010-0085-5.

Johnson, D.L., Maguire, K.L., Anderson, D.R., McGrath, S.P., 2004. Enhanced dissipation of chrysene in planted soil: the impact of a rhizobial inoculum. Soil Biol. Biochem. 36, 33–38. Available from: https://doi.org/10.1016/j.soilbio.2003.07.004.

Kelishadi, R., 2012. Environmental pollution: health effects and operational implications for pollutants removal. J. Environ. Public Health 2012.

Kirkby, M.J., 2016. Water in the critical zone: soil, water and life from profile to planet. SOIL 2, 631–645. Available from: https://doi.org/10.5194/soil-2-631-2016.

Klamerus-Iwan, A., Błońska, E., Lasota, J., et al., 2015. Influence of oil contamination on physical and biological properties of forest soil after chainsaw use. Water Air Soil. Pollut. 226. Available from: https://doi.org/10.1007/s11270-015-2649-2.

Koshlaf, E., Ball, A.S., 2017. Soil bioremediation approaches for petroleum hydrocarbon polluted environments. AIMS Microbiol. 3, 25–49. Available from: https://doi.org/10.3934/microbiol.2017.1.25.

Kudoyarova, G., Arkhipova, T., Korshunova, T., et al., 2019. Phytohormone mediation of interactions between plants and non-symbiotic growth promoting bacteria under edaphic stresses. Front. Plant. Sci. 10.

Kuiper, I., Kravchenko, L.V., Bloemberg, G.V., Lugtenberg, B.J.J., 2002. *Pseudomonas putida* strain PCL1444, selected for efficient root colonization and naphthalene degradation, effectively utilizes root exudate components. Mol. Plant-Microbe Interact. 15, 734–741. Available from: https://doi.org/10.1094/MPMI.2002.15.7.734.

Lebeau, T., Braud, A., Jézéquel, K., 2008. Performance of bioaugmentation-assisted phytoextraction applied to metal contaminated soils: a review. Environ. Pollut. 153, 497–522.

Li, J.H., Gao, Y., Wu, S.C., et al., 2008. Physiological and biochemical responses of rice (*Oryza sativa* L.) to phenanthrene and pyrene. Int. J. Phytoremediat. 10, 106–118. Available from: https://doi.org/10.1080/15226510801913587.

Mamy, L., Patureau, D., Barriuso, E., et al., 2015. Prediction of the fate of organic compounds in the environment from their molecular properties: a review. Crit. Rev. Environ. Sci. Technol. 45, 1277–1377. Available from: https://doi.org/10.1080/10643389.2014.955627.

Mobin, M., Abbas, Z.K., Alamri, S.A., 2018. Fertilizers and their contaminants in soils, surface and groundwater. Encycl. Anthr. 5, 225–240. Available from: https://doi.org/10.1016/B978-0-12-809665-9.09888-8.

Mulligan, C.N., Yong, R.N., 2004. Natural attenuation of contaminated soils. Environ. Int. 30, 587–601.

Muratova, A.Y., Bondarenkova, A.D., Panchenko, L.V., Turkovskaya, O.V., 2010. Use of integrated phytoremediation for cleaning-up of oil-sludge-contaminated soil. Appl. Biochem. Microbiol. 46, 789–794. Available from: https://doi.org/10.1134/S0003683810080090.

Muratova, A., Golubev, S., Wittenmayer, L., et al., 2009. Effect of the polycyclic aromatic hydrocarbon phenanthrene on root exudation of *Sorghum bicolor* (L.) Moench. Environ. Exp. Bot. 66, 514–521. Available from: https://doi.org/10.1016/j.envexpbot.2009.03.001.

Muratova, A.Y., Turkovskaya, O.V., Antonyuk, L.P., et al., 2005. Oil-oxidizing potential of associative rhizobacteria of the genus *Azospirillum*. Microbiology 74, 210–215. Available from: https://doi.org/10.1007/s11021-005-0053-4.

Nikolopoulou, M., Kalogerakis, N., 2009. Biostimulation strategies for fresh and chronically polluted marine environments with petroleum hydrocarbons. J. Chem. Technol. Biotechnol. 84, 802–807. Available from: https://doi.org/10.1002/jctb.2182.

Ojuederie, O., Babalola, O., 2017. Microbial and plant-assisted bioremediation of heavy metal polluted environments: a review. Int. J. Environ. Res. Public. Health 14, 1504. Available from: https://doi.org/10.3390/ijerph14121504.

Olanrewaju, O.S., Glick, B.R., Babalola, O.O., 2017. Mechanisms of action of plant growth promoting bacteria. World J. Microbiol. Biotechnol. 33.

Ortíz-Castro, R., Contreras-Cornejo, H.A., Macías-Rodríguez, L., López-Bucio, J., 2009. The role of microbial signals in plant growth and development. Plant. Signal. Behav. 4, 701–712. Available from: https://doi.org/10.4161/psb.4.8.9047.

Ossai, I.C., Ahmed, A., Hassan, A., Hamid, F.S., 2020. Remediation of soil and water contaminated with petroleum hydrocarbon: a review. Environ. Technol. Innov. 17.

Panno, M.T., Morelli, I.S., Engelen, B., Berthe-Corti, L., 2005. Effect of petrochemical sludge concentrations on microbial communities during soil bioremediation. FEMS Microbiol. Ecol. 53, 305–316. Available from: https://doi.org/10.1016/j.femsec.2005.01.014.

Pichtel, J., 2016. Oil and gas production wastewater: soil contamination and pollution prevention. Appl. Environ. Soil. Sci. 2016.

Reed, M.L.E., Glick, B.R., 2005. Growth of canola (*Brassica napus*) in the presence of plant growth-promoting bacteria and either copper or polycyclic aromatic hydrocarbons. Can. J. Microbiol. 51, 1061–1069. Available from: https://doi.org/10.1139/w05-094.

Reichenauer, T.G., Germida, J.J., 2008. Phytoremediation of organic contaminants in soil and groundwater. ChemSusChem 1, 708–717. Available from: https://doi.org/10.1002/cssc.200800125.

Rockström, J., Williams, J., Daily, G., et al., 2017. Sustainable intensification of agriculture for human prosperity and global sustainability. Ambio 46, 4–17. Available from: https://doi.org/10.1007/s13280-016-0793-6.

Sarvade, S., Upadhyay, V.B., Kumar, M., et al., 2019. Soil and water conservation techniques for sustainable agriculture. Sustainable Agriculture, Forest and Environmental Management. Springer, Singapore, pp. 133–188.

Sharma, S., Tiwari, S., Hasan, A., et al., 2018. Recent advances in conventional and contemporary methods for remediation of heavy metal-contaminated soils. 3 Biotech. 8.

Sheng, X.F., Gong, J.X., 2006. Increased degradation of phenanthrene in soil by *Pseudomonas* sp. GF3 in the presence of wheat. Soil Biol. Biochem. 38, 2587–2592. Available from: https://doi.org/10.1016/j.soilbio.2006.03.014.

Singh, P., Jain, R., Srivastava, N., et al., 2017. Current and emerging trends in bioremediation of petrochemical waste: a review. Crit. Rev. Environ. Sci. Technol. 47, 155–201.

Tahri, N., Bahafid, W., Sayel, H., El Ghachtouli, N., 2013. Biodegradation: involved microorganisms and genetically engineered microorganisms. Biodegradation-Life of Science. InTech.

Tchounwou, P.B., Yedjou, C.G., Patlolla, A.K., Sutton, D.J., 2012. Heavy metal toxicity and the environment. EXS 101, 133–164.

Teng, Y., Shen, Y., Luo, Y., et al., 2011. Influence of *Rhizobium meliloti* on phytoremediation of polycyclic aromatic hydrocarbons by alfalfa in an aged contaminated soil. J. Hazard. Mater. 186, 1271–1276. Available from: https://doi.org/10.1016/j.jhazmat.2010.11.126.

Thijs, S., Weyens, N., Gkorezis, P., Vangronsveld, J., 2016. Plant-endophyte partnerships to assist petroleum hydrocarbon remediation. Consequences of Microbial Interactions with Hydrocarbons, Oils, and Lipids: Biodegradation and Bioremediation. Springer International Publishing, pp. 1–34.

Tsukanova, K.A., Chebotar, V., Meyer, J.J.M., Bibikova, T.N., 2017. Effect of plant growth-promoting Rhizobacteria on plant hormone homeostasis. South Afr. J. Bot. 113, 91–102.

Tyagi, M., da Fonseca, M.M.R., de Carvalho, C.C.C.R., 2011. Bioaugmentation and biostimulation strategies to improve the effectiveness of bioremediation processes. Biodegradation 22, 231–241.

Varjani, S.J., 2017. Microbial degradation of petroleum hydrocarbons. Bioresour. Technol. 223, 277–286.

Villegas, L.G.C., Mashhadi, N., Chen, M., et al., 2016. A short review of techniques for phenol removal from wastewater. Curr. Pollut. Rep. 2, 157–167.

Weyens, N., Croes, S., Dupae, J., et al., 2010a. Endophytic bacteria improve phytoremediation of Ni and TCE co-contamination. Environ. Pollut. 158, 2422–2427. Available from: https://doi.org/10.1016/j.envpol.2010.04.004.

Weyens, N., Monchy, S., Vangronsveld, J., et al., 2010b. Plant-microbe partnerships. Handbook of Hydrocarbon and Lipid Microbiology. Springer Berlin Heidelberg, pp. 2545–2574.

Xiao, M., Zytner, R.G., 2019. The effect of age on petroleum hydrocarbon contaminants in soil for bioventing remediation. Bioremediat J. 23, 311–325. Available from: https://doi.org/10.1080/10889868.2019.1671306.

Xu, X., Liu, W., Tian, S., et al., 2018. Petroleum hydrocarbon-degrading bacteria for the remediation of oil pollution under aerobic conditions: a perspective analysis. Front. Microbiol. 9. Available from: https://doi.org/10.3389/fmicb.2018.02885.

Yang, S.Z., Jin, H.J., Wei, Z., et al., 2009. Bioremediation of oil spills in cold environments: a review. Pedosphere 19, 371–381. Available from: https://doi.org/10.1016/S1002-0160(09)60128-4.

Yan, J., Wang, L., Fu, P.P., Yu, H., 2004. Photomutagenicity of 16 polycyclic aromatic hydrocarbons from the US EPA priority pollutant list. Mutat. Res. – Genet. Toxicol. Environ. Mutagen. 557, 99–108. Available from: https://doi.org/10.1016/j.mrgentox.2003.10.004.

Yousaf, S., Ripka, K., Reichenauer, T.G., et al., 2010. Hydrocarbon degradation and plant colonization by selected bacterial strains isolated from Italian ryegrass and birdsfoot trefoil. J. Appl. Microbiol. 109, 1389–1401. Available from: https://doi.org/10.1111/j.1365-2672.2010.04768.x.

Yu, H., 2002. Environmental carcinogenic polycyclic aromatic hydrocarbons: photochemistry and phototoxicity. J. Environ. Sci. Heal. C: Environ. Carcinog. Ecotoxicol. Rev. 20, 149–183.

Yu, X.Z., Wu, S.C., Wu, F.Y., Wong, M.H., 2011. Enhanced dissipation of PAHs from soil using mycorrhizal ryegrass and PAH-degrading bacteria. J. Hazard. Mater. 186, 1206–1217. Available from: https://doi.org/10.1016/j.jhazmat.2010.11.116.

Zhang, B., Zhang, L., Zhang, X., 2019. Bioremediation of petroleum hydrocarbon-contaminated soil by petroleum-degrading bacteria immobilized on biochar. RSC Adv. 9, 35304–35311. Available from: https://doi.org/10.1039/c9ra06726d.

Zhao, X.Q., Shen, R.F., 2018. Aluminum–nitrogen interactions in the soil–plant system. Front. Plant. Sci. 9.

Chapter 9

Biosurfactant producing microbes for clean-up of soil contaminants

Sandeep Kumar Singh[1], Manoj Kumar Singh[2], Hariom Verma[1], Prem Pratap Singh[1], Anand Vikram Singh[1], Kumari Rashmi[3] and Ajay Kumar[4]

[1]*Centre for Advanced Study in Botany, Banaras Hindu University, Varanasi, India,* [2]*Chemical Division, Intellectual Property Office, Delhi, India,* [3]*L.N.Mishra Institute, Patna, India,* [4]*Agriculture Research Organization, Ministry of Agriculture and Rural Development, Volcani Center, Rishon LeZion, Israel*

9.1 Introduction

Currently a range of organic and inorganic contaminants deposited in the environment due to rapid industrialization or anthropogenic behavior. These environmental contaminants severally affect the texture and productivity of plant and soil, as well health and surrounding of human beings. Traditionally, to remediate these environmental contaminants various physiochemical methods such as landfilling, incineration at high temperature are the common ones but their expenses and toxic residual products still serious concern for environment management. In this regards, bioremediation emerges as alternative and sustainable approach for remediation and contaminants management (Gupta et al., 2020). In bioremediation, various microbes or microbial products have been used for the contaminants management (Singh et al., 2019, 2020). Biosurfactant is also a microbial products currently utilized in remediation of toxic, hydrophobic contaminants in economic and effective ways. The term biosurfactant is utilized for the microbial-derived surfactant molecules, surfactants are an amphipathic compounds, made up of hydrophilic and hydrophobic constituents, the hydrophilic components are generally ionic (cationic, anionic, and amphoteric) in nature and mainly constituted of amino acids, ammonium salts, ester sulfonates, lipids, whereas hydrophobic groups are made up of long hydrocarbon chains. The amphipathic nature of this biological molecules, lower the interfacial surface tension and separate the compounds in two phases. As in the case of oil−water interface, surfactants lower the interfacial tension of oil−water interface or surface tension of water, which helps in separation and mass transport of hydrocarbons from oil phase into the aqueous phase (Silva et al., 2014; Cerqueira et al., 2011). The average molecular mass of the biosurfactants ranges from 500 to 1500 Da (Bognolo, 1999) and their effectivity has been measured by critical micelle concentration (CMC) values and it normally ranges from 1 to 2000 mg. However, interfacial (oil/water) and surface tensions are in the range of 1−30 mN/m, respectively (Bognolo, 1999; Santos et al., 2016).

Petroleum-derived surfactant are commonly utilized during remediation of range of contaminants which are immiscible in nature and having low solubility, but long-term effect on the environment and low degradation potential, higher toxicity, as well as difficult management strategies made petro-derived surfactant not so popular. However, the microbial derived biosurfactant are easy to manage in terms less toxicity, high degradation potential, and eco-friendly mitigation process (Vijayakumar and Saravanan, 2015) in addition structural diversity and better adaptability under various stress conditions, including temperature, pH, and salinity, these biosurfactants drawn attention to various researchers and industries for the exploration of various aspects of microbial surfactants.

9.2 Microbes and biosurfactants

Currently microbial (bacteria, fungi, and cyanobacteria) biosurfactant emerges as a better alternative against traditional physiochemical method in terms of management of range of environmental contaminants (Fig. 9.1). Researchers continuously exploring novel microbial strains that have capability to produce biosurfactant molecules with specific properties

Microbe Mediated Remediation of Environmental Contaminants. DOI: https://doi.org/10.1016/B978-0-12-821199-1.00009-2

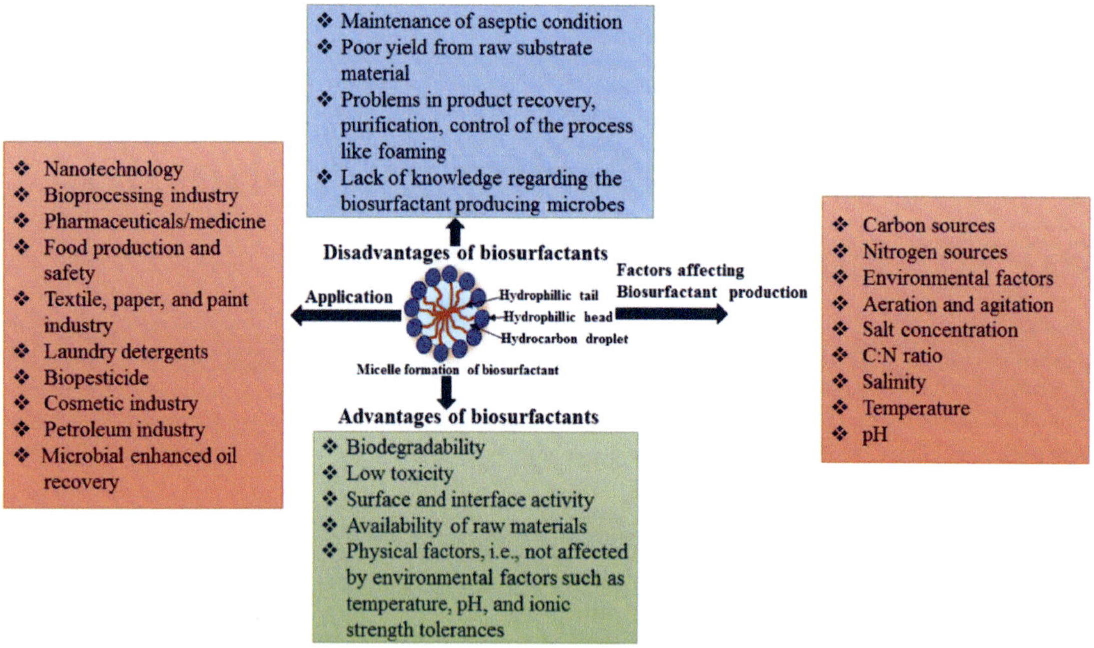

FIGURE 9.1 An overview of biosurfactant in environmental contamination management.

such as low CMC, high emulsifying power, lesser toxicity, and environment friendly nature. The CMC of any surfactants defines as a concentration, above which formation of micelles started in the solution (Silva et al., 2014; Deleu and Paquot, 2004; Marchant et al., 2014; Chakraborty and Das, 2014).The microbial derived biosurfactant are diverse in nature, having structural complexity with specific surface activities and it varies from species to species (Campos et al., 2013). Now there is a range of microbial strains including, bacteria, fungi, and cyanobacteria are reported as biosurfactant producer. Silva et al. (2014) reported genera of *Bacillus* and *Pseudomonas* are most studied and diverse biosurfactant synthesizer bacteria genera; however, these biosurfactant are not frequently used in food industries due to their possible pathogenic natures (Shepherd et al., 1995). In case of yeast the *Candida* genera and the species especially *C. bombicola* and *C. lipolytica* are most commonly known biosurfactants (Campos et al., 2013).

9.3 Factors affecting biosurfactant production

Biosurfactants can be survived or worked at diverse range of temperature and pH and also up to 10% saline conditions. Meanwhile the less toxicity and high degrading potential of biosurfactant are main reason to use as potential agent for mitigation of environmental contamination. Biosurfactant molecules constituted of different functional group and each constitutional functional group have specific properties and help in detoxifying the nature of contaminants or also in the emulsification or deemulsifications of the industrial pollutants or emulsions (Velikonja and Kosaric, 1993; Cirigliano and Carman, 1985). Emulsion is a form of heterogeneous system in which one liquid types having immiscible in nature dispersed in another liquid. The stability of emulsion is very low, but the addition of biosurfactant enhances their stability even up to month or year (Velikonja and Kosaric, 1993).

The production of biosurfactant depends upon numerous factors in which the nature of the substrate, nutrient source, growth conditions such as temperature, pH, speed, and duration of rotation, incubation temperature are the some prime factors that influenced the amount and nature of biosurfactant production.

During production source of carbon is one of the most important factors that influence the amount of biosurfactant production. In a study Kim et al. (1997) reported, combined application of glucose and vegetable oils had showed better yield enhancement in comparison to singly used vegetable oil and glucose as a source of carbon. However, application of soybean oil with the glucose as a carbon source showed better yield production than the vegetable oil and glucose. Similarly, Sarubbo et al. (2007) reported in case of biosurfactant production from species mixed application of canola oil and glucose at 10% of yielded 8 g/L of sophorolipids from *C. lipolytica*.

Nitrogen is also an important supplement and utilized through a range of inorganic or organic sources during the production of biosurfactants. Mulligan and Gibbs (2004) reported nitrates, ammonium, and amino acids as a nitrogen sources during the production of rhamnolipids from *Pseudomonas aeruginosa* strain and lipids act as rate determining state instead of glucose molecules. However, during sophorose synthesis from strains of *Torulopsis bombicola* and *C. bombicola*, lipids/urea and yeast extract full fill the nitrogen requirement during (Deshpande and Daniels, 1995). Similarly, during production of mannosylerythritol lipid using *Candida* sp. ammonium nitrate and yeast extract used as nitrogen source (Sarubbo et al., 2007; Rufino et al., 2007; Kitamoto et al., 2001).

Generally optimum temperature is the ultimate requirement for biosurfactant production and it is strains specific, varies, with the strains. Generally, optimum temperature for biosurfactant production is room temperature or in between 25°C and 30°C; however, in case of fungal strain optimum temperature is 30°C (Cooper and Paddock, 1984; Deshpande and Daniels, 1995; Kim et al., 1999; Rufino et al., 2008). The optimum pH is also an important factor during biosurfactant production. Various authors reported the importance of pH factor and it also varies with the nature of strains, *Candida glabrata*, required pH of 5.7 for the optimum growth; however, *Candida* sp. required pH of 7.8, whereas pH 6.0 was required for the growth of *Candida batistae* species (Kim et al., 1999; Asfora Sarubbo et al., 2006; Konishi et al., 2008). Similarly the incubation time during the biosurfactant production is also an important factor and also depends upon the microbial strains as, for example, *C. bombicola*, 7, 8, 10 days of incubation for optimum yield production (Felse et al., 2007; Cavalero and Cooper, 2003).

9.4 Biosurfactants and hydrocarbons degradation/remediation

The deposition of hydrocarbon or hydrocarbon residual products in the soil or aquatic ecosystem causes severe effect on health, productivity, as well as life quality of living organism. The remediation of hydrocarbon contaminants from the soil or aquatic ecosystem in sustainable ways is one the most challenging task for the environmentalist or researcher due to limited oxidative–reductive chemicals and also lower bioavailability potential for microbial use. In this regard application of microbial based biosurfactant emerges as a new approach for the management of various hydrocarbon-based environmental pollutants. The action mechanism of biosurfactant is to immobilize or enhance substrate bioavailability for the microbial strains via enhancing interaction between contaminant molecules and surfaces of bacterial cell (Mulligan and Gibbs, 2004). The biosurfactant molecules reduce interfacial tensions and enhanced surface area of insoluble compounds that results in the enhancement of mobility, solubility, emulsification, and bioavailability of hydrocarbons (Nguyen et al., 2008; Deziel et al., 1996; Bai et al., 1997; Rahman et al., 2003; Urum and Pekdemir, 2004; Nievas et al., 2008).

The mobilization of pollutants depends upon CMC concentration of biosurfactant, lower concentration of CMC, and reduces interfacial force between biosurfactants with soil/oil system; however, at higher CMC, biosurfactant enhanced the formation of micelles. Inside the micelles, the hydrophobic groups or ends associated with each other, whereas the hydrophilic groups comes out to the surface, which helps in solubilization of hydrophobic contaminants (Pacwa-Płociniczak et al., 2011; Urum and Pekdemir, 2004).

Sometimes molecular weight of biosurfactant also a factor for remediation of organic contaminants. Biosurfactant of higher molecular weight proven as an efficient emulsifying agent and stimulate bioremediation of oil substances from environment contaminants. In a study, Franzetti et al. (2010) describe details mechanism of interactions and modulation between microorganism, biosurfactant, and hydrocarbons contaminants and reported higher hydrophobicity of the cell interacts the beneficial microbes and contact with oil droplets as well as hydrocarbons; however, lower hydrophobicity of cell permits adhesion to micelles or emulsified the oils. During hydrocarbon remediation, three main mechanism has been followed first the access to water solubilized contaminants, the direct contact of microbial cell surface with the oil droplets. Biosurfactant change the accession modes of hydrocarbons during growth stages of microorganisms (Franzetti et al., 2009, 2010).

In a study, Cameotra and Singh (2009) reported details mechanism of alkane mitigation by the use of rhamnolipids synthesized by *P. aeruginosa* and reported that biosurfactant increases the bioavailability of hydrocarbon to the bacterial cell by dispersing hexadecane into microdroplets and enhances biodegradation of hydrocarbons.

There are various report present that showed effective potential of microbial synthesized biosurfactant in the mitigation of hydrocarbon-based organic contaminants (Deziel et al., 1996; Rahman et al., 2003; Inakollu et al., 2004). Reddy et al. (2010) reported the effectivity of *Brevibacterium* sp. PDM-3 synthesized biosurfactant against the remediation of polyromantic hydrocarbon contaminants, during study found effectivity of surfactant up to 93.92% against phenanthrene along with some other contaminant such as anthracene and fluorine. Similarly, Obayori et al. (2009) reported biodegradation potential of biosurfactant produced by *Pseudomonas* sp. LP1 against crude oil and diesel contamination and found excellent remediation potential up to 92.34% of crude oil and up to 95.29% of diesel oil.

During bioremediation of hydrocarbon contaminants, biosurfactant microbes can directly applied on the contaminant sites. In this microbiological phenomenon, there is no need of detection of active site of the chemicals compounds and stability or effectivity of biosurfactant is found very high in the culture medium (Pacwa-Płociniczak et al., 2011). In a study, Das and Mukherjee (2007) reported the biosurfactant potential of *Bacillus subtilis* and *P. aeruginosa* strains and their effectiveness against the remediation of crude petroleum–oil hydrocarbons.

Kumari et al. (2012) isolated strain *Pseudomonas* sp. BP10 and *Rhodococcus* sp. NJ2 from the crude oil contaminated site and both the strain had potential capability to degrade different fractions of crude oil. The strain BP10 degraded more than 60% of petroleum hydrocarbons; however, strain *Rhodococcus* sp. NJ2 degraded 49.5% after incubating in minimal salt media with 2% of crude oil for 30 days under optimum conditions. Both the strains have different degradation potential and found differences in biosurfactant production, catabolizing enzymes, cell hydrophobicity. On comparing enzyme production, both the strains produce catechol 2,3-dioxygenase and the rate of production in *Pseudomonas* sp. BP10 is more in comparison to *Rhodococcus* sp. NJ2, and these enzymes have significant role in in crude oil degradation. Parthipan et al. (2017) reported about the biosurfactant synthesizing strain *B. subtilis* A1 and their activity against remediation of hydrocarbon present in crude oil contamination. Furthermore, author reported *B. subtilis* strains completely degraded hydrocarbon contaminants that are made up of, that is, C10–C14, while 97% of degradation observed in the contaminants of C15–C19 compounds and overall 87% of degradation were found for all the oil contaminants after 7 days of incubation. In another study Parthipan et al. (2017) reported bioremediation potential of enzymes secreted by strains *Pseudomonas stutzeri* NA3 and *Acinetobacter baumannii* MN3 against crude oil contamination and during study found 86% remediation efficiency of mixed consortia followed by *P. stutzeri* NA3 (84%) after incubation for short period of time.

9.5 Future directions and concluding remarks

In the recent last few years biosurfactant has been emerged as one of the potential resource for the bioremediation of various toxic environmental contaminants such as petrochemicals, polysaturated hydrocarbons, heavy metals, or the industrial effluents. From last few decades' rapid industrialization, adversely affect the environments with their residual toxic products, which ultimately affect the surrounding environments and human health. The traditional approach of environment contamination management is not so efficient because of higher cost and toxic residual products. Biosurfactant-based bioremediation is an alternative and effective approach for the contaminants management especially hydrophobic hydrocarbon-based pollutants, petroleum, oil spills, etc. But the limited productions of biosurfactant is still a matter of concern, and this may arises mainly due to the unexplored and less known microorganism, resources, and the traditional fermenters. In this context, it is necessary to evaluate and understand the limiting factors of culturing conditions to overcome and enhanced the rate of production.

References

Asfora Sarubbo, L., Moura de Luna, J., de Campos-Takaki, G.M., 2006. Production and stability studies of the bioemulsifier obtained from a new strain of *Candida glabrata* UCP 1002. Electron. J. Biotechnol. 9.

Bai, G., Brusseau, M.L., Miller, R.M., 1997. Biosurfactant-enhanced removal of residual hydrocarbon from soil. J. Contam. Hydrol. 25, 157–170.

Bognolo, G., 1999. Biosurfactants as emulsifying agents for hydrocarbons. Colloids Surf. A: Physicochem. Eng. Aspects 152.

Cameotra, S.S., Singh, P., 2009. Synthesis of rhamnolipid biosurfactant and mode of hexadecane uptake by *Pseudomonas* species. Microb. Cell Fact. 8, 16.

Campos, J.M., Montenegro Stamford, T.L., Sarubbo, L.A., de Luna, J.M., Rufino, R.D., Banat, I.M., 2013. Microbial biosurfactants as additives for food industries. Biotechnol. Prog. 29, 1097–1108.

Cavalero, D.A., Cooper, D.G., 2003. The effect of medium composition on the structure and physical state of sophorolipids produced by *Candida bombicola* ATCC 22214. J. Biotechnol. 103, 31–41.

Cerqueira, V.S., Hollenbach, E.B., Maboni, F., Vainstein, M.H., Camargo, F.A., Maria do Carmo, R.P., et al., 2011. Biodegradation potential of oily sludge by pure and mixed bacterial cultures. Bioresour. Technol. 102, 11003–11010.

Chakraborty, J., Das, S., 2014. Biosurfactant-based bioremediation of toxic metals. Microbial Biodegradation and Bioremediation. Elsevier, pp. 167–201.

Cirigliano, M.C., Carman, G.M., 1985. Purification and characterization of liposan, a bioemulsifier from *Candida lipolytica*. Appl. Environ. Microbiol. 50, 846–850.

Cooper, D.G., Paddock, D.A., 1984. Production of a biosurfactant from *Torulopsis bombicola*. Appl. Environ. Microbiol. 47, 173–176.

Das, K., Mukherjee, A.K., 2007. Crude petroleum-oil biodegradation efficiency of *Bacillus subtilis* and *Pseudomonas aeruginosa* strains isolated from a petroleum-oil contaminated soil from North-East India. Bioresour. Technol. 98, 1339–1345.

Deleu, M., Paquot, M., 2004. From renewable vegetables resources to microorganisms: new trends in surfactants. C. R. Chim. 7, 641–646.

Deshpande, M., Daniels, L., 1995. Evaluation of sophorolipid biosurfactant production by *Candida bombicola* using animal fat. Bioresour. Technol. 54, 143–150.

Deziel, E., Paquette, G., Villemur, R., Lepine, F., Bisaillon, J., 1996. Biosurfactant production by a soil *Pseudomonas* strain growing on polycyclic aromatic hydrocarbons. Appl. Environ. Microbiol. 62, 1908–1912.

Felse, P.A., Shah, V., Chan, J., Rao, K.J., Gross, R.A., 2007. Sophorolipid biosynthesis by *Candida bombicola* from industrial fatty acid residues. Enzyme Microb. Technol. 40, 316–323.

Franzetti, A., Caredda, P., Ruggeri, C., La Colla, P., Tamburini, E., Papacchini, M., et al., 2009. Potential applications of surface active compounds by *Gordonia* sp. strain BS29 in soil remediation technologies. Chemosphere. 75, 801–807.

Franzetti, A., Gandolfi, I., Bestetti, G., Smyth, T.J., Banat, I.M., 2010. Production and applications of trehalose lipid biosurfactants. Eur. J. Lipid Sci. Technol. 112, 617–627.

Gupta, A., Singh, S.K., Singh, V.K., Singh, M.K., Modi, A., Zhimo, V.Y., et al., 2020. Endophytic microbe approaches in bioremediation of organic pollutants. Microbial Endophytes. Woodhead Publishing, pp. 157–174.

Inakollu, S., Hung, H.C., Shreve, G.S., 2004. Biosurfactant enhancement of microbial degradation of various structural classes of hydrocarbon in mixed waste systems. Environ. Eng. Sci. 21, 463–469.

Kim, S.Y., Oh, D.K., Lee, K.H., Kim, J.H., 1997. Effect of soybean oil and glucose on sophorose lipid fermentation by *Torulopsis bombicola* in continuous culture. Appl. Microbiol. Biotechnol. 48, 23–26.

Kim, H.S., Yoon, B.D., Choung, D.H., Oh, H.M., Katsuragi, T., Tani, Y., 1999. Characterization of a biosurfactant, mannosylerythritol lipid produced from *Candida* sp. SY16. Appl. Microbiol. Biotechnol. 52, 713–721.

Kitamoto, D., Ikegami, T., Suzuki, G.T., Sasaki, A., Takeyama, Y.I., Idemoto, Y., et al., 2001. Microbial conversion of *n*-alkanes into glycolipid biosurfactants, mannosylerythritol lipids, by Pseudozyma (*Candida antarctica*). Biotechnol. Lett. 23, 1709–1714.

Konishi, M., Fukuoka, T., Morita, T., Imura, T., Kitamoto, D., 2008. Production of new types of sophorolipids by *Candida batistae*. J. Oleo Sci. 57, 359–369.

Kumari, B., Singh, S.N., Singh, D.P., 2012. Characterization of two biosurfactant producing strains in crude oil degradation. Process Biochem. 47, 2463–2471.

Marchant, R., Funston, S., Uzoigwe, C., Rahman, P.K.S.M., Banat, I.M., 2014. Production of biosurfactants from nonpathogenic bacteria. In: Biosurfactants: Production and Utilization-Processes, Technologies, and Economics, CRC Press, pp. 73-82.

Mulligan, C.N., Gibbs, B.F., 2004. Types, production and applications of biosurfactants. Proc. Indian Natl. Sci. Acad. B 70, 31–56.

Nguyen, T.T., Youssef, N.H., McInerney, M.J., Sabatini, D.A., 2008. Rhamnolipid biosurfactant mixtures for environmental remediation. Water Res. 42, 1735–1743.

Nievas, M.L., Commendatore, M.G., Esteves, J.L., Bucalá, V., 2008. Biodegradation pattern of hydrocarbons from a fuel oil-type complex residue by an emulsifier-producing microbial consortium. J. Hazard. Mater. 154, 96–104.

Obayori, O.S., Ilori, M.O., Adebusoye, S.A., Oyetibo, G.O., Omotayo, A.E., Amund, O.O., 2009. Degradation of hydrocarbons and biosurfactant production by *Pseudomonas* sp. strain LP1. World J. Microbiol. Biotechnol. 25, 1615–1623.

Pacwa-Płociniczak, M., Płaza, G.A., Piotrowska-Seget, Z., Cameotra, S.S., 2011. Environmental applications of biosurfactants: recent advances. Int. J. Mol. Sci. 12, 633–654.

Parthipan, P., Preetham, E., Machuca, L.L., Rahman, P.K., Murugan, K., Rajasekar, A., 2017. Biosurfactant and degradative enzymes mediated crude oil degradation by bacterium *Bacillus subtilis* A1. Front. Microbiol. 8, 193.

Rahman, K.S.M., Rahman, T.J., Lakshmanaperumalsamy, P., Marchant, R., Banat, I.M., 2003. The potential of bacterial isolates for emulsification with a range of hydrocarbons. Acta Biotechnol. 23, 335–345.

Reddy, M.S., Naresh, B., Leela, T., Prashanthi, M., Madhusudhan, N.C., Dhanasri, G., et al., 2010. Biodegradation of phenanthrene with biosurfactant production by a new strain of *Brevibacillus* sp. Bioresour. Technol. 101, 7980–7983.

Rufino, R.D., Sarubbo, L.A., Campos-Takaki, G.M., 2007. Enhancement of stability of biosurfactant produced by *Candida lipolytica* using industrial residue as substrate. World J. Microbiol. Biotechnol. 23, 729–734.

Rufino, R.D., Sarubbo, L.A., Neto, B.B., Campos-Takaki, G.M., 2008. Experimental design for the production of tensio-active agent by *Candida lipolytica*. J. Ind. Microbiol. Biotechnol. 35, 907–914.

Santos, D.K.F., Rufino, R.D., Luna, J.M., Santos, V.A., Sarubbo, L.A., 2016. Biosurfactants: multifunctional biomolecules of the 21st century. Int. J. Mol. Sci. 17, 401.

Sarubbo, L.A., Farias, C.B., Campos-Takaki, G.M., 2007. Co-utilization of canola oil and glucose on the production of a surfactant by *Candida lipolytica*. Curr. Microbiol. 54, 68–73.

Shepherd, R., Rockey, J., Sutherland, I.W., Roller, S., 1995. Novel bioemulsifiers from microorganisms for use in foods. J. Biotechnol. 40, 207–217.

Silva, R.D.C.F., Almeida, D.G., Rufino, R.D., Luna, J.M., Santos, V.A., Sarubbo, L.A., 2014. Applications of biosurfactants in the petroleum industry and the remediation of oil spills. Int. J. Mol. Sci. 15, 12523–12542.

Singh, V.K., Singh, M., Singh, S.K., Kumar, C., Kumar, A., 2019. Sustainable agricultural practices using beneficial fungi under changing climate scenario. Climate Change and Agricultural Ecosystems. Woodhead Publishing, pp. 25–42.

Singh, D., Singh, S.K., Singh, V.K., Gupta, A., Aamir, M., Kumar, A., 2020. Plant growth-promoting bacteria and their role in environmental management. Abatement of Environmental Pollutants. Elsevier, pp. 161–175.

Urum, K., Pekdemir, T., 2004. Evaluation of biosurfactants for crude oil contaminated soil washing. Chemosphere. 57, 1139–1150.

Velikonja, J., Kosaric, N., 1993. Biosurfactants in Food Applications. Surfactant Science Series. CRC Press, p. 419.

Vijayakumar, S., Saravanan, V., 2015. Biosurfactants-types, sources and applications. Res. J. Microbiol. 10, 181–192.

Chapter 10

Rhizoremediation for efficient management of inorganic and organic contaminants in soil

Pratibha Vyas[1,*], Simranjeet Singh[2,3,4,*], Vaishali Dhaka[2], Daljeet Singh Dhanjal[2] and Joginder Singh[5]

[1]Department of Microbiology, College of Basic Sciences and Humanities, Punjab Agricultural University, Ludhiana, India, [2]Department of Biotechnology, Lovely Professional University, Jalandhar, India, [3]Punjab Biotechnology Incubators, Mohali, India, [4]Regional Advanced Water Testing Laboratory, Mohali, India, [5]Department of Microbiology, School of Bioengineering and Biosciences, Lovely Professional University, Jalandhar, India

10.1 Introduction

The major emitters or repositories of chemicals are the urban areas due to the increasing industrialized global economy. Pesticides, metals, petroleum hydrocarbons, halogenated hydrocarbons, and polycyclic aromatic hydrocarbons (PAHs) are the common contaminants present in the soil (CCME, 2001). The PAHs have been found in different environmental samples, that is, in air, water, soil, oil, sediments, foodstuffs, and tars (Lijinsky, 1991). The heavy metals (cadmium, lead, and mercury) are a serious concern as they have contaminated most superfund sites (Peters, 1999). The bacteria associated to plant (rhizobacteria) contributed in the biodegradation of such organic compounds that are toxic in nature and enhance the phytoremediation (Dowling and Doty, 2009).

Rhizoremediation is an efficient strategy to reduce the contaminants present in the soil (Kuiper et al., 2004). It is the combination of both phytoremediation and bioaugmentation. The pollutant degradation via phytoremediation involves the plant's enzymes, whereas, during bioaugmentation, the indigenous population of microbes establishes the degradation. The microbial community present in the rhizosphere region of the plants, which contributes to the degradation of pollutants during natural vegetation or phytoremediation on a contaminated site. Rhizoremediation is known as the contribution of these rhizomicrobial species involved in the degradation process (Anderson et al., 1993; Schwab and Banks, 1994). An increase in the number of microbes along with the metabolic activity is the main reason behind the pollutant's degradation. By inoculating the *Lewia* sp., which is an endophytic fungus in the rhizospheric region of *Festuca arundinacea*, enhances the rhizoremediation of hydrocarbons (Cruz-Hernández et al., 2013). The nutrients released from the roots stimulate the microbial activity by creating the nutrient-rich medium enriched with amino acids, sugars, and organic acids (Vančura and Hovadik, 1965). The populations of *Verrucomicrobia* and *Actinobacteria* were found increased in the soil contaminants which corresponds to an increase in the number of β-proteobacteria *Denitratisoma*, indicating that this genus might be important for rhizoremediation (Kawasaki et al., 2012).

10.2 Rhizosphere: a novel habitat for microorganisms

The plant roots surrounding zone is known as rhizosphere, which is a habitat of several organisms (Raaijmakers et al., 2009; Hinsinger et al., 2009). The different types of microorganisms are found in this region, namely, fungi, archaea, oomycetes, and bacteria (Bonkowski et al., 2009). The rhizosphere microbiome utilizes the nutrients that are released by the plant. These exudates (rhizodeposits) are the major driving force that regulates the microbial activity and diversity on roots. These microorganisms promote the growth of the plant as well as protect the plant from the attack of a

*Equal contribution

Microbe Mediated Remediation of Environmental Contaminants. DOI: https://doi.org/10.1016/B978-0-12-821199-1.00010-9

pathogen by ways of different mechanisms (Lugtenberg and Kamilova, 2009). It includes the rhizoremediation, disease control, biofertilization, and control of abiotic stress. The mycorrhizal fungi responsible for the translocation of nutrients and minerals from soil to plant (Adeleke et al., 2012; Gianinazzi et al., 2010; Johnson and Graham, 2013), for suppression of pathogens related to soilborne diseases (Whipps, 2001; Pozo and Azcón-Aguilar, 2007), and for generating the stable aggregates of soil (Degens et al., 1996; Miller and Jastrow, 2000) are important symbiont (Salvioli and Bonfante, 2013; Johnson and Graham, 2013). The various types of nitrogen-fixing bacteria have been discovered other than *Bradyrhizobium* and *Rhizobium* (Gaby and Buckley, 2011; Zehr et al., 2003). In soil, iron is an abundant element that is consumed by the rhizosphere microorganisms. It exists in the insoluble ferric oxide state under the conditions of neutral to alkaline due to which this element is not available for the growth of microbes. To overcome the scarcity of iron, bacteria use a variety of mechanisms by siderophore secretions (Hider and Kong, 2010; Andrews et al., 2003). The species of rhizobacterial are mostly organotrophs, which means that they gain energy by assimilating the organic compounds (Aldén et al., 2001).

10.3 Soil contaminants

10.3.1 Inorganic

The industrial waste containing metals gives rise to inorganic contaminants in the soil. Arsenic fluids, along with the sulfur dioxide, are the two major contaminants that are produced by the activity of industries (Richardson et al., 2006). The accumulation of arsenic, nickel, copper, lead, mercury, and cadmium elements occurs in the soil through the wastewater of industries or mine. The lead element is also present in the smoke of automobiles, which gets immersed with the soil particles, and it is also toxic to the plants (Van Zorge, 1996). Due to the high mobility of salts (sodium chloride), the ions of salt are traveled from the effluents of soap and detergent industries to the soil. It increases the concentration of sodium ion in the soil, which damages the normal growth of the plant because of the excess salinity (Bergmann, 1992). de Souza Silva and Fay (2012) demonstrated that salinity affects the microbial community present in the soil. The extensive usage of agrochemical products such as pesticides, herbicides, and fertilizers contaminates the rhizospheric region (Saha et al., 2010). The heavy metals affect the soil in many ways like a diminution of soil microbes, decrease in the productivity of crops, and contamination of the food (Rooney et al., 2007; Chaudri et al., 1993; Åkerblom et al., 2007).

10.3.2 Organic

The several types of organic pollutants, namely, polychlorinated biphenyls (PCBs), pesticides, surfactants, polychlorinated dibenzofurans, biocides, flame retardants, polychlorinated dibenzo-*p*-dioxins, and pharmaceuticals are ended up in soil either intentionally or unintentionally by human activities (Schaeffer et al., 2016). The majorly found organic contaminants are as follows:

- insecticides—captan, benomyl, endrin, heptachlor, and endosulfan (Valentin et al., 2013);
- oil hydrocarbons—alkenes, cycloalkanes, and alkanes (Stokes et al., 2005);
- pesticides and herbicides—bifenox, alachlor, atrazine, and acetochlor (Andreu and Picó, 2004);
- monomeric aromatic hydrocarbons—xylene, toluene, benzene, and ethylbenzene (Gennadiev and Tsibart, 2013); and
- fungicides—penconazole, metalaxyl, procymidone, and lindane (Bermúdez-Couso et al., 2007).

Some of the abovementioned organic pollutants are bioaccumulative, toxic, and persistent (Wang et al., 2016). They can be transferred to the soil directly or indirectly in many ways. The products related to personal care or human pharmaceuticals might get released via irrigation with waste effluents or sewage sludge (Tijani et al., 2016). Moreover, these pollutants utilized to soil surface could be leached quickly to groundwater (Arias-Estévez et al., 2008).

10.4 Rhizoremediation

The removal of pollutants from soil using the techniques of landfilling and incineration is very expensive. An alternative in situ remediation approach, known as rhizoremediation, involves the mutualistic relationship between microorganisms and plants for degrading organic and inorganic pollutants in the root zone. Rhizosphere provides a rich source of nutrients in the form of root exudates, thereby, stimulating the survivability and activities of bacteria leading to efficient degradation of soil pollutants. The rhizosphere is a dynamic microenvironment where microorganisms have access to a large supply of carbon and energy sources (Clegg and Murray, 2002). In addition, to provide the essential nutrients to sustain the diverse microbial community in the rhizosphere, plant root exudates also contain chemical signals influencing the rhizospheric interactions. Root exudates possessing amino acids, enzymes, sugars, low molecular weight carbohydrates aliphatic and aromatic compounds are released in the soil ecosystem, leads an enhancement in microbial activity and degradation of soil pollutants (Burken and Schnoor, 1996). The plant root system helps to spread bacteria in the soil and penetrate them in the impermeable soil layers. When inoculated on plant seed, the pollutant-degrading bacteria help in improving the efficiency of rhizoremediation or phytoremediation.

Rhizoremediation, an emerging green technology using microorganisms, is an effective, eco-friendly, and inexpensive means to remediate polluted soils. It has been reported to be more cost-effective technique compared to mechanical and chemical methods of removing hazardous compounds from the environments (Pradhan et al., 1998). Many studies have shown that the degradation of pollutants is much higher in the rhizosphere compared to nonrhizospheres or nonvegetated soils. In a study, it was found that the degradation of pyrene was 74% in vegetated soil as compared to about 40% degradation in nonvegetated soils (Liste and Alexander, 2000).

Selection of a suitable plant is an important aspect during the design of a rhizoremediation technology since high pollutant levels prevent any vegetation. There is a possibility of the presence of a high concentration of pollutant-degrading bacteria in the rhizosphere of such plants able to thrive under pollutants. Herein, the physical structure of the roots, including root width, surface-to-volume ratio, total biomass, and surface area, plays an important role in the successful implementation of rhizoremediation. Plant root systems bring about the dissolution of pollutants in the rhizosphere by taking up a large amount of water from their surroundings (Erickson, 1997).

During rhizoremediation the organic pollutants, including PAHs, pesticides, and herbicides, are converted to degradable compounds, whereas the inorganic compounds or heavy metals, including cadmium, copper, lead, tin, and zinc, are converted from one oxidation state to another. Different mechanisms used by rhizospheric microorganisms include the production of biosurfactants able to solubilize hydrophobic pollutants, siderophore production for the acquisition of heavy metals, biofilm formation, humification, and production of different acids. The process of rhizoremediation is affected by different physical, chemical, and biological factors such as pH, temperature, nature of pollutants, condition of soils, and native microflora. During the process, plant–microbe interactions play a major role that includes root colonization by bacteria, carrying out catabolic activity, and effect of the different factors on the interaction.

10.4.1 Management of organic contaminants

10.4.1.1 Polycyclic aromatic hydrocarbons and polychlorinated biphenyl

PAHs and PCB are persistent pollutants of great environmental concern due to their toxic, mutagenic, and carcinogenic properties and thus cause environmental pollution worldwide. Due to their chemical structures, insolubility in water, and to adsorb to soil components, they are persistent in the environment (Chaineau et al., 2000). In addition to PAHs, half of the world's organic pollutants are the chlorinated organic compounds (Tiedje et al., 1993). Most significant means of decomposition of these compounds is the biological degradation. Davies and Evans (1964) for the first time demonstrated the metabolic pathway and enzymatic reactions in pseudomonads leading to the decomposition of naphthalene. Since then, many studies have reported the ability of rhizospheric microorganisms to degrade such compounds (Table 10.1). Degradation of low-molecular-weight PAHs has been studied extensively. Single bacterial species or consortium containing two or more bacterial species can either totally degrade or transform PAHs depending on the bacterial population and environmental conditions (Juhasz et al., 2000; Cuypers et al., 2002).

Bacteria generally carry out the degradation of PAHs under aerobic conditions using either monooxygenase or dioxygenase enzymes. The enzyme dioxygenase is a multicomponent consisting of ferredoxin, reductase, and terminal oxygenase subunits (Mallick et al., 2011). Hydroxylation of the aromatic ring with the help of dioxygenase leading to the formation of *cis*-dihydrodiol rearomatized with the help of dehydrogenase to a diol intermediate is the first step in the aerobic degradation of PAH by bacteria. Further, the diol intermediates can be cleaved by dioxygenases, leading to the formation of intermediates that are converted to Krebs cycle intermediates (Cerniglia, 1992; Mueller et al., 1996; Mallick et al., 2011). Bacteria also have the ability to degrade PAH either using cytochrome P450–mediated pathway leading to the production of *trans*-dihydrodiols or anaerobically under nitrate-reducing conditions (Moody et al., 2004; Foght, 2008; Carmona et al., 2009).

Majority of the PAHs are highly hydrophobic, so their metabolization by bacteria takes place via water phase depending upon their solubility (Bouwer and Zehnder, 1993). Therefore, due to the sorption by soil particles, these pollutants are not easily available to bacteria for degradation leading to their long-term persistence in soil. Many bacteria produce surface-active agents known as biosurfactants, with the ability to lower the surface tension between interfaces

TABLE 10.1 Rhizoremediation of organic pollutants.

Plant	Microorganism	Pollutant	Reference
Medicago sativa	*Rhizobium meliloti*	PAHs and PCBs	Ahmad et al. (1997)
Hordeum vulgare	*Burkholderia cepacia*	2,4-Dichlorophenoxyacetic acid	Jacobsen (1997)
Wheat	*Pseudomonas fluorescens*	Trichloroethylene	Yee et al. (1998)
Zinnia angustifolia	*Chrysobacterium indologenes* and *P. fluorescens*	Mefenoxam	Pai et al. (2001)
Spartina alterniflora, Distichlis spicata, Sporobolus airoide, Juncus gerardi	*Paenibacillus naphthalenovorans*	Naphthalene	Daane et al. (2002)
Alfalfa	*Comamonas* sp.	4-Chloronitrobenzene	Liu et al. (2007)
Pinus nigra, Salix caprea	*Rhodococcus, Luteibacter,* and *Williamsia*	Polychlorinated biphenyl	Leigh et al. (2006)
Barmultra	*Pseudomonas putida*	Naphthalene	Kuiper et al. (2002)
Rhizosphere	*Pseudomonas chlororaphis*	Naphthalene	Siunova et al. (2007)
Rhizosphere	*Sphingobium chlorophenolicum*	Pentachlorophenol	Dams et al. (2007)
	Sphingomonas sp.	Naphthalene, acenaphthene, phenanthrene, anthracene, and fluoranthene	Baboshin et al. (2008)
Corn	*Sphingomonas* sp.	Lindane	Böltner et al. (2008)
Populus deltoides	*Kurthia* sp. and *Bacillus circulans*	Naphthalene and anthracene	Bisht et al. (2010)
Cymbopogon ambiguus, Brachiaria decumbens, and *Microlaena stipoides*	Lipase- and hydrocarbon-producing bacteria	Hydrocarbons	Gaskin and Bentham (2010)
Withania somnifera	*Staphylococcus cohnii* subspecies *urealyticus*	Lindane	Abhilash et al. (2011)
Zea mays	*Gordonia* sp.	Diesel oil	Hong et al. (2011)
Z. mays	*P. putida*	Naphthalene	Fernández et al. (2012)
Pennisetum pedicellatum	*Stenotrophomonas maltophilia*	Cypermethrin	Dubey and Fulekar (2013)
Phaseolus vulgaris	*Rhizobium*	Phenanthrene and benzo(*a*) pyrene	Yessica et al. (2013)
Jatropha curcas	Rhizospheric bacteria	Lindane	Abhilash et al. (2013)
Indian mustard, alfalfa	Hydrocarbon-degrading bacterial consortia	Diesel oil	Graj et al. (2013)
M. sativa L.	*Ensifer meliloti*	Phenanthrene	Muratova et al. (2014)
Spirodela polyrhiza	*Pseudomonas* sp. *Cupriavidus* sp., and *Rhodococcus* sp.	Nitrophenols	Kristanti et al. (2014)
M. sativa L.	*Pseudomonas kunmingensis, E. meliloti, Rhizobium petrolearium,* and *Stenotrophomonas* sp.	Phenanthrene	Muratova et al. (2015)
Wheat	Bacterial *nidA* gene	Phenanthrene and pyrene	Shahsavari et al. (2015)
Legumes	*R. petrolearium*	Phenanthrene	Huang et al. (2016)
Phragmites australis and *Typha angustifolia*	*Pseudomonas* sp. and *Alcaligenes* sp.	Hexachlorobenzene	Zhang et al. (2017)

(*Continued*)

TABLE 10.1 (Continued)

Plant	Microorganism	Pollutant	Reference
Lens culinaris	*P. fluorescens*	Sulfonylureas	Rainbird et al. (2018)
Acorus calamus, Typha latifolia, and *Phragmites karka*	*Pseudomonas* sp.	Atrazine	James et al. (2018)
Pennisetum americanum (L.)	Rhizospheric bacterial community	Atrazine	Cao et al. (2018)
Elytrigia repens	*Rhodococcus* and *Acinetobacter*	Naphthalene, anthracene, and phenanthrene	Belovezhets et al. (2020)

and form micelles (Neu, 1996). *Pseudomonas aeruginosa* and *Bacillus* spp. strains have been reported to produce biosurfactants (Pornsunthorntawee et al., 2008; Diab and Din, 2013).

Many studies have reported that rhizoremediation of PAHs or PCBs is more effective than the remediation in the bulk soil. In a study carried out in the soils polluted with aliphatic hydrocarbons and PAHS, the plate counts and respiration rates of the soils were higher in the ryegrass rhizosphere than in the bulk soil (Gunther et al., 1996). Degradation of pyrene was higher in the rhizosphere and, on the addition of organic acids found in the root exudates, pyrene degradation increased (Reilley et al., 1996). Nichols et al. (1997) reported that bacteria able to grow in medium supplemented with a mixture of PAHs also had the capability of degrading PAHs in the rhizospheres of bluegrass and alfalfa.

PAH catabolic genes in bacteria are located on plasmids generally along with the regulatory genes, and with the genes responsible for chemotaxis toward PAH, in some cases (Grimm and Harwood, 1997). Due to the presence of the catabolic plasmids, the indigenous bacteria becomes adaptable to PAH pollutants. Naphthalene-degrading bacteria are present in the rhizosphere of different plants, and a number of reports have mentioned the ability of rhizospheric bacteria to degrade naphthalene (Table 10.1). In *Pseudomonas putida* G7 the naphthalene catabolic genes on the plasmid NAH7 is reported (Simon et al., 1993). Bacteria belonging to *Pseudomonas*, *Sphingomonas*, *Sphingobium*, and *Novosphingobium* have been reported to degrade a large number of xenobiotics, including high-molecular-weight PAHs (Table 10.1).

In addition to naphthalene, a number of studies have reported the ability of different Gram-negative and Gram-positive bacteria to degrade phenanthrene (Table 10.1). *Staphylococcus* sp. strain PN/Y has been reported to degrade phenanthrene through the deoxygenation followed by metacleavage of phenanthrene-1,2-diol, leading to the formation of metabolic intermediate 2-hydroxyl-naphthoic acid (Mallick et al., 2007). Likewise, *Ochrobactrum* sp. isolated by enrichment technique from the municipal waste-contaminated soil sample has been studied for the assimilation of phenanthrene (Ghosal et al., 2010). The strain also had the ability to degrade phenanthrene leading to the formation of 2-hydroxy-1-naphthoic acid, salicylic acid, and catechol. *Ochrobactrum* sp. was able to utilize 2-hydroxy-1-naphthoic acid and salicylic acid as sole carbon substrates. Salicylic acid was metabolized to catechol, which was further degraded to 2-hydroxy muconoaldehyde acid by catechol 2,3-dioxygenase, whereas 2-hydroxy-1-naphthoic acid was metabolized by a ferric-dependent metacleavage dioxygenase.

Many lower molecular weight PAHs, including acenaphthene, anthracene, acenaphthylene, and fluorene, have also been found to be in high quantities in PAH-polluted sites. Different bacterial species with the ability to use these PAHs as sole carbon and energy sources, thereby, metabolizing these pollutants have been reported (Table 10.1). In a study, *Acinetobacter* sp. isolated from municipal waste-contaminated soil sample using acenaphthene as carbon and energy source has been reported to degrade acenaphthene through 1-acenaphthenol, naphthalene-1,8-dicarboxylic acid, 1-naphthoic acid, salicylic acid, and *cis, cis* muconic acid (Ghosal et al., 2013).

10.4.1.2 Pesticides and herbicides

Pesticides and herbicides have the ability to persist in the environment and, therefore, cause concerns for human health worldwide. Traditional physicochemical methods to treat the persistent pesticides and herbicides are costly and inefficient, thereby, making it necessary to search for novel remediation technologies. Rhizoremediation while utilizing microorganisms and plants to degrade pollutants has come up as a sound technology for in situ remediations of various pollutants. The rhizoremediation involves the transformation or conversion of complex forms of pollutants into nonhazardous forms. It is very important to understand the various mechanisms used by microorganisms for the degradation of pollutants for the successful implementation of the technology. Production of pesticide-degrading enzymes in the plant rhizosphere by microorganisms has been reported (Huang et al., 2018). *Pseudomonas* sp. utilized atrazine as a sole carbon source, whereby, three enzymes, AtzA, AtzB, and AtzC, were found to be involved in its degradation. The enzyme AtzA was a key enzyme involved in the hydrolytic dechlorination of atrazine to hydroxyl atrazine, which is further converted *N*-isopropyl cyanuric amide by the action of enzyme AtzB. *N*-Isopropyl cyanuric amide is converted into cyanuric acid and isopropylamine by the enzyme AtzC, which is further converted to CO_2 and NH_3 (De Souza et al., 1996; Wackett et al., 2002).

Various studies have reported the rhizoremediation of pesticides and herbicides (Table 10.1). In a study, degradation rates of butachlor were studied in the rhizosphere and nonrhizosphere soils. The degradation of butachlor was found to be higher in the rhizosphere of wheat compared to nonrhizosphere (Yu et al., 2003). Microorganisms were found to play an important role during the degradation of aldicarb in the rhizosphere of corn, mung bean, and cowpea (Sun et al., 2004). Higher degradation of simazine was observed in rhizospheric soils, mainly in the inoculated soils as compared to bulk soils (Liao and Xie, 2018). Likewise, in a study, plant species were selected for the rhizoremediation of three pesticides, chlorpyrifos, cypermethrin, and fenvalerate. Enhanced remediation of chlorpyrifos was shown in association with grass species *Pennisetum pedicellatum* (Dubey and Fulekar, 2011, 2012).

Many bacteria, including *Burkholderia cepacia*, *Chryseobacterium indologenes*, *Pseudomonas fluorescens*, *Sphingomonas* sp., *Staphylococcus cohnii*, and *Stenotrophomonas maltophilia*, have been reported to be involved in the rhizoremediation of pesticides and herbicides (Table 10.1). Pure cultures of *P. fluorescens* and *Chryseobacterium indologenes* were reported to degrade the fungicide mefenoxam completely to free acids in the rhizosphere of *Zinnia angustifolia* (Pai et al., 2001). *Pseudomonas nitroreducens* was identified as a potential degrader of chlorpyrifos in the rhizosphere of ryegrass (Korade and Fulekar, 2009). The potential of *S. cohnii* in association with *Withania somnifera* has been studied for rhizoremediation of lindane (Abhilash et al., 2011). The inoculation of *S. cohnii* enhanced the lindane uptake ability of *W. somnifera*. Recently, *Pseudomonas* sp. has been shown to degrade atrazine associated with the plants, *Acorus calamus*, *Typha latifolia*, and *Phragmites karka* (James et al., 2018).

10.4.2 Management of inorganic contaminants

Among the inorganic contaminants, pollution of agricultural soils due to heavy metals originating from agricultural and industrial sources is one of the major environmental problems. Plants adsorb heavy metals, which enter the food chain, causing health risks. Heavy metals lead to the deterioration of plant quality and yield and also the microbial community in soil and rhizosphere. In addition, abiotic stress caused by different heavy metals also affects the plant growth as well as growth and metabolism of beneficial microorganisms present in the soil. Heavy metals, including cadmium, lead, and mercury, are among the most toxic pollutants according to the priority list of hazardous substances by 2003 Comprehensive Environmental Response, Compensation, and Liability Act (commonly known as Superfund). Many approaches have been proved to be effective in reducing the harmful effects of these pollutants from soil. However, rhizoremediation seems to be a pragmatic technique to remove these pollutants. During rhizoremediation, plants are prevented from accumulating heavy metals due to the action of bacteria, including plant growth–promoting rhizobacteria that are able to detoxify or accumulate these pollutants. The rhizosphere of plants growing in metal-contaminated environments harbors bacterial species that are either metal resistant or are able to bioaccumulate/bioadsorb/sequester the metals or have the ability to produce enzymes or polysaccharides able to transform the toxic forms of metals to nontoxic one. Bacteria residing in the rhizosphere of plants alleviate metal toxicity and increase the bioavailability of heavy metals through different mechanisms, including the production of plant growth–promoting substances, siderophores, and phytochelators. Bacterial siderophores are low-molecular-weight metal chelating agents regulating the availability of iron in the rhizosphere of plants.

Various heavy metals, including arsenic, lead, cadmium, chromium, mercury, cobalt, copper, nickel, and zinc, have been reported to be detoxifying by the action of microorganisms in the plant rhizospheres (Table 10.2). Bacteria, including *Aeromonas aquariorum*, *Arthrobacter rhombi*, *Bacillus* sp., *Clavibacter xyli*, *Enterobacter aerogenes*, *Gordonia* sp., *Kocuria flava*, *Microbacterium arabinogalactanolyticum*, *P. putida*, *P. fluorescens*, *P. aeruginosa*, *Pseudomonas composti*, *Rhizobium mongolense*, *Variovorax paradoxus*, and *Serratia* sp., have been reported to be involved in the remediation of heavy metals. *Pseudomonas maltophilia* was found to reduce the toxic and mobile form of chromium (Cr [6 +]) to immobile and nontoxic form Cr [3 +] and in addition also reduced the mobility of mercury, lead, and cadmium (Blake et al., 1993). Inoculation of *Helianthus annuus* roots with *P. putida* has been shown to increase the accumulation of cadmium in the test plant and a reduction in its phytotoxicity (Wu et al., 2006). Bacteria belonging to *Arthrobacter* and *Ochrobactrum* have been identified in the rhizosphere of *Agrostis tenuis* growing in arsenic-contaminated soil, facilitating the uptake of the metal (Chopra et al., 2007). Likewise, in another study, it was found that copper-resistant bacterial communities in the rhizosphere of *Phragmites* sp. were more in comparison to bulk soil

TABLE 10.2 Rhizoremediation of inorganic pollutants.

Plant	Bacteria	Pollutant	Reference
Mung bean	*Pseudomonas putida*	Cadmium	Tripathi et al. (2005)
Alfalfa	*Pseudomonas fluorescens*	Arsenic	Ryan et al. (2007)
Alyssum murale	*Arthrobacter rhombi, Clavibacter xyli, Microbacterium arabinogalactanolyticum, Rhizobium ongolense*, and *Variovorax paradoxus*	Arsenate, cadmium, chromium, zinc, mercury, lead, cobalt, copper, and nickel	Abou-Shanab et al. (2007)
Black gram	*Pseudomonas aeruginosa*	Cadmium	Ganesan (2008)
Mustard and pumpkin	*P. aeruginosa*	Cadmium	Sinha and Mukherjee (2008)
Echinochloa crus-galli	*Serratia* sp.	Lead	Koo and Cho (2010)
Zea mays	*Gordonia* sp. and *Serratia* sp.	Copper, cadmium, and lead	Hong et al. (2010)
Z. mays and *Helianthus annuus*	*Pseudomonas* spp.	Zinc, cesium, lead, arsenate, and mercury	Li and Ramakrishna (2011)
Spartina densiflora	*Aeromonas aquariorum, Pseudomonas composti*, and *Bacillus* sp.	Arsenic, nickel, cobalt, zinc, copper, lead, and mercury	Andrades-Moreno et al. (2014)
Brassica napus	*Pantoea* sp.	Chromium	Ontañon et al. (2014)
Sedum alfredii, Festuca arundinacea	*Bacillus cereus*	Lead, cadmium, and zinc	Lu and Zhang (2014)
Sedum plumbizincicola	*Bacillus* sp.	Lead, cadmium, and zinc	Ma et al. (2015)
Wheat	*Dietzia maris* and *Lysinibacillus* sp.	Cadmium	Gusain et al. (2017)
Mangroves	*Kocuria flava* and *Bacillus vietnamensis*	Arsenic	Mallick et al. (2018)
Groundnut	*Bacillus* and *Halobacillus*	Zinc, aluminum, and lead	Banik et al. (2018)
Rice	*Enterobacter aerogenes*	Cadmium	Pramanik et al. (2018)
Rice	Cadmium-tolerant bacterial strain KDML105	Arsenic	Thongnok et al. (2018)
Rice	*Enterobacter* sp.	Cadmium	Mitra et al. (2018)
B. napus L.	Unidentified bacterial community	Copper, cadmium, and zinc	Lacalle et al. (2018)
Vetiveria zizanioides	*Aeromonas* sp.	Cadmium	Itusha et al. (2019)

and were able to produce exopolymers helping in the detoxification of the metal (Huang et al., 2007). Arsenic immobilizing halophilic *K. flava* and *Bacillus vietnamensis* from mangrove rhizosphere has also been reported (Mallick et al., 2018). Recently, biofilm-forming cadmium-resistant *Aeromonas* sp. has been reported to bioaugment cadmium in the rhizosphere of *Vetiveria zizanioides* (Itusha et al., 2019).

10.5 Challenges and future prospects

Rhizoremediation is an emerging eco-friendly and effective technology using microorganisms to remediate polluted environments. Several studies have reported the restoration of polluted sites; however, different mechanisms used by microorganisms during pollutant degradation need to be investigated in detail for the successful execution of the technology. Biosurfactant-producing bacteria are considered a valuable tool for the improvement of rhizoremediation; their mode of action needs to be studied. Many studies reported the selection of one plant–bacterium combination for treating a pollutant, indicating that similar bacterial–plant combinations can be selected for other pollutants. Since standard methods to study the degradation of organic pollutants are available but not for inorganics, there is a challenge in studying in situ heavy metal transformation in polluted systems. Rhizoremediation has the potential to become a promising future technology with a deep insight into sequencing techniques and technological advancement.

References

Abhilash, P.C., Singh, B., Srivastava, P., Schaeffer, A., Singh, N., 2013. Remediation of lindane by *Jatropha curcas* L: utilization of multipurpose species for rhizoremediation. Biomass Bioenergy 51, 189–193.

Abhilash, P.C., Srivastava, S., Srivastava, P., Singh, B., Jafri, A., Singh, N., 2011. Influence of rhizospheric microbial inoculation and tolerant plant species on the rhizoremediation of lindane. Environ. Exp. Bot. 74, 127–130.

Abou-Shanab, R.A.I., Van Berkum, P., Angle, J.S., 2007. Heavy metal resistance and genotypic analysis of metal resistance genes in gram-positive and gram-negative bacteria present in Ni-rich serpentine soil and in the rhizosphere of *Alyssum murale*. Chemosphere 68 (2), 360–367.

Adeleke, R.A., Cloete, T.E., Bertrand, A., Khasa, D.P., 2012. Iron ore weathering potentials of ectomycorrhizal plants. Mycorrhiza 22 (7), 535–544.

Ahmad, D., Mehmannavaz, R., Damaj, M., 1997. Isolation and characterization of symbiotic N2-fixing *Rhizobium meliloti* from soils contaminated with aromatic and chloroaromatic hydrocarbons: PAHs and PCBs. Int. Biodeterior. Biodegrad. 39 (1), 33–43.

Åkerblom, S., Bååth, E., Bringmark, L., Bringmark, E., 2007. Experimentally induced effects of heavy metal on microbial activity and community structure of forest mor layers. Biol. Fertil. Soils 44 (1), 79–91.

Aldén, L., Demoling, F., Bååth, E., 2001. Rapid method of determining factors limiting bacterial growth in soil. Appl. Environ. Microbiol. 67 (4), 1830–1838.

Anderson, T.A., Guthrie, E.A., Walton, B.T., 1993. Bioremediation in the rhizosphere. Environ. Sci. Technol. 27 (13), 2630–2636.

Andrades-Moreno, L., Del Castillo, I., Parra, R., Doukkali, B., Redondo-Gómez, S., Pérez-Palacios, P., et al., 2014. Prospecting metal-resistant plant-growth promoting rhizobacteria for rhizoremediation of metal contaminated estuaries using *Spartina densiflora*. Environ. Sci. Pollut. Res. 21 (5), 3713–3721.

Andreu, V., Picó, Y., 2004. Determination of pesticides and their degradation products in soil: critical review and comparison of methods. TrAC Trends Anal. Chem. 23 (10–11), 772–789.

Andrews, S.C., Robinson, A.K., Rodríguez-Quiñones, F., 2003. Bacterial iron homeostasis. FEMS Microbiol. Rev. 27 (2–3), 215–237.

Arias-Estévez, M., López-Periago, E., Martínez-Carballo, E., Simal-Gándara, J., Mejuto, J.C., García-Río, L., 2008. The mobility and degradation of pesticides in soils and the pollution of groundwater resources. Agric. Ecosyst. Environ. 123 (4), 247–260.

Baboshin, M., Akimov, V., Baskunov, B., Born, T.L., Khan, S.U., Golovleva, L., 2008. Conversion of polycyclic aromatic hydrocarbons by *Sphingomonas* sp. VKM B-2434. Biodegradation 19 (4), 567–576.

Banik, A., Pandya, P., Patel, B., Rathod, C., Dangar, M., 2018. Characterization of halotolerant, pigmented, plant growth promoting bacteria of groundnut rhizosphere and its in-vitro evaluation of plant-microbe protocooperation to withstand salinity and metal stress. Sci. Total. Environ. 630, 231–242.

Belovezhets, L.A., Tretyakova, M.S., Markova, Y.A., Levchuk, A.A., 2020. Use of rhizosphere microorganisms for bioremediation of oil contaminated soils. In: IOP Conference Series: Earth and Environmental Science, vol. 408, No. 1, IOP Publishing, p. 012086.

Bergmann, W., 1992. Nutritional Disorders of Plants - Development, Visual and Analytical Diagnosis. Gustav Fischer Verlag, Jena, Stuttgart.

Bermúdez-Couso, A., Arias-Estévez, M., Nóvoa-Muñoz, J.C., López-Periago, E., Soto-González, B., Simal-Gándara, J., 2007. Seasonal distributions of fungicides in soils and sediments of a small river basin partially devoted to vineyards. Water Res. 41 (19), 4515–4525.

Bisht, S., Pandey, P., Sood, A., Sharma, S., Bisht, N.S., 2010. Biodegradation of naphthalene and anthracene by chemo-tactically active rhizobacteria of *Populus deltoides*. Braz. J. Microbiol. 41 (4), 922–930.

Blake, R.C., Choate, D.M., Bardhan, S., Revis, N., Barton, L.L., Zocco, T.G., 1993. Chemical transformation of toxic metals by a *Pseudomonas* strain from a toxic waste site. Environ. Toxicol. Chem. 12 (8), 1365–1376.

Böltner, D., Godoy, P., Muñoz-Rojas, J., Duque, E., Moreno-Morillas, S., Sánchez, L., et al., 2008. Rhizoremediation of lindane by root-colonizing *Sphingomonas*. Microb. Biotechnol. 1 (1), 87–93.

Bonkowski, M., Villenave, C., Griffiths, B., 2009. Rhizosphere fauna: the functional and structural diversity of intimate interactions of soil fauna with plant roots. Plant Soil 321 (1–2), 213–233.

Bouwer, E.J., Zehnder, A.J., 1993. Bioremediation of organic compounds—putting microbial metabolism to work. Trends Biotechnol. 11 (8), 360–367.

Burken, J.G., Schnoor, J.L., 1996. Phytoremediation: plant uptake of atrazine and role of roots exudates. J. Environ. Eng. 122, 958–963.

Cao, B., Zhang, Y., Wang, Z., Li, M., Yang, F., Jiang, D., et al., 2018. Insight into the variation of bacterial structure in atrazine-contaminated soil regulating by potential phytoremediator: *Pennisetum americanum* (L.) K. Schum. Front. Microbiol. 9, 864.

Carmona, M., Zamarro, M.T., Blazquez, B., Durante-Rodriguez, G., Juarez, J.F., Valderrama, J.A., et al., 2009. Anaerobic catabolism of aromatic compounds: a genetic and genomic view. Microbiol. Mol. Biol. Rev. 73, 71–133. Available from: https://doi.org/10.1128/MMBR.00021-08.

CCME, 2001. Canada-Wide Standards for Petroleum Hydrocarbons (PHCs) in Soil—Technical Supplement.

Cerniglia, C.E., 1992. Biodegradation of polycyclic aromatic hydrocarbons. Biodegradation 3, 351–368. Available from: https://doi.org/10.1007/BF00129093.

Chaineau, C.H., Morel, J.L., Oudot, J., 2000. Biodegradation of fuel oil hydrocarbons in the rhizosphere of maize. J. Environ. Qual. 29, 549–578.

Chaudri, A.M., McGrath, S.P., Giller, K.E., Rietz, E., Sauerbeck, D.R., 1993. Enumeration of indigenous *Rhizobium leguminosarum* biovar *trifolii* in soils previously treated with metal-contaminated sewage sludge. Soil Biol. Biochem. 25 (3), 301–309.

Chopra, B.K., Bhat, S., Mikheenko, I.P., Xu, Z., Yang, Y., Luo, X., et al., 2007. The characteristics of rhizosphere microbes associated with plants in arsenic-contaminated soils from cattle dip sites. Sci. Total. Environ. 378 (3), 331–342.

Clegg, C., Murray, P., 2002. Soil Microbial Ecology and Plant Root Interactions. IGER Innovations.

Cruz-Hernández, A., Tomasini-Campocosio, A., Pérez-Flores, L.J., Fernández-Perrino, F.J., Gutiérrez-Rojas, M., 2013. Inoculation of seed-borne fungus in the rhizosphere of *Festuca arundinacea* promotes hydrocarbon removal and pyrene accumulation in roots. Plant Soil 362 (1–2), 261–270.

Cuypers, C., Pancras, T., Grotenhuis, T., Rulkens, W., 2002. The estimation of PAH bioavailability in contaminated sediments using hydroxypropyl-ß-cyclodextrin and triton *x*-100 extraction techniques. J. Chem. 46, 1235–1245.

Daane, L.L., Harjono, I., Barns, S.M., Launen, L.A., Palleron, N.J., Häggblom, M.M., 2002. PAH-degradation by *Paenibacillus* spp. and description of *Paenibacillus naphthalenovorans* sp. nov., a naphthalene-degrading bacterium from the rhizosphere of salt marsh plants. Int. J. Syst. Evol. Microbiol. 52 (1), 131–139.

Dams, R.I., Paton, G.I., Killham, K., 2007. Rhizoremediation of pentachlorophenol by *Sphingobium chlorophenolicum* ATCC 39723. Chemosphere 68 (5), 864–870.

Davies, J.I., Evans, W.C., 1964. Oxidative metabolism of naphthalene by soil pseudomonads. Biochem. J. 91, 251–261.

de Souza Silva, C.M.M., Fay, E.F., 2012. Effect of salinity on soil microorganisms. Soil Health Land Use Manage. 10, 177–198.

De Souza, M.L., Sadowsky, M.J., Wackett, L.P., 1996. Atrazine chlorohydrolase from Pseudomonas sp. strain ADP: gene sequence, enzyme purification, and protein characterization. J. Bacteriol. 178 (16), 4894–4900.

Degens, B.P., Sparling, G.P., Abbott, L.K., Takeuchi, Y., Murakami, M., Nakajima, N., et al., 1996. Increasing the length of hyphae in a sandy soil increases the amount of water-stable aggregates2407421 Appl. Soil Ecol. 3 (2), 149–159.

Diab, A., Din, G.E., 2013. Application of the biosurfactants produced by *Bacillus* spp. (SH 20 and SH 26) and *P. aeruginosa* SH 29 isolated from the rhizosphere soil of an Egyptian salt marsh plant for the cleaning of oil-contaminataed vessels and enhancing the biodegradation of oily sludge. Afr. J. Environ. Sci. Technol. 7 (7), 671–679.

Dowling, D.N., Doty, S.L., 2009. Improving phytoremediation through biotechnology. Curr. Opin. Biotechnol. 20 (2), 204–206.

Dubey, K.K., Fulekar, M.H., 2011. Effect of pesticides on the seed germination of *Cenchrus setigerus* and *Pennisetum pedicellatum* as monocropping and co-cropping system: implications for rhizospheric bioremediation. Rom. Biotech. Lett. 16, 5908–5918.

Dubey, K.K., Fulekar, M.H., 2012. Chlorpyrifos bioremediation in *Pennisetum* rhizosphere by a novel potential degrader *Stenotrophomonas maltophilia* MHF ENV20. World J. Microbiol. Biotechnol. 28 (4), 1715–1725.

Dubey, K.K., Fulekar, M.H., 2013. Investigation of potential rhizospheric isolate for cypermethrin degradation. 3 Biotech 3 (1), 33–43.

Erickson, M.D., 1997. Analytical Chemistry of PCBs. CRC Press.

Fernández M., Niqui-Arroyo J.L., Conde S., Ramos J.L. and Duque E., 2012. Enhanced tolerance to naphthalene and enhanced rhizoremediation performance for *Pseudomonas putida* KT2440 via the NAH7 catabolic plasmid, *Appl. Environ. Microbiol*, 78 (15), 5104–5110.

Foght, J., 2008. Anaerobic biodegradation of aromatic hydrocarbons: pathways and prospects. J. Mol. Microbiol. Biotechnol. 15, 93–120. Available from: https://doi.org/10.1159/000121324.

Gaby, J.C., Buckley, D.H., 2011. A global census of nitrogenase diversity. Environ. Microbiol. 13 (7), 1790–1799.

Ganesan, V., 2008. Rhizoremediation of cadmium soil using a cadmium-resistant plant growth-promoting rhizopseudomonad. Curr. Microbiol. 56 (4), 403–407.

Gaskin, S.E., Bentham, R.H., 2010. Rhizoremediation of hydrocarbon contaminated soil using Australian native grasses. Sci. Total Environ. 408 (17), 3683–3688.

Gennadiev, A.N., Tsibart, A.S., 2013. Pyrogenic polycyclic aromatic hydrocarbons in soils of reserved and anthropogenically modified areas: factors and features of accumulation. Eurasian Soil Sci. 46 (1), 28–36.

Ghosal D., Chakraborty J., Khara P. and Dutta T.K., 2010. Degradation of phenanthrene via meta-cleavage of 2-hydroxy-1-naphthoic acid by *Ochrobactrum* sp. strain PWTJD. *FEMS Microbiol Lett.* 313 (2), 103–110.

Ghosal D., Dutta A., Chakraborty J., Basu S. and Dutta T.K., 2013. Characterization of the metabolic pathway involved in assimilation of acenaphthene in *Acinetobacter* sp. strain AGAT-W, *Res. Microbiol.* 164 (2), 155–163.

Gianinazzi, S., Gollotte, A., Binet, M.N., van Tuinen, D., Redecker, D., Wipf, D., 2010. Agroecology: the key role of arbuscular mycorrhizas in ecosystem services. Mycorrhiza 20 (8), 519–530.

Graj, W., Lisiecki, P., Szulc, A., Chrzanowski, Ł., Wojtera-Kwiczor, J., 2013. Bioaugmentation with petroleum-degrading consortia has a selective growth-promoting impact on crop plants germinated in diesel oil-contaminated soil. Water Air Soil Pollut. 224 (9), 1676.

Grimm, A.C., Harwood, C.S., 1997. Chemotaxis of *Pseudomonas* spp. to the polyaromatic hydrocarbon naphthalene. Appl. Environ. Microbiol. 63 (10), 4111–4115.

Gunther, T.H., Dornberger, U., Fritsche, W., 1996. Effect of ryegrass on biodegradation of hydrocarbons in soil. J. Chem. 33, 203–215.

Gusain, P., Paliwal, R., Singh, V., 2017. Rhizoremediation of cadmium-contaminated soil associated with hydroxamate siderophores isolated from Cd-resistant plant growth–promoting *Dietzia maris* and *Lysinibacillus* strains. Int. J. Phytorem. 19 (3), 290–299.

Hider, R.C., Kong, X., 2010. Chemistry and biology of siderophores. Nat. Prod. Rep. 27 (5), 637–657.

Hinsinger, P., Bengough, A.G., Vetterlein, D., Young, I.M., 2009. Rhizosphere: biophysics, biogeochemistry and ecological relevance. Plant Soil 321 (1–2), 117–152.

Hong, S.H., Koo, S.Y., Kim, S.H., Ryu, H.W., Lee, I.S., Cho, K.S., 2010. Rhizoremediation of petroleum and heavy metal-contaminated soil using rhizobacteria and *Zea mays*. Microbiol. Biotechnol. Lett. 38 (3), 329–334.

Hong, S.H., Ryu, H., Kim, J., Cho, K.S., 2011. Rhizoremediation of diesel-contaminated soil using the plant growth-promoting rhizobacterium *Gordonia* sp. S2RP-17. Biodegradation 22 (3), 593–601.

Huang, A., Teplitski, M., Rathinasabapathi, B., Ma, L., 2007. Characterization of rhizosphere bacteria of the arsenic hyperaccumulator Chinese Brake Fern (*Pteris vittata* L.). In: The ASA-CSSA-SSSA International Annual Meetings (Nov 4–8, 2007). <http://a-c-s.confex.com/a-c-s/2007am/techprogram/P33676.HTM>.

Huang, X., Shi, J., Cui, C., Yin, H., Zhang, R., Ma, X., et al., 2016. Biodegradation of phenanthrene by *Rhizobium petrolearium* SL-1. J. Appl. Microbiol. 121 (6), 1616–1626.

Huang, Y., Xiao, L., Li, F., Xiao, M., Lin, D., Long, X., et al., 2018. Microbial degradation of pesticide residues and an emphasis on the degradation of cypermethrin and 3-phenoxy benzoic acid: a review. Molecules 23 (9), 2313.

Itusha, A., Osborne, W.J., Vaithilingam, M., 2019. Enhanced uptake of Cd by biofilm forming Cd resistant plant growth promoting bacteria bioaugmented to the rhizosphere of *Vetiveria zizanioides*. Int. J. Phytorem. 21 (5), 487–495.

Jacobsen, C.S., 1997. Plant protection and rhizosphere colonization of barley by seed inoculated herbicide degrading *Burkholderia* (*Pseudomonas*) *cepacia* DBO1 (pRO101) in 2,4-D contaminated soil. Plant Soil 189 (1), 139–144.

James, A., Singh, D.K., Khankhane, P.J., 2018. Enhanced atrazine removal by hydrophyte–bacterium associations and in vitro screening of the isolates for their plant growth-promoting potential. Int. J. Phytorem. 20 (2), 89–97.

Johnson, N.C., Graham, J.H., 2013. The continuum concept remains a useful framework for studying mycorrhizal functioning. Plant Soil 363 (1–2), 411–419.

Juhasz, A.L., Britz, M.L., Stanley, G.A., 2000. Evaluation of a creosote based medium for the growth and preparation of a PAH-degrading bacterial community for bioaugmentation. J. Ind. Microbiol. Biotechnol. 24, 272–284.

Kawasaki, A., Watson, E.R., Kertesz, M.A., 2012. Indirect effects of polycyclic aromatic hydrocarbon contamination on microbial communities in legume and grass rhizospheres. Plant Soil 358 (1–2), 169–182.

Koo, S.Y., Cho, K.S., 2010. Characterization of *Serratia* sp. K1RP-49 for application to the rhizoremediation of heavy metals. Survival and Sustainability. Springer, Berlin, Heidelberg, pp. 3–13.

Korade, D.L., Fulekar, M.H., 2009. Rhizosphere remediation of chlorpyrifos in mycorrhizospheric soil using ryegrass. J. Hazard. Mater. 172 (2–3), 1344–1350.

Kristanti, R.A., Toyama, T., Hadibarata, T., Tanaka, Y., Mori, K., 2014. Sustainable removal of nitrophenols by rhizoremediation using four strains of bacteria and giant duckweed (*Spirodela polyrhiza*). Water Air Soil Pollut. 225 (4), 1928.

Kuiper, I., Kravchenko, L.V., Bloemberg, G.V., Lugtenberg, B.J., 2002. *Pseudomonas putida* strain PCL1444, selected for efficient root colonization and naphthalene degradation, effectively utilizes root exudate components. Mol. Plant-Microbe Interact. 15 (7), 734–741.

Kuiper, I., Lagendijk, E.L., Bloemberg, G.V., Lugtenberg, B.J., 2004. Rhizoremediation: a beneficial plant-microbe interaction. Mol. Plant-Microbe Interact. 17 (1), 6–15.

Lacalle, R.G., Gómez-Sagasti, M.T., Artetxe, U., Garbisu, C., Becerril, J.M., 2018. *Brassica napus* has a key role in the recovery of the health of soils contaminated with metals and diesel by rhizoremediation. Sci. Total Environ. 618, 347–356.

Leigh, M.B., Prouzová, P., Macková, M., Macek, T., Nagle, D.P., Fletcher, J.S., 2006. Polychlorinated biphenyl (PCB)-degrading bacteria associated with trees in a PCB-contaminated site. Appl. Environ. Microbiol. 72 (4), 2331–2342.

Li, K., Ramakrishna, W., 2011. Effect of multiple metal resistant bacteria from contaminated lake sediments on metal accumulation and plant growth. J. Hazard. Mater. 189 (1–2), 531–539.

Liao, M., Xie, X., 2018. Effects of combination of plant and microorganism on degradation of simazine in soil. J. Environ. Sci. 20 (2), 195–198.

Lijinsky, W., 1991. The formation and occurrence of polynuclear aromatic hydrocarbons associated with food. Mutat. Res./Genet. Toxicol. 259 (3-4), 251–261.

Liste, H.H., Alexander, M., 2000. Plant- promoted pyrene degradation in soil. J. Chem. 40, 7–10.

Liu, L., Jiang, C.Y., Liu, X.Y., Wu, J.F., Han, J.G., Liu, S.J., 2007. Plant–microbe association for rhizoremediation of chloronitroaromatic pollutants with *Comamonas* sp. strain CNB-1. Environ. Microbiol. 9 (2), 465–473.

Lu, M., Zhang, Z.Z., 2014. Phytoremediation of soil co-contaminated with heavy metals and deca-BDE by co-planting of *Sedum alfredii* with tall fescue associated with *Bacillus cereus* JP12. Plant Soil 382 (1–2), 89–102.

Lugtenberg, B., Kamilova, F., 2009. Plant-growth-promoting rhizobacteria. Annu. Rev. Microbiol. 63, 541–556.

Ma, Y., Oliveira, R.S., Wu, L., Luo, Y., Rajkumar, M., Rocha, I., et al., 2015. Inoculation with metal-mobilizing plant-growth-promoting rhizobacterium *Bacillus* sp. SC2b and its role in rhizoremediation. J. Toxicol. Environ. Health, A 78 (13–14), 931–944.

Mallick, I., Bhattacharyya, C., Mukherji, S., Dey, D., Sarkar, S.C., Mukhopadhyay, U.K., et al., 2018. Effective rhizoinoculation and biofilm formation by arsenic immobilizing halophilic plant growth promoting bacteria (PGPB) isolated from mangrove rhizosphere: a step towards arsenic rhizoremediation. Sci. Total Environ. 610, 1239–1250.

Mallick, S., Chakraborty, J., Dutta, T.K., 2011. Role of oxygenases in guiding diverse metabolic pathways in the bacterial degradation of low-molecular weight polycyclic aromatic hydrocarbons: a review. Crit. Rev. Microbiol. 37, 64–90. Available from: https://doi.org/10.3109/1040841X.2010.512268.

Mallick S., Chatterjee S. and Dutta T.K., 2007. A novel degradation pathway in the assimilation of phenanthrene by *Staphylococcus* sp. strain PN/Y via meta-cleavage of 2-hydroxy-1-naphthoic acid: formation of trans-2, 3-dioxo-5-(2′-hydroxyphenyl)-pent-4-enoic acid. *Microbiol.* 153 (7), 2104–2115.

Miller, R.M., Jastrow, J.D., 2000. Mycorrhizal fungi influence soil structure. Arbuscular Mycorrhizas: Physiology and Function. Springer, Dordrecht, pp. 3–18.

Mitra, S., Pramanik, K., Sarkar, A., Ghosh, P.K., Soren, T., Maiti, T.K., 2018. Bioaccumulation of cadmium by *Enterobacter* sp. and enhancement of rice seedling growth under cadmium stress. Ecotoxicol. Environ. Saf. 156, 183–196.

Moody, J.D., Freeman, J.P., Fu, P.P., Cerniglia, C.E., 2004. Degradation of benzo[*a*]pyrene by *Mycobacterium vanbaalenii* PYR-1. Appl. Environ. Microbiol. 70, 340–345. Available from: https://doi.org/10.1128/AEM.70.1.340-345.2004.

Mueller, J.G., Cerniglia, C.E., Pritchard, P.H., 1996. Bioremediation of environments contaminated by polycyclic aromatic hydrocarbons. In: Crawford, R.L., Crawford, L.D. (Eds.), Bioremediation: Principles and Applications. Cambridge University Press, Cambridge, pp. 125–194.

Muratova, A., Dubrovskaya, E., Golubev, S., Grinev, V., Chernyshova, M., Turkovskaya, O., 2015. The coupling of the plant and microbial catabolisms of phenanthrene in the rhizosphere of *Medicago sativa.* J. Plant. Physiol. 188, 1–8.

Muratova, A., Pozdnyakova, N., Makarov, O., Baboshin, M., Baskunov, B., Myasoedova, N., et al., 2014. Degradation of phenanthrene by the rhizobacterium *Ensifer meliloti*. Biodegradation 25 (6), 787–795.

Neu, T.R., 1996. Significance of bacterial surface-active compounds in interaction of bacteria with interfaces. Microbiol. Rev. 60 (1), 151.

Nichols, T.D., Wolf, D.C., Rogers, H.B., Beyrouty, C.A., Reynolds, C.M., 1997. Rhizosphere microbial populations in contaminated soils. Water Air Soil Pollut. 95, 165–178.

Ontañon, O.M., González, P.S., Ambrosio, L.F., Paisio, C.E., Agostini, E., 2014. Rhizoremediation of phenol and chromium by the synergistic combination of a native bacterial strain and Brassica napus hairy roots. Int. Biodeterior. Biodegrad. 88, 192–198.

Pai, S.G., Riley, M.B., Camper, N.D., 2001. Microbial degradation of mefenoxam in rhizosphere of *Zinnia angustifolia.* Chemosphere 44 (4), 577–582.

Peters, R.W., 1999. Chelant extraction of heavy metals from contaminated soils. J. Hazard. Mater. 66 (1–2), 151–210.

Pornsunthorntawee, O., Wongpanit, P., Chavadej, S., Abe, M., Rujiravanit, R., 2008. Structural and physicochemical characterization of crude biosurfactant produced by *Pseudomonas aeruginosa* SP4 isolated from petroleum-contaminated soil. Bioresour. Technol. 99 (6), 1589–1595.

Pozo, M.J., Azcón-Aguilar, C., 2007. Unraveling mycorrhiza-induced resistance. Curr. Opin. Plant Biol. 10 (4), 393–398.

Pradhan, S.P., Conrad, J.R., Paterek, J.R., Srivastava, V.J., 1998. Potential of phytoremediation for treatment of PAH S in soil at MGP sites. J. Soil Contam. 7, 467–480.

Pramanik, K., Mitra, S., Sarkar, A., Maiti, T.K., 2018. Alleviation of phytotoxic effects of cadmium on rice seedlings by cadmium resistant PGPR strain *Enterobacter aerogenes* MCC 3092. J. Hazard. Mater. 351, 317–329.

Raaijmakers, J.M., Paulitz, T.C., Steinberg, C., Alabouvette, C., Moënne-Loccoz, Y., 2009. The rhizosphere: a playground and battlefield for soilborne pathogens and beneficial microorganisms. Plant Soil 321 (1–2), 341–361.

Rainbird, B., Bentham, R.H., Soole, K.L., 2018. Rhizoremediation of residual sulfonylurea herbicides in agricultural soils using *Lens culinaris* and a commercial supplement. Int. J. Phytorem. 20 (2), 104–113.

Reilley, K.A., Banks, M.K., Schab, A.P., 1996. Dissipation of polycyclic aromatic hydrocarbons in the rhizosphere. J. Environ. Qual. 25, 212–219.

Richardson, G.M., Bright, D.A., Dodd, M., 2006. Do current standards of practice in Canada measure what is relevant to human exposure at contaminated sites? II: Oral bioaccessibility of contaminants in soil. Hum. Ecol. Risk Assess. 12 (3), 606–616.

Rooney, C.P., Zhao, F.J., McGrath, S.P., 2007. Phytotoxicity of nickel in a range of European soils: influence of soil properties, Ni solubility and speciation. Environ. Pollut. 145 (2), 596–605.

Ryan, R.P., Ryan, D., Dowling, D.N., 2007. Plant protection by the recombinant, root-colonizing *Pseudomonas fluorescens* F113rifPCB strain expressing arsenic resistance: improving rhizoremediation. Lett. Appl. Microbiol. 45 (6), 668–674.

Saha, J.K., Panwar, N., Singh, M.V., 2010. An assessment of municipal solid waste compost quality produced in different cities of India in the perspective of developing quality control indices. Waste Manage. 30 (2), 192–201.

Salvioli, A., Bonfante, P., 2013. Systems biology and "omics" tools: a cooperation for next-generation mycorrhizal studies. Plant Sci. 203, 107–114.

Schaeffer, A., Amelung, W., Hollert, H., Kaestner, M., Kandeler, E., Kruse, J., et al., 2016. The impact of chemical pollution on the resilience of soils under multiple stresses: a conceptual framework for future research. Sci. Total Environ. 568, 1076–1085.

Schwab, A.P., Banks, M.K.,1994. Biologically mediated dissipation of polyaromatic hydrocarbons in the root zone. In: Bioremediation through Rhizosphere Technology, Anderson, T., et al. (Eds.) ACS Symposium Series; American Chemical Society: Washington, DC, pp 132–141.

Shahsavari, E., Adetutu, E.M., Taha, M., Ball, A.S., 2015. Rhizoremediation of phenanthrene and pyrene contaminated soil using wheat. J. Environ. Manage. 155, 171–176.

Simon, M.J., Osslund, T.D., Saunders, R., Ensley, B.D., Suggs, S., Harcourt, A., et al., 1993. Sequences of genes encoding naphthalene dioxygenase in *Pseudomonas putida* strains G7 and NCIB 9816-4. Gene 127 (1), 31–37.

Sinha, S., Mukherjee, S.K., 2008. Cadmium–induced siderophore production by a high Cd-resistant bacterial strain relieved Cd toxicity in plants through root colonization. Curr. Microbiol. 56 (1), 55–60.

Siunova, T.V., Anokhina, T.O., Mashukova, A.V., Kochetkov, V.V., Boronin, A.M., 2007. Rhizosphere strain of *Pseudomonas chlororaphis* capable of degrading naphthalene in the presence of cobalt/nickel. Microbiology 76 (2), 182–188.

Stokes, J.D., Paton, G.I., Semple, K.T., 2005. Fate of hydrophobic compounds in the soil environment. Soil Use Manage. 21, 475–486.

Sun, H., Xu, J., Yang, S., Liu, G., Dai, S., 2004. Plant uptake of aldicarb from contaminated soil and its enhanced degradation in the rhizosphere. Chemosphere 54 (4), 569–574.

Thongnok, S., Siripornadulsil, W., Siripornadulsil, S., 2018. Mitigation of arsenic toxicity and accumulation in hydroponically grown rice seedlings by co-inoculation with arsenite-oxidizing and cadmium-tolerant bacteria. Ecotoxicol. Environ. Saf. 162, 591–602.

Tiedje, J.M., Quensen, J.F., Sanford, J.C., Schimel, J.P., Boyd, S.A., 1993. Microbial reductive dechlorination of PCBS. J. Biodegrad. 4, 231–240.

Tijani, J.O., Fatoba, O.O., Babajide, O.O., Petrik, L.F., 2016. Pharmaceuticals, endocrine disruptors, personal care products, nanomaterials and perfluorinated pollutants: a review. Environ. Chem. Lett. 14 (1), 27–49.

Tripathi, M., Munot, H.P., Shouche, Y., Meyer, J.M., Goel, R., 2005. Isolation and functional characterization of siderophore-producing lead-and cadmium-resistant *Pseudomonas putida* KNP9. Curr. Microbiol. 50 (5), 233–237.

Valentin, L., Nousiainen, A., Mikkonen, A., 2013. Introduction to organic contaminants in soil: concepts and risks. Emerging Organic Contaminants in Sludges. Springer, Berlin, Heidelberg, pp. 1–29.

Van Zorge, J.A., 1996. Exposure to mixtures of chemical substances: is there a need for regulations? Food Chem. Toxicol. 34 (11–12), 1033–1036.

Vančura, V., Hovadik, A., 1965. Root exudates of plants: II. Composition of root exudates of some vegetables. Plant Soil 21–32.

Wackett, L., Sadowsky, M., Martinez, B., Shapir, N., 2002. Biodegradation of atrazine and related *s*-triazine compounds: from enzymes to field studies. Appl. Microbiol. Biotechnol. 58 (1), 39–45.

Wang, X., Sun, D., Yao, T., 2016. Climate change and global cycling of persistent organic pollutants: a critical review. Sci. China Earth Sci. 59 (10), 1899–1911.

Whipps, J.M., 2001. Ecological and biotechnological considerations in enhancing disease biocontrol. NATO Sci. Ser., I: Life Behav. Sci. 339, 43–51.

Wu, C.H., Wood, T.K., Mulchandani, A., Chen, W., 2006. Engineering plant-microbe symbiosis for rhizoremediation of heavy metals. Appl. Environ. Microbiol. 72 (2), 1129–1134.

Yee, D.C., Maynard, J.A., Wood, T.K., 1998. Rhizoremediation of trichloroethylene by a recombinant, root-colonizing *Pseudomonas fluorescens* strain expressing toluene ortho-monooxygenase constitutively. Appl. Environ. Microbiol. 64 (1), 112–118.

Yessica, G.P., Alejandro, A., Ronald, F.C., José, A.J., Esperanza, M.R., Samuel, C.S.J., et al., 2013. Tolerance, growth and degradation of phenanthrene and benzo [a] pyrene by *Rhizobium tropici* CIAT 899 in liquid culture medium. Appl. Soil Ecol. 63, 105–111.

Yu, Y.L., Chen, Y.X., Luo, Y.M., Pan, X.D., He, Y.F., Wong, M.H., 2003. Rapid degradation of butachlor in wheat rhizosphere soil. Chemosphere 50 (6), 771–774.

Zehr, J.P., Jenkins, B.D., Short, S.M., Steward, G.F., 2003. Nitrogenase gene diversity and microbial community structure: a cross-system comparison. Environ. Microbiol. 5 (7), 539–554.

Zhang, C., Wang, B., Dai, X., Li, S., Lu, G., Zhou, Y., 2017. Structure and function of the bacterial communities during rhizoremediation of hexachlorobenzene in constructed wetlands. Environ. Sci. Pollut. Res. 24 (12), 11483–11492.

Chapter 11

Metal-oxidizing microbes and potential application in bioremediation

Vipin Kumar Singh[1], Anand Kumar Chaudhari[1], Kin Israel R. Notarte[2], Ajay Kumar[3], Rishikesh Singh[4] and Rahul Bhadouria[5]

[1]*Department of Botany, Centre of Advanced Study in Botany, Institute of Science, Banaras Hindu University, Varanasi, India,* [2]*Faculty of Medicine and Surgery, University of Santo Tomas, Manila, Philippines,* [3]*Agriculture Research Organization, Ministry of Agriculture and Rural Development, Volcani Center, Rishon LeZion, Israel,* [4]*Institute of Environment and Sustainable Development, Banaras Hindu University, Varanasi, India,* [5]*Department of Botany, University of Delhi, New Delhi, India*

11.1 Introduction

The revolution in metal industries worldwide and anthropogenic activities has added significantly large amount of metals into the natural environment leading to hazardous impact on diverse organisms (Manoj et al., 2020). Remediation of metal contaminants using physicochemical techniques is expensive, ecologically unsafe, and secondary product generating calling for the judicious application of biological agents. The microbes inhabiting in metal-polluted soil and water, during the long course of evolution, have evolved the intricate mechanism of metal oxidation by employing cellular processes. The potential ability of microorganisms in oxidizing metals also possesses advantageous application in extraction of heavy metals (Wu et al., 2020), thereby serving as a foundation for contaminant treatment (Aziz et al., 2020). Sometimes, the metals used by these microbes are utilized as electron donor to derive energy during the respiratory processes (Rawlings, 2005). For instance, under natural environmental conditions ferrous form of iron is rapidly oxidized to ferric iron and offers a strong substrate for its oxidation. Since the electrode potential of ferrous/ferric under acidic condition is high enough, only oxygen molecule with the sufficient availability of protons can be preferred as electron acceptor and subsequent formation of water molecule. Therefore ferrous iron can serve as electron donor only under aerobic environment. The process of iron oxidation occurs spontaneously if the pH is not very acidic. However, under strong acidic conditions, acidophilic microbes, especially bacteria, catalyze the oxidation of ferrous iron eventually resulting into formation of biogenic iron oxides. The oxidation of manganese [Mn(II)] to produce insoluble manganese oxides [Mn(III) and Mn(IV)] is also evident in different bacterial and fungal species (Miyata et al., 2007; Martínez-Ruiz et al., 2020) directing their utilization for toxic substance remediation. Microbially facilitated Mn (II) oxidation under circumneutral environment has been recorded multifold greater as compared to abiotically catalyzed processes (Tebo et al., 2004) and thereby indicates the promising potential in generation of various oxides and hydroxides under aqueous and terrestrial environment. The formation of insoluble manganese oxides has environmentally important contribution in regulating the biogeochemical cycling of multiple elements occurring in the natural ecosystem (Tebo et al., 2004; Meng et al., 2009). Moreover, biogenic manganese oxides can sequester organic as well as inorganic metal contaminants of hazardous nature (Tran et al., 2018; He et al., 2020). The arsenite [As(III)]−oxidizing bacteria are mostly documented in arsenic-enriched water and soil samples. The conversion of arsenite to arsenate [As (V)] facilitated by *aio* genes (Silver and Phung, 2005) is considered as a mechanism of detoxification. Oxidative transformation of As(III) to more adsorptive As(V) by microbiological activities is quite helpful in treatment of arsenic-laden water and wastewater (Bagade et al., 2019). The process of As(III) oxidation linked with reduction of nitrate causing reduced bioavailability of metalloid has also been demonstrated as a mechanism of alleviation of arsenic toxicity (Li et al., 2020a,b). The presence of arsenite-oxidizing microorganism in aqueous and terrestrial environment is intimately linked with immobilization (Zhang et al., 2020) biogeochemical cycling and abundance of different arsenic species. Oxidation of comparatively highly toxic form of antimony Sb (III) to Sb (V) via microbiological actions offers an

Microbe Mediated Remediation of Environmental Contaminants. DOI: https://doi.org/10.1016/B978-0-12-821199-1.00011-0

important mechanism of detoxification (Li et al., 2013). The oxidation of antimony is associated with the formation of mineral oxides displaying strong adsorption characteristics and hence may affect the biogeochemical cycling (Terry et al., 2015; Loni et al., 2020) not only of antimony but also of other metals and metalloids. The bacterial species showing resistance, tolerance, and potential to oxidize higher concentrations of antimony can be successfully employed for bioremediation.

The microbial community occurring in aquatic and terrestrial environment has differing potential in oxidation of noxious metals and metalloids. The formation of different minerals of metals by metal-oxidizing microbes has promising potential in removal of hazardous metals and metalloids to an acceptable level recommended by regulatory authorities. Further, oxidation and fate of any metal in an environment is largely determined by different environmental factors and microbial species involved. The present chapter has discussed about different metal-oxidizing microbes, particularly bacteria and potential opportunities of metal oxidizers in bioremediation.

11.2 Iron-oxidizing microbes and application in metal and other contaminant removal

Iron found in the Earth is a biologically important metal element required for myriads of biochemical processes and exists naturally as ferrous (Fe^{2+}) and ferric (Fe^{3+}) form. The oxidation state of iron under prevailing environmental conditions is regulated considerably by redox status, pH, temperature, nutritional composition, and microbial communities surviving therein. Oxidation of iron occurs in most of the aqueous environments having elemental or reduced form of iron, that is, ferrous iron (Eggerichs et al., 2020). Microorganisms catalyzing the iron oxidation obtain energy required for their multiplication with the involvement of either oxygen or nitrate as electron acceptor (Emerson et al., 2010). The microbial iron oxidation has been categorized into four distinct types as (1) aerobic Fe^{2+} oxidation under acidic environment, (2) microaerobic Fe^{2+} oxidation under neutral condition, (3) anoxic phototropic Fe^{2+} oxidation under neutral situation, and (4) nitrate-dependent Fe^{2+} oxidation performed under neutral and anoxic situations (Blake et al., 1992; Widdel et al., 1993; Neubauer et al., 2002; Sobolev and Roden, 2002). Studies on removal of heavy metals from contaminated soil by indigenous iron-oxidizing bacteria are demonstrated by Zagury et al. (1994). The isolated bacteria (*Acidithiobacillus ferrooxidans*) acclimated at pH 4 and at room temperature were observed to lower the pH of soil solution below 2.5 and increased the oxidation reduction potential ranging from 500 to 600 mV. The addition of $(NH_4)_2SO_4$ and K_2HPO_4 augmented the pace of iron oxidation by iron oxidizers. The percent removal of zinc and manganese was higher as compared to copper. Interestingly, the removal of metal by iron-oxidizing microflora was affected by metal speciation as well as soil types taken in study. Role of iron-oxidizing bacteria *Leptothrix ochracea* and *Gallionella ferruginea* in elimination of iron and manganese from groundwater is described by Katsoyiannis and Zouboulis (2004). The oxidation of iron resulted into formation of amorphous iron minerals, further speeding up the oxidation rate by multiple orders. Removal of iron followed first-order reaction kinetics with half-life of 0.9 min demonstrating rapid removal and in that way offers an economical and nature-friendly approach because of no requirement of chemicals for bioremediation of groundwater contaminants. The study pertaining to involvement of iron-oxidizing bacteria in sequestration of metals from sewage sludge is presented by Blais et al. (1993). The oxidation of iron by acclimated microbes reduced the pH and increased the oxidation reduction potential of sewage sludge. The order of metal removal from sludge by iron-oxidizing bacteria *Thiobacillus fenooxidans* (now *A. ferrooxidans*) was found as manganese > zinc > nickel > cadmium > copper > lead > chromium and not affected considerably by sludge form used during experimental investigation. Another batch study directed toward the elimination of metals from sewage sludge using indigenous iron-oxidizing bacteria was conducted by Xiang et al. (2000). Addition of iron in form of ferrous sulfate in inoculated sets led to the decrease in pH of sludge ranging from 2.0 to 2.5. After 16 days of treatment with iron oxidizers, considerable reduction in contents of metals such as copper, zinc, chromium, nickel, and lead present in sewage sludge reaching to agriculturally acceptable limits was recorded, supporting the employment of biological agents for metal contaminant remediation. Contribution of microaerophilic iron-oxidizing bacteria in removal of arsenic is recently reported (Tong et al., 2019). The ferric oxyhydroxides produced after remarkable biological activity of iron oxidizers with strong adsorption affinity for arsenic were responsible for elimination and adsorption of arsenite and arsenate. Treatment with iron oxidizers reduced the arsenic level to 4.8 μg/L from 600 μg/L. The generation of iron oxide mineral in due course led to the conversion of most of the arsenite into arsenate supporting the fact of arsenic oxidation followed by immobilization. The study also indicated the presence of arsenite oxidase gene referring to the capability of microbe-dependent arsenite conversion into arsenate. Oxidation of iron by iron-oxidizing bacteria *Gallionella* and *Leptothrix* culminating into the generation of so-called biogenic iron oxide has immense potential in removal of phosphate from solution phase (Buliauskaitė et al., 2020). *Gallionella* sp. was described more efficient in removal of phosphate as compared to *Leptothrix* sp. The biologically derived iron oxides displayed better phosphate sequestration performance as compared to chemically synthesized ones suggesting their implications in constructed treatment system for recovery of phosphate. Their study concluded the major role of adsorption and precipitation in phosphate removal by biogenic iron oxides. Groundwater contaminated with higher concentrations of iron and manganese can be effectively treated by iron-oxidizing bacteria (Aziz et al., 2020). The iron-oxidizing bacteria isolated from groundwater and grown on filter modulated the characteristic of limestone and removed 81.72% and 83.63% of iron and manganese, respectively, from contaminated groundwater within 2 h as compared to filter applied alone. The study revealed the enhancement effect of treatment time on iron and manganese oxidation by iron oxidizers. The formation of insoluble and oxidized minerals of iron and manganese on limestone filter further assisted the oxidation and adsorption of soluble forms of iron and manganese. Elimination of heavy metals (copper, zinc, chromium, and nickel) from electroplating sludge by *A. ferrooxidans* under low-voltage stimulus is recently documented (Wu et al., 2020) and thus paved the way for industrial-scale metal contaminant treatment. The potential iron-oxidizing bacteria *A. ferrooxidans* catalyzing the formation of crystalline biogenic tooeleite (iron–arsenic-based mineral) are considered to remove approximately 95% of As(III) under laboratory condition having pH 2.0–3.0 and iron concentration nearly 11 g/L (Li et al., 2020a,b). The study offered basic mechanisms underlying biogenic mineral formation, thereby providing an important clue for the development of treatment system exploiting rapidly multiplying iron-oxidizing bacteria for decontamination of As-affected environment.

11.3 Manganese-oxidizing microbes and application in metals and other contaminants removal

Manganese (Mn), an important micronutrient occurring in oxidation states between −3 and +7, is required by most of the life forms. The manganese existing in oxidation states +2, +3, and +4 is biologically important form (Brouwers et al., 2000). The reduced form Mn(II) is soluble and can be oxidized to unstable Mn(III) and insoluble Mn(IV) form by microbiological activities. Among different microorganisms explored, Mn(II)-oxidizing bacteria are of immense importance because of potential feature in removal of variety of heavy metals and organics. The amorphous nature of biogenically produced Mn oxides rendering vacant positions is the key factor responsible for adsorption of inorganic and organic contaminants. Manganese-oxidizing bacteria are involved in oxidation of Mn(II) leading to the formation of oxides of Mn(III) and Mn(IV) and most commonly found in habitats such as water transport pipes and aquifers enriched with Mn(II). Presence of Mn(II) in water beyond the recommended limit poses a health hazards on humans and animals. Studies are available indicating the Mn(II) removal in full-scale biofiltration system predominantly by the action of manganese-oxidizing bacteria leading to formation of manganese oxides and sorption of available Mn(II) rather than contributed maximally by the well-known oxidizer *Leptothrix* as previously considered (Burger et al., 2008). The biogenic manganese oxides as better adsorbent for cadmium as compared to synthetic manganese oxide (todorokite) in aquatic environment under varying pH conditions is presented by Meng et al. (2009). The biogenic manganese oxide synthesized by *Bacillus* sp. WH4 recovered from Fe–Mn nodules was in nano-range. Under given conditions the cadmium adsorption potential of biogenic manganese oxide was determined as 2.04, whereas for synthetic one the removal potential was 0.69 mmol/g, implying better sequestration characteristics of biologically derived manganese oxides and important role in cycling of toxic metals. Ability of manganese-oxidizing bacteria in efficient degradation of 17α-ethinylestradiol at small concentrations is recognized (Sabirova et al., 2008). The oxidative degradation of test xenoestrogen was carried out by oxidized forms Mn(III) and Mn(IV) produced after biological oxidation of Mn(II) mediated by *Leptothrix discophora* and different strains of *Pseudomonas putida*. The biological degradation product of 17α-ethinylestradiol was devoid of estrogenic potential, suggesting the employment of Mn(II)-oxidizing bacterial species in remediation of estrogen-contaminated environment. The ability of manganese-oxidizing bacteria *Acinetobacter* isolated from contaminated groundwater in removal of arsenic from synthetic solution containing 5 ppm As(III) at neutral pH is described by Singh et al. (2016). The manganese-encrusted bacteria upon immobilization in sodium alginate were more effective in arsenic sequestration in comparison to free cells. Isolation of manganese-resistant forms and contribution of acclimated manganese-oxidizing bacteria *P. putida* NRRL B-14878 in elimination of micropollutant paracetamol is clearly evident (Tepe and Cömert, 2019). Paracetamol at a concentration of 2 ppm was completely removed in 9 days at neutral pH by using bacteriogenic manganese oxide at a dose of 5 g/L thus offering opportunities in remediation of organic contaminants. The optimal conditions of pH, paracetamol concentration, and content of Mn oxide for maximum removal as determined by response surface methodology technique were found as 6.81, 9.82 ppm, and 6.36 g/L, respectively. Role of bacteria *Nocardia*, *Comamonas*, *Pseudomonas*, *Hyphomicrobium*, and *Mycobacterium* in catalyzing Mn(II) oxidation followed by manganese removal from synthetic manganese–enriched water in open bioreactors system is illustrated by Shoiful et al. (2020). Different substrates used for growth and multiplication of Mn(II)-oxidizing bacteria were equally effective and minor difference in Mn(II) sequestration was observed. The bioreactor fed with methanol had maximum removal of Mn(II) as well as consistency in performance, while one fed with K medium had least Mn(II) removal rate.

The application of manganese-oxidizing fungus *Acremonium strictum* strain KR21-2 in oxidation of Cr(III) and its subsequent sequestration under acidic and aerobic condition is recently presented by Suzuki et al. (2020). The biogenically synthesized manganese oxides further catalyzed the oxidation of reduced Cr(III) and facilitated their sorption. Under anaerobic condition the oxidation of Cr(III) by biogenic manganese oxide was very much reduced and caused the considerable release of Mn(II). However, repeated treatment under anaerobic environment caused increased elimination of Cr(III). The biogenic manganese oxides thus generated by microbiological action offer an ecologically sound technique for removing different ionic forms of chromium present in contaminated sites. The contribution of Mn(II)-oxidizing fungi *Cladosporium halotolerans* and *Hypocrea jecorina* in removal of Mn(II) is currently described (Mota et al., 2020). *C. halotolerans* removed approximately all of the Mn(II) present in the medium likely due to the activity of extracellular multicopper oxidase. A total of 50% Mn(II) was sequestered by *H. jecorina* biomass within 21 days of treatment and the removal was ascribed as the resultant of MnO_2 formation in the medium followed by adsorption of available Mn(II). The isolation of fungal strains equipped with Mn(II)-oxidizing activity has promising potential in treatment of water and wastewater containing excess of Mn(II).

11.4 Arsenite-oxidizing microbes and application in metalloids removal

Presence of metalloid arsenic in water (Singh and Singh, 2015, 2018) and soil beyond the safety limits is a major challenge worldwide. In aqueous environment, arsenic exists as arsenite [As(III)] and arsenate [As(V)] under reducing and oxidizing environmental conditions, respectively. For humans, As(III) is nearly sixty times more toxic as compared to As(V) (Neff, 1997). In water, As(III) occurs in neutral form, whereas As(V) as oxyanion; therefore considerable differences in mobilization properties are clearly evident. Being charged, the oxidized form of arsenic is much easier to remove in comparison to reduced form of arsenic. Therefore oxidation of arsenite to arsenate is carried out prior to elimination from contaminated environment. In contrast to the widely employed chemical oxidation method for removal of arsenic, biological methods are preferred due to their ecologically safe nature and absence of expensive chemicals leading to minimal environmental perturbations. The process of arsenite oxidation by bacteria is well recognized (Oremland and Stolz, 2003; Rhine et al., 2005) and can be successfully applied for remediation of arsenic-contaminated environment. Depending upon nutritional requirement, the arsenite-oxidizing bacteria (AsOB) can be classified as heterotrophic (Ike et al., 2008) and chemolithoautotrophic bacteria (Rhine et al., 2007). Study on AsOB *Microbacterium lacticum* possessing the capability of oxidizing 50 mmol/L responsible for sequestration of arsenic from water is presented by Mokashi and Paknikar (2002). The bacterially oxidized arsenic was easily removed by using $FeCl_3$, charcoal, and zero valent iron. The study offered the opportunity for designing arsenic treatment facilities based on integrated biological and chemical treatment methods. The bacterium isolated from sewage and supported on broken brick parts, under column condition, fed with methanol catalyzed the oxidation of arsenic at neural pH and room temperature. Potential role of AsOB *Thiomonas arsenivorans* in arsenic removal in a reactor system is reported by Wan et al. (2010). Reactor containing zero valent iron and inoculated with *T. arsenivorans* was able to remove 45% of total arsenic content in 1 h retention time and was not stable in nature because of immediate As(III) oxidation followed by sorption onto zero valent iron and oxidized form. The oxidation of arsenic by bacteria was linked with the axis length of the reactor regulating the distribution of bacterial cells. Exploration of AsOB communities present in groundwater causing biotransformation of arsenic species followed by its removal in biofiltration reactor system is described by Crognale et al. (2019) and provided a clue for the development of household filtration unit. The study investigated the effect of different parameters, including As(III)-to-As(V) ratio, initial As(III) content, filling substances utilized, and volume of filter, on process of biological arsenite oxidation. Biofiltration reactor unit containing coarse sand as filling material displayed oxidation reaching upto 90% within 3 h for initial arsenic concentration 100 ppb. The study revealed increase in arsenite-oxidizing genes during the oxidation of arsenic in biofiltration unit. Arsenic oxidation activity of bacteria isolated from ore mine is recently demonstrated by Thul et al. (2019). Among different isolates, six bacterial strains were able to withstand high arsenic content together with the As(III) oxidation as confirmed through silver nitrate assay and HPLC-ICP-MS. Analysis of isolates based on 16S rDNA sequences led to characterization of bacteria belonging to *Bacillus*, *Pseudomonas*, *Paenibacillus*, *Enterobacter*, and *Ochrobactrum*. Further studies on arsenic biotransformation coupled with the arsenic removal potential either by using the same bacterial isolate or suitable adsorbent could serve as an important step toward the mitigation of arsenic-induced toxicity. Characterization of bacterial strain *Bacillus firmus* L-148 isolated from lake having tolerance to different heavy metals with potential in rapid arsenic oxidation even in the presence of minimal nutritional sources and exploration in bioremediation practices is recently recognized (Bagade et al., 2020). The isolate was found tolerant to 3 M arsenic and was able to catalyze the oxidation of 75 mM As(III) in 2 weeks. The oxidation capability of As(III) even in the presence of heavy metals together with maximal oxidation efficiency at alkaline pH revealed its promising exploitation for bioremediation of contaminated water under changing natural conditions. As(III) oxidation reaching upto 77% within 15 days was observed for a synthetic medium with As(III) concentration 75 mM and supplemented with sodium acetate and ferrous iron. The experimental investigation demonstrated the varying effects of multimetals, including iron as well as carbon sources on bacterial growth and oxidation rate of As(III).

11.5 Antimony-oxidizing microbes and application in metalloids removal

Antimony (Sb) contamination resulting from anthropogenic activities, including industrial mining and smelting, has considerably deteriorated the quality of natural ecosystem (Sun et al., 2020). Contamination of soil and water by Sb, eventually gaining access to human food chain, is an alarming condition in many countries around the globe. Under natural environmental condition the metalloid antimony (Sb) occurs as antimonite [Sb(III)] and antimonate [Sb(V)]. The existence of two different oxidation states is determined largely by prevailing natural condition. Sb shares several geochemical characteristics with As and is thought to have similar effects on microorganisms. For instance, similar to arsenic, the oxygenic environment favors the presence of oxidized form Sb(V), whereas reducing conditions favor the dominance of Sb(III) (Hockmann et al., 2014), hence modulates their mobilization in a given aqueous system. Further, reduced form of Sb is more toxic as compared to oxidized form. Therefore, remediation of Sb-contaminated sites involving the oxidation in order to transform highly toxic form of Sb into less/nontoxic form is the basic strategy of detoxification. Since the oxidation processes relying on chemical methods are much expensive, amenable to create secondary chemical hazards and contaminate the nearby environment, the biological agent–based oxidation is considered much safe. Antimony-resistant bacteria obtained from sediment samples directing the heterotrophic oxidation of antimony is reported by Nguyen and Lee (2015). The Sb(III)-oxidizing bacteria falling in beta-Proteobacteria were *Cupriavidus* sp. NL4 and *Comamonas* sp. NL11 and those falling in the division of gamma-Proteobacteria were identified as *Acinetobacter* sp. NL1, *Acinetobacter* sp. NL12, *Pseudomonas* sp. NL2, *Pseudomonas* sp. NL5, *Pseudomonas* sp. NL6, and *Pseudomonas* sp. NL10. The isolate identified as *Cupriavidus* sp. NL4 was the most efficient oxidizer and observed to oxidize 100 μM of Sb(III) absolutely in 500 h. Other isolates were also able to oxidize the Sb(III) under similar conditions; however, complete oxidation could not be achieved even after increasing the treatment duration. The study provided important insights into involvement of diverse bacterial species in detoxification and cycling of Sb (III). Indigenous bacteria isolated from mine soil participating in aerobic Sb(III) oxidation is recently documented (Li et al., 2018). Three bacteria resistant to high Sb(III) concentrations were identified as *Cupriavidus*, *Moraxella*, and *Bacillus*. The strain *Moraxella* sp. S2 isolated from mine soil having highest Sb(III) concentration showed highest minimum inhibitory concentration; however, it was marked as least effective in catalyzing Sb(III) oxidation and could not absolutely convert 50 μM of Sb(III) to Sb(V) in 15 days. Interestingly, *Bacillus* sp. S3 exhibiting minimal resistance to Sb(III) was found to aerobically transform 100 μM Sb(III) within a time frame of 2 days. The isolation of indigenous bacteria from mining areas holds a promising future in remediation of heavily metal-contaminated sites. Bacterially catalyzed anaerobic oxidation of Sb linked with Fe oxidation and denitrification by facultative anaerobic bacterium *Sinorhizobium* sp. GW3 is presented by Li et al. (2019). The experimental analysis demonstrated oxidation of both Fe and Sb by isolated bacteria using nitrate as electron acceptor under in vitro anaerobic condition. The anaerobic oxidation of Sb(III) was characteristically increased by the presence of Fe, Sb, and nitrate, leading to the synthesis of ferric iron–based minerals accelerating non-biological oxidation of Sb and further sequestration/precipitation therein, and provides background information for understanding the geochemical cycling of metalloids under anaerobic condition. The peculiar anaerobic biological oxidation potential of both Fe and Sb has immense potential in the development of treatment system for remediation of Sb-contaminated groundwater as well as other sites having reducing condition.

11.6 Conclusion

Metal pollutants originating from different industrial activities are severely contaminating the water and soil posing harmful impact on human health and environment. The concentration of toxic metals and metalloids beyond the recommended limits has deteriorated the natural soil characteristics and microbial communities thereby limiting agricultural productivity. The metal decontamination strategies using chemicals have several disadvantages calling for the development of biological methods for remediation of contaminated sites. Metal- and metalloid-contaminated habitats are expected to support the growth and multiplication of metal-resistant microbial forms catalyzing the process of metal oxidation. The biological activities of metal-oxidizing microbes convert reduced form of toxic metals into oxidized and less toxic form through the process known as detoxification. In addition, the oxidation of some of the metals may result into the synthesis of mineral oxides with efficient adsorption potential for variety of metals and metalloids. Most importantly, the formation of oxidized forms of minerals regulates the mobilization and therefore geochemical cycling of diverse metals as well as organic materials in the vicinity. Thus metal-oxidizing microbes resistant to higher concentrations of metals and possessing the ability to rapidly acclimatize under changing natural environmental conditions hold promising application in treatment of sites contaminated with metals, metalloids, and toxic organics.

Acknowledgments

Authors are thankful to Head, Department of Botany, Banaras Hindu University, Institute of Environment and Sustainable Development (IESD), Banaras Hindu University, and University Grants Commission (UGC) for providing necessary facilities.

References

Aziz, H.A., Tajarudin, H.A., Wei, T.H.L., Alazaiza, M.Y.D., 2020. Iron and manganese removal from groundwater using limestone filter with iron-oxidized bacteria. Int. J. Environ. Sci. Technol. 17, 2667–2680.

Bagade, A., Nandre, V., Ghosh, S., Battu, S., Haram, S., Giri, A., et al., 2019. Rapid and efficient sequestration of arsenic from contaminated water using hypertolerant *Bacillus* L-148 sp.: a two-step process. Green Chem. 21 (9), 2245–2251.

Bagade, A., Nandre, V., Paul, D., Patil, Y., Sharma, N., Giri, A., et al., 2020. Characterisation of hyper tolerant *Bacillus firmus* L-148 for arsenic oxidation. Environ. Pollut. 261, 114124.

Blais, J.F., Tyagi, R.D., Auclair, J.C., 1993. Metals removal from sewage sludge by indigenous iron-oxidizing bacteria. J. Environ. Sci. Health, A 28 (2), 443–467.

Blake, R., Shute, E.A., Waskovsky, J., Harrison, A.P. J.R., 1992. Respiratory components in acidophilic bacteria that respire on iron. Geomicrobiol. J. 10 (3–4), 173–192.

Brouwers, G.J., Vijgenboom, E., Corstjens, P.L.A.M., De Vrind, J.P.M., De Vrind-De Jong, E.W., 2000. Bacterial Mn^{2+} oxidizing systems and multicopper oxidases: an overview of mechanisms and functions. Geomicrobiol. J. 17 (1), 1–24.

Buliauskaitė, R., Wilfert, P., Suresh Kumar, P., de Vet, W.W., Witkamp, G.J., Korving, L., et al., 2020. Biogenic iron oxides for phosphate removal. Environ. Technol. 41 (2), 260–266.

Burger, M.S., Krentz, C.A., Mercer, S.S., Gagnon, G.A., 2008. Manganese removal and occurrence of manganese oxidizing bacteria in full-scale biofilters. J. Water Supply Res. Technol. AQUA 57 (5), 351–359.

Crognale, S., Casentini, B., Amalfitano, S., Fazi, S., Petruccioli, M., Rossetti, S., 2019. Biological As (III) oxidation in biofilters by using native groundwater microorganisms. Sci. Total Environ. 651, 93–102.

Eggerichs, T., Wiegand, M., Neumann, K., Opel, O., Thronicker, O., Szewzyk, U., 2020. Growth of iron-oxidizing bacteria *Gallionella ferruginea* and *Leptothrix cholodnii* in oligotrophic environments: Ca, Mg, and C as limiting factors and *G. ferruginea* necromass as C-source. Geomicrobiol. J. 37 (2), 190–199.

Emerson, D., Fleming, E.J., McBeth, J.M., 2010. Iron-oxidizing bacteria: an environmental and genomic perspective. Annu. Rev. Microbiol. 64, 561–583.

He, Z., Zhang, Q., Wei, Z., Zhu, Y., Pan, X., 2020. Simultaneous removal of As (III) and Cu (II) from real bottom ash leachates by manganese-oxidizing aerobic granular sludge: performance and mechanisms. Sci. Total Environ. 700, 134510.

Hockmann, K., Lenz, M., Tandy, S., Nachtegaal, M., Janousch, M., Schulin, R., 2014. Release of antimony from contaminated soil induced by redox changes. J. Hazard. Mater. 275, 215–221.

Ike, M., Miyazaki, T., Yamamoto, N., Sei, K., Soda, S., 2008. Removal of arsenic from groundwater by arsenite-oxidizing bacteria. Water Sci. Technol. 58 (5), 1095–1100.

Katsoyiannis, I.A., Zouboulis, A.I., 2004. Biological treatment of Mn (II) and Fe (II) containing groundwater: kinetic considerations and product characterization. Water Res. 38 (7), 1922–1932.

Li, J., Wang, Q., Zhang, S., Qin, D., Wang, G., 2013. Phylogenetic and genome analyses of antimony-oxidizing bacteria isolated from antimony mined soil. Int. Biodeterior. Biodegrad. 76, 76–80.

Li, J., Yu, H., Wu, X., Shen, L., Liu, Y., Qiu, G., et al., 2018. Novel hyper antimony-oxidizing bacteria isolated from contaminated mine soils in China. Geomicrobiol. J. 35 (8), 713–720.

Li J., Zhang Y., Zheng S., Liu F., Wang, G., 2019. Anaerobic bacterial immobilization and removal of toxic Sb (III) coupled with Fe (II)/Sb (III) oxidation and denitrification. Front. Microbiol. 10, 360.

Li, Q., Zhang, M., Yang, J., Liu, Q., Zhang, G., Liao, Q., et al., 2020a. Formation and stability of biogenic tooeleite during Fe (II) oxidation by *Acidithiobacillus ferrooxidans*. Mater. Sci. Eng.: C. 111, 110755. Available from: https://doi.org/10.1016/j.msec.2020.110755.

Li, X., Qiao, J., Li, S., Häggblom, M.M., Li, F., Hu, M., 2020b. Bacterial communities and functional genes stimulated during anaerobic arsenite oxidation and nitrate reduction in a paddy soil. Environ. Sci. Technol. 54 (4), 2172–2181.

Loni, P.C., Wu, M., Wang, W., Wang, H., Ma, L., Liu, C., et al., 2020. Mechanism of microbial dissolution and oxidation of antimony in stibnite under ambient conditions. J. Hazard. Mater. 385, 121561.

Manoj, S.R., Karthik, C., Kadirvelu, K., Arulselvi, P.I., Shanmugasundaram, T., Bruno, B., et al., 2020. Understanding the molecular mechanisms for the enhanced phytoremediation of heavy metals through plant growth promoting rhizobacteria: a review. J. Environ. Manage. 254, 109779.

Martínez-Ruiz, E.B., Cooper, M., Fastner, J., Szewzyk, U., 2020. Manganese-oxidizing bacteria isolated from natural and technical systems remove cylindrospermopsin. Chemosphere 238, 124625.

Meng, Y.T., Zheng, Y.M., Zhang, L.M., He, J.Z., 2009. Biogenic Mn oxides for effective adsorption of Cd from aquatic environment. Environ. Pollut. 157 (8–9), 2577–2583.

Miyata, N., Tani, Y., Sakata, M., Iwahori, K., 2007. Microbial manganese oxide formation and interaction with toxic metal ions. J. Biosci. Bioeng. 104 (1), 1–8.

Mokashi, S.A., Paknikar, K.M., 2002. Arsenic (III) oxidizing *Microbacterium lacticum* and its use in the treatment of arsenic contaminated groundwater. Lett. Appl. Microbiol. 34 (4), 258–262.

Mota, E.A., Felestrino, É.B., Leão, V.A., Guerra-Sá, R., 2020. Manganese (II) removal from aqueous solutions by *Cladosporium halotolerans* and *Hypocrea jecorina*. Biotechnol. Rep. e00431.

Neff, J.M., 1997. Ecotoxicology of arsenic in the marine environment. Environ. Toxicol. Chem. 16 (5), 917–927.

Neubauer, S.C., Emerson, D., Megonigal, J.P., 2002. Life at the energetic edge: kinetics of circumneutral iron oxidation by lithotrophic iron-oxidizing bacteria isolated from the wetland-plant rhizosphere. Appl. Environ. Microbiol. 68 (8), 3988–3995.

Nguyen, V.K., Lee, J.U., 2015. Antimony-oxidizing bacteria isolated from antimony-contaminated sediment—a phylogenetic study. Geomicrobiol. J. 32 (1), 50–58.

Oremland, R.S., Stolz, J.F., 2003. The ecology of arsenic. Science 300 (5621), 939–944

Rawlings, D.E., 2005. Characteristics and adaptability of iron-and sulfur-oxidizing microorganisms used for the recovery of metals from minerals and their concentrates. Microb. Cell Fact. 4 (1), 13.

Rhine, E.D., Chadhain, S.N., Zylstra, G.J., Young, L.Y., 2007. The arsenite oxidase genes (aroAB) in novel chemoautotrophic arsenite oxidizers. Biochem. Biophys. Res. Commun. 354 (3), 662–667.

Rhine, E.D., Garcia-Dominguez, E., Phelps, C.D., Young, L.Y., 2005. Environmental microbes can speciate and cycle arsenic. Environ. Sci. Technol. 39 (24), 9569–9573.

Sabirova, J.S., Cloetens, L.F.F., Vanhaecke, L., Forrez, I., Verstraete, W., Boon, N., 2008. Manganese-oxidizing bacteria mediate the degradation of 17α-ethinylestradiol. Microb. Biotechnol. 1 (6), 507–512.

Shoiful, A., Ohta, T., Kambara, H., Matsushita, S., Kindaichi, T., Ozaki, N., et al., 2020. Multiple organic substrates support Mn (II) removal with enrichment of Mn (II)-oxidizing bacteria. J. Environ. Manage. 259, 109771.

Silver, S., Phung, L.T., 2005. Genes and enzymes involved in bacterial oxidation and reduction of inorganic arsenic. Appl. Environ. Microbiol. 71 (2), 599–608.

Singh, A.L., Singh, V.K., 2015. Arsenic contamination in ground water of Ballia, Uttar Pradesh state, India. J. Appl. Geochem. 17 (1), 78–85.

Singh, A.L., Singh, V.K., 2018. Assessment of groundwater quality of Ballia district, Uttar Pradesh, India, with reference to arsenic contamination using multivariate statistical analysis. Appl. Water Sci. 8 (3), 95.

Singh, A.L., Singh, V.K., Yadav, A., 2016. Arsenic sequestration by manganese-oxidizing *Acinetobacter* sp. Indian J. Biotechnol. 15, 525–530.

Sobolev, D., Roden, E.E., 2002. Evidence for rapid microscale bacterial redox cycling of iron in circumneutral environments. Antonie van Leeuwenhoek 81 (1–4), 587–597.

Sun, X., Kong, T., Xu, R., Li, B., Sun, W., 2020. Comparative characterization of microbial communities that inhabit arsenic-rich and antimony-rich contaminated sites: responses to two different contamination conditions. Environ. Pollut. 260, 114052.

Suzuki, R., Tani, Y., Naitou, H., Miyata, N., Tanaka, K., 2020. Sequestration and oxidation of Cr (III) by fungal Mn oxides with Mn (II) oxidizing activity. Catalysts 10 (1), 44.

Tebo, B.M., Bargar, J.R., Clement, B.G., Dick, G.J., Murray, K.J., Parker, D., et al., 2004. Biogenic manganese oxides: properties and mechanisms of formation. Annu. Rev. Earth Planet. Sci. 32, 287–328.

Tepe, O., Cömert, S., 2019. Enhanced removal of paracetamol using biogenic manganese oxides produced by *Pseudomonas putida* B-14878 and process optimization by RSM. Water Air Soil Pollut. 230 (12), 309.

Terry, L.R., Kulp, T.R., Wiatrowski, H., Miller, L.G., Oremland, R.S., 2015. Microbiological oxidation of antimony (III) with oxygen or nitrate by bacteria isolated from contaminated mine sediments. Appl. Environ. Microbiol. 81 (24), 8478–8488.

Thul, S.T., Nigam, B., Tiwari, S., Pandey, R.A., 2019. Arsenite-oxidation performance of microbes from abandoned iron ore mine. Indian J. Biotechnol. 18, 34–41.

Tong, H., Liu, C., Hao, L., Swanner, E.D., Chen, M., Li, F., et al., 2019. Biological Fe (II) and As (III) oxidation immobilizes arsenic in micro-oxic environments. Geochim. Cosmochim. Acta 265, 96–108.

Tran, T.N., Kim, D.G., Ko, S.O., 2018. Synergistic effects of biogenic manganese oxide and Mn (II)-oxidizing bacterium *Pseudomonas putida* strain MnB1 on the degradation of 17 α-ethinylestradiol. J. Hazard. Mater. 344, 350–359.

Wan, J., Klein, J., Simon, S., Joulian, C., Dictor, M.C., Deluchat, V., et al., 2010. AsIII oxidation by *Thiomonas arsenivorans* in up-flow fixed-bed reactors coupled to As sequestration onto zero-valent iron-coated sand. Water Res. 44 (17), 5098–5108.

Widdel, F., Schnell, S., Heising, S., Ehrenreich, A., Assmus, B., Schink, B., 1993. Ferrous iron oxidation by anoxygenic phototrophic bacteria. Nature 362 (6423), 834–836.

Wu, P., Zhang, L.J., Lin, C.B., Xie, X.X., Yong, X.Y., Wu, X.Y., et al., 2020. Extracting heavy metals from electroplating sludge by acid and bioelectrical leaching using *Acidithiobacillus ferrooxidans*. Hydrometallurgy 191, 105225.

Xiang, L., Chan, L.C., Wong, J.W.C., 2000. Removal of heavy metals from anaerobically digested sewage sludge by isolated indigenous iron-oxidizing bacteria. Chemosphere 41 (1–2), 283–287.

Zagury, G.J., Narasiah, K.S., Tyagi, R.D., 1994. Adaptation of indigenous iron-oxidizing bacteria for bioleaching of heavy metals in contaminated soils. Environ. Technol. 15 (6), 517–530.

Zhang, D.R., Chen, H.R., Xia, J.L., Nie, Z.Y., Fan, X.L., Liu, H.C., et al., 2020. Humic acid promotes arsenopyrite bio-oxidation and arsenic immobilization. J. Hazard. Mater. 384, 121359.

Further reading

Wang, G., Li, J., Zhang, Y., Zheng, S., Liu, F., 2019. Anaerobic bacterial immobilization and removal of toxic Sb (III) coupled with Fe (II)/Sb (III) oxidation and denitrification. Front. Microbiol. 10, 360.

Chapter 12

Extremophilic microbes and their application in bioremediation of environmental contaminants

Bhupendra Koul, Roni Chaudhary and Pooja Taak

Department of Biotechnology, School of Bioengineering and Biosciences, Lovely Professional University, Phagwara, India

12.1 Introduction

The industrial and technological advancements and activities have resulted in the problem of accumulation of toxic wastes and pollution of soil, water, and air (Tabak et al., 2005; Azubuike et al., 2016). Various anthropogenetic activities such as smelting, industrial waste disposal, accidental oil spillage, petroleum refining, automobile exhausts, mining, burning of fossil fuels, R&D activities produce organic and inorganic contaminants in the environment (Sar et al., 2013; Hazen et al., 2016; Singh et al., 2020). Natural activities such as volcanic eruption, mineral rock weathering also result in environmental contamination which affects the flora and fauna of both water and land ecosystems (Giovanella et al., 2019). Conventional techniques of pollution remediation are less reproducible, involve high operational cost, and may produce residual potential contaminants (Chen et al., 2015; Giovanella et al., 2019). On the contrary, biological remediation techniques have achieved more attention due to their feasibility, cost-effectiveness, and environment friendliness (Sar et al., 2013; Alegbeleye et al., 2017; Giovanella et al., 2019; Gupta et al., 2020).

Bioremediation strategies deploy microorganisms (bacteria, archaea, and fungi) to reduce, eliminate, and modify complex toxic pollutants from soil and water into simpler nontoxic degradable ones (Tabak et al., 2005; Giovanella et al., 2019; Rajkumari et al., 2019; Singh et al., 2019). The microorganisms exhibit rapid growth rate, have variable metabolic needs, and can be genetically manipulated. These properties make them useful for effective bioremediation of environmental contaminants.

Various extremophilic microbes that inhabit extreme environmental conditions including acidic or alkaline pH, higher or low temperature, and higher salt concentration have evolved various adaptation strategies for their survival (Peeples, 2014). Based on their different habitat characteristics extremophiles can be grouped as (1) thermophiles, (2) mesophiles, (3) halophiles, (4) psychrophiles, and (5) xerophiles. To adapt themselves in unfavorable conditions, extremophiles have developed different physiological, enzymatic, and catalytic characteristics; therefore they can be harnessed for the remediation of environmental pollutants (Chino et al., 2011; Ramos et al., 2011; Sar et al., 2013; Raddadi et al., 2015; Giovanella et al., 2019). Various extremophilic thermophiles (*Pyrobaculum aerophilum*) can remediate arsenite toxic heavy metals while the halophiles such as *Haloferax gibbonsii*, *Haloarcula marismortui*, and *Haloferax mediterranei* remediate chlorates and nitrates from polluted soils (Stepanov et al., 2014; Fox et al., 2016; Wang et al., 2019; Miralles-Robledillo et al., 2019).

In this chapter, bioremediation of environmental contaminants such as hydrocarbon pollutants, polycyclic aromatic hydrocarbon (PAH), and halogenated hydrocarbon by various extremophiles such as psychrophiles, halophiles, haloalkaliphiles, haloacidophiles, thermophiles, and mesophiles is discussed.

12.2 Environmental contaminants

Crude oil and petroleum-based products contain various hydrocarbons such as aromatic compounds (30%), alkanes (60%), and several inorganic components such as sulfur (S), nitrogen (N), and oxygen (O) (Hazen et al., 2016). Nearly

Microbe Mediated Remediation of Environmental Contaminants. DOI: https://doi.org/10.1016/B978-0-12-821199-1.00012-2

40% of the total energy consumption of the world is accomplished through crude oil (Brown and Skipsey, 1986). However, the processing, refining, storage, and transport of petroleum-based products leads to accidental oil spillage, which cause extensive damage to air, land, and water ecosystem (Holliger et al., 1997; Rajkumari et al., 2019). There are specific reports of accidental oil spillage from across the world that has proved to be detrimental for both flora and fauna of soil and water ecosystems: (1) Amoco Cadiz in Brittany coast (1978); (2) Atlantic express in Trinidad and Tobago (1979); (3) Arabian gulf in Kuwait (2002); (4) deepwater horizon oil spill in the Gulf of Mexico (2010); (5) keystone pipeline in the United States Menno (2016); (6) Ennore oil spill in India (2017); and (7) Bunker oil spill in Hong Kong, China (2019) (Swannell et al., 1996; Burgherr, 2007; Issa and Vempatti, 2018; Kostka et al., 2011; Mazoomdaar, 2017; Bland, 2018; Mike, 2019; Martin and Eddie, 2019). The National Oil Spill Detection and Response Agency (NOSDRA), Nigeria has recorded 9343 cases of oil spills in the last 10 years. Ballast water discharged (approximately 7.5 million L/h) by the ships (cruise, cargo, large tankers, and bulk cargo) have negative impact (ecological and economic damage) on marine environment (oceanic and shordine) along with serious human health issues (Prasad et al., 2013; Nguyen and Dong, 2018; Rajkumari et al., 2019; Giovanella et al., 2019). Various hydrocarbon pollutants result from oil spills; PAHs originate from industrial effluents and pyrolysis of organic matter; and halogenated hydrocarbon results from agricultural use of pesticides and herbicides (Kennish et al., 2017). Hydrocarbon contamination in soil decreases the agricultural land productivity and also water bodies that are used for fishing (Dabbs et al., 1996; Odokuma et al., 2012; McGenity et al., 2012; Rajkumari et al., 2019). PAHs are one of the primary pollutants that may cause immunogenic, toxigenic, mutagenic, and carcinogenic effects on both lower and higher organisms (Prasad et al., 2013; Varjani, 2017; Giovanella et al., 2019). They may cause damage to kidney, liver, lung, skin, and eyes (Varjani, 2017). The damage caused by the earthquake to Fukushima Nuclear Power Plant on March 11, 2011, released a large amount of radioactive material (xenon, tellurium, cesium, strontium, iodine, etc.) into the environment (Chino et al., 2011; Stohl et al., 2012; Iwamoto and Minoda, 2018). This radioactive contamination can be also remediated by extremophilic microorganisms.

Arsenite is the most toxic metal present in ample amount in sediments and groundwater, which affects the quality of water resources by contaminating fluvial waters. Thermophilic archaea such as *P. aerophilum* and *Pyrobaculum aersenaticum* can oxidize, intracellularly bioaccumulate and methylate arsenic species, and can be useful in remediation of arsenic-contaminated soil and water (Jebelli et al., 2017; Serrano et al., 2017). Chlorates and nitrates bioremediation are also done by extremophilic organisms such as *Halophiles*, *Haloferax gibbonsii*, *H. marismortui*, and *H. mediterranei* (Stepanov et al., 2014; Fox et al., 2016; Wang et al., 2019; Miralles-Robledillo et al., 2019). Textiles dyes are wastes released from the textile industries which can be decolorized and detoxified by mesophilic microorganisms (Khandegar and Saroha, 2013). All these aforementioned environmental contaminants show toxigenic and carcinogenic effects in the environment which can be degraded by the extremozymes (oxidoreductases, ligninolytic, and catalase enzyme) present in the extremophiles.

12.3 Extremophiles: adaptation mechanisms

The extremophiles are grouped according to the chemical or physical parameters of the ecological conditions in which they live and show their optimum growth. Hence, based on these characterizing parameters they can be divide into (1) hyperthermophiles that live at temperature greater than 80°C; (2) thermophiles that survive at temperatures ranging from 60°C to 80°C; and (3) psychrophiles that survive at temperatures below 15°C. The extremophiles can also be classified according to their pH requirement for survival: (1) alkaliphiles (surviving at $pH > 9$) or (2) acidophiles (surviving at $pH < 3$) (Chandrasekaran and Kumar, 2010; Di Donato et al., 2019). There are other terminologies such as halophiles—bacterial species that live in high saline (high NaCl) environmental condition; microanaerobes—that survive in total absence of oxygen or in its low concentration; xerophiles—that survive and reproduce in low water availability; radioresistant—those microorganisms which live under high radiation level; barophiles—those microorganisms that thrive under high pressure.

Some extremophiles can simultaneously resist more than one extreme environmental conditions and can be defined as poly-extremophiles, for example, thermoacidophiles bacteria that survive at both low pH and high temperature; some halophilic species can tolerate both alkaline pH and high salt concentration; or finally, some radioresistant species that tolerate other unfavorable conditions such as very low temperature, absence of water, and vacuum (Rothschild and Mancinelli, 2001; Caruso et al., 2018a). In the marine ecosystem, various types of extremophiles can be found that survive in more than one extremity such as extreme temperature, hypersalinity, and high pressure. Some extreme conditions include the hydrothermal vents, ocean depths, Antarctic or arctic regions, cold sea, the black smokers, and hypersaline lakes. In the last few decades, various biotechnologically useful molecules such as enzymes (also called extremozymes) have been reported and extracted from the extremophiles (Bolhuis and Cretoiu, 2016; Di Donato et al., 2019). These unique survival capabilities/strategies of the extremophiles make them useful candidates for bioremediation of polluted soil and water (Fig. 12.1). Table 12.1 shows research reports on extremophilic microorganism-mediated remediation of polluted soil and water.

As already mentioned extremophiles are microorganisms that are capable of surviving in various extreme conditions where life is not possible, various mechanisms such as alteration in DNA repair system, cell permeability, and protein folding are involved (Orellana et al., 2018; Rajkumari et al., 2019). Acidophiles are microbes that grow optimally at pH < 5 and are classified into groups, as extreme acidophiles that survive at pH 3 or less and moderate extremophiles which survive at pH between 3 and 5 (Johnson, 2007; Johnson, 2008; Orellana et al., 2018). Three mechanisms are evolved in acidophiles for adaptation in acidic conditions: (1) first mechanism includes proton flux system that involves active transport of proton to maintain pH equivalence in both intracellular and extracellular environment. Various transport system F0–F1 ATP synthase system, antiporter, and symporter system are also involved in this mechanism. This mechanism has been reported in various acidophiles such as *Thermoplasma acidophilum*, *Leptosprillium ferrooxidans* (Michels and Bakker, 1985; Christel et al., 2018; Orellana et al., 2018); (2) the second mechanism that involves inhibition mechanism where proton influx in inhibited by efflux of K+ ions to create a positive membrane potential

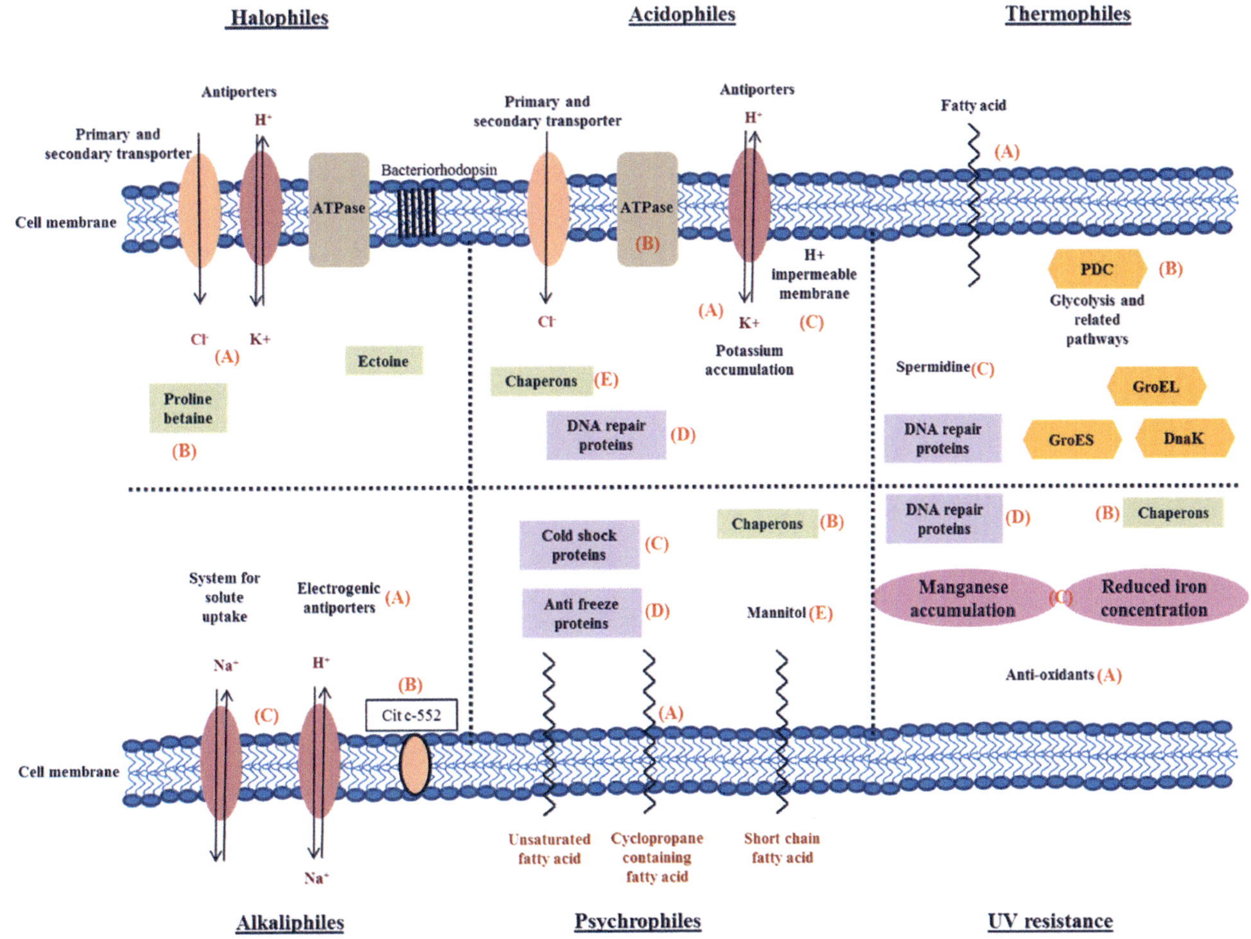

FIGURE 12.1 Molecular mechanisms of extremophiles that are useful for bioremediation of polluted soils and waters: [I] *Halophiles*: (A) High salt-in mechanism (i) potassium uptake into cells by concerted action of bacteriorhodopsin and ATP synthase, (ii) chloride transporters and (B) low-salt strategy (i) synthesis or uptake of osmoprotectants (proline-betaine, ectoine) to maintain osmotic pressure; [II] *Acidophiles*: (A) potassium antiporter, (B) ATP synthase, (C) proton impermeable membrane, (D) DNA-repair proteins, and (E) chaperones; [III] *Thermophiles*: (A) long chain dicarboxylic fatty acids, (B) upregulated glycolysis proteins and (C) polyamines (spermidine); [IV] *Alkaliphiles*: (A) electrogenic antiporters for proton accumulation, (B) Cytochrome *c*-552 enhanced terminal oxidation function by electron and H^+ accumulation and (C) Na^+-solute uptake system; [V] *Psychrophiles*: (A) cyclopropane containing fatty acids and short chain fatty acids, (B) chaperones, (C) cold shock proteins, (D) antifreeze proteins and (E) manniol or other solutes in cytoplasm as cryo-protectants to prevent protein aggregation; and [VI] *UV resistance*: (A) antioxidants, (B) chaperones, (C) manganese accumulation and reduced iron levels and (D) DNA-repair proteins.

TABLE 12.1 Reports on extremophilic microorganism-mediated remediation of polluted soil and water.

Microbe(s)	Pollutant	Soil/water	Reference(s)
Pseudomonas BI7 and B18	Naphthalene, C5–C12 *n*-alkanes, and toluene	Soil	Whyte et al. (1997)
Pseudomonas sp.	Phenol	Soil	Selvaratnam et al. (1997)
Acinetobacter ADH-1	Cyclohexane, crude oil, hexadecane, aromatic gas–oil, hydrogenated gas–oil, dodecane, and kerosene	Soil	MacCormack and Fraile (1997)
Arthrobacter protophormiae	Hexadecane	Soil	Pruthi and Cameotra (1997)
Acidiphilium facilis, Acidocella sp. SRS 1	Anthracene, toluene, phenanthrene, and naphthalene	Coal pile storage area	Stapleton et al. (1998)
Pseudomonas Sag-50G and Cam-1	Biphenyl	Soil	Master and Mohn (1998)
Nocardioides sp. strain KP7	PAH	Soil	Saito et al. (2000)
Bacillus thermoleovorans Hamburg 2	Naphthalene	Compost	Annweiler et al. (2000)
Rhodococcus	Pristine, C6–C20 *n*-alkanes, and JP8 jet fuel	Soil	Bej et al. (2000)
Pseudomonas DhA-91	Dodecane, Jet A-1 jet fuel, and octane	Soil	Yu et al. (2000)
Sphingomonas	Pristine, Jet A-1 jet fuel, and dodecane	Soil	
Pseudomonas DhA-91	Dodecane, Jet A-1 jet fuel, and octane	Soil	
Pseudomonas IpA-92	Toluene	Soil	
Pseudomonas IpA-93	Benzene and toluene	Soil	
Pseudomonas Ant 5	JP8 jet fuel	Soil	Aislabie et al. (2000)
Pseudomonas Ant 7	2-Methyl naphthalene, *p*-xylene, 1,2,4-trimethyl benzene naphthalene, 1-methyl naphthalene, and JP8 jet fuel	Soil	
Pseudomonas 7/22	*m*- and *p*-Xylene, JP8 jet fuel, 1,2,4-trimethyl benzene, and toluene	Soil	
Sphingomonas Ant 20	Phenanthrene, 1-methyl naphthalene, and JP8 jet fuel	Soil	
Pseudomonas Ps 8	Pristine, hexadecane and Jet A-1 fuel	Soil	Thomassin-Lacroix et al. (2001)
Rhodococcus Rho10	Dodecane and Jet A-1 fuel	Soil	
Cycloclasticus sp.	PAH	Soil	Kasai et al. (2002); Alegbeleye et al. (2017)
Sphingomonas sp.	Aromatic hydrocarbon	Soil	Baraniecki et al. (2002)
Pseudomonas 5B	Hexane and JP8 jet fuel	Soil	Eckford et al. (2002)
Pseudomonas PK4	Hexadecane, dodecane, and pyrene	Soil	Eriksson et al. (2002)
Pseudomonas 30-3	JP8 jet fuel, C8–C13 *n*-alkanes, toluene, *m*- and *p*-xylene, 1,2,4-trimethyl benzene	Soil	Panicker et al. (2002)
Sphingomonas Ant 17	2-Ethylnaphthalene, *m*-xylene, 1-methyl naphthalene, dimethylnaphthalene, fluorene, JP8 jet fuel, 2-methyl naphthalene, and phenanthrene	Soil	Baraniecki et al. (2002)
Pseudomonas K319	Pyrene	Soil	Eriksson et al. (2002)

(*Continued*)

TABLE 12.1 (Continued)

Microbe(s)	Pollutant	Soil/water	Reference(s)
Pseudomonas 5A	*m*-Xylene, JP-8, toluene, and benzene	Soil	Eckford et al. (2002)
Bordetella sp. and *Pseudomonas* sp.	Cadmium, cobalt and nickel	Soil	Abou-Shanab et al. (2003)
Rhodococcus erythropolis MS11	Di-, tri-, and tetrachlorobenzenes	–	Rapp and Gabriel-Jurgens (2003)
Pseudomonas sp. C3211	Dibenzofuran	–	Jensen et al. (2003)
Rhodococcus 43/02	Pristine, JP5 jet fuel, hexadecane and dodecane	Soil	Saul et al. (2005)
Sphingomonas 43/03	Phenanthrene	Soil	
Sphingomonas 44/02	Phenanthrene	Soil	
Hydrogenophaga taenospiralis IA3	Mono-, di-, tri-, and tetrachlorobiphenyls	–	Lambo and Patel (2006a, 2006b), Lambo and Patel (2007)
Pseudoalteromonas sp.	Di-, tri-, tetra-, and pentachlorobiphenyls	–	Michaud et al. (2007)
Psychrobacter sp.	Di-, tri-, tetra-, and pentachlorobiphenyls	–	
Arthrobacter sp.	Di-, tri-, tetra-, and pentachlorobiphenyls	–	
Rhodococcus erythropolis S-7	3-Chlorobenzoate	–	Qi et al. (2007)
Shewanella sp. Asc-3	Arsenic precipitate	Soil	Demergasso et al. (2007)
Nocardia otitidiscaviarum TSH1	PAHs, *n*-alkanes, and phenol	Soil	Zeinali et al. (2007)
Staphylococcus aureus	Chromate	Soil	Aguilar-Barajas et al. (2008)
Halomonas nitroreducens	Selenite	Soil	González-Domenech et al. (2008)
Pseudomonas, *Rhodococcus erythropolis*, *Hydrogenophaga*, *Arthrobacter*, and *Actinobacteria*	Xylene, toluene, and benzene	Ground water	Fahy et al. (2008)
Idiomarina baltica DSM 15154T, *Alcanivorax venustensis* ISO4T, *Halomonas ventosae* Al12T, *Alcanivorax borkumensis*, *Bacillus oshimensis* JCM, *Alcanivorax dieselolei* B-5T, *Bacillus litoralis* DSM 16303T, and *Bacillus licheniformis* ATCC 14580T	Alkanes	Soil	Wang et al. (2010)
Anoxybacillus flavithermus, *Anoxybacillus kamchatkensis*, and *Anoxybacillus pushchinoensis*	Reactive Black 5	Hot springs	Deive et al. (2010)
Enterobacter cloacae	Total petroleum hydrocarbon	Oilfield	Hua et al. (2010)
Gamma-proteobacterium	Alkane	Mediterranean Sea	Grossi et al. (2010)
Bacillus cereus and *Bacillus subtilis*	Cadmium, cobalt, and zinc	Soil	Abdelatey et al. (2011)
Haloferax sp. CL47	Benzanthracene, naphthalene, pyrene, phenanthrene, and anthracene	Soil	Bonfá et al. (2011)
Pseudomonas toyotomiensis	*n*-Alkanes	Water	Hirota et al. (2011)
Sphingomonas indicum B90A	g-Hexachlorocyclohexane	–	Zheng et al. (2011)
Sphingomonas japonicum UT26	g-Hexachlorocyclohexane	–	

(*Continued*)

TABLE 12.1 (Continued)

Microbe(s)	Pollutant	Soil/water	Reference(s)
Haloferax sulfurifontis, Halomonas salina, Haloferax mucosum, Halomonas organivorans, Halobacterium piscisalsi, Chromohalobacter salexigens, Halobacterium salinarum, Marinobacter sedimentalis, Haloferax lucentense, Marinobacter flavimaris, Bacillus oleronius, Exiguobacterium homiense, Salinivibrio siamensis, Pseudomonas sp., and *Stenotrophomonas maltophilia*	Hydrocarbons	Soil	Al-Mailem et al. (2013)
A. flavithermus	Acid Black 48 and Reactive Black 5	Hot springs	Álvarez et al. (2013)
Aquiflexum sp. DL6	Azo dyes (Fast Red E and amaranth)	Soil	Misal et al. (2013)
Cryptococcus sp.	Copper, zinc, and lead	Water	Singh et al. (2013)
Bacillus safensis (CFA-06)	Aromatic compounds	Soil	–
Acidocella aromatica	Chromium(VI)	Soil	Masaki et al. (2015)
Microbacterium sp.	Arsenic	Soil	Mandakovic et al. (2015)
Pseudomonas, Chryseobacterium, and *Bacillus*	Crude oil	Soil	Wang et al. (2015)
Bacillus sp. L10, *Bacillus flexus* strain NBN2, *Bacillus cytotoxicus* NVH 391-98 and *B. cereus* strain AGP-03	Direct red 31 and direct blue 151	Soil	Lalnunhlimi and Veenagayathri (2016)
Corynebacterium variabile HRJ4	Pyrene, *n*-alkanes (C16, C18, C19, C26, C28), and naphthalene	Soil	Zhang et al. (2016a,b)
Geobacillus stearothermophilus strain A-2	Aromatic hydrocarbons and *n*-alkanes	Soil	Zhou et al. (2016)
Enterobacter hormaechei and *E. cloacae*	Crude oil	Soil	Ghoreishi et al. (2017)
Exiguobacterium sp.	Arsenic	Soil	Castro Severyn et al. (2017)
Pseudomonas balearica, Halomonas alkaliphila, and *Kocuria turfanesis*	Toluene and benzene	Petrochemical wastewater	Ahmadi et al. (2017)
Bacillus thuringiensis SH24 and *Bacillus amyloliquefaciens* SH20	Crude oil	Soil	Barakat et al. (2017)
Bacillus clausii P6, *Bacillus aryabhattai* P1, *B. cereus* P5, and *Bacillus* sp. P4	Direct Orange-34 and direct Red-81	Soil	Rathod and Pathak (2018)
Gracilibacter, Clostridium, Desulfotomaculum, Proteiniborus, and *Lynsinibacillus*	Zinc	Water	Hwang and Jho (2018)
Ruminococcaceae, Veillonellaceae, Bacillaceae, and *Clostridiaceae*	Sulfate, acidic, iron, and heavy metal-rich soil	Soil	Gupta et al. (2018)
Oleispira antarctica	Hydrocarbon	Antarctic seawater	Marietou et al. (2018)
A. dieselolei B-5, *Pseudomonas stutzeri* DEH130, *Psychromonas ingrahamii,* and *Paracoccus* sp. DEH99	Halogenated pollutants	Soil	Zhang et al. (2013),), Novak et al. (2013), Zhang et al. (2013), Li and Shao (2014) and Nikolaivits et al. (2017)
Pseudomonas aeruginosa	Inorganic mercury	Soil	De et al. (2008) and Dash and Das (2012)

(Christel et al., 2018; Orellana et al., 2018). Various tetrapetric lipids and genes are also involved in stability of nucleic acids and intracellular components of cells; and (3) the third mechanism which also increases the acid tolerance capacity of acidophiles by modifying DNA repair system and proteins (chaperones) in acidophiles (Orellana et al., 2018). Other extremophilic microorganisms such as halophiles that optimally grown in saline habitats have also developed two strategies for their adaptation in extreme environments: (1) first mechanism involves potassium and chloride uptake by various symporter and antiporter systems that maintain equivalence between intracellular salt concentration and extracellular environment (Oren et al., 1992; Margesin and Schinner, 2001; Orellana et al., 2018); and (2) another mechanism involves production of betaine and ectoine to maintain osmotic equilibrium that prevents cellular dehydration and desiccation (Oren, 1999; Orellana et al., 2018). Alkaliphiles microorganisms are extremophiles that show optimal growth in alkaline habitats ($pH > 9$). These microorganisms maintain their thermodynamic stability at hypersalinity condition by maintaining their proton and solute influx in cell due to alteration in ion distribution by symporter and antiporter systems (Krulwich, 1995; Krulwich et al., 1998; Matsuno and Yumoto, 2015; Orellana et al., 2018). The psychrophiles too exhibit special features for their survival at very low temperatures: (1) increased synthesis of unsaturated cyclopropane-containing fatty acids and short-chain fatty acids in membranes that maintains membrane fluidity (D'Amico et al., 2006); (2) high synthesis of coldshock proteins and chaperones to protect the synthesis of RNA and proteins (Godin-Roulling et al., 2015); (3) synthesis of antifreeze proteins that bind to ice-crystals and bring thermal hysteresis; transport of mannitol to stabilize the cytoplasmatic environment and prevent ice-formation (Baraúna et al., 2017; Muñoz et al., 2017); and (4) accumulation of mannitol and other compatible solutes as cryo-protectants to prevent cell damage by UV radiation.

12.4 Bioremediation by psychrophiles

Psychrophiles belong to extremophilic archaea or bacteria (that are cold-loving) and are also termed as psychrotolerant that survive at low temperatures. They optimally grow at 15°C and maximally grow at 20°C. It is estimated that 71% of the Earth's surface is accounted by marine region and 14% is a polar region. *Methanogenium frigidum* is the first and only truly psychrophilic methanogen archaeon isolated from Antarctica ice lake (Franzmann et al., 1997). Taxonomically, in both Archaea and Eucarya cold-loving microbes are found; they occur solitary or in association with sponges. They are distributed in various genera such as *Exiguobacterium*, *Halobacillus*, *Colwellia*, *Arthrobacter*, *Gelidibacter*, *Halomonas*, *Hyphomonas*, *Listeria*, *Methanogenium*, *Marinobacter*, *Planococcus*, *Moritella*, *Glaciecola*, *Pseudomonas*, *Psychroserpens*, *Pseudoalteromonas*, *Psychroflexus*, *Psychrobacter*, *Sphingomonas*, and *Shewanella* (Poli et al., 2017; Caruso et al., 2018b; Di Donato et al., 2019).

Various pollutants such as chlorinated hydrocarbons, PAHs, heavy metals, and pesticides are released in the environment through various routes such as mining for oil extraction, accidental oil spills, and industrial activities (Sivaperumal et al., 2017; Di Donato et al., 2019). Various extremozymes, oxidoreductases, and laccases have been proved to be useful for bioremediation and detoxification of these pollutants. In the marine environment, the main causes of hydrocarbons contamination are oil spills and natural oil seepage in marine environments (Rothschild and Mancinelli, 2001; Bolhuis and Cretoiu, 2016; Di Donato et al., 2019). Bioremediation of these polluting agents through extremozymes of marine origin can be attributed to the alkane hydroxylases which are nonhem diiron (AlkA, AlkB) and the protein belong to CYP153 family 53 which is cytochrome P450 (Di Donato et al., 2019). Nocardioides sp. strain KP7 isolated from Kuwait beach, releases a dioxygenase enzyme which is used for the degradation of aromatic compound (phenanthrene) after an oil spill accidents or oil spillage (Saito et al., 2000; Wang et al., 2010).

The primary pollutants which are released by the textile industry are dyes that can be degraded by oxidoreductases produced by psychrophiles. From Campos Basin (Brazil) petroleum bacterial species *Bacillus safensis* (CFA-06) has been isolated that produces a catalase and three oxidoreductases. Two of the enzymes show a very higher efficiency in the rate of degradation of petroleum contaminant. Another crucial oxidoreductase enzyme produced by marine bacterial species is laccase that catalyzes the oxidation of both nonphenolic and phenolic aromatic compounds (Fang et al., 2012).

12.5 Bioremediation by haloalkaliphiles/halophiles

Haloalkaliphiles/halophiles belong to the class of extremophiles adapted to hypersalinity or alkaline environments and harsh conditions. Halophiles optimally grow in the pH range of 9 with high salinity (~33% NaCl w/v) (Horikoshi, 2008; Oren et al., 2010; Singh et al., 2012). This unique capability of these organisms makes them suitable for fundamental research as well as for their application. Moderately halophiles and haloalkaliphiles have enzymes that are stable at higher salt concentrations including polymer degrading enzymes (cellulases, amylases, and chitinases) and hydrolases (proteases, lipases, nucleases, phosphatases). All of these find applications in various industries that manufacture paper, food products, pharmaceuticals, detergent, and perform pulp waste treatment (Singh et al., 2012). Haloalkaliphilic organisms producing biopolymer degrading enzymes have the potential to treat oilfield wastes rich in salts in extremely hot environments. Nonextremophiles are not adapted to perform efficiently in such conditions and fail to remove organic pollutants in such extreme environments. But Halophilic microorganisms such as *Staphylococcus* sp. MN-005, and *Bacillus naphthovorans* strain MN-003. *Mycobacterium vanbaalenii* PYR-1 and *Neptunomonas naphthovorans* NAG-2 N-126 are having different metabolic activities and are adapted to saline localities (Hedlund et al., 1999; Zhuang et al., 2002, 2003; Singh et al., 2012). Hence, these extremophiles are promising candidates for remediation of saline effluents and hypersaline environment.

12.6 Bioremediation by thermophiles

Thermophilic microorganisms are grouped into three classes: (1) moderately thermophilic—which survive at temperature of 45°C; (2) extreme thermophilic—which are capable of surviving at between 70°C and 80°C; and (3) hyperthermophilic—microorganisms that exhibiting optimum growth at 80°C (Charlier and Droogmans, 2005; Sar et al., 2013). Thermophiles found in various genera of bacterial class like green nonsulfur bacteria, cyanobacteria, purple sulfur bacteria, *Clostridium*, *Thiobacillus*, lactic acid bacteria, Thermus, Spirochetes, and other numerous genera. The most thermophilic microorganisms genera present in archaea domain Thermoplasma, Sulfolobus, Methanogens, Pyrococcus, and Pyrolobus (Rothschild and Mancinelli, 2001; Gadd, 2010; Sar et al., 2013). These thermophilic microorganisms evolved some metal resistance mechanisms and adaptation strategies for their survival in unfavorable conditions (Miroshnichenko, 2004). At higher temperature thermostability and adaptability of thermophilic microorganisms depend on the presence of thermophilic proteins, modification of enzymes by adding hydrogen bond, special amino acids, and ion pair and various other mechanisms (Li and Shao, 2000; Shen and Shen, 2010; Rajkumari et al., 2019). Thermoadaptation mechanisms can be attributed to the presence of various lipids in cell membrane including increased amounts of saturated fatty acids and acyl chain length, presence of iso-C15 fatty acids in cyclization and branching, which results in sustaining ideal membrane structure and its stability at higher temperature (Sprott et al., 1997; Rothschild and Mancinelli, 2001; Charlier and Droogmans, 2005; Koga, 2012; Rajkumari et al., 2019). Thermophilic proteins (core and peripheral) are stabilized by reducing nonpolar residues and increasing the amount of bulky hydrophobic residues. Their flexibility at room temperature is decreased by increased ion-pair content. Other factors such as introduction of metal-binding sites, multiple substitution of amino acid, and addition of negatively charged amino acid also increase the protein stability (Daniel and Cowan, 2000; Rothschild and Mancinelli, 2001; Kumar and Nussinov, 2001; Santos and Da Costa, 2002; Sar et al., 2013). Majorly, thermostability of nucleic acid in thermophilic archaea is achieved by the presence of polyamines, DNA-binding proteins, high intracellular concentration of salts, reverse gyrase, or histones. Thermostability of ribosomal and transfer RNA depends on high G + C content (Galtier et al., 1999; Rothschild and Mancinelli, 2001; Miroshnichenko, 2004; Sar et al., 2013). Genome analysis of thermophilic microorganisms has revealed the presence of special DNA repair enzymes that are detected at high temperatures which belong to DNA metabolic enzymes having special properties (Koulis et al., 1996; Makarova et al., 2002). The unique structure of cell wall and adaptation mechanisms makes them suitable candidates for remediation of metal contaminated environments (Sar et al., 2013).

The physical methods of bioremediation such as bioaccumulation and biosorption are the most widely used approaches for sequestration of toxic metals (Barkay and Schaefer, 2001; Tabak et al., 2005; Gadd, 2010; Sar et al., 2013). Bioaccumulation is the metabolism-dependent intracellular accumulation of metals while biosorption refers to metabolism-independent passive sequestration of metals by live cells. *Geobacillus thermodenitrificans* isolated from Damodar River, India are being used for bioremediation of heavy metals such as arsenite, chromium, nitrate from industrial wastewater, by biosorption. Their metal-binding efficiency has also been studied by using both industrial wastewater and synthetic metal solutions (Chatterjee et al., 2010). Bioremediation of metals is done by oxidative and reductive transformation that affects its various properties such as mobility, toxicity and solubility, which have found great possible applications. Various environmental contaminants such as radionuclides and toxic metal are degraded by direct enzymatic reduction where metal oxidized forms is used as terminal electron acceptor and this process results in precipitation (less bioavailable) of metals (Valls and De Lorenzo, 2002; Tabak et al., 2005). *Pseudomonas aeruginosa* AP02-1 is thermophilic biosurfactant–producing bacteria which degrades environmental contaminant within 7 days of its incubation. It degrades 99% of crude oil (v/v) and 2% (v/v) diesel oil at 45°C. Another thermophilic bacterium *Geobacillus thermoleovorans* T80, degrades *n*-hexadecane twofold faster at high temperature: 91% at 60°C and 48.5%

at 18°C (Perfumo et al., 2006; Rajkumari et al., 2019). Three *Geobacillus* sp. that were isolated from compost were able to degrade phenanthrene at 60°C (Viamajala et al., 2007; Rajkumari et al., 2019). The alkane monooxygenase gene (*alk*B) presents in *Aeribacillus* and *Geobacillus* thermophilic bacteria has been known to degrade environmental contaminants (Van Beilen and Funhoff, 2007; Nie et al., 2014; Tourova et al., 2016; Rajkumari et al., 2019). In *G. thermoleovorans* T70, *alk*B gene expression induced by *n*-hexadecane was able to reduce *n*-alkanes, at an optimal growth temperature of 55°C (Marchant et al., 2006; Rajkumari et al., 2019). From Dagang petroleum reservoir, *Geobacillus stearothermophilus* A-2 was isolated at high temperature of 73°C, which degrade aromatic hydrocarbons (phenanthrene, naphthalene, and fluorene), long-chain *n*-alkanes and also showed highest degradation rate of 89.83% (Zhou et al., 2016; Rajkumari et al., 2019). The polyaromatic and aliphatic hydrocarbons can be degraded by *Thermus brockii* bacteria at a high temperature of 60°C–70°C. This bacterium also showed a high degradation rate of hexadecane (1000 mg/h) and pyrene (40 mg/h) at a pH range of 6–7 and a high temperature of 70°C (Feitkenhauer and Markl, 2003; Rajkumari et al., 2019).

12.7 Conclusion

Extremophilic microorganisms are invaluable to the scientific community because of their biotechnological applications. These extremophilic microorganisms are being utilized for detoxification or degradation of environmental contaminants such as PAHs, halogenated hydrocarbons, and crude oil as they were discovered from polluted area at extreme pH, temperature, pressure, and salinity. This highlights the physical, metabolic, and genetic apparatus of extremophiles and shows their capabilities to degrade various contaminants at extreme conditions. Some extremophiles are also called polyextremophiles as they adapt to more than one extreme condition that prevails altogether and decontaminate pollutants. The exceptional performance of extremophilic bacteria and archaea in the remediation of soil and water contaminated with both inorganic and organic pollutants need a deeper understanding about their mechanism, their tolerance mechanism to extreme condition and enzymes related to it. In this context, both international and national scientific community should meticulously utilize extremophilic flora and also facilitate sample collection from extreme environmental conditions to discover novel candidates which can be deployed in the bioremediation of environmental contaminants.

References

Abdelatey, L.M., Khalil, W.K.B., Ali, T.H., Mahrous, K.F., 2011. Heavy metal resistance and gene expression analysis of metal resistance genes in Gram-positive and Gram-negative bacteria present in Egyptian soils. J. Appl. Sci. Environ. Sanit. 6, 201–211.

Abou-Shanab, R.I., Delorme, T.A., Angle, J.S., Chaney, R.L., Ghanem, K., Moawad, H., et al., 2003. Phenotypic characterization of microbes in the rhizosphere of *Alyssum murale*. Int. J. Phytorem. 5, 367–380.

Aguilar-Barajas, E., Paluscio, E., Cervantes, C., Rensing, C., 2008. Expression of chromate resistance genes from *Shewanella* sp. strain ANA-3 in *Escherichia coli*. FEMS Microbiol. Lett. 285, 97–100.

Ahmadi, M., Jorfi, S., Kujlu, R., Ghafari, S., Soltani, R.D.C., Haghighifard, N.J., 2017. A novel salt-tolerant bacterial consortium for biodegradation of saline and recalcitrant petrochemical wastewater. J. Environ. Manage. 191, 198–208.

Aislabie, J., Foght, J., Saul, D., 2000. Aromatic-hydrocarbon degrading bacteria isolated from soil near Scott Base, Antarctica. Polar Biol. 23, 183–188.

Al-Mailem, D.M., Eliyas, M., Radwan, S.S., 2013. Bioremediation of oily hypersaline soil and water via potassium and magnesium amendment. Can. J. Microbiol. 59 (12), 837–844.

Alegbeleye, O.O., Opeolu, B.O., Jackson, V.A., 2017. Polycyclic aromatic hydrocarbons: a critical review of environmental occurrence and bioremediation. Environ. Manage 60, 758–783.

Álvarez, M.S., Moscoso, F., Rodríguez, A., Sanromán, M.A., Deive, F.J., 2013. Novel physic-biological treatment for the remediation of textile dyes-containing industrial effluents. Bioresour. Technol. 146, 689–695.

Annweiler, E., Richnow, H.H., Antranikian, G., Hebenbrock, S., Garms, C., Franke, S., et al., 2000. Naphthalene degradation and incorporation of naphthalene-derived carbon into biomass by the thermophile *Bacillus thermoleovorans*. Appl. Environ. Microbiol. 66.

Azubuike, C.C., Chikere, C.B., Okpokwasili, G.C., 2016. Bioremediation techniques-classification based on site of application: principles, advantages, limitations and prospects. World J. Microbiol. Biotechnol. 32 (11), 180.

Barakat, K.M., Hassan, S.W.M., Darwesh, O.M., 2017. Biosurfactant production by haloalkaliphilic *Bacillus* strains isolated from Red Sea, Egypt. Egypt. J. Aquat. Res. 43, 205–211.

Baraniecki, C.A., Aislabie, J., Foght, J.M., 2002. Characterization of *Sphingomonas* sp. Ant 17, an aromatic hydrocarbon-degrading bacterium isolated from Antarctic soil. Microb. Ecol. 43, 44–54.

Baraúna, R.A., Freitas, D.Y., Pinheiro, J.C., Folador, A.R.C., Silva, A., 2017. A proteomic perspective on the bacterial adaptation to cold: integrating OMICs data of the psychrotrophic bacterium *Exiguobacterium antarcticum* B7. Proteomes 5, E9.

Barkay, T., Schaefer, J., 2001. Metal and radionuclide bioremediation: issues, considerations and potentials. Curr. Opin. Microbiol. 4 (3), 318–323.

Bej, A.K., Saul, D.J., Aislabie, J., 2000. Cold tolerant alkane-degrading *Rhodococcus* species from Antarctica. Polar Biol. 23, 100–105.

Bland, B., 2018. Rescuers battle toxic oil blaze off China coast. Financial Times .

Bolhuis, H., Cretoiu, M.S., 2016. What is so Special about Marine Microorganisms? Introduction to the marine microbiome—from diversity to biotechnological potential, The Marine Microbiome: An Untapped Source of Biodiversity and Biotechnological, vol. 1. Springer, Cham, pp. 3–20.

Bonfá, M.R.L., Grossman, M.J., Mellado, E., Durrant, L.R., 2011. Biodegradation of aromatic hydrocarbons by Haloarchaea and their use for the reduction of the chemical oxygen demand of hypersaline petroleum produced water. Chemo 84, 1671–1676.

Brown, G.C., Skipsey, E., 1986. Energy Resources, Geology, Supply and Demand. Open University Press, Milton Keynes.

Burgherr, P., 2007. In depth analysis of accidental oil spills from tankers in the context of global spill trends from all sources. J. Hazard. Mater. 140, 245–256.

Caruso, C., Rizzo, C., Mangano, S., Poli, A., Di Donato, P., Finore, I., et al., 2018a. Production and biotechnological potential of extracellular polymeric substances from sponge associated Antarctic bacteria. Appl. Environ. Microbiol. 84 (4), 1617–1624.

Caruso, C., Rizzo, C., Mangano, S., Poli, A., Di Donato, P., Nicolaus, B., et al., 2018b. First evidence of extracellular polymeric substance production by a cold adapted *Marinobacter* isolate from Antarctic seawater. Antarct. Sci. (in press).

Castro Severyn, J., Remonsellez, F., Valenzuela, S.L., Salinas, C., Fortt, J., Aguilar, P., 2017. Comparative genomics analysis of a new *Exiguobacterium* strain from Salar de Huasco reveals a repertoire of stress related genes and arsenic resistance. Front. Microbiol. 8, 456.

Chandrasekaran, M., Kumar, S.R., 2010. Marine microbial enzymes. In: Werner, H., Roken, S. (Eds.), Biotechnology, vol. 9. EOLSS, Paris, pp. 47–49.

Charlier, D., Droogmans, L., 2005. Microbial life at high temperature, the challenges, the strategies. Cell. Mol. Life Sci. 62 (24).

Chatterjee, S.K., Bhattacharjee, I., Chandra, G., 2010. Biosorption of heavy metals from industrial waste water by *Geobacillus thermodenitrificans*. J. Hazard. Mater. 175 (1–3), 117–125.

Chen, H., Teng, Y., Lu, S., Wang, Y., Wang, J., 2015. Contamination features and health risk of soil heavy metals in China. Sci. Total Environ. 512–513, 143–153.

Chino, M., Nakayama, H., Nagai, H., 2011. Preliminary estimation of release amount of 131I and 137Cs accidentally discharged from the Fukushima Daiichi nuclear power plant into the atmosphere. J. Nucl. Sci. Technol. 48, 1129–1134.

Christel, S., Herold, M., Bellenberg, S., El Hajjami, M., Buetti-Dinh, A., Pivkin, I.V., 2018. Multi-omics reveals the lifestyle of the acidophilic, mineral-oxidizing model species *Leptospirillum ferriphilum*^{T}. Appl. Environ. Microbiol. 84.

D'Amico, S., Collins, T., Marx, J.C., Feller, G., Gerday, C., 2006. Psychrophilic microorganisms: challenges for life. EMBO Rep. 7, 385–389.

Daniel, R.M., Cowan, D.A., 2000. Biomolecular stability and life at high temperatures. Cell. Mol. Life Sci. (CMLS) 57 (2), 250–264.

Dabbs, W.C., 1996. Oil production and environmental damage. Cell. Mol. Life Sci. (CMLS) 8, 250–264. http//www.american.edu.TED/hpl.htm.

Dash, H.R., Das, S., 2012. Bioremediation of mercury and importance of bacterial mer genes. Int. Biodeterior. Biodegrad. 75, 207–213.

De, J., Ramaiah, N., Vardanyan, L., 2008. Detoxification of toxic heavy metals by marine bacteria highly resistant to mercury. Mar. Biotechnol. 10, 471–477.

Deive, F.J., Domínguez, A., Barrio, T., Moscoso, F., Morán, P., Longo, M.A., et al., 2010. Decolorization of dye Reactive Black 5 by newly isolated thermophilic microorganisms from geothermal sites in Galicia (Spain). J. Hazard. Mater. 182, 735–742.

Demergasso, C.S., Chong, D.G., Escudero, G.L., Mur, J.J.P., Pedrós-Alió, C., 2007. Microbial precipitation of arsenic sulfides in Andean salt flats. Geomicrobiol. J. 24, 111–123.

Di Donato, P., Buono, A., Poli, A., Finore, I., Abbamondi, G.R., Nicolaus, B., et al., 2019. Exploring marine environments for the identification of extremophiles and their enzymes for sustainable and green bioprocesses. Sustainability 11 (1), 149.

Eckford, R., Cook, F.D., Saul, D., Aislabie, J., Foght, J., 2002. Free living nitrogen–fixing bacteria from fuel-contaminated Antarctic soils. Appl. Environ. Microbiol. 68, 5181–5185.

Eriksson, M., Dalhammar, G., Mohn, W.W., 2002. Bacterial growth and biofilm production on pyrene. FEMS Microbiol. Ecol. 40, 21–27.

Fahy, A., Ball, A.S., Lethbridge, G., Timmis, K.N., McGenity, T.J., 2008. Isolation of alkali-tolerant benzene-degrading bacteria from a contaminated aquifer. Lett. Appl. Microbiol. 47 (1), 60–66.

Fang, Z., Li, T., Chang, F., Zhou, P., Fang, W., Hong, Y., et al., 2012. A new marine bacterial laccase with chloride-enhancing, alkaline-dependent activity and dye decolorization ability. Bioresour. Technol. 111, 36–41.

Feitkenhauer, H., Markl, H., 2003. Biodegradation of aliphatic and aromatic hydrocarbons at high temperatures. Water Sci. Technol. 47, 123–130.

Fox, S., Bruner, T., Oren, Y., Gilron, J., Ronen, Z., 2016. Concurrent microbial reduction of high concentrations of nitrate and perchlorate in an ion exchange membrane bioreactor. Biotechnol. Bioeng. 113, 1881–1891.

Franzmann, P.D., Liu, Y., Balkwill, D.L., Aldrich, H.C., De Macario, E.C., Boone, D.R., 1997. *Methanogenium frigidum* sp. nov.; a psychrophilic, H2-using methanogen from Ace Lake, Antarctica. Int. J. Syst. Bacteriol. 47, 1068–1072.

Gadd, G.M., 2010. Metals, minerals and microbes: geomicrobiology and bioremediation. Microbiology 156 (3), 609–643.

Galtier, N., Tourasse, N., Gouy, M., 1999. A nonhyperthermophilic common ancestor to extant life forms. Science 283 (5399), 220–221.

Ghoreishi, G., Alemzadeh, A., Mojarrad, M., Djavaheri, M., 2017. Bioremediation capability and characterization of bacteria isolated from petroleum contaminated soils in Iran. Sustain. Environ. Res. 27 (4), 195–202.

Giovanella, P., Vieira, G.A., Otero, I.V.R., Pellizzer, E., de Jesus Fontes, B., Sette, L.D., 2019. Metal and organic pollutants bioremediation by extremophile microorganisms. J. Hazard. Mater. 121024.

Godin-Roulling, A., Schmidpeter, P.A., Schmid, F.X., Feller, G., 2015. Functional adaptations of the bacterial chaperone trigger factor to extreme environmental temperatures. Environ. Microbiol. 17, 2407–2420.

González-Domenech, C.M., Béjar, V., Martínez-Checa, F., Quesada, E., 2008. *Halomonas nitroreducens* sp. nov., a novel nitrate-and nitrite reducing species. Int. J. Syst. Evol. Microbiol. 58, 872–876.

Grossi, V., Yakimov, M.M., Al Ali, B., Tapilatu, Y., Cuny, P., Goutx, M., 2010. Hydrostatic pressure affects membrane and storage lipid compositions of the piezotolerant hydrocarbon-degrading *Marinobacter hydrocarbonoclasticus* strain 5. Environ. Microbiol. 12, 2020–2033.

Gupta, A., Dutta, A., Sarkar, J., Panigrahi, M.K., Sar, P., 2018. Low-abundance members of the firmicutes facilitate bioremediation of soil impacted by highly acidic mine drainage from the Malanjkhand Copper Project, India. Front. Microbiol. 9, 2882.

Gupta, A., Singh, S.K., Singh, V.K., Singh, M.K., Modi, A., Zhimo, V.Y., et al., 2020. Endophytic microbe approaches in bioremediation of organic pollutants. In: Kumar, A., Singh, V.K. (Eds.), Microbial Endophytes: Prospects of Sustainable Agriculture. Elsevier, pp. 157–174.

Hedlund, B.P., Geiselbrecht, A.D., Bair, T.J., Staley, J.T., 1999. Polycyclic aromatic hydrocarbon degradation by a new marine bacterium, *Neptunomonas naphthovorans* gen. nov., sp. nov. Appl. Environ. Microbiol. 65 (1), 251–259.

Hazen, T.C., Prince, R.C., Mahmoudi, N., 2016. Marine oil biodegradation. Environ. Sci. Technol 50, 2121–2129. Available from: https://doi.org/10.1021/acs.est.5b03333.

Hirota, K., Yamahira, K., Nakajima, K., Nodasaka, Y., Okuyama, H., Yumoto, I., 2011. *Pseudomonas toyotomiensis* sp. nov., a psychrotolerant facultative alkaliphile that utilizes hydrocarbons. Int. J. Syst. Evol. Microbiol. 61 (8), 1842–1848.

Holliger, C., Gaspard, S., Glod, G., 1997. Contaminated environments in the subsurface and bioremediation: organic contaminants. FEMS Microbiol. Rev. 20 (3–4), 517–523.

Horikoshi, K., 2008. Past, present and future of extremophiles. *Extremophiles*, *12*(1), 1.

Hua, X., Wang, J., Wu, Z., Zhang, H., Li, H., Xing, X., et al., 2010. A salt tolerant Enterobacter cloacae mutant for bioaugmentation of petroleum-and salt-contaminated soil. Biochem. Eng. J. 49 (2), 201–206.

Hwang, S.K., Jho, E.H., 2018. Heavy metal and sulfate removal from sulfate-rich synthetic mine drainages using sulfate reducing bacteria. Sci. Total. Environ. 635, 1308–1316.

Issa, N., Vempatti, S., 2018. Oil spills in the Arabian Gulf: a case study and environmental review. Environ. Nat. Resour. Res. 8 (2).

Iwamoto, K., Minoda, A., 2018. Bioremediation of biophilic radionuclides by algae. Algae. Intech Open.

Jebelli, M.A., Maleki, A., Amoozegar, M.A., Kalantar, E., Shahmoradi, B., Gharibi, F., 2017. Isolation and identification of indigenous prokaryotic bacteria from arsenic-contaminated water resources and their impact on arsenic transformation. Ecotoxicol. Environ. Saf. 140, 170–176.

Jensen, A., Finster, K., Karlson, U., 2003. Degradation of carbazole, dibenzothiophene, and dibenzofuran at low temperature by *Pseudomonas* sp. strain C3211. Environ. Toxicol. Chem. 22, 730–735.

Johnson, D.B., 2007. Physiology and ecology of acidophilic microorganisms. In: Gerday, G., Glansdorff, N. (Eds.), Physiology and Biochemistry of Extremophiles. AMS Press, Washington, DC, pp. 275–270.

Johnson, D.B., 2008. Biodiversity and interactions of acidophiles: key to understanding and optimizing microbial processing of ores and concentrates. Trans. Nonferrous Met. Soc. China 18, 1367–1373.

Kasai, Y., Kishira, H., Harayama, S., 2002. Bacteria belonging to the genus *Cycloclasticus* play a primary role in the degradation of aromatic hydrocarbons released in a marine environment. Appln Environ. Microbiol. 68, 5625–5633.

Koga, Y., 2012. Thermal adaptation of the archaeal and bacterial lipid membranes. Archaea 789652.

Khandegar, V., Saroha, A.K., 2013. Electrocoagulation for the treatment of textile industry effluent—a review. J Environ Manage 128, 949–963.

Kostka, J.E., Prakash, O., Overholt, W.A., Green, S.J., Freyer, G., Canion, A., et al., 2011. Hydrocarbon-degrading bacteria and the bacterial community response in Gulf of Mexico beach sands impacted by the deepwater horizon oil spill. Appl. Environ. Microbiol. 77 (22), 7962–7974.

Koulis, A., Cowan, D.A., Pearl, L.H., Savva, R., 1996. Uracil-DNA glycosylase activities in hyperthermophilic microorganisms. FEMS Microbiol. Lett. 143 (2–3), 267–271.

Krulwich, T.A., 1995. Alkaliphiles: 'basic' molecular problems of pH tolerance and bioenergetics. Mol. Microbiol. 15, 403–410.

Krulwich, T.A., Ito, M., Hicks, D.B., Gilmour, R., Guffanti, A.A., 1998. pH homeostasis and ATP synthesis: studies of two processes that necessitate inward proton translocation in extremely alkaliphilic *Bacillus* species. Extremophiles 2, 217–222.

Kumar, S., Nussinov, R., 2001. How do thermophilic proteins deal with heat? Cell. Mol. Life Sci. (CMLS) 58 (9), 1216–1233.

Lalnunhlimi, S., Veenagayathri, K., 2016. Decolorization of azo dyes (Direct Blue 151 and Direct Red 31) by moderately alkaliphilic bacterial consortium. Braz. J. Microbiol. 47, 39–46.

Lambo, A., Patel, T., 2006a. Isolation and characterization of a biphenyl-utilizing psychrotrophic bacterium, *Hydrogenophaga taeniospiralis* IA3A, that cometabolize dichlorobiphenyls and polychlorinated biphenyl congeners in Aroclor 1221. J. Basic. Microbiol. 46, 94–107.

Lambo, A., Patel, T., 2006b. Cometabolic degradation of polychlorinated biphenyls at low temperature by psychrotolerant bacterium *Hydrogenophaga* sp. IA3-A. Curr. Microbiol. 53, 48–52.

Lambo, A., Patel, T., 2007. Biodegradation of polychlorinated biphenyls in Aroclor 1232 and production of metabolites from 2, 4, 40-trichlorobiphenyl at low temperature by psychrotolerant *Hydrogenophaga* sp. strain IA3-A. J. Appl. Microbiol. 102, 1318–1329.

Li, H.Y., Shao, J.Y., 2000. The thermophilic mechanism of thermophilic bacteria. Chin. Bull. Life Sci. 12 (1), 30–33.

Li, A., Shao, Z., 2014. Biochemical characterization of a haloalkane dehalogenase DadB from *Alcanivorax dieselolei* B-5. PLoS One 9, e89144.

MacCormack, W.P., Fraile, E.R., 1997. Characterization of a hydrocarbon-degrading psychrotrophic Antarctic bacterium. Ant. Sci. 9, 150–155.

Makarova, K.S., Aravind, L., Grishin, N.V., Rogozin, I.B., Koonin, E.V., 2002. A DNA repair system specific for thermophilic Archaea and bacteria predicted by genomic context analysis. Nucleic Acids Res. 30 (2), 482–496.

Mandakovic, D., Cabrera, P., Pulgar, R., Maldonado, J., Aravena, P., Latorre, M., 2015. Complete genome sequence of *Microbacterium* sp. CGR1, bacterium tolerant to wide abiotic conditions isolated from the Atacama Desert. J. Biotechnol. 216, 149–150.

Marchant, R., Sharkey, F.H., Banat, S.I., Rahman, T.J., 2006. The degradation of *n*-hexadecane in soil by thermophilic geobacilli. FEMS Microbiol. Ecol. 56 (1), 44–54.

Margesin, R., Schinner, F., 2001. Bioremediation (natural attenuation and biostimulation) of diesel-oil-contaminated soil in an alpine glacier skiing area. Appl. Environ. Microbiol. 67, 3127–3133.

Marietou, A., Chastain, R., Beulig, F., Scoma, A., Hazen, T.C., Bartlett, D.H., 2018. The effect of hydrostatic pressure on enrichments of hydrocarbon degrading microbes from the Gulf of Mexico following the deepwater Horizon oil spill. Front. Microbiol. 21, 415–427.

Martin, L., Eddie, O., 2019. We cannot swim, we cannot eat': Solomon Islands struggle with nation's worst oil spill. Guardian.

Masaki, Y., Hirajima, T., Sasaki, K., Okibe, N., 2015. Bioreduction and immobilization of hexavalent chromium by the extremely acidophilic Fe(III)-reducing bacterium *Acidocella aromatica* strain PFBC. Extremophiles 19 (2), 495–503.

Master, E.R., Mohn, W.W., 1998. Psychrotolerant bacteria isolated from Arctic soil that degrade polychlorinated biphenyls at low temperatures. Appl. Environ. Microbiol. 64, 4823–4829.

Matsuno, T., Yumoto, I., 2015. Bioenergetics and the role of soluble cytochromes c for alkaline adaptation in gram-negative alkaliphilic *Pseudomonas*. Biomed. Res. Int. 2015, 847945.

Mazoomdaar, J., 2017. Oil spill: we're well prepared on paper but sluggish response made preparedness a joke. Indian Express 3 Feburary.

Michaud, L., Di Marco, G., Bruni, V., Lo Giudice, A., 2007. Biodegradative potential and characterization of psychrotolerant polychlorinated biphenyl-degrading marine bacteria isolated from a coastal station in the Terra Nova Bay (Ross Sea, Antarctica). Mar. Pollut. Bull. 54, 1754–1761.

McGenity, T.J., Folwell, B.D., McKew, B.A., Sanni, G.O., 2012. Marine crude-oil biodegradation: a central role for interspecies interactions. Aquat. Biosyst. 8 (1), 10.

Michels, M., Bakker, E.P., 1985. Generation of a large, protonophoresensitive proton motive force and pH difference in the acidophilic bacteria *Thermoplsma acidophilum* and *Bacillus acidocaldarius*. J. Bacteriol. 161, 231–237.

Mike, S., 2019. Bunker fuel spilled from Maersk ship at port of Hong Kong. G Captain .

Miralles-Robledillo, J.M., Torregrosa-Crespo, J., Martínez-Espinosa, R.M., Pire, C., 2019. DMSO reductase family: phylogenetics and applications of extremophiles. Int. J. Mol. Sci. 20 (13), 3349.

Miroshnichenko, M.L., 2004. Thermophilic microbial communities of deep-sea hydrothermal vents. Microbiology 73 (1), 1–13.

Misal, S.A., Lingojwar, D.P., Gawai, K.R., 2013. Properties of NAD (P) H azoreductase from alkaliphilic red bacteria *Aquiflexum* sp. DL6. Protein J. 32, 601–608.

Muñoz, P.A., Márquez, S.L., González-Nilo, F.D., Márquez-Miranda, V., Blamey, J.M., 2017. Structure and application of antifreeze proteins from Antarctic bacteria. Microb. Cell Fact. 16, 138.

Nguyen, X.P., Dong, V.H., 2018. A report of the impacts of pollutants on maritime operation in Vietnam. Eur. J. Eng. Res. Sci. 3 (10), 120–125.

Nie, Y., Chi, C.Q., Fang, H., Liang, J.L., Lu, S.L., Lai, G.L., et al., 2014. Diverse alkane hydroxylase genes in microorganisms and environments. Sci. Rep. 4, 4968.

Nikolaivits, E., Dimarogona, M., Fokialakis, N., Topakas, E., 2017. Marine-derived biocatalysts: importance, accessing, and application in aromatic pollutant bioremediation. Front. Microbiol. 8, 265.

Novak, H.R., Sayer, C., Panning, J., Littlechild, J.A., 2013. Characterisation of an L-haloacid dehalogenase from the marine psychrophile *Psychromonas ingrahamii* with potential industrial application. Mar. Biotechnol. 15, 695–705.

Orellana, R., Macaya, C., Bravo, G., Dorochesi, F., Cumsille, A., Valencia, R., et al., 2018. Living at the frontiers of life: extremophiles in Chile and their potential for bioremediation. Front. Microbiol. 9, 2309.

Odokuma, L.O., Akponah, E., 2012. Effect of nutrient supplementation on biodegradation and metal uptake by three bacteria in crude oil impacted fresh and brackish waters of the Niger Delta. J. ***Cell*** Anim. Biol. 4 (1), 001–018.

Oren, A., 1999. Bioenergetic aspects of halophilism. Microbiol. Mol. Biol. Rev. 63, 334–348.

Oren, A., Gurevich, P., Azachi, M., Hents, Y., 1992. Microbial degradation of pollutants at high salt concentrations. Biodegradation. 3, 387–398.

Oren, A., 2010. Industrial and environmental applications of halophilic microorganisms. Environ. Tech. 31, 825–834.

Panicker, G., Aislabie, J., Saul, D., Bej, A.K., 2002. Cold tolerance of *Pseudomonas* sp. 30–3 isolated from oil-contaminated soil, Antarctica. Polar Biol. 25, 5–11.

Peeples, T.L., 2014. Bioremediation using extremophiles. Microb. Biodegrad. Biorem. 251–268.

Perfumo, A., Banat, I.M., Marchant, R., 2006. The use of thermophilic bacteria in accelerated hydrocarbon bioremediation. WIT Trans. Ecol. Environ. 88, 67–77.

Poli, A., Finore, I., Romano, I., Gioiello, A., Lama, L., Nicolaus, B., 2017. Microbial diversity in extreme marine habitats and their biomolecules. Microorganisms 5, 25.

Prasad, A.S.A., Rao, K.V.B., 2013. Aerobic biodegradation of Azo dye by *Bacillus cohnii* MTCC 3616; an obligately alkaliphilic bacterium and toxicity evaluation of metabolites by different bioassay systems. Appl. Microbiol. Biotechnol. 97, 7469–7481.

Pruthi, V., Cameotra, S.S., 1997. Production and properties of a biosurfactant synthesized by *Arthrobacter protophormiae*—an Antarctic strain. World J. Microbiol. Biotechnol. 13, 137–139.

Qi, Y., Zhao, L., Ojekunle, Z., Tan, X., 2007. Isolation and preliminary characterization of a 3-chlorobenzoate degrading bacteria. J. Environ. Sci. 19, 332–337.

Raddadi, N., Cherif, A., Daffonchio, D., Neifar, M., Fava, F., 2015. Biotechnological applications of extremophiles, extremozymes and extremolytes. Appl. Microbiol. Biotechnol. 99, 7907–7913.

Rajkumari, J., Bhuyan, B., Das, N., Pandey, P., 2019. Environmental applications of microbial extremophiles in the degradation of petroleum hydrocarbons in extreme environments. Environ. Sustainability 1–18.

Ramos, J.-L., Marqués, S., van Dillewijn, P., Espinosa-Urgel, M., Segura, A., Duque, E., et al., 2011. Laboratory research aimed at closing the gaps in microbial bioremediation. Trends Biotechnol. 29, 641–647.

Rapp, P., Gabriel-Jurgens, L., 2003. Degradation of alkanes and highly chlorinated benzenes, and production of biosurfactants, by a psychrophilic *Rhodococcus* sp. and genetic characterization of its chlorobenzene dioxygenase. Microbiology 149, 2879–2890.

Rathod, M.G., Pathak, A.P., 2018. Efficient decolorization of textile dyes by alkaline protease producing bacterial consortia. Indian J. Geo Mar. Sci. 47, 1468–1477.

Rothschild, L.J., Mancinelli, R.L., 2001. Life in extreme environments. Nature 409 (6823), 1092–1101.

Saito, A., Iwabuchi, T., Harayama, S., 2000. A novel phenanthrene dioxygenase from Nocardioides sp. strain KP7: expression in *Escherichia coli*. J. Bacteriol. 182, 2134–2141.

Santos, H., Da Costa, M.S., 2002. Compatible solutes of organisms that live in hot saline environments. Environ. Microbiol. 4 (9), 501–509.

Sar, P., Kazy, S.K., Paul, D., Sarkar, A., 2013. Metal bioremediation by thermophilic microorganisms. Thermophilic Microbes Environ. Ind. Biotechnol. 171–201.

Saul, D.J., Aislabie, J., Brown, C.E., Harris, L., Foght, J.M., 2005. Hydrocarbon contamination changes the bacterial diversity of soil from around Scott Base, Antarctica. FEMS Microbiol. Ecol. 53, 141–155.

Selvaratnam, S., Schoedel, B.A., McFarland, B.L., Kulpa, C.F., 1997. Application of the polymerase chain reaction (PCR) and reverse transcriptase/PCR for determining the fate of phenoldegrading *Pseudomonas putida* ATCC 11172 in a bioaugmented sequencing batch reactor. Appl. Microbiol. Biotechnol. 47, 236–240.

Serrano, J., Leiva, E., Serrano, J., Leiva, E., 2017. Removal of arsenic using acid/metal-tolerant sulfate reducing bacteria: a new approach for bioremediation of high-arsenic acid mine waters. Water 9, 994.

Shen, J.S., Shen, B., 2010. Thermostability mechanism of thermophilic enzyme. J. Microbiol. 2 (30), 80–85.

Singh, S.P., Raval, V.H., Purohit, M.K., Thumar, J.T., Gohel, S.D., Pandey, S., et al., 2012. Haloalkaliphilic bacteria and actinobacteria from the saline habitats: new opportunities for biocatalysis and bioremediation. Microorg. Environ. Manage. 415–429.

Singh, P., Raghukumar, C., Parvatkar, R.R., Mascarenhas-Pereira, M.B.L., 2013. Heavy metal tolerance in the psychrotolerant *Cryptococcus* sp. isolated from deep-sea sediments of the Central Indian Basin. Yeast 148, 442–456.

Singh, V.K., Singh, M., Singh, S.K., et al., 2019. Sustainable agricultural practices using beneficial fungi under changing climate scenario. In: Choudhary, K.K., Kumar, A., Singh, A.K. (Eds.), Climate Change and Agricultural Ecosystems. Woodhead Publishing, pp. 25–42.

Singh, P., Singh, V.K., Singh, R., Borthakur, A., Madhav, S., Ahamad, A., et al., 2020. Bioremediation: a sustainable approach for management of environmental contaminants. In: Singh, P., Kumar, A., Borthakur, A. (Eds.), Abatement of Environmental Pollutants. Elsevier, pp. 1–23.

Sivaperumal, P., Kamala, K., Rajaram, R., 2017. Bioremediation of industrialwaste through enzyme producing marine microorganisms. Adv. Food Nutr. Res. 80, 165–179.

Sprott, G.D., Agnew, B.J., Patel, G.B., 1997. Structural features of ether lipids in the archaeobacterial thermophiles *Pyrococcus furiosus*, *Methanopyrus kandleri*, *Methanothermus fervidus*, and *Sulfolobus acidocaldarius*. Can. J. Microbiol. 43 (5), 467–476.

Stapleton, R.D., Savage, D.C., Sayler, G.S., Stacey, G., 1998. Biodegradation of aromatic hydrocarbons in an extremely acidic environment. Appl. Environ. Microbiol. 64 (11), 4180–4184.

Stepanov, V.G., Xiao, Y., Tran, Q., Rojas, M., Willson, R.C., Fofanov, Y., et al., 2014. The presence of nitrate dramatically changed the predominant microbial community in perchlorate degrading cultures under saline conditions. BMC Microbiol. 14, 225.

Stohl, A., Seibert, P., Wotawa, G., Arnold, D., Burkhart, J.F., Eckhardt, S., 2012. Xenon-133 and caesium-137 release into the atmosphere from the Fukushima Daiichi nuclear power plant: Determination of the source term, atmospheric dispersion, and deposition. Atmos. Chem. Phys. 12, 2313–2343.

Swannell, R.P.J., Lee, K., McDonagh, M., 1996. Field evaluations of marine oil spill bioremediation. Microbiol. Rev. 60, 342.

Tabak, H.H., Lens, P., van Hullebusch, E.D., Dejonghe, W., 2005. Developments in bioremediation of soils and sediments polluted with metals and radionuclides—1. Microbial processes and mechanisms affecting bioremediation of metal contamination and influencing metal toxicity and transport. Rev. Environ. Sci. Biotechnol. 4 (3), 115–156.

Thomassin-Lacroix, E.J.M., Yu, Z., Eriksson, M., Reimer, K.J., Mohn, W.W., 2001. DNA-based and culture-based characterization of hydrocarbon-degrading consortium enriched from Arctic soil. Can. J. Microbiol. 47, 1107–1115.

Tourova, T.P., Sokolova, D.S., Semenova, E.M., Shumkova, E.S., Korshunova, A.V., Babich, T.L., et al., 2016. Detection of *n*-alkane biodegradation genes alkB and ladA in thermophilic hydrocarbon-oxidizing bacteria of the genera *Aeribacillus* and *Geobacillus*. Microbiology 85, 693–707.

Valls, M., De Lorenzo, V., 2002. Exploiting the genetic and biochemical capacities of bacteria for the remediation of heavy metal pollution. FEMS Microbiol. Rev. 26 (4), 327–338.

Van Beilen, J.B., Funhoff, E.G., 2007. Alkane hydroxylases involved in microbial alkane degradation. Appl. Microbiol. Biotechnol. 74, 13–21.

Varjani, S.J., 2017. Microbial degradation of petroleum hydrocarbons. Bioresour. Technol. 223, 277–286.

Viamajala, S., Peyton, B.M., Richards, L.A., Petersen, J.N., 2007. Solubilization, solution equilibria, and biodegradation of PAH's under thermophilic conditions. Chemosphere 66 (6), 1094–1106.

Wang, L., Cho, D.W., Tsang, D.C., Cao, X., Hou, D., Shen, Z., et al., 2019. Green remediation of As and Pb contaminated soil using cement-free clay-based stabilization/solidification. Environ. **Int.** 66 (126), 336–345.
Wang, L., Wang, W., Lai, Q., Shao, Z., 2010. Gene diversity of CYP153A and AlkB alkane hydroxylases in oil-degrading bacteria isolated from the Atlantic Ocean. Environ. Microbiol. 12, 1230–1242.
Wang, J., Wang, J., Zhang, Z., Li, Y., Zhang, B., Zhanga, Z., et al., 2015. Cold-adapted bacteria for bioremediation of crude oil-contaminated soil. J. Chem. Technol. Biotechnol. 91 (8), 2286–2297.
Whyte, L.G., Bourbonniere, L., Greer, C.W., 1997. Biodegradation of petroleum hydrocarbons by psychrotrophic *Pseudomonas* strains possessing both alkane (alk) and naphthalene (nah) catabolic pathways. Appl. Environ. Microbiol. 63, 3719–3723.
Yu, Z., Stewart, G.R., Mohn, W.W., 2000. Apparent contradiction: psychrotolerant bacteria from hydrocarbon-contaminated Arctic tundra soils that degrade diterpenoids synthesized by trees. Appl. Environ. Microbiol. 66, 5148–5154.
Zeinali, M., Vossoughi, M., Ardestani, S.K., 2007. Characterization of a moderate thermophilic Nocardia species able to grow on polycyclic aromatic hydrocarbons. Lett. Appl. Microbiol. 45 (6), 622–628.
Zhang, J., Cao, X., Xin, Y., Xue, S., Zhang, W., 2013. Purification and characterization of a dehalogenase from *Pseudomonas stutzeri* DEH130 isolated from the marine sponge *Hymeniacidon perlevis*. World J. Microbiol. Biotechnol. 29, 1791–1799.
Zhang, H., Tang, J., Wang, L., Liu, J., Gurav, R.G., Sun, K., 2016a. A novel bioremediation strategy for petroleum hydrocarbon pollutants using salt tolerant *Corynebacterium variabile* HRJ4 and biochar. J. Environ. Sci. 47, 7–13.
Zhang, X., Feng, X., Tao, J., Ma, L., Xiao, Y., Liang, Y., et al., 2016b. Comparative genomics of the extreme acidophile *Acidithiobacillus thiooxidans* reveals intraspecific divergence and niche adaptation. Int. J. Mol. Sci. 17.
Zheng, G., Selvam, A., Wong, J., 2011. Rapid degradation of lindane (*g*-hexachlorocyclohexane) at low temperature by Sphingobium strains. Int. Biodeterior. Biodegrad. 65, 612–618.
Zhou, J., Li, G., Xie, J., Cui, X., Dai, X., Tian, H., et al., 2016. A novel bioemulsifier from *Geobacillus stearothermophilus* A-2 and its potential application in microbial enhanced oil recovery. RSC Adv. 98.
Zhuang, W.Q., Tay, J.H., Maszenan, A., Tay, S., 2002. *Bacillus naphthovorans* sp. nov. from oil-contaminated tropical marine sediments and its role in naphthalene biodegradation. Appl. Microbiol. Biotechnol. 58 (4), 547–554.
Zhuang, W.Q., Tay, J.H., Maszenan, A.M., Krumholz, L.R., Tay, S.L., 2003. Importance of Gram-positive naphthalene-degrading bacteria in oil-contaminated tropical marine sediments. Lett. Appl. Microbiol. 36 (4), 251–257.

Chapter 13

Microbes as natural degraders of polychlorinated biphenyls

Paramdeep Kaur, Deepanshu Monga and Baljinder Singh
Department of Biotechnology, Panjab University, Chandigarh, India

13.1 Introduction

Polychlorinated biphenyls (PCBs) are a group of chlorinated aromatic compounds classified as one of the persistent organic pollutants (POPs), which is a class of hazardous pollutants with serious health effects on living organisms. PCBs are produced by catalytic chlorination of biphenyls yielding up to 209 products called congeners. Individual PCB congeners differ in the number and positions of chlorine substituents (refer to Figs. 13.1 and 13.2). PCB congeners with equal number of chlorine atoms are known as homologs, and the homologs with chlorine atoms located on different positions are called isomers (Borja et al., 2005). The individual chlorine substitutions are further categorized into singly or doubly flanked and unflanked chlorines, referring to whether the neighboring positions carry chlorine substitutions or not, respectively (Wiegel and Wu, 2000). PCB congeners have also been classified on the basis of structure into two categories, coplanar (dioxin-like) and noncoplanar (nondioxin-like) PCBs. PCBs in which ortho position of the biphenyl rings substituted with no chlorines have a coplanar biphenyl orientation in solution, and PCBs with ortho position substituted with one to four chlorines have noncoplanar biphenyl orientation. The dioxin-like PCBs are considered to be more toxic in nature.

In 1881 PCB was synthesized for the first time in the laboratory and extensively used between the 1930s and 1980s for various industrial and commercial purposes due to their excellent chemical and physical properties such as chemical inertness, thermal stability, resistance to flame, and high dielectric constant. They have been used in transformer oils, plasticizers, hydraulic fluids, surface coatings, fire retardants, adhesives, pesticides, carbonless copy paper, inks, dyes, and waxes. The commercial PCB mixtures consisting of about 60–90 PCB congeners have been sold under various trade names in different countries such as Aroclor in the United States, Canada, and United Kingdom; Phenoclor in France and Spain; and Kanechlor in Japan. But due to recognition of the toxic effects of PCBs, they were eventually banned in most of the countries by the late 1970s.

PCBs are a great public concern because of their potential toxicity to humans and wildlife and their property to get bioaccumulated. They are frequently detected in almost all environmental media such as air, water, sediments, soil, and also in living organisms. The movement of PCBs by "grasshopper effect" (volatizing from soil to air in warm weather and falling back to the Earth miles away where the temperature is cool) has made them ubiquitous (Schmidt, 2010). The properties of PCBs such as lipophilicity, poor water solubility, and their resistance to biodegradation promote their

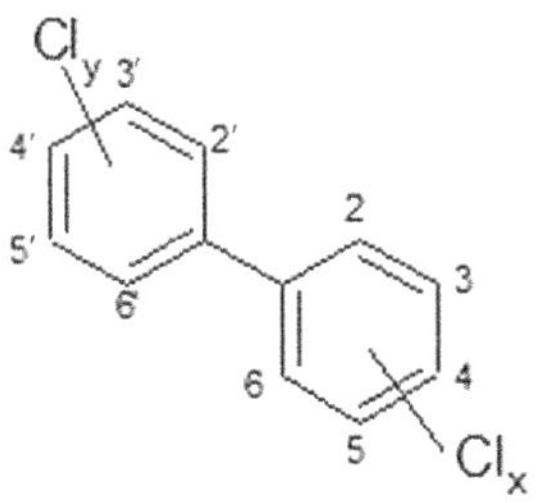

FIGURE 13.1 Basic structure of PCB. *PCB*, Polychlorinated biphenyls.

Microbe Mediated Remediation of Environmental Contaminants. DOI: https://doi.org/10.1016/B978-0-12-821199-1.00013-4

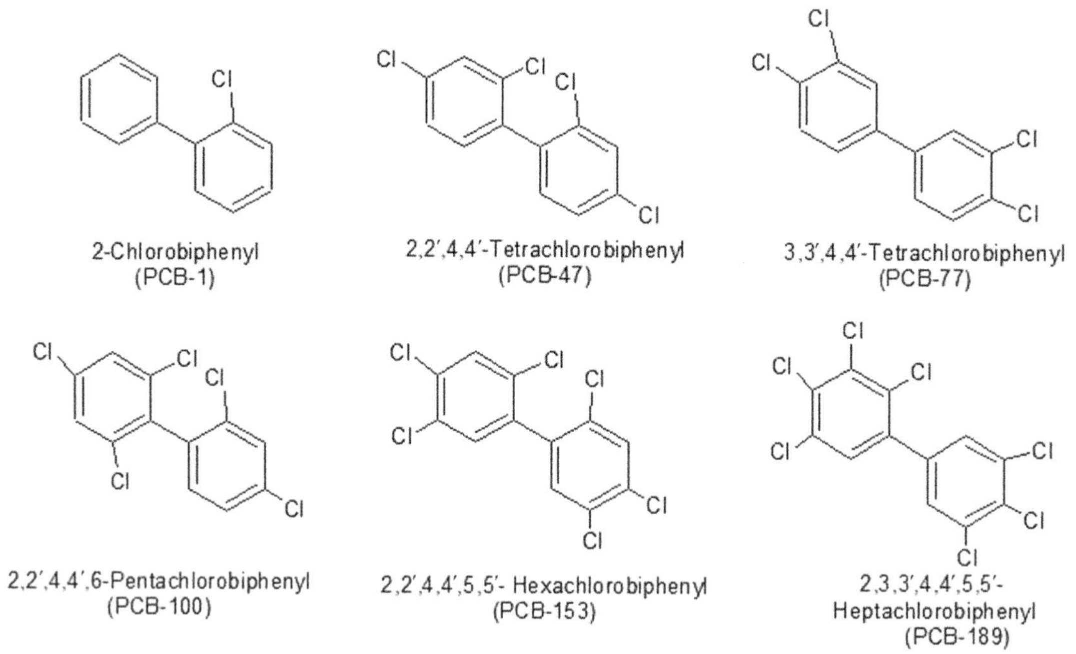

FIGURE 13.2 Examples of various PCB congeners. *PCB*, Polychlorinated biphenyls.

accumulation and biomagnification in living organisms. The populations exposed to high PCB concentrations are known to show various detrimental health effects due to the toxic nature of PCB.

Presently, PCBs are considered as one of the most hazardous pollutants in the world. Various regulatory measures have also been commenced with the aim of eradicating PCBs from the environment. Great Lakes Water Quality Agreement was signed by both Canada and the United States to control PCB pollution in the great lakes. The global treaty Stockholm Convention published a list of 12 POPs (among including PCBs) in May 1995, which should be eliminated from the environment. Throughout the years, this list was revised and updated for PCB and other chemicals, called persistent organic pollutants. It aims to phase out the use of PCB in equipment by 2025 and ensure elimination of PCB by 2028 (http://www.pops.int).

Regardless of all regulatory efforts, PCBs are still quite persistent, thus continuing to be one of the topmost public health concerns. There are various ongoing studies on the degradation of PCBs by physical, chemical, or biological methods. Bioremediation of PCBs by microbes is being extensively explored due to their advantages over the traditional methods of degradation. Various bacteria and fungi have been found to be capable of transforming PCBs to lesser toxic substances or completely degrading some specific PCB congeners.

13.2 Toxicity of polychlorinated biphenyls

Currently, PCBs are a serious threat to the environment due to their detrimental effects on both terrestrial and aquatic life. Properties of PCBs such as high soil adsorption capacity, low water solubility, stability in environment, and bioaccumulation in animal tissues make their exposure to the living beings more prevalent and chronic.

In humans and other animals, PCBs have been proved to cause neurological, reproductive, endocrine, cardiovascular, immunological, and systemic adverse effects. PCBs have also been associated with hypertension, diabetes, and obesity, which are all related to the development of cardiovascular disease. Epidemiological studies showed that serum PCB levels were significantly higher among individuals with hypertension (Christensen, White et al., 2011). Also dioxin-like PCBs are more capable of causing hypertension as compared to nondioxin-like PCBs, which also depends upon the chlorine substitution (Everett et al., 2007).

PCB exposure increases the risk of type 2 diabetes. PCBs also increase gestational diabetes risk by causing insulin resistance. PCB 138 and PCB 180 were associated with increasing 2-h glucose levels and PCB 180 was involved in increasing immunoreactive insulin levels in expecting mothers (Arrebola et al., 2015).

Many of the PCB congeners are neurotoxic. PCBs have been associated with decrease in the dopamine levels leading to altered dopamine signaling. Hypothyroidism is also an implication of PCB exposure. PCBs also induce the production of reactive oxygen species (ROS) and reactive nitrogen species causing oxidative stress in the cells, which leads to cell damage or cell death (Pessah et al., 2019). The adverse effects of the PCBs on the reproductive system have been extensively studied. At high concentrations Aroclor 1242 exposure results in detrimental effects to Leydig cell homeostasis. It also leads to impaired steroidogenesis, especially testosterone biosynthesis (Aydin et al., 2017). Aroclor 1254 has been reported to affect fertilization potential of sperms by reducing sperm motility, increasing ROS production, and causing general mitochondrial dysfunction (Jiang et al., 2017). High PCB level exposure is also associated with heightened risk of endometriosis in women (Hoffman et al., 2007). PCBs are also carcinogenic in nature. They have been known to induce and promote cancers associated with liver, intestine, biliary duct, lip, stomach, breast, and brain. They are also linked to Hodgkin's and non-Hodgkin's lymphoma, multiple myeloma, and cancers of the hematopoietic system (Faroon et al., 2001). Epidemiological studies have proved PCB exposure, which causes many other health conditions as well such as clinical hepatitis, skin lesions, and liver damage.

In aquatic organisms, biomagnification of PCBs within aquatic food chain has been observed. Aquatic organisms also act as source of PCBs to humans when taken as food. The aquatic organisms themselves suffer from adverse consequences of PCB contact as well. In juvenile zebrafish, PCB 77 exposure during the gonadal differentiation decreased vitellogenesis, which lasted up to adulthood resulting in abnormal female gonad development, ultimately leading to decreased reproductive efficiency(Quintaneiro et al., 2019). Recently, a study reported decrease in the killer whale populations and the risk of their extinction by high toxicity effects of PCB on the calves because of PCB's transfer from mother to offspring (Schnitzler et al., 2019).

Plants show moderate-to-high toxic effects when exposed to PCBs depending on the congener and the plant species. High exposure of PCBs decreases plant growth by affecting the height, weight, and water usage. It also affects photosynthesis and respiration (Strek and Weber, 1982). Plants can take up PCBs and can metabolize them to some extent. They transform PCBs into their hydroxy derivatives, which are considered to be more toxic than PCBs. The toxicity of parent PCB (2,5-dichlorobiphenyl) was less as compared to its three hydroxy derivatives, 2′-OH-, 3′-OH-, and 4′-OH-2,5-dichlorobiphenyls toward the model plant, *Arabidopsis thaliana*, resulting in severe iron deficiency by interfering with the brassinosteroid pathway (Subramanian et al., 2019).

13.3 Microbial degradation of polychlorinated biphenyls

The persistence and toxicity of PCBs makes their efficient degradation important to eliminate them from environment. The traditional methods of degradation such as solvent extraction, landfarming, and thermal desorption are available but there are some limitations associated with them. They are (1) expensive, (2) laborious, (3) not suitable for treatment of large volumes of contaminated material. Microbial degradation represents an effective, cost-competitive, and environment-friendly method of remediation (Singh et al., 2020; Gupta et al., 2020). Many bacteria and fungi have been reported to successfully degrade PCBs. *Anabaena* PD-1 isolated from PCB-contaminated paddy soils is the only cyanobacterium capable of degrading PCB congeners (Zhang et al., 2015).

The efficiency of biodegradation depends on several factors such as structure of pollutant; solubility; pollutant concentration; physical and environmental parameters (oxygen content, temperature, and pH); and biological parameters (such as the presence, type, and number of microorganisms) (Borja et al., 2005). Lower chlorinated PCBs are more readily degraded by microorganisms because of their higher aqueous solubility.

Biodegradation is also affected by the pollutant concentration. In general, a low pollutant concentration may fail to induce the degrading machinery of competent organisms. In contrast, a very high concentration may be toxic to the organisms. The uses of microorganisms, both anaerobic and aerobic, are known to degrade PCBs in soil systems or aquatic environments but by different mechanisms. The environment of anaerobes is suitable to reductive transformations, which leads to formation of dechlorinated compound, whereas aerobic microorganisms perform oxidative transformation resulting in mineralization of the compound.

Temperature and pH also play major roles in determining suitable conditions for PCB degradation. Different temperature and pH conditions favor different bacterial strains over the others and also affect their degradation skills. In Woods Pond sediment, pH values ranging between 5.0 and 8.0 were considered suitable for PCB dechlorination and variations in temperature affected the lag time preceding the dechlorination of 2,4,6-CB along with the preferred route of its dechlorination. Para dechlorination was favored at 20°C, whereas ortho dechlorination dominated at 15°C (Wiegel and Wu, 2000).

13.3.1 Bacterial polychlorinated biphenyl degradation

There are two types of biologically mediated processes for the degradation of PCBs by bacteria: anaerobic reductive dechlorination and aerobic oxidative degradation.

13.3.1.1 Anaerobic reductive dechlorination

Anaerobic reductive dechlorination involves removal of chlorine atoms from the biphenyl rings of higher chlorinated PCB congeners resulting in accumulation of lower chlorinated congeners. Dechlorination reduces the toxicity of PCBs and makes them more available for aerobic degradation. In this process PCBs act as alternative electron acceptors in anaerobic respiration and the hydrogen atoms replace chlorine atoms by removal of chloride ions ($R-Cl + 2e- + H+ \rightarrow R-H + Cl-$) (Elangovan et al., 2019). Reductive dehalogenase (RDase) enzymes are used by the bacteria to catalyze the dehalogenation reaction but RDase genes that participate in PCB dechlorination have not been clearly stated (Hug et al., 2013).

Many bacteria have been associated with anaerobic PCB dechlorination. These include anaerobic *Chloroflexi bacterium* o-17, DF-1, *Dehalococcoides mccartyi*, *Dehalogenemonas*, and *Dehalobacter*. Microbial dehalogenation of PCB was first reported in sediments of the upper Hudson River. The PCB dehalogenation was detected from the changes in PCB isomer and homolog concentrations in the various PCB-contaminated samples (Brown et al., 1987). This was followed by several evidences of anaerobic dehalogenation in the sediment and soil slurries (Quensen et al., 1988). *Dehalococcoides ethenogenes* strain 195 (now *D. mccartyi* 195) was the first pure bacterial strain isolated, which was capable of performing reductive dechlorination of two PCB congeners chlorinated on a single ring (Fennell et al., 2004).

The presence of sediment or sediment substitutes is considered important to maintain dechlorination activity of bacteria but it interferes with the isolation and enrichment of bacteria. Sediment-free cultures reported till date are few but they are important for studying organisms' nutritional requirements, ecophysiology, and reductive dehalogenase genes involved in PCB dichlorination. Sediment-free JN cultures propagating *Dehalococcoides* under defined conditions were found to be capable of dechlorinating Aroclor 1260 by targeting flanked or doubly flanked metachorines (Bedard et al., 2007). Another sediment-free culture AD14, consisting of *Dehalococcoides* and *Dehalobacter* dechlorinated the commercial PCB mixture Aroclor 1260, mainly by removing flanked para- and doubly flanked metachlorines (Wang and He, 2013a,b). Microcosms including *Dehalogenimonas* and *Dehalococcoides* species have also been successfully designed (Wang and He, 2013a,b). A highly efficient sediment-free culture GD1, with *Dehalococcoides* playing a major role in reductive dechlorination has been developed, which dechlorinated hexa- and hepta-CBs in Aroclor 1260 with dechlorination efficiencies of 78.3% and 65.5%, respectively (Liang and Wang, 2017).

Mostly meta and/or para chlorines are removed to generate primarily ortho-substituted chlorobiphenyls, but ortho dechlorination of several PCB congeners has also been reported. Mono- or dichlorobiphenyls are the common products of reductive dechlorination but there are examples of complete PCB dechlorination resulting in biphenyl compounds. Conversion of 2,4,6-trichlorobiphenyl (Williams, 1994), 3,4,3′,4′-CB (Rhee et al., 1993), and 2,3,4,5,6-pentachlorobiphenyl (Natarajan et al., 1996) to biphenyl has been reported. Dechlorination of 2,3,4,5,6-pentachlorobiphenyl includes removal of all chlorine atoms from the biphenyl ring (refer to Fig. 13.3) (Natarajan et al., 1996).

Chloroflexi o-17 isolated from estuarine sediments of the Baltimore Harbor was the first bacteria reported to dechlorinate a PCB congener 2,3,5,6-pentachlorobiphenyl in the ortho positions (Cutter et al., 2001). Another phylogenetically

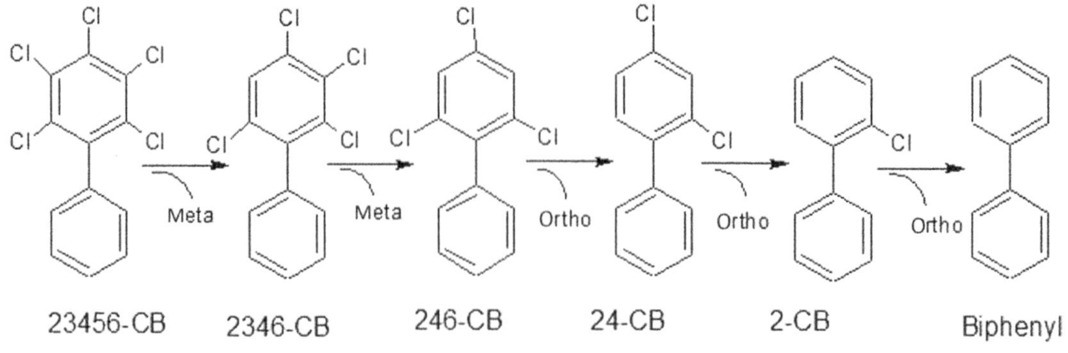

FIGURE 13.3 Pathway for anaerobic degradation of PCBs. *PCB*, Polychlorinated biphenyls.

similar bacterium, that is, DF-1, was identified in cultures enriched from estuarine sediments of the Charleston Harbor as the dechlorinator of 2,3,4,5 CB, 3,4,5-chlorobiphenyl and 2,3,4-CB congeners in the doubly flanked meta and para positions (Wu et al., 2002).

In situ remediation using PCB dehalogenating bacteria has not been widely adopted due to the slow process, low bioavailable concentrations of PCBs, and frequently low abundances of indigenous dehalorespiring microorganisms. The efficiency of in situ remediation can be improved by using bioaugmentation and biostimulation approaches. Use of biofilms for bioaugmentation can be an efficient approach. DF1 biofilm bacteria adsorbed on coal activated carbon, bone biochar, polyoxymethylene, and sand microcosms showed 73%, 93%, 100%, and 83% degradation efficiencies, respectively (Capozzi et al., 2019). Interspecies synergistic interactions were found between *D. mccartyi* strains (CG3 and SG1) and with other beneficial nondechlorinating microorganisms such as *Methanosarcina* and *Desulfovibrio*, which may play roles in mediating acetate and H2 sources for *D. mccartyi* and can enhance the degradation activity (Wang et al., 2019).

Microbial fuel cell is an advanced method of converting organic pollutants in sediment and simultaneously generating electricity. The use of surfactant along with sediment microbial fuel cell (SMFC) led to the highest removal efficiency (43.26% after 60 days) and produced the maximum power output (Xu et al., 2015). Another report described that the combination of nanoscale zerovalent iron and SMFC increases the degradation of PCBs in sediment and the electricity production. It was found that the combined application provided higher PCBs removal efficiency (37.55% ± 1.11%) and produced a higher power density (Wu et al., 2019).

13.3.1.2 Aerobic oxidative degradation

The oxidative aerobic biodegradation of PCB is suitable for the degradation of less chlorinated PCB congeners. A large number of bacteria capable of aerobic PCB degradation have been identified (refer to Table 13.1). They include Gram-negative *Pseudomonas, Alcaligenes, Burkholderia, Comamonas, Sphingomonas, Ralstonia, Cupriavidus, Achromobacter, Acidovorax, Nocardia, Castellaniella*, and *Acinetobacter* strains, and Gram-positive *Rhodococcus, Corynebacterium*, and *Bacillus* strains. *Achromobacter* was the first bacteria capable of degrading dichlorobiphenyl and monochlorobiphenyl to chlorobenzoic acid (Ahmed and Focht, 1973).

Breakdown of PCB molecules with more than one chlorine atom proceeds by cometabolic process, which means that the microorganisms involved in the process use an additional carbon source other than PCBs. Biphenyl and monochlorobiphenyl are the common enrichment factors used. Biphenyl also acts as an inductor of degrading enzymes. Use of readily available sources of carbon such as glucose, acetate, and pyruvate also increases the rate of metabolism. PCB-degrading abilities of four bacterial strains *Alcaligenes xylosoxidans, Pseudomonas stutzeri, Ochrobactrum anthropi*, and *Pseudomonas veronii* increased when they were cometabolically grown on glucose or biphenyl or both (Murínová et al., 2014). *Sinorhizobium meliloti* NM showed increased efficiency of PCB degradation in the presence of variable enrichment factors such as mannitol, succinate, phenolic acids (caffeic acid, coumalic acid), 4-hydroxybenzoic acid, and a surfactant (Tween 20) (Wang et al., 2016). It has been found that particular plant species cultivated can also change the soil microbial communities in favor of PCB biodegradation (Terzaghi et al., 2019). Mostly oxidative PCB degradation is limited to less chlorinated PCB congeners (biphenyls with five or fewer chlorines). But *Enterobacter* (LY 402) is an efficient strain that can degrade both the higher and lower chlorinated PCBs under aerobic conditions (Cao et al., 2011). Aerobic oxidative destruction involves two clusters of genes. The first one is responsible for the transformation of PCB congeners to chlorobenzoic acid and aliphatic acids and the second cluster is responsible for the degradation of chlorobenzoic acid. The degradation process can be divided into two pathways—the biphenyl upper pathway and the biphenyl lower pathway (refer to Fig. 13.4) (Elangovan et al., 2019; Borja et al., 2005).

The biphenyl upper pathway

This pathway is involved in the conversion of PCB congeners to chlorobenzoic acid.

This aerobic degradation is carried out by enzymes of the *bph* gene cluster. The *bph* genes are present on bacterial chromosomes, plasmids, and transposons. The upper pathway consists of the following four steps:

1. conversion of chlorobiphenyl to 2,3-dihydroxy-1-phenylcyclohexa-4,6-diene (dihydrodiol) by introducing molecular oxygen at the 2,3 position of the nonchlorinated or the lesser chlorinated ring of the PCB catalyzed by biphenyl2,3-dioxygenases (*bphA*);
2. dehydrogenation of dihydrodiol (chlorinated) to 2,3-dihydroxybiphenyl (chlorinated) catalyzed by *cis*-2,3-dihydro-2,3-dihydroxybiphenyl dehydrogenases (*bphB*);

TABLE 13.1 Bacteria involved in aerobic polychlorinated biphenyls (PCB) degradation.

Microorganism	Type of PCB	Isolation from or source	Enrichment factor for cometabolism	Pathway of degradation	Efficiency of degradation	Reference
Rhodococcus sp. *Strain RHA1*	PCB 48/PCB mixture of Kanechlor 200, 300, 400, and 500	Gamma-hexachlorocyclohexane-contaminated soil in Japan	Biphenyl	Both upper and lower pathway	80%–100% transformation of PCB 48/45 of the 62 components of Kanechor mixture degraded	Seto et al. (1995)
Alcaligenes xylosoxidans	DELOR 103 mixture	PCB-contaminated soil (area surrounding the former producer of PCBs Chemko Strázske)	Glucose (5 g/L) or biphenyl (10 mg/L)	–	55% and 46% PCB depletion, respectively	Murínová et al. (2014)
Pseudomonas stutzeri	DELOR 103 mixture	PCB-contaminated soil (Chemko Strázske)	Glucose (5 g/L) and biphenyl (10 mg/L)	–	27%	Murínová et al. (2014)
Ochrobactrum anthropi	DELOR 103 mixture	Sediment from Strázsky canal	Glucose (5 g/L)	–	33%	Murínová et al. (2014)
Pseudomonas veronii	DELOR 103 mixture	Sediment from Strázsky canal	Glucose (5 g/L) and biphenyl (10 mg/L)	–	40%	Murínová et al. (2014)
Sinorhizobium meliloti NM	*PCB 77*	Agricultural Culture Collection of China	Mannitol/succinate/phenolic acids (caffeic acid, coumalic acid)/4-hydroxybenzoic acid/surfactant (Tween 20)	–	High efficiency 91.93(±5.89)% at low PCB levels (0.25 mg/L). Addition of enrichment factors increases efficiency	Wang et al. (2016)
Lysinibacillus macrolides DSM54T	2-CB and 2,4-diCB	Contaminated soil sample	–	Biphenyl upper pathway and degradation of 4-chlorobenzoic acid	80%	Samadi et al. (2019)
Bacillus firmus NBRC15306T	2-CB and 2,4-diCB	Contaminated soil sample	–	Biphenyl upper pathway and degradation of 4-CBA	80% in 96 h	Samadi et al. (2019)
Enterobacter sp. LY402	PCB mixtures (Aroclors)	PCB-contaminated soil from China (Wenling, Zhejiang)	–	Biphenyl upper pathway and CBA degradation	75.5% in 6 days	Jia et al. (2008)
Castellaniella sp. SPC4	3,3′,4,4′-Tetrachlorobiphenyl (PCB 77)	PCB-contaminated soil samples from e-waste recycling town Taizhou area of Zhejiang Province	Glucose	Biphenyl upper pathway and mineralization pathway of 3,4-dichlorobenzoate	43.2% in degradation within 48 h	Su et al. (2019)
Sphingobium fuliginis HC3	various PCB congeners	Soil samples e-waste recycling area in Taizhou	Glucose, sodium acetate, tryptone, and yeast extract	Biphenyl upper pathway and CBA/benzoate degradation	Mono-, di-, and tri-PCBs significantly degraded, 4-CBA and benzoate nearly completely degraded; tryptone and yeast extract give 97% biphenyl degradation	Hu et al. (2015)

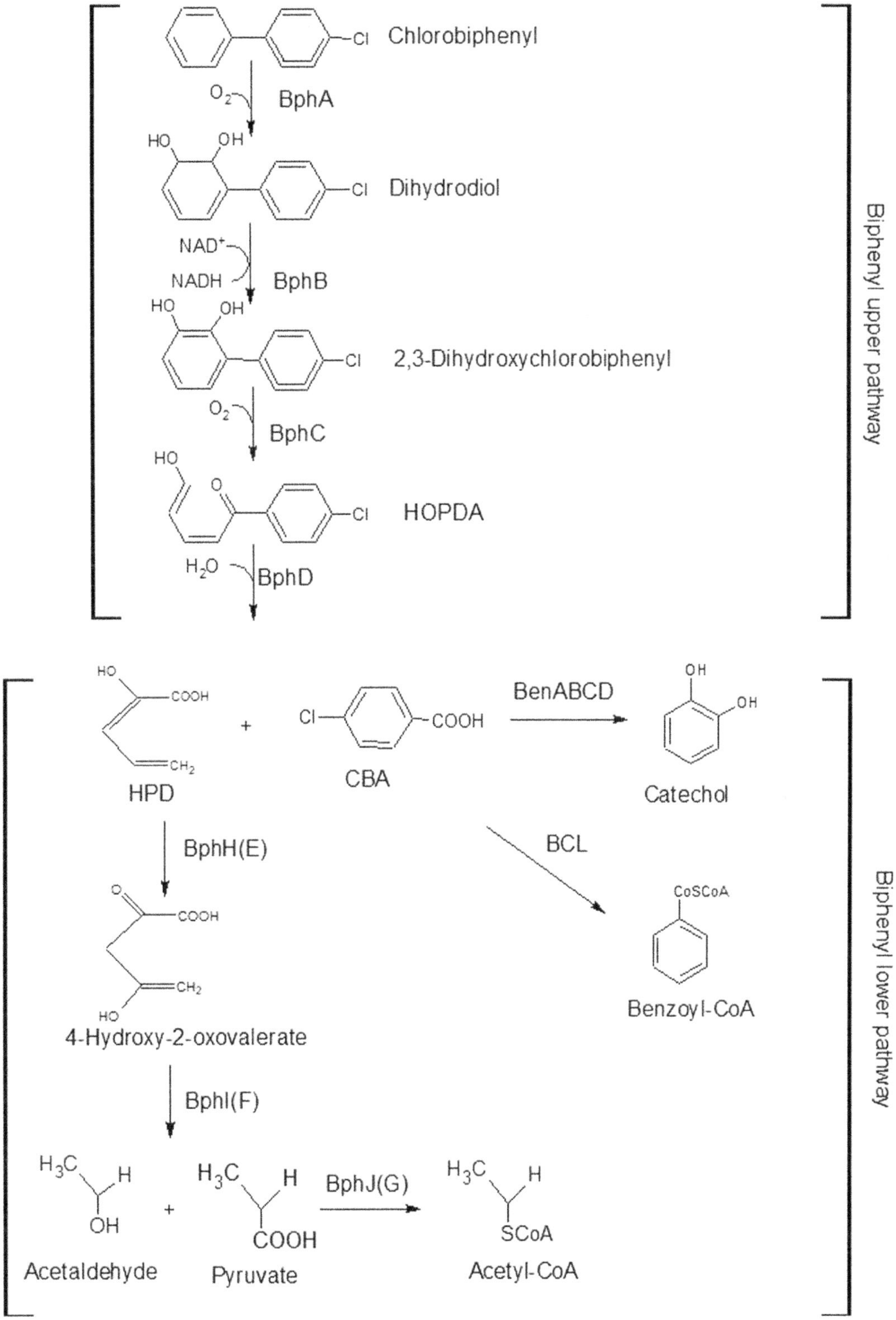

FIGURE 13.4 Upper and lower pathways of aerobic degradation.

3. cleavage of 2,3-dihydroxy biphenyl at the 1,2 position to form metacleavage product HOPDA (2-hydroxy-6-oxo-6-phenylhexa-2,4-dienoic acid) catalyzed by 2,3-dihydroxybiphenyl-1,2-dioxygenases (*bphC*); and
4. hydroxylation HOPDA to chlorobenzoic acid (CBA) and 2-hydroxypenta-2,4-dienoate (HPD) catalyzed by 2-hydroxy-6-phenyl-6-oxohexa-2,4-dieneoate (HOPDA) hydrolases (*bphD*).

The biphenyl lower pathway

The products of the upper biphenyl pathway (HPD and CBA) act as the substrates of the lower pathway. HPD undergoes further metabolism catalyzed by the products of *bphEFG* or *bphHIJ* genes. Genes important for the conversion of aliphatic end product of the biphenyl upper pathway to Krebs cycle intermediates in the *bph* locus of *Pseudomonas* sp. LB400 had already been discovered between the *bphC* and *bphD locus* (Hofer et al., 1994). HPD is converted to 4-hydroxy-2-oxo-valerate by (*bphH(E)*) encoded enzyme 2-hydroxypenta-2,4-dienoate hydratase followed by conversion to acetaldehyde and pyruvic acid by 4-hydroxy-2-oxovalerate aldolase of *BphI(F)* gene. Finally, an acylating acetaldehyde dehydrogenase (*bphJ(G)*) acts upon acetaldehyde and converts it into acetyl-CoA, which enters the Krebs cycle. The other product of the upper pathway CBA is also further metabolized to form lesser toxic products and the breakdown of CBA may vary among different microorganisms. CBA is converted to benzoyl-CoA by benzoate-CoA ligase (BCL) followed by conversion to 2,3-dihydroxydihydrobenzoyl CoA by benzoyl-CoA dioxygenase. This is followed by nonoxygenolytic cleavage of the aromatic ring and the beta-oxidation-like pathway of the ring cleavage product. This pathway produces 3-hydroxyadipyl CoA and 3-ketoadipyl CoA and then finally is converted to succinyl-CoA and acetyl-CoA.

CBA could also be converted to catechol by the enzyme benzoate 1,2-dioxygenase and benzoate dihydrodiol dehydrogenase.

13.3.1.3 Combined anaerobic and aerobic degradation

Combination of anaerobic dechlorination and aerobic oxidation is a promising bioremediation strategy to achieve complete degradation. Two types of combined anaerobic dechlorination and aerobic oxidation have been reported in PCB bioremediation: two-stage (or sequential) anaerobic–aerobic degradation and concurrent (or alternating) anaerobic–aerobic degradation. Two-stage degradation of Aroclor 1260–contaminated soil mixed with organic waste (2:3 ratio) exposed to 70 days of anaerobic composting followed by 28 days of aerobic composting resulted in 25.0% reduction in total PCBs (Long et al., 2015). Bioaugmentation of laboratory mesocosms containing Aroclor-contaminated sediment from Baltimore Harbor with anaerobic halorespiring *Dehalobium chlorocoercia* DF1 and aerobic *Burkholderia xenovorans* LB400 added concurrently with granulated activated carbon (GAC) as a delivery system has been determined. The greatest effect was seen in the mesocosm bioaugmented with both DF1 and LB400 together, which resulted in an 80% decrease by mass of PCBs (Payne et al., 2013). Utilizing facultative microbial strains capable of degradation of PCBs under both anaerobic and aerobic conditions instead of separate obligate aerobic and anaerobic microorganisms provides an efficient and effective alternative to commonly used bioaugmentation methods. Four facultative anaerobic bacteria discovered from soil and sediments and identified as *Achromobacter* sp. NP03, *Ochrobactrum* sp. NP04, *Lysinibacillus* sp. NP05, and *Pseudomonas* sp. NP06 were found to be capable of degrading Aroclor 1260 (50 mg/L) under aerobic, anaerobic, and two-stage anaerobic–aerobic conditions (Pathiraja et al., 2019b). Aroclor 1260 degradation studied under alternating (AN) and two-stage (TS) anaerobic–aerobic conditions by facultative anaerobic bacterial consortium consisting of *Achromobacter* sp. NP03, *Lysinibacillus* sp., and Ochrobactrum sp. NP04 reported 49.2% ± 2.5% total reduction of the original 50 mg/L Aroclor 1260 mixture after 2 weeks in AN treatment as compared to only 24.44% ± 2.46% in TS treatment. The overall results suggested that anaerobic and aerobic conditions achieved better PCB degradation by repeated exposure to short periods of alternative anaerobic and aerobic conditions rather than the conventional two-stage anaerobic–aerobic conditions (Pathiraja et al., 2019a).

13.3.2 Polychlorinated biphenyl degradation by fungi

Fungi have proved to be excellent candidates for degradation of PCBs and show features, which give them an edge over bacterial degradation. They are filamentous and therefore the hyphae can penetrate the environmental matrices, which makes the environmental pollutants more accessible to them. They employ both intracellular and extracellular enzymes in PCB degradation unlike bacteria that utilize enzymes that are usually intracellular, which makes PCB intake a rate-limiting step in bacteria (Chun et al., 2019). The major fungal enzymes involved in PCB degradation have been found to be intracellular cytochrome P450 monooxygenases and extracellular peroxidases such as lignin peroxidase (LiP), manganese peroxidase (MnP), versatile peroxidases (VP), and phenol oxidases such as laccase, but the pathways through which these PCBs are degraded are not completely elucidated due to complexity of fungal enzyme systems (Čvančarová et al., 2012). Bioremediation of PCBs using fungi, mostly white-rot causing basidiomycetes are exhibited to be an extremely effective approach. Various white-rot fungi, including *Phanerochaete chrysosporium, Trametes versicolor, Lentinus edodes, Grifola frondosa, Pycnoporus cinnabarinus, Coriolopsis polyzona, Phlebia brevispora, Irpex lacteus, Bjerkandera adusta, Phanerochaete magnoliae*, and *Pleurotus ostreatus*, can successfully orchestrate PCB removal. Degradation of PCBs by a fungi was first reported in *P. chrysosporium*, which degraded [U-14C]-Aroclor 1254 with the release of 14CO2 and water-soluble degradation products (Eaton, 1985). White-rot fungi are capable of degrading highly chlorinated PCB congeners quite efficiently. Biodegradation of four PCB congeners, 4,4′-dichlorobiphenyl, 2,3′,4′,5-tetrachlorobiphenyl, 2,2′,4,5,5′-pentachlorobiphenyl, and 2,2′,4,4′,5,5′-hexachlorobiphenyl by *Ceriporia* sp. ZLY-2010 showed that highly chlorinated PCB congeners were degraded more efficiently, with 2,2′,4,4′,5,5′-hexachlorobiphenyl being degraded with 40.05% efficacy (Hong et al., 2012). *P. ostreatus*, breaks down penta- and hexa-chlorinated biphenyls with 98.4% and 99.6% efficiency in complex and mineral media, respectively, and is the most efficient PCB-degrading fungi so far (Čvančarová et al., 2012). Recently a highly efficient white-rot fungi *Pleurotus sajor-caju* LBM 105 capable of removing more than 90% of PCBs after 35 days has been reported (Sadañoski et al., 2019). Many ascomycetes have also been identified with the potential of PCB degradation such as *Thermothelomyces thermophila, Thermothelomyces heterothallica, Thermoascus crustaceus, Fusarium solani, Doratomyces nanus, Doratomyces purpureofuscus, Doratomyces verrucisporus, Myceliophthora thermophila, Phoma eupyrena*, which may prove to be promising bioremediation candidates in future (Périgon et al., 2019; Sage et al., 2014).

13.4 Conclusion

There has been great progress in the studies associated with bacterial and fungal PCB degradation, which reflect their potential for use to minimize the PCB-related toxicity effects on environment. Bacterial degradation has been most widely studied but still pathways of degradation associated with many strains need further clarification. Establishment of pure cultures of anaerobic bacteria is required for understanding the conditions required for degradation and the pathway-related details. As compared to the studies of bacterial degradation only a few articles have been published about fungi. White-rot fungi hold a great potential in PCB degradation, but the mechanisms of degradation and the factors limiting the microbe activity are not completely confirmed. Although various studies tried to elucidate the mechanism of degradation from the transformation products, further explanation about the activity of enzymes is still required (Stella et al., 2017; Čvančarová et al., 2012).

References

Ahmed, M., Focht, D.D., 1973. Degradation of polychlorinated biphenyls by two species of Achromobacter. Can. J. Microbiol. 19, 47–52.

Arrebola, J.P., González-Jiménez, A., Fornieles-González, C., Artacho-Cordón, F., Olea, N., Escobar-Jiménez, F., et al., 2015. Relationship between serum concentrations of persistent organic pollutants and markers of insulin resistance in a cohort of women with a history of gestational diabetes mellitus. Environ. Res. 136, 435–440.

Aydin, Y., Erkan, M.J.T., Health, I., 2017. The toxic effects of polychlorinated biphenyl (Aroclor 1242) on Tm3 Leydig cells. Toxicol. Ind. Health 33, 636–645.

Bedard, D.L., Ritalahti, K.M., Löffler, F.E., 2007. The *Dehalococcoides* population in sediment-free mixed cultures metabolically dechlorinates the commercial polychlorinated biphenyl mixture Aroclor 1260. Appl. Environ. Microbiol. 73, 2513–2521.

Borja, J., Taleon, D.M., Auresenia, J., Gallardo, S., 2005. Polychlorinated biphenyls and their biodegradation. Process Biochem. 40, 1999–2013.

Cao, Y.-M., Xu, L., Jia, L.-Y.J.N.B., 2011. Analysis of PCBs degradation abilities of biphenyl dioxygenase derived from *Enterobacter* sp. LY402. Mol. Simul. 29, 90–98.

Capozzi, S.L., Bodenreider, C., Prieto, A., Payne, R.B., Sowers, K.R., Kjellerup, B.V., 2019. Colonization and growth of dehalorespiring biofilms on carbonaceous sorptive amendments. Biofouling 35, 50–58.

Chun, S.C., Muthu, M., Hasan, N., Tasneem, S., Gopal, J., 2019. Mycoremediation of PCBs by *Pleurotus ostreatus*: possibilities and prospects. Appl. Sci. 9, 4185.

Cutter, L.A., Watts, J.E., Sowers, K.R., May, H.D., 2001. Identification of a microorganism that links its growth to the reductive dechlorination of 2,3,5,6-chlorobiphenyl. Environ. Microbiol. 3, 699–709.

Čvančarová, M., Křesinová, Z., Filipová, A., Covino, S., Cajthaml, T., 2012. Biodegradation of PCBs by ligninolytic fungi and characterization of the degradation products. Chemosphere 88, 1317–1323.

Eaton, D.C., 1985. Mineralization of polychlorinated biphenyls by *Phanerochaete chrysosporium*: a ligninolytic fungus. Enzyme Microb. Technol. 7, 194–196.

Elangovan, S., Pandian, S.B.S., Geetha, S., Joshi, S.J., 2019. Polychlorinated Biphenyls (PCBs): Environmental Fate, Challenges and Bioremediation. Microbial Metabolism of Xenobiotic Compounds. Springer.

Everett, C.J., Frithsen, I.L., Diaz, V.A., Koopman, R.J., Simpson, W.M., Mainous, A.G., 2007. Association of a polychlorinated dibenzo-*p*-dioxin, a polychlorinated biphenyl, and DDT with diabetes in the 1999–2002 National Health and Nutrition Examination Survey. Environ. Res. 103, 413–418.

Faroon, O.M., Keith, S., Jones, D., De Rosa, C., 2001. Carcinogenic effects of polychlorinated biphenyls. Toxicol. Ind. Health 17, 41–62.

Fennell, D.E., Nijenhuis, I., Wilson, S.F., Zinder, S.H., Häggblom, M.M., 2004. *Dehalococcoides ethenogenes* strain 195 reductively dechlorinates diverse chlorinated aromatic pollutants. Environ. Sci. Technol. 38, 2075–2081.

Gupta, A., Singh, S.K., Singh, V.K., Singh, M.K., Modi, A., Zhimo, V.Y., et al., 2020. Endophytic microbe approaches in bioremediation of organic pollutants. In: Kumar, A., Kumar Singh, V. (Eds.), Microbial endophytes: Prospects of Sustainable Agriculture. Elsevier, pp. 157–174.

Hofer, B., Backhaus, S., Timmis, K.N.J.G., 1994. The biphenyl/polychlorinated biphenyl-degradation locus (bph) of *Pseudomonas* sp. LB400 encodes four additional. Metab. Enzymes 144, 9–16.

Hoffman, C.S., Small, C.M., Blanck, H.M., Tolbert, P., Rubin, C., Marcus, M., 2007. Endometriosis among women exposed to polybrominated biphenyls. Ann. Epidemiol. 17, 503–510.

Hong, C.-Y., Gwak, K.-S., Lee, S.-Y., Kim, S.-H., Lee, S.-M., Kwon, M., et al., 2012. Biodegradation of PCB congeners by white rot fungus, Ceriporia sp. ZLY-2010, and analysis of metabolites. J. Environ. Sci. Health A: Tox. Hazard. Subst. Environ. Eng 47, 1878–1888.

Hu, J., Qian, M., Zhang, Q., Cui, J., Yu, C., Su, X., et al., 2015. *Sphingobium fuliginis* HC3: a novel and robust isolated biphenyl-and polychlorinated biphenyls-degrading bacterium without dead-end intermediates accumulation. PLoS One 10, e0122740.

Hug, L.A., Maphosa, F., Leys, D., Löffler, F.E., Smidt, H., Edwards, E.A., 2013. Overview of organohalide-respiring bacteria and a proposal for a classification system for reductive dehalogenases. Philos. Trans. R. Soc. Lond. B: Biol. Sci 368, 20120322.

Jia, L.-Y., Zheng, A.-P., Xu, L., Huang, X.-D., Zhang, Q., Yang, F.-L.J.J.M.B., 2008. Isolation and characterization of comprehensive polychlorinated biphenyl degrading bacterium, *Enterobacter* sp. LY402. J. Microbiol. Biotechnol. 18, 952–957.

Jiang, L.-G., Cheng, L.-Y., Kong, S.-H., Yang, Y., Shen, Y.-J., Chen, C., et al., 2017. Toxic effects of polychlorinated biphenyls (Aroclor 1254) on human sperm motility. Asian J. Androl. 19, 561.

Liang, D., Wang, S., 2017. Development and characterization of an anaerobic microcosm for reductive dechlorination of PCBs. Front. Environ. Sci. Eng. 11, 2.

Long, Y.-Y., Fang, Y., Zhang, C., Du, Y., Shentu, J., Shen, D.-S., 2015. Degradation of polychlorinated biphenyls by sequential anaerobic–aerobic composting. Water Air Soil Pollut 226, 44.

Murínová, S., Dercová, K., Dudášová, H., 2014. Degradation of polychlorinated biphenyls (PCBs) by four bacterial isolates obtained from the PCB-contaminated soil and PCB-contaminated sediment. Int. Biodeterioration Biodegradation 91, 52–59.

Natarajan, M., Wu, W.-M., Nye, J., Wang, H., Bhatnagar, L., Jain, M., 1996. Dechlorination of polychlorinated biphenyl congeners by an anaerobic microbial consortium. Appl. Microbiol. Biotechnol. 46, 673–677.

Pathiraja, G., Egodawatta, P., Goonetilleke, A., Te'o, V.S., 2019a. Effective degradation of polychlorinated biphenyls by a facultative anaerobic bacterial consortium using alternating anaerobic aerobic treatments. Sci. Total Environ. 659, 507–514.

Pathiraja, G., Egodawatta, P., Goonetilleke, A., Te'o, V.S., 2019b. Solubilization and degradation of polychlorinated biphenyls (PCBs) by naturally occurring facultative anaerobic bacteria. Sci. Total Environ. 651, 2197–2207.

Payne, R.B., Fagervold, S.K., May, H.D., Sowers, K.R., 2013. Remediation of polychlorinated biphenyl impacted sediment by concurrent bioaugmentation with anaerobic halorespiring and aerobic degrading bacteria. Environ. Sci. Technol. 47, 3807–3815.

Périgon, S., Massier, M., Germain, J., Binet, M.-N., Legay, N., Mouhamadou, B.J.E.S., et al., 2019. Metabolic adaptation of fungal strains in response to contamination by polychlorinated biphenyls. Environ. Sci. Pollut. Res. Int. 26, 14943–14950.

Pessah, I.N., Lein, P.J., Seegal, R.F., Sagiv, S.K., 2019. Neurotoxicity of polychlorinated biphenyls and related organohalogens. Acta Neuropathol. 138, 1–25.

Quensen, J.F., Tiedje, J.M., Boyd, S.A., 1988. Reductive dechlorination of polychlorinated biphenyls by anaerobic microorganisms from sediments. Science 242, 752–754.

Quintaneiro, C., Soares, A.M., Costa, D., Monteiro, M.S., 2019. Effects of PCB-77 in adult zebrafish after exposure during early life stages. J. Environ. Sci. Health A: Tox. Hazard. Subst. Environ. Eng 54, 478–483.

Rhee, G.Y., Sokol, R.C., Bethoney, C.M., Bush, B., 1993. A long-term study of anaerobic dechlorination of PCB congeners by sediment microorganisms: pathways and mass balance. Environ. Toxicol. Chem. 12, 1829–1834.

Sadañoski, M.A., Benítez, S.F., Fonseca, M.I., Velázquez, J.E., Zapata, P.D., Levin, L.N., et al., 2019. Mycoremediation of high concentrations of polychlorinated biphenyls with *Pleurotus sajor-caju* LBM 105 as an effective and cheap treatment. J. Environ. Chem. Eng. 7, 103453.

Sage, L., Périgon, S., Faure, M., Gaignaire, C., Abdelghafour, M., Mehu, J., et al., 2014. Autochthonous ascomycetes in depollution of polychlorinated biphenyls contaminated soil and sediment. Chemosphere 110, 62–69.

Samadi, A., Sharifi, H., Ghobadi Nejad, Z., Yaghmaei, S.J.I.J.O.E., 2019. Biodegradation of polychlorinated biphenyls by and isolated from contaminated soil. Int. J. Eng 32, 628–633.

Schmidt, C., 2010. How PCBs are like grasshoppers. ACS Publ. 44 (8), 2752.

Schnitzler, J.G., Reckendorf, A., Pinzone, M., Autenrieth, M., Tiedemann, R., Covaci, A., et al., 2019. Supporting evidence for PCB pollution threatening global killer whale population. Aquat. Toxicol. 206, 102–104.

Seto, M., Kimbara, K., Shimura, M., Hatta, T., Fukuda, M., Yano, K.J.A.E.M., 1995. A novel transformation of polychlorinated biphenyls by *Rhodococcus* sp. Strain RHA1. Appl. Environ. Microbiol 61, 3353–3358.

Singh, P., Singh, V.K., Singh, R., Borthakur, A., Madhav, S., Ahamad, A., et al., 2020. Bioremediation: a sustainable approach for management of environmental contaminants. In: Singh, P., Kumar, A., Borthakur, A. (Eds.), Abatement of Environmental Pollutants. Elsevier, pp. 1–23.

Stella, T., Covino, S., Čvančarová, M., Filipová, A., Petruccioli, M., D'annibale, A., et al., 2017. Bioremediation of long-term PCB-contaminated soil by white-rot fungi. J. Hazard. Mater. 324, 701–710.

Strek, H., Weber, J.B., 1982. Behaviour of polychlorinated biphenyls (PCBs) in soils and plants. Environ. Pollut. Ser. A, Ecol. Biol. 28, 291–312.

Su, X., Li, S., Cai, J., Xiao, Y., Tao, L., Hashmi, M.Z., et al., 2019. Aerobic degradation of 3, 3′, 4, 4′-tetrachlorobiphenyl by a resuscitated strain *Castellaniella* sp. SPC4. Kinetics Model. Pathw. Biodegrad. 688, 917–925.

Subramanian, S., Tehrani, R., Van Aken, B., 2019. Transcriptomic response of *Arabidopsis thaliana* exposed to hydroxylated polychlorinated biphenyls (OH-PCBs). Int. J. Phytoremediation 21, 52–59.

Terzaghi, E., Vergani, L., Mapelli, F., Borin, S., Raspa, G., Zanardini, E., et al., 2019. Rhizoremediation of weathered PCBs in a heavily contaminated agricultural soil: results of a biostimulation trial in semi field conditions. Sci. Total Environ. 686, 484–496.

Wang, S., He, J., 2013a. Phylogenetically distinct bacteria involve extensive dechlorination of Aroclor 1260 in sediment-free cultures. PLoS One 8, e59178.

Wang, S., He, J., 2013b. Dechlorination of commercial PCBs and other multiple halogenated compounds by a sediment-free culture containing *Dehalococcoides* and *Dehalobacter*. Environ. Sci. Technol. 47, 10526–10534.

Wang, X., Teng, Y., Luo, Y., Dick, R.P., 2016. Biodegradation of 3,3′,4,4′-tetrachlorobiphenyl by *Sinorhizobium meliloti* NM. Bioresour. Technol. 201, 261–268.

Wang, S., Chen, C., Zhao, S., He, J., 2019. Microbial synergistic interactions for reductive dechlorination of polychlorinated biphenyls. Sci. Total Environ. 666, 368–376.

Wiegel, J., Wu, Q.J.F.M.E., 2000. Microbial reductive dehalogenation of polychlorinated biphenyls. FEMS Microbiol. Ecol 32, 1–15.

Williams, W.A., 1994. Microbial reductive dechlorination of trichlorobiphenyls in anaerobic sediment slurries. Environ. Sci. Technol. 28, 630–635.

Wu, Q., Watts, J.E., Sowers, K.R., May, H.D., 2002. Identification of a bacterium that specifically catalyzes the reductive dechlorination of polychlorinated biphenyls with doubly flanked chlorines. Appl. Environ. Microbiol. 68, 807–812.

Wu, M., Xu, X., Lu, K., Li, X., 2019. Effects of the presence of nanoscale zero-valent iron on the degradation of polychlorinated biphenyls and total organic carbon by sediment microbial fuel cell. Sci. Total Environ. 656, 39–44.

Xu, X., Zhao, Q., Wu, M., 2015. Improved biodegradation of total organic carbon and polychlorinated biphenyls for electricity generation by sediment microbial fuel cell and surfactant addition. RSC Adv. 5, 62534–62538.

Zhang, H., Jiang, X., Lu, L., Xiao, W., 2015. Biodegradation of polychlorinated biphenyls (PCBs) by the novel identified cyanobacterium Anabaena PD-1. PLoS One 10, e0131450.

Chapter 14

Perspectives in desulfurization of coal using microbes

Vipin Kumar Singh[1], Rishikesh Singh[2], Ajay Kumar[3], Rahul Bhadouria[4] and Kin Israel Notarte[5]

[1]*Department of Botany, Centre of Advanced Study in Botany, Institute of Science, Banaras Hindu University, Varanasi, India,* [2]*Institute of Environment and Sustainable Development, Banaras Hindu University, Varanasi, India,* [3]*Agriculture Research Organization, Ministry of Agriculture and Rural Development, Volcani Center, Rishon LeZion, Israel,* [4]*Department of Botany, University of Delhi, New Delhi, India,* [5]*Faculty of Medicine and Surgery, University of Santo Tomas, Manila, Philippines*

14.1 Introduction

Coal is one of the cheapest, reliable, heterogeneous, noncrystalline, and naturally abundant nonrenewable major fossil fuel employed for energy production around the globe. The gradual loss of good quality coal (Tang et al., 2018a) as well as release of large amount of SO_2 into atmosphere during industrial burning culminating into atmospheric pollution (Makgato and Chirwa, 2020), acid rain, and eventually negative consequences on natural environment (Medunić et al., 2016; Xu et al., 2019) has compelled the fuel scientists to develop the coal beneficiation strategies for efficient sulfur reduction. Coal has been classified as anthracite, bituminous, subbituminous, and lignite (Speight, 2015). The coal categorization as "high rank" and "low rank" is considerably determined by the coal characteristics such as amount of carbon, volatile matter, and heat content. The sulfur present in coal may be either of organic or of inorganic (pyritic and sulfide) nature and modulates energy production (Acharya et al., 2005). The removal of organic sulfur from low-quality coal is considered very much difficult (Xia, 2018). High-content sulfur in different coal types has limited its industrial applications and requires the minimization of sulfur content by suitable methods. The process of removing sulfur from coal is termed as desulfurization and can be achieved by various physical, chemical, and biological methods.

Some of the commonly used physical processes for desulfurization such as flotation, flocculation, gravity separation, sedimentation, ultrasonication, microwave irradiation, and pyrolysis are lesser in cost but proved little effective in the elimination of organic sulfur (Shen et al., 2011). The chemical treatment includes the application of different acids and bases allowing either oxidation or reduction for rapid removal of sulfur from coal matrices. Treatment with sodium borohydride (Li et al., 2010), potassium hydroxide, hydrochloric acid (Anwar et al., 2020), nitric acid (Karaca et al., 2003; Ahmad et al., 2019), subcritical water (Giray et al., 2019), and supercritical liquids (Park et al., 2007) such as methanol and ethanol is well documented. The physicochemical processes are expensive, may produce secondary pollutants leading to environmental degradation, difficult in application, and not ecologically sound. In contrast, the biological agent–based sulfur removal from coal (biodesulfurization) is economical, environmentally safe, results into minimal release of SO_2 into atmospheric environment, and even able to improve the overall calorific value. Interestingly, the process of biodesulfurization as a clean coal technology has been found much effective in sulfur elimination from low rank coal (Restrepo et al., 2015). Among various biological agents used for removing sulfur from coal, bacteria (Makgato and Chirwa, 2020), actinomycetes (Omori et al., 1992; Dahlberg et al., 1993), and fungi (Sabar et al., 2019) have been reported for successful elimination of sulfur from various coal types.

The important step during the removal of sulfur from coal involves oxidation of sulfur moiety, and the breakage of C–C as well as C–S bond (Gupta et al., 2005) as largely known from the investigations on model organic sulfur compound dibenzothiophene (DBT). The process of sulfur removal may be of aerobic or anaerobic nature (Soleimani et al., 2007). Noteworthy, the process of desulfurization may be facilitated by autotrophic or heterotrophic microorganisms (Xu et al., 2019). The enzymes contributing in desulfurization have been identified as monooxygense, desulfinase, and FMN-oxidoreductase and expressed as "*DszC* and *DszA*," "*DszB*," *DszE*, respectively (Martín-Cabello et al., 2020).

Microbe Mediated Remediation of Environmental Contaminants. DOI: https://doi.org/10.1016/B978-0-12-821199-1.00014-6

Since, the sulfur removal from coal based on living organisms is affected largely by environmental factors, optimization of different parameters at laboratory scale is a prerequisite step in obtaining clean coal at industrial scale. Temperature, pH, moisture content, composition of nutritional media, coal types, particle size, pulp density, and microorganisms used are important factors to be considered while trying to minimize the level of sulfur in various coal types. The assessment of cost benefit ratio of biodesulfurization process as compared to other physicochemical methods would be helpful in better exploration of the process at industrial scale. In present chapter, we have tried to briefly discuss the different processes for sulfur removal from coal with emphasis on biological desulfurization, different microbes performing sulfur removal, mechanism of biodesulfurization, effect of different factors on sulfur elimination, and future perspectives.

14.2 Physicochemical methods for coal desulfurization

The process of physical removal of sulfur from coal may be of particular interest for some coal types. Some of the important physical methods for sulfur elimination from coal include gravity separation, sedimentation, flocculation, microwave irradiation, ultrasonication, flotation, and so on. The process of physical elimination of sulfur from coal is rapid, economical, very often does not modulate coal attributes, and sometimes involves the application of hazardous chemicals. However, the physical methods are not much effective in removing organic sulfur present in coal matrices (Wang et al., 2016). Coal desulfurization based on application of different chemicals is rapid (Anwar et al., 2020) and in most of the cases are able to eliminate both inorganic and organic sulfur contents. But, the treatment with chemicals at high temperature and pressure has been described to be associated with the formation of secondary products and may result in loss of calorific value. Further, the release of large amount of SO_2 during chemical processes needs further treatment to reduce the atmospheric pollution and makes the process less feasible and more expensive at industrial scale (Meshram et al., 2015).

Desulfurization using chemical approaches deploys treatment with chemical substances such as hydrochloric acid, sulfuric acid, sodium hydroxide, potassium permanganate, hydrogen peroxide, cupric chloride, and nitric acid. The integrated application of physical and chemical methods (Luo and Wang, 2013) has also been reported by different workers and in few cases is proved more effective as compared to single treatment method applied. Some of the important physicochemical process for removal of sulfur from coal is presented in Table 14.1.

14.3 Biological coal desulfurization

The role of microorganisms falling in the category of autotrophs, heterotrophs, mixotrophs, acidophiles, and thermophiles in the elimination of sulfur from coal is attractive from environmental point of view. So far, numbers of bacteria and fungi frequently contributing in removal of sulfur from different coal types are described. Biological processes for desulfurization as a suitable ecofriendly alternative of clean coal technology are independent of chemical application, economical, able to sequester finely scattered sulfur compounds, and, most notably, do not produce the SO_2 into atmosphere, hence those are thought to have negligible impact on environmental homoeostasis. Furthermore, the employment of microbes for removing different forms of sulfur from coal may be advantageous as compared to physical and chemical processes (Kargi, 1986; Gunam et al., 2006; He et al., 2012) because of low temperature requirement and minimal changes in some coal characteristics, important from industrial view point. The biodesulfurization process can be successfully applied under mild laboratory conditions to remove both inorganic (pyrite, sulfate, and sulfide) and organic sulfur (dibenzothiphene and benzothiophene) present in coal. Interestingly, during the process of biodesulfurization, carbon and sulfur present in coal are used as energy source for microbial growth and multiplication. In contrast to easy removal of inorganic sulfur rendered by loose association of sulfur compounds such as pyrite, organic sulfur elimination by microbiological processes is quite difficult probably due to covalent linkage with coal matrices (Aytar et al., 2008) and varies among microbes used.

In general, the biological sulfur removal from typical organic sulfur compounds of coal like DBT is achieved by oxidative breakage of C–S and C–C bond present in ring responsible for synthesis of degradation products such as DBT-5-oxide, DBT-5-dioxide, benzoate, and sulfate (Seo, 2012). However, the so-called 4S pathway of desulfurization without involvement of cleavage of aromatic ring structure is also recognized. On the other hand, the removal of inorganic sulfur mostly present as iron pyrite involves oxidation of reduced sulfur by microbes interacting with coal minerals. The probable pathway of pyrite oxidation in a sequential manner results into derivation of thiosulfate followed by tetrathionates/polythionates and ultimately sulfate as the end product (Schippers et al., 1996). Very often, in the case of biodesulfurization of pyritic sulfur, formation of mineral such as jarosite responsible for reduced rate of sulfur elimination is clearly evident. This reduction in sulfur removal is mainly because of restricted interaction of microorganism

TABLE 14.1 An overview of different physicochemical processes of sulfur removal from coal.

Studied process	Investigated coal	Country	Remarks	Reference
Physical process				
Microwave irradiation	Bituminous	China	Nearly absolute removal of inorganic sulfur was achieved by exposure to irradiation for 100 s followed by acid treatment.	Weng and Wang (1992)
Microwave irradiation coupled with $Fe_2(SO_4)_3$ oxidation	–	China	Under optimized conditions the process removed 62.40% of total sulfur, increased calorific value, and augmented features of coal.	Luo and Wang (2013)
Microwave	Bituminous	China	The technique was able to reduce 14.4% pyrite sulfur and 4% organic sulfur. Slight reduction in calorific value was observed.	Xu and Tao (2015)
Microwave oxidation	–	China	The treatment caused reduction in sulfur content and ash content up to 48.38% and 42.51%, respectively.	Huang et al. (2019)
Ultrasonication combined with microwave	–	China	The study indicated synergistic action of ultrasonication and microwave on sulfur removal. Maximum reduction in pyritic sulfur was 85% as compared to 22% removal of organic sulfur.	Tang et al. (2019)
Ultrasonication	Brazilian coal	Brazil	The application of ultrasonication resulted into reduction of total sulfur, organic sulfur, pyritic sulfur, and sulfate sulfur up to 70%–77%, 67%–81%, 48%–60%, 91%–96%, respectively, under set conditions.	Saikia et al. (2016)
Ultrasonication	Sulfur rich Indian coal	India	The investigation revealed 83.92% removal of pyritic sulfur, 12.50% reduction in sulfate sulfur, and 18.80% elimination of total sulfur under defined conditions.	Saikia et al. (2014)
Flotation	Lignite	Turkey	In second part of experiment, total sulfur removed was 56.25%. The Aegean sea water was most effective medium in flotation for total sulfur elimination.	Temel et al. (2010)
Column froth flotation	Lignite and bituminous	Turkey	Column floatation showed better performance in sulfur removal as compared to conventional flotation.	Demirbaş (2002)
Flotation	Fine coal	Turkey	Pyritic sulfur and sulfate sulfur was removed to a level of 90.04% and 81.45%, respectively, under predefined conditions.	Sahinoglu (2018)
Float sink separation	Lignite	Turkey	The process was not much effective in removing sulfur from low rank coal.	Çetinkaya and Bayat (2020)
Gravity separation	Lignite	Turkey	Only little changes (17.46%) in total sulfur content was recorded.	Özbakir et al. (2017)
Gravity separation integrated with pyrolysis and Fe powder	Low rank coal	China	The process effectively declined organic sulfur content from 2.0% to 0.33%.	Xia et al. (2018)
Chemical process				
Nitric acid treatment	Lignite	Turkey	The pyritic sulfur removal was affected by particle size, nitric oxide concentration, and stirring rate. Higher rate of pyrite sulfur was observed for coal with smaller particle size, increased with increase in initial acid concentration and temperature.	Karaca et al. (2003)
Sequential KOH and HCl treatment	Low grade	Pakistan	The highest sulfur removal noted from coal was 74.2%. Increase in treatment time and temperature improved desulfurization.	Anwar et al. (2020)
Leaching by HNO_3	–	Iran	Sulfur removal efficiency was expected to reach 86.96% under optimized temperature (80°C), acid concentration (28%), and treatment time (84 min) as determined by response surface methodology.	Shahraki et al. (2018)
Flotation combined with acid treatment	Mezino coal	Iran	Highest total sulfur removal (75.4%) was found with 30% acid concentration.	Alam et al. (2009)
Cupric chloride treatment	Lignite	Pakistan	At 150°C the highest total sulfur and organic sulfur removal observed was 50% and 56%, respectively, while maximum pyritic sulfur elimination equivalent to 66% was noticed at 200°C.	Khan et al. (2014)
NaOH/HCl treatment	High sulfur coal	India	16% NaOH and 10% HCl treatment to coal was able to eliminate 10% organic sulfur and 30%–33% total sulfur.	Mukherjee and Borthakur (2001)
Solvent extraction coupled with alkali treatment	Oxidized Indian coal	India	Maximum 95% inorganic sulfur and 31% organic sulfur elimination was reported in the investigation.	Baruah and Khare (2007)
Acid/base leaching integrated with oxidation	Lakhra coal	Pakistan	The combined action of acid and alkali treatment together with catalytic oxidation led to sulfur reduction up to 83%.	Ahmad et al. (2019)
Hydrogen peroxide (H_2O_2) treatment	Subbituminous coal	India	H_2O_2 at a concentration of 15% was effective in reducing coal sulfur in pyritic state to 76%, sulfate form to 70% and organic sulfur to 5%.	Mukherjee et al. (2001)
Sodium hypochlorite and sodium hydroxide treatment	Pittsburgh No. 8 and Illinois No. 6 coal	United States	Total 70% pyritic sulfur removal was noted after treatment with 0.4 molar sodium hypochlorite and similar concentration of sodium hydroxide at 90°C for Pittsburgh No. 8 coal. In case of Illinois No. 6 coal, the reduction in organic sulfur from coal was determined as 37.8% after treatment with sodium hypochlorite followed by alkaline hydrolysis by NaOH.	Li and Cho (2005)

with coal minerals such as pyrite present in coal caused by heavy deposition of jarosite on surface. Usually, biodesulfurization leads to transformation of reduced sulfur into forms that are soluble, non-hazardous, and amenable to removal by conventional washing (Acharya et al., 2004). Biodesulfurization is reported to be assisted by enzymes such as monooxygenase and desulfinase synthesized by so-called genes *dszABC* (Oldfield et al., 1997; Gunam et al., 2006). The efficiency of biodesulfurization by particular microbes is determined by variables such as pH, temperature, humidity, composition of nutritional media, coal characteristics, content of sulfur, treatment time, shaking and static condition, and particle size of coal.

14.3.1 Role of bacteria in removal of sulfur from coal

The presence of aromatic sulfur compounds and highly hydrophobic nature of coal makes it resistant to microbiological desulfurization. Nevertheless, in nature, diverse bacteria associated with the desulfurization are described. These bacterial species may be utilized as an ecofriendly tool for management of atmospheric sulfur contamination resulting from burning of fossil fuels such as coal. The bacteria having rapid multiplication rate, ability to survive under adverse environmental conditions such as tolerance to increased concentrations of heavy metals present in coal matrices, are considered effective means for sulfur removal from coal. Desulfurization of coal by microbiological activities is a wet procedure (Dhawan and Sharma, 2019). The microbiological desulfurization by iron oxidizers may be of direct or indirect nature and occurs in a concurrent manner. The direct action during inorganic sulfur rejection encompasses oxidation of pyrite by interacting microbe, whereas during the process of indirect action, the ferric iron produced by bacterial metabolism catalyzes the oxidation of pyrite (Tripathi et al., 2019).

To date, bacterial genera such as *Bacillus*, *Acidithiobacillus*, *Mycobacterium*, *Sulfolobus*, *Corynebacterium*, *Leptospirillum*, *Arthrobacter*, *Sphingomonas*, *Burkholderia*, *Rhodococcus*, *Pseudomonas*, *Desulfovibrio*, *Serratia*, and *Stenotrophomonas* have been reported to contribute considerably in the removal of inorganic and organic sulfur from coal and representative sulfur compounds generally encountered in coal matrices. Bacterial action on coal leads to changes in surface characteristics without any negative effect on energy production ability. Such changes in coal matrix can be easily observed through electron microscopy, infrared spectroscopy, X-ray diffraction, X-ray photoelectron spectroscopy and X-ray absorption near edge spectroscopy, and elemental analysis.

Application of *Acidithiobacillus ferrooxidans* in catalyzing the removal of sulfur from coal is presented recently (Liu et al., 2020). The desulfurization study conducted in three batches in medium M9K with and without iron was shown to affect the desulfurization ability of taken microorganism. In different sets with iron containing media the removal of pyritic and total sulfur was observed as 55.6% and 10.0%, 77.1% and 16.1%, and 86.5% and 28.2%, respectively. The supplementation of iron-free M9K media improved the desulfurization potential of the said bacterium; however, higher amount had negative impact on microbiological sulfur removal phenomenon. The reduction in desulfurization was also ascribed to the presence of compounds of calcium such as $CaCO_3$ and CaO. The experimental study performed by Makgato and Chirwa (2020) have shown even positive impact on calorific value (19% enhancement) of biologically desulfurized coal assisted by bacteria isolated from coal. A total of 70.3% pyrite form, 51.0% organic form, and 43.8% sulfide form of sulfur was successfully eliminated after bacterial treatment. The treatment was also effective in minimizing ash content by 33%. In conclusion, total sulfur removed by biodesulfurization was determined as 65.4%. A study on biodesulfurization of Indonesian coal by five bacteria is described currently by Arifin et al. (2020). The maximum percent total sulfur removal observed was 19.61% and facilitated by *Citrobacter murliniae*; however, it was least effective in minimizing the ash content.

The bacterial consortium could be considered for better results in view of total sulfur as well as ash removal. *Thiobacillus* induced removal of sulfur from petcoke obtained from Gujarat, India is represented recently (Tripathi et al., 2019). The microbial action of acidiophilic bacterium was able to remove more than 92% of total sulfur after treatment for 4 weeks. The bacterial consortium comprising of *Thiobacillus thiooxidans* and *Thiobacillus ferrooxidans* was demonstrated to be more effective in desulfurization as compared to pure culture of single bacterial species used. In order to produce energy without negative impact on ecosystem, microbe facilitated reduction in different sulfur forms present as impurity in various coal types is an industrially important and environmentally relevant process for its safe utilization. Etemadifar et al. (2019) have presented the potential of growing and resting cells of different strains of *Rhodococcus erythropolis* and *Rhodococcus qingshengii* in sulfur exclusion. In comparison to resting cells, actively multiplying cells of strain PD1 had maximum 46% sulfur elimination efficiency in media devoid of any carbon source. Further, the resting as well as multiplying cells of strain PD1 was more active in sulfur removal as compared to other strains taken under investigation, implying suitability for industrial coal remediation practices. In addition, all the selected strains were capable of DBT desulfurization via 4S pathway. The pilot scale experiment may be helpful in industrializing the process for efficient sulfur removal. The detailed investigation of the process under laboratory condition would be helpful in designing the industrial system to combat the problem of coal burning induced atmospheric sulfur-di-oxide (SO_2) pollution. Some of the important bacterial species contributing in desulfurization is provided in Table 14.2.

TABLE 14.2 List of bacterial species participating in desulfurization.

Bacterial species	Country	Remarks	Reference
Acidithiobacillus ferrooxidans	China	Maximum pyritic sulfur removal was 92% and that for total sulfur was recorded as 31.6%.	Liu et al. (2020)
Isolated from coal; microorganism details not mentioned	Republic of South Africa	The bacterial action removed total sulfur by 65.4%, decreased ash content, and improved calorific value.	Makgato and Chirwa (2020)
Citrobacter murliniae, Dietzia psychralcaliphila, Pseudomonas aeruginosa, Alcaligenes faecalis, Bacillus altitudini	Indonesia	Among different bacteria, *C. murliniae* was most effective in eliminating total sulfur content by 19.61%. The treatment was also helpful in alleviating ash content.	Arifin et al. (2020)
A. ferrooxidans and *A. thiooxidans*	Colombia	Sulfate removal by bacterial consortium increased in presence of cysteine up to 60 ppm and was efficient as compared to *A. ferrooxidans* alone.	Pineda and Godoy (2019)
Thiobacillus ferrooxidans	Korea	Sulfur removal from anthracite coal in bioreactor system ranging from 90% to 95% was expected in 15–20 days and tended to increase after introducing CO_2 enriched air even at higher density.	Ryu et al. (1993)
Bacterial isolates from acid mine drainage water and coal; microbe not identified	Spain	For different coal types the percent total sulfur removal and pyritic sulfur reduction was determined as 7.8%–57.7% and 11.7%–91.4%, respectively.	Rubiera et al. (1999)
T. ferrooxidans	India	After 25 days of bacterial treatment, maximum sulfate released from coal containing 9.11% of total sulfur was determined as 225 mg/g. The desulfurization was affected by changes in pH, media composition, and ferrous sulfate.	Mannivannan et al. (1994)
T. ferrooxidans	Japan	Pyrite sulfur content in microbial flotation column reduced to 3.9% from initial 11%. The microbially driven long-column flotation was able to minimize the pyrite sulfur content up to 1.8% from the feed coal sample having 11% pyrite sulfur.	Ohmura and Saiki (1994)
Thiobacillus and *Leptospirillum*	Spain	Maximum 34% removal of total sulfur and 48.6% of pyritic sulfur from column packed coal was attained after 6 weeks of bacterial treatment.	Morán et al. (1997)
T. ferrooxidans	India	Percent total sulfur removal ranged from 11.46% to 17.34%, sulfate removal varied from 61.04% to 72.04%, pyritic sulfur removal in the range of 33.33% to 82.60%, and organic sulfur decreased in the range of 1.40% to 6.73% during bacterial treatment (6–26 days) of different coal types.	Chandra and Mishra (1988)
T. ferrooxidans	Japan	Direct microbiological oxidation of pyrite caused 82% reduction in sulfur during 14 days treatment at 30°C and pH 2.0 in bioreactor system.	Konishi et al. (1993)
Pseudomonas sp. NP22 and *A. ferrooxidans*	China	The tested bacterium *Pseudomonas* sp. NP22 was effective in removing 46% of sulfur present in	Liu et al. (2017b)

(Continued)

TABLE 14.2 (Continued)

Bacterial species	Country	Remarks	Reference
		lignite coal. Application of both the bacteria minimized the sulfur content from 3.23% to 2.76%. The bacterial treatment enhanced the calorific value.	
Rhodococcus erythropolis DSM 44308	Turkey	Treatment of lignite coal with *R. erythropolis* DSM 44308 in the presence of surfactant caused decrease in sulfur content by 44.6% in shake flask investigation. In bioreactor system with capacity of 1 and 10 L, the process minimized sulfur level to 49.2% and 51.2%, respectively. The process integration with ultrasonication had only 6% increment in sulfur reduction.	Mishra et al. (2018)
Bacillus sp., *Pseudomonas* sp., *P. aeruginosa*, *Pseudomonas putida; Pseudomonas stutzeri*, *Pseudomonas monteilii*, *Pseudomonas rhizosphaerae*, and *Pseudomonas alcaligenes*	South Africa	Biodesulfurization efficiency of bacteria in terms of total sulfur elimination was in between 23.6% and 65.4% for coal with different particle size.	Makgato and Chirwa-Nkhalambayausi (2019)
Agrobacterium MC501, *Xanthomonas* MC701, *Corynebacterium* sp. MC401, and *Corynebacterium* sp. MC402,	Spain	The isolates were capable to degrade the derivative of DBT and other sulfur-based compounds into hydroxybiphenyl and sulfate. The mixed culture released higher content of sulfate in medium at cell density $1-2 \times 10^8$ cells/mL. 61 μM sulfate was generated after the degradation of DBT sulfone having concentration 231 μM.	Constantí et al. (1996)
Rhodococcus rhodochrous	Turkey	The 75 h treatment of lignite coal by selected bacterium resulted into 30.2% and 27.1% elimination of total and organic sulfur, respectively.	Bozdemir et al. (1996)
Sulfolobus solfataricus ATCC 35091	Bulgaria and Belgium	Bacterial treatment of chemically demineralized and depyritized coal was responsible for reduction in total sulfur by 16.9% and organic sulfur by 18.3%.	Gonsalvesh et al. (2012)
P. putida	Bulgaria	The lignite coal desulfurization by integrated chemical treatment and bacterial metabolic action reduced total, pyritic, and organic sulfur by 71.0%, 90.6%, and 49.4%, respectively.	Gonsalvesh et al. (2013)
A. ferrooxidans and *A. thiooxidans*	Colombia	Approximately 60% of pyrite present in coal was oxidized biologically in a large reactor after 4 days treatment under controlled condition having pH 1.8 and 4 mg/L oxygen.	Caicedo et al. (2012)
Rhodococcus spp. (Eu-32)	Pakistan	Under optimized set of conditions, a maximum of 40% total sulfur and 60% organic sulfur was demonstrated to be eliminated after 15 days bacterial treatment.	Akhtar et al. (2016)
A. ferrooxidans and *A. thiooxidans*	Colombia	Media supplementation with chloride salts favored more than 70% oxidation of pyrite sulfur present in coal without any impact of pulp density.	Caicedo Pineda and Márquez Godoy (2013)
R. erythropolis SX-12 and *A. ferrooxidans* GF	China	Two step treatment with bacteria for 1 month led to decrease in total sulfur content by 65.23%.	Hu et al. (2019)
R. rhodochrous ATCC 53968	Turkey	The reactor-based investigation under defined conditions having 8% inoculum density, pH 8.0 and 20 mM carbon source concentration revealed maximum decrease in total sulfur equivalent to 21.4% and that for organic sulfur as 9.2% from lignite coal rendered by bacterial metabolism. Further treatment for 8 days under similar conditions reduced the organic sulfur content by 10.2%.	Bayram et al. (2002)

14.3.2 Role of fungi in removal of sulfur from coal

The application of fungi in management of environmental contaminants has received considerable attention globally (Daccò et al., 2020). Fungi are the eukaryotic organisms with diverse metabolic activities responsible for the degradation of various organic and inorganic substrates by large array of extracellularly secreted enzymes and cytochromes (Van Hamme et al., 2003) even in the conditions with very low moisture and nutrient availability. Moreover, fungi can survive under dry environmental conditions by the so-called crptobiosis (Morgan and Watkinson, 1994). Fungal enzymes such as lignin peroxidase, manganese peroxidase, and laccase possess the broad range selectivity offering promising potential to mineralize the variety of hazardous xenobiotics and sulfur-based minerals (Ofori-Sarpong et al., 2013), making them more applicable industrially as compared to bacterial systems (Etemadifer et al., 2006; Husaini et al., 2008). In addition, favorable growth in acidic pH and release of various organic acids (Ruijter et al., 2002; Mancera-López et al., 2008) responsible for degradation of contaminants advocates their superiority over prokaryotic system. So far, different fungi have been reported to participate in sulfur elimination, thus serving as a nature friendly candidate to combat the global menace.

The fungi, as an effective means of clean coal technology, can be utilized singly, in combination with other fungi as well as bacteria to improve the desulfurization efficiency. Fungi belonging to genera *Aspergillus*, *Alternaria*, *Fusarium*, *Trametes*, *Phanerochaete*, *Exophiala*, *Gordonia*, *Cunninghamella*, *Pleurotus*, *Stachybotrys*, and *Rhodosporidium* have been documented for their application in sulfur removal from coal and organic sulfur compounds typically present in coal. The selection of particular fungus for desulfurization purposes is based on its ability to withstand harsh environmental conditions, better growth, and rapid sulfur rejection from desired coal samples. Most importantly, like bacterial systems, the desulfurization by fungi is affected by parameters such as pH, temperature, nutritional composition, aeration, shaking speed, coal particle size, and pulp density. The fungi mediated biodegradation pathway of organic sulfur components such as DBT follows the typical 4S pathway or another route leading to cleavage of C−S bond (Elmi et al., 2015). In other words the desulfurization process by microbes proceeds either without destruction of C−C bond or with destruction of molecular arrangement present in sulfur bearing substrate (Soleimani et al., 2007). During coal biodesulfurization, the carbon and sulfur present in coal or model organic sulfur substrates are utilized by attacking fungi eventually resulting into generation of different metabolites in order to survive and successfully multiply in that environment. The process of desulfurization by using fungal resources can be improved much by investigating diverse enzyme pools both intracellular and extracellular, search for newer candidate genes participating in sulfur elimination, process integration with currently available physicochemical methods, and application of genetic engineering principles so as to introduce more than one degradation pathway in single fungus. List of some fungi involved in sulfur removal are presented in Table 14.3.

14.4 Biological mechanisms of coal desulfurization

The employment of microbial processes for desulfurization of coal is an ecofriendly alternative of clean coal technology and may be aerobic or anaerobic in nature. The removal of inorganic sulfur including pyrite form is comparatively much easier as compared to organic sulfur present in coal. The important factors restricting the removal of organic sulfur from coal lies into the (1) absence of suitable analytical techniques for the measurement of organic sulfur content in coal and (2) sparse information pertaining to biological mechanism of organic sulfur removal (ElSawy and Gray, 1991). The mechanisms of sulfur removal from organic and inorganic constituents of coal are quite different. So far as the removal of sulfur from organic sulfur components of coal is concerned, three different mechanisms identified as (1) destructive, (2) anaerobic, and (3) oxidative sulfur elimination pathways are mentioned in literature (Soleimani et al., 2007). For detailed investigation of biodesulfurization mechanism, DBT has long been used as model compound (Martín-Cabello et al., 2020). To date, numbers of bacterial and fungal species efficient in degradation of C−S bond to facilitate the removal of sulfur from different substrates so as to use the derived sulfur products for growth have been presented (Kilbane, 2017).

The first desulfurization pathway referred to as "Kodama pathway" consisting of three distinct steps of hydroxylation, ring breakage, and hydrolysis characteristically employs pools of different cellular enzymes attacking on carbon atom of phenyl ring present in DBT. The second pathway documented in *Desulfovibrio desulfuricans* M6 leading to 42% removal of sulfur from DBT under anaerobic condition is proposed by Kim et al. (1990), but the process could not be exploited at industrial scale treatment because of limitations associated with the maintenance of anaerobic environment. The third pathway supposed to be existing in most of the sulfur-removing microbes known as 4S pathway and originally proposed by Kilbane (1990) involves oxidative removal of sulfur from organic compounds such as DBT

TABLE 14.3 List of important fungal genera responsible for desulfurization.

Fungal species	Country	Remarks	Reference
Alternaria sp.	Turkey	Treatment of lignite coal with *Alternaria* sp. in drum reactor system led to drop in pyritic sulfur by 17% and organic sulfur by 19%.	Şener et al. (2018)
Phanerochaeta chrysosporium ME446	Bulgaria and Belgium	Fungal treatment of chemically demineralized and depyritized coal was responsible for reduction in total sulfur by 24.2% and organic sulfur by 23.8%.	Gonsalvesh et al. (2012)
Fusarium oxysporum FE and *Exophiala spinifera* FM	Iran	The growing cells of *E. spinifera* FM favoring the creation of acidic condition induced the exclusion of sulfate sulfur, whereas non-dividing cells of *F. oxysporum* FE promoted the elimination of organic sulfur. Non-dividing cells of *F. oxysporum* FE facilitated 34.21% release of sulfur. The dividing as well as non-dividing cells of *F. oxysporum* FE showed greater efficiency in catalyzing sulfur loss as compared to *E. spinifera* FM.	Etemadzadeh et al. (2016)
Aspergillus sp.	India	The indigenously isolated fungus was capable in reducing 70%–80% of total sulfur content from coal having size −74 μm within 10 days treatment.	Acharya et al. (2005)
Trametes versicolor ATCC 200801 and *Phanerochaete chrysosporium* ME 446	Turkey	*T. versicolor* showed better efficiency in sulfur removal (40%) under conditions having pulp density 5%, pH 6 and stirring rate 125 rpm during 6 days treatment at 35°C. The fungal treatment did not alter the overall calorific value of coal.	Aytar et al. (2008)
Alternaria sp.	Turkey	The fungus isolated from lignite coal samples represented the removal of total sulfur, organic sulfur, and sulfate sulfur by 52%, 38%, and 51%, respectively, after 12 days of treatment under optimized conditions with pH 4, 2% inoculums density, and 1% pulp density.	Aytar et al. (2014)
P. chrysosporium	China	The fungal treatment released 33% sulfur from pyrite. The dissolution of pyrite resulted from the combined action of fungal enzymes, H_2O_2, Fe^{3+}, and organic acids.	Yang et al. (2018)
The fungus name is not indicated	China	The fungus directed desulfurization of coal leading to 45% minimization of total sulfur and 54.87% reduction in organic sulfur content after 2 days treatment at pH 2, 45°C temperature, 10% coal particles having size 100 μm.	He et al. (2005)
White rot fungi	China	Under optimized conditions having 200 mesh size of coal particles, 0.9 g inoculum and 10% density, maximum 35.77% of total sulfur was eliminated after 5 days fungal treatment.	Zhang et al. (2008)
T. versicolor ATCC200801, *P. chrysosporium* ME446, *Pleurotus sajor*-Caju	Bulgaria	Nearly 13% maximum reduction in organic sulfur due to fungal activity was registered for previously bacterially treated sub-bituminous coal. Maximum reduction in total sulfur by fungal action was determined as 25%.	Gonsalvesh et al. (2008)
Trametes trogii UAMH 8156, *Trametes hirsuta* UAMH 8165, *P. chrysosporium* ATCC 24725, *T. versicolor* IFO 30340 and *Tyromyces palustris* IFO 30339	Canada	Maximum 80% oxidation of sulfur compound DBS by fungal action was recorded.	Van Hamme et al. (2003)
Agrocybe aegerita and *Coprinellus radians*	Germany	Absolute oxidation of DBT by *A. aegerita* into 8 biodegradation products, (major components being DBT sulfoxide, DBT sulfone) was encountered after treatment with fungi for 16 days. In contrast, *C. radians* transformed only 60% of initial DBT content.	Aranda et al. (2009)

(*Continued*)

TABLE 14.3 (Continued)

Fungal species	Country	Remarks	Reference
E. spinifera strain FM	Iran	Nearly 100% desulfurization of DBT within 1 week by action of fungus *E. spinifera* was demonstrated under conditions having 30°C temperature and stirring rate 180 rpm.	Elmi et al. (2015)
Rhodosporidium toruloides strain DBVPG 6662	Italy	Transformation of 27% of DBT (10 μM) as a means of desulfurization within 36 h was observed. The supplementation of medium with orimulsion led to nearly 70% of DBT conversion within 15 days treatment.	Baldi et al. (2003)
Paecilomyces sp. TLi	United States	Specific sulfur oxidation of DBT to dihydroxybiphenyl as a suitable pathway of desulfurization is explored. Further transformation of metabolites was not found.	Faison et al. (1991)
Gordonia sp. IITR100	India	Six to eight days of incubation with fungus was found effective in completely reducing dibenzyl sulfide and dibenzyl sulfone. Conversion of dibenzyl sulfide into metabolic products such as sulfoxide, DBS sulfone, and benzoic acid was markedly registered.	Ahmad et al. (2015)
Cunninghamella elegans	United States of America	*C. elegans* showed roughly absolute transformation of DBT into DBT sulphoxide during 24 h treatment duration and was supposed to be facilitated by P450 as extracellular filtrate was unable to catalyze the conversion.	Schlenk et al. (1994)
Pleurotus ostreatus	Israel	The fungal metabolic action offered by cytochrome P-450 monooxygenase converted 84% of DBT used in experimental investigation. The conversion was noticed at aliphatic moiety.	Bezalel et al. (1996)
C. elegans	Brazil	Complete transformation based on new pathway catalyzed by fungus leading to desulfurization of DBT was documented within 24 h at initial substrate concentration 2 mM.	de Souza et al. (2017)
Stachybotrys bisbyi TUSb1	Saudi Arabia	The desulfurization potential of fungus for initial 0.3 mM DBT concentration was around 100% subsequent to 7 days incubation at 30°C temperature and 180 rpm stirring rate. Hydroxybiphenyl was ascribed as the resultant of orderly oxidation of sulfur followed by breakage of C–S bond.	Gherbawy et al. (2016)
Coriolus versicolor	Japan	The lignin degrading fungus showed potential ability in degrading 4-methyldibenzothiophene indicating application in desulfurization. The primary step consisted of oxidation of sulfur moiety resulting into formation of oxide and initial compound after 10 days treatment was reduced to 78%.	Ichinose et al. (1999)
Stachybotrys sp. WS4	Iran	The percent desulfurization of two different crude oil by fungus at 30°C ranged from 64.8% to 76% within 72–144 h incubation.	Torkamani et al. (2008)
Aspergillus flavus	Nigeria	The immobilized *A. flavus* spores having 50 g weight eliminated maximum 94.7% sulfur from crude oil during 7 days treatment.	Adegunlola et al. (2012)
C. elegans	United States of America	After incubation for 1 week the fungus metabolized DBT, reducing its initial concentration from 90 to 0.72 ppm (99.20%).	Crawford and Gupta (1990)

DBS, Dibenzyl sulphide; *DBT*, dibenzothiophene.

(Martínez et al., 2017). This process of desulfurization leads to the conversion of DBT sulfur into sulfoxide, sulfone, sulfinate, and finally hydroxybiphenyl by degradation of C—S bond facilitated by microbial enzymes monooxygenases (Piddington et al., 1995), flavin reductase, and desulfinase. Eventually the sulfur of DBT is transported to sulfate. Thus the destruction of C—S bond present in thiophene ring is mandatory for desulfurization. The oxidative desulfurization is considered to be less feasible industrially due to limited process effectiveness and oxidation of coal by oxygen during the process (Tang et al., 2018b). Van Afferden et al. (1990) have proposed another pathway of aerobic sulfur elimination from model organic compound DBT by using *Brevibacterium* sp. and concluded the eventual conversion of sulfur into sulfate. The metabolic products of biodesulfurization of DBT as characterized through ultraviolet spectroscopy and gas chromatography-mass spectroscopy (GC—MS) analysis were recorded as DBT-5-oxide, DBT-5-dioxide, and benzoate. The complete mineralization of DBT by bacterial action culminated into formation of water and carbon dioxide (CO_2).

The organic sulfur occurring as thiols, thiophene, and sulfide serves as an important framework constituent of coal. The elimination of organic sulfur from coal hence could be achieved by cleaving C—S bond. The detailed mechanism of biodesulfurization can be revealed through the assessment of sulfur removal from representative organic sulfur compounds like DBT and benzyl methyl sulfide (BMS) (Van Afferden et al., 1993). While working on biodesulfurization of organic substances, the investigations of Van Afferden et al. (1993) using *Brevibacterium* sp. DO and *Pseudomonas aeruginosa* OSl demonstrated existence of two different mechanisms broadly classified as carbon regulation and sulfur regulation mechanism responsible for sulfur removal. The study concluded similar mechanism of sulfur removal from selected organic compound. The primary step toward the desulfurization involves the oxidation of sulfur into respective sulfone followed by breakage of covalently linked C—S bond through the synthesis of unstable oxidized sulfur hemimercaptal resulting from the oxidation of carbon near to sulfur. The enzymes participating in process was recognized as sulfoxygenases. The formation of sulfinic acid as an intermediary substance was observed during the C—S bond cleavage in both organic sulfur compounds. For DBT, sulfinic acid was ultimately hydrolyzed to give rise to sulfite and benzoate; however, in the case of BMS the desulfurization commenced by oxidation of sulfonic acid.

Metagenomics can provide important insights into genes directing the desulfurization of dibenzothiphene (Abbasian et al., 2016). The application of metagenomics approach to decipher the operon and associated regulatory process taking part in removal of sulfur from DBT is reported by Martín-Cabello et al. (2020). One of the obtained vectors harbored four closely associated genes responsible for the expression of monooxygense, desulfinase, and FMN-oxidoreductase denoted as "*DszC* and *DszA*," "*DszB*," "*DszE*," respectively. Transcription of operon *dszEABC* rendering *Escherichia coli* the potential to metabolize DBT as only sulfur substrate was illustrated to be operated by promoter and activator. Still, very little knowledge regarding the transcriptional control of genes involved in sulfur removal from model sulfur containing compound such as dibenzothiphene is documented in literature. The detailed information on genetic controls would provide the foundation for development of genetically engineered microbes for rapid removal of sulfur from coal matrices.

14.5 Effect of environmental factors on coal biodesulfurization

The coal desulfurization by microbial processes is affected by different environmental factors. Some of the important parameters affecting the removal of sulfur from coal includes pH, temperature, coal characteristics, media composition, use of surfactants used and of utmost importance the biological agents used. Study of effect of different factors on sulfur removal from coal using sulfur oxidizer *T. ferrooxidans* is demonstrated by Mannivannan et al. (1994). The optimum pH for desulfurization was recorded at pH 2.2 and the percent removal of sulfur was not influenced by different medium used. The process of desulfurization was induced after the supplementation of ferrous iron in the form of ferrous sulfate in the medium as revealed through the measurement of sulfate present in medium and reduction in sulfur content of coal samples used. Effect of treatment time, coal pulp density, and size of coal matrices on sulfur elimination by bacterium *Sinomonas flava* is reported by Mishra et al. (2014). Maximum percent removal of sulfur was observed at 15 days of incubation period with 2% pulp density and 150 μm size of coal matrix. The process of bacterial coal desulfurization led to increase in calorific value showing advantage over conventional treatment methods. The optimum conditions of treatment time, pH, temperature, and coal pulp density for coal desulfurization by *Aspergillus* sp. was denoted as 8 days, 5.0, 35°C, and 5%, respectively (Liu et al., 2017a). The isolated fungus was able to remove the total sulfur to a level of 2.63% from an initial content of 4.97% indicating 47.08% reduction in total sulfur content. The simultaneous application of *Aspergillus* sp. with *T. ferrooxidans* was more effective in desulfurization causing decline in total sulfur content up to 59%. Impact of addition of iron-less M9K medium on coal desulfurization employing *A. ferrooxidans* is recently explored (Liu et al., 2020). Addition of iron free medium up to certain extent was shown to increase the efficiency of bacterium in sulfur removal from coal. The inhibition of sulfur removal beyond a limit was ascribed the resultant of jarosite and gypsum formation limiting pyrite oxidation, thereby reduced sulfur elimination. Overall, the extent of sulfur removal from coal using microorganism is affected significantly by various environmental factors indicating the need of process optimization for specific coal types and microorganisms used for successful application at commercial scale.

14.6 Conclusion

Coal desulfurization currently being practiced at industrial scale involves different physical and chemical methods. Very little commercialization has been made so far regarding the application of microbial resources for the management of atmospheric SO_2 contamination resulting from indiscriminate and huge coal burning across the globe. Three different mechanisms of coal biodesulfurization facilitated by microbial enzymes leading to release of sulfur from coal are presented. Various microorganisms such as bacteria, fungi, and actinomycetes have promising potential in removing sulfur from different coal types. The removal of sulfur from coal, however, is influenced by numbers of environmental factors such as pH, metal content, the proportion of organic and inorganic sulfur content, temperature, moisture content, nutritional composition of media, shaking and static condition, microorganisms involved, and most importantly the nature of coal itself. The identification of genes governing the expression of enzymes participating in coal desulfurization could be helpful in enhancing the microbial efficiency of sulfur removal from coal under natural environmental condition.

14.7 Future perspectives

The implication of different microorganisms in catalyzing diverse biological process can be harnessed for the management of contaminated environments. Microbes occurring in diverse ecological habitats with efficiency in coal desulfurization have great opportunities in clean coal technologies. The search for newer microorganisms, especially bacterial and fungal species possessing the ability to withstand adverse environmental conditions and quickly adapt under given habitat, could help to solve the atmospheric SO_2 pollution resulting from coal combustion. Detailed investigations pertaining to enzymes involved and intermediate products generated during coal biodesulfurization should be conducted in order to elucidate the intricate pathway of coal desulfurization. The study in perspectives of metagenomics may provide important insights into novel genes and their products taking part in desulfurization of coal and could be helpful for designing engineered microbes for rapid elimination of both organic and inorganic sulfur from coal matrices. Further, the complete genome sequencing of presently known microorganisms involved in desulfurization may be a breakthrough step for industrialization of microbe directed sulfur removal process not only for different coal types but also for petroleum wastes.

References

Abbasian, F., Lockington, R., Megharaj, M., Naidu, R., 2016. Identification of a new operon involved in desulfurization of dibenzothiophenes using a metagenomic study and cloning and functional analysis of the genes. Enzyme Microb. Technol. 87, 24–28.

Acharya, C., Sukla, L.B., Misra, V.N., 2004. Biodepyritisation of coal. J. Chem. Technol. Biotechnol. 79 (1), 1–12.

Acharya, C., Sukla, L.B., Misra, V.N., 2005. Biological elimination of sulphur from high sulphur coal by *Aspergillus*-like fungi. Fuel 84 (12–13), 1597–1600.

Adegunlola, G.A., Oloke, J.K., Majolagbe, O.N., Adebayo, E.A., Adegunlola, C.O., Adewoyin, A.G., et al., 2012. Microbial desulphurization of crude oil using *Aspergillus flavus*. Eur. J. Exp. Biol. 2 (2), 400–403.

Ahmad, A., Chauhan, A.K., Kushwaha, H.N., Javed, S., Kumar, A., 2015. Preferential desulfurization of dibenzyl sulfide by an isolated *Gordonia* sp. IITR100. 3 Biotech. 5 (3), 237–243.

Ahmad, W., Ahmad, I., Ahmad, R., Ullah, Z., Ibrahim, M., 2019. Desulfurization of Lakhra coal by combined leaching and catalytic oxidation techniques. Int. J. Coal Prep. Util. 17, 1–17.

Akhtar, N., Ghauri, M.A., Akhtar, K., 2016. Exploring coal biodesulfurization potential of a novel organic sulfur metabolizing *Rhodococcus* spp. (Eu-32)—a case study. Geomicrobiol. J. 33 (6), 468–472.

Alam, H.G., Moghaddam, A.Z., Omidkhah, M.R., 2009. The influence of process parameters on desulfurization of Mezino coal by HNO_3/HCl leaching. Fuel Process. Technol. 90 (1), 1–7.

Anwar, F., Nasir, R., Maqsood, K., Suleman, H., Rehman, F., Ali, A., et al., 2020. Desulfurization and optimization of low-grade local coal by sequential KOH and HCl treatment. J. Sulfur Chem. 41 (1), 44–56.

Aranda, E., Kinne, M., Kluge, M., Ullrich, R., Hofrichter, M., 2009. Conversion of dibenzothiophene by the mushrooms *Agrocybe aegerita* and *Coprinellus radians* and their extracellular peroxygenases. Appl. Microbiol. Biotechnol. 82 (6), 1057.

Arifin, M., Sanwani, E., Chaerun, S.K., 2020. Removal of sulfur and ash from Indonesian coal by indigenous mixotrophic bacteria. Curr. Res. Biosci. Biotechnol. 1 (2), 71–76.

Aytar, P., Sam, M., Çabuk, A., 2008. Microbial desulphurization of Turkish lignites by white rot fungi. Energy Fuels 22 (2), 1196–1199.

Aytar, P., Aksoy, D.O., Toptas, Y., Çabuk, A., Koca, S., Koca, H., 2014. Isolation and characterization of native microorganism from Turkish lignite and usability at fungal desulphurization. Fuel 116, 634–641.

Baldi, F., Pepi, M., Fava, F., 2003. Growth of *Rhodosporidium toruloides* strain DBVPG 6662 on dibenzothiophene crystals and orimulsion. Appl. Environ. Microbiol. 69 (8), 4689–4696.

Baruah, B.P., Khare, P., 2007. Desulfurization of oxidized Indian coals with solvent extraction and alkali treatment. Energy Fuels 21 (4), 2156–2164.

Bayram, Z., Bozdemir, T., Durusoy, T., Yürüm, Y., 2002. Biodesulfurization of Mengen lignite with *Rhodoccocus rhodochrous*: effects of lignite concentration and retreatment. Energy Sources 24 (7), 625–631.

Bezalel, L., Hadar, Y., Fu, P.P., Freeman, J.P., Cerniglia, C.E., 1996. Initial oxidation products in the metabolism of pyrene, anthracene, fluorene, and dibenzothiophene by the white rot fungus *Pleurotus ostreatus*. Appl. Environ. Microbiol. 62 (7), 2554–2559.

Bozdemir, T.Ö., Durusoy, T., Erincin, E., Yürüm, Y., 1996. Biodesulfurization of Turkish lignites: 1. Optimization of the growth parameters of *Rhodococcus rhodochrous*, a sulfur-removing bacterium. Fuel 75 (13), 1596–1600.

Caicedo, G., Prada, M., Pelaez, H., Moreno, C., Marquez, M., 2012. Evaluation of a coal biodesulfurization process (semi-continuous mode) on the pilot plant level. Dyna 79 (174), 114–118.

Caicedo Pineda, G.A., Márquez Godoy, M.A., 2013. Effect of chloride salts on biodesulfurization process of a colombian coal. Rev. Fac. Ing., Univ. Antioquia 68, 115–123.

Çetinkaya, Z., Bayat, O., 2020. Upgrading low-rank coals (Çan, Çanakkale/Turkey) by float-sink separation in dense media. Energy Sources, A: Recover. Util. Environ. Eff. 42 (1), 113–120.

Chandra, D., Mishra, A.K., 1988. Desulfurization of coal by bacterial means. Resour. Conserv. Recycl. 1 (3–4), 293–308.

Constantí, M., Giralt, J., Bordons, A., 1996. Degradation and desulfurization of dibenzothiophene sulfone and other sulfur compounds by *Agrobacterium* MC501 and a mixed culture. Enzyme Microb. Technol. 19 (3), 214–219.

Crawford, D.L., Gupta, R.K., 1990. Oxidation of dibenzothiophene by *Cunninghamella elegans*. Curr. Microbiol. 21 (4), 229–231.

Daccò, C., Girometta, C., Asemoloye, M.D., Carpani, G., Picco, A.M., Tosi, S., 2020. Key fungal degradation patterns, enzymes and their applications for the removal of aliphatic hydrocarbons in polluted soils: a review. Int. Biodeterior. Biodegrad. 147, 104866.

Dahlberg, M.D., Rohrer, R.L., Fauth, D.J., Sprecher, R., Olson, G.J., 1993. Biodesulfurization of dibenzothiophene sulfone by *Arthrobacter* sp. and studies with oxidized Illinois No. 6 coal. Fuel 72 (12), 1645–1649.

de Souza, P.M., e Silva, T.A.D.L., Lima, M.A.B., de Oliveira Franco, L., Schwartz, M., da Silva, P.H., et al., 2017. Reduction in the sulfur content of fossil fuels by *Cunninghamella elegans* (UCP 0596) to dibenzothiophene compound. Recent Insights in Petroleum Science and Engineering. IntechOpen.

Demirbaş, A., 2002. Demineralization and desulfurization of coals via column froth flotation and different methods. Energy Convers. Manage. 43 (7), 885–895.

Dhawan, H., Sharma, D.K., 2019. Advances in the chemical leaching (inorgano-leaching), bio-leaching and desulphurisation of coals. Int. J. Coal Sci. Technol. 1–15.

Elmi, F., Etemadifar, Z., Emtiazi, G., 2015. A novel metabolite (1,3-benzenediol, 5-hexyl) production by *Exophiala spinifera* strain FM through dibenzothiophene desulfurization. World J. Microbiol. Biotechnol. 31 (5), 813–821.

ElSawy, A., Gray, D., 1991. A critical review of biodesulphurization systems for removal of organic sulphur from coal. Fuel 70 (5), 591–594.

Etemadifer, Z., Giti, E., Sharareh, P., 2006. Removal of dibenzothiophene, biphenyl and phenol from waste by *Trichosporon* sp. Sci. Res. Essay 1 (3), 72–76.

Etemadifar, Z., Etemadzadeh, S.S., Emtiazi, G., 2019. A novel approach for bioleaching of sulfur, iron, and silica impurities from coal by growing and resting cells of *Rhodococcus* spp. Geomicrobiol. J. 36 (2), 123–129.

Etemadzadeh, S.S., Emtiazi, G., Etemadifar, Z., 2016. Heterotrophic bioleaching of sulfur, iron, and silicon impurities from coal by *Fusarium oxysporum* FE and *Exophiala spinifera* FM with growing and resting cells. Curr. Microbiol. 72 (6), 707–715.

Faison, B.D., Clark, T.M., Lewis, S.N., Ma, C.Y., Sharkey, D.M., Woodward, C.A., 1991. Degradation of organic sulfur compounds by a coal-solubilizing fungus. Appl. Biochem. Biotechnol. 28 (1), 237–251.

Gherbawy, Y., El-Deeb, B., Hazzani, A.A., Maher, A., Shehata, A., 2016. Mycobiota of oil-contaminated soil samples and their abilities for dibenzothiophene desulfurization. Geomicrobiol. J. 33 (7), 618–624.

Giray, E.S., Hüyükpınar, T., Sönmez, Ö., 2019. Leaching of coals with subcritical water. Int. J. Coal Prep. Util. 1–11. https://doi.org/10.1080/19392699.2019.1649664.

Gonsalvesh, L., Marinov, S.P., Stefanova, M., Yürüm, Y., Dumanli, A.G., Dinler-Doganay, G., et al., 2008. Biodesulphurized subbituminous coal by different fungi and bacteria studied by reductive pyrolysis. Part 1: Initial coal. Fuel 87 (12), 2533–2543.

Gonsalvesh, L., Marinov, S.P., Stefanova, M., Carleer, R., Yperman, J., 2012. Organic sulphur alterations in biodesulphurized low rank coals. Fuel 97, 489–503.

Gonsalvesh, L., Marinov, S.P., Stefanova, M., Carleer, R., Yperman, J., 2013. Biodesulphurized low rank coal: Maritza east lignite and its "humus-like" byproduct. Fuel 103, 1039–1050.

Gunam, I.B.W., Yaku, Y., Hirano, M., Yamamura, K., Tomita, F., Sone, T., et al., 2006. Biodesulfurization of alkylated forms of dibenzothiophene and benzothiophene by sphingomonas subarctica T7b. J. Biosci. Bioeng. 101 (4), 322–327.

Gupta, N., Roychoudhury, P.K., Deb, J.K., 2005. Biotechnology of desulfurization of diesel: prospects and challenges. Appl. Microbiol. Biotechnol. 66 (4), 356–366.

He, D.W., Jin, Y., Chai, L.Y., Huang, R., Peng, B., Wang, Y.Y., 2005. Effect and influence factors of sulfur removal in coal with fungus. J. Cent. South Univ. Technol. 12 (1), 98–101.

He, H., Hong, F.F., Tao, X.X., Li, L., Ma, C.Y., Zhao, Y.D., 2012. Biodesulfurization of coal with *Acidithiobacillus caldus* and analysis of the interfacial interaction between cells and pyrite. Fuel Process. Technol. 101, 73–77.

Hu, T., Yang, Y., Zhang, M., Gao, Y., Cheng, Q., Ji, H., 2019. Biodesulfurization of coal using *Rhodococcus erythropolis* SX-12 and *Acidithiobacillus ferrooxidans* GF: A two-step approach. Energy Sci. Eng. 7 (1), 162–169.

Huang, Y., Fu, J., Zhang, Y., Zhao, D., Wang, X., Yuan, L., 2019. Desulfurization and removal ash of high sulfur coal by microwave oxidation treatment. Energy Sources, A: Recover. Util. Environ. Eff. 1–9.

Husaini, A., Roslan, H.A., Hii, K.S.Y., Ang, C.H., 2008. Biodegradation of aliphatic hydrocarbon by indigenous fungi isolated from used motor oil contaminated sites. World J. Microbiol. Biotechnol. 24 (12), 2789–2797.

Ichinose, H., Wariishi, H., Tanaka, H., 1999. Bioconversion of recalcitrant 4-methyldibenzothiophene to water-extractable products using lignin-degrading basidiomycete *Coriolus versicolor*. Biotechnol. Prog. 15 (4), 706–714.

Karaca, S., Akyürek, M., Bayrakçeken, S., 2003. The removal of pyritic sulfur from Aşkale lignite in aqueous suspension by nitric acid. Fuel Process. Technol. 80 (1), 1–8.

Kargi, F., 1986. Microbial methods for desulfurization of coal. Trends Biotechnol. 4 (11), 293–297.

Khan, M.H., Sibtain, S.F., Khan, A.R., 2014. Desulfurization and deashing of lignitic coal with cupric chloride and behaviors of copper. Int. J. Coal Prep. Util. 34 (2), 59–64.

Kilbane II, J.J., 1990. Sulfur-specific microbial metabolism of organic compounds. Resour. Conserv. Recycl. 3 (2–3), 69–79.

Kilbane II, J.J., 2017. Biodesulfurization: how to make it work? Arab. J. Sci. Eng. 42 (1), 1–9.

Kim, H.Y., Kim, T.S., Kim, B.H., 1990. Degradation of organic sulfur compounds and the reduction of dibenzothiophene to biphenyl and hydrogen sulfide by *Desulfovibrio desulfuricans* M6. Biotechnol. Lett. 12 (10), 761–764.

Konishi, Y., Kawamura, T., Asai, S., 1993. Removal of inorganic sulfur from coal by *Thiobacillus ferrooxidans*. J. Chem. Eng. Jpn. 26 (1), 83–88.

Li, W., Cho, E.H., 2005. Coal desulfurization with sodium hypochlorite. Energy Fuels 19 (2), 499–507.

Li, Z., Sun, T., Jia, J., 2010. An extremely rapid, convenient and mild coal desulfurization new process: sodium borohydride reduction. Fuel Process. Technol. 91 (9), 1162–1167.

Liu, T., Hou, J.H., Peng, Y.L., 2017a. Biodesulfurization from the high sulfur coal with a newly isolated native bacterium, *Aspergillus* sp. DP06. Environ. Prog. Sustain. Energy 36 (2), 595–599.

Liu, T., Hou, J.H., Peng, Y.L., 2017b. Effect of a newly isolated native bacteria, *Pseudomonas* sp. NP22 on desulfurization of the low-rank lignite. Int. J. Miner. Process. 162, 6–11.

Liu, F., Lei, Y., Shi, J., Zhou, L., Wu, Z., Dong, Y., et al., 2020. Effect of microbial nutrients supply on coal bio-desulfurization. J. Hazard. Mater. 384, 121324.

Luo, D.C., Wang, W., 2013. Study on desulfurization of coal by ferric sulfate oxidation in combination with microwave pretreatment. Miner. Eng. Res. 28, 70–74.

Makgato, S.S., Chirwa-Nkhalambayausi, E.M., 2019. Modelling performance evaluation of microbial desulphurization of waterberg steam coal in CSTR. Chem. Eng. 76, 145–150.

Makgato, S.S., Chirwa, E.M.N., 2020. The desulphurization potential of waterberg steam coal using bacteria isolated from coal: The SO_2 emissions control technique. J. Clean. Prod. 121051. Available from: https://doi.org/10.1016/j.jclepro.2020.121051.

Mancera-López, M.E., Esparza-García, F., Chávez-Gómez, B., Rodríguez-Vázquez, R., Saucedo-Castaneda, G., Barrera-Cortés, J., 2008. Bioremediation of an aged hydrocarbon-contaminated soil by a combined system of biostimulation–bioaugmentation with filamentous fungi. Int. Biodeterior. Biodegrad. 61 (2), 151–160.

Mannivannan, T., Sandhya, S., Pandey, R.A., 1994. Microbial desulfurization of coal by chaemoautotrophic *Thiobacillus ferrooxidans*—an iron mine isolate. J. Environ. Sci. Health A 29 (10), 2045–2061.

Martín-Cabello, G., Terrón-González, L., Ferrer, M., Santero, E., 2020. Identification of a complete dibenzothiophene biodesulfurization operon and its regulator by functional metagenomics. Environ. Microbiol. 22 (1), 91–106.

Martínez, I., Mohamed, M.E.S., Santos, V.E., García, J.L., García-Ochoa, F., Díaz, E., 2017. Metabolic and process engineering for biodesulfurization in Gram-negative bacteria. J. Biotechnol. 262, 47–55.

Medunić, G., Ahel, M., Mihalić, I.B., Srček, V.G., Kopjar, N., Fiket, Ž., et al., 2016. Toxic airborne S, PAH, and trace element legacy of the superhigh-organic-sulphur Raša coal combustion: Cytotoxicity and genotoxicity assessment of soil and ash. Sci. Total. Environ. 566, 306–319.

Meshram, P., Purohit, B.K., Sinha, M.K., Sahu, S.K., Pandey, B.D., 2015. Demineralization of low grade coal—a review. Renew. Sustain. Energy Rev. 41, 745–761.

Mishra, S., Panda, P.P., Pradhan, N., Satapathy, D., Subudhi, U., Biswal, S.K., et al., 2014. Effect of native bacteria *Sinomonas flava* 1C and *Acidithiobacillus ferrooxidans* on desulphurization of Meghalaya coal and its combustion properties. Fuel 117, 415–421.

Mishra, S., Akcil, A., Panda, S., Tuncuk, A., 2018. Effect of Span-80 and ultrasonication on biodesulphurization of lignite by *Rhodococcus erythropolis*: lab to semi-pilot scale tests. Miner. Eng. 119, 183–190.

Morgan, P., Watkinson, R.J., 1994. Biodegradation of components of petroleum. In: Ratledge, C. (Ed.), Biochemistry of Microbial Degradation. Kluwer Academic Publishers, The Netherlands, pp. 1–31.

Morán, A., Aller, A., Cara, J., Martinez, O., Encinas, J.P., Gómez, E., 1997. Microbiological desulfurization of column-packed coal. Fuel Process. Technol. 52 (1–3), 155–164.

Mukherjee, S., Borthakur, P.C., 2001. Chemical demineralization/desulphurization of high sulphur coal using sodium hydroxide and acid solutions. Fuel 80 (14), 2037–2040.

Mukherjee, S., Mahiuddin, S., Borthakur, P.C., 2001. Demineralization and desulfurization of subbituminous coal with hydrogen peroxide. Energy Fuels 15 (6), 1418–1424.

Ofori-Sarpong, G., Osseo-Asare, K., Tien, M., 2013. Mycohydrometallurgy: Biotransformation of double refractory gold ores by the fungus, *Phanerochaete chrysosporium*. Hydrometallurgy 137, 38–44.

Ohmura, N., Saiki, H., 1994. Desulfurization of coal by microbial column flotation. Biotechnol. Bioeng. 44 (1), 125–131.

Oldfield, C., Pogrebinsky, O., Simmonds, J., Olson, E., Kulpa, C.F., 1997. Elucidation of the metabolic pathway for dibenzothiophene desulphurization by *Rhodococcus* sp. strain IGTS8 (ATCC 53968). Microbiology 143, 2961–2973.

Omori, T., Monna, L.I.S.A., Saiki, Y.U.K.O., Kodama, T.O.H.R.U., 1992. Desulfurization of dibenzothiophene by *Corynebacterium* sp. strain SY1. Appl. Environ. Microbiol. 58 (3), 911–915.

Özbakir, O., Koltka, S., Sabah, E., 2017. Modeling and optimization of fine coal beneficiation by hydrocyclone and multi-gravity separation to produce fine lignite clean coal. Part. Sci. Technol. 35 (6), 712–722.

Park, J.H., Joung, Y.O., Do Park, S., 2007. Sulfur removal from coal with supercritical fluid treatment. Korean J. Chem. Eng. 24 (2), 314–318.

Piddington, C.S., Kovacevich, B.R., Rambosek, J., 1995. Sequence and molecular characterization of a DNA region encoding the dibenzothiophene desulfurization operon of *Rhodococcus* sp. strain IGTS8. Appl. Environ. Microbiol. 61 (2), 468–475.

Pineda, G.A.C., Godoy, M.A.M., 2019. Effect of *Acidithiobacillus thiooxidans*-cysteine interactions on pyrite biooxidation by *Acidithiobacillus ferrooxidans* in the presence of coal compounds. Braz. J. Chem. Eng. 36 (2), 681–692.

Restrepo, Á., Bazzo, E., Miyake, R., 2015. A life cycle assessment of the Brazilian coal used for electric power generation. J. Clean. Prod. 92, 179–186.

Rubiera, F., Arenillas, A., Martinez, O., Moran, A., Fuente, E., Pis, J.J., 1999. Biodesulfurization of coals of different rank: effect on combustion behavior. Environ. Sci. Technol. 33 (3), 476–481.

Ruijter, G.J.G., Kubicek, C.P., Visser, J., 2002. Production of organic acids by fungi. In: Esser, K., Bennett, J.W., Osiewacz volume, H.D. (Eds.), The Mycota. Industrial Applications, vol. X. Springer-Verlag Berlin, Heidelberg, pp. 213–230.

Ryu, H.W., Chang, Y.K., Kim, S.D., 1993. Microbial coal desulfurization in an airlift bioreactor by sulfur-oxidizing bacterium *Thiobacillus ferrooxidans*. Fuel Process. Technol. 36 (1–3), 267–275.

Sabar, M.A., Ali, M.I., Fatima, N., Malik, A.Y., Jamal, A., Farman, M., et al., 2019. Degradation of low rank coal by *Rhizopus oryzae* isolated from a Pakistani coal mine and its enhanced releases of organic substances. Fuel 253, 257–265.

Sahinoglu, E., 2018. Cleaning of high pyritic sulfur fine coal via flotation. Adv. Powder Technol. 29 (7), 1703–1712.

Saikia, B.K., Dutta, A.M., Saikia, L., Ahmed, S., Baruah, B.P., 2014. Ultrasonic assisted cleaning of high sulphur Indian coals in water and mixed alkali. Fuel Process. Technol. 123, 107–113.

Saikia, B.K., Dalmora, A.C., Choudhury, R., Das, T., Taffarel, S.R., Silva, L.F., 2016. Effective removal of sulfur components from Brazilian power-coals by ultrasonication (40 kHz) in presence of H_2O_2. Ultrason. Sonochem. 32, 147–157.

Schippers, A., Jozsa, P., Sand, W., 1996. Sulfur chemistry in bacterial leaching of pyrite. Appl. Environ. Microbiol. 62 (9), 3424–3431.

Schlenk, D., Bevers, R.J., Vertino, A.M., Cerniglia, C.E., 1994. P450 catalysed S-oxidation of dibenzothiophene by *Cunninghamella elegans*. Xenobiotica 24 (11), 1077–1083.

Şener, B., Aksoy, D.Ö., Çelik, P.A., Toptaş, Y., Koca, S., Koca, H., et al., 2018. Fungal treatment of lignites with higher ash and sulphur contents using drum type reactor. Hydrometallurgy 182, 64–74.

Seo, J.S., 2012. Multiple pathways in the degradation of dibenzothiophene by *Mycobacterium aromativorans* strain JS19b1 T. J. Korean Soc. Appl. Biol. Chem. 55 (5), 613–618.

Shahraki, S., Karamoozian, M., Azizi, A., 2018. Desulfurization of coal by HNO_3 leaching: Optimization of influential factors using Box–Behnken design. J. Min. Environ. 9 (3), 657–665.

Shen, Y., Sun, T., Jia, J., 2011. Novel desulfurization method of sodium borohydride reduction for coal water slurry. Energy Fuels 25 (7), 2963–2967.

Soleimani, M., Bassi, A., Margaritis, A., 2007. Biodesulfurization of refractory organic sulfur compounds in fossil fuels. Biotechnol. Adv. 25 (6), 570–596.

Speight, J.G., 2015. Handbook of Coal Analysis. John Wiley & Sons, pp. 1–21, Chapter 1.

Tang, L., Chen, S., Wang, S., Tao, X., He, H., Zheng, L., et al., 2018a. Heterocyclic sulfur removal of coal based on potassium tert-butoxide and hydrosilane system. Fuel Process. Technol. 177, 194–199.

Tang, L., Wang, S., Zhu, X., Guan, Y., Chen, S., Tao, X., et al., 2018b. Feasibility study of reduction removal of thiophene sulfur in coal. Fuel 234, 1367–1372.

Tang, L., Fan, H., Chen, S., Tao, X., He, H., Zhu, X., 2019. Investigation on the synergistic mechanism of coal desulfurization by ultrasonic with microwave. Energy Sources, A: Recover. Util. Environ. Eff. 1–10.

Temel, H.A., Bozkurt, V., Ayhan, F.D., 2010. Desulfurization and deashing of Adiyaman-Gölbaşi lignite by flotation. Energy Sources, A: Recover. Util. Environ. Eff. 32 (8), 727–743.

Torkamani, S., Shayegan, J., Yaghmaei, S., Alemzadeh, I., 2008. Study of the first isolated fungus capable of heavy crude oil biodesulfurization. Ind. Eng. Chem. Res. 47 (19), 7476–7482.

Tripathi, N., Singh, R.S., Hills, C.D., 2019. Microbial removal of sulphur from petroleum coke (petcoke). Fuel 235, 1501–1505.

Van Afferden, M., Schacht, S., Klein, J., Trüper, H.G., 1990. Degradation of dibenzothiophene by *Brevibacterium* sp. DO. Arch. Microbiol. 153 (4), 324–328.

Van Afferden, M., Tappe, D., Beyer, M., Trüper, H.G., Klein, J., 1993. Biochemical mechanisms for the desulfurization of coal-relevant organic sulfur compounds. Fuel 72 (12), 1635–1643.

Van Hamme, J.D., Wong, E.T., Dettman, H., Gray, M.R., Pickard, M.A., 2003. Dibenzyl sulfide metabolism by white rot fungi. Appl. Environ. Microbiol. 69 (2), 1320–1324.

Wang, M., Fu, C., Chang, L., Xie, K., 2016. Effects of flotation on the release behavior of sulfur and nitrogen during coal pyrolysis. Energy Sources, A: Recover. Util. Environ. Eff. 38 (3), 347–353.

Weng, S., Wang, J., 1992. Exploration on the mechanism of coal desulfurization using microwave irradiation/acid washing method. Fuel Process. Technol. 31 (3), 233–240.

Xia, W., 2018. A novel and effective method for removing organic sulfur from low rank coal. J. Clean. Prod. 172, 2708–2710.

Xia, W., Li, Y., He, W., Peng, Y., 2018. Desulfurization of low rank coal co-pyrolysis with reduced iron powder followed by dry magnetic separation. J. Clean. Prod. 204, 525–531.

Xu, N., Tao, X., 2015. Changes in sulfur form during coal desulfurization with microwave: Effect on coal properties. Int. J. Min. Sci. Technol. 25 (3), 435–438.

Xu, J., Liu, X., Song, C., Du, Z., Wang, F., Luo, J., et al., 2019. Biodesulfurization of high sulfur coal from Shanxi: Optimization of the desulfurization parameters of three kinds of bacteria. Energy Sources, A: Recover. Util. Environ. Eff. 1–19.

Yang, H.Y., Qian, L.I.U., Chen, G.B., Tong, L.L., Auwalu, A.L.I., 2018. Bio-dissolution of pyrite by *Phanerochaete chrysosporium*. Trans. Nonferrous Met. Soc. China 28 (4), 766–774.

Zhang, D.M., Xie, Q.L., Zhang, P., Zhang, C.F., 2008. Experimental studies on coal desulphurization using white rot fungi. Energy Technol. Manage. (1), 26.

Chapter 15

Microbially synthesized nanomaterials for remediation of contaminated soil and water environment

Sahadevan Neethu, Sebastian Jose Midhun, E.K. Radhakrishnan and Mathew Jyothis
School of Biosciences, Mahatma Gandhi University, Kottayam, India

15.1 Introduction

Environmental pollution is undeniably one of the major problems that affect worldwide. The exponential growth of human population piloted the abuse of nature and natural resources. The industrial revolution has resulted in the production of numerous synthetic or natural waste by-products resulted in the prolonged contamination of the environment. Environmental Protection Agency (EPA) reported that the properly disposed wastes share a low percentage, approximately 10%, out of all wastes (Guerra et al., 2018). Environmental contaminations by heavy metals from industrial as well as anthropogenic activities have caused irreparable damage to the soil as well as aquatic ecosystems. Particulate matter, pesticides, mining and smelting of ores, herbicides, fertilizers, toxic gases, oil spills, sewage, and effluent from storage batteries are just a few examples of the many concerning environmental contaminants. Lead, chromium, uranium, mercury, zinc, selenium, cadmium, arsenic, silver, nickel, and gold are the major metals and metalloids that most commonly found in the environment. These metals are highly toxic and are not biodegradable in nature (Igiri et al., 2018). The environmental pollution is being increasing in an alarming rate and even detectable rates of contamination can be seen in the remote areas of ocean waters. In addition, large-scale contamination of the Rhine River; conifer forest and aquatic life deterioration in the Northeastern United States, Canada, and Europe; oil spills reported in Exxon Valdez; the Union Carbide disaster in Bhopal; or the Chernobyl accident and Fukushima reactor in Japan; and the recent crisis of crude oil pollution reported from the gulf of Mexico are some of the examples of serious episodes of environmental destruction or disaster (Sardrood et al., 2013). Hence, there is a need to seek new methods in developing approaches to minimize and eliminate environmental contaminants from the soil and water ecosystem.

The recent past has witnessed a sudden growth in applying nanotechnology in the field of biology. The first long-circulating polymeric nanoparticle with improved bioavailability was reported by Gref et al. (1994) and Bruchez et al. (1998). The authors showed the first description of semiconductor quantum dots as fluorescent biological labels. Chan and Nie (1998) reported the application of quantum dots bioconjugate for biological detection (Chan and Nie, 1998). Cui et al. (2001) reported the use of nanowires and nanosensors for the detection of biological and chemical molecules (Cui et al., 2001). During the last 20th century, achievements in the field of nanotechnology have been enormous. Nanotechnology has applications in drug discovery (Pantarotto et al., 2003), fluorescent labeling (Bruchez et al., 1998), detection of proteins (Nam et al., 2003), and pathogen detection (Edelstein et al., 2000). Moreover, due to their unique surface chemistry, nanoparticles can be functionalized or grafted with functional groups that can target specific molecules of interest such as pollutants. In addition, the intentional tuning of the properties of nanoparticles such as shape, size, porosity, and chemical composition can grant significant advantages over conventional methods for addressing environmental contamination (Pandey, 2018). The functionalization of nanoparticles on a scaffold responsible for targeting contaminant molecules will help to increase the efficacy and selectivity of the material. Hence, the application of nanotechnology in bioremediation has opened vast opportunities for utilizing nanomaterials or devices for better fruitful outcomes. The use of microbially generated nanoparticles in bioremediation is still in its starting stage, though

Microbe Mediated Remediation of Environmental Contaminants. DOI: https://doi.org/10.1016/B978-0-12-821199-1.00015-8

the present book chapter is designed to provide a general overview of "nanobioremediation" using microbially generated nanoparticles.

15.2 A brief account of water and soil contamination

Water is known as a universal solvent; hence, water bodies are particularly vulnerable to pollution. Water pollution is the contamination of any water body, including streams, lakes, rivers, oceans, and aquifer, by chemicals or microorganisms, thus degrading the water quality, rendering it to toxic.

15.2.1 Different types of water pollution

15.2.1.1 Groundwater

Groundwater is one of the important natural resources and it is the water found in the cracks, crevices, and porous spaces of the underground. Groundwater gets polluted when contaminants from pesticides as well as fertilizers leached from landfills or septic systems and contaminate the underground aquifer rendering it unsafe. The groundwater once it gets contaminated, it remains so for several decades. The contamination of the groundwater can easily spread to streams, lakes, and ocean even far from the original polluting source (Sasakova et al., 2018).

15.2.1.2 Surface water

Surface water is the water found everywhere. It fills the "all blue bits" on the world map, including the saltwater of ocean and the freshwater in rivers, lakes, and streams. Unfortunately, nearly half of the streams and rivers and more than one-third of lakes are polluted and unfit for fishing, drinking, and swimming, according to the recent surveys of US EPA. Nutrient pollution such as with phosphate and nitrates is the lead cause of contamination of freshwater sources. Municipal waste, industrial waste, and random junks that individual dumps also contribute their fair share of toxins to the surface water (Wang et al., 2013).

15.2.1.3 Ocean water

A percentage of 80 of marine pollution originates from land. Contaminants such as nutrients, chemicals, and heavy metals are carried from factories, cities, industrial areas, into the estuaries and bays and finally reach into the ocean. Another source of contamination is oil spills, leaks, and carbon pollution from the air. The fossil fuels are not only polluting the air but the oceans too. The sea absorbs as much as a quarter of all man-made carbon emissions, which changes the pH of surface water, leading to acidification. The contamination of the ocean also affects its inhabitants. The aquatic creatures such as mussels, coral, clams, and oysters require calcium carbonate. One of the important aftereffects of acidification of seawater is the decrease in the concentration of calcium carbonate, which eventually affects all the aquatic creatures (Gesamp, 2009).

15.2.2 Status of soil contamination

Soil pollution is the contamination of soil by toxic chemicals. Soil contamination occurs either due to human activities or by natural processes. The excess amount of chemicals such as pesticides, ammonia, herbicides, petroleum hydrocarbons, nitrate, lead, mercury, and naphthalene adversely affects the soil and ultimately reduces the quality of the soil.

15.2.2.1 Causes of soil pollution

One of the major causes of soil contamination is pesticides. They are synthetic toxic chemicals that definitely kill different types of pests and insects causing damage to agriculture. But it has many ecological repercussions. Pesticides are generally nonbiodegradable and insoluble in water. Hence, these chemicals will not gradually decompose and keep on accumulating in the soil. It may cause several metabolic as well as physiological problems in humans as the concentration of these chemicals getting increase when it transfers from lower to higher trophic level through the food chain. Excessive use of inorganic nitrogen fertilizers also leads to acidification of soil and contaminates the agricultural soil (Cachada et al., 2018).

The incorrect way of waste disposal from industries results in heavy contamination and acidification of the soil. Disposal of plastics, cans, and other solid waste also falls into the category of soil pollution. Dumping of electrical goods such as batteries causes an adverse effect on the soil due to the presence of harmful chemicals. Lithium present in batteries can cause leaching of soil (Medina-Perez et al., 2019). Lack of proper industrial as well as residential waste disposal creates excessive damage to the soil. Leaking of sewage systems affects soil quality and causes soil pollution by changing its chemical composition. Altogether, the soil contamination finally results in inferior crop quality, harmful effects on human health, contamination of water sources, and a negative impact on ecosystem and biodiversity (Cachada et al., 2018).

15.3 Bioremediation

The term "bioremediation" originated from the two words "bios" and "to remediate" These two words that mean "life" and "to solve the problem," respectively, give meaning to the word "bioremediation" (Sardrood et al., 2013). Bioremediation is the use of biological organisms such as bacteria, fungi, protists, or enzymes to eliminate serious environmental problems such as soil and water contamination. In this method living organisms are used agents to neutralize different environmental pollutants. In other words, it is the cleaning up of the contaminated environment by exploiting metabolic activities of microorganisms and converting these contaminants into its lesser harmful state by mineralizing or converting into microbial biomass or generation of carbon dioxide and water. It is necessary to put emphasis on the confusion between the meanings of bioremediation and biodegradation. Bioremediation is a process or a technique that includes biodegradation as its one of the key mechanisms. The most common bioremediation techniques include bioventing, bioreactor, bioleaching, bioaugmentation, composting, land farming, biostimulation, phytoremediation, and rhizofiltration (Sardrood et al., 2013; Azubuike et al., 2016).

There are two approaches to bioremediation. In one method, suitable temperature, oxygen, and nutrients are used to enhance the indigenous microbial population living at the contaminated site. In the other method, exogenous microbes are added to the affected site to degrade the contaminants, but it is a less common method. Therefore bioremediation applications fall mainly into two broad categories based on the sites where wastes are confiscated: in situ and ex situ methods (Fig. 15.1).

15.3.1 In situ bioremediation

In situ bioremediation is the treating of polluted substances in contaminated soil and water system at the site of pollution. It is a superior method because there is no excavation in in situ bioremediation. Therefore in situ bioremediation accompanied by little or no disturbances to the normal soil structure. When compared to ex situ bioremediation, it is less expensive due to no expense for the excavation process. However, the major concern is the cost of design and onsite installation of some sophisticated instruments to improve microbial activities during bioremediation. It uses GRAS (generally regarded as safe) microorganisms to remove the contaminants. The microorganisms used for this process have some encouraging chemotaxis toward the contaminants. This quality of organisms facilitates the likelihood of bioremediation even in the bioremediate shut points.

Another important benefit of in situ bioremediation relies on the possibility of simultaneous management of contaminated water and soil. Moreover, in situ bioremediation has been successfully utilized to treat dyes, chlorinated solvents, heavy metals, hydrocarbons, and polluted sites. Finally, a successful in situ bioremediation requires important environmental conditions such as nutrient availability, the status of electron acceptor, moisture content, pH, and temperature.

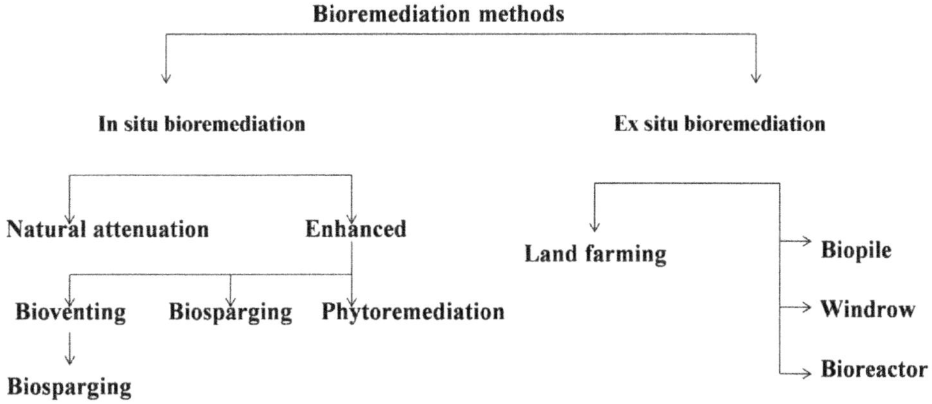

FIGURE 15.1 Bioremediation techniques.

However, in situ bioremediation is slow in action compared to other counteractive processes and directs to major seasonal variations in the activity of microbes due to the straight exposure to uncontrolled environmental factors (Sardrood et al., 2013; Azubuike et al., 2016).

15.3.2 Ex situ bioremediation

Ex situ bioremediation involves the treatment of contaminated soil or groundwater excavating from the polluted sites and transporting them to another site for bioremediation. Although ex situ bioremediation technique requires the removal of contaminated soil prior to the treatment, the method is faster and usually able to treat a wider range of contaminants. Ex situ bioremediation can occur in the following two ways.

15.3.2.1 Solid-phase bioremediation

Solid-phase bioremediation treats the contaminated soil in an aboveground treatment center. This type of treatment is easy to maintain but it needs a lot of space and longer periods compared to slurry-phase bioremediation. Solid-phase bioremediation consists of land farming, soil biopiles, and composting. This system aims to bioremediate wastes of organic nature and challenging industrial as well as domestic effluents and municipal solid waste.

15.3.2.2 Slurry-phase bioremediation

In this method, contaminated soil is kept in a large bioreactor tanks and mixed with water and other additives, which mingled together to bring the indigenous microorganisms in close contact with the contaminated soil. Then oxygen and essential nutrients are added, so as to generate an ideal environment for the breakdown of contaminants. After entire process the water is separated from the soil and the soil is tested. The solid waste is either replaced in the environment or processed little more to decontaminate the remaining pollutants (Sardrood et al., 2013).

15.3.3 The advantages and limitations of bioremediation

Bioremediation is a recent technology developed for the eco-friendly removal and detoxification of environmental contaminants, even though the microbes have been using in the transformation as well as the treatment of waste for centuries. It is completely natural with almost no harmful side effects and therefore it is considered to be an accepted method for the handling of soil contamination. Bioremediation can be implemented in an integrated manner united with other treatments regimes. Furthermore, natural bioremediants are present and that can be utilized in vast varieties of contaminants. Many dangerous as well as detrimental agents could be transformed into harmless products, which will eliminate the chance of future liability in the handling and removal of contaminated material. Another important merit is most of the bioremediation processes carried out in situ with no dangerous transport, for instance, land to water and water to air. The complete destruction of a pollutant can be carried out in one environmental medium. This could stop the channel of hazardous contaminants from one bionetwork to another. Bioremediation can be efficiently done on-site without making prominent interruptions on regular activities, which in turn eradicates the potential threat to human health as well as to the environment that may arise during transportation. Finally, bioremediation methods are a little expensive when compared with other technologies used for cleanup procedures.

However, bioremediation suffers few limitations. Bioremediation is applicable only to the compounds that can be degraded biologically. Some of the compounds are not easily undergo complete and rapid degradation. A successful bioremediation process requires important site factors such as the presence of metabolically capable microbial population, appropriate level of nutrients and contaminants, and suitable environmental growth factors. Each biological process is often specific. Thus pilot studies could not be easily extrapolated into full-scale on-site operations. Another important factor is that some of the bacteria and fungi are capable of operating in a multitude of pollutants. So far organisms that are omnivorous to remove large variety of natural chemicals hardly exist in nature. Moreover, bioremediation is time-consuming and slow to establish its effectiveness. Another drawback of bioremediation is that the application of additives to increase the bioreactivity of microorganisms may sometimes intervene the niches of other organisms present in the same surroundings. Nutrient imbalance followed by bioremediation may obstruct biodegradation. Insufficiency of potassium, phosphorus, nitrogen, and sulfur could hinder the degradation of hydrocarbons in the environment, which eventually affects the flora and fauna of the environment.

These drawbacks may be bypassed through a few solutions. The breakthrough findings in microbial species diversity and its synchronous use directly or indirectly increase the bioreactivity of the bioremediate. Genetic manipulation is another opportunity to obtain new microbial populations with enhanced bioremediation activity. Another resolution could be the adding up of stimulants or enhancers to the formulation and/or environment to biochemically build up certain bioremediation pathways in the bioremediate. And finally the uses of nanoparticles for bioremediation, which will not only improve the microbial activity of specific waste and toxic materials but also reduce the overall cost and time consumption.

15.4 Nanobioremediation

Bioremediation is a considerably green, clean, and sustainable solution for the treatment of contaminated materials. However, bio-based techniques at higher contaminant concentrations toxic to the organisms involved for the purpose. In the recent past, scientists have intensively explored the concealed avenue of "biomimetic nanotechnology." The field of nanobiotechnology intersects the knowledge from all branches of science and technology and involves all living organisms. The scope of nanobiotechnology has higher potential and can extensively be used to address global environmental challenges. Thus integration of nanobiotechnology with bioremediation possibly achieves effective, efficient, and sustainable solutions for a clean environment (Tripathi et al., 2018).

15.4.1 Uniqueness of nanoparticles

As reported by the British Standard Institute (BSI) (2007), the nanomaterial is "any material having one or more dimension in the nanoscale (1–100 nm)." Nanosized materials exhibit unique and unusual properties due to the change in volume-to-surface ratio and quantum effects in the nanosize range (Royal Society, 2004). Their properties cannot be described by the classical laws of physics. The nanorange is an intermediate stage between the bulk state and the molecular or atomic state (Suresh, 2012). At the same time, "nanotechnology" is the study of manipulation and manufacture of ultrasmall structures by controlling size and shape at the nanometer scale (Royal Society, 2004). Nanotechnology is a multidisciplinary science that covers a diverse area of research and technology in chemistry, physics, and biology (Whitesides, 2005), evolved in the late 1960s (Karpagam et al., 2011). Now, this specific field of applied nanoscience has hailed great interest and enthusiasm among scientists due to multitudes of applications it has in various fields of science.

15.4.2 Significance of nanoparticles in bioremediation

There are different reasons for the use of nanoparticles in bioremediation. The pronounced effect of the physical properties of nanomaterial is due to its large surface area or surface energy–reduced imperfections (Suresh, 2012). The large "surface-to-volume" ratio brings about changes both in chemical and physical properties of nanoparticles such as its catalytic activity; mechanical, biological, and optical properties; and melting point compared to their bulk counterparts (Bogunia-Kubik and Sugisaka, 2002). As a result, a large amount of material comes in contact with the surrounding areas and this affects its reactivity. Nanoparticles exhibit the quantum effect. Hence, less activation energy is required to make the chemical reactions feasible. Another property of nanoparticle is its surface plasmon resonance (SPR). The SPR arises due to the coherent oscillation of the valence band electrons in the nanoparticles (Mulvaney, 2002). Hence, it can be used for the detection of toxic materials. Different metallic and nonmetallic nanoparticles exhibit different sizes and shapes. Their broad ranges of applications are based on their size- and shape-dependent characteristics. Hence, nanoparticles can diffuse into a contamination zone where microparticles cannot reach, and they have high reactivity to redox susceptible contaminants. Nurmi et al. (2005) reported that oxide-coated Fe^0 can form weak and outer sphere complexes with contaminants such as carbon tetrachloride (Nurmi et al., 2005). The oxide coating increases the reactivity and through electron transfer, carbon tetrachloride can be broken down into carbon monoxide and methane.

The last decade witnessed a rapid growth of nanotechnology-based industries due to the introduction of several newly fabricated nanomaterials. The demand for novel nanomaterial is increasing at a faster rate. Most commonly used nanomaterials are majorly based on five classes: metal-based, carbonaceous nanomaterial, semiconductor materials, zerovalent nanometals such as silver and iron, and finally nanopolymers such as dendrimers (Kroto et al., 1985). The studies on the ability of nanoparticle to control the pollution production are in progress and it would be the most revolutionary field in environmental biotechnology in the upcoming decades.

15.4.3 Methods of synthesis of nanoparticles

"Top-down" and "bottom-up" methods are the two main strategies used for the creation of nanomaterials. In bottom-up approach, there is self-assembly of small particles, and the nanostructures are created from smaller building blocks. The bottom-up approach has the advantage to synthesize nanostructures with homogeneous chemical composition and has fewer defects. In the top-down approach the starting bulk material is reduced into nanosize by chemical or physical methods while retaining the original integrity. The main limitation with the top-down approach is the imperfection of surface structure and significant crystallographic damage to the synthesized nanomaterial. These imperfections ultimately raise some extra challenges in the device design and fabrication. Both approaches are possible in physical, chemical, and biological methods of synthesis (Wang and Xia, 2004) (Fig. 15.2).

15.4.3.1 Physical method of synthesis of nanoparticles

The important physical method of nanoparticle synthesis is plasma catalysis, pyrolysis, laser ablation, sputtering, and evaporation–condensation. Pingali et al. (2005) reported the synthesis of uniform-sized nanoparticles by spray pyrolysis (Pingali et al., 2005). Kruis et al. (2002) reported the synthesis of nanoparticles of different materials, including Ag, PbS, and Au using condensation techniques in the vacuum chamber (Kruis et al., 2002). Li et al. (1999) reported the synthesis of crystalline silver nanoparticles using sputtering induced by an electron beam (Li et al., 1999).

Gold nanoparticles were synthesized successfully in liquid polyethylene glycol using sputter deposition (Hatakeyama et al., 2011). Mafune et al. (2000) synthesized nanoparticles with a plasma plume by laser ablation (Mafune et al., 2000). But the high energy needed for the synthesis is the major drawback of this method. The major demerits of physical methods are the high cost of production, low production rate, and high energy consumption.

15.4.3.2 Chemical method of synthesis

Chemical reduction, photochemical reactions, and electrochemical techniques are the common chemical methods used for nanoparticle synthesis. The nanoparticles are synthesized in a liquid medium containing various reactants, including reducing agents such as sodium borohydride, methoxy polyethylene glycol, potassium bitartrate, and hydrazine. The agglomeration of particles during the synthesis is prevented by the addition of stabilizing agents such as polyvinylpyrrolidone and sodium dodecyl benzyl sulfate. The major limitations of the chemical methods are the contamination from precursor chemicals, the generation of hazardous by-products, and the use of toxic solvents. In addition, the toxicity of the particles and the environmental toxicity are also accounted for its demerits (Yu et al., 2008).

15.4.3.3 Biological synthesis of nanoparticles

Alternative methods or modifications of methods have been developed to overcome the toxicity associated with the chemical synthesis of nanoparticles. The "green chemistry" approaches have been evolved as major alternatives for the synthesis of commercially viable nanoparticles of noble metals. Green synthesis is defined as the use of environmentally compatible materials such as fungi, bacteria, and plants in the synthesis of nanoparticles. The biological synthesis of nanoparticles is preferred over chemical and physical means because of the rapidity of synthesis, less toxicity, cost-effectiveness, better control over size and shape, generation of nonhazardous waste, absence of contamination from precursor chemicals, eco-friendly nature, and importantly, the lack of costly downstream processing required for producing the nanoparticles. Hence, nanoparticles from biological extracts tentatively offer a route for large-scale production of commercially attractive nanoparticles (Thakkar et al., 2010) (Fig. 15.3).

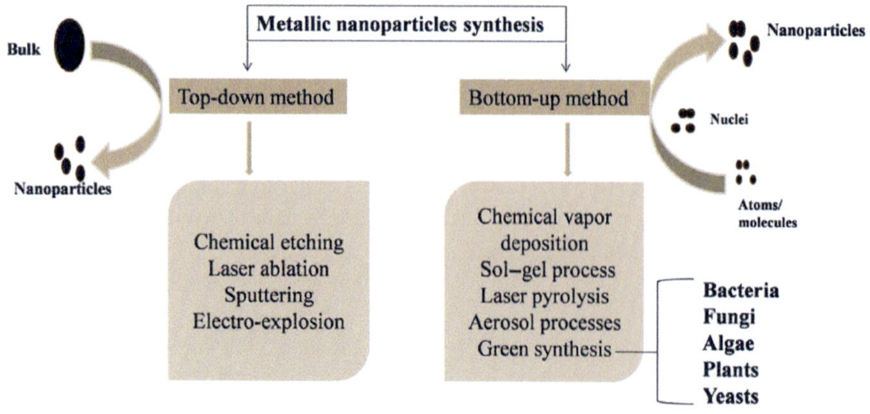

FIGURE 15.2 Synthesis of nanoparticles by top-down and bottom-up approaches.

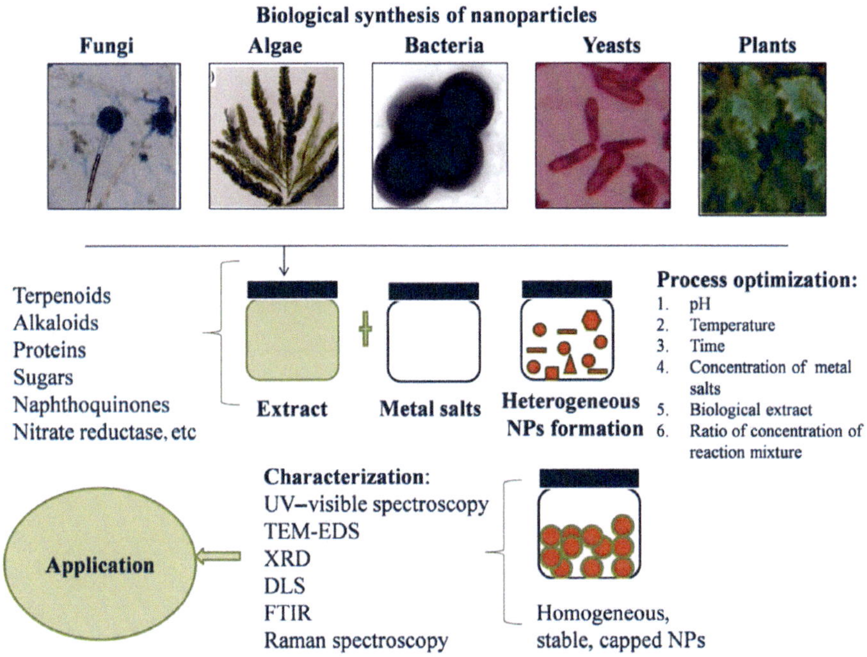

FIGURE 15.3 Schematic representation of the procedure for green synthesis of silver nanoparticles using various biological extracts.

Recently living extracts have been widely utilized by researchers for the synthesis of nanoparticles. Several reports suggest that the molecules existing in the plant extracts and microbial cells such as terpenoids, carbonyl groups, phenolic, amines, flavonoids, amides, amines, proteins, alkaloids and pigments to have the ability to favor the synthesis of nanoparticles (Kumar et al., 2015). In this regard, it is worth studying the biologically synthesized nanoparticles in detail.

15.4.4 Biosynthesis of nanoparticles using microorganisms

Microorganisms have the potential to reduce metal ions leading to the synthesis of nanoscale materials. Microorganisms such as bacteria, fungi, actinomycetes, and, cyanobacteria have been reported for their intrinsic ability to make inorganic nanoparticles of gold, silver, silicon, gypsum, calcium, iron, and lead both intra- and extracellularly (Asmathunisha and Kathiresan, 2013). Extracellular production of the nanoparticle is preferred over intracellular production of nanoparticles due to the easiness of extracting the final product. Tedious processes such as treatment with detergents and ultrasonication can be avoided in extracellular production (Sharma et al., 2007). Many microorganisms have the potential to produce metallic nanoparticles with characteristic features similar to nanomaterials synthesized by chemical and physical methods (Huang et al., 2009). Right now, microbially generated nanomaterial synthesis is restricted to metals, metal sulfides, and some metal oxides such as Fe_3O_4 and TiO_2. It is expected that the other metal oxides can also be produced by means of hydrolytic activity of the microorganisms (Mubarakali et al., 2012). As of now, numerous microorganisms from algae (Rajeshkumar et al., 2013) to bacteria (Iravani, 2014) have been reported to synthesize inorganic materials either intracellularly or extracellularly and thus have the potential to be used as eco-friendly nanosystems.

15.4.4.1 Biosynthesis of nanoparticles by bacteria

Bacteria have a special affinity for metals and are capable of mobilizing and immobilizing metals. Hence, bacteria are considered to be potential "nano-factories" for the synthesis of nanoparticles such as silver, gold, palladium, platinum, titanium, titanium dioxide, cadmium sulfide, and magnetite. Optimization of environmental conditions such as temperature, pH, light, the concentration of metal salts, and the ratio of the concentration of metal salts to bacterial extract, the properties of synthesized nanoparticles can be easily manipulated. The extracellular syntheses of nanoparticles by bacteria lead to the production of nanoparticles of size range between 100 and 200 nm in a relatively pure state. Purification of synthesized nanoparticles can be achieved by filtration. Certain surface antigens such as S-layer proteins and special metal-binding abilities make them useful for technical applications in nanotechnology and further in bioremediation (Yadav et al., 2017) (Table 15.1).

TABLE 15.1 Metal nanoparticles synthesized by bacteria.

Bacteria	Product	Extracellular/ intracellular	Size	References
Multicellular prokaryotes	Fe_3S_4	Intracellular	N/A	Lefevre et al. (2010)
Magnetotactic bacterium	Probably polyphosphate	Extracellular	N/A	Lefevre et al. (2010)
Pseudomonas aeruginosa	Au	Cell wall surface	15 ± 30	Husseiny et al. (2007)
Rhodopseudomonas capsulata	Au	Extracellular	10–20	Bai et al. (2009)
Shewanella algae	Au	Intracellular	10–20	Konishi et al. (2007b)
Brevibacterium casei	Au/Ag	Intracellular	10–50	Kalishwaralal et al. (2010)
Bacillus licheniformis	Ag	Extracellular	50	Kalimuthu et al. (2008)
Corynebacterium glutamicum	Ag		5–50	Sneha et al. (2010)
Ureibacillus thermosphaericus	Ag	Extracellular	50 ± 70	Juibari et al. (2011)
Bacillus cereus	Ag	Intracellular	4–5	Ganesh Babu and Gunasekaran (2009)
S. algae	Pt	Intracellular	5	Konishi et al. (2007a)
Enterobacter	Hg	Intracellular	2 ± 5	Sinha and Khare (2011)
Shewanella sp.	Se	Extracellular	181 ± 40	Lee et al. (2007)
Escherichia coli	CdTe	Extracellular	2 ± 3.2	Bao et al. (2010)
Pyrobaculum islandicum	U(VI), Tc(VII), Cr(VI), Co(III), Mn (IV)	Extracellular	N/A	Kashefi and Lovley (2000)
Desulfovibrio desulfuricans	Pd	Extracellular	50	Lloyd et al. (1998)
Shewanella oneidensis	Fe_3O_4	Extracellular	40 ± 50	Gref et al. (1994)
S. oneidensis MR1	Fe_3O_3	Intracellular	30 ± 43	Bose et al. (2009)
Lactobacillus sp.	$BaTiO_3$	Extracellular	20 ± 80	Jha and Prasad (2010)
Lactobacillus sp.	TiO_2	Extracellular	8 ± 35	Jha et al. (2009a)
Lactobacillus	CdS	Intracellular	4.9 ± 0.2	Prasad and Jha (2010)
Sulfate-reducing bacteria	FeS	Extracellular	2	Watson et al. (1999)
Desulfobacteriaceae	CdS	Intracellular	2–5	Lloyd et al. (1998)
E. coli	CdS	Intracellular	2–5	Sweeney et al. (2004)
Rhodopseudomonas palustris	CdS	Intracellular	8	Bai et al. (2009)
B. casei	Polyhydroxybutyrate	Intracellular	100–125	Ram Kumar Pandian et al. (2009)

15.4.4.2 Biosynthesis of nanoparticles by fungi

Among the microorganisms, fungi are more efficient for nanoscale metal particle synthesis. Fungi grow fast; their culturing and maintenance in the laboratory conditions are very simple. Fungi are more convenient to be handled and fabricated in downstream processing compared with other microorganisms (Boroumand Moghaddam et al., 2015). Fungi are known to produce more bioactive substances compared to bacteria. Hence, fungi can be more suitable for large-scale production of biosynthesized nanoparticles (Madhurima et al., 2017). Boroumand Moghaddam et al. (2015) reported that many active substances secreted by fungi played important roles as reducing agents as well as capping agents during the biosynthesis of nanoparticles (Boroumand Moghaddam et al., 2015). Fungi are more considered to be suitable for large-scale production of biosynthesized nanoparticles because of their immense bioactive substance production.

15.4.5 The advantages and limitations of microbially synthesized nanoparticles

It should be noted that the synthesis of nanoparticles by microorganisms satisfies the three main criteria of green synthesis protocol. These are the use of water as the solvent instead of organic solvents, environmentally benign reducing agent, and nontoxic material for the stabilization of the nanoparticles. Replacing organic solvents by water during microbial synthesis improves biocompatibility and applicability of the synthesized particles (Saxena et al., 2014). The nanoparticles for biomedical applications should be devoid of metal toxicity. The synthesized nanoparticles are coated by a layer of biomolecules that are absorbed from the reaction mixture. The corona of biomolecules around AgNPs makes it less toxic, and hence eco-friendly and biocompatible. It is hypothesized that these coronas provide additional efficacy of AgNPs over bare nanoparticles (Singh et al., 2017). Moreover, unlike physicochemical synthesis, the biological synthesis of nanoparticles proceeds even without a strong ligand in the solution. In microbial synthesis the metallic ions directly interact with the biomolecules present in the microbial extract through electrostatic interactions. Hence, using fungi and bacteria as a creator for nanomaterial has gained considerable attention in recent years (Madhurima et al., 2017).

However, microbial synthesis of nanoparticles is in the early stages of advance. Consequently, the creation of valuable nanoparticles with the desired size remains a challenge. A wide variety of biomolecules such as proteins, carbohydrates, and nucleic acids present in the microbial extract can act as capping or reducing agents. The stronger reducing agents produce smaller nanoparticles, whereas the weak reducing agents usually produce larger nanoparticles. Therefore the microbially synthesized nanoparticle brings an overall heterogeneity to the synthesized nanoparticles (Zhao et al., 2017).

Another major challenge is the influence of reaction conditions such as temperature, pH, incubation time, and concentration of precursor metal salts on the morphology and size of synthesized particles. In the case of fungi the compositions of biomolecules depend on the environmental habitat where it grows. Thus reproducing the exact morphological characteristics by mycosynthesis is more challenging. At the same time, these limitations make further studies more relevant (Zhao et al., 2017).

Microbial synthesis of nanoparticles is comparatively slower than phytosynthesis. Phytosynthesis is a simple and rapid process and it does not involve microbial cultures and its maintenance or associated health risks. The major advantage of phytosynthesis is its simplicity in using living plants for reducing metal salts to its nanoparticles. In phytosynthesis, nanoparticles are synthesized extracellularly and the conditions required for the synthesis can be easily controlled so as to obtain nanoparticles with desired size, shape, and stability without the use of any capping or stabilizing agents (Yadav et al., 2017).

15.5 Nanobioremediation of soil and groundwater

Nanoparticles can be used for the bioremediation of contaminated soil and groundwater. Different treatment methods such as photocatalytic oxidation, disinfection, adsorption, separation, filtration, and bioremediation have been developed to combat water pollution. Nanoparticles exhibit high absorption and interaction–reaction capability and are able to form colloids by mixing with aqueous suspensions. Nanoparticles can penetrate deeper layers (Pandey, 2018). Hence, nanoparticles are a suitable candidate for the treatment of ground or wastewater treatment. Paknikar et al. (2005) reported that nanoscale calcium peroxide can be used to clean up oil spills and nanoscale zerovalent iron (ZVI) for soil and groundwater remediation (Paknikar et al., 2005). Ex situ remediation such as soil washing and pump-and-treat operations and in situ remediation, including chemical oxidation, thermal treatment, and use of reactive barriers with iron, are the conventional methods of remediation.

15.5.1 Biogenesis of nanoparticles and metal detoxification

The microorganism can grow and survive at higher metal ion concentrations. Microorganism plays an important role in the bioremediation of metals through their reduction to metal nanoparticles. The exposure to extreme environmental conditions forcing the microorganisms to develop a specific defense mechanism to combat the metal toxicity and its subsequent reduction of metal ions to metal nanoparticles. The toxicity of the metal ion is eliminated or reduced during the reduction process by changing the redox potential and subsequent precipitation of the metals intra- or extracellularly in the microorganisms. This is the basis of microbially generated nanobioremediation (Cameotra and Dhanjal, 2010).

Dissimilatory metal-reducing bacteria (DMRB) are a group of microorganisms that are capable of using metals or metalloids as terminal electron acceptor in anaerobic respiration (Cameotra and Dhanjal, 2010). These DMRB bacteria play a very important role in the reduction of iron and other metalloids in the Earth's crust in natural environment. The DMRB mediated reduction has an important role in biogeochemical cycles of these metals (Nealson, 1997). Hence, there have been many attempts of exploiting the power of DMRB for various applications. One of the notable applications of these bacteria is bioremediation of environmental contaminants. The biosynthesis of nanoparticles using DMRB usually occurs in water under environmentally benign conditions. Hence, it has become an attractive research focus in green nanobiotechnology (Cameotra and Dhanjal, 2010).

Among the DMRB, *Shewanella oneidensis* has attracted great interest. It is the most often used DMRB in biosynthesis of nanoparticles. *S. oneidensis* is able to reduce several metal ions and the mechanism behind the synthesis of metal nanoparticles is well characterized. *S. oneidensis* was used to synthesis diverse nanoparticles. However, the synthesis of palladium nanoparticles with great potentials has applications in the treatment of environmental pollutants (Lovley, 1993).

15.5.1.1 Remediation of organohalogen-contaminated groundwater

The groundwater is frequently contaminated by halogenated substances such as chlorinated solvents and pesticides. Hexachlorocyclohexane (HCH) is a recalcitrant chlorinated organic compound that is widely distributed in the soil and aquatic systems due to its wide spread use as a commercial pesticide (Hennebel et al., 2011a). The HCH is a mixture of five isomers: a-HCH (60%–70%), b-HCH (5%–12%), c-HCH (10%–15%), d-HCH (6%–10%), and e-HCH (3%–4%), and these isomers are formed by the chlorination of benzene. The c-HCH isomer is also known as lindane. It is the isomer with highest pesticidal activity. Fractional crystallization and concentration are used for subsequent treatment for technical HCH. However, these processes are inefficient and produce 6–10 t of other isomers for each tons of lindane. The resulting rest products are often have been dumped near the production site leading to the high contamination of soil (Hennebel et al., 2011a, 2012). Several strategies are developed to remediate the HCH-contaminated groundwater. The treatments of contaminated groundwater with bioaugmentation using specialized microorganisms such as aerobic (Mertens et al., 2006) and anaerobic bacteria (Jagnow et al., 1977) and fungi (Tu, 1976) or adsorption on materials such as activated zeolites, carbon, silica, and bark are the most commonly used remediation methods. Elliott et al. (2009) have been studied the promising use of ZVI nanoparticles for the reductive dechlorination of indane at laboratory scale (Elliott et al., 2009). In an another study conducted by Joo and Zhao (2008), the increased reactivity of ZVI–Pd bimetallic nanoparticles on indane compared to ZVI nanoparticles is reported (Joo and Zhao, 2008).

Hennebel et al. (2011a) developed a new method to produce a nanopalladium catalyst by precipitating palladium on a bacterium (*S. oneidensis*) in order to get the nanoparticles, that is known as biopad. Biogenic palladium nanoparticles (biopad) can be considered as very promising for full-scale environmental purposes. The metal palladium (Pd) is ubiquitous in nature. Pd can very effectively catalyzed dehalogenation reactions under ambient conditions. Hence, one of the most straight forward applications of biopad is dehalogenation of organic contaminants from contaminated groundwater and soil (Hennebel et al., 2011a).

S. oneidensis can be charged with H* radicals by adding different substrates such as acetate, hydrogen, and formate as electron donors in a bioreductive assay containing Pd(II). Bacterium can reduce Pd(II), when an electron donor is provided and subsequently precipitate it as Pd nanocrystals on their cell wall and in their periplasmic space. In this process, bacteria serve as a reducing, stabilizing, and capping agent. Moreover, it also improves the activity of Pd. When this charged Pd(0) deposited *S. oneidensis* cells are brought in contact with chlorinated compounds, the H* radicals on the Pd(0) can catalytically react with pentachlorophenol in the contaminated water, resulting in the removal of chlorine molecule from the chlorinated compounds (Cameotra and Dhanjal, 2010). However, the retaining of Pd is of utmost important during these processes because its release in the aquatic environment is undesirable. But the magnetical property of biopad nanoparticles or in combination of Pd with biogenic magnetite can surpass this problem. Fig. 15.4 shows a schematic representation of dehalogenation of groundwater using biopad Pd nanoparticles synthesized by *S. oneidensis* as shown by Hennebel et al. (2011a).

15.5.1.2 Treatment of contaminated soil and sediments

The soil and sediments contaminated with adsorbed organohalogens are very difficult to bioremediate. One of the most persistent groups of contaminants in these matrices is polychlorobifenyls (PCBs). PCB cause adverse health effects along the entire food chain. The PCBs are hydrophobic in nature; hence, the removal of PCB from contaminated soil is

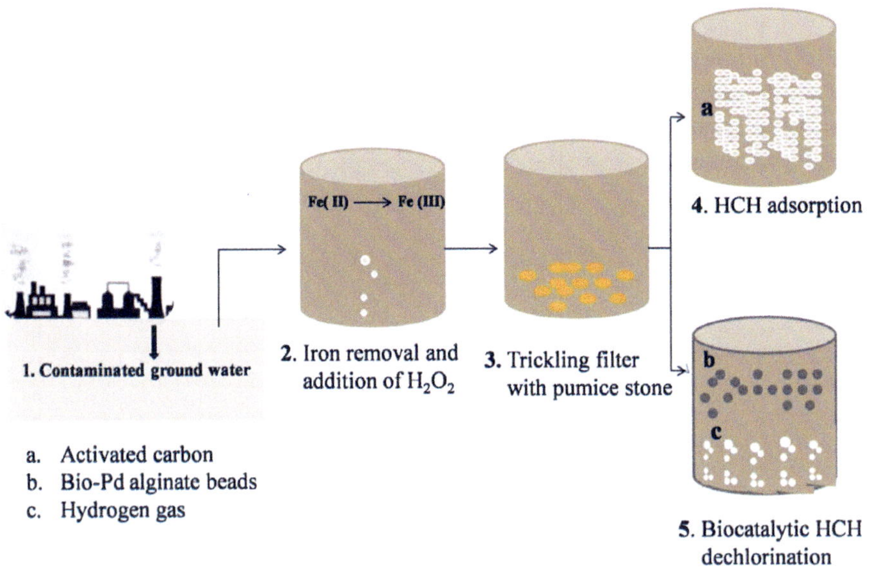

FIGURE 15.4 Schematic representation of treatment steps in the bioremediation of HCH using biopad Pd nanoparticles synthesized by *Shewanella oneidensis* (1) contaminated groundwater with HCH and chlorobenzenes; (2) aeration: iron removal by oxidation or precipitation of Fe(II) to Fe(III) and subsequent addition of H_2O_2, pH corrected with NaOH; (3) trickling filter with pumice stone, here dehalogenating bacteria (e.g., *S. oneidensis*) forms biofilms on pumice stones; (4) removal of HCH by adsorption on activated carbon; (5) dechlorination of HCH by biopad Pd nanoparticles encapsulated in alginate beads. *HCH*, Hexachlorocyclohexane.

very important. One and most common method for the removal of PCBs is incineration, but it is costly. An alternative method is the aerobic and anaerobic biodegradation of PCBs. This technique is preferred for the in situ treatment of contaminated lake, river, or deep sea sediments, which are difficult to excavate. However, the process takes months to years for completion. Under this scenario a biopad catalyzed dehalogenation is an effective alternative, because the PCBs are degraded under mild reaction conditions (Hennebel et al., 2012). The ability of both biopad Pd nanoparticles and chemically produced Pd nanoparticles for the dechlorination of PCBs has already been extensively studied (Mohn and Tiedje, 1992; Hennebel et al., 2011a; Agarwal et al., 2009). The resulting reaction products are biphenyl and lower chlorinated PCBs. Both products have been shown to be more biodegradable and less toxic.

15.5.1.3 *Microbial synthesis of arsenic nanoparticles*

Arsenic is a very toxic metalloid. It affects almost all living organisms as well as enzyme systems and a common cause of lung and skin cancers. Although there are numerous reports of the isolation of microorganisms and their detoxification of toxic arsenic compounds from arsenic-contaminated soil, only a few reports on the synthesis of arsenic nanoparticles by these microorganisms. Jiang et al. (2009) reported that the metal-reducing bacteria *Shewanella* sp. HN-41 synthesizes arsenic-sulfide (As-S) nanotubes with a size range of 20–100 nm with a length of ~30 nm (Jiang et al., 2009). During the synthesis of As-S nanotubes, initially the reduction of As(V) and $S_2O_3^{2-}$ takes place, which leads to the formation of amorphous As_2S_3 and finally the chalcogenide minerals, realgar (AsS) and duranusite (As_4S), were formed.

15.5.1.4 *Selenium and tellurium nanoparticles*

The recent past has witnessed selenium (Se) toxicity in surface water as well as in groundwater. It has become a major problem worldwide. The direct or indirect consumption of selenium is highly fatal due to its toxicity to human health. Selenium is an essential trace element in the glutathione peroxidase enzyme system, which gives protection to the intracellular structure from oxidative damage. However, excessive selenium exposure or consumption leads to selenium toxicity. Similar to arsenic, selenium also adversely affects the sulfhydryl group of amino acids. Selenosis results from chronic oral exposure to selenium. Here are few examples of bacteria that can reduce selenium into its nanoparticles and thus reduce its metal toxicity. Oremland et al. (2004) and Baesman et al. (2007) reported the reduction of selenite and tellurite to elemental selenium and tellurium by two anaerobic bacteria such as *Bacillus selenireducens* and *Sulfurospirillum barnesii* (Baesman et al., 2007; Oremland et al., 2004). The selenium nanoparticles were synthesized extracellularly as granules and consist of stable, uniform nanospheres with a diameter of ~300 nm monoclinic crystalline structures. Although in tellurium, initially nanorods with a size of ~10 nm were synthesized by *B. selenireducens*, these particles were clustered together to form large rosettes (~1000 nm). The other bacterium *S. barnesii* formed irregularly shaped nanoparticles that coalesce to form large aggregates.

TABLE 15.2 Biosynthesis of metal nanoparticles using fungi and other microorganisms.

Microorganisms	Product	Extracellular/intracellular	Morphology of nanoparticles	Size (nm)	References
Phanerochaete chrysosporium	Ag	Extracellular	Pyramidal	50–200	Vigneshwaran et al. (2006)
Aspergillus fumigatus	Ag	Extracellular	Spherical, triangular	5–25	Gudikandula et al. (2017)
Neurospora crassa	Ag/Au, Au	Intra/extracellular	Spherical	20–50, 32	Castro-Longoria et al. (2011)
Aspergillus niger	Ag	Extracellular	Spherical	3–30	Kandasamy et al. (2010)
Aspergillus flavus	Ag	Cell wall surface	Spherical	8.92	Gudikandula et al. (2017)
Yeast	Au/Ag	Extracellular	Irregular polygonal	9–25	Zheng et al. (2010)
A. niger	Ag	Extracellular	Spherical	5–35	Jaidev and Narasimha (2010)
A. fumigatus	Ag	Extracellular	Spherical hexagonal	15–45	Alani et al. (2012)
Aspergillus clavatus	Ag	Extracellular	Mostly spherical	10–25	Verma et al. (2010)
Penicillium fellutanum	Ag	Extracellular	Mostly spherical	5–25	Kathiresan et al. (2009)
Penicillium strain J3	Ag	Extracellular	Spherical	10–100	Maliszewska et al. (2009)
Penicillium brevicompactum	Ag	Extracellular	Spherical	17–58	Shaligram et al. (2009)
Rhizoctonia stolonifer	Ag	Extracellular	Spherical	5–50	Afreen et al. (2011)
Verticillium sp.	Ag	Intracellular	Highly variable	25	Mukherjee et al. (2001)
Fusarium oxysporum	Ag	Extracellular	Spherical	5–15	Ahmad et al. (2017)
F. oxysporum	Ag	Extracellular	Spherical	20–50	Duran et al. (2005)
Fusarium acuminatum	Ag	Extracellular	Spherical	4–50	Pierrat et al. (2008)
Trichoderma viride	Ag	Extracellular	Rod-like	2–4, 10–40	Mohammed Fayaz et al. (2009a)
T. viride	Ag	Extracellular	Nanocrystalline	5–40	Mohammed Fayaz et al. (2009b)
Trichoderma asperellum	Ag	Extracellular	Spherical	13–18	Dey et al. (2008)
Pleurotus sajorcaju	Ag	Extracellular	Spherical	30.5	Vigneshwaran et al. (2007)
Volvariella volvacea	Ag	Extracellular	Spherical	15	Thakkar et al. (2010)
Cladosporium cladosporioides	Ag	Extracellular	Spherical	60–80	Balaji et al. (2009)
Coriolus versicolor	Ag	Extracellular	Spherical	25–75	Thakkar et al. (2010)
Phoma glomerata	Ag	Intracellular	Spherical	444–491	Birla et al. (2009)
Saccharomyces boulardii	Ag	Extracellular	Spherical rods and triangles	60–80	Kaler et al. (2013)
Xanthomonas oryzae	Ag	Extracellular	Spherical	3–10	Saxena et al. (2014)
Cylindrocladium floridanum	Ag	Extracellular	Spherical	14.8	Sanghi and Verma (2009a)

Beauveria bassiana	Ag	Extracellular	Spherical	25	Banu and Balasubramanian (2014)
Sargassum wightii	Au	Extracellular	planar	8–12	Singaravelu et al. (2007)
F. oxysporum	Au	Extracellular	Multishaped	20–40	Zhang et al. (2011)
Fusarium semitectum	Au	Extracellular	Multishaped	18–80	Sawle et al. (2008)
A. fumigatus	Au	Extracellular	Triangles	17.7–26	Gupta and Bector (2013)
A. flavus	Au	Extracellular	Hexagons, spherical	20–25	Gupta and Bector (2013)
Plectonema boryanum	Au	Intracellular	Cubic	10–25	Lengke et al. (2006a)
A. niger	Au	Extracellular	Polydispersed	10–20	Xie et al. (2007)
F. oxysporum	Au	Extracellular	Spherical	25	Mandal et al. (2006)
P. boryanum UTEX485	Au	Extracellular	Octahedral	10 nm–6 μm	Lengke et al. (2006b)
F. oxysporum	Au	Extracellular	Spherical	2–50	Mukherjee et al. (2002)
Rhizopus oryzae	Au	Cell surface	Nanocrystalline	10	Das et al. (2009a)
Candida utilis	Au	Intracellular	N/A	N/A	Gericke and Pinches (2006a)
Aureobasidium pullulans	Au	Intracellular	Spherical	29	Zhang et al. (2011)
Helminthosporium solani	Au	Extracellular	Polydispersed	2–70	Kumar et al. (2007)
N. crassa	Au	Extracellular	Spherical	32	Castro-Longoria et al. (2012)
P. brevicompactum	Au	Extracellular	Spherical	10–50	Mishra et al. (2011)
Yarrowia lipolytica	Au	Extracellular	Triangles	15	Agnihotri et al. (2009)
Penicillium aurantiogriseum	Au	Extracellular	Spherical	153	Honary et al. (2013)
Penicillium citrinum	Au	Extracellular	Quasihexagonal	172	Saxena et al. (2014)
Phaenerochaete chrysosporium	Au	Extracellular	Pyramidal	50–200	Vigneshwaran et al. (2006)
Penicillium waksmani	Au	Extracellular	Spheres and rods	160	Saxena et al. (2014)
Verticillium sp.	Au	Cell wall	Spherical	20 ± 8	Hulkoti and Taranath (2014)
Verticillium luteoalbum	Au	Cytoplasmic membrane	Spheres and rods	~10	Gericke and Pinches (2006b)
C. floridanum	Au	Extracellular	Spherical	19.5	Narayanan and Sakthivel (2013)
V. volvacea	Au	Extracellular	Spherical	20–150	Philip (2009)
P. chrysosporium	Au	Extracellular	Spherical	10–100	Sanghi et al. (2011)
Sclerotium rolfsii	Au	Extracellular	Triangles, decahedral, hexagonal, rods	25	Narayanan and Sakthivel (2011)
Colletotrichum sp.	Au	Extracellular	Spherical	8–40	Shankar et al. (2003)
Candida albicans	Au	Intracellular	Spherical	20–40 60–80	Chauhan et al. (2011)

(Continued)

TABLE 15.2 (Continued)

Microorganisms	Product	Extracellular/intracellular	Morphology of nanoparticles	Size (nm)	References
A. fumigatus	Zinc oxide	Extracellular	Triangular	1.2–6.8	Raliya et al. (2015)
Yeast cells	Magnetite	Extracellular	Worm hole–like	N/A	Zhou et al. (2009a)
Yeast cells	$FePO_4$	Extracellular	Nanopowders	N/A	Zhou et al. (2009b)
Saccharomyces cerevisiae	Sb_2O_3	Intracellular	Spherical	2–10	Jha et al. (2009b)
Aspergillus tubingensis	Calcium phosphate	Extracellular	Cubo-octahedral	28.2	Tarafdar et al. (2012)
Aspergillus oryzae	Ferric chloride	Extracellular	Quasispherical	10–24.6	Raliya et al. (2015)
V. volvaceae	Silver–gold alloy	Extracellular	Spherical	20–150	Tarafdar and Raliya (2013)
A. flavus TFR7	Titanium dioxide	Extracellular	Rectangular–triangular–spherical	12–15	Hulkoti and Taranath (2014)
F. oxysporum	Titanium dioxide	Extracellular	Spherical	6–13	Sanyal et al. (2005)
F. oxysporum	Cadmium	Extracellular	Hexagonal, pentagonal rectangles	9–15	Kumar et al. (2008)
F. oxysporum	Barium titanate	Extracellular, intracellular	Irregular	70–18010–100	Bansal et al. (2006)
F. oxysporum f. sp. *lycopersici*	Platinum	Intracellular	Spherical	10–100	Riddin et al. (2006)
Fusarium sp.	Zinc	Surface of mycelia	Spherical	20.5	You et al. (2010)
Aspergillus versicolor	Mercury	Extracellular	Spherical	2.5–5.5	Das et al. (2009b)
C. versicolor	CdS	Extracellular	Spherical	100–200	Sanghi and Verma (2009b)
Yeast	CdS	Extracellular	Spherical	4.9 ± 0.2	Prasad and Jha (2010)
Schizosaccharomyces pombe	CdS	Intracellular	Hexagonal lattice	11.5	Filipe et al. (2001)
Candida glabrata	CdS	Intracellular	Hexagonal lattice	2	Dameron et al. (1989)
S. cerevisiae	Magnetite	Extracellular	Spherical	100	Saxena et al. (2014)
F. oxysporum	$PbCO_3$, $CdCO_3$	Extracellular	Spherical	120–200	Rautaray et al. (2004)
F. oxysporum	$SrCO_3$	Extracellular	Needle-like	10–50	Rautaray et al. (2004)
F. oxysporum	CdSe ·	Extracellular	Spherical	9–15	Kumar et al. (2007)
Yeast	$Zn_3(PO_4)_2$	Extracellular	Butterfly-like structure	10–80	Kumar et al. (2007)

15.5.1.5 Cadmium nanoparticles

Heavy metals are important pollutants in the water system. Among this, cadmium is a toxic heavy metal. It is widely used in Ni–Cd batteries, discarded electronic devices, and the coloration of plastics, and this heavy metal is frequently released into the soil as well as to the groundwater resulting in serious health problems. The exposure to cadmium for a very long period results in the well-known disease, itai–itai or ouch–ouch disease. This disease was first reported in postmenopausal women in Japan and the area where they resided was contaminated with cadmium. Till now, only a few reports on the synthesis of cadmium nanoparticles by bacteria were reported. The first report of the bacterial synthesis of cadmium nanoparticles came from Sweeney et al. (2004). They showed that the important facultative anaerobic bacteria *Escherichia coli* synthesize cadmium sulfide semiconductor nanocrystals when incubated with cadmium chloride and sodium sulfide. The results showed that various physiological parameters of the reaction conditions, under which the synthesis took place much, affect the formation of nanoparticles. However, the synthesis was increased 20-fold when *E. coli* at the stationary phase of their growth.

In another report, the Gram-positive bacteria *Clostridium thermoaceticum* also synthesize cadmium sulfide nanoparticles extra- and intracellularly when exposed to cadmium chloride, where cysteine hydrochloride was used as the source of sulfide in the growth medium (Sweeney et al., 2004). Bai et al. (2009) explored that the incubation of biomass of photosynthetic bacterium *Rhodopseudomonas palustris* along with cadmium sulfate solution resulted in the color change of reaction mixture to a yellow from 8 h onward. This was a prima facie for the synthesis of cadmium sulfide nanoparticles. The crystalline structure of the synthesized nanoparticles was confirmed by X-ray diffraction studies (Bai et al., 2009). These results showed that the described bacterial species could be used as a powerful arsenal to combat the cadmium pollution of the soil and groundwater.

15.5.1.6 Platinum nanoparticles

Platinum is once considered as a nondangerous metal. However, platinum group elements are frequently released into the environment from catalytic converters during driving and this metal is largely deposited in the roadsides. Platinum metal is also highly used in fuel cells. Recent research showed that platinum can dissolve in water and are absorbed by plants and animals and these result in several health effects such as allergic reaction to the skin and mucous membrane; cancer; DNA alterations; damage to organs such as kidneys, intestines, and bone marrow. Iron-reducing enzyme from *Shewanella algae* can be used to convert platinum metals into metal nanoparticles as reported by Konishi et al. (2007a). The results showed that *S. algae* reduced aqueous $PtCl_6^{2-}$ ions into elemental platinum at room temperature using lactate as an electron acceptor. The synthesis of nanoparticles was carried out within 60 min at neutral pH (Konishi et al., 2007a).

15.5.1.7 Gold and silver nanoparticles

Elemental gold and silver are nontoxic in nature. However, the soluble compounds of gold and its cyanide salts such as potassium gold cyanide used in gold electroplating are highly toxic to human health. The recent past has witnessed the reports of the synthesis of gold and silver nanoparticles by several microorganisms (Table 15.2).

The size, shape, and surface chemistry of nanoparticles are not yet optimal. Hence, the application of nanobiotechnology in bioremediation and the technological development in this field are still in their infant stage. One of the important limitations of nanobioremediation is the release of nanomaterials into the environment. The mankind could be exposed to these nanoparticles through dermal contact, inhalation, ingestion, or injection. In order to confirm the safety of using fabricated nanostructures, detailed research is needed to know the nanotoxicity on the biological systems.

15.6 Concluding remarks

Nanotechnology has emerged as a powerful tool to make the environment clean. Nanoparticles can detect the pollutants, control the release of pollutants, and, interestingly, can remediate the contaminated soil and aquatic system by in situ and ex situ approaches. Several nanomaterials such as ZVI, silver, gold, titanium, and quantum dots have been used as nanobioremediants. The use of microbially generated nanoparticles in bioremediation gives an added advantage of green and clean nanotechnology for environmental cleanup. Soil microbiota may have the potential to biosynthesize nanoparticles and further can be exploited in bioremediation. Even though, nanobioremediation is a feasible and interesting technology, the health risks as well as the safety concerns associated with the large-scale production and usage are not fully studied. It has become necessary to bloom up the nascent field of green nanotechnology and find innovative methods for eco-friendly microbial synthesis of nanoparticles in order to keep the Earth more clean and green.

References

Afreen, Rathod, V., Ranganath, E., 2011. Synthesis of monodispersed silver nanoparticles by *Rhizopus stolonifer* and its antibacterial activity against MDR strains of *Pseudomonas aeruginosa* from burnt patients. Int. J. Environ. Sci. 1 (7), 1830–1840.

Agarwal, S., Al-Abed, S.R., Dionysiou, D.D., 2009. A feasibility study on Pd/Mg application in historically contaminated sediments and PCB spiked substrates. J. Hazard. Mater. 172 (2–3), 1156–1162.

Agnihotri, M., Joshi, S., Kumar, A.R., Zinjarde, S., Kulkarni, S., 2009. Biosynthesis of gold nanoparticles by the tropical marine yeast *Yarrowia lipolytica* NCIM 3589. Mater. Lett. 63 (15), 1231–1234.

Ahmad, A., Mukherjee, P., Senapati, S., Mandal, D., Khan, M.I., Sastry, M., 2017. Extracellular biosynthesis of silver nanoparticles using the fungus *Fusarium oxysporum*. Nature 28, 313–318.

Alani, F., Moo-Young, M., Anderson, W., 2012. Biosynthesis of silver nanoparticles by a new strain of *Streptomyces* sp. compared with *Aspergillus fumigatus*. World J. Microbiol. Biotechnol. 28 (3), 1081–1086.

Asmathunisha, N., Kathiresan, K., 2013. A review on biosynthesis of nanoparticles by marine organisms. Colloids Surf., B: Biointerfaces 103, 283–287.

Azubuike, C.C., Chikere, C.B., Okpokwasili, G.C., 2016. Bioremediation techniques–classification based on site of application: principles, advantages, limitations and prospects. World J. Microbiol. Biotechnol. 32, 11.

Baesman, S.M., Bullen, T.D., Dewald, J., Zhang, D., Curran, S., Islam, F.S., et al., 2007. Formation of tellurium nanocrystals during anaerobic growth of bacteria that use Te oxyanions as respiratory electron acceptors. Appl. Environ. Microbiol. 73 (7), 2135–2143.

Bai, H.J., Zhang, Z.M., Guo, Y., Yang, G.E., 2009. Biosynthesis of cadmium sulfide nanoparticles by photosynthetic bacteria *Rhodopseudomonas palustris*. Colloids Surf., B: Biointerfaces 70 (1), 142–146.

Balaji, D.S., Basavaraja, S., Deshpande, R., Mahesh, D.B., Prabhakar, B.K., Venkataraman, A., 2009. Extracellular biosynthesis of functionalized silver nanoparticles by strains of *Cladosporium cladosporioides* fungus. Colloids Surf., B: Biointerfaces 68 (1), 88–92.

Bansal, V., Poddar, P., Ahmad, A., Sastry, M., 2006. Room-temperature biosynthesis of ferroelectric barium titanate nanoparticles. J. Am. Chem. Soc. 128 (36), 11958–11963.

Banu, A.N., Balasubramanian, C., 2014. Myco-synthesis of silver nanoparticles using *Beauveria bassiana* against dengue vector, *Aedes aegypti* (Diptera: Culicidae). Parasitol. Res. 113 (8), 2869–2877.

Bao, H., Lu, Z., Cui, X., Qiao, Y., Guo, J., Anderson, J.M., et al., 2010. Extracellular microbial synthesis of biocompatible CdTe quantum dots. Acta Biomater. 6 (9), 3534–3541.

Birla, S.S., Tiwari, V.V., Gade, A.K., Ingle, A.P., Yadav, A.P., Rai, M.K., 2009. Fabrication of silver nanoparticles by *Phoma glomerata* and its combined effect against *Escherichia coli*, *Pseudomonas aeruginosa* and *Staphylococcus aureus*. Lett. Appl. Microbiol. 48 (2), 173–179.

Bogunia-Kubik, K., Sugisaka, M., 2002. From molecular biology to nanotechnology and nanomedicine. Biosystems 65 (2–3), 123–138.

Boroumand Moghaddam, A., Azizi, S., Namvar, F., Moniri, M., Mohamad, R., Md. Tahir, P., 2015. Nanoparticles biosynthesized by fungi and yeast: a review of their preparation, properties, and medical applications. Molecules 20 (9), 16540–16565.

Bose, S., Hochella, M.F., Gorby, Y.A., Kennedy, D.W., McCready, D.E., Madden, A.S., et al., 2009. Bioreduction of hematite nanoparticles by the dissimilatory iron reducing bacterium *Shewanella oneidensis* MR-1. Geochim. Cosmochim. Acta 73 (4), 962–976.

British Standard. 2007. 2: 2007: Business Continuity Management-Part 2: Specification. BS25999, British Standard Institute (BSI) Std.

Bruchez, M., Moronne, M., Gin, P., Weiss, S., Alivisatos, A.P., 1998. Semiconductor nanocrystals as fluorescent biological labels. Science (80) 281 (5385), 2013–2016.

Cachada, A., Rocha-Santos, T., Duarte, A.C., 2018. Soil and pollution: an introduction to the main issues. In: Soil Pollution. Academic Press, 1–28.

Cameotra, S., Dhanjal, S., 2010. Bioremediation technology. Environ. Nanotechnol. Nanopart. Bioremed. Toxic. 1, 348–374.

Castro-Longoria, E., Vilchis-Nestor, A.R., Avalos-Borja, M., 2011. Biosynthesis of silver, gold and bimetallic nanoparticles using the filamentous fungus *Neurospora crassa*. Colloids Surf., B: Biointerfaces 83 (1), 42–48.

Castro-Longoria, E., Moreno-Velasquez, S.D., Vilchis-Nestor, A.R., Arenas-Berumen, E., Avalos-Borja, M., 2012. Production of platinum nanoparticles and nanoaggregates using *Neurospora crassa*. J. Microbiol. Biotechnol. 22 (7), 1000–1004.

Chan, W.C.W., Nie, S., 1998. Quantum dot bioconjugates for ultrasensitive nonisotopic detection. Science (80) 281 (5385), 2016–2018.

Chauhan, A., Zubair, S., Tufail, S., Sherwani, A., Sajid, M., Raman, S.C., et al., 2011. Fungus-mediated biological synthesis of gold nanoparticles: potential in detection of liver cancer. Int. J. Nanomed. 6, 2305–2319.

Cui, Y., Wei, Q., Park, H., Lieber, C.M., 2001. Nanowire nanosensors for highly sensitive and selective detection of biological and chemical species. Science 293, 1289–1292.

Dameron, C.T., Reeses, R.N., Mehra, R.K., Kortan, A.R., Carrol, P.J., Steigerwald, M.L., et al., 1989. Biosynthesis of cadmium sulphide quantum semiconductor crystallites. Nature 338, 596–597.

Das, S.K., Das, A.R., Guha, A.K., 2009a. Gold nanoparticles: microbial synthesis and application in water hygiene management. Langmuir 25 (14), 8192–8199.

Das, S.K., Das, A.R., Guha, A.K., 2009b. Adsorption behavior of mercury on functionalized *Aspergillus versicolor* mycelia: atomic force microscopic study. Langmuir 25 (1), 360–366.

Dey, G.K., Mandal, B.P., Roy, M., Kale, S.P., Ghatak, J., Tyagi, A.K., et al., 2008. Green synthesis of highly stabilized nanocrystalline silver particles by a non-pathogenic and agriculturally important fungus *T. asperellum*. Nanotechnology 19 (7), 075103.

Duran, N., Marcato, P.D., Alves, O.L., De Souza, G.I.H., Esposito, E., 2005. Mechanistic aspects of biosynthesis of silver nanoparticles by several *Fusarium oxysporum* strains. J. Nanobiotechnol. 3, 1–7.

Edelstein, R.L., Tamanaha, C.R., Sheehan, P.E., Miller, M.M., Baselt, D.R., Whitman, L.J., et al., 2000. The BARC biosensor applied to the detection of biological warfare agents. Biosens. Bioelectron. 14 (10–11), 805–813.

Elliott, D.W., Lien, H.L., Zhang, W.X., 2009. Degradation of lindane by zero-valent iron nanoparticles. J. Environ. Eng. 135 (5), 317–324.

Filipe, C.D.M., Daigger, G.T., Grady, C.P.L., 2001. Stoichiometry and kinetics of acetate uptake under anaerobic conditions by an enriched culture of phosphorus-accumulating organisms at different pHs. Biotechnol. Bioeng. 76 (1), 32–43.

Ganesh Babu, M.M., Gunasekaran, P., 2009. Production and structural characterization of crystalline silver nanoparticles from *Bacillus cereus* isolate. Colloids Surf., B: Biointerfaces 74 (1), 191–195.

Gericke, M., Pinches, A., 2006a. Biological synthesis of metal nanoparticles. Hydrometallurgy 83 (1–4), 132–140.

Gericke, M., Pinches, A., 2006b. Microbial production of gold nanoparticles. Gold Bull. 39 (1), 22–28.

GESAMP, 2009. Pollution in the Open Oceans: A Review of Assessments and Related Studies. p. 64.

Gref, R., Minamitake, Y., Peracchia, M.T., Trubetskoy, V., Torchilin, V., Langer, R., 1994. Biodegradable long-circulating polymeric nanospheres. Science (80) 263 (5153), 1600–1603.

Gudikandula, K., Vadapally, P., Singara Charya, M.A., 2017. Biogenic synthesis of silver nanoparticles from white rot fungi: their characterization and antibacterial studies. OpenNano 2, 64–78.

Guerra, F.D., Attia, M.F., Whitehead, D.C., Alexis, F., 2018. Nanotechnology for environmental remediation: materials and applications. Molecules 23 (7), 1–23.

Gupta, S., Bector, S., 2013. Biosynthesis of extracellular and intracellular gold nanoparticles by *Aspergillus fumigatus* and *A. flavus*. Antonie van Leeuwenhoek 103 (5), 1113–1123.

Hatakeyama, Y., Morita, T., Takahashi, S., Onishi, K., Nishikawa, K., 2011. Synthesis of gold nanoparticles in liquid polyethylene glycol by sputter deposition and temperature effects on their size and shape. J. Phys. Chem. C 115 (8), 3279–3285.

Hennebel, T., Simoen, H., Verhagen, P., de Windt, W., Dick, J., Weise, C., et al., 2011a. Biocatalytic dechlorination of hexachlorocyclohexane by immobilized bio-Pd in a pilot scale fluidized bed reactor. Environ. Chem. Lett. 9 (3), 417–422.

Hennebel, T., De Corte, S., Verstraete, W., Boon, N., 2012. Microbial production and environmental applications of Pd nanoparticles for treatment of halogenated compounds. Curr. Opin. Biotechnol. 23 (4), 555–561.

Honary, S., Gharaei-Fathabad, E., Barabadi, H., Naghibi, F., 2013. Fungus-mediated synthesis of gold nanoparticles: a novel biological approach to nanoparticle synthesis. J. Nanosci. Nanotechnol. 13 (2), 1427–1430.

Huang, X., Neretina, S., El-Sayed, M.A., 2009. Gold nanorods: from synthesis and properties to biological and biomedical applications. Adv. Mater. 21 (48), 4880–4910.

Hulkoti, N.I., Taranath, T.C., 2014. Biosynthesis of nanoparticles using microbes—a review. Colloids Surf., B: Biointerfaces 121, 474–483.

Husseiny, M.I., El-Aziz, M.A., Badr, Y., Mahmoud, M.A., 2007. Biosynthesis of gold nanoparticles using *Pseudomonas aeruginosa*. Spectrochim. Acta, A: Mol. Biomol. Spectrosc. 67 (3–4), 1003–1006.

Igiri, B., Okoduwa, S., Idoko, G., Akabuogu, E., Adeyi, I.E., 2018. Toxicity and bioremediation of heavy metals contaminated ecosystem from tannery wastewater: a review. J. Toxicol. 7, 609–616.

Iravani, S., 2014. Bacteria in nanoparticle synthesis: current status and future prospects. Int. Scholarly Res. Not. 2014, 359316.

Jagnow, G., Haider, K., Ellwardt, P.C., 1977. Anaerobic dechlorination and degradation of hexachlorocyclohexane isomers by anaerobic and facultative anaerobic bacteria. Arch. Microbiol. 115 (3), 285–292.

Jaidev, L.R., Narasimha, G., 2010. Fungal mediated biosynthesis of silver nanoparticles, characterization and antimicrobial activity. Colloids Surf., B: Biointerfaces 81 (2), 430–433.

Jha, A.K., Prasad, K., 2010. Ferroelectric $BaTiO_3$ nanoparticles: biosynthesis and characterization. Colloids Surf., B: Biointerfaces 75 (1), 330–334.

Jha, A.K., Prasad, K., Kulkarni, A.R., 2009a. Synthesis of TiO_2 nanoparticles using microorganisms. Colloids Surf., B: Biointerfaces 71 (2), 226–229.

Jha, A.K., Prasad, K., Prasad, K., 2009b. A green low-cost biosynthesis of Sb_2O_3 nanoparticles. Biochem. Eng. J. 43 (3), 303–306.

Jiang, S., Lee, J.H., Kim, M.G., Myung, N.V., Fredrickson, J.K., Sadowsky, M.J., et al., 2009. Biogenic formation of As-S nanotubes by diverse *Shewanella* strains. Appl. Environ. Microbiol. 75 (21), 6896–6899.

Joo, S.H., Zhao, D., 2008. Destruction of lindane and atrazine using stabilized iron nanoparticles under aerobic and anaerobic conditions: effects of catalyst and stabilizer. Chemosphere 70 (3), 418–425.

Juibari, M.M., Abbasalizadeh, S., Jouzani, G.S., Noruzi, M., 2011. Intensified biosynthesis of silver nanoparticles using a native extremophilic *Ureibacillus thermosphaericus* strain. Mater. Lett. 65 (6), 1014–1017.

Kaler, A., Jain, S., Banerjee, U.C., 2013. Green and rapid synthesis of anticancerous silver nanoparticles by Saccharomyces boulardii and insight into mechanism of nanoparticle synthesis. Biomed. Res. Int. 2013, 1–8.

Kalimuthu, K., Suresh Babu, R., Venkataraman, D., Bilal, M., Gurunathan, S., 2008. Biosynthesis of silver nanocrystals by *Bacillus licheniformis*. Colloids Surf., B: Biointerfaces 65 (1), 150–153.

Kalishwaralal, K., Deepak, V., Ram Kumar Pandian, S.B., Kottaisamy, M., BarathManiKanth, S., Kartikeyan, B., et al., 2010. Biosynthesis of silver and gold nanoparticles using *Brevibacterium casei*. Colloids Surf., B: Biointerfaces 77 (2), 257–262.

Kandasamy, S., Pathmanaban, S., Kathiresan, K., Alikunhi, N.M., Nabikhan, A., 2010. Analysis of antimicrobial silver nanoparticles synthesized by coastal strains of *Escherichia coli* and *Aspergillus niger*. Can. J. Microbiol. 56 (12), 1050–1059.

Karpagam, R., Gopalakrishnan, S., Natarajan, M., Ramesh Babu, B., 2011. Mapping of nanoscience and nanotechnology research in India: a scientometric analysis, 1990—2009. Scientometrics 89 (2), 501–522.

Kashefi, K., Lovley, D.R., 2000. Reduction of Fe(III), Mn(IV), and toxic metals at 100°C by *Pyrobaculum islandicum*. Appl. Environ. Microbiol. 66 (3), 1050–1056.

Kathiresan, K., Manivannan, S., Nabeel, M.A., Dhivya, B., 2009. Studies on silver nanoparticles synthesized by a marine fungus, *Penicillium fellutanum* isolated from coastal mangrove sediment. Colloids Surf., B: Biointerfaces 71 (1), 133–137.

Konishi, Y., Ohno, K., Saitoh, N., Nomura, T., Nagamine, S., Hishida, H., et al., 2007a. Bioreductive deposition of platinum nanoparticles on the bacterium *Shewanella algae*. J. Biotechnol. 128 (3), 648–653.

Konishi, Y., Tsukiyama, T., Tachimi, T., Saitoh, N., Nomura, T., Nagamine, S., 2007b. Microbial deposition of gold nanoparticles by the metal-reducing bacterium *Shewanella algae*. Electrochim. Acta 53 (1), 186–192.

Kroto, H.W., Heath, J.R., O'Brien, S.C., 1985. RFC & RES. C 60: buckminster-fullerene. Nature 318, 162–164.

Kruis, F.E., Fissan, H., Rellinghaus, B., 2002. Sintering and evaporation characteristics of gas-phase synthesis of size-selected PbS nanoparticles. Mater. Sci. Eng., B 69–70, 329–334.

Kumar, S.A., Ansary, A.A., Abroad, A., Khan, M.I., 2007. Extracellular biosynthesis of CdSe quantum dots by the fungus, *Fusarium oxysporum*. J. Biomed. Nanotechnol. 3 (2), 190–194.

Kumar, S.A., Peter, Y.A., Nadeau, J.L., 2008. Facile biosynthesis, separation and conjugation of gold nanoparticles to doxorubicin. Nanotechnology 19 (49).

Kumar, K.R., Nattuthurai, N., Gopinath, P., Mariappan, T., 2015. Synthesis of eco-friendly silver nanoparticles from *Morinda tinctoria* leaf extract and its larvicidal activity against *Culex quinquefasciatus*. Parasitol. Res.

Lee, J.H., Han, J., Choi, H., Hur, H.G., 2007. Effects of temperature and dissolved oxygen on Se(IV) removal and Se(0) precipitation by *Shewanella* sp. HN-41. Chemosphere 68 (10), 1898–1905.

Lefevre, C.T., Abreu, F., Lins, U., Bazylinski, D.A., 2010. Nonmagnetotactic multicellular prokaryotes from low-saline, nonmarine aquatic environments and their unusual negative phototactic behavior. Appl. Environ. Microbiol. 76 (10), 3220–3227.

Lengke, M.F., Fleet, M.E., Southam, G., 2006a. Morphology of gold nanoparticles synthesized by filamentous cyanobacteria from gold(I)-thiosulfate and gold(III)-chloride complexes. Langmuir 22 (6), 2780–2787.

Lengke, M.F., Ravel, B., Fleet, M.E., Wanger, G., Gordon, R.A., Southam, G., 2006b. Mechanisms of gold bioaccumulation by filamentous cyanobacteria from gold(III)-chloride complex. Environ. Sci. Technol. 40 (20), 6304–6309.

Li, Y., Duan, X., Qian, Y., Yang, L., Liao, H., 1999. Nanocrystalline silver particles: synthesis, agglomeration, and sputtering induced by electron beam. J. Colloid Interface Sci. 209 (2), 347–349.

Lloyd, J.R., Yong, P., Macaskie, L.E., 1998. Enzymatic recovery of elemental palladium by using sulfate-reducing bacteria. Appl. Environ. Microbiol. 64 (11), 4607–4609.

Lovley, D.R., 1993. Dissimilatory metal reduction. Annu. Rev. Microbiol. 47, 236–290.

Madhurima, B., Mahanty, S., Chaudhuri, P., 2017 Fungi mediated biosynthesis of nanoparticles and application in metal sequestration. In: Das, S., Dash, H.R. (Eds.), Handbook of Metal-Microbe Interactions and Bioremediation. CRC Press, Taylor & Francis Group, NY, pp. 423–434.

Mafune, F., Kohno, J.-y, Takeda, Y., Kondow, T., 2000. Structure and stability of silver nanoparticles in aqueous solution produced by laser ablation. J. Phys. Chem. B 104 (35), 8333–8337.

Maliszewska, I., Szewczyk, K., Waszak, K., 2009. Biological synthesis of silver nanoparticles. J. Phys. Conf. Ser. 146.

Mandal, D., Bolander, M.E., Mukhopadhyay, D., Sarkar, G., Mukherjee, P., 2006. The use of microorganisms for the formation of metal nanoparticles and their application. Appl. Microbiol. Biotechnol. 69 (5), 485–492.

Medina-Perez, G., Fernandez-Luqueno, F., Vazquez-Nunez, E., Lopez-Valdez, F., Prieto-Mendez, J., Madariaga-Navarrete, A., 2019. Remediating polluted soils using nanotechnologies: environmental benefits and risks. Pol. J. Environ. Stud. 28 (3), 1013–1030.

Mertens, B., Boon, N., Verstraete, W., 2006. Slow-release inoculation allows sustained biodegradation of gamma-hexachlorocyclohexane. Appl. Environ. Microbiol. 72 (1), 622–627.

Mishra, A., Tripathy, S.K., Wahab, R., Jeong, S.H., Hwang, I., Yang, Y.B., et al., 2011. Microbial synthesis of gold nanoparticles using the fungus *Penicillium brevicompactum* and their cytotoxic effects against mouse mayo blast cancer C_2C_{12} cells. Appl. Microbiol. Biotechnol. 92 (3), 617–630.

Mohammed Fayaz, A., Balaji, K., Girilal, M., Kalaichelvan, P.T., Venkatesan, R., 2009a. Mycobased synthesis of silver nanoparticles and their incorporation into sodium alginate films for vegetable and fruit preservation. J. Agric. Food Chem 57 (14), 6246–6252.

Mohammed Fayaz, A., Balaji, K., Kalaichelvan, P.T., Venkatesan, R., 2009b. Fungal based synthesis of silver nanoparticles—an effect of temperature on the size of particles. Colloids Surf., B: Biointerfaces 74 (1), 123–126.

Mohn, W.W., Tiedje, J.M., 1992. Microbial reductive dehalogenation. Microbiol. Rev. 56 (3), 482–507.

Mubarakali, D., Gopinath, V., Rameshbabu, N., Thajuddin, N., 2012. Synthesis and characterization of CdS nanoparticles using C-phycoerythrin from the marine cyanobacteria. Mater. Lett. 74, 8–11.

Mukherjee, P., Ahmad, A., Mandal, D., Senapati, S., Sainkar, S.R., Khan, M.I., et al., 2001. Bioreduction of $AuCl_4^-$ ions by the fungus, *Verticillium* sp. and surface trapping of the gold nanoparticles formed. Angew. Chemie. Int. Ed. 40 (19), 3585.

Mukherjee, P., Senapati, S., Mandal, D., Ahmad, A., Islam Khan, M., Kumar, R., et al., 2002. Extracellular synthesis of gold nanoparticles by the fungus *Fusarium oxysporum*. ChemBioChem. 5, 461–463.

Mulvaney, P., 2002. Surface plasmon spectroscopy of nanosized metal particles. Langmuir 12 (3), 788–800.

Nam, J.M., Thaxton, C.S., Mirkin, C.A., 2003. Nanoparticle-based bio-bar codes for the ultrasensitive detection of proteins. Science (80) 301 (5641), 1884–1886.

Narayanan, K.B., Sakthivel, N., 2011. Facile green synthesis of gold nanostructures by NADPH-dependent enzyme from the extract of *Sclerotium rolfsii*. Colloids Surf., A: Physicochem. Eng. Asp. 380 (1–3), 156–161.

Narayanan, K.B., Sakthivel, N., 2013. Mycocrystallization of gold ions by the fungus *Cylindrocladium floridanum*. World J. Microbiol. Biotechnol. 29 (11), 2207–2211.

Nealson, K.H., 1997. Sediment bacteria: who's there, what are they doing, and what's new? Annu. Rev. Earth Planet. Sci. 25 (1), 403–434.

Nurmi, J.T., Tratnyek, P.G., Amonette, J.E., Pecher, K., Wang, C., Linehan, J.C., et al., 2005. Characterization and properties of metallic iron nanoparticles: spectroscopy, electrochemistry, kinetics. Environ. Sci. Technol. 39, 1221–1230.

Oremland, R.S., Herbel, M.J., Blum, J.S., Langley, S., Beveridge, T.J., Ajayan, P.M., et al., 2004. Structural and spectral features of selenium nanospheres produced by Se-respiring bacteria. Appl. Environ. Microbiol. 70 (1), 52–60.

Paknikar, K.M., Nagpal, V., Pethkar, A.V., Rajwade, J.M., 2005. Degradation of lindane from aqueous solutions using iron sulfide nanoparticles stabilized by biopolymers. Sci. Technol. Adv. Mater 6 (3–4 Spec. Iss.), 370–374.

Pandey, G., 2018. Prospects of nanobioremediation in environmental cleanup. Orient. J. Chem. 34 (6), 2838–2850.

Pantarotto, D., Partidos, C.D., Hoebeke, J., Brown, F., Kramer, E., Briand, Jean-Paul, et al., 2003. Immunization with peptide-functionalized carbon nanotubes enhances virus-specific neutralizing antibody responses. Chem. Biol. 10, 961–966.

Philip, D., 2009. Biosynthesis of Au, Ag and Au–Ag nanoparticles using edible mushroom extract. Spectrochim. Acta, A: Mol. Biomol. Spectrosc. 73 (2), 374–381.

Pierrat, S., Rai, M., Gade, A., Ingle, A., Sonnichsen, C., 2008. Mycosynthesis of silver nanoparticles using the fungus *Fusarium acuminatum* and its activity against some human pathogenic bacteria. Curr. Nanosci. 4 (2), 141–144.

Pingali, K.C., Rockstraw, D.A., Deng, S., 2005. Silver nanoparticles from ultrasonic spray pyrolysis of aqueous silver nitrate. Aerosol. Sci. Technol. 39 (10), 1010–1014.

Prasad, K., Jha, A.K., 2010. Biosynthesis of CdS nanoparticles: an improved green and rapid procedure. J. Colloid Interface Sci. 342 (1), 68–72.

Rajeshkumar, S., Malarkodi, C., Vanaja, M., Gnanajobitha, G., Paulkumar, K., Kannan, C., et al., 2013. Antibacterial activity of algae mediated synthesis of gold nanoparticles from turbinaria conoides. Der Pharma Chem. 5 (2), 224–229.

Raliya, R., Biswas, P., Tarafdar, J.C., 2015. TiO_2 nanoparticle biosynthesis and its physiological effect on mung bean (*Vigna radiata* L.). Biotechnol. Rep. 5 (1), 22–26.

Ram Kumar Pandian, S., Deepak, V., Kalishwaralal, K., Muniyandi, J., Rameshkumar, N., Gurunathan, S., 2009. Synthesis of PHB nanoparticles from optimized medium utilizing dairy industrial waste using *Brevibacterium casei* SRKP2: a green chemistry approach. Colloids Surf., B: Biointerfaces 74 (1), 266–273.

Rautaray, D., Sanyal, A., Adyanthaya, S.D., Ahmad, A., Sastry, M., 2004. Biological synthesis of strontium carbonate crystals using the fungus *Fusarium oxysporum*. Langmuir 20 (16), 6827–6833.

Riddin, T.L., Gericke, M., Whiteley, C.G., 2006. Analysis of the inter- and extracellular formation of platinum nanoparticles by *Fusarium oxysporum* f. sp. *lycopersici* using response surface methodology. Nanotechnology 17 (14), 3482–3489.

Royal Society, 2004. Nanoscience and nanotechnologies: opportunities and uncertainties. Neuroradiology 46, 618–618. Available from: <http://royalsociety.org/policy/publications/2004/nanoscience-nanotechnologies/>.

Sanghi, R., Verma, P., Puri, S., 2011. Enzymatic formation of gold nanoparticles using *Phanerochaete Chrysosporium*. Adv. Chem. Eng. Sci. 01 (03), 154–162.

Sanghi, R., Verma, P., 2009a. Biomimetic synthesis and characterisation of protein capped silver nanoparticles. Bioresour. Technol. 100 (1), 501–504.

Sanghi, R., Verma, P., 2009b. A facile green extracellular biosynthesis of CdS nanoparticles by immobilized fungus. Chem. Eng. J. 155 (3), 886–891.

Sanyal, A., Ahmad, A., Bharde, A., Sastry, M., Bansal, V., Ahire, K., et al., 2005. Fungus-mediated biosynthesis of silica and titania particles. J. Mater. Chem. 15 (26), 2583.

Sardrood, B., Goltapeh, E., Verma, A., 2013. An introduction to bioremediation. Soil Biol. 32, 203–226 (Chapter 1).

Sasakova, N., Gregova, G., Takacova, D., Mojzisova, J., Papajova, I., Venglovsky, J., et al., 2018. Pollution of surface and ground water by sources related to agricultural activities. Front. Sustain. Food Syst. 2.

Sawle, B.D., Salimath, B., Deshpande, R., Bedre, M.D., Prabhakar, B.K., Venkataraman, A., 2008. Biosynthesis and stabilization of Au and Au-Ag alloy nanoparticles by fungus, *Fusarium semitectum*. Sci. Technol. Adv. Mater. 9 (3).

Saxena, J., Sharma, M., Gupta, S., 2014. Emerging role of fungi in nanoparticle synthesis and their applications. World J. Pharm. Sci. 3 (9), 2278–4357 (Internet).

Shaligram, N.S., Bule, M., Bhambure, R., Singhal, R.S., Singh, S.K., Szakacs, G., et al., 2009. Biosynthesis of silver nanoparticles using aqueous extract from the compactin producing fungal strain. Process Biochem. 44 (8), 939–943.

Shankar, S.S., Ahmad, A., Pasricha, R., Sastry, M., 2003. Bioreduction of chloroaurate ions by geranium leaves and its endophytic fungus yields gold nanoparticles of different shapes. J. Mater. Chem. 13 (7), 1822–1826.

Sharma, N.C., Sahi, S.V., Nath, S., Parsons, J.G., Gardea-Torresdey, J.L., 2007. Synthesis of plant-mediated gold nanoparticles and catalytic role of biomatrix-embedded nanomaterials. Environ. Sci. Technol. 2007 41 (14), 5137–5142.

Singaravelu, G., Arockiamary, J.S., Kumar, V.G., Govindaraju, K., 2007. A novel extracellular synthesis of monodisperse gold nanoparticles using marine alga, *Sargassum wightii* Greville. Colloids Surf., B: Biointerfaces 57 (1), 97–101.

Singh, T., Jyoti, K., Patnaik, A., Singh, A., Chauhan, R., Chandel, S.S., 2017. Biosynthesis, characterization and antibacterial activity of silver nanoparticles using an endophytic fungal supernatant of *Raphanus sativus*. J. Genet. Eng. Biotechnol. 15 (1), 31–39 (Internet).

Sinha, A., Khare, S.K., 2011. Mercury bioaccumulation and simultaneous nanoparticle synthesis by *Enterobacter* sp. cells. Bioresour. Technol. 102 (5), 4281–4284.

Sneha, K., Sathishkumar, M., Mao, J., Kwak, I.S., Yun, Y.S., 2010. Corynebacterium glutamicum-mediated crystallization of silver ions through sorption and reduction processes. Chem. Eng. J. 162 (3), 989–996.

Suresh, A.K., 2012. Introduction to nanocrystallites, properties, synthesis, characterizations, and potential applications. In: L. L. B. (Ed.), Metallic Nanocrystallites and Their Interaction With Microbial Systems. Springer, pp. 1–23.

Sweeney, R., Mao, C., Gao, X., Burt, L., Belcher, A., Geotgiou, G.I.B., 2004. Bacterial biosynthesis of cadmium sulfide nanocrystals. Chem. Biol. 11, 1553–1559.

Tarafdar, J.C., Raliya, R., Rathore, I., 2012. Microbial synthesis of phosphorous nanoparticle from tri-calcium phosphate using *Aspergillus tubingensis* TFR-5. J. Bionanosci. 6 (2), 84–89.

Tarafdar, J.C., Raliya, R., 2013. Rapid, low-cost, and ecofriendly approach for iron nanoparticle synthesis using *Aspergillus oryzae* TFR9. J. Nanopart. 2013.

Thakkar, K.N., Mhatre, S.S., Parikh, R.Y., 2010. Biological synthesis of metallic nanoparticles. Nanomed. Nanotechnol. Biol. Med. 6 (2), 257–262.

Tripathi S., Sanjeevi R., Anuradha J., Chauhan D.S., Rathoure A.K., 2018. Nano-bioremediation: nanotechnology and bioremediation. In: Biostimulation Remediation Technologies for Groundwater Contaminants. IGI Global, pp. 202–219.

Tu, C.M., 1976. Utilization and degradation of lindane by soil microorganisms. Arch. Microbiol. 108 (3), 259–263.

Verma, V.C., Kharwar, R.N., Gange, A.C., 2010. Biosynthesis of antimicrobial silver nanoparticles by the endophytic fungus *Aspergillus clavatus*. Nanomedicine 5 (1), 33–40.

Vigneshwaran, N., Kathe, A.A., Varadarajan, P.V., Nachane, R.P., Balasubramanya, R.H., 2006. Biomimetics of silver nanoparticles by white rot fungus, *Phanerochaete chrysosporium*. Colloids Surf., B: Biointerfaces 53 (1), 55–59.

Vigneshwaran, N., Kathe, A.A., Varadarajan, P.V., Nachane, R.P., Balasubramanya, R.H., 2007. Silver-protein (core-shell) nanoparticle production using spent mushroom substrate. Langmuir 23 (13), 7113–7117.

Wang, Q., Li, S., Jia, P., Qi, C., Ding, F., 2013. A review of surface water quality models. Sci. World J. 2013, 231768.

Wang, Y., Xia, Y., 2004. Bottom-up and top-down approaches to the synthesis of monodispersed spherical colloids of low melting-point metals. Nanoletters 4 (10), 2047–2050.

Watson, J.H.P., Ellwood, D.C., Soper, A.K., Charnock, J., 1999. Nanosized strongly-magnetic bacterially-produced iron sulfide materials. J. Magn. Magn. Mater. 203 (1–3), 69–72.

Whitesides, G.M., 2005. Nanoscience, nanotechnology, and chemistry. Small 1 (2), 172–179.

Xie, J., Lee, J.Y., Wang, D.I.C., Ting, Y.P., 2007. High-yield synthesis of complex gold nanostructures in a fungal system. J. Phys. Chem. C 111 (45), 16858–16865.

Yadav, K.K., Singh, J.K., Gupta, N., Kumar, V., 2017. A review of nanobioremediation technologies for environmental cleanup: a novel biological approach. J. Mater. Environ. Sci. 8 (2), 740–757.

You, Y., Choi, S., Kim, H.J., Kamala-Kannan, S., Oh, B.-T., Velmurugan, P., et al., 2010. Removal of zinc by live, dead, and dried biomass of *Fusarium* spp. isolated from the abandoned-metal mine in South Korea and its perspective of producing nanocrystals. J. Hazard. Mater. 182 (1–3), 317–324.

Yu, C.H., Tam, K., Tsang, E.S.C., 2008. Chapter 5 Chemical methods for preparation of nanoparticles in solution. Handb. Met. Phys. 5, 113–141.

Zhang, X., He, X., Wang, K., Yang, X., 2011. Different active biomolecules involved in biosynthesis of gold nanoparticles by three fungus species. J. Biomed. Nanotechnol. 7 (2), 245–254.

Zhao, X., Zhou, L., Rajoka, M.S.R., Yan, L., Jiang, C., Shao, D., et al., 2017. Fungal silver nanoparticles: synthesis, application and challenges. Crit. Rev. Biotechnol 38 (6), 817–835 (Internet).

Zheng, D., Hu, C., Gan, T., Dang, X., Hu, S., 2010. Preparation and application of a novel vanillin sensor based on biosynthesis of Au-Ag alloy nanoparticles. Sens. Actuators, B: Chem. 148 (1), 247–252.

Zhou, W., He, W., Zhang, X., Yan, S., Sun, X., Tian, X., et al., 2009a. Biosynthesis of iron phosphate nanopowders. Powder Technol. 194 (1–2), 106–108.

Zhou, W., He, W., Zhong, S., Wang, Y., Zhao, H., Li, Z., et al., 2009b. Biosynthesis and magnetic properties of mesoporous Fe_3O_4 composites. J. Magn. Magn. Mater. 321 (8), 1025–1028.

Chapter 16

Microbial systems as a source of novel genes for enhanced phytoremediation of contaminated soils

Sonal Srivastava[1,2], Vandana Anand[1,2], Pallavi Singh[1], Aditi Roy[3], Sushree Pallavi[1], Vidisha Bist[1,2], Jasvinder Kaur[1], Subhi Srivastava[4], Ratna Katiyar[5] and Suchi Srivastava[1,2]

[1]*Division of Microbial Technology, CSIR-NBRI, Lucknow, India,* [2]*Academy of Scientific and Innovative Research, AcSIR, Ghaziabad, India,* [3]*Division of Environmental Sciences, CSIR-NBRI, Lucknow, India,* [4]*Plant Molecular Biology Division, CSIR-NBRI, Lucknow, India,* [5]*Botany Department, Lucknow University, Lucknow, India*

16.1 Introduction

Soil is a complex mixture of minerals, organic material, water, and various life-forms. It serves as a key to sustain life on Earth by supporting agriculture and maintaining ecological balance. In its original state, soil was an uncontaminated substance covering the Earth. However, natural soil environment gets contaminated by the presence of harmful chemicals, namely, industrial activity, agricultural chemicals, or improper disposal of waste. Contamination of soil is considered as one of the major environmental concern because of their mixed nature, nondegradable nature, long-term persistence, and difficulties to remediate. The most common chemicals involved in soil contamination based on the intensity of their occurrence are petroleum hydrocarbons, polycyclic aromatic hydrocarbons (PAHs), pesticides, and heavy metals. Various natural processes and technological advancements result in the excess release of toxic compounds into the biosphere and keep on increasing with time due to human activities including excessive use of pesticides/unsafe agricultural practices, rapid urbanization, industrialization, and deforestation. The excessive presence of different agrochemicals and heavy metals in soil led to their absorption in crops and their entry into food chain. Contamination of soil not only affects the soil diversity and quality but also poses a severe effect on agricultural production and human health and becomes a serious problem worldwide due to geologic and anthropogenic activities (Kong and Glick, 2017; Necibi and Mzoughi, 2017).

Contamination of soil with toxic metals is mainly caused by the burning of fossil fuels, manufactured waste, and the indiscriminate use of chemicals leading to contamination with PAHs. Low concentrations of certain transition metals such as cobalt, copper, nickel, and zinc are essential for many cellular processes of bacteria. However, higher concentrations of these metals often are cytotoxic. Other heavy metals, including lead, cadmium, mercury, silver, and chromium, have no known beneficial effects on bacterial cells and are toxic even at low concentrations (Nies, 2004). Natural occurrence of these metals and metalloids (Pb, Ni, and As) are very low; however, their emission due to anthropogenic activities is shown to be several times higher than its natural emissions. Increased acidification due to excess fertilization leads to enhanced solubility of these toxic metals in soil and water. These metals and metalloids enter into soils and water bodies causing contamination via many processes including atmospheric deposition from industrial activities; waste disposal, industrial byproducts; utilization of fertilizers, lime, or agrochemicals.

Use of pesticides and agrochemicals (herbicides, insecticides, fungicides, nematicides, fertilizers) became dominant in worldwide agriculture systems for a notable increase in crop yield and food production. Since 1940 by the introduction of synthetic insecticides, organo-chlorine (OCl) was the first insecticide to be used for pest management (Gill and Garg, 2014). The worldwide consumption of pesticide has reached to about 2 million tons per year and India contributes to about 4% of world's total pesticide use with an average pesticide usage rate of 0.5 kg/ha. The excess use of pesticide leads to its persistence in soil and water and raising the risk for living organisms and humans due to lack of their

Microbe Mediated Remediation of Environmental Contaminants. DOI: https://doi.org/10.1016/B978-0-12-821199-1.00016-X

selectivity. There have been many cases of human intoxication due to poising effect of these synthetic chemicals on human health. It was estimated that excessive exposure and inappropriate use of toxic chemicals kill ~355,000 people every year. (Carvalho, 2017; World Health Organization, 2012; World Health Organization, 2017).

Different remediation approaches have been used to remediate the soil from this inorganic and organic pollutant in order to have safe food for growing population. The conventional methods for the remediation of contaminated sites commonly involve the physical and chemical methods with some limitations such as physical remediation strategies are expensive, nonspecific, and often make the soil unsuitable for agriculture and other uses by disturbing the microenvironment by generating huge amount of toxic waste. Therefore it is imperative to develop agro-biotechnological approaches for decontamination of soil and this includes more eco-friendly and sustainable approaches such as phytoremediation, microbe-assisted phytoremediation, also known as bioremediation to remove contaminants from polluted sites (Mosa et al., 2016).

Phytoremediation refers to a diverse collection of plant-based technologies that use either naturally occurring (hyperaccumulators) or genetically engineered (GE) plants to clean contaminated environments (Flathman and Lanza, 1998) and contain necessary genetic, biochemical, and physiological characteristics for the removal of contaminant from soil and water. Phytoremediation is a cost-effective, green-clean technology with long-term applicability for the cleaning up of contaminated sites, but this process requires longer time to remove contaminants. Therefore microbe-assisted bioremediation, which involves root-colonizing microbes for the removal of organic and inorganic pollutants, has been used (Pilon-Smits, 2005). This technology involves the use of plants (phytoremediation), plant–microbe interactions (rhizoremediation), and microbial communities involving stimulation of viable native microbial population (biostimulation), artificial introduction of viable population (bioaugmentation), bioaccumulation (live cells), and use of dead microbial biomass (biosorption) to clean up the hazardous material from contaminated sites, thereby keeping ecosystem in its original form (Ojuederie and Babalola, 2017). The development of such strategy acts as an alternative tool for the abatement of a contaminated medium and is important for environmental conservation and human health. Bioremediation is a simple process and it can be applied over large areas and more eco-friendly is and inexpensive than other conventional methods (Barzanti et al., 2007). In the present scenario, deployment of GE organisms is one of the interesting alternative techniques for the removal of toxic waste from soil (Pandey et al., 2018) due to the limitation of bioremediation. Many naturally or genetically modified microorganisms have the ability to degrade, transform, or chelate various toxic chemicals and provide better strategies to combat environmental pollution (Dixit et al., 2015). Therefore the development of a new genetic tool through transformation of microbial- and plant-based gene provides advancement for improved remediation strategies.

16.2 Different soil contaminants and their impact on human health

Different soil contaminants profoundly present in soil and are known to have detrimental effects on plant and human health mainly include heavy metals [Cd, Cr(VI), Pb, Co, Cu, Pb, Ni, Se, Zn], polychlorinated biphenyls (PCBs), PAHs, chlorinated pesticides and organophosphate insecticides (e.g., parathion), and excess of nutrients (nitrate, ammonium, phosphate) used as chemical fertilizers. However, some sites are contaminated with other classes of contaminants such as explosives, radionuclides, chlorinated solvents, and petroleum hydrocarbons. Some of the commonly present contaminants in soil are as detailed in the following subsections.

16.2.1 Heavy metals

Heavy metals are the naturally occurring trace elements that cause environmental contamination mainly through industrial, agricultural, atmospheric, pharmaceutical, and domestic effluents. These metals show their toxicity even at their low concentrations such as arsenic, lead, cadmium, mercury, chromium, zinc, selenium. Increasing contamination due to heavy metals in our resources is currently an area of concern as a huge part of it results from anthropogenic activities such as industrial discharge, mining, and smelting operations and various domestic and agricultural uses of metals without any adequate treatment (Tchounwou et al., 2012). Plants require several heavy metals for normal growth; however, excess accumulation of these metals in plants severely affects their metabolism. These heavy metals are accumulated in plants due to their excessive presence in soil and water bodies, causing problem of phytotoxicity and nutrient deficiency (Prasad, 2008; Hall, 2002; Adams et al., 2013). They affect the soil and water quality with limited biodegradation (Anyasi and Atagana, 2011; Siddiquee et al., 2015). Presence of heavy metal inhibits soil respiration, nitrogen mineralization, and nitrification (Sobolev and Begonia, 2008), resulting in reduction of cultivable land (Oves et al., 2016). Heavy metals are transferred through food chain via irrigation with industrial wastewater and thus cause high risk to human health (Mahmood and Malik, 2014). Heavy metals bioaccumulation causes toxic effect in humans when consumed above the bio-recommended limits that could be toxic (acute, chronic, or subchronic), neurotoxic, carcinogenic, mutagenic, or teratogenic. When the volatile form of heavy metals is inhaled at their toxic levels, specific signs and symptoms occur, which include gastrointestinal diseases, diarrhea, stomatitis, tremor, hemoglobinuria, ataxia, paralysis, vomiting and convulsion, depression, and pneumonia (McCluggage, 1991). Toxicity of different heavy metals has been discussed in the following subsections.

16.2.1.1 Arsenic

Arsenic is a ubiquitous naturally occurring metalloid and is widely distributed in Earth's crust ranging from trace levels up to hundreds of milligrams per kilogram. Naturally, it occurs in all living matters including rocks, soil, water, air, plants, and animals. Most of the arsenic released in the environment is through natural activities, namely, the erosion of rocks, forest fires, and volcanic activities. Arsenic contamination includes erosion of local rocks, industrial effluents, various commercial processes, and combustion of fossil fuels. Arsenic concentration in the environment is also influenced by both natural processes, such as weathering, mineral leaching, and biological activity, as well as anthropogenic activities, such as mining, combustion of fossil fuels, and its use in pesticides, herbicides, and wood preservation (Bhattacharya et al., 2002). Its concentration in soil ranges from 0.1 to 55 mg/kg with an average baseline concentration of 7.2 mg/kg.

Arsenic is also present as contaminants in aquatic and marine water bodies and the level of arsenic is lowest in seawater, higher in rivers and lakes, and highest in underground water in areas containing volcanic rock or arsenic-rich minerals. Freshwater systems also contain arsenic contamination, with higher concentrations present in groundwaters than the surface waters (Smedley and Kinniburgh, 2013). Groundwater supply wells in the United States were reported to have arsenic concentration exceeding 10 μg/L (Welch et al., 2000). Other regions as Bangladesh and the Xinjiang province of China have severe arsenic contamination in groundwater with more than 30% of groundwater arsenic concentrations ranging from 50 to 1860 μg/L (Bhattacharya et al., 2002). Arsenic contamination in freshwater causes its accumulation in aquatic biota and thus enters the food chain. The total arsenic concentrations observed in the tissues of shellfish ranges from 0.2 to 125.9 μg/g dry weight (Liao et al., 2005).

Arsenic speciation and toxicity in the environment

Arsenic is present in environment in both inorganic (arsenate and arsenite) and organic forms (monomethylarsonic acid and dimethylarsonic acid) where inorganic forms are more toxic than organic. Arsenic is immobile under toxic conditions and neutral pH due to sorption or coprecipitation with metal oxides through surface complexation reactions and the formation of inner-sphere complexes (Bauer and Blodau, 2006). Although theoretically concentration of arsenite is insignificant due to its thermodynamic instability, it is detected both under anoxic and oxic conditions through natural and man-made sources (Rhine et al., 2006; Smedley and Kinniburgh, 2013).

The toxicity of arsenic depends on the several factors such as their physical and gaseous state, particle size, cell adsorption and elimination rate, the nature of chemical substituent's in the toxic compound (Anderson and Cook, 2004), and on the basis of their toxicity level, different forms of arsenic are ordered as follows: volatile arsenic > inorganic arsenite > organic arsenite > inorganic arsenate > organic arsenate > free arsenic (Mandal and Suzuki, 2002). Arsenic accumulation leads to low production of adenosine triphosphate (ATP), it coagulates protein, forms complexes with coenzymes, and inhibits the production of ATP during respiration [Institute of Environmental Conservation and Research (INECAR), 2000]. Arsenic toxicity leads to a disorder similar to Guillain–Barre syndrome, an antiimmune disorder causing nerve inflammation and muscle weakness [Kantor, 2006; National Institute of Neurological Disorders and Stroke (NINDS), 2007].

16.2.1.2 Lead

Lead is reported as the second major hazardous metal after arsenic due to its frequency of occurrence, toxicity, and human exposure by Agency for Toxic Substances and Disease registry (ATSDR) (2005). Environmental lead contamination occurs due to consumption of food and water, mining and smelting of lead associated ores, automobile exhausts, lead-containing paints, diesel fuels and explosives, and effluents from battery industries (Verma and Dubey, 2003; Sharma and Dubey, 2005). Lead contamination also leads to severe plant toxicity and the results are reduced growth, blackening of root system, chlorosis, reduced photosynthesis rate, and mineral uptake by disturbing the membrane stability and permeability (Emamverdian et al., 2015). Lead contamination affects the development of brain and nervous system of young children. Lead poisoning in humans causes inhibition of the hemoglobin synthesis; dysfunctioning of kidneys, joints, and reproductive systems, cardiovascular system; and acute and chronic damage to the central nervous system and peripheral nervous system (World Health Organization, 2017).

16.2.1.3 Mercury

Mercury has been reported as one of the hazardous metalloids by the Agency for Toxic Substances and Disease Registry (ATSDR) (2007) due to its toxicity, mobility, and long-time residence in the atmosphere. Mercury exerts its toxic effects when it accumulates in human beings, animals, and plants. Humans are exposed to different forms of Hg and several health risks have been reported such as nervous system damage, dental amalgams, neurological disorder in newborns and child. In plants the higher concentrations of metal accumulation have shown to be toxic and affect various physical and biochemical processes such as chlorophyll and carotenoid synthesis, growth, nodulation for legume crops (Feng et al., 2010).

16.2.1.4 Uranium

Uranium (Ur) is a naturally occurring radioactive element that is present in nearly all rocks and soils (Du Preez, 1989; Beach, 1984). It may be redistributed in the environment by both anthropogenic and natural processes. The three primary industrial processes that cause this redistribution are mining, milling, and processing of uranium ores or uranium end products. Uranium can undergo oxidation–reduction reactions in the environment or microbial reactions to form complexes with organic matter (Premuzic et al., 1985). Since all three forms of the naturally occurring uranium isotopes have very long half-lives, that is why the only mechanism for decreasing activity of uranium is radioactive decay (Beach, 1984). Chronic exposure of uranium radionuclide in drinking water causes potential health risk to animals including human (Blantz et al., 1985). The principal sites for uranium deposition are kidney, liver and bones, where it accumulates and leads to chemical and radioactive effect. The severity of Ur toxicity in human beings depends on the route of exposure, particle solubility, contact time, and route of elimination [Agency for Toxic Substances and Disease Registry (ATSDR), 2007].

16.2.1.5 Nickel

Nickel (Ni) is a major environmental contaminant due to increasing industrial activities, burning of fossil fuels, incorrect disposal of sewage sludges, excessive use of manure, and chemical-based fertilizers and pesticide (Lešková et al., 2019). It is required in very low amounts by most of the plant species (ranging between 0.01 and 10 μg Ni per g dry weight) (Marschner, 2011), but above the threshold limit, it affects plant growth and other physiological and biochemical processes such as disturbed water balance, inhibition of photosynthesis and increased oxidative stress (Pandey and Sharma, 2002; Gajewska and Skłodowska, 2007).

16.2.1.6 Cadmium

Cadmium usually occurs as a toxic metal and its concentration in water and soil is increasing day by day due to both natural and anthropogenic activities. The probable source of cadmium transfer to plants is through both natural (coal combustion, volcanic activities, steel production, etc.) and industrial (mining, textile, smelting, etc.) sources causing their discharge to agricultural fields (Mahajan and Kaushal, 2018). Cadmium being toxic at a very low level causes several bone disorders, namely, osteomalacia, osteoporosis, and spontaneous fractures, affects reproductive and excretory systems, increased blood pressure, and myocardial dysfunctions. High exposure also leads to obstructive lung disease like cadmium pneumonitis, characterized by chest pain, cough with foamy and bloody sputum, and death of the lining of the lung tissues.

16.2.2 Polycyclic aromatic hydrocarbons

PAHs are a group of organic compounds with two or more fused aromatic rings (Lee et al., 2010). It is a predominant environmental pollutant that occurs naturally or due to anthropogenic activities (Abdel-Shafy and Mansour, 2016; Singh and Haritash, 2019). Naturally, PAHs are formed through forest fires, volcanic actions, synthesis of bacteria and algae, petroleum discharge, corrosion of sedimentary rocks containing petroleum products, and decomposition of vegetative wastes (Gitipour et al., 2018), whereas anthropogenic sources of PAHs include incomplete combustion, automotive emissions, smoke from wood-burning stoves, jet aircraft exhausts, cigarette and cigar smoke, petroleum product spills, sewage sludge, and waste materials (Haritash and Kaushik, 2016). They are highly stable contaminants demonstrating recalcitrant nature of soil and difficult to degrade as compared to other organic contaminants due to their hydrophobic nature (Gan et al., 2009; Singh and Haritash, 2019). Their commercial products are used in pharmaceuticals, agricultural products, photographic products, thermosetting plastics, lubricating materials, and other chemical industries (Singh and Haritash, 2019). Among different PAHs; naphthalene, anthracene, and phenanthrene belong to low molecular weight, which consists of two or three benzene rings (Abdel-Shafy and Mansour, 2016), whereas pyrene, chrysene, and benzo(a)pyrene are high molecular weight PAHs composed of more than three benzene rings (García-Alonso et al., 2008). The presence of PAH affects all life forms showing its toxicity, mutagenicity, and carcinogenicity effects by altering soil characteristics (particle size, porosity, and water holding capacity and affecting the microbial diversity). Binding of PAHs with soil depends on soil properties, fine grain of silts and clay easily bind with it because of higher surface area and the hydrophobic nature (Magi et al., 2002; Smith et al., 2006; García-Alonso et al., 2008; Saba et al., 2011; Mao and Guan, 2016). Acidic soil that is rich in humic acid and fulvic acid binds preferably with PAHs resulting in higher and persistent contamination (Saba et al., 2011).

Different PAHs have been identified to cause adverse health effects to humans especially cancer (Sreelal and Jayanthi, 2017, Abed et al., 2015). The most significant health effect to be expected is an excess risk of lung cancer through their inhalation (Pandey et al., 2012). The major exposure to PAHs is from breathing ambient and indoor air, eating contaminated food, smoking cigarettes or breathing smoke from open fireplaces (Kiamarsi et al., 2019). Some crops, such as wheat, rye, and lentils, may synthesize PAHs or absorb these *via* water, air, or soil (Saba et al., 2012). Water can also contain certain amounts of PAHs due to leaching of chemicals from the soil into water or from industrial discharge and marine accidental oil leakage during shipment (Breedveld and Sparrevik, 2000). Due to high lipophilicity, their bioavailability after ingestion and inhalation is significant, particularly in organs that are rich in adipose tissue (Gitipour et al., 2018). Several health concerns occur due to their long-term exposure such as decreased immune function, vision problems, kidney and liver damage (e.g., jaundice), and respiratory diseases such as asthma and pulmonary dysfunction (Singh and Haritash, 2019). Therefore, there is need to remove these contaminants from polluted soil site in order to have safer food and better health for growing population.

16.2.3 Hexachlorocyclohexane

Another major contaminant known as hexachlorocyclohexane (HCH) is commercially produced as chlorinated benzene, consisting of mainly five stable isomers of HCH—α (60%–70%), β (5%–12%), γ (10%–12%), δ (6%–12%), and ε (3%–4%) (Lal et al., 2010)—and are used as pesticides since Green Revolution. It is a white crystalline, solid, highly stable in light, air, heat, carbon dioxide, and acids which has been declared as a persistent organic pollutant (Vijgen et al., 2011; Abhilash and Singh, 2008). The predominant contamination of HCH presumably takes place from three sources: (1) industrial production site, (2) application in field, and (3) abandoned sites and obsolete chemicals. Presence of high dose of HCH at the dump site and its continuous discharge into water, soil, animal milk, etc. poses a serious environmental threat. The presence of HCH isomers in such elevated concentrations in soil could cause long-term eco-toxicological damage to the inhabitants. γ-HCH (lindane) also known as gammaxene having insecticidal property has various uses as agricultural insecticide, fumigant for pest control and for pharmaceutical treatment of pets (Nayyar and Lal, 2016). Due to lipophilic nature, HCH easily accumulates in animal bodies and has been considered as potential carcinogen (Pal et al., 2005). The most probable route of lindane exposure in humans is ingestion of food-containing insecticides. The α-, β-, and δ-HCH isomers are more problematic than α-HCH (lindane) due to the presence of major portion of technical HCH, whereas β- and δ-isomers are considered as more stable than the γ-HCH (Quintero et al., 2005). The stability of PAHs depends on the presence of different chlorine atom orientation because it provides the active sites for enzymatic degradation, and it is generally found that γ- and α-HCH are more easily biodegradable than the δ-isomer due to presence of more equatorial chlorine atoms (Salam and Das, 2012). Lindane volatilization gets affected by soil moisture content facilitated by the capillary action (Paraiba and Spadotto, 2002; Chen et al., 2015).

16.3 Tool for detoxification of contaminated soil

Remediation of different soil contaminant is a global concern due to adverse risk to public health (Mao and Guan, 2016). Remediation techniques involve the removal or alteration of the contaminant through various physical, chemical, thermal, and biological means either in situ or ex situ conditions (Chaineau et al., 2005). Among different approaches, phytoremediation has been considered as an emerging technology that involves the potential use of selected plant species to remove and degrade the contaminants from the polluted soils.

16.3.1 Phytoremediation

The term "phytoremediation" is derived from Greek (*phyto*) and Latin words (*remedium*), which stands for restoring balance using plants. The process depends on the natural ability of plants to bioaccumulate, degrade harmful pollutants in the soil, air, and water from ancient time (Chaney et al., 1997). Plants that are used for phytoremediation are the hyperaccumulators, which accumulate particular trace elements of metal and metalloid in their living tissues up to 1%–2% of total dry weight (Van der Ent et al., 2013; Shrivastava et al., 2019). Due to the unique properties of the hyperaccumulator plants, they have been extensively used for extreme adaptations toward heavy metal stresses. Some extremely important properties of hyperaccumulator plants are high growth efficiency with better above-ground biomass, well-distributed root system, high accumulation, and tolerance ability of heavy metals (Cu, Ni, Cd, Ni, Zn, As, Mn, Fe, Co, U, Cr, and Pb) based on their bioavailability (Prasad, 2003), detoxification to the metals and nonmetals, and better translocation to the shoot system from the root system. Three types of plants reported for waste-depuration processes are Graminae in buffer zones, trees such as poplar or willow in riparian zones, and aquatic plants due to high evapotranspiration capacities. Approximately 721 plant species belonging to 52 families have been identified as hyperaccumulator as per the July 2017 database records. Highest number of plant species (523) have been reported for Ni, 53 for copper, 42 for cobalt, 1 for chromium, 42 for manganese, 20 for zinc, 41 for selenium, 7 for cadmium, 5 for arsenic, and 8 for lead for phytoremediation purposes. These plants chiefly belong to family Brassicaceae (83 species) and Phyllanthaceae (59 species). Table 16.1 represents the family and genera of plants categorized as hyperaccumulator for heavy metals (Reeves et al., 2018). Phytoremediation applicable to large areas of low-to-moderate surface soil (0–3 ft.) can use several species at same site, simultaneously or subsequently, to remove more than one contaminant. Despite various procedures for remediation of contaminants, phytoremediation via hyperaccumulator has been proved to be an effective tool.

The hyperaccumulators such as *Solanum nigrum*, *Populus* spp., *Salix calodendron*, *Arabis paniculata*, and *Siegesbeckia orientalis* are potential candidates for the removal of cadmium (Zhang et al., 2013a). The potential hyperaccumulators for Pb, Ni, Cu, and As are *Cannabis sativa* and *Zea mays*, since they can accumulate high concentration of these metals in their tissues and also have higher biomass production (Seleiman et al., 2013). *Alyssum murale* is the well-known hyperaccumulator of nickel, accumulates Ni in exceedingly high level up to 2% (W/W) (Reeves et al., 2018). *Alyssum bertolonii* and *A. murale* have an increased proportion of bacteria with metal resistance (Barzanti et al.,

TABLE 16.1 Important families and genera of hyperaccumulator plants.

Elements	Families	Genera	References
Arsenic (As)	Pteridaceae	*Pteris, Pityrogramma*	Reeves et al. (2018)
Cadmium (Cd)	Brassicaceae, Crassulaceae	*Noccaea, Sedum, Brassica*	Ehsan et al. (2014)
Copper (Cu)	Asteraceae, Commelinaceae, Fabaceae, Lamiaceae Linderniaceae, Malvaceae	*Anisopappus, Commelina*	Kim and Lee (2005)
Cobalt (Co)	Asteraceae, Lamiaceae, Linderniaceae, Orobanchaceae, Phyllanthaceae	*Anisopappus, Phyllanthus, Presicaria*	Lee et al. (2010)
Manganese (Mn)	Myrtaceae, Celastraceae, Proteaceae	*Gossia, Denhamia, Virotia, Cichorium intybus, Zea mays, Medicago sativa*	Poniedziałek et al. (2005)
Nickel (Ni)	Asteraceae, Brassicaceae, Buxaceae, Cunoniaceae, Phyllanthaceae, Salicaceae, Violaceae	*Alyssum, Buxus, Berkheya, Glochidion, Geissois, Homalium, Hybanthus, Phyllanthus, Leucocroton, Senecio, Xylosma*	Abou-Shanab et al. (2007)
Zinc (Zn)	Brassicaceae, Crassulaceae	*Arabidopsis, Noccaea, Sedum*	Gasic and Korban (2007)
Lead (Pb)	Brassicaceae	*Noccaea*	Luo et al. (2019)
Selenium (Se)	Fabaceae	*Astragalus, Stanleya*	Lindblom et al. (2013)
Thallium (Tl)	Brassicaceae	*Biscutella, Iberis*	Leblanc et al. (1999)

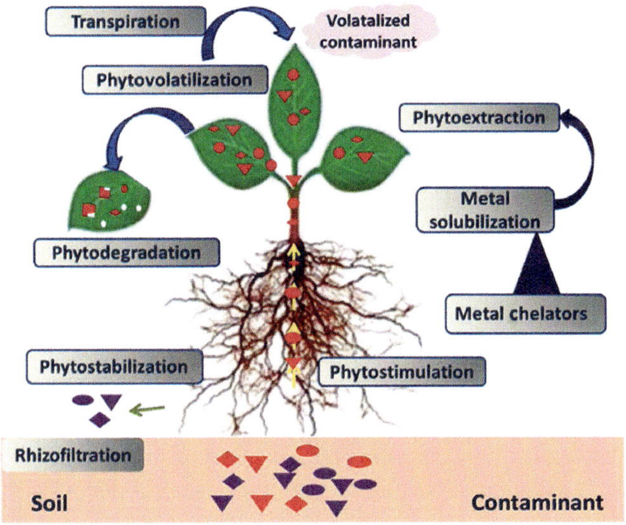

FIGURE 16.1 Schematic representation of different phytoextraction processes.

2007). Two Ni-hyperaccumulating plant species *A. murale* has received the most attention due to their remarkable ability of Ni extraction and accumulation from soils (Abou-Shanab et al., 2003). *Arabidopsis halleri* is a naturally selected zinc and cadmium tolerant Zn hyperaccumulator (Becher et al., 2004).

Among different processes for degradation of PAHs, phytoremediation has been considered as an appropriate in situ technique. Different types of plants and grasses are found to degrade different PAHs such as tall fescue (*Festuca arundinacea*), ryegrass (*Lolium multiflorum*, *Lolium perenne*), orchard grass (*Dactylis glomerata*), red fescue (*Festuca rubra*), yellow sweet clover (*Melilotus officinalis*), birdsfoot-trefoil (*Lotus corniculatus*), red clover (*Trifolium pratense*), and white clover (*Trifolium repens*) (Smith et al., 2006; Rezek et al., 2008). Enhanced removal of phenanthrene and anthracene in soils by phytoremediation with *Brassica rapa* has been reported (Cameselle and Gouveia, 2019). The mechanism of phytoremediation involves five basic techniques: (1) rhizofiltration, (2) phytoextraction, (3) phytotransformation, (4) phytostimulation, and (5) phytostabilization. The details of each process are discussed in the following subsections (Fig. 16.1).

16.3.1.1 Rhizofiltration

Rhizofiltration, a hydroponic-based green technology, implements the use of aquatic and land plants for water treatment by absorbing, concentrating, precipitating heavy metals from wastewater. Heavy metals such as lead, cadmium, cobalt, copper, uranium, arsenic are too low for efficient removal through conventional methods but can be efficiently removed through this technique. The method is cost-effective and sustainable as it involves the use of plant roots to absorb and concentrate heavy metals from polluted aqueous streams. It is a plant-based technique where hydroponically grown plants' roots act as filters to remove the metal contaminants by absorbing and precipitating the chemicals from aquatic environments (López-Chuken, 2012). Root surface provides efficient area for the absorption of the metals or contaminants (copper, lead, cadmium, zinc, chromium, nickel) into the root tissues. Both aquatic and terrestrial plants are used for rhizofiltration. A few examples of aquatic plants usually engaged in rhizofiltration are water hyacinth (*Eichhornia crassipes*, Mahmood et al., 2010); duckweed (*Lemna minor*, Hou et al., 2007), whereas terrestrial plants are tobacco (*Nicotiana tabacum*), sunflower (*Helianthus annuus*), spinach (*Spinacia oleracea*), rye (*Secale cereale*), and Indian mustard (*Brassica juncea*). Terrestrial plants are widely preferred for rhizofiltration as they have long, fibrous root system with fast growth and increased surface area (López Chuken et al., 2010). Eapen et al., (2003) developed the hairy root cultures of *Brassica juncea* and *Chenopodium amaranticolor* through genetic transformation of *Agrobacterium rhizogenes* and their study showed that *B. juncea* could take up 20-30% of Uranium while *Chenopodium amaranticolor* followed a slow trend, uptake of 13% from the solution. Thereby, depicted that hairy roots could remove uranium from the aqueous solution over short period of time.

16.3.1.2 Phytodegradation

Phytodegradation is the process that involves the reductive and oxidative enzymes released through the root zone for the degradation of the organic contaminants. Phytodegradation, also referred as rhizodegradation, involves the process of biodegradation by the rhizospheric bacteria. The organic contaminants are taken up by the root through metabolic process and metabolized into a less toxic form. Many different compounds and classes of hazardous compounds can be removed by this method, including bisphenol A (BPA), a volatile organic compound reported to be a groundwater contaminant by the use of *Azolla filiculoides*, which can remove about 60%–90% (Zazouli et al, 2014). Poplar tree (*Populus* spp.) has been used successfully in phytodegradation of toxic organic compounds.

Phytodegradation is also referred as phytotransformation where the organic contaminants are degraded or transformed into the simpler compounds inside the plant tissue, which in turn permits the plant growth. Remediation of the contaminants through phytotransformation directly depends on the uptake of contaminants into the plant root tissues and their efficient removal by the process of transpiration.

16.3.1.3 Phytoextraction

Plants have been used as the sustainable strategy for the remedial of contaminants present in the soil and water. Phytoextraction is the procedure where the plants are able to accumulate contaminants above ground. The two promising factors that determine the efficiency of phytoextraction is the ability of plant to hyperaccumulate the metals and the biomass production (Lombi et al., 2000). The hyperaccumulators should have the following properties: (1) tolerance to high concentrations of metals, (2) accumulation of high concentration of metals in their aerial tissues, (3) rapid growth of the plants, (4) high biomass production, (5) profuse root system, and (6) easy to cultivate and easily harvestable plants. Phytoextraction basically involves the repeated harvesting of the biomass of the plants in order to lower the concentration of the contaminants present in the soil. Phytoextraction can exist in two different forms: continuous and induced process. Continuous process of phytoextraction can be described as the ability of the plants to accumulate the soil contaminants (toxic metals) steadily into their biomass; an induced process is use of chemicals to increase the bioavailability of the contaminants to the soil. The process of phytoextraction can only be successful if there is high biomass of plant species and the bioavailability of metals for the plant uptake. Hence, the efficiency of phytoextraction can be improved by the use of fast-growing plants that subsequently removes the contaminants by harvesting. These plants can accumulate the metals or metalloids without any toxic effect. The hyperaccumulating plants can contain more than 1% of metal in their biomass; if the accumulated heavy metals are translocated up to the shoot, then the traditional farming can be used for harvesting in order to prevent dead leaf and decomposed matter of plants to disperse in the soil. The metals can be recovered from the soil by incinerating the plants through the process of phytomining.

Phytomining restores the mined-degraded lands by planting hyperaccumulator plant species. A number of hyperaccumulators endemic to the metalliferous soil are responsible for the accumulation of metals such as nickel, cobalt, copper, and lead to the incomparable amount. Despite various procedures for remediation of contaminants, phytoremediation via hyperaccumulator has been proved to be an effective tool. The hyperaccumulators such as *S. nigrum*, *Populus* spp., *S. calodendron*, *A. paniculata*, and *S. orientalis* are the potential candidates for removal of cadmium (Zhang et al., 2013b). The potential hyperaccumulator for Pb, Ni, Cu, and As are *C. sativa* and *Z. mays*, since they can accumulate high concentration of these metals in their tissues and also have higher biomass production (Seleiman et al., 2013). *A. murale* is the well-known hyperaccumulator of nickel, accumulates Ni in exceedingly high level up to 2% (W/W) (Reeves et al., 2000). *A. bertolonii* Desv. were isolated from roots, stems, and leaves of *A. bertolonii* (Brassicaceae), potential helper plant growth, and Ni hyperaccumulator (Barzanti et al., 2007).

16.3.1.4 Phytostabilization

Phytostabilization is the process that retains the contaminants in the soil and stabilizes it in the rhizospheric regions. It reduces the mobility of heavy metals in the soil and immobilizes them in the roots near the soil, thus reducing the off-site contaminants. The process involves the reduction of the contaminants through the transpiration and elongation of root growth. The contaminants are reduced through the process of leaching and creating an aerobic environment in the root zone. Phytostabilization, also referred as phytosequestration, involves absorption by roots, adsorption on the root surface, and the sequestration, precipitation, and immobilization by the exudates released in the rhizospheric regions. The pollutant source containing the organic matters associated with the soil is degraded with the help of microbes present in the soil (Bolan et al., 2011). The low fertility soil can be stabilized by the immobilization of metals or metalloids in combination with the plants such as *Agrostis tenuis* cv. Goginan and cv. Parys and *F. rubra* cv. Merlin, which are characterized as high metal-tolerant grass species. Phytostabilization requires regular monitoring of the soil in order to check the immobilization of the soil contaminants.

16.3.1.5 Phytovolatilization

Plants transport variety of environmental pollutants, volatile compounds from atmosphere. This process involves the uptake of volatile compounds from plant roots and releases into the atmosphere through the process of evapotranspiration either in the form of same volatile compounds or their byproducts. Phytovolatilization has been divided into two types— direct and indirect. Direct phytovolatilization is the volatilization of organic compounds from the stems and leaves; indirect phytovolatilization is volatilization from the soil due to involvement of root activities (Limmer and Burken, 2016).

16.3.1.6 Advantages and limitations of phytoremediation

Phytoremediation is the process to clean up contaminated soil and groundwater, using plants and plant-associated microbes with their natural abilities to take up, accumulate, and/or degrade constituents of their soil and water environments (Sharma et al., 2018). The foremost advantages reported for phytoremediation as compared to traditional remediation technologies comprise the generation of less secondary wastes, and low environmental disturbances. The advantages of phytoremediation also include decline in dispersed contaminants and dust by wind treatment, decline in runoff contaminants, decline in leached soil contaminants.

The disadvantages of phytoremediation include the long time period that usually requires several growing seasons (3–5 years) of the plants especially crop plants. In addition, heavy farming tools and machineries are required for large-scale cleanup operations, which are usually distant from the contaminated urban areas (Mudhoo, 2011). The other limitations involve the dependency of phytoremediation on the local climate, depth of the treatment zone as it is limited to the shallow soil, tolerance level of plants for contaminants and other toxins, efficiency for remediating varied contaminants and the possibility of these plants entering in the food chains through animal consumption.

16.3.2 Bioremediation

Microbe-based bioremediation is an attractive and novel technology of using microbial communities involving stimulation of viable native microbial population (biostimulation), artificial introduction of viable population (bioaugmentation), bioaccumulation (live cells), and use of dead microbial biomass (biosorption) to clean up the contaminated sites. Microbial survival in polluted soils depends on intrinsic biochemical and structural properties, physiological, and/or genetic adaptation including morphological, changes of cells, as well as environmental modifications of metal speciation (Wuertz, 1997). Metal resistance of microbes involves expolymer binding, siderophore, and biosurfactant complexation, metal precipitation, and metal-dependent mechanism of metal resistance (Nies, 2004). These mechanisms may be encoded by chromosomal genes, or plasmids conferring resistance (Cervantes and Gutierrez-Corona, 1994; Wuertz, 1997). The basic principles of bioremediation involve reducing the solubility of environmental contaminants by changing pH, the redox reactions, and adsorption of contaminants from polluted environment (Jain and Arnepalli, 2019).

16.3.2.1 Heavy metals

Microorganisms adopt different mechanisms to interact and survive in the presence of inorganic metals. Various mechanisms evolved by microbes to survive against metal toxicity are biotransformation, extrusion, use of enzymes, exopolysaccharide production (Dixit et al., 2015), and synthesis of metallothioneins (MTs). In response to metals in the environment, microorganisms have developed ingenious mechanisms of metal resistance and detoxification. The mechanism involves several procedures, together with electrostatic interaction, ion exchange, precipitation, redox process, and surface complexation (Yang et al., 2015). The major mechanical means to resist heavy metals by microorganism are metal oxidation, methylation, enzymatic degradation, metal complexion, metal-ligand degradation, metal efflux pumps, demethylation, intracellular and extracellular metal sequestration, exclusion by permeability barrier, and production of metal chelators such as MTs and biosurfactants (Ramasamy and Banu, 2007). A metal-tolerant *Rahnella* sp. with plant growth-promoting attributes from roots of Mn-hyperaccumulator *Polygonum pubescens* has been found to tolerate very high Cd, Pb, and Zn and effectively solubilized $CdCO_3$, $PbCO_3$, and $Zn_3(PO_4)_2$ (He et al., 2013). Endophytic bacteria identified as *Bacillus* sp. from the roots of Alnus firma (metal hyperaccumulator plant) exhibit bioremoval of Pb with an ability to produce siderophores and indoleacetic acid; however, they do not harbor responsible genes (Shin et al., 2012). Different endophytes, resistant to Zn/Cd and mostly affiliated with *Proteobacteria*, *Actinobacteria*, and *Bacteroidetes/Chlorobi* with different PGP traits, and enhanced metal uptake in roots and/or in leaves were also reported (Kuffner et al., 2010). Two bacterial isolates identified as *Pseudomonas fluorescens* and *Microbacterium* sp. exhibiting multiple heavy metal resistance characteristics have also been reported by Sheng et al. (2008).

16.3.2.2 Polyaromatic hydrocarbons

Polyaromatic hydrocarbons (PAHs) become bioavailable when they are either in dissolved or the vapor phase. Naturally occurring soil microorganisms, that is, aerobic and anaerobic microorganisms belonging to genera *Pseudomonas, Alcanivorax, Microbulbifer, Mycobacteria, Sphingomonas, Micrococcus, Alcaligenes, Ralstonia, Paenibacillus, Bacillus, Aeromonas, Xanthomonas, Arthrobacter, Acinetobacter, Corynebacterium, Enterobacter*, and others are reported for their degradation (Haritash and Kaushik, 2016). For the fungal bioremediation of PAH contaminated soil, *Aspergillus* and white-rot fungi, such as *Phanerochaete chrysosporium, Bjerkandera adusta*, and *Pleurotus ostreatus, Basidiomycetes P. ostreatus* and *Irpex lacteus* have been documented for efficient biodegradation of PAHs (Haritash and Kaushik, 2016). Byss et al. (2008) reported 55%–67% PAH removal in *P. ostreatus* treatments and 27%–36% PAH removal in *I. lacteus* treatments. Several studies have described various PAH catabolic genes encoding for PAH catabolic enzymes. Monooxygenases and PAH ring-hydroxylating dioxygenase enzymes were identified in *Novosphingobium pentaromativorans* proteome analysis which is involved in the degradation of PAHs (Lyu et al., 2014). *Serratia marcesencs* L-11 strain produce lipases and aromatic ring cleavage enzyme (catechol 1,2-dioxygenase) which can degrade phenanthrene, anthracene, fuorene, and pyrene (Pandey et al., 2012). Various PAH catabolic enzymes such as salicylate hydroxylase, 2-carboxybenzaldehyde dehydrogenase, catechol 1,2-dioxygenase, and catechol 2,3-dioxygenase are differentially expressed in *Pseudomonas aeruginosa* PSA5 and *Rhodococcus* sp. NJ2 which were investigated for B(*a*)P degradation. Fungal strain of *Scopulariopsis brevicaulis* PZ-4 isolated from a PAH contaminated soil was found to have the ability to degrade PAHs up to 77% (Mao and Guan, 2016). Fungal isolate, *Penicillium* sp. 06, was effective at oxidizing a range of PAH in petroleum-contaminated soils. This isolate also has the ability to oxidize more than 89% of the phenanthrene and 75% of the acenaphthene, fuorene, and fuoranthene after 30 days of incubation (Zheng and Obbard, 2003). Studies have shown 90% of fluorene degradation in a soil slurry system by augmenting with *Absidia cylindrospora* a fungal isolate after 288 hours (Garon et al., 2004). This technique could also be beneficial for other in-situ as well as ex situ remediation processes with respect to favorable environmental conditions that affect the biodegradation.

16.3.2.3 Hexachlorocyclohexane

The rate of biodegradation of HCH isomers depends on their solubility, volatility, and persistence in the soil (Girish and Kunhi, 2013). Among different HCH isomers degradation of γ-HCH was faster under anoxic conditions. The α- and γ-isomers degrade more rapidly than β- and δ-HCH (Phillips et al., 2005). The key reaction associated with HCH biodegradation is dehalogenation, in which chloride is replaced by hydrogen atom or by a hydroxyl group resulting in intermediates of lower toxicity (Raina et al., 2008). Intermediate metabolites of HCH degradation as pentachlorocyclohexane (PCCH), tetrachlorocyclohexene (TCCH), tri-, di-, and monochlorobenzenes were observed during degradation (Quintero et al., 2005). HCH isomers can be biodegraded to less toxic compounds under both aerobic and anaerobic conditions as evident by enrichment culture (Quintero et al., 2005). The first HCH-degrading microorganism to be isolated was *Clostridium sphenoides* UQM780, which anaerobically reduce the concentration of lindane in minimal salts media (Philips et al., 2005). From this bacterial degradation, the intermediate was γ-3,4,5,6-TCCH (Endo et al., 2007; Philips et al., 2005). The degradation process by *C. sphenoides* was postulated to be the reductive dechlorination and the first product was PCCH, which is highly unstable (Chen et al, 2015). Similarly, through reductive dechlorination, PCCH was detected as a metabolite of γ-HCH degradation in a mixed culture containing a large number of *Clostridium* sp. (Phillips et al., 2005) and TCCH and benzene were intermediates in degradation of γ-HCH by another anaerobic mixed culture (Chen et al., 2015). Formation of small amounts of tri- and tetrachlorinated benzenes from anaerobic biodegradation of γ-HCH has also been reported (Chen et al., 2015).

Several studies have also reported various levels of degradation of the four major HCH isomers under aerobic conditions. Most aerobic bacteria belonged to the family Sphingomonadaceae, for example, *Sphingobium japonicum* UT26, *Sphingobium indicum* B90A, and *Sphingobium francense* Sp+ (Pal et al., 2005; Lal et al., 2010). Strain UT26 utilizes γ-HCH as a sole source of carbon and energy. These include *C. sphenoides* UQM780, *Clostridium rectum* S-17, and several other *Clostridium* sp., two species of *Bacillus*, several *Pseudomonas* and *Sphingomonas* species, a *Pandoraea* sp., *Citrobacter freundii*, and *Rhodanobacter lindaniclasticus* (Phillips et al., 2005). Under anaerobic conditions, microbial degradation of HCH isomers includes *C. rectum, C. sphenoides, Clostridium butyricum, Clostridium pasteurianum, C. freundii, Desulfovibrio gigas, Desulfovibrio africanus, Desulfococcus multivorans*, and a *Dehalobacter* sp. bacterial strains (Lal et al., 2010). Raina et al. (2008) have extensively investigated for the degradation and characterization of metabolites, genes, and enzymes of the bacterial strains UT26 and B90A and their further application for bioremediation-producing PCCH as end product (Philips et al., 2005). Manickam et al. (2007) isolated a *Xanthomonas* sp. ICH12 strain capable of γ-HCH degradation (Fig. 16.2).

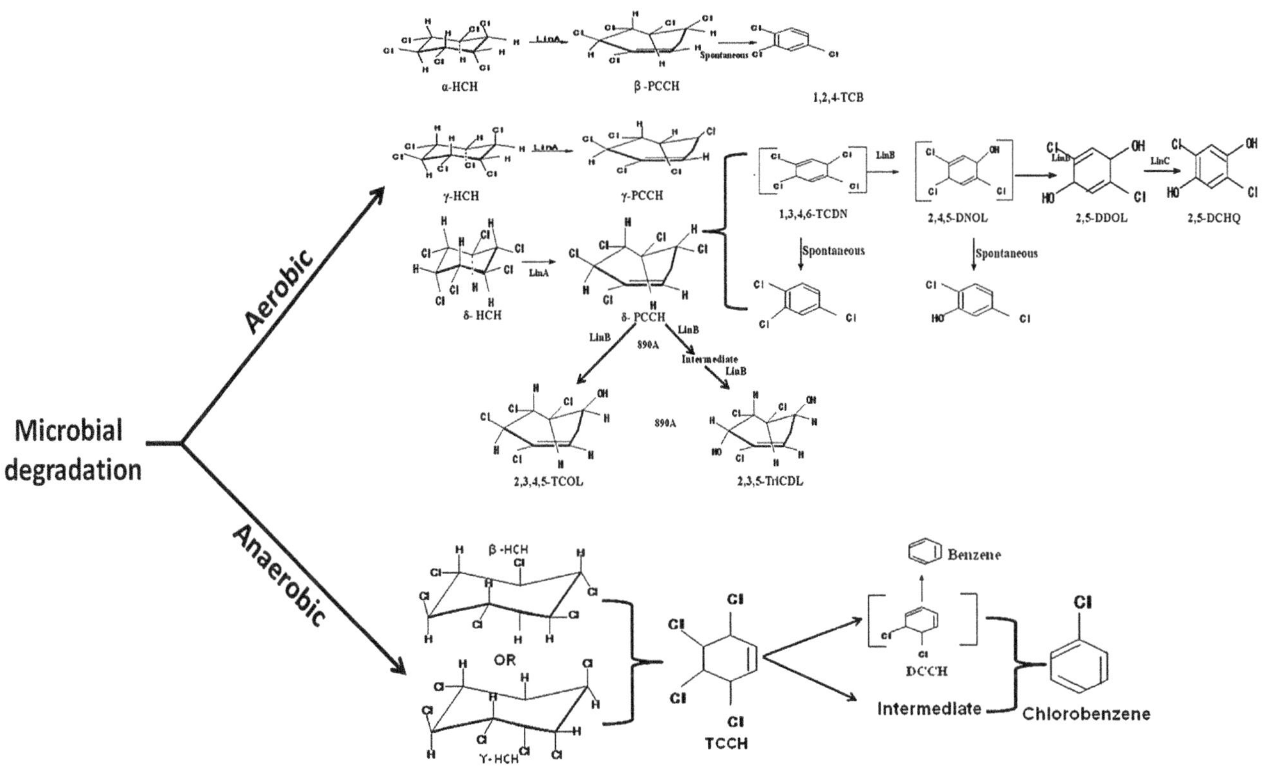

FIGURE 16.2 Bioremediation processes of degradation of HCH isomers. *HCH*, Hexachlorocyclohexane.

16.4 Plant–microbe interaction as detoxification strategy

Microorganisms are the integral part of soil and affect the plants' growth both positively and negatively. They are also known to affect the plant remediation property of heavy metal and other contaminants because of their higher tolerance mechanism (Schlegel et al., 1991). Soil microorganisms can endure the metal toxicity by affecting mobility of the metal and availability to the plant and exhibit natural traits to confront with contaminants and contribute to the establishment of a plant–rhizosphere microbial community, by altering soil pH, release of chelating substances for iron source (siderophores), and solubilizes the minerals that present in soil as insoluble form and other oxidation/reduction mechanism (Gadd, 2000).

Microbe-mediated phytoremediation has been considered as a cost-effective and a more promising tool for the remediation of various contaminants. Phytoremediation using microbe to enhance extraction (phytoextraction) or stabilization (phytostabilization) of metals has been proven to be a reliable and sustainable approach for the removal of metals and restoration of soil quality (Hao et al., 2015). However, interactions between plants, microbe, and heavy metals may greatly affect the efficiency of phytoremediation. The availability of soil heavy metals to plants can be influenced by soil factors, including plant root exudates and microbial activities (Ma et al., 2016). For survival and adaptation under metal-stressed conditions, plant growth–promoting rhizobacteria (PGPR) has developed a range of mechanisms by which they can immobilize, mobilize, or transform heavy metals, thereby rendering them inactive and nontoxic for their own development (Nies, 2004) and resulting in either enhanced or repressed metal transfer from soil to plant (Tak et al., 2013). Plants produce a wide range of sugar, organic acid, vitamins, and minerals, which serve as the rich source of nutrient to the microbial population. Bioremediation, an alternative approach for reclamation of environment contaminated with the heavy metals, novel and multidisciplinary approach that involves microbial community involvement through the process of biostimulation and bioaugmentation.

16.4.1 Biostimulation

It is the process of environmental modification through addition of nutrients such as nitrogen, phosphorous, carbon, organic biostimulants, and oxygen to stimulate the activity of contaminant degrading indigenous microorganisms. The addition of these stimulants allows the microbes to synthesize necessary enzymes for the degradation of contaminants.

The addition of nutrients stimulates PAH degradation rate in the topsoil (Breedveld and Sparrevik, 2000). Chaineau et al. (2005) reported that the biodegradation efficiency has increased up to 62% due to the addition of nutrients in crude oil–contaminated soil. The findings of Zucchi et al. (2003) showed 39.5% reduction of total hydrocarbon content through biostimulation of hydrocarbon-degrading bacterial community with mineral nutrient and surfactant solution in crude oil–contaminated soil. Similarly, Abed et al. (2015) investigated that addition of NH_4CL and NaH_2PO_4 as nitrogen and phosphorus sources for biostimulation of oil-contaminated desert soil has increased the removal efficiency by 20%. The biomass of *Spirulina platensis* was found to be the most effective biostimulant for diesel removal, as reported by Decesaro et al. (2017).

16.4.2 Bioaugmentation

Bioagumentation is a widely applied technique for supporting the bioremediation, using the microbial populations with the specialized capability for degradation of the contaminants. This technique involves the addition of indigenous or exogenous microorganisms for remediating any contaminated site. Studies have been done to illustrate the effect of bioaugmentation both under the aerobic and anaerobic conditions. Factors such as pollutant availability, survival of microbial community, and their enzymatic catabolic activity play an important role for bioaugmentation (Kiamarsi et al., 2019). Inoculation of bacterial endophyte *Pseudomonas* sp. able to produce biosurfactants, siderophore, and organic acids has improved the Fe and P uptake in Cd-hyperaccumulator *S. nigrum*, thereby potentially increasing the biomass of plant (Chen et al., 2014).

16.4.3 Microbial traits for enhanced bioremediation

Microbes colonize the rhizospheric roots of plants that improve plant growth and productivity, enhance soil fertility often termed as PGPR. They showed mechanism against different biotic and abiotic stresses or heavy metal toxicity/contaminants. PGPRs can directly or indirectly modulate plant growth through various mechanisms. Direct mechanism may include fixation of nitrogen (Soares et al., 2014), synthesis of various phytohormones (auxins, cytokinins, gibberellins, etc.), enzymes (Cheng et al., 2007), solubilization of phosphate in plants and reduction of ethylene levels in plants by 1-aminocyclopropane-1-carboxylic acid deaminase (Panhwar et al., 2012), whereas indirect mechanisms include the production of cell wall lytic enzymes and the production of plant's defense mechanisms against various stresses (Weyens et al., 2009). Actively growing microbes are more colonized in rhizospheric soil of hyperaccumulating plant as well as other crop plants. The root exudates enhance the metal uptake through phytosiderophores which are released by the symbiotically attached root microbes (Nautiyal, 1999). The exact relationships of root exudates and microbial community are not known (Lombi et al., 2000); however, recent study suggested that root exudates from *Thlaspi* do not enhance mobilization of metals in soil (Zhao et al., 2001), in addition to simple and complex sugars and growth regulators, different classes of primary and secondary compounds including amino acids, organic acids, phenolic acids, flavonoids, fatty acids, enzymes, steroids, alkaloids, and vitamins (Uren, 2000). The microbes associated with plants produce various enzymes to detoxify and mineralize PAHs, namely, monooxygenase, dioxygenase, dehydrogenase, hydrolase, peroxidases, and dehalogenase which have the ability to transform or degrade aromatic contaminants with respect to their concentration, bioavailability and environmental factors (Campos et al., 2008; Sreelal and Jayanthi, 2017) (Fig. 16.3).

16.5 Microbial genes as potential biotechnological strategies for phytoremediation

All plants contain a complex metal homeostatic network for the detoxification of excess or nonessential metals by evolving their metal homeostasis network throughout the entire life cycle. The candidates for metal tolerance or hyperaccumulation are members of the cation diffusion facilitator and zinc regulated transporter, iron-regulated transporters, and ZIP families of transition metal membrane transport proteins. In *Arabidopsis thaliana*, expression of ZIP family genes AtZIP1, AtZIP3, AtzIP4 are induced under conditions of Zn depletion. The AtZIP1 and AtZIP3 have been shown to mediate cellular Zn uptake across the plasma membrane in *Arabidopsis*. In *Thlaspi caerulescens* a Zn hyperaccumulator, endogenously overexpressed Zn transporter gene Tc ZNT1, showed sixfold higher expression (Gisbert et al., 2003).

Generally hyperaccumulators are low-biomass/slow-growing plants so it is advisable to incorporate their metal-binding proteins/peptides or genes into high biomass and fast-growing plants. For example, overexpression of selenocysteine methyltransferase (SMT), gene from Se hyperaccumulator *Astragalus bisulcatus* in Indian mustard showed enhanced Se tolerance and produced three- to sevenfold more biomass than wild type and threefold longer root lengths (LeDuc et al., 2004). Transgenic *A. thaliana* plants expressing SMT isolated from the Se-hyperaccumulator *A. bisulcatus* accumulated

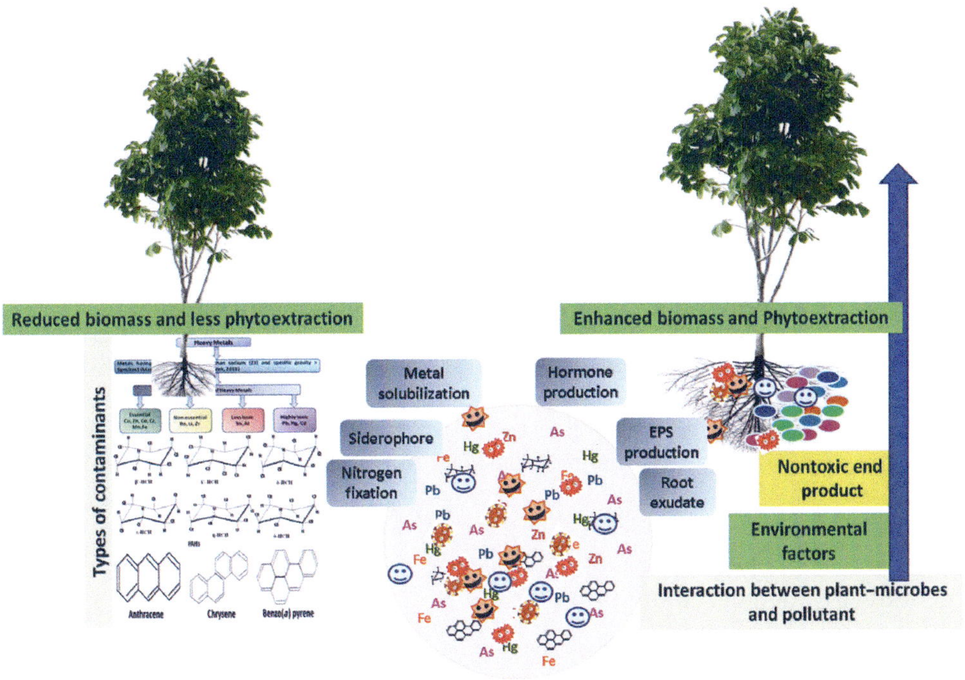

FIGURE 16.3 Schematic representation of microbe-assisted enhanced phytoremediation.

methylselenocysteine and contained up to eightfold higher Se concentrations than wild-type plants (Ellis et al., 2004). *N. tabacum* calmodulin-binding protein (NtCBP4), structural homolog of vertebrate, and invertebrate K^+ and to nonselective cation channels have been recently reported from barley and *Arabidopsis* that can modulate plant tolerance to heavy metals. Arazi et al. (1999) first reported a plant protein that modulates plant tolerance or accumulation of Pb^{2+}. The subcellular localization of NtCBP4 and the phenotype of transgenic plants showed higher overexpression of this protein and they exhibited improved tolerance to Ni^{2+} and hypersensitivity to Pb^{2+}, respectively. Finally, their findings highlighted the involvement of NtCBP4 in metal uptake across the plant plasma membrane and also proved it as a useful tool for implementation of selective ion tolerance in crops (Arazi et al., 1999). Another approach for improved phytoextraction through plants is the overexpression of metal-catalyzing enzymes, for example, ATP sulfurylase (APS), which facilitates the reduction of selenate to selenite (LeDuc et al., 2006). Wangeline et al. (2004) evaluated the overexpression of APS gene in transgenic Indian mustard to tolerate and accumulate 12 metals, i.e. arsenic (As), cadmium (Cd), chromium (Cr), copper (Cu), mercury (Hg), manganese (Mn), molybdenum (Mo), nickel (Ni), lead (Pb), vanadium (V), tungsten (W) and zinc (Zn) in its shoots when exposed to metal polluted site and it accumulates As(III), As(V), Hg and a 2.5-fold higher concentration of molybdenum and vanadium than its wild type (WT). Similarly, grasses are known for better metal extraction and one of the best examples for this is *Nicotiana glauca*, reported to tolerate a wide range of environmental contaminants, namely, Cd, Cu, Pb, Zn in its wild type form. But when it was genetically modified, the overexpression of TaPCS1, a wheat gene encoding phytochelatin (PC) synthase, increased the plants' tolerance to cadmium and developed roots up to 160% longer than nontransformed plants (Gisbert et al., 2003).

Plants, being deep rooted in soil, undergo constant interaction between root system and microorganisms which raises the need for full mechanistic understanding of microbial genes and the transgenic approaches. Biotechnological application of transfer of foreign gene from microorganism into the plants enhances the phytoremediation of heavy metals and organic compounds. These plant interacting microbial species produce various metabolites and organic compounds for the transformation and detoxification of soil. The use of transgenic plants was a limiting technology but the need for low-cost remediation strategy and plant-transformation protocols have proved the use of transgenic plants and associated bacteria to be the efficient environmental-friendly technology for the remediation of soil pollutants.

16.5.1 Microbial genes for higher tolerance

Microorganism contains specific gene(s) with unique metabolic characteristics for enhancing the efficacy of decontamination. Engineered plants containing novel traits from contaminant-resistant microorganisms such as bacterial, yeast, fungi have been demonstrated for the enhancement of phytoremediation. The use of these tools makes it possible for some plants to take up xenobiotics or high levels of heavy metals and then survives or helps the soil biology. *Microbacterium arabinogalactanolyticum* AY509226, the genes identified, designates high homology to the *czc*D, *chr*B, *ncc*A, and *mer* genes of *Ralstonia metallidurans* CH34. These genes are located on the bacterial chromosome or on the plasmids pMOL28 or pMOL30 that encodes the resistance to Cd, Zn, and Co (Mergeay et al., 2009). *A. murale* has a significant ability to hyperaccumulate Ni from Ni-rich soils, due to the presence of Ni, Hg, Zn, and Cr resistance genes in both Gram-positive and -negative isolates obtained from *A. murale* rhizospheric zone (Abou-Shanab et al., 2007).

The nickel–cobalt–cadmium resistance genes are found in plasmid pTOM9 of *Alcaligenes xylosoxidans* 31A which are located on a 14.5 kb BamHI fragment (Shaik et al., 2008). The two distinct Ni resistance loci *ncc* and *nre* have been identified for high- and low-mediating Ni resistance, respectively. Metal-resistant bacterium *Alcaligenes eutrophus* CH34 (strain AE104) represents high-level resistance of combined Ni, Cd, and Co due to the presence of *ncc* locus, while *Escherichia coli* do not have *ncc* locus. Locus *nre* shows low level of Ni resistance in *A. eutrophus* as well as in *E. coli*. Grass et al. (2005) found the involvement of single *nreB* gene in nre operon of *Achromobacter xylosoxidans 31A* for nickel resistance. However, the *ncc* operon represents the seven open reading frames *nccYXHCBAN*, which shows momentous similarity with proteins encoded by the metal-resistance loci *cnr (cnrYXHCBA)* and *czc (czcRCBAD)* of *A. eutrophus* CH34 (Schmidt and Schlegel, 1994).

16.5.2 Development of genetically engineered microbes

Removal of contaminant from polluted sites by GE microbes is more efficient than other conventional methods due to presence of specific characteristics such as presence of specific enzyme or gene to degrade certain kind of pollutant, such as PCBs and trichloroethylene (TCE) (Divya and Kumar, 2011). Most of the genes for pesticide degradation are present on plasmid and organophosphorous-degrading (*opd*) gene was first reported in *Pseudomonas diminuta*. Although in *Agrobacterium radiobacter* the *opd*A was reported to be present in chromosome. de la Pena Mattozzi et al. (2006) constructed a genetically modified strain of *Pseudomonas putida* to degrade paraxon, an organophosphorous compound. Liu et al (2006) developed recombinant *Sphingomonas* sp. CDS-1 for the degradation of other organophosphate and carbamates. The main classes of metal chelators include PCs and MTs, are metal-binding peptides, and are capable to enhance metal-binding capabilities of microorganisms. For this, genes from these proteins have been cloned in *E. coli* from plants and fungi. Deng et al. (2007) developed GE *E. coli* JM109 for bioaccumulation of Cd^{2+} from heavy metals polluted site. Similarly, various GE bacteria such as *Mesorhizobium huakuii*, *P. putida*, and *Caulobacter crescentus* are reported for the enhanced accumulation of Cd^{2+} by the expression of fusion proteins for PC, metal-binding peptide (EC20), and RsaA-6His, respectively (Sarma and Prasad, 2019). Treatment of cauliflower (*Brassica oleracea* var. *botrytis*) with $CdCl_2$, suggested the presence of a Cd-inducible PC and its involvement in Cd-tolerance of cauliflower (Hasegawa et al., 1997). The overexpression of the MT protein, namely, glutathione *S*-transferase fusion protein and nixA-encoded membrane transport in bacteria exhibited high affinity to accumulate large amount of Ni^{2+} (Singh et al., 2011).

Use of arsenite *S*-adenosylmethionine methyltransferase gene (*ars*M) cloned in *E. coli* from *Rhodopseudomonas palustris* which methylates inorganic As to less toxic and volatile trimethylarsine has been reported (Liu et al., 2011). Expression of another metalloregulatory protein *ArsR* in *E. coli* also showed high affinity to remove As from the contaminated sites (Kostal et al., 2004). The detoxification and volatilization of highly toxic elemental mercury are a unique example for the successful phytoremediation based on genetic engineering approach. Another strategy for enhancing the bioremediation is to develop hybrid strain by conjugation of appropriate organism or by introduction of specific gene. This approach has been applied for the degradation of highly toxic trihalopropanes and the hybrid gene was formed by combining the constitutive promoter and translational signals from a fungal *gpd* gene (encoding dehydrogenase) with the structural region of the bacterial *opd* gene (Divya and Kumar, 2011).

Recently, a new class of metal-binding peptides known as synthetic phytochelatin (PCs)with the repetitive metal-binding motif $(GluCys)_nGly$ were shown to have improved Cd^{2+} binding capability. Overexpression of such metal binding in bacterial cells resulted in enhanced Hg^{2+} accumulation and thus offers a promising strategy for the development of microbe-based biosorbents (Bae et al., 2001). Another approach is to express both naturally occurring metal-binding peptides such as MTs and synthetic peptides such as synthetic PCs on the surface of bacterial cells for improved uptake and biosorption of mercury. However, one major problem associated with these cysteine-rich peptides is their lack of specificity. Development of resistance against heavy metals by expression of associated proteins, for example, MerR protein resulted in controlled expression of enzymes responsible for mercury detoxification (Bae et al., 2003).

TABLE 16.2 Genes manipulated in transgenic plants for improved metal tolerance and phytoextraction potential.

	Pollutants	Transgenic plants	Gene source	Foreign gene or DNA	References
Inorganic pollutants (heavy metals)	As	*Brassica napus*	*Enterobacter cloacae*	UW4 ACC deaminase	Nie et al. (2002)
	Cd	*Brassica juncea*	*Escherichia coli*	*gshI, gshII*	Zhu et al. (1999a, 1999b)
	Cd, Pb	*Arabidopsis thaliana*	*Saccharomyces cerevisiae*	*YCFI*	Song et al. (2003)
	Hg(II)	*A. thaliana*	Bacteria	*merA*	Heaton et al. (1998)
	Hg(II)	*Liriodendron tulipifera*	Bacteria	*merA*	Rugh et al. (1998)
	MeHg	*A. thaliana*	*E. coli*	*merApe9*	Rugh et al. (1996)
	MeHg	*A. thaliana*	Bacteria	*merBpe*	Bizily et al. (1999)
	MeHg	*A. thaliana*	Bacteria	*merA, merB*	Bizily et al. (2000)
	MeHg	*A. thaliana*	Bacteria	*merB*	Bizily et al. (2003)
Organic pollutants	PMA	*Nicotiana tabacum*	Bacteria	*merA, merB*	Ruiz et al. (2003)
	ACET, MET	*Populus*	Bacteria	γ- glutamyl synthetase	Gullner et al. (2001)
	PCP	*N. tabacum*	*Coriolus versicolor*	*MnP* Manganese peroxidase	Iimura et al. (2002)
	Cd, Cu, Pb	*B. juncea*	*E. coli*	gshI, gshII	Bennett et al. (2003)
	RDX	*A. thaliana*	*Rhodococcus rhodochrous*	Cytochrome P450	Rylott et al. (2006)
	TNT	*N. tabacum, A. thaliana. Hybrid aspen (Populus tremula X Populus tremuloides)*	*E. cloacae, E. coli, Pseudomonas putida*	Nitroreductase (*nfsI, nfsA, pnrA*)	Hannink et al. (2001, 2007), Kurumata et al. (2005), Van Dillewijn et al. (2008)
	GNT, TNT	*N. tabacum*	*E. cloacae*	*onr*	French et al. (1999)
	RDX	*A. thaliana*	*R. rhodochrous*	*XplA* and *XplB*	Jackson et al. (2007)
	Sulfonylurea	*N. tabacum*	*Streptomyces griseolus*	*CYP105A1*	O'Keefe et al. (1994)
	PCBs	*N. tabacum*	PCB-degrading bacteria	*bphc*	Chrastilova et al.,2007
	Atrazine	*Alfalfa N. tabacum*	*E. coli*	atzA	Wang et al. (2005)
	Protoporphyrinogen IX oxidase	*Oryza sativa*	*Bacillus subtilis*	*Protox*	Jung et al. (2008)
	OPH	*N. tabacum*	*Pseudomonas pseudoalcaligenes*	*ophc2*	Wang et al. (2008)
	Fungal laccase	*N. tabacum*	*C. versicolor*	*LAC*	Sonoki et al. (2005)
	PCB	*N. tabacum*	*Burkholderia xenovorans*	BphA	Mohammadi et al. (2007)
	Haloalkane dehydrogenase	*N. tabacum*	*Terrabacter* sp.	*DhaA*	Uchida et al. (2005)
	Aromatic-cleaving extradiol dioxygenase	*A. thaliana*	*Terrabacter* sp.	*DbfB*	Uchida et al. (2005)
	Biphenyl catabolic enzymes	*N. tabacum*	*Pandoraea pnomenusa*	*bphC*	Francova et al. (2003), Novakova et al. (2009)

ACET, Acetochlor; *BphA*, biphenyl dioxygenase gene; *GTN*, glycerol trinitrate; *GTN*, glyceroltrinitrate; *MeHg*, methyl mercury; *MET*, metolachlor; *OPH*, organophosphorous hydrolase; *PCB*, polychlorinated biphenyl; *PCP*, pentachlorophenol; *RDX*, hexahydro-1,3,5-trinitro-1,3,5-triazine; *TNT*, trinitrotoluene; *TNT*, trinitrotoluene.

16.5.3 Transgenics plants using microbial genes

The early study illustrates the examples of expression of yeast MT gene in tobacco plants for higher tolerance to Cd (Misra and Gedamu, 1989). MTs are metal-binding proteins that provide heavy metal tolerance especially to copper and cadmium. Thomas et al. (2003) performed the expression of yeast MT (CUP 1) gene in tobacco plant for the phytoextraction of the Cu and Cd. The transformed plants were subjected to both copper stress and nonstress conditions for 7 days and showed Cu accumulation seven times higher in the older leaves under stress condition. Dhankher et al. (2002) incorporated bacterial genes for arsenic reductase and *g*-glutamylcysteine synthase together in *Arabidopsis* to increase its As(V) tolerance and hyperaccumulation. For mercury removal, dicistronic construct was developed that contained the bacterial mercury detoxification genes merA and merB under the control of the *Arabidopsis* Actin2 promoter and terminator. The resulted construct was introduced into *Agrobacterium* competent cells (Rahman et al., 2008). Rahman et al. (2008) also confirmed the effectiveness of GE *Agrobacterium* for Hg detoxification by germinating tobacco (*N. tabacum* var. Gold leaf) seeds on MS media. The results showed that transgenic plants are resistant to both Phenyl mercuric acetate (PMA) and $HgCl_2$. For the removal of uranium, hairy root cultures of *B. juncea* and *C. amaranticolor* were developed by genetic transformation using *Agrobacterium rhizogenes* (Eapen et al., 2003). Other microbial genes used for development of transgenics for remediation of contaminated soils are enlisted in Table 16.2.

16.6 Conclusion

Since conventional phytoremediation methods are not effective for removal of higher concentration of soil contaminants, most of the known hyperaccumulating plant species are selective, their growth rate is usually slow, produce less biomass and most of them can be used in their natural habitats only. Therefore, to increase its effectiveness, plant and microbe-based phytoremediation associated biotechnological approaches may be an emerging technology. Transferring knowledge about the application of such approaches to agricultural practices would be a step toward removal of environmental contaminants. Depending on different strategies, transgenic plants can be engineered through transfer or overexpression of gene for metal uptake. Genes from hyperaccumulating microbes offer good opportunities that can be employed for genetic modification to reduce a higher concentration of contamination from environment after getting regulatory approvals.

References

Abdel-Shafy, H.I., Mansour, M.S., 2016. A review on polycyclic aromatic hydrocarbons: source, environmental impact, effect on human health and remediation. Egypt. J. Pet. 25 (1), 107–123.

Abed, R.M., Al-Kharusi, S., Al-Hinai, M., 2015. Effect of biostimulation, temperature and salinity on respiration activities and bacterial community composition in an oil polluted desert soil. Int. Biodeterior. Biodegrad. 98, 43–52.

Abhilash, P.C., Singh, N., 2008. Distribution of hexachlorocyclohexane isomers in soil samples from a small scale industrial area of Lucknow, North India, associated with lindane production. Chemosphere 73 (6), 1011–1015.

Abou-Shanab, R.A., Angle, J.S., Delorme, T.A., Chaney, R.L., Van Berkum, P., Moawad, H., et al., 2003. Rhizobacterial effects on nickel extraction from soil and uptake by *Alyssum murale*. N. Phytol. 158 (1), 219.

Abou-Shanab, R.A.I., Van Berkum, P., Angle, J.S., 2007. Heavy metal resistance and genotypic analysis of metal resistance genes in gram-positive and gram-negative bacteria present in Ni-rich serpentine soil and in the rhizosphere of *Alyssum murale*. Chemosphere 68 (2), 360–367.

Adams, A., Raman, A., Hodgkins, D., 2013. How do the plants used in phytoremediation in constructed wetlands, a sustainable remediation strategy, perform in heavy-metal-contaminated mine sites? Water Environ. J. 27 (3), 373–386.

Agency for Toxic Substances and Disease Registry (ATSDR), 2007. 2007 CERCLA Priority List of Hazardous Substances.

Anderson, C.R., Cook, G.M., 2004. Isolation and characterization of arsenate-reducing bacteria from arsenic-contaminated sites in New Zealand. Curr. Microbiol. 48 (5), 341–347.

Anyasi, R.O., Atagana, H.L., 2011. Biological remediation of polychlorinated biphenyls (PCB) in the environment by microorganisms and plants. Afr. J. Biotechnol. 10 (82), 18916–18938.

Arazi, T., Sunkar, R., Kaplan, B., Fromm, H., 1999. A tobacco plasma membrane calmodulin-binding transporter confers Ni^{2+} tolerance and Pb^{2+} hypersensitivity in transgenic plants. Plant J. 20 (2), 171–182.

ATSDR, U., 2005. Toxicological profile for lead. US Department of Health and Human Services. Public Health Service, Atlanta, USA, p. 582.

Bae, W., Mehra, R.K., Mulchandani, A., Chen, W., 2001. Genetic engineering of *Escherichia coli* for enhanced uptake and bioaccumulation of mercury. Appl. Environ. Microbiol. 67 (11), 5335–5338.

Bae, W., Wu, C.H., Kostal, J., Mulchandani, A., Chen, W., 2003. Enhanced mercury biosorption by bacterial cells with surface-displayed MerR. Appl. Environ. Microbiol. 69 (6), 3176–3180.

Barzanti, R., Ozino, F., Bazzicalupo, M., Gabbrielli, R., Galardi, F., Gonnelli, C., et al., 2007. Isolation and characterization of endophytic bacteria from the nickel hyperaccumulator plant *Alyssum bertolonii*. Microb. Ecol. 53 (2), 306–316.

Bauer, M., Blodau, C., 2006. Mobilization of arsenic by dissolved organic matter from iron oxides, soils and sediments. Sci. Total Environ. 354 (2–3), 179–190.

Beach, L., 1984. Exposure From the Uranium Series with Emphasis on Radon and Its Daughters, Report No. 77. National Council on Radiation Protection and Measurements Bethesda, NCRP, p. 131.

Becher, M., Talke, I.N., Krall, L., Krämer, U., 2004. Cross-species microarray transcript profiling reveals high constitutive expression of metal homeostasis genes in shoots of the zinc hyperaccumulator *Arabidopsis halleri*. Plant J. 37 (2), 251–268.

Bennett, L.E., Burkhead, J.L., Hale, K.L., Terry, N., Pilon, M., Pilon-Smits, E.A.H., 2003. Analysis of transgenic Indian mustard plants for phytoremediation of metal-contaminated mine tailings. J. Environ. Qual. 32, 432–440.

Bhattacharya, P., Frisbie, S.H., Smith, E., Naidu, R., Jacks, G., Sarkar, B., 2002. Arsenic in the environment: a global perspective. Handbook of Heavy Metals in the Environment. Marcell Dekker Inc., New York, pp. 147–215.

Bizily, S.P., Kim, T., Kandasamy, M.K., Meagher, R.B., 2003. Subcellular targeting of methylmercury lyase enhances its specific activity for organic mercury detoxification in plants. Plant Physiol. 131 (2), 463–471.

Bizily, S.P., Rugh, C.L., Meagher, R.B., 2000. Phytodetoxification of hazardous organomercurials by genetically engineered plants. Nat. Biotechnol. 18 (2), 213–217.

Bizily, S.P., Rugh, C.L., Summers, A.O., Meagher, R.B., 1999. Phytoremediation of methylmercury pollution: merB expression in *Arabidopsis thaliana* confers resistance to organomercurials. Proc. Natl. Acad. Sci. U.S.A. 96 (12), 6808–6813.

Blantz, R.C., Pelayo, J.C., Gushwa, L.C., Myers, R.R., Evan, A.P., 1985. Functional basis for the glomerular alterations in uranyl nitrate acute renal failure. Kidney Int. 28 (5), 733–743.

Bolan, N.S., Park, J.H., Robinson, B., Naidu, R., Huh, K.Y., 2011. Phytostabilization: a green approach to contaminant containment. Adv. Agron. 112, 145–204.

Breedveld, G.D., Sparrevik, M., 2000. Nutrient-limited biodegradation of PAH in various soil strata at a creosote contaminated site. Biodegradation 11 (6), 391–399.

Byss, M., Elhottová, D., Tříska, J., Baldrian, P., 2008. Fungal bioremediation of the creosote-contaminated soil: influence of *Pleurotus ostreatus* and *Irpex lacteus* on polycyclic aromatic hydrocarbons removal and soil microbial community composition in the laboratory-scale study. Chemosphere 73 (9), 1518–1523.

Cameselle, C., Gouveia, S., 2019. Phytoremediation of mixed contaminated soil enhanced with electric current. J. Hazard. Mater. 361, 95–102.

Campos, V.M., Merino, I., Casado, R., Pacios, L.F., Gómez, L., 2008. Phytoremediation of organic pollutants. Span. J. Agric. Res. 1, 38–47.

Carvalho, F.P., 2017. Pesticides, environment, and food safety. Food Energy Secur. 6 (2), 48–60.

Cervantes, C., Gutierrez-Corona, F., 1994. Copper resistance mechanisms in bacteria and fungi. FEMS Microbiol. Rev. 14 (2), 121–137.

Chaineau, C.H., Rougeux, G., Yepremian, C., Oudot, J., 2005. Effects of nutrient concentration on the biodegradation of crude oil and associated microbial populations in the soil. Soil Biol. Biochem. 37 (8), 1490–1497.

Chaney, R.L., Malik, M., Li, Y.M., Brown, S.L., Brewer, E.P., Angle, J.S., et al., 1997. Phytoremediation of soil metals. Curr. Opin. Biotechnol. 8 (3), 279–284.

Chen, H., Gao, B., Wang, S., Fang, J., 2015. Microbial Degradation of Hexachlorocyclohexane (HCH) Pesticides. Adv. Biodegrad. Biorem. Ind. Waste 195–224.

Chen, L., Luo, S., Li, X., Wan, Y., Chen, J., Liu, C., 2014. Interaction of Cd-hyperaccumulator *Solanum nigrum* L. and functional endophyte *Pseudomonas* sp. Lk9 on soil heavy metals uptake. Soil Biol. Biochem. 68, 300–308.

Cheng, Y., Qin, G., Dai, X., Zhao, Y., 2007. NPY1, a BTB-NPH3-like protein, plays a critical role in auxin-regulated organogenesis in *Arabidopsis*. Proc. Natl. Acad. Sci. U.S.A. 104 (47), 18825–18829.

Chrastilova, Z., Mackova, M., Novakova, M., Macek, T., Szekeres, M., 2007. Transgenic plants for effective phytoremediation of persistent toxic organic pollutants present in the environment. J. Biotechnol. 2 (131), S38.

de la Pena Mattozzi, M., Tehara, S.K., Hong, T., Keasling, J.D., 2006. Mineralization of paraoxon and its use as a sole C and P source by a rationally designed catabolic pathway in *Pseudomonas putida*. Appl. Environ. Microbiol. 72 (10), 6699–6706.

Decesaro, A., Rampel, A., Machado, T.S., Thomé, A., Reddy, K., Margarites, A.C., et al., 2017. Bioremediation of soil contaminated with diesel and biodiesel fuel using biostimulation with microalgae biomass. J. Environ. Eng. 143 (4), 04016091.

Deng, X., Yi, X.E., Liu, G., 2007. Cadmium removal from aqueous solution by gene-modified *Escherichia coli* JM109. J. Hazard. Mater. 139 (2), 340–344.

Dhankher, O.P., Li, Y., Rosen, B.P., Shi, J., Salt, D., Senecoff, J.F., et al., 2002. Engineering tolerance and hyperaccumulation of arsenic in plants by combining arsenate reductase and γ-glutamylcysteine synthetase expression. Nat. Biotechnol. 20 (11), 1140.

Divya, B., Kumar, M.D., 2011. Plant-microbe interaction with enhanced bioremediation. Res. J. Biotechnol. 6 (1), 72–79.

Dixit, R., Malaviya, D., Pandiyan, K., Singh, U.B., Sahu, A., Shukla, R., et al., 2015. Bioremediation of heavy metals from soil and aquatic environment: an overview of principles and criteria of fundamental processes. Sustainability 7 (2), 2189–2212.

Du Preez, J.G.H., 1989. A review of the industrial processes involving uranium-from the ore to the reactor. Radiat. Prot. Dosim. 26 (1–4), 7–13.

Eapen, S., Suseelan, K.N., Tivarekar, S., Kotwal, S.A., Mitra, R., 2003. Potential for rhizofiltration of uranium using hairy root cultures of *Brassica juncea* and *Chenopodium amaranticolor*. Environ. Res. 91 (2), 127–133.

Ehsan, S., Ali, S., Noureen, S., Mahmood, K., Farid, M., Ishaque, W., et al., 2014. Citric acid assisted phytoremediation of cadmium by *Brassica napus* L. Ecotoxicol. Environ. Saf. 106, 164–172.

Ellis, D.R., Sors, T.G., Brunk, D.G., Albrecht, C., Orser, C., Lahner, B., et al., 2004. Production of Se-methylselenocysteine in transgenic plants expressing selenocysteine methyltransferase. BMC Plant Biol. 4 (1), 1.

Emamverdian, A., Ding, Y., Mokhberdoran, F., Xie, Y., 2015. Heavy metal stress and some mechanisms of plant defense response. Sci. World J. 2015.

Endo, R., Ohtsubo, Y., Tsuda, M., Nagata, Y., 2007. Identification and characterization of genes encoding a putative ABC-type transporter essential for utilization of γ-hexachlorocyclohexane in *Sphingobium japonicum* UT26. J. Bacteriol. 189 (10), 3712–3720.

Feng, J., Shi, Q., Wang, X., Wei, M., Yang, F., Xu, H., 2010. Silicon supplementation ameliorated the inhibition of photosynthesis and nitrate metabolism by cadmium (Cd) toxicity in *Cucumis sativus* L. Sci. Hortic. 123 (4), 521–530.

Flathman, P.E., Lanza, G.R., 1998. Phytoremediation: current views on an emerging green technology. J. Soil Contam. 7 (4), 415–432.

Francova, K., Sura, M., Macek, T., Szekeres, M., Bancos, S., Demnerova, K., et al., 2003. Preparation of plants containing bacterial enzyme for degradation of polychlorinated biphenyls. Fresenius Environ. Bull. 12 (3), 309–313.

French, C.E., Rosser, S.J., Davies, G.J., Nicklin, S., Bruce, N.C., 1999. Biodegradation of explosives by transgenic plants expressing pentaerythritol tetranitrate reductase. Nat. Biotechnol. 17 (5), 491–494.

Gadd, G.M., 2000. Bioremedial potential of microbial mechanisms of metal mobilization and immobilization. Curr. Opin. Biotechnol. 11 (3), 271–279.

Gajewska, E., Skłodowska, M., 2007. Relations between tocopherol, chlorophyll and lipid peroxides contents in shoots of Ni-treated wheat. J. Plant Physiol. 164 (3), 364–366.

Gan, S., Lau, E.V., Ng, H.K., 2009. Remediation of soils contaminated with polycyclic aromatic hydrocarbons (PAHs). J. Hazard. Mater. 172 (2–3), 532–549.

García-Alonso, S., Pérez-Pastor, R.M., Sevillano-Castano, M.L., Escolano, O., García-Frutos, F.J., 2008. Influence of particle size on the quality of PAH concentration measurements in a contaminated soil. Polycyclic Aromat. Compd. 28 (1), 67–83.

Garon, D., Sage, L., Wouessidjewe, D., Seigle-Murandi, F., 2004. Enhanced degradation of fluorene in soil slurry by *Absidia cylindrospora* and maltosyl-cyclodextrin. Chemosphere 56 (2), 159–166.

Gasic, K., Korban, S.S., 2007. Expression of *Arabidopsis* phytochelatin synthase in Indian mustard (*Brassica juncea*) plants enhances tolerance for Cd and Zn. Planta 225 (5), 1277–1285.

Gill, H.K., Garg, H., 2014. Pesticides: environmental impacts and management strategies. Pesticides—Toxic Aspects 8, 187.

Girish, K., Kunhi, A.M., 2013. Microbial degradation of gamma-hexachlorocyclohexane (lindane). Afr. J. Microbiol. Res. 7 (17), 1635–1643.

Gisbert, C., Ros, R., De Haro, A., Walker, D.J., Bernal, M.P., Serrano, R., et al., 2003. A plant genetically modified that accumulates Pb is especially promising for phytoremediation. Biochem. Biophys. Res. Commun. 303 (2), 440–445.

Gitipour, S., Sorial, G.A., Ghasemi, S., Bazyari, M., 2018. Treatment technologies for PAH-contaminated sites: a critical review. Environ. Monit. Assess. 190 (9), 546.

Grass, G., Fricke, B., Nies, D.H., 2005. Control of expression of a periplasmic nickel efflux pump by periplasmic nickel concentrations. Biometals 18 (4), 437–448.

Gullner, G., Kömives, T., Rennenberg, H., 2001. Enhanced tolerance of transgenic poplar plants overexpressing γ-glutamylcysteine synthetase towards chloroacetanilide herbicides. J. Exp. Bot. 52 (358), 971–979.

Hall, A.H., 2002. Chronic arsenic poisoning. Toxicol. Lett. 128 (1–3), 69–72.

Hannink, N., Rosser, S.J., French, C.E., Basran, A., Murray, J.A., Nicklin, S., et al., 2001. Phytodetoxification of TNT by transgenic plants expressing a bacterial nitroreductase. Nat. Biotechnol. 19 (12), 1168–1172.

Hannink, N.K., Subramanian, M., Rosser, S.J., Basran, A., Murray, J.A., Shanks, J.V., et al., 2007. Enhanced transformation of TNT by tobacco plants expressing a bacterial nitroreductase. Int. J. Phytorem. 9 (5), 385–401.

Hao, X., Xie, P., Zhu, Y.G., Taghavi, S., Wei, G., Rensing, C., 2015. Copper tolerance mechanisms of *Mesorhizobium amorphae* and its role in aiding phytostabilization by *Robinia pseudoacacia* in copper contaminated soil. Environ Sci Tech 49 (4), 2328–2340.

Haritash, A.K., Kaushik, C.P., 2016. Degradation of low molecular weight polycyclic aromatic hydrocarbons by microorganisms isolated from contaminated soil. Int. J. Environ. Sci. 6 (5), 808–819.

Hasegawa, I., Terada, E., Sunairi, M., Wakita, H., Shinmachi, F., Noguchi, A., et al., 1997. Genetic improvement of heavy metal tolerance in plants by transfer of the yeast metallothionein gene (CUP1). Plant Nutrition for Sustainable Food Production and Environment. Springer, Dordrecht, pp. 391–395.

He, H., Ye, Z., Yang, D., Yan, J., Xiao, L., Zhong, T., et al., 2013. Characterization of endophytic *Rahnella* sp. JN6 from *Polygonum pubescens* and its potential in promoting growth and Cd, Pb, Zn uptake by *Brassica napus*. Chemosphere 90 (6), 1960–1965.

Heaton, A.C., Rugh, C.L., Wang, N.J., Meagher, R.B., 1998. Phytoremediation of mercury-and methylmercury-polluted soils using genetically engineered plants. J. Soil Contam. 7 (4), 497–509.

Hou, W., Chen, X., Song, G., Wang, Q., Chang, C.C., 2007. Effects of copper and cadmium on heavy metal polluted waterbody restoration by duckweed (*Lemna minor*). Plant Physiol. Biochem. 45 (1), 62–69.

Iimura, Y., Ikeda, S., Sonoki, T., Hayakawa, T., Kajita, S., Kimbara, K., et al., 2002. Expression of a gene for Mn-peroxidase from *Coriolus versicolor* in transgenic tobacco generates potential tools for phytoremediation. App. Microbial. Biotechnol. 59 (2–3), 246–251.

Institute of Environmental Conservation and Research (INECAR), 2000. Position Paper Against Mining in Rapu-Rapu. INECAR, Ateneo de Naga University, Philippines. <www.adnu.edu.ph/Institutes/Inecar/pospaper1.asp>.

Jackson, R.G., Rylott, E.L., Fournier, D., Hawari, J., Bruce, N.C., 2007. Exploring the biochemical properties and remediation applications of the unusual explosive-degrading P450 system XplA/B. Proc. Natl. Acad. Sci. U.S.A. 104 (43), 16822–16827.

Jain, S., Arnepalli, D.N., 2019. Biominerlisation as a remediation technique: A critical review. Geotechnical Characterisation and Geoenvironmental Engineering. Springer, Singapore, pp. 155–162.

Jung, S., Lee, H.J., Lee, Y., Kang, K., Kim, Y.S., Grimm, B., et al., 2008. Toxic tetrapyrrole accumulation in protoporphyrinogen IX oxidase-overexpressing transgenic rice plants. Plant Mol. Biol. 67 (5), 535–546.

Kantor, D., 2006. Guillain–Barre Syndrome: The Medical Encyclopedia. National Library of Medicine and National Institute of Health, Bethesda, MD.

Kiamarsi, Z., Soleimani, M., Nezami, A., Kafi, M., 2019. Biodegradation of n-alkanes and polycyclic aromatic hydrocarbons using novel indigenous bacteria isolated from contaminated soils. Int. J. Environ. Sci. Technol. 16 (11), 6805–6816.

Kim, S.H., Lee, I.S., 2005. Phytoremediation of Cu-contaminated Soil and Water by *Commelina communis*. Korean J. Ecol. 28 (1), 7–13.

Kong, Z., Glick, B.R., 2017. The role of bacteria in phytoremediation. Appl. Bioeng. 327–353.

Kostal, J., Mulchandani, A., Chen, W., 2004. Affinity purification of plasmid DNA by temperature-triggered precipitation. Biotechnol. Bioeng. 85 (3), 293–297.

Kuffner, M., De Maria, S., Puschenreiter, M., Fallmann, K., Wieshammer, G., Gorfer, M., et al., 2010. Culturable bacteria from Zn- and Cd-accumulating *Salix caprea* with differential effects on plant growth and heavy metal availability. J. App. Microbiol. 108 (4), 1471–1484.

Kurumata, M., Takahashi, M., Sakamoto, A., Ramos, J.L., Nepovim, A., Vanek, T., et al., 2005. Tolerance to, and uptake and degradation of 2,4,6-trinitrotoluene (TNT) are enhanced by the expression of a bacterial nitroreductase gene in *Arabidopsis thaliana*. Z. Naturforsch. 60 (3–4), 272–278.

Lal, R., Pandey, G., Sharma, P., Kumari, K., Malhotra, S., Pandey, R., et al., 2010. Biochemistry of microbial degradation of hexachlorocyclohexane and prospects for bioremediation. Microbiol. Mol. Biol. Rev. 74 (1), 58–80.

Leblanc, M., Petit, D., Deram, A., Robinson, B.H., Brooks, R.R., 1999. The phytomining and environmental significance of hyperaccumulation of thallium by *Iberis intermedia* from southern France. Econ. Geol. 94 (1), 109–113.

LeDuc, D.L., AbdelSamie, M., Móntes-Bayon, M., Wu, C.P., Reisinger, S.J., Terry, N., 2006. Overexpressing both ATP sulfurylase and selenocysteine methyltransferase enhances selenium phytoremediation traits in Indian mustard. Environ. Pollut. 144 (1), 70–76.

LeDuc, D.L., Tarun, A.S., Montes-Bayon, M., Meija, J., Malit, M.F., Wu, C.P., et al., 2004. Overexpression of selenocysteine methyltransferase in Arabidopsis and Indian mustard increases selenium tolerance and accumulation. Plant Physiol. 135 (1), 377–383.

Lee, E.H., Cho, K.S., Kim, J., 2010. Comparative study of rhizobacterial community structure of plant species in oil-contaminated soil. J. Microbiol. Biotechnol. 20 (9), 1339–1347.

Lešková, A., Zvarík, M., Araya, T., Giehl, R.F., 2019. Nickel toxicity targets cell wall-related processes and PIN2-mediated auxin transport to inhibit root elongation and gravitropic responses in Arabidopsis. Plant Cell Physiol. 61 (3), 519–535.

Liao, X.Y., Chen, T.B., Xie, H., Liu, Y.R., 2005. Soil As contamination and its risk assessment in areas near the industrial districts of Chenzhou City, Southern China. Environ. Int. 31 (6), 791–798.

Limmer, M., Burken, J., 2016. Phytovolatilization of organic contaminants. Environ. Sci. Technol. 50 (13), 6632–6643.

Lindblom, S.D., Valdez-Barillas, J.R., Fakra, S.C., Marcus, M.A., Wangeline, A.L., Pilon-Smits, E.A., 2013. Influence of microbial associations on selenium localization and speciation in roots of *Astragalus* and *Stanleya* hyperaccumulators. Environ. Exp. Bot. 88, 33–42.

Liu, S., Zhang, F., Chen, J., Sun, G., 2011. Arsenic removal from contaminated soil via biovolatilization by genetically engineered bacteria under laboratory conditions. J. Environ. Sci. 23 (9), 1544–1550.

Liu, Z., Hong, Q., Xu, J.H., Jun, W., Li, S.P., 2006. Construction of a genetically engineered microorganism for degrading organophosphate and carbamate pesticides. Int. Biodeterior. Biodegrad. 58 (2), 65–69.

Lombi, E., Zhao, F.J., Dunham, S.J., McGrath, S.P., 2000. Cadmium accumulation in populations of *Thlaspi caerulescens* and *Thlaspi goesingense*. N. Phytol. 145 (1), 11–20.

López Chuken, U.J., Young, S.D., Guzmán Mar, J.L., 2010. Evaluating a 'biotic ligand model' applied to chloride enhanced Cd uptake by *Brassica juncea* from nutrient solution at constant Cd^{2+} activity. Environ. Technol. 31 (3), 307–318.

López-Chuken, U.J., 2012. Hydroponics and environmental clean-up. Hydroponics–a Standard Methodology for Plant Biological Researches. IntechOpen, p. 181.

Luo, J., Xing, X., Qi, S., Wu, J., Gu, X.S., 2019. Comparing the risk of metal leaching in phytoremediation using *Noccaea caerulescens* with or without electric field. Chemosphere 216, 661–668.

Lyu, Y., Zheng, W., Zheng, T., Tian, Y., 2014. Biodegradation of polycyclic aromatic hydrocarbons by *Novosphingobium pentaromativorans* US6-1. PLoS One 9 (7).

Ma, X., Yu, G., Li, X., Wang, T., Wu, D., Olsson, K.S., et al., 2016. Interfacial control of *Dzyaloshinskii-Moriya* interaction in heavy metal/ferromagnetic metal thin film hetero-structures. Phys. Rev. B 94 (18), 180–408.

Magi, E., Bianco, R., Ianni, C., Di Carro, M., 2002. Distribution of polycyclic aromatic hydrocarbons in the sediments of the Adriatic Sea. Environ. Pollut. 119 (1), 91–98.

Mahajan, P., Kaushal, J., 2018. Role of phytoremediation in reducing cadmium toxicity in soil and water. J. Toxicol. 2018.

Mahmood, A., Malik, R.N., 2014. Human health risk assessment of heavy metals via consumption of contaminated vegetables collected from different irrigation sources in Lahore, Pakistan. Arabian J. Chem. 7 (1), 91–99.

Mahmood, T., Malik, S.A., Hussain, S.T., 2010. Biosorption and recovery of heavy metals from aqueous solutions by *Eichhornia crassipes* (water hyacinth) ash. Bioresources 5 (2), 1244–1256.

Mandal, B.K., Suzuki, K.T., 2002. Arsenic round the world: a review. Talanta 58 (1), 201–235.

Manickam, N., Misra, R., Mayilraj, S., 2007. A novel pathway for the biodegradation of γ-hexachlorocyclohexane by a *Xanthomonas* sp. strain ICH12. J. Appl. Microbiol. 102 (6), 1468–1478.

Mao, J., Guan, W., 2016. Fungal degradation of polycyclic aromatic hydrocarbons (PAHs) by *Scopulariopsis brevicaulis* and its application in bioremediation of PAH-contaminated soil. Acta Agric. Scand. Sect B 66 (5), 399–405.

Marschner, H., 2011. Marschner's Mineral Nutrition of Higher Plants. Academic Press.

McCluggage, D., 1991. Heavy Metal Poisoning. NCS Magazine, published by the Bird Hospital, CO, USA, https://www.cockatiels.org/articles/Diseases/metals.html.

Mergeay, M., Monchy, S., Janssen, P., Van Houdt, R., Leys, N., 2009. Megaplasmids in *Cupriavidus* genus and metal resistance. Microbial Megaplasmids. Springer, Berlin, Heidelberg, pp. 209–238.

Misra, S., Gedamu, L., 1989. Heavy metal tolerant transgenic *Brassica napus* L. and *Nicotiana tabacum* L. plants. Theor. Appl. Genet. 78 (2), 161–168.

Mohammadi, M., Chalavi, V., Novakova Sura, M., Laliberté, J.F., Sylvestre, M., 2007. Expression of bacterial biphenyl-chlorobiphenyl dioxygenase genes in tobacco plants. Biotechnol. Bioeng. 97 (3), 496–505.

Mosa, K.A., Saadoun, I., Kumar, K., Helmy, M., Dhankher, O.P., 2016. Potential biotechnological strategies for the cleanup of heavy metals and metalloids. Front. Plant Sci. 7, 303.

Mudhoo, A., 2011. Phytoremediation of cadmium: a green approach. Green. Chem. Environ. Remediation 661–698.

National Institute of Neurological Disorders and Stroke (NINDS), 2007. Guillain-Barre Syndrome, Guillain-Barre Syndrome Fact Sheet.

Nayyar, N., Lal, R., 2016. Hexachlorocyclohexane contamination and solutions: brief history and beyond. Emerging model to study evolution of catabolic genes and pathways. J. Bioremed. Biodegrad. 7, 338.

Nautiyal, C.S., 1999. An efficient microbiological growth medium for screening phosphate solubilizing microorganisms. FEMS Microbiol Let 170 (1), 265–270.

Necibi, M., Mzoughi, N., 2017. The Distribution of Organic and Inorganic Pollutants in Marine Environment, Nova Science Publishers, New York.

Nie, L., Shah, S., Rashid, A., Burd, G.I., Dixon, D.G., Glick, B.R., 2002. Phytoremediation of arsenate contaminated soil by transgenic canola and the plan growth-promoting bacterium *Enterobacter cloacae* CAL2. Plant Physiol. Biochem. 40 (4), 355–361.

Nies, D.H., 2004. Essential and toxic effects of elements on microorganisms. In: Anke, K., Ihnat, M., Stoeppler, M. (Eds.), Metals And Their Compounds In The Environment. Wiley, Weinheim, Part II.1.

Novakova, M., Mackova, M., Chrastilova, Z., Viktorova, J., Szekeres, M., Demnerova, K., et al., 2009. Cloning the bacterial bphC gene into *Nicotiana tabacum* to improve the efficiency of PCB phytoremediation. Biotechnol. Bioeng. 102 (1), 29–37.

Ojuederie, O.B., Babalola, O.O., 2017. Microbial and plant-assisted bioremediation of heavy metal polluted environments: a review. Int. J. Environ. Res. Public Health 14 (12), 1504.

O'Keefe, D.P., Tepperman, J.M., Dean, C., Leto, K.J., Erbes, D.L., Odell, J.T., 1994. Plant expression of a bacterial cytochrome P450 that catalyzes activation of a sulfonylurea pro-herbicide. Plant Physiol. 105 (2), 473–482.

Oves, M., Saghir Khan, M., Huda Qari, A., Nadeen Felemban, M., Almeelbi, T., 2016. Heavy metals: biological importance and detoxification strategies. J. Bioremed. Biodegrad. 7 (2), 1–15.

Pal, R., Bala, S., Dadhwal, M., Kumar, M., Dhingra, G., Prakash, O., et al., 2005. Hexachlorocyclohexane-degrading bacterial strains *Sphingomonas paucimobilis* B90A, UT26 and Sp + , having similar lin genes, represent three distinct species, *Sphingobium indicum* sp. nov., *Sphingobium japonicum* sp. nov. and *Sphingobium francense* sp. nov., and reclassification of [*Sphingomonas*] *chungbukensis* as *Sphingobium chungbukense* comb. nov. Int. J. Syst. Evol. Microbiol. 55 (5), 1965–1972.

Pandey, A., Tripathi, P.H., Pandey, S.C., Pathak, V.M., Nailwal, T.K., 2018. Removal of toxic pollutants from soil using microbial biotechnology. Microbial Biotechnology in Environmental Monitoring and Cleanup. IGI Global, pp. 86–105.

Pandey, A.K., Chaudhary, P., Singh, S.B., Arora, A., Kumar, K., Chaudhry, S., et al., 2012. Deciphering the traits associated with PAH degradation by a novel *Serratia marcesencs* L-11 strain. J. Environ. Sci. Health A 47 (5), 755–765.

Pandey, N., Sharma, C.P., 2002. Effect of heavy metals Co^{2+}, Ni^{2+} and Cd^{2+} on growth and metabolism of cabbage. Plant Sci. 163 (4), 753–758.

Panhwar, Q.A., Othman, R., Rahman, Z.A., Meon, S., Ismail, M.R., 2012. Isolation and characterization of phosphate-solubilizing bacteria from aerobic rice. Afr. J. Biotechnol. 11 (11), 2711–2719.

Paraiba, L.C., Spadotto, C.A., 2002. Soil temperature effect in calculating attenuation and retardation factors. Chemosphere 48 (9), 905–912.

Phillips, T.M., Seech, A.G., Lee, H., Trevors, J.T., 2005. Biodegradation of hexachlorocyclohexane (HCH) by microorganisms. Biodegradation 16 (4), 363–392.

Pilon-Smits, E., 2005. Phytoremediation. Annu. Rev. Plant Biol. 56, 15–39.

Poniedziałek, M., Sękara, A., Ciura, J., Jędrszczyk, E., 2005. Nickel and manganese accumulation and distribution in organs of nine crops. Folia Hortic. 17 (1), 11–22.

Prasad, M.N.V., 2003. Phytoremediation of metal-polluted ecosystems: hype for commercialization. Russ. J. Plant Physiol. 50 (5), 686–701.

Prasad, M.N.V., 2008. Trace Elements as Contaminants and Nutrients: Consequences in Ecosystems and Human Health. John Wiley & Sons.

Premuzic, E.T., Francis, A.J., Lin, M., Schubert, J., 1985. Induced formation of chelating agents by *Pseudomonas aeruginosa* grown in presence of thorium and uranium. Arch. Environ. Contam. Toxicol. 14 (6), 759–768.

Quintero, J.C., Moreira, M.T., Feijoo, G., Lema, J.M., 2005. Anaerobic degradation of hexachlorocyclohexane isomers in liquid and soil slurry systems. Chemosphere 61 (4), 528–536.

Rahman, R.A., Abou-Shanab, R.A., Moawad, H., 2008. Mercury detoxification using genetic engineered *Nicotiana tabacum.* Global Nest J. 10, 432–438.

Raina, V., Suar, M., Singh, A., Prakash, O., Dadhwal, M., Gupta, S.K., et al., 2008. Enhanced biodegradation of hexachlorocyclohexane (HCH) in contaminated soils via inoculation with *Sphingobium indicum* B90A. Biodegradation 19 (1), 27–40.

Ramasamy, K., Banu, S.P., 2007. Bioremediation of metals: microbial processes and techniques. Environmental Bioremediation Technologies. Springer, Berlin, Heidelberg, pp. 173–187.

Reeves, R.D., Baker, A.J.M., Raskin, I., Ensley, B.D., 2000. Phytoremediation of toxic metals. Using Plants to Clean Up the Environment. John Wiley & Sons, New York, pp. 193–229.

Reeves, R.D., Baker, A.J., Jaffré, T., Erskine, P.D., Echevarria, G., van der Ent, A., 2018. A global database for plants that hyperaccumulate metal and metalloid trace elements. N. Phytol. 218 (2), 407–411.

Rezek, J., in der Wiesche, C., Mackova, M., Zadrazil, F., Macek, T., 2008. The effect of ryegrass (*Lolium perenne*) on decrease of PAH content in long term contaminated soil. Chemosphere 70 (9), 1603–1608.

Rhine, E.D., Phelps, C.D., Young, L.Y., 2006. Anaerobic arsenite oxidation by novel denitrifying isolates. Environ. Microbiol. 8 (5), 899–908.

Rugh, C.L., Senecoff, J.F., Meagher, R.B., Merkle, S.A., 1998. Development of transgenic yellow poplar for mercury phytoremediation. Nat. Biotechnol. 16, 925–928.

Rugh, C.L., Wilde, H.D., Stack, N.M., Thompson, D.M., Summers, A.O., Meagher, R.B., 1996. Mercuric ion reduction and resistance in transgenic *Arabidopsis thaliana* plants expressing a modified bacterial merA gene. Proc. Natl. Acad. Sci. U.S.A. 93 (8), 3182–3187.

Ruiz, O.N., Hussein, H.S., Terry, N., Daniell, H., 2003. Phytoremediation of organomercurial compounds via chloroplast genetic engineering. Plant Physiol. 132 (3), 1344–1352.

Rylott, E.L., Jackson, R.G., Edwards, J., Womack, G.L., Seth-Smith, H.M., Rathbone, D.A., et al., 2006. An explosive-degrading cytochrome P450 activity and its targeted application for the phytoremediation of RDX. Nat. Biotechnol. 24 (2), 216–219.

Saba, B., Hashmi, I., Awan, M.A., Nasir, H., Khan, S.J., 2012. Distribution, toxicity level, and concentration of polycyclic aromatic hydrocarbons (PAHs) in surface soil and groundwater of Rawalpindi, Pakistan. Desalin Water Treat 49 (1-3), 240–247.

Saba, B., Rafique, U., Hashmi, I., 2011. Adsorption kinetics of anthracene and phenanthrene in different soils of Attock Refinery Limited (ARL) Rawalpindi, Pakistan. Desalin. Water Treat. 30 (1–3), 333–338.

Salam, J.A., Das, N., 2012. Remediation of lindane from environment—an overview. Int. J. Adv. Biol. Res. 2, 9–15.

Sarma, H., Prasad, M.N.V., 2019. Metabolic engineering of rhizobacteria associated with plants for remediation of toxic metals and metalloids. Transgenic Plant Technology for Remediation of Toxic Metals and Metalloids. Academic Press, pp. 299–318.

Schlegel, H.G., Cosson, J.P., Baker, A.J.M., 1991. Nickel hyperaccumulating plants provide a niche for nickel resistant bacteria. Bot. Acta 104 (1), 18–25.

Schmidt, T., Schlegel, H.G., 1994. Combined nickel-cobalt-cadmium resistance encoded by the ncc locus of *Alcaligenes xylosoxidans* 31A. J. Bacteriol. Res. 176 (22), 7045–7054.

Seleiman, M.F., Santanen, A., Jaakkola, S., Ekholm, P., Hartikainen, H., Stoddard, F.L., et al., 2013. Biomass yield and quality of bioenergy crops grown with synthetic and organic fertilizers. Biomass Bioenergy 59, 477–485.

Shaik, R., Pillay, B., Pillay, D., 2008. Genetic sequences derived from suppression subtractive hybridization analysis provides insight into their possible roles in *Xanthomonas albilineans*. Afr. J. Biotechnol. 7, 12.

Sharma, P., Dubey, R.S., 2005. Lead toxicity in plants. Braz. J. Plant Physiol. 17 (1), 35–52.

Sharma, S., Tiwari, S., Hasan, A., Saxena, V., Pandey, L.M., 2018. Recent advances in conventional and contemporary methods for remediation of heavy metal-contaminated soils. Biotechnology 8 (4), 216.

Sheng, X.F., Xia, J.J., Jiang, C.Y., He, L.Y., Qian, M., 2008. Characterization of heavy metal-resistant endophytic bacteria from rape (*Brassica napus*) roots and their potential in promoting the growth and lead accumulation of rape. Environ. Pollut. 156 (3), 1164–1170.

Shin, M.N., Shim, J., You, Y., Myung, H., Bang, K.S., Cho, M., et al., 2012. Characterization of lead resistant endophytic *Bacillus* sp. MN3-4 and its potential for promoting lead accumulation in metal hyperaccumulator *Alnus firma*. J. Hazard. Mater. 199, 314–320.

Shrivastava, M., Khandelwal, A., Srivastava, S., 2019. Heavy metal hyperaccumulator plants: the resource to understand the extreme adaptations of plants towards heavy metals. Plant-Metal Interactions. Springer, Cham, pp. 79–97.

Siddiquee, S., Rovina, K., Azad, S.A., Naher, L., Suryani, S., Chaikaew, P., 2015. Heavy metal contaminants removal from wastewater using the potential filamentous fungi biomass: a review. J. Microb. Biochem. Technol. 7 (6), 384–395.

Singh, S.K., Haritash, A.K., 2019. Polycyclic aromatic hydrocarbons: soil pollution and remediation. Int. J. Environ. Sci. 1–24.

Singh, S.K., Iizuka, Y., Xu, Q., 2011. Nickel-palladium nanoparticle catalyzed hydrogen generation from hydrous hydrazine for chemical hydrogen storage. Int. J. Hydrogen Energy 36 (18), 11794–11801.

Smedley, P.L., Kinniburgh, D.G., 2013. Arsenic in groundwater and the environment. Essentials of Medical Geology. Springer, Dordrecht, pp. 279–310.

Smith, M.J., Flowers, T.H., Duncan, H.J., Alder, J., 2006. Effects of polycyclic aromatic hydrocarbons on germination and subsequent growth of grasses and legumes in freshly contaminated soil and soil with aged PAHs residues. Environ. Pollut. 141 (3), 519–525.

Soares, B.L., Ferreira, P.A.A., Oliveira-Longatti, S.M.D., Marra, L.M., Rufini, M., Andrade, M.J.B.D., et al., 2014. Cowpea symbiotic efficiency, pH and aluminum tolerance in nitrogen-fixing bacteria. Sci. Agricola 71 (3), 171–180.

Sobolev, D., Begonia, M., 2008. Effects of heavy metal contamination upon soil microbes: lead-induced changes in general and denitrifying microbial communities as evidenced by molecular markers. Int. J. Environ. Res. Public Health 5 (5), 450–456.

Song, W.Y., Sohn, E.J., Martinoia, E., Lee, Y.J., Yang, Y.Y., Jasinski, M., et al., 2003. Engineering tolerance and accumulation of lead and cadmium in transgenic plants. Nat. Biotechnol. 21 (8), 914–919.

Sonoki, T., Kajita, S., Ikeda, S., Uesugi, M., Tatsumi, K., Katayama, Y., et al., 2005. Transgenic tobacco expressing fungal laccase promotes the detoxification of environmental pollutants. Appl. Microbiol. Biotechnol 67 (1), 138–142.

Sreelal, G., Jayanthi, R., 2017. Review on phytoremediation technology for removal of soil contaminant. Indian J. Sci. Res. 14 (1), 127–130.

Tak, H.I., Ahmad, F., Babalola, O.O., 2013. Advances in the application of plant growth-promoting rhizobacteria in phytoremediation of heavy metals. Rev. Environ. Contam. T 223, 33–52.

Tchounwou, P.B., Yedjou, C.G., Patlolla, A.K., Sutton, D.J., 2012. Heavy metal toxicity and the environment. Molecular, Clinical and Environmental Toxicology. Springer, Basel, pp. 133–164.

Thomas, J.C., Davies, E.C., Malick, F.K., Endreszl, C., Williams, C.R., Abbas, M., et al., 2003. Yeast metallothionein in transgenic tobacco promotes copper uptake from contaminated soils. Biotechnol. Prog. 19 (2), 273–280.

Uchida, E., Ouchi, T., Suzuki, Y., Yoshida, T., Habe, H., Yamaguchi, I., et al., 2005. Secretion of bacterial xenobiotic-degrading enzymes from transgenic plants by an apoplastic expressional system: an applicability for phytoremediation. Environ. Sci. Technol. 39 (19), 7671–7677.

Uren, N.C., 2000. Types, amounts, and possible functions of compounds released into the rhizosphere by soil-grown plants. The Rhizosphere. CRC Press, pp. 35–56.

Van der Ent, A., Baker, A.J., Reeves, R.D., Pollard, A.J., Schat, H., 2013. Hyperaccumulators of metal and metalloid trace elements: facts and fiction. Plant Soil 362 (1–2), 319–334.

Van Dillewijn, P., Couselo, J.L., Corredoira, E., Delgado, A., Wittich, R.M., Ballester, A., et al., 2008. Bioremediation of 2, 4, 6-trinitrotoluene by bacterial nitroreductase expressing transgenic aspen. Environ. Sci. Technol. 42 (19), 7405–7410.

Verma, S., Dubey, R.S., 2003. Lead toxicity induces lipid peroxidation and alters the activities of antioxidant enzymes in growing rice plants. Plant Sci. 164 (4), 645–655.

Vijgen, J., Abhilash, P.C., Li, Y.F., Lal, R., Forter, M., Torres, J., et al., 2011. Hexachlorocyclohexane (HCH) as new Stockholm Convention POPs—a global perspective on the management of lindane and its waste isomers. Environ. Sci. Pollut 18 (2), 152–162.

Wang, L., Samac, D.A., Shapir, N., Wackett, L.P., Vance, C.P., Olszewski, N.E., et al., 2005. Biodegradation of atrazine in transgenic plants expressing a modified bacterial atrazine chlorohydrolase (atzA) gene. Plant Biotechnol. J 3 (5), 475–486.

Wang, X., Wu, N., Guo, J., Chu, X., Tian, J., Yao, B., et al., 2008. Phytodegradation of organophosphorus compounds by transgenic plants expressing a bacterial organophosphorus hydrolase. Biochem. Biophys. Res. Commun. 365 (3), 453–458.

Wangeline, A.L., Burkhead, J.L., Hale, K.L., Lindblom, S.D., Terry, N., Pilon, M., et al., 2004. Overexpression of ATP sulfurylase in Indian mustard. J. Environ. Qual. 33 (1), 54–60.

Welch, A.H., Westjohn, D.B., Helsel, D.R., Wanty, R.B., 2000. Arsenic in ground water of the United States: occurrence and geochemistry. Groundwater 38 (4), 589–604.

Weyens, N., van der Lelie, D., Taghavi, S., Newman, L., Vangronsveld, J., 2009. Exploiting plant–microbe partnerships to improve biomass production and remediation. Trends Biotechnol. 27 (10), 591–598.

World Health Organization, 2012.). Guidelines for Procuring Public Health Pesticides (No. WHO/HTM/NTD/WHOPES/2012.4. WHO, Geneva.

World Health Organization, 2017. Lead poisoning and health fact sheet. updated August 2017.

Wuertz, S., 1997. The impact of heavy metals on soil microbial communities and their activities.—p. 607-642 (No. COLECCION GENERAL/631.46 M689m). En: Modern Soil Microbiology. Marcel Dekker, New York, 1997.

Yang, T., Chen, M.L., Wang, J.H., 2015. Genetic and chemical modification of cells for selective separation and analysis of heavy metals of biological or environmental significance. Trends Anal. Chem. 66, 90–102.

Zazouli, M.A., Mahdavi, Y., Bazrafshan, E., Balarak, D., 2014. Phytodegradation potential of bisphenolA from aqueous solution by *Azolla filiculoides*. J. Environ. Health Sci. 12 (1), 66.

Zhang, S., Lin, H., Deng, L., Gong, G., Jia, Y., Xu, X., et al., 2013a. Cadmium tolerance and accumulation characteristics of *Siegesbeckia orientalis* L. Ecol. Eng. 51, 133–139.

Zhang, Y., Liu, J., Zhou, Y., Gong, T., Wang, J., Ge, Y., 2013b. Enhanced phytoremediation of mixed heavy metal (mercury)–organic pollutants (trichloroethylene) with transgenic alfalfa co-expressing glutathione S-transferase and human P450 2E1. J. Hazard. Mater. 260, 1100–1107.

Zhao, F.J., Hamon, R.E., McLaughlin, M.J., 2001. Root exudates of the hyperaccumulator Thlaspi caerulescens do not enhance metal mobilization. N. Phytol. 151 (3), 613–620.

Zheng, Z., Obbard, J.P., 2003. Oxidation of polycyclic aromatic hydrocarbons by fungal isolates from an oil contaminated refinery soil. Environ. Sci. Pollut 10 (3), 173–176.

Zhu, Y.L., Pilon-Smits, E.A., Jouanin, L., Terry, N., 1999a. Overexpression of glutathione synthetase in Indian mustard enhances cadmium accumulation and tolerance. Plant Physiol. 119 (1), 73–80.

Zhu, Y.L., Pilon-Smits, E.A., Tarun, A.S., Weber, S.U., Jouanin, L., Terry, N., 1999b. Cadmium tolerance and accumulation in Indian mustard is enhanced by overexpressing γ-glutamylcysteine synthetase. Plant Physiol. 121 (4), 1169–1177.

Zucchi, M., Angiolini, L., Borin, S., Brusetti, L., Dietrich, N., Gigliotti, C., et al., 2003. Response of bacterial community during bioremediation of an oil polluted soil. J. Appl. Microbiol. 94 (2), 248–257.

Chapter 17

Bioremediation potential of methylotrophic bacteria

Deen Dayal Giri[1], Sandeep Kumar Singh[2], Ajay Giri[2], Himanshu Dwivedi[1] and Ajay Kumar[3]
[1]*Department of Botany, Maharaj Singh College, Saharanpur, India,* [2]*Department of Botany, Centre of Advance Study in Science, Institute of Science, Banaras Hindu University, Varanasi, India,* [3]*Agriculture Research Organization, Ministry of Agriculture and Rural Development, Volcani Center, Rishon LeZion, Israel*

17.1 Introduction

The bacteria utilizing methane as the sole source of carbon and energy are called methanotrophs whereas other related bacteria utilizing reduced one carbon or multi-carbon compounds with no carbon−carbon bond like methanol and other methylated compounds are called methylotrophs. Methylotrophs were first reported by Söhngen (1906), but its various details such as widespread distribution, taxonomic diversity, and physiological diversity were described quite later (Colby and Dalton, 1979; Hanson, 1980; Quayle and Ferenci, 1978). The methylotrophs are widely distributed on phylloplane, phyllosphere, rhizosphere of the different plant species. Reliable stable isotope probing showed their presence in the rhizosphere (Taubert et al., 2016). In their habitat, these bacteria perform variety of important functions including minimizing global warming effect by assimilating radiatively active greenhouse gas methane from their environment. Depending upon the ease of utilizing the carbon substrates they can be divided into three subgroups (Kolb and Stacheter, 2013; Iguchi et al., 2015). First, the specific obligate methylotrophs utilize only single carbon compounds; second, limited facultative methylotrophs utilize a limited variety of complex compounds of carbon, whereas third, less-restricted facultative methylotrophs utilize a wide range of complicated carbon compounds (Jenkins et al., 1987). *Methylovorus*, *Methylophilus*, *Methylobacillus* of β-proteobacteria are examples of restricted facultative methylotrophs and genus *Methylobacterium* of *α-proteobacteria* is less-restricted facultative methylotrophs (Govorukhina and Trotsenko, 1991; Yordy and Weaver, 1977; Jenkins et al., 1987; Urakami and Komagata, 1986). Their ability to utilize different organic compounds (Giri et al., 2013a), tolerance to harsh nutrient-limited spoiled soiled (Giri et al., 2013b), and nonpathogenic nature makes them useful in commercial application in agriculture as seed coat bioinoculants for boosting seed germination (Meena et al., 2012) as well as application in bioremediation of spoiled soils. They were readily increased their number in spoiled coalmine soils from 10^4 to 10^6 after 12 years of revegetation and helped in restoring the normal biological activity in these soils (Giri et al., 2013b).

17.2 Mechanism of carbon assimilation

The methanotrophs oxidize C_1 compounds to methanol with the help of enzyme methane monooxygenase, whereas in methylotrophs, this methanol or other organic derivatives are further converted to formaldehyde, a key compound to the catabolism and anabolism of this group of bacteria. Formaldehyde in part is being oxidized to CO_2 in catabolism for energy production pathway and part being absorbed for carbon in the cell via two specific pathways, the ribulose monophosphate pathway or the serine pathway. Farmaldehyde is important in understanding methylotrophic metabolism. Due to its cytotoxicity, methylotrophs adapt some strategies for its intracellular accumulation (Vorholt, 2002, Yurimoto et al., 2005). The first way is trapping of formaldehyde with cofactors such as tetrahydrofolate, tetrahydromethanopterin, glutathione, or mycothiol before its oxidation to CO_2. The other alternative is its trapping on the sugar phosphates as the first reaction in the C_1 assimilation RuMP pathways in methylotrophic bacteria. The RuMP pathway works as a profoundly proficient framework for catching free HCHO at generally low concentrations (Dijkhuizen et al., 1992).

Microbe Mediated Remediation of Environmental Contaminants. DOI: https://doi.org/10.1016/B978-0-12-821199-1.00017-1

Bacteria such as *Methylomonas*, *Methylocaldum*, *Methylosphaera*, *Methylomicrobium*, and *Methylobacter* assimilate formaldehyde produced from the oxidation of CH_4 via CH_3OH intermediate by RuMP pathway (Anthony, 1986).

Aldol condensation of HCHO with ribulose 5P (5C) produces a 6C compound D-arabino-3-hexulose 6P with the help of enzyme 3-hexulose-6-phosphate synthase (HPS). The D-arabino-3-hexulose 6P is isomerized to fructose-6P by enzyme 6-phospho-3-hexuloisomerase (PHI). The genes for the enzyme HPS and PHI has been cloned from an obligate methylotroph *Methylomonas aminofaciens* (Yanase et al., 1996, Sakai et al., 1999), facultative methylotroph *Mycobacterium gastri* MB19 (Mitsui et al., 2000), and thermotolerant methylotroph, *Bacillus brevis* S1 (Yurimoto et al., 2002).

The fructose 1,6-bisP is metabolized to pyruvate through the glycolytic pathway. The acceptor of HCHO is regenerated from $C_6H_{13}O_9P$ and glyceraldehyde-3P through a combination of transketolase, transaldolase, and several reactions for pentose phosphate isomerizations (regeneration stage). There are two elective stages for the cleavage of $C_6H_{13}O_9P$ to C_3 compounds and two phases for the regeneration of $C_5H_{11}O_8P$. The combination of two cleavage and two regeneration stages, thus, leads to a total of 4 possible variants of the RuMP pathway (Kato et al., 2006).

17.2.1 Serine pathway

Bacteria such as *Methylocystis* and *Methylosinus* assimilate formaldehyde by relatively less-productive pathway serine pathway (Hanson and Hanson, 1996). The pathway starts with the build-up of methylene tetrahydrofolate and glycine which synthesize serine. This 3C compound then undergoes a series of transformations to phosphoenolpyruvate, which is carboxylated to form malate. By completing the cycle, malate is converted into malylCoA that cleaved into two 2C compounds acetyl CoA and glyoxalate. In microbes having serine cycle, transformation of acetyl CoA to glyoxylate is not clear. The usual path could be via enzyme isocitrate lyase and found in a few strains only (Chistoserdova and Lidstrom, 1996). Most of the genes encoding enzymes involved in the serine cycle have been cloned and sequenced from *Methylobacterium extorquens* AM1 or *Hyphomicrobium methylovorum* GM2 (Chistoserdova and Lidstrom, 1997; Tanaka et al., 1997). In every C_3 compound produced via the serine cycle, 2C are derived from HCHO and 1C originates from CO_2 (Lidstrom, 2006). The methylotrophic carbon assimilation pathways are depicted in Fig. 17.1.

17.3 Bioremediation of simple organic compounds

The methylotrophs and their enzymes are apparently well suited for their application in processes of biodegradation and bioremediation because they oxidize CH_3OH to HCHO and $HCOO^-$ without formation of undesirable by-products (Jakobsen, 2006; Koopman et al., 2009; Qiu, 2014). Their high phenotypic plasticity allows them to occupy different extreme habitats of soil/sediment and water in addition to well-suited inhabitant on plants and animals (Fedorov et al., 2011; Dourado et al., 2015). The various habitats mentioned previously are occupied by phylogenetically diverse methylotrophs adapted to the specific environmental challenges by tuning their metabolism. Such niches could be used for selection of methylotrophs for developing of desired technologies in bioremediation (Chistoserdova, 2018).

17.3.1 Bioremediation of methylated amines

Methylamine generally presents in environment due to degradation of nitrogenous compounds (Anthony, 1982, Barrett and Kwan, 1985, Burg and Ferraris, 2008). Bellion et al. (1980) described an uptake system for methylamine in a methylotrophic bacterium *Pseudomonas* sp. strain MA by growing this bacteria on methylamine as the sole source of C and N. Methylamine was the sole N, but not C in *Pseudomonas* sp. strain MA (Bellion and Wayland, 1982). Bacteria such as *Methylopila helvetica* sp. and *Methylobacterium dichloromethanicum* utilized methylamine (Doronina et al., 2000). The C_1-utilizing bacteria effectively removed methylated amines (Schaefer et al., 2010; Nanda et al., 2012). Methylamine is utilized in methylamine dehydrogenase pathway, methylamine oxidase pathway, and a pathway having methylated glutamate (GMA—γ-glutamylmethylamide—and NMG—*N*-methylglutamate) (Anthony, 1982). The genes for GMA synthetase, NMG synthase, and NMG dehydrogenase have been identified in methylotrophic bacteria *Methyloversatilis universalis* (Latypova et al., 2010).

A total of 12 bacterial isolates grow on trimethylamine, methylamine, dimethylamine as the source of carbon and energy. In these, only 2 isolates that grow on methylamines/methanol are called obligate methylotrophs and rest of the 10 isolates growing on a variety of C_1 organic substrates are called facultative methylotrophs. The obligate methylotrophs oxidized trimethylamine by pathways involving enzyme trimethylamine dehydrogenase and phenazine

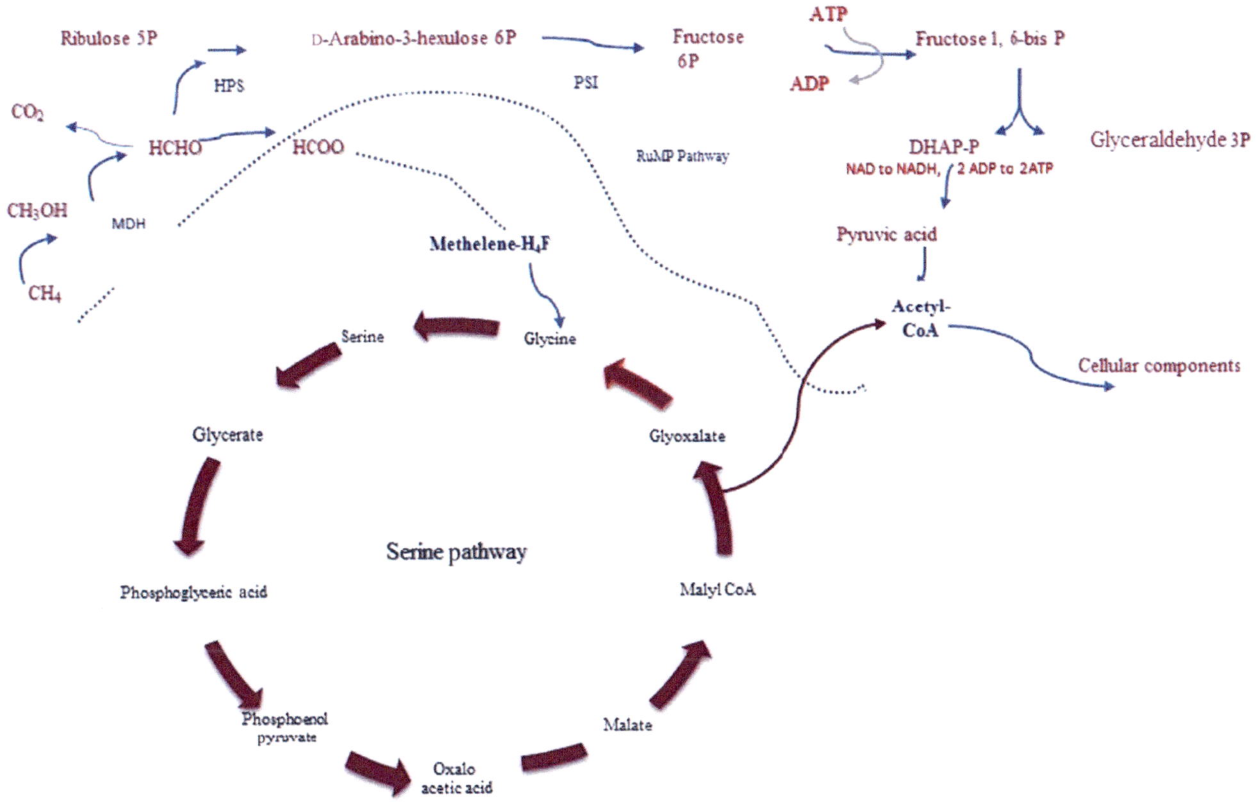

FIGURE 17.1 A generalized pathway for carbon assimilation in methylotrophic bacteria. *DHAP*, Dihydroxy acetone P; *enzyme MMO*, methane monooxygenase in methanotrophs; *HPS*, D-arabino-3-huxulose 6P synthase; *MDH*, methanol dehydrogenase; *PSI*, 6-phospho-3-hexuloisomerase.

methosulfate as a primary hydrogen acceptor, whereas facultative methylotrophs used enzyme trimethylamine monooxygenase and trimethylamine N-oxide demethylase in the oxidation of trimethylamine (Colby and Zatman, 1973).

17.3.2 Bioremediation of methylated sulfur compounds

Monomethyl sulfate (MS) is a by-product of large-scale reaction synthesizing agrochemical and dye stuffs. It is produced during methylation of methyl-accepting group such as hydroxyl, mercapto, amino, or imino group in organic syntheses using dimethyl sulfate (DMS) as the methyl donor. The MS remained in mother liquors is generally useless as it is not recycled or remethylated to produce DMS. The disposal of MS by hydrolysis can create security problems. So, the only alternative is biodegradation of mother liquor using microorganism. Most of the previously reported methylotrophs were not able to degrade MS, but later MS-degrading methylotrophs were isolated by selective enrichment in MV7 agar from sewage sludge and industrial sewage (Ghisalba and Kuenzi, 1983). In the last decade, methylotrophs capable of utilizing sulfur-containing compounds such as dimethylsulfide, dimethyldisulfide, methanesulfonic acid have been reported (Schaefer et al., 2010; Nanda et al., 2012).

17.3.3 Bioremediation of halogenated methane

Methylotrophic bacteria are capable to utilize the various halogenated compounds such as chloro, bromo, and iodomethanes (Schaefer et al., 2007; Flanagan, 1998). The utilization ability of halogenated compounds develops in methylotrophs in long-term selective pressure of a specific C_1 substrate (Goodwin et al., 2001; McDonald, 2002).

17.3.3.1 Chloromethane (CM CH_3Cl)

Chloromethane is a volatile compound responsible for stratospheric ozone destruction (Harper et al., 2003, Tsai, 2017). Global emissions of chloromethane are about 2.0 Tg/year (Carpenter et al., 2014). Its terrestrial emissions are mainly through plants and white rot fungi in soil (Harper, 2000, Keppler et al., 2005; Schäfer et al., 2007). This CH_3Cl is produced from methoxy groups of lignin and pectin in plant cell wall during plant biomass burning (Keppler et al., 2004). Global forest soils consume naturally about 1.0 Tg/year CH_3Cl (Yoshida et al., 2004, Carpenter et al., 2014). However, its global budget is uncertain in respect to the nature of the biotic sink and spatiotemporal variation at the regional and global scale (Redeker and Kalin, 2012).

17.3.3.2 Dichloromethane (DCM $CHCl_2$)

Several chlorinated derivatives of CH_4, $CH_3{-}CH_3$, and $CH_2 = CH_2$ were shown to be degraded by mixed bacterial cultures enriched on CH_4 as carbon substrate (Fogel et al., 1986, Janssen et al., 1987). Some aerobic dichloromethane (DCM)-utilizers are *Methylophilus leisingerii* DM11 (Doronina and Trotsenko, 1994), *Methylorhabdus multivorans* DM13 (Doronina et al., 1995), *Paracoccus methylutens* DM12 (Doronina et al., 1998), *M. helvetica* sp., *M. dichloromethanicum* sp. (Doronina et al., 2000).

Methylotrophic genera *Methylobacterium* sp. strain DM4 and *Methylophilus* sp. strain DM11 can grow with DCM as the sole source of carbon and energy by virtue of homologous glutathione-dependent DCM dehalogenases (Gisi et al., 1998). *Hyphomicrobium* and *Methylobacterium* dehalogenate $CHCl_2$ to HCHO by a glutathione transferase reaction catalyzed by the DCM dehalogenase (Kohler-Staub and Leisinger, 1985). DCM dehalogenase was found to be a product of the gene dcmA in methylotrophs (Bader and Leisinger, 1994). Glutathione 5-transferases function was not well known (Vuilleumier, 1997), this inducible enzyme catalyzes thiolytic-substitutive dehalogenation in facultative methylotrophs. DCM is converted to HCHO and Cl^- via 5-chloromethylgutathione intermediate. This enzyme is also present in some other bacteria genera such as *Pseudomonas* sp., *Hyphomicrobium* sp., and *Methylobacterium* sp. (Bhatt et al., 2007).

17.3.3.3 Trichloroethylene

Trichloroethylene (TCE) is a noninflammable industrial solvent for many organic compounds and is commonly used as degreaser. It is contaminated and very harmful if drank. Tetrachloroethene ($Cl_2C = CCl_2$) and trichloroethene ($ClCH = CCl_2$) are the primary contaminants, whereas DCE ($ClCH = CHCl$) and vinyl chloride ($CH_2 = CHCl$) are toxic by-products of their incomplete bacterial degradation. Vinyl chloride degrades more rapidly than DCE and TCE, whereas CH_2Cl_2 oxidizes faster than $CHCl_3$ (McCarty, 1988).

Methylococcus capsulatus (Bath) and *Methylosinus* sp. were able to degrade TCE (Hanson and Hanson (1996). *Methylosinus trichosporium* oxidized different chlorinated hydrocarbons such as $ClCH = CHCl$, CH_3CHCl_2, and $CH_2 = CHCl$ (Fox et al., 1990). TCE degradation by *Methylosinus* sp. was more rapid than ammonia monooxygenase-containing nitrifying bacteria (Arciero et al., 1989), toluene 2(or 4)-monooxygenase-containing *Pseudomonas* sp. (Folsom et al., 1990, Winter et al., 1989), and propane monooxygenase *Mycobacterium* sp. (Wackett and Gibson, 1988). Thus the pure culture as well as methylotrophic consortia efficiently degraded TCE (Little et al., 1988; Fernandes et al., 2009). The indigenous-dechlorinating consortia present in the contaminated groundwater could be used as a starting point for future bioremediation activities (Kotik et al., 2013).

17.3.3.4 tert-Butyl alcohol

tert-Butyl alcohol (TBA) is produced by the degradation of methyl *tert*-butyl ether, ethyl *tert*-butyl ether as well as *tert*-amyl alcohol methylotrophic bacteria *Burkholderia cepacia* CIP 1−2052 isolated from an activated sludge that was able to use TBA (Piveteau et al., 2001).

17.3.3.5 Brominated hydrocarbons

Brominated halocarbons are responsible for the destruction of atmospheric ozone in the troposphere and lower stratosphere (Montzka and Reimann, 2011). Atmospheric life of $CHBr_3$ is 24 days and CH_2Br_2 123 days as per WMO (World Meteorological Organization) (2011). Abiotic anoxic hydrolysis under strong sunlight $CHBr_3$ is transformed CH_2Br_2 (Betterton et al., 1995).

Brominated halocarbons are heterogeneously distributed in ocean (Yamamoto et al., 2001). However, their oceanic production, degradation, and transformation of brominated halocarbons are not well known. In ocean, they are biogenically produced by microorganism (Quack and Wallace, 2003), phytoplankton, and macroalgae (Sturges et al., 1993). In the Atlantic Ocean, CH_2Br_2 production was correlated with the abundance of cyanobacteria *Synechococcus*, *picoeukaryotes*, and heterotrophic bacteria (Liu et al., 2013). When $CHBr_3$ added to subarctic surface seawater at eastern coast of

Hokkaido, there developed a bacterial community in response to $CHBr_3$ and/or CH_3OH that plays a major role in dehalogenation reaction (Kataoka et al., 2019). In early exponential phase, marine bacteria such as *Phaeobacter gallaeciensis*, *Roseobacter denitrificans*, and *Rhodobacter vinaykumarii* were able to convert $CHBr_3$ to CH_2Br_2 (Ichikawa et al., 2015), but no further debromination takes place by these bacteria. Thus these bacteria may act as new sinks for $CHBr_3$ and new sources for CH_2Br_2 marine environments.

17.3.4 Bioremediation of complex xenobiotics

Xenobiotics are those chemical compounds that are very much similar in characteristics to naturally occurring compounds supporting life forms. So, when these are present in the environment, these affect numerous life forms in different ways. These compounds are eminently toxic and may impede with ordinary growth and developmental of different life forms (Mansuy, 2013). Xenobiotics are difficult to process and they may be gathered in creatures of subordinate trophic level in food chain and passed to creatures of higher trophic level (i.e., biomagnified). However, major form of xenobiotics present in soil are distinctive forms of pesticides due to their substantial use in agriculture (Aktar et al., 2009).

In many reports, it was mentioned that enormous number of methylotrophs fit for degrading assortment of xenobiotics. Large et al. (1972) reported oxygen incorporation in trimethylamine to develop trimethylamine N-oxide when grown *Pseudomonas aminovorans* and *Hyphomicrobium vulgare* on trimethylamine. Microbial enzymes catalyzing the demethylation of C_3H_9NO yielding dimethylamine and formaldehyde have been seen in *Bacillus* PM6 and *P. aminovorans* (Large, 1971; Myers and Zatman, 1971).

17.3.5 Polyaromatic compounds

Methylotrophic bacteria can degrade large number of aromatic pollutants due to broad substrate specificity of enzyme MMO (Jechorek et al., 2003). They can degrade aromatic and heterocyclic compounds, for example, *M. helvetica* sp. and *M. dichloromethanicum* sp. utilizing variety of polycarbon compounds as the C and N source. It was studied for their potential use in bioremediation (Doronina et al., 2000). *Methylobacterium populi* destructed polyaromatic hydrocarbons (Ventorino, 2014). Some methylotrophs were able to destruct toxic explosives (TNT: 2,4,6-trinitrotoluene; RDX: hexahydro-1,3,5-trinitro-1,3,5-triazene; HMX octahydro-1,3,5,7-tetranitro-1,3,5-tetrazocine) (Van Aken et al., 2004). The *Methylobacterium* sp. GPE1 was able to degrade carbazole, heterocyclic constituent in smoke of tobacco (Pasternak and Kołwzan, 2013). Bacterial consortia with methylotrophs efficiently removed phenol (Khongkhaem et al., 2011) and also common soil pollutants such as methyl-*tert*-butyl ether (Müller et al., 2008; Zhang et al., 2008) and TCE (Fernandes et al., 2009).

Genomic data can predict methylotrophic lifestyle basing on the concept of minimal metabolic modules (Chistoserdova, 2011) is constantly expanding our understanding about the distribution and environmental roles of this group of bacteria. Genomic data of polyethylene–terephthalate degrading bacteria *Ideonella sakaiensis* 201-F6 (Yoshida et al., 2016) suggest that it can also employ autotrophic methylotrophy.

17.3.6 Methylotrophs consortium treating industrial effluent

Industrial effluents contain large quantity of toxic pollutants depending upon the industry type. Bioremediation of the industrial effluent remains difficult. To remediate the industrial effluents, methylotrophs were isolated from the variety of soils and sediments of polluted and pristine sites to withstand high level of pollutants (De Marco et al., 2004). The mixed methylotrophic bacterial consortium was used as microbial seed that effectively remediated the common industrial effluent from fertilizer chemical and pesticide industry. It lowered COD range from 950 to 2000 mg/mL to below detection limit in batch reactor (60–96 hours, 6 L) and continuous reactor (9–15 days, 6 L) (Hingurao and Nerurkar, 2018).

17.3.7 Biodegradation of hydrocarbons in oil spill

In the Gulf of Mexico an incidence of oil spill was reported on April 20, 2010 that continued for a period over 12 weeks (Reddy et al., 2012). The estimated release of total crude oil and gas was about 7 lakhs tonnes. This deepwater oil plume entrained below 1000–1300 m and spanned in large dimension of $35 \times 2 \times 0.2$ km (Camilli et al., 2010; Diercks et al., 2010).

17.3.8 Biodegradable polymers production

Petroleum-based plastics are notorious for polluting a range of soil and water habitats across the globe and this is the matter of concern for the scientific communities to replace it with the biodegradable plastic. The biocompatible and biodegradable polymers polyhydroxyalkanoates (PHA) could replace nondegradable petroleum-based plastics (Shah et al., 2008; Gao et al., 2011). The PHAs can be produced from the organic compounds such as saccharides, *n*-alkanes, *n*-alcohols, and gases by archaebacteria and bacteria (Verlinden et al., 2007). The methylotrophic bacteria can biosynthesize PHAs using methanol as substrate (Orita et al., 2014, Wang et al., 2014). The bacteria *M. extorquens* growing on methanol produced about 149 g/L polyhydroxybutyrate, a natural polymer (Suzuki et al., 1986).

17.4 Conclusion

Considering the abovementioned literature regarding the methylotrophs could be promising in combating several environmental issues. They could be grown on low cost and renewable substrate methanol for synthesizing various chemical leading to methanol-based bioeconomy. The synthesis will improve with the advancement of technical improvement in biological, metabolic, and protein engineering. They could minimize the volatile organic compounds, remediate industry effluents and oil spill, and synthesize biodegradable substitute of polymers for a better sustainable future.

References

Aktar, W., Sengupta, D., Chowdhury, A., 2009. Impact of pesticides use in agriculture: their benefits and hazards. Interdisciplin. Toxicol. 2, 1–12.

Anthony, C., 1982. The Biochemistry of Methylotrophs. Academic Press, London.

Anthony, C., 1986. Bacterial oxidation of methane and methanol. Adv. Microb. Physiol. 27, 113–210.

Arciero, D., Vannelli, T., Logan, M., Hopper, A.B., 1989. Degradation of trichloroethylene by the ammonia-oxidizing bacterium *Nitrosomonas europaea*. Biochem. Biophys. Res. Commun. 159, 640–643.

Bader, R., Leisinger, T., 1994. Isolation and characterization of the *Methylophilus* sp. strain DM11 gene encoding dichloromethane dehalogenase/glutathione *S*-transferase. J. Bacteriol. 176, 3466–3473.

Barrett, E.L., Kwan, H.S., 1985. Bacterial reduction of trimethylamineoxide. Annu. Rev. Microbiol. 39, 131–149.

Bellion, E., Wayland, L., 1982. Methylamine uptake in *Pseudomonas* species strain MA: utilization of methylamine as the sole nitrogen source. J. Bacteriol. 149, 395–398.

Bellion, E., Khan, M.Y., Romano, M.J., 1980. Transport of methylamine by *Pseudomonas* sp. MA. J. Bacteriol. 142, 786–790.

Betterton, E.A., Arnold, R.G., Kuhler, R.J., Santo, G.A., 1995. Reductive dehalogenation of bromoform in aqueous solution. Environ. Health Perspect. 103, 89–91.

Bhatt, P., Kumar, M.S., Mudliar, S., Chakrabarti, T., 2007. Biodegradation of chlorinated compounds—a review. Crit. Rev. Environ. Sci. Tech. 37, 165–198.

Burg, M.B., Ferraris, J.D., 2008. Intracellular organic osmolytes: function and regulation. J. Biol. Chem. 283, 7309–7313.

Camilli, R., Reddy, C.M., Yoerger, D.R., Van Mooy, B.A., Jakuba, M.V., Kinsey, J.C., et al., 2010. Tracking hydrocarbon plume transport and biodegradation at Deepwater Horizon. Sci 330, 201–204.

Carpenter, L.J., Reimann, S., Burkholder, J.B., Clerbaux, C., Hall, B.D., Hossaini, R., et al., 2014. Chapter 1, Ozone-depleting substances (ODSs) and other gases of interest to the Montreal Protocol. In: Scientific Assessment of Ozone Depletion: Global Ozone Research and Monitoring Project, Report No. 55. World Meteorological Organization, Geneva.

Chistoserdova, L., 2011. Modularity of methylotrophy, revisited. Environ. Microbiol. 13, 2603–2622.

Chistoserdova, L., 2018. Applications of methylotrophs: can single carbon be harnessed for biotechnology? Curr. Opin. Biotechnol. 50, 189–194.

Chistoserdova, L.V., Lidstrom, M.E., 1996. Molecular characterization of a chromosomal region involved in the oxidation of acetyl-CoA to glyoxylate in the isocitrate-lyase-negative methylotroph *Meihylobacterium extorquens* AMI. Microbiology 142, 1459–1468.

Chistoserdova, L., Lidstrom, M.E., 1997. Molecular and mutational analysis of a DNA region separating two methylotrophy gene clusters in *Methylobacterium extorquens* AMI. Microbiology 143, 1729–1736.

Colby, J., Dalton, H., 1979. Characterization of the second prosthetic group of the flavoenzyme NADH-acceptor reductase (component C) of the methane monooxygenase from *Methylococcus capsulatus* (Bath). Biochem. J. 177, 903–908.

Colby, J., Zatman, L.J., 1973. Trimethylamine metabolism in obligate and facultative methylotrophs. Biochem. J. 132, 101–112.

De Marco, P., Pacheco, C.C., Figueiredo, A.R., Moradas-Ferreira, P., 2004. Novel pollutant-resistant methylotrophic bacteria for use in bioremediation. FEMS Microbiol. Lett. 234, 75–80.

Diercks, A.R., Highsmith, R.C., Asper, V.L., Joung, D., Zhou, Z., Guo, L., et al., 2010. Characterization of subsurface polycyclic aromatic hydrocarbons at the Deepwater Horizon site. Geophys. Res. Lett. 37, 20.

Dijkhuizen, L., Levering, P.R., De Vries, G.E., 1992. The physiology and biochemistry of aerobic methanol-utilizing gram-negative and gram-positive bacteria. In: Murrell, J.C., Dalton, H. (Eds.), Methane and Methanol Oxidizers. Plenum Press, New York, pp. 149–181.

Doronina, N.V., Trotsenko, Y.A., 1994. Methylophilus leisingerii sp. nov., a new species of restricted facultatively methylotrophic bacteria. Microbiology (English translation of Mikrobiologiia) 63, 298–302.

Doronina, N.V., Braus-Stromeyer, S.A., Leisinger, T., Trotsenko, Y.A., 1995. Isolation and characterization of a new facultatively methylotrophic bacterium: description of *Methylorhabdus multivorans*, gen. nov., sp. nov. Syst. Appl. Microbiol. 18, 92–98.

Doronina, N.V., Trotsenko, Y.A., Krausova, V.I., Suzina, N.E., 1998. *Paracoccus methylutens* sp. nov.—a new aerobic facultatively methylotrophic bacterium utilizing dichloromethane. Syst. Appl. Microbiol. 21, 230–236.

Doronina, N.V., Trotsenko, Y.A., Tourova, T.P., Kuznetsov, B.B., Leisinger, T., 2000. *Methylopila helvetica* sp. nov. and *Methylobacterium dichloromethanicum* sp. nov.—novel aerobic facultatively methylotrophic bacteria utilizing dichloromethane. Syst. Appl. Microbiol. 23, 210–218.

Dourado, M.N., Aparecida Camargo Neves, A., Santos, D.S., Araújo, W.L., 2015. Biotechnological and agronomic potential of endophytic pink-pigmented methylotrophic Methylobacterium spp. BioMed. Res. Int. 2015.

Fedorov, D.N., Doronina, N.V., Trotsenko, Y.A., 2011. Phytosymbiosis of aerobic methylobacteria: new facts and views. Microbiology 80, 443–454.

Fernandes, V.C., Albergaria, J.T., Oliva-Teles, T., Delerue-Matos, C., De Marco, P., 2009. Dual augmentation for aerobic bioremediation of MTBE and TCE pollution in heavy metal-contaminated soil. Biodegradation 20, 375–382.

Flanagan, W.P., 1998. Biodegradation of dichloromethane in a granular activated carbon fluidized-bed reactor. Water Environ. Res. 70, 60–66.

Fogel, M.M., Taddeo, A.R., Fogel, S., 1986. Biodegradation of chlorinated ethenes by a methane-utilizing mixed culture. Appl. Environ. Microbiol. 51, 720–724.

Folsom, B.R., Chapman, P.J., Pritchard, P.H., 1990. Phenol and trichloroethylene degradation by *Pseudomonas cepacia* G4: kinetics and interactions between substrates. Appl. Environ. Microbiol. 56, 1279–1285.

Fox, B.G., Borneman, J.G., Wackett, L.P., Lipscomb, J.D., 1990. Haloalkene oxidation by the soluble methane monooxygenase from *Methylosinus trichosporium* OB3b: mechanistic and environmental implications. Biochemistry 29, 6419–6427.

Gao, X., Chen, J.C., Wu, Q., Chen, G.Q., 2011. Polyhydroxyalkanoates as a source of chemicals, polymers, and biofuels. Curr. Opin. Biotechnol. 22, 768–774.

Ghisalba, O., Kuenzi, M., 1983. Biodegradation and utilization of monomethyl sulfate by specialized methylotrophs. Experientia 39, 1257–1263.

Gisi, D., Willi, L., Traber, H., Leisinger, T., Vuilleumier, S., 1998. Effects of bacterial host and dichloromethane dehalogenase on the competitiveness of methylotrophic bacteria growing with dichloromethane. Appl. Environ. Microbiol. 64, 1194–1202.

Giri, D.D., Shukla, P.N., Singh, R., Kumar, A., Pandey, K.D., 2013a. Substrate utilization of stress tolerant methylotrophs isolated from revegetated heavy metal polluted coalmine spoil. World J. Microbiol. Biotechnol. 29, 635–643.

Giri, D.D., Kumar, A., Shukla, P.N., Singh, R., Singh, P.K., Pandey, K.D., 2013b. Salt stress tolerance of methylotrophic bacteria *Methylophilus* sp. and *Methylobacterium* sp. isolated from coal mine spoils. Pol. J. Microbiol. 62, 273–280.

Goodwin, K.D., Varner, R.K., Crill, P.M., Oremland, R.S., 2001. Consumption of tropospheric levels of methyl bromide by C1 compound-utilizing bacteria and comparison to saturation kinetics. Appl. Environ. Microbiol. 67, 5437–5443.

Govorukhina, N.I., Trotsenko, Y.A., 1991. *Methylovorus*, a new genus of restricted facultatively methylotrophic bacteria. Int. J. Syst. Bacteriol. 41, 158–162.

Hanson, R.S., 1980. Ecology and diversity of methylotrophic organisms. Adv. Appl. Microbiol. 26, 3–39.

Hanson, R.S., Hanson, T.E., 1996. Methanotrophic bacteria. Microbiol. Mol. Biol. Rev. 60, 439–471.

Harper, D.B., 2000. The global chloromethane cycle: biosynthesis, bio-degradation and metabolic role. Nat. Prod. Rep. 17, 337–348.

Harper, D.B., Hamilton, J.T.G., Ducrocq, V., Kennedy, J.T., Downey, A., Kalin, R.M., 2003. The distinctive isotopic signature of plant-derived chloromethane: possible application in constraining the atmospheric chloromethane budget. Chemosphere. 52, 433–436.

Hingurao, K., Nerurkar, A., 2018. A novel methylotrophic bacterial consortium for treatment of industrial effluents. Appl. Biochem. Biotechnol. 185 (3), 691–704.

Ichikawa, K., Kurihara, M., Tamegai, H., Hashimoto, S., 2015. Decomposition of brominated organic halogens by cultures of marine proteobacteria: *Phaeobacter*, *Roseobacter*, and *Rhodobacter*. Mar. Chem. 176, 133–141.

Iguchi, H., Yurimoto, H., Sakai, Y., 2015. Interactions of methylotrophs with plants and other heterotrophic bacteria. Microorganisms 3, 137–151.

Jakobsen, O.M., 2006. Upregulated transcription of plasmid and chromosomal ribulose monophosphate pathway genes is critical for methanol assimilation rate and methanol tolerance in the methylotrophic bacterium *Bacillus methanolicus*. J. Bacteriol. 188, 3063–3072.

Janssen, D.B., Keuning, S., Witholt, B., 1987. Involvement of a quinoprotein alcohol dehydrogenase and an NAD-dependent aldehyde dehydrogenase in 2-chloroethanol metabolism in *Xanthobacter autotrophicus* GJ10. Microbiology 133, 85–92.

Jechorek, M., Wendlandt, K.D., Beck, M., 2003. Cometabolic degradation of chlorinated aromatic compounds. J. Biotechnol. 102, 93–98.

Jenkins, O., Byrom, D., Jones, D., 1987. *Methylophilus*: a new genus of methanol-utilizing bacteria. Int. J. Syst. Bacteriol. 37, 446–448.

Kataoka, T., Ooki, A., Nomura, D., 2019. Production of dibromomethane and changes in the bacterial community in bromoform-enriched seawater. Microbes Environ. 34, 215–218.

Kato, N., Yurimoto, H., Thauer, R.K., 2006. The physiological role of the ribulose monophosphate pathway in bacteria and archaea. Biosci. Biotechnol. Biochem. 70, 10–21.

Keppler, F., Kalin, R.M., Harper, D.B., McRoberts, W.C., Hamilton, J.T.G., 2004. Carbon isotope anomaly in the major plant C1 pool and its global biogeochemical implications. Biogeoscience 1, 123–131.

Keppler, F., Harper, D.B., Röckmann, T., Moore, R.M., Hamilton, J.T.G., 2005. New insight into the atmospheric chloromethane budget gained using stable carbon isotope ratios. Atmos. Chem. Phys. 5, 2403–2411.

Khongkhaem, P., Intasiri, A., Luepromchai, E., 2011. Silica-immobilized *Methylobacterium* sp. NP3 and *Acinetobacter* sp. PK1 degrade high concentrations of phenol. Lett. Appl. Microbiol. 52, 448–455.

Kohler-Staub, D., Leisinger, T., 1985. Dichloromethane dehalogenase of *Hyphomicrobium* sp. strain DM2. J. Bacteriol. 162, 676–681.

Kolb, S., Stacheter, A., 2013. Prerequisites for amplicon pyrosequencing of microbial methanol utilizers in the environment. Front. Microbiol. 4, 268.

Koopman, F.W., De Winde, J.H., Ruijssenaars, H.J., 2009. C1 compounds as auxiliary substrates for engineered *Pseudomonas putida* S12. Appl. Microbiol. Biotechnol. 83, 705–713.

Kotik, M., Davidová, A., Voříšková, J., Baldrian, P., 2013. Bacterial communities in tetrachloroethene-polluted groundwaters: a case study. Sci. Total Environ. 454, 517–527.

Large, P.J., 1971. Non-oxidative demethylation of trimethylamine N-oxide by *Pseudomonas aminovorans*. FEBS Lett. 18, 297–300.

Large, P.J., Boulton, C.A., Crabbe, M.J., 1972. The reduced nicotinamide-adenine dinucleotide phosphate-and oxygen-dependent N-oxygenation of trimethylamine by *Pseudomonas aminovorans*. Biochem. J. 128, 137P.

Latypova, E., Yang, S., Wang, Y.S., Wang, T., Chavkin, T.A., Hackett, M., et al., 2010. Genetics of the glutamate-mediated methylamine utilization pathway in the facultative methylotrophic beta-proteobacterium *Methyloversatilis universalis* FAM5. Mol. Microbiol. 75, 426–439.

Lidstrom, M.E., 2006. Aerobic methylotrophic prokaryotes. In: Balows, A., Truper, H.G., Dworkin, M., Harder, W., Schleifer, K.-H. (Eds.), The Prokaryotes. Springer-Verlag, New York, pp. 618–634.

Little, C.D., Palumbo, A.V., Herbes, S.E., Lidstrom, M.E., Tyndall, R.L., Gilmer, P.J., 1988. Trichloroethylene biodegradation by a methane-oxidizing bacterium. Appl. Environ. Microbiol. 54, 951–956.

Liu, Y., Yvon-Lewis, S.A., Thornton, D.C.O., Butler, J.H., Bianchi, T.S., Campbell, L., et al., 2013. Spatial and temporal distributions of bromoform and dibromomethane in the Atlantic Ocean and their relationship with photosynthetic biomass. J. Geophys. Res. Oceans 118, 3950–3965.

Mansuy, D., 2013. Metabolism of xenobiotics: beneficial and adverse effects. Biol. Aujourd'hui 207, 33–37.

McCarty, P.L., 1988. Bioengineering issues related to in situ remediation of contaminated soils and groundwater. Environmental Biotechnology. Springer, Boston, MA, pp. 143–162.

McDonald, I.R., 2002. A review of bacterial methyl halide degradation: biochemistry, genetics and molecular ecology. Environ. Microbiol. 4, 193–203.

Meena, K.K., Kumar, M., Kalyuzhnaya, M.G., Yandigeri, M.S., Singh, D.P., Saxena, A.K., et al., 2012. Epiphytic pink-pigmented methylotrophic bacteria enhance germination and seedling growth of wheat (*Triticum aestivum*) by producing phytohormone. Antonie van Leeuwenhoek 101, 777–786.

Mitsui, R., Sakai, Y., Yasueda, H., Kato, N., 2000. A novel operon encoding formaldehyde fixation: the ribulose monophosphate pathway in the Gram-positive facultative methylotrophic bacterium *Mycobacterium gastri* MB19. J. Bacteriol. 184, 944–948.

Montzka, S.A., Reimann, S., 2011. Ozone-depleting substances (ODSs) and related chemicals. In: Ennis, C.A. (Ed.), Global Ozone Research and Monitoring Project-Report, vol. 52. World Meteorological Organization, Geneva, pp. 1–108.

Müller, R.H., Rohwerder, T., Harms, H., 2008. Degradation of fuel oxygenates and their main intermediates by *Aquincola tertiaricarbonis* L108. Microbiology 154, 1414–1421.

Myers, P.A., Zatman, L.J., 1971. The metabolism of trimethylamine N-oxide by *Bacillus* PM6. Biochem. J. 121, 10P.

Nanda, S., Sarangi, P.K., Abraham, J., 2012. Microbial biofiltration technology for odour abatement: an introductory review. J. Soil. Sci. Environ. Manage. 3, 28–35.

Orita, I., Nishikawa, K., Nakamura, S., Fukui, T., 2014. Biosynthesis of polyhydroxy-alkanoate copolymers from methanol by *Methylobacterium extorquens* AM1 and the engineered strains under cobalt-deficient conditions. Appl. Microbiol. Biotechnol. 98, 3715–3725.

Pasternak, G., Kołwzan, B., 2013. Surface tension and toxicity changes during biodegradation of carbazole by newly isolated methylotrophic strain *Methylobacterium* sp. GPE1. Int. Biodeterior. Biodegrad. 84, 143–149.

Piveteau, P., Fayolle, F., Vandecasteele, J.P., Monot, F., 2001. Biodegradation of *tert*-butyl alcohol and related xenobiotics by a methylotrophic bacterial isolate. Appl. Microbiol. Biotechnol. 55, 369–373.

Qiu, L.W., 2014. Formaldehyde biodegradation by immobilized *Methylobacterium* sp. XJLW cells in a three phase fluidized bed reactor. Bioprocess Biosyst. Eng. 37, 1377–1384.

Quack, B., Wallace, D.W.R., 2003. Air-sea flux of bromoform: controls, rates, and implications. Global Biogeochem. Cycles 17, 1023.

Quayle, J.R., Ferenci, T., 1978. Evolutionary aspects of autotrophy. Microbiol. Rev. 42, 251–273.

Reddy, C.M., Arey, J.S., Seewald, J.S., Sylva, S.P., Lemkau, K.L., Nelson, R.K., et al., 2012. Composition and fate of gas and oil released to the water column during the Deepwater Horizon oil spill. Proc. Natl. Acad. Sci. U.S.A. 109, 20229–20234.

Redeker, K.R., Kalin, R.M., 2012. Methyl chloride isotopic signatures from Irish forest soils and a comparison between abiotic and biogenic methyl halide soil fluxes. Global Change Biol. 18, 1453–1467.

Sakai, Y., Mitsui, R., Katayama, Y., Yanase, H., Kato, N., 1999. Organization of the genes involved in the ribulose monophosphate pathway in an obligate methylotrophic bacterium, *Methylomonas aminofaciens* 77a. FEMS Microbiol. Lett. 176, 125–130.

Schaefer, J.K., Miller, L.G., Oremland, R.S., Murrell, J.C., 2007. Bacterial cycling of methyl halides. Adv. Appl. Microbiol. 61, 307–346.

Schaefer, H., Myronova, N., Boden, R., 2010. Microbial degradation of dimethylsulfide and related C1-sulfur compounds: organisms and pathways controlling fluxes of sulfur in the biosphere. J. Exp. Bot. 61, 315–334.

Schäfer, H., Miller, L.G., Oremland, R.S., Murrell, J.C., 2007. Bacterial cycling of methyl halides. Adv. Appl. Microbiol. 61, 307–346.

Shah, A.A., Hasan, F., Hameed, A., Ahmed, S., 2008. Biological degradation of plastics: a comprehensive review. Biotechnol. Adv. 26, 246–265.

Söhngen, N.L., 1906. Über bakterien, welche methan als kohlenstoffnahrung und energiequelle gebrauchen. Zentrabl. Bakteriol. Parasitenk Infektionskr. 15, 513–517.

Sturges, W.T., Sullivan, C.W., Schnell, R.C., Heidt, L.E., Pollock, W.H., 1993. Bromoalkane production by Antarctic Ice Algae. Tellus B 45, 120–126.

Suzuki, T., Yamane, T., Shimizu, S., 1986. Kinetics and effect of nitrogen-source feeding on production of poly-beta-hydroxybutyric acid by fed-batch culture. Appl. Microbiol. Biotechnol. 24, 366–369.

Tanaka, Y., Yoshida, T., Watanabe, K., Izumi, Y., Mitsunaga, T., 1997. Cloning and analysis of methanol oxidation genes in the methylotroph *Hyphomicrobium methylovorum* GM2. FEMS Microbiol. Lett. 154, 397–401.

Taubert, M., Grob, C., Howat, A.M., Burns, O.J., Chen, Y., Neufeld, J.D., et al., 2016. Analysis of active methylotrophic communities: when DNA-SIP meets high-throughput technologies. Microbial Environmental Genomics (MEG). Humana Press, New York, pp. 235–255.

Tsai, W.T., 2017. Fate of chloromethanes in the atmospheric environment: implications for human health, ozone formation and depletion, and global warming impacts. Toxics 5 (4), 23.

Urakami, T., Komagata, K., 1986. Emendation of *Methylobacillus yordy* and *weaver* 1977, a genus for methanol-utilizing bacteria. Int. J. Syst. Bacteriol. 36, 502–511.

Van Aken, B., Yoon, J.M., Schnoor, J.L., 2004. Biodegradation of nitro-substituted explosives 2,4,6-trinitrotoluene, hexahydro-1,3,5-trinitro-1,3,5-triazine, and octahydro-1,3,5,7-tetranitro-1,3,5-tetrazocine by a phytosymbiotic *Methylobacterium* sp. associated with poplar tissues (*Populus deltoides* × *nigra* DN34). Appl. Environ. Microbiol. 70, 508–517.

Ventorino, V., 2014. *Methylobacterium populi* VP2: plant growth promoting bacterium isolated from a highly polluted environment for polycyclic aromatic hydrocarbon (PAH) biodegradation. Sci. World J. 1–12.

Verlinden, R.A., Hill, D.J., Kenward, M.A., Williams, C.D., Radecka, I., 2007. Bacterial synthesis of biodegradable polyhydroxyalkanoates. J. Appl. Microbiol. 102, 1437–1449.

Vorholt, J.A., 2002. Cofactor-dependent pathway of formaldehyde oxidation in methylotrophic bacteria. Arch. Microbiol. 178, 239–249.

Vuilleumier, S., 1997. Bacterial glutathione *S*-transferases: what are they good for? J. Bacteriol. 179, 1431.

Wackett, L.P., Gibson, D.T., 1988. Degradation of trichloroethylene by toluene dioxygenase in whole-cell studies with *Pseudomonas putida* F1. Appl. Environ. Microbiol. 54, 1703–1708.

Wang, Y., Yin, J., Chen, G.Q., 2014. Polyhydroxyalkanoates, challenges and opportunities. Curr. Opin. Biotechnol. 30, 59–65.

Winter, R.B., Yen, K.M., Ensley, B.D., 1989. Efficient degradation of trichloroethylene by a recombinant *Escherichia coli*. Biotechnol. 7, 282.

WMO (World Meteorological Organization), 2011. Scientific Assessment of Ozone Depletion: 2010. WMO, Geneva.

Yamamoto, H., Yokouchi, Y., Otsuki, A., Itoh, H., 2001. Depth profiles of volatile halogenated hydrocarbons in seawater in the Bay of Bengal. Chemosphere 45, 371–377.

Yanase, H., Ikeyama, K., Mitsui, R., Ra, S., Kita, K., Sakai, Y., et al., 1996. Cloning and sequence of the gene encoding 3-hexulose-6-phosphate synthase from the methylotrophic bacterium, *Methylomonas aminofaciens* 77a, and its expression in *Escherichia coli*. FEMS Microbiol. Lett. 135, 201–205.

Yordy, J.R., Weaver, T.L., 1977. *Methylobacillus*: a new genus of obligate methylotrophic bacteria. Int. J. Syst. Bacteriol. 27, 247–255.

Yoshida, Y., Wang, Y., Zeng, T., Yantosca, R., 2004. A three-dimensional global model study of atmospheric methyl chloride budget and distributions. J. Geophys. Res. Atmos. 109 (D24).

Yoshida, S., Hiraga, K., Takehana, T., Taniguchi, I., Yamaji, H., Maeda, Y., et al., 2016. A bacterium that degrades and assimilates poly(ethylene terephthalate). Science 351, 1196–1199.

Yurimoto, H., Hirai, R., Yasueda, H., Mitsui, R., Sakai, Y., Kato, N., 2002. The ribulose monophosphate pathway operon encoding formaldehyde fixation in a thermotolerant methylotroph, *Bacillus brevis* S1. FEMS Microbiol. Lett. 214, 189–193.

Yurimoto, H., Kato, N., Sakai, Y., 2005. Assimilation, dissimilation, and detoxification of formaldehyde, a central metabolic intermediate of methylotrophic metabolism. Chem. Rec. 5, 367–375.

Zhang, L.L., Chen, J.M., Fang, F., 2008. Biodegradation of methyl *t*-butyl ether by aerobic granules under a cosubstrate condition. Appl. Microbiol. Biotechnol. 78, 543–550.

Chapter 18

Advancement in bioinformatics and microarray-based technologies for genome sequence analysis and its application in bioremediation of soil and water pollutants

Jyoti Singh[1], Madhuri Gupta[2], Kaushalendra Kumar Singh[3], Ajay Kumar[4], Deepanker Yadav[4], Wang Wenjing[5] and Prashant Kumar Singh[6]

[1]Department of Botany, Institute of Sciences, Banaras Hindu University, Varanasi, India, [2]Sardar Vallabhbhai Patel University of Agriculture and Technology, Meerut, India, [3]Department of Zoology, Pachhunga University College, Mizoram University, Aizawl, India, [4]Agriculture Research Organization, Ministry of Agriculture and Rural Development, Volcani Centre, Rishon LeZion, Israel, [5]State Key Laboratory of Cotton Biology, Henan Key Laboratory of Plant Stress Biology, School of Life Science, Henan University, Kaifeng, China, [6]Department of Biotechnology, Pachhunga University College, Mizoram University, Aizawl, India

18.1 Introduction

Modern agriculture practices in combination with industrialization have dramatically enforcing world toward environmental pollution in the environment, which imposes negative impact on normal ecosystem functioning (Megharaj et al., 2011). Burning of fossil fuels (coal, petroleum, diesel, and hydrocarbons) and application of fertilizers, pesticides, insecticides, and herbicides are the major players contributing to environmental pollution. Most of the fertilizers and biocides used in agriculture are readily water-soluble and runoffs from the contaminated sites. Frequent cycling by volatilization and condensation these contaminants also precepted in the form of rain, fog, and snow (Dubus et al., 2000). Moreover, on migration, these pollutants leach to soil or as dust and contribute further to the natural ecosystem's pollution (Lopez-Errasquin and Vazquez, 2003). Therefore our soil, air, and water resources are heavily polluted with various pollutants that prime worsening of soil and water quality. Notwithstanding above, by leaching and runoff these pollutants also contaminate our surface and underground water. These pollutants negatively affect the human health and also hazardous environment. Furthermore, currently, chemical pesticides are frequently used in agriculture to enhance crop productivity worldwide and are also a major factor influencing natural environmental. Bioremediation is an effective, economical, and eco-friendly way to check out these environmental hazards by natural degradation mediated by microorganisms and/or plants (Pierzynski et al., 1994). Nowadays various techniques such as genomics, transcriptomics, proteomics, metabolomics, lipidomics, phenomics, bioinformatics, and sequencing are employed to identify and understand the mechanism of natural bioremediation in different environments (Chakraborty et al., 2012). The microarray-based technologies in combination with bioinformatics are useful in documentation of unprecedented microbes and their mechanism of bioremediation and to uncover the gene and their products in a cell involved in the process. Metagenomics and whole-genome sequencing have generated novel possibilities to explore the microorganism mediated remediation. The present chapters high spot the role of microarray-based technologies and bioinformatics to understand the bioremediation and future perspectives.

Microbe Mediated Remediation of Environmental Contaminants. DOI: https://doi.org/10.1016/B978-0-12-821199-1.00018-3

18.2 Mechanism of bioremediation

Microorganism and/or plant-mediated degradation of soil water contaminants is known as bioremediation. The process comprises (1) plants or microorganisms, (2) biodegradation of contaminants, and (3) a bioreactor having microbes and/or plants with potential contaminants. Most of the microbes utilize carbon as a source of energy and degrade the contaminants. Microbes not only degrade contaminants in both ex situ and in situ. In ex-situ bioremediation, before treatment, the contaminated soil excavated or groundwater extracted, while excavation or extraction not required in situ remediations. Consequently, the polluted soil or groundwater assists as bioreactor. A huge number of microbes have been identified as a potential candidate to remediate almost all kinds of environmental contaminants (Table 18.1).

As many pollutants released into the environment; therefore, a diverse microbe is required for efficient bioremediation (Table 18.1). Although a large number of database sequences are available for such microorganisms, still a huge number of microbes are yet to be sequenced for their genomes.

18.3 Genome sequencing

The complete genome of some microorganisms with a broad range of base pairs has been successfully sequenced, including both prokaryotic (example) and eukaryotic (e.g., human) organisms. Analysis of sequence deals with the

TABLE 18.1 Microbes utilized for bioremediation.

S. no.	Pollutant	Organism	References
1.	Atrazine	*Pseudomonas* sp. (ADP)	Newcombe and Crowley (1999)
2.	Alkyl benzene sulfonate polycyclic aromatics	*Alcaligenes* spp.	Lal and Khanna (1996)
3.	TNT	*Methanococcus* sp.	Boopathy and Kulpa (1994)
4.	PCB formaldehyde	*Candida tropicalis*	Ijah (1998)
5.	Halogenated hydrocarbon	*Corynebacterium* spp.	Jogdand (1995)
6.	Polycyclic hydrocarbon	*Xanthomas* sp.	Jogdand (1995)
7.	Chlorpyrifos	*Enterobacter* strain B-14	Singh et al. (2004)
8.	DBT	*Rhizobium meliloti*	Frassinetti et al. (1998)
9.	RDX	*Acetobacterium paludosum*	Sherburne et al. (2005)
		Clostridium acetobutylicum	Zhang and Hughes (2003)
10.	PAHs	*Pseudomonas* sp.	Arun et al. (2008)
11.	Phenanthrene, PAH	*Agrobacterium, Bacillus*	Aitken et al. (1998)
12.	PCB	*Rhodococcus* sp.	Chung et al. (1994)
13.	Endosulfan	*Bacillus* sp.	Mohamed et al. (2011)
14.	Chlorpyrifos	*Enterobacter* sp.	Niti et al. (2013)
15.	Malathion	*Arthrobacter* sp.	Hassan et al. (2016)
16.	Chlorpyrifos and methyl parathion	*Acinetobacter* sp., *Pseudomonas* sp.	Ravi et al. (2015)
17.	Heavy metals, lead, mercury, and nickel	*Saccharomyces cerevisiae*	Chen and Wang (2007), Infante et al. (2014)
18.	Cobalt, copper, chromium	*Lysinibacillus sphaericus* CBAM5	Peña-Montenegro et al. (2015)
19.	Cadmium	*Aspergillus* sp., *Microsporum* sp., *Cladosporium* sp.	Mohammadian Fazli et al. (2015)
20.	Monocyclic aromatic hydrocarbons	*Pseudomonas putida*	Safiyanu et al. (2015)
21.	Fe^{3+}, U^{6+}	*Geobacter* spp.	Mirlahiji and Eisazadeh (2014)

DBT, Dibenzothiophene; *PCB*, polychlorinated biphenyl; *PAH*, polycyclic aromatic hydrocarbon; *RDX*, hexahydro-1,3,5-trinitro-1,3,5-triazine; *TNT*, 2,4,6-trinitrotoluene.

structure, features, function, evolution, etc. information that can be conferred through DNA sequences with the help of various methodologies and algorithms such as biological databases, sequence alignment. Computation of whole genome is a challenging task especially in eukaryotes with a large quantity of junk DNA in the form of introns, repetitive sequences, or nontranscribed region whereas coding region, that is, exon in confined areas.

Complementary base pairing of DNA is the fundamental principle of sequencing technique. Maxam and Gilbert (1977) were the first to develop the method utilized for sequencing. However, this technique got replaced later by prominently used advanced Sanger sequencing technique (Sanger et al., 1977). The sequencing mixture contains dNTPs, ddNTPs, primer, and DNA polymerase utilizes ssDNA fragment for the formation of complementary strand whose synthesis gets stopped after the attachment of fluorescently labeled ddNTPs as they lack 3′OH. All the four ddNTPs are marked by different fluorophores which are observed by the detector after the passage of laser beam. This first-generation sequencing machine can utilize approximately 800 nucleotides for sequencing. Therefore shotgun strategy came into existence for breaking longer strands into utilizable shorter fragments. Several overlapping and variable-sized sequenced pieces are gathered which are aligned and analyzed by computer algorithms resulting in the whole sequence.

The era of sequence analysis prevailed from 1969 to 1977 first with the residue correlation interaction of tRNA sequences which ultimately led to the 3D structure of tRNA (Levitt, 2001; Levitt, 1969). Alignment of two strands with the help of a computer algorithm was reported by Needleman and Wunsch (1970). Continuous improvement in techniques and methodologies for obtaining nucleotide sequence led to the decoding of the complete bacteriophage genome (Sanger et al., 1977). RNA molecule was first sequenced by Robert Holley and his team in Cornell University (Holley et al., 1965). Along with the involvement of advanced bioinformatics tools, Sanger sequencing technique was the backbone of the human genome project (Bishop, 1998; DeAngelis and Batzer, 2001).

Sequence alignment is the most efficient method for analysis and understanding of sample sequences in respect of their functions or metabolism they are involved in. The basis of sequence analysis is to gain information about conserved sequences, regulatory regions, and certain function-specific genes to nurture the research and discoveries in the field of basic, clinical, and other areas of sciences. There are two significant sequence methodologies alignment; a pairwise alignment that associates two sequences at a time with two main algorithms whereas multiple sequences are compared at a time by multiple sequence alignment. Number of sequences are successfully aligning the sequences with the help of numerous pairwise and multiple sequences alignment tools such as Dot plots, BLAST, ClustalW, MAFFT, T-Coffee, MUSCLE, PROBCONS for comparison and identification of homologous sequences which is measured for statistical significance (Durbin et al., 1998; Pedersendagger et al., 2000). Continuously new genes are discovered in the partially or uncharacterized DNA sequences mostly during the sequencing project. At the initial stages, DNA sequencing technologies were working with short DNA sequences collected by breaking long strands; sequencing them and reconstructing them by obtained information from DNA fragments. Advancement in both molecular and computational biology has increased the attraction for the whole-genome sequence which led to the sequencing of more than 40 organisms including a wide range from 450 to 100,000 genes (Fleischmann et al., 1995).

Any of the biological samples such as epithelial cells, hair, bone marrow, plant leaves, seeds even holding very ancient or less amount of DNA can be utilized as a sample for whole-genome sequencing. The evolution of technologies resulted in the whole-genome sequencing of *Haemophilus influenza* in 1995 which was followed by certain bacteria and archaea due to the small genome through the application of shotgun technique (Fleischmann et al., 1995). *Saccharomyces cerevisiae*, a unicellular model organism with ~12 Mb genome size, was the first eukaryotic organism sequenced in 1996. This is further proceeded by whole-genome sequencing of the *Caenorhabditis elegans*, a nematode worm as first multicellular eukaryote and animal in 1998 through the help of large DNA clones obtained from DNA libraries (The *C. elegans* Sequencing Consortium, 1998; Alberts, 2008). The shortest human autosome, that is, human chromosome 22 was entirely sequenced and published in 1999, whereas whole-genome draft was released in 2001 (Dunham, 1999). The *Drosophila melanogaster* (fruit fly) a model organism was the first insect and the second invertebrate with *Arabidopsis thaliana* genome was fully sequenced by the year 2000 (Adams et al., 2000). By 2002 the complete genome of *Mus musculus*, a laboratory mouse, was sequenced, and till now the number is increasing exponentially (Venter et al., 2001; Waterston et al., 2002).

A broad coverage of objectives is upheld by sequence analysis that includes similarity search for homologous sequences, inferring the features intrinsic to the specific sequences, search of variations and differences such as single nucleotide polymorphism (SNP) or point mutation for identification of marker, evolution and diversity study of sequence in respect of same organisms with nearest neighbors, studying molecular structures with reference of sequences and many more with a nonending list. With the advancement of sequencing techniques from manual to automated and rapidly, the whole genome of organisms comprising both mitochondrial and chloroplast DNA can be sequenced in a single time, which comes under entire genome sequencing (Durbin et al., 1998; Levitt, 2001).

The continuous increase in whole-genome databases such as DDBJ, GenBank, and EMBL through advanced techniques enhances the search for genes coding RNA products, proteins, etc. For example, Entrez nucleotide database retrieves all complete and partial genomes from the primary databases and compile them for different information such as a single gene, chromosomal genes. Application of modified algorithms and techniques for similarity searches and accordingly assigning a function to the sequences enhance the understanding of biology. The initial and most crucial step just ahead of sequencing is gene finding or prediction, that is, identification of gene coding for desired mechanism or products (Baker et al., 2000; Benson et al., 2000; Okayama et al., 1998; Schuler et al., 1996).

Nowadays, experiments analyzing expression level for a particular gene through quantification of produced mRNA or converted cDNA due to rapid degradation of mRNA or protein products are new sources for procuring genomic data. This technique is being considered more promising as in general, genomic information within a cell is approximately same, whereas expression of genes is regulated by different factors such as their position, developmental stage, environmental effects. Expression analysis of all transcriptome reveals that approximately one-third of genome express at a particular time (Holstege et al., 1998). Microarray measuring the relative amount of mRNA in form of cDNA, serial analysis of gene expression (SAGE), and Affinity matrix GeneChip inferring absolute level of expressed products are three primary techniques involved in former analysis, among which microarray lead in the case of humans (Eisen and Brown, 1999; Duggan et al., 1999; Lipshutz et al., 1999). Expression analysis mainly focuses on yeast and human genome as till day no central depository has been mentioned. For yeast, gene chip technology is being utilized by Church, Younge, and Samsung datasets, whereas microarray is employed by diauxic shift, Stanford cell cycle, and deletion mutation datasets. Numerous projects such as lymphoma and leukemia molecular profiling, molecular portraits of breast tumors, etc. procure data regarding cancer and tumor cells through expression analysis from application of microarray techniques (Roth et al., 1998; Jelinsky and Samson, 1999; Cho et al., 1998; DeRisi et al., 1997; Winzeler et al., 1999; Perou et al., 2000; Holstege et al., 1998).

18.4 Application of microarray technique in genome sequence analysis

Understanding the genes' functions and its functional characterization is among the significant postgenomics challenge due to recent dramatic advancement of genetic technologies, providing substantial knowledge for a better understanding of genome–environment interactions, diagnosis, and their treatments. For the genes function and regulatory assignment in molecular networks, several strategies such as transcriptomics, proteomics, metabolomics are implemented because the gene expression of a cell determines its function, phenotype, and response to the environment. Therefore the analysis of gene expression is mandatory for providing information about regulatory mechanisms, metabolic pathways, and cellular functions. However, in researches done at the genomic level, DNA microarrays have provided a powerful tool that performs important cell and molecular biology as well as several clinical diagnostic assays including analysis of inter- and intraindividual genomic variation within a species.

Conventional expression profiling strategies such as reverse transcriptase–polymerase chain reaction (RT–PCR), Northern blot, reverse Northern blot, Western blot, enzyme-linked immunosorbent assay, in situ hybridization, and immunohistochemistry are optimized for single gene analysis. DNA microarrays have been developed as a high-throughput, efficient, and comprehensive strategy, for the simultaneous measurement of all the genes, or a primary genome encoded defined subset (Schena et al., 1995, 1996). The DNA microarray technology is primarily based on the immobilization of known DNA sequences (also known as probes) on a solid surface and their hybridization with a complementary DNA or a given target sequence. The analysis of DNA microarray represents one of the significant advances leading to the development of proteomics and functional genomics; advancement in microarray-based approaches enables the essential forms of genomic variations to be detected deletions, insertions, rearrangements, amplifications, and base-pair changes likewise.

Several methodologies such as differential display PCR, Northern blots (Taniguchi et al., 2001), quantitative RT–PCR, SAGE (Yamamoto et al., 2001; Bertelsen and Velculescu, 1998), and TIGR Orthologous Gene Alignments (TOGAs) (Lee et al., 2002; Sogayar et al., 2004) are used alongside microarrays.

18.5 Advancement in microarray technology

18.5.1 From the discovery of DNA to modern DNA array

The understanding of the molecular basis of healthy and unhealthy, that is, diseased tissues has always been a significant challenge for research and development since ages. However, following the discovery of the DNA molecule

(Watson and Crick, 1953) and labeling of nucleic acid molecules by hybridization over solid support (Southern, 1975), the field of array germinated.

As when several techniques such as differential gene display (DD-PCR), RNase protection assay (Liang and Pardee, 1992), and SAGE (Velculescu et al., 1995) were developed, the study of some tens of genes became possible. Afterward, the increasing knowledge of the entire genome sequences along with the Human Genome Project challenged scientists with the demand for the detection of expression levels of more than one gene in a single experiment. This led to the first genomic filter arrays that allow for the clone libraries screening, based on a correlation between clones and hybridization signals. These filters permitted the study of hundreds of genes in a single experiment.

Due to the advancement of research tools in the last two decades, traditional Southern- and Northern-blotting techniques were replaced by microarray technology (Tarca et al., 2006). Genomic microarrays commenced with the use of fluorescence-based detection with nonporous materials (glass and silicon), photolithography, automation, robotic spotters, and bioinformatics. The introduction genomic microarrays allowed simultaneous and efficient quantitative gene expression (Naidu and Suneetha, 2012). They are DNA microarray—the most significant, meaningful, and best-developed technology of all types of microarrays (Chalah, 2016).

18.5.2 Modern DNA array

DNA array technology had progressed from the late 1990s as when the methods such as production and fluorescent detection were adapted. The first cDNA microarray was pioneered in 1995, which allowed the detection of gene expression patterns and it has undergone significant advancement (Naidu and Suneetha, 2012). Again in the same year, the first cDNA microarray was pioneered, which also allow the detection of gene expression patterns. In cDNA microarray extracted RNA reverse transcribed into cDNA by incorporating fluorescent labels (cyanine 3 and cyanine 5) and hybridized simultaneously into an array (Lander et al., 2001). The fluorescent signal recorded with a laser scanner and the signal intensity ratios indicates differential gene expression levels (Schena et al., 1995).

Following the DNA microarray protein microarray (Wang et al., 2004; Freeman et al., 2001) was developed for the identification of protein functional analysis, screening and production of antibodies (Liljedahl et al., 2003) and recombinant proteins (Amin et. al., 2004), discovery of proteins implicated in disease (Waring et al., 2002; Pennie et al., 2004), protein–protein interaction screening, enzyme–substrate and DNA–protein interactions (Sebat, 2004; Weiss, 2008). High-density human protein array provided stability and structure of human genes (Perry, 2007); however, the protein microarrays are not as successful as the DNA microarray (Lupski, 2007).

Cell-based arrays recently emerged as handy tools for the functional genomics, which employs engineered living cell matrices to express a protein of interests and it facilitates genome-scale studies for identification of protein function (Sebat, 2007). Transfection microarray has also been another living cell array identified as an alternative to protein microarrays for pharmacogenomics and recognized as an expression system for the gene products identification (Sjoblom, 2006). Recently, a siRNA transfected cell microarray has been developed for large-scale, functional genomics (Wood, 2007). An array of small organic compounds is called chemical microarrays, which was designed for analysis of chemical libraries. These libraries widely analyze the proteins' interaction with organic compounds (Wood, 2007). Various technologies have been implemented with a leg on each side of macromolecular microarrays to cell arrays.

In general, the DNA microarray works through the following order:

1. *Target:* a previously marked DNA sample engrossed in a solution;
2. *Labels:* are markers, which are positioned over the target DNAs, and changes the color of the medium by showing a specific color;
3. *Probes:* are DNA sequences placed on a solid support corresponding to the particular genes and serve as a tool for determination of the degree of gene expression; and
4. *Solid support:* the support media with which DNA probe anchored.

Though various protocols and microarray are known, however, the necessary microarray experiment steps have been shown in Fig. 18.1. It starts with the isolation of RNA from biological samples. The application of fluorescent tag with the nucleic acid and the hybridization of this labeled sample (target) with a microarray probe for a period following to its excess of probe washed off.

The excitation and scanning of the microarray plate under laser light and analysis of data are generated by using the appropriate software. A brief description of the three basic types of DNA microarray experiments and their applications are provided in Table 18.2.

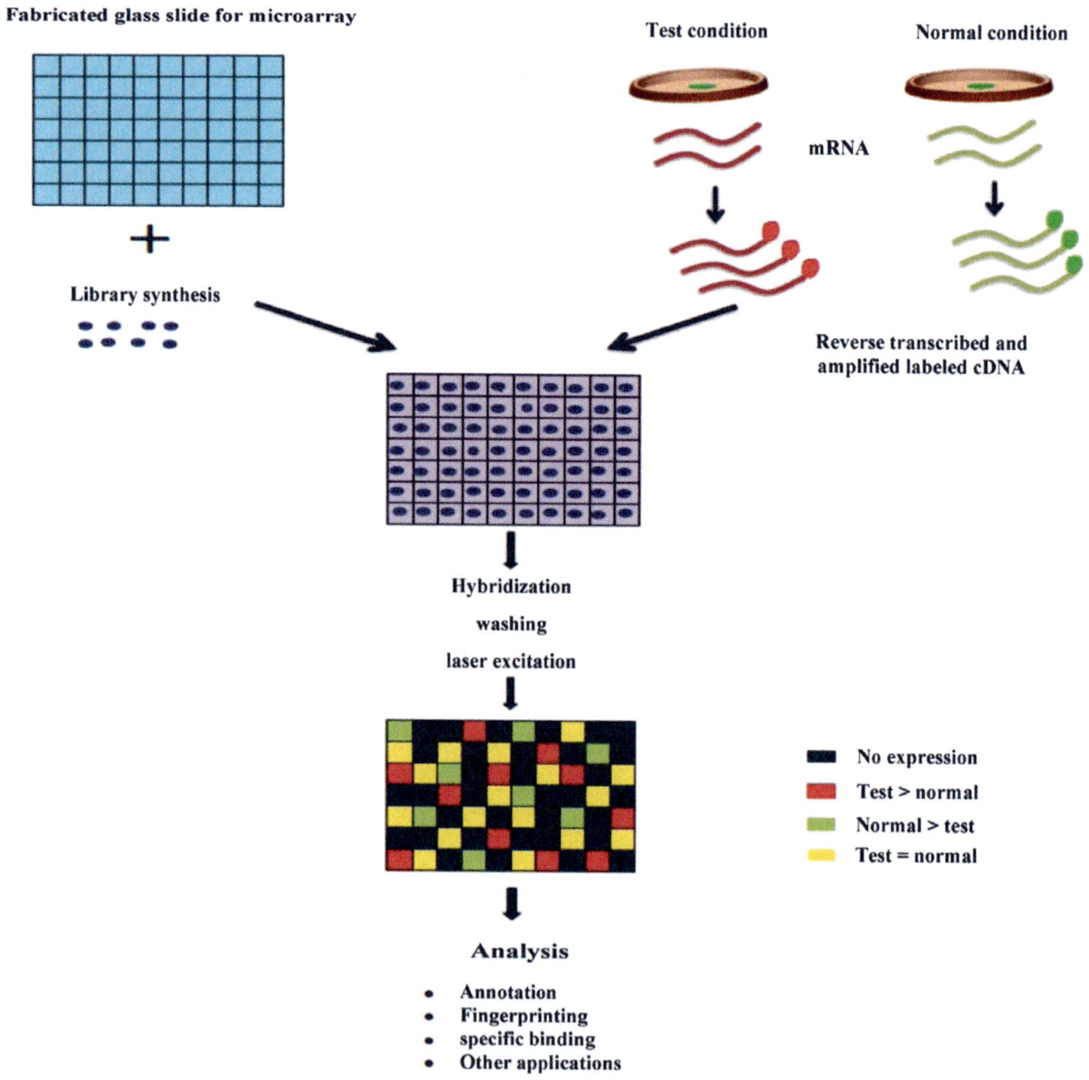

FIGURE 18.1 The schematic diagram use in microarray-based expression analysis.

TABLE 18.2 Types of DNA microarrays.

Type of microarray	Description	Utilization
Gene expression analysis	Gene expression analysis	Study of developmental stages, development of drugs, pharmacogenomics, disease progression tracking, response to biotic and abiotic stresses, epigenetics, epigenomics
CGH	A cytogenetic way for analyzing the variations in ploidy for detection of chromosomal aberrations	classification of Tumors, Risk assessment, Disease diagnosis, Evolutionary relatedness healthy and diseased tissues analysis
Polymorphism/ Mutation analysis	polymorphisms/mutations detection	Development of drugs, SNPs, risk assessment of diseases, species identification, genotyping, population genetics, and somatic/germline mutation

CGH, Comparative genomic hybridization; *SNPs*, single nucleotide polymorphisms.

18.6 Applications of microarray technologies for genome sequencing

Several methods have been introduced for genome sequencing which includes ultra-thin gel technology (Bharaj et al., 1998) and capillary electrophoresis (Carrilho, 2000). However, these technologies still have limitations. At present, the most efficient way to perform massively parallel sequencing includes hybridization sequencing on microarrays (McKenzie et al, 1998; Kruglyak, 1998). For sequencing, this method comprises immobilization of the target in a microarray format, hybridized with an extensive set of short, labeled probes (e.g., 10,000 probes that are each seven nucleotides long), and then the examination hybridization pattern and original DNA sequence computation (Drmanac et al., 1998). Another approach involves immobilization of thousands of short probes (18–20 nucleotides) in a microarray and their hybridization with the short probes of the target of interest, previously labeled fluorescent probe. One can then easily predict the interest DNA sequence in one analysis by observing hybridization pattern (Pease et al., 1994). The GeneChip sequencing and other tools provide a conceptual solution for the endurance of sequence variability, over large genomic areas, with reasonable ease, speed, and cost. These methods are proving very promising and will become routine in coming years.

Microarrays preferably suited against a small sample size for thousands of genes analysis and following to various discoveries, different processes and methods are being included to take the resulting genes of interest. These screened genes used for screening of thousands of samples, tissues, or cell lines for further investigation of gene function for their involvement in diseases.

18.7 Single nucleotide polymorphisms

In 1998 for genotyping of different samples, SNP array was developed (Wang et al., 1998) and subsequent dramatic improvement in this technology designated it as one of the most potent genomic analysis tools. The first SNP array contained more than 500 loci, and SNPs present within a sample amplified by multiplex PCR, in which primer pairs from many different loci were combined in a single reaction (Wang et al., 1998). The DNA amplified was then hybridized on the SNP array to detect the genotyping of several SNPs in the sample. SNP array analysis is also applicable in Cancer Research, and it can be easily adjusted for DNA methylation studies by comparing the presence and absence of DNA sequences among the methylation sensitive and nonsensitive enzymes.

SNP arrays although were initially developed for genomic analysis, however later adopted for epigenetic as well as gene expression analysis. The whole-genome genotyping SNP array technology possesses potential to include almost all existing SNPs into a single experiment. The genetic abnormalities resolution was selected by high-density SNP array analysis more sufficient than for direct PCR and sequencing analysis. However, this technique also has some of its limitations similar to the other DNA array tools. SNP array is unable to detect inversions, whole-genome ploidy changes, and balanced chromosomal translocations deprived of combination with other techniques.

18.8 Clinical diagnostics

SNP-based microarray technology enables the detection of copy number neutral regions and copy number variations (CNVs) that are absent of heterozygosity throughout the genome in clinical diagnostics. The SNP-based microarray analysis enhances the detection of abnormalities in human pregnancy and has made possible the identification of diseases at an early gestational age (Steigrad, 2003; Benirschke and Kaufmann, 1995).

18.9 Forensic detection

There is an increase in interest for SNP typing in the forensic field, not only for their usefulness for defining various chromosomal or mitochondrial DNA haplogroups or for geographical origin of samples but also for the autosomal SNPs. The autosomal SNPs have attracted the audience due to the inherent advantages in paternity testing as well as analysis of degraded samples by using short amplicons.

18.10 Pharmacogenomics

Xenobiotics or pharmaceutical agents induced or repressed gene expression profiling can be done by cDNA microarray and in pharmacogenomics (Wang et al., 2004). Microarray genotyping system has been successfully developed in pharmacogenetics for the development of antihypertensive drug response (Liljedahl et al., 2003; Freeman et al., 2001; Amin et al., 2002). This technique also provides a reliable basis for prediction of an individual's response to herbal drugs. Furthermore, this technology also applied in various areas of biology, including toxicology. Recently, microarray-based technology has also been employed for the understanding toxicant mechanism of action, known as Toxicogenomic (Pennie, 2002; Waring et al., 2002).

18.11 Variation in the human genome

With the increasing range of genomic diversity-related questions in human microarrays, SNP genotyping is providing an essential means of discovering new genomic variations methods and assaying their genome-wide frequency in the human genome. Among the wide range there are some of the applications based on microarray-based approaches for studying the diversity of human genome:

- *Nature of structural variation study.* The recent discoveries based on the widespread CNV in the human genome have exploded a new interest in structural genomic variation, which is amenable to detection using array comparative genome hybridization (Sebat, 2004; Weiss, 2008).
- *To study human evolution by identification of the role of gene amplification.* Recently, it has been advocated that the CNV at the *AMY1* locus (salivary amylase) has been under selective pressure through human history (Perry, 2007). The selection for CNV for a particular locus can be a general theme in the human genome.
- *To study the relationship between structural variation and human disorders.* Gene amplifications or deletions have been identified as a significant cause of human diseases (Lupski,2007). However, recently by employing genome-scale approaches, it has been demonstrated that the high frequency of de novo CNV and its role in various diseases like autism (Sebat, 2007).
- *To study the cancers associated with genomic changes.* Typically, somatic point mutations related to genome-wide studies related to diseases in humans use Sanger sequencing method (Sjoblom, 2006). Microarray-based tools possess the potential for addressing the various genomics events associated efficiently with cancers.
- *To study the human's insertional variation extents:* The human genome mostly constituted with mobile elements. However, a lack of information about the mobile elements sequence difference across the human population: therefore, by using CNV, one can quickly identify and compare the variation among the mobile elements in the human society (Wood et al., 2007).

18.12 Expression analysis of plants and microbes through microarray

To characterize and quantify the daunting uncultivated biodiversity of environmental microbes with functional abilities and their linkage to the ecosystem are a tedious task; therefore there is an urgent need of high-throughput technology (Wagner et al., 2007). Microarray technologies application allows us to study the single nucleotide variations in the sequences, expression of a particular gene at one specific event, the concentration of resultant protein products, the presence of various chemical and biological molecules, etc. on a large scale (Stears et al., 2003; Stoughton, 2005). Most of the sequence-based information can be inferred through whole-genome sequencing with a high cost as a limitation, whereas with the properties of automation, miniaturization, parallelism for extensive data and analysis of the entire genome, microarray becomes a crucial element for functional genomics (Zhou and Thompson, 2002; Aharoni and Vorst, 2001).

Advanced technologies can be applied to study the expression analysis of 99% unculturable soil microorganism through the utilization of RNA extraction, microarray, qRT–PCR, RT–PCR, stable isotopic probes, and presence of reporter genes. It comes up with information on resultant mechanisms and cellular activity of microbiome against environmental effects such as pH, moisture, salinity, temperature, oxygen availability, nutrients, the presence of pollutants (Saleh-Lakha et al., 2005). Mining of nucleic acid from the contaminated or affected sites along with their amplification by the use of polymerase chain various profiling techniques can utilize reaction through tagged probes for the estimation of the population, genes responsible for bioremediation, phylogenetic study relative to ecological niches, etc. (Malik et al., 2008). A massive list of parameters such as fruit ripening, phytochrome A level, nitrate assimilation, circadian clock, seed development, and many more biological functions can be analyzed by using microarray as a tool (Aharoni and Vorst, 2001). The presence of methanotrophs involved in reducing the methane emission at landfill sites can be detected through the microarray technique by targeting methane monooxygenase (pmoA) gene (Stralis-Pavese et al., 2004).

Microarray as miniaturization can confront an epidemic or crisis due to the presence of biodefence potential as it is involved in the preparation of medicines or vaccines against numerous diseases by a quantitative check for nucleic acids, proteins, antibodies, carbohydrates, and other products included in the defense mechanisms. It provides the information regarding the expression during intentional experiments for diseased conditions, disease subphenotypes, predicts disease progression, gene disruptions, and other parameters (Stoughton, 2005; Uttamchandani et al., 2009; Murray-Rust, 1994; Friend, 1999). Microbes present in the rhizospheric periphery of plants or at crucial environmental sites can be easily and rapidly characterized by utilizing genomic DNA microarray employing open reading frames (ORFs), functional protein-coding genes, genes involved in physiological responses, or genes coding virulence factors as probes (Cummings and Relman, 2000). It can be utilized by the breeders to obtain a gene cascade or a metabolic pathway involved in conferring more efficiency, precise improvement and resistance to plants against abiotic stresses such as salinity, drought, high temperature as an in silico genotype can be designed by segregating the information about allele at all the loci of a population (Kathiresan et al., 2006; Varshney et al., 2005). Species, specific oligonucleotide probes of 17–25 base pairs, can be utilized to target the variable 16 S rRNA regions present within compost forming microbial communities (Franke-Whittle et al., 2005). For example, to trace the metabolic pathway of metal-reducing bacteria, full GeoChip containing 24,243 oligonucleotides (50 mer) probes have been developed for studying biogeochemical processes involved in environmental management, restoration, climate change, agriculture, etc. It covers more than 10,000 genes involved in degradation of contaminant, metal reduction, resistance, carbon, nitrogen, sulfur, and phosphorus cycle (Gentry et al., 2007). In the same context, nine genes were found to be overexpressed in *Candida albicans* during the exposure of nitric oxide and provide the resistance against the toxic effects (Hromatka et al., 2005).

18.13 Microarray for proteomics

Proteins are the molecular drivers of the biological systems that lead to comprehensive knowledge about complex pathways and processes supporting the living world. Due to the development of different forms of microarray, we can also trace the presence and quantify the amount of translated protein products, posttranslational modification, protein–protein interaction, functions, and other processes at the particular event through DNA–protein hybridization. For in situ conversion of DNA microarray into protein microarray containing a variety of proteins, membrane filter sandwich technique was employed in which protein synthesized on one DNA template arrayed slide and is transferred through membrane filter on capturing agent coated slide (He et al., 2008). In an alternative, robust, cost-effective, and mostly applied process, biomolecule libraries (such as a peptide, small molecules, and protein libraries) are synthesized and deposited onto glass slide robotically which can be efficiently fabricated for large-scale screening. Peptide libraries are one of the best and reproducible sources for protein microarray due to solid-phase peptide synthesis which reduce the effort for the preparation of synthetic compounds and also lead to convenient fabrication of microarray which is a limitation for delicate protein library (Merrifield, 1985). For the spotted microarray development after the completion of library synthesis, immobilization is a crucial step that is achieved by the robotic deposition of biomolecules in the form of small droplets on a solid support which gets linked to surface either covalently, noncovalently, or specific to the particular site. Microarray proteomics used to study the function and kinetics of proteins or enzymes in the form of absorbance, fluorescence, or luminescence in response to the appropriate substrate (Chen et al., 2003). It can also be used to analyze physiological activities of proteins (such as histone peptide microarray) through mapping of their ligand-binding specificities concerning their substrate (Gurard-Levin et al., 2010). Protein–RNA interactions can also be deduced through amphilic peptide array that binds to fluorescently labeled hairpin RNAs (Pai et al., 2012).

18.14 ChIP-sequencing

ChIP-sequencing (ChIP-seq) is the most common technique that utilizes chromatin immunoprecipitation with DNA sequencing to study the protein–DNA interaction through the global or precise binding sites. It is mainly employed to study the phenotypic alteration due to various transcription factors, modification in expression due to the presence of proteins, epigenetic changes, and modifications during diseased state (Johnson and Mortazavi, 2007). It has also been applied to isolate the DNA sequences directly interact with structural proteins, transcriptional machinery, polymerases, or transcription factors and give rise to in vivo library of target DNA sites. It employs protein-specific antibody to enrich DNA–protein complex attached with oligonucleotide adapter with simultaneous massive and high-resolution sequencing either using genome sequencer (ChIP-SEQ) or hybridization to a microarray (ChIP-CHIP) followed by amplification and purification. Approximately 1000 clonal copies can be analyzed through cluster amplification of adapter-ligated ChIP DNA fragments on solid flow cell substrate which is sequenced base by base through a genome

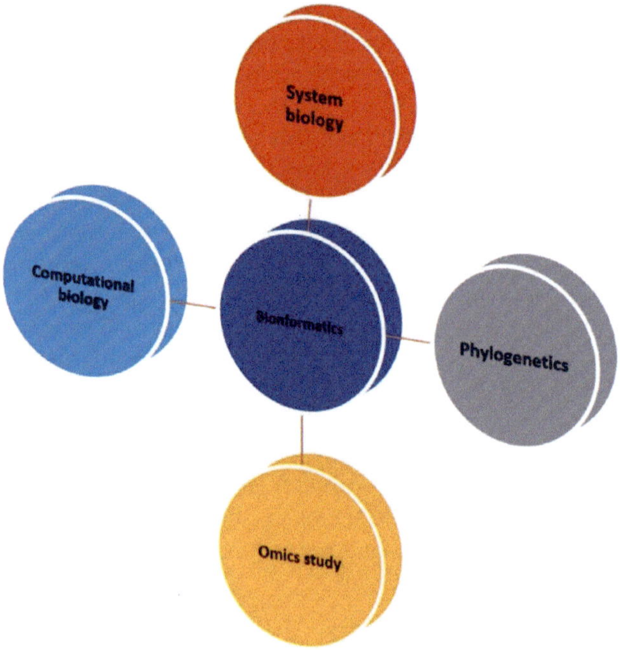

FIGURE 18.2 Flow diagram for interdisciplinary studies with bioinformatics.

analyzing the program. Size of the genome, depth of sequence run, and distribution of target factors regulate the performance of this technique which is not hindered by the spacing of predetermined probes (Kaufmann et al., 2010; Chen et al., 2017; Jothi, 2008).

A rich cDNA microarray using RNA gel blot analysis, upholding 1700 cDNAs obtained from the library has been prepared for the screening of abiotic stresses (drought, cold, abscisic acid and high salinity) tolerant genes which up came with unreported 57 novel genes in the plant (Rabbani et al., 2003). Numerous species–specific quantitative chip kits are available for plants such as Abcam's ChIP Kit—Plants (ab117137) for 24–48 chip reactions with operational duration of 6 hours and maintain its specificity through both qualitative and quantitative PCR, DNA sequencing, Southern blot and microarray (Baker, 2015). Some computational methods have been developed to analyze the enormous resultant datasets and to predict the DNA binding sites from ChIP-seq read count data through peak call method. For example, MACS is the most commonly used the technique to evaluate the size of ChIP-Seq tags, whereas ChIPDiff and ODIN identify significant differences in signals of two ChIP-seq from distinct biological situations (Allhoff et al., 2014; Zhang et al., 2008) (Fig. 18.2).

18.15 Role of microarray-based technologies and bioinformatics in bioremediation

The objective of such studies to develop and validate user-friendly, portable, microarray-based monitoring systems for microbial community structure and dynamics in groundwater and soil. The field-portable microarray studies by using the TruArray designed to investigate the bioremediation efficiency of microbes and enzymatic mechanism of remediation. Furthermore, the system biology is also applied for the bioremediation in documentation and depiction of microbes and the molecular processes involved in the process. Many bioremediation reactions are catalyzed by different enzymes and mechanisms that need to be understood by gene expression analysis and various computational tools such as PhyloChip, or sequencing, and 16S rRNA clone library technique employed for identification of microbial community structure involved in bioremediation. Likewise, for RNA GeoChip and RNA-seq, for proteins/enzymes various mass spectroscopy (MS) techniques, and for metabolites (metabolomics) gas chromatography (GC) and high-performance liquid chromatography (HPLC) coupled with MS, nuclear magnetic resonance, matrix-assisted laser desorption/ionization, and desorption electrospray ionization are employed to uncover the process of bioremediation. For constant monitoring of electron donors, electron acceptors, limiting nutrients, and hydrology these mechanisms should be known. Therefore it is imperative to comprehend the on-site bioremediation activities to exploit. Microarray, quantitative PCR, phylogenetics, enzyme probes, proteomics, metabolomics, and bioinformatics could be employed as monitoring tool to record terminal electron donors and electron acceptors in bioremediation process in an ecosystem and hence offer a better picture to understand the role of microbes in bioremediation.

TABLE 18.3 Bioinformatics tool/ software applied for bioremediation.

Tools	Database	Features/description	Web address/URL	Reference(s)
Gene prediction tools	FgenesB	Bacterial operon and gene	prediction http://linux1.softberry.com	Koski et al. (2005)
	GeneMark	Gene prediction in archaea, bacteria, and metagenomes	http://opal.biology.gatech.edu/GeneMark	Vallenet et al. (2009)
	Glimmer	Microbial gene finding system	www.cbcb.umd/software/glimmer	Moriya et al. (2007)
	REGANOR	Gene prediction server and database	www.cebitec.unibielefeld.de	Pellegrini et al. (1999)
Annotation pipeline	KAAS	KEGG automatic server annotation	www.genome.jp/tools/kaas	Zhang et al. (2011)
	Blast2GO	Annotation and sequence analysis tool	http://www.blast2go.com	–
	RAST	Rapid annotation using subsystem technology	www.nmpdr.org/FIG/wikiiew.cgi	Romualdi et al. (2005)
	BASys	Bacterial annotation system	http://basys.cal/basyscgi/submit.pi	Aziz et al. (2008)
Database and resources	NCBI	GenBank, RefSeq, TPA and PDB, databanks for storage and downloadable genomic information	http://www.ncbi.nlm.nih.gov	–
	EMBL	Nucleotide sequence database	http://www.ebi.ac.uk/embl	–
	KEGG	An integrated database resource, provides genomic, chemical, and systemic information	http://www.kegg.jp	–
	UniProt	Protein resource and functional information	http://www.uniprot.org	–
	InterProScan	Protein sequence analysis and classification	https://www.ebi.ac.uk/interpro	–
	MetaRouter	Bioinformatics for bioremediation	http://pdg.cnb.uam.es/MetaRouter	Pazos et al. (2005)
	STRING	Protein interaction network	https://string-db.org/	–
	CD database	Protein conserved domain	https://www.ncbi.nlm.nih.gov/Structure/cdd/cdd.shtml	Marchler-Bauer et al. (2013)

The decontamination of hazardous compounds (such as heavy metals and radioactive nuclei) is achieved by changing the redox state by using appropriate electron donors known as biotransformation (Suzuki et al., 2005). Therefore the microbial respiration, enzyme activity as well as metabolism deliver a healthier insight of bioremediation. Enzymes or nucleic acid-based probes could be exploited to decipher the bioremediation pathways and such probes very effective (Beller et al., 2002; Hazen et al., 2009; Bell et al., 2011). Oligonucleotide microarrays probes have been utilized to decipher the genes involved in biodegradation and/or biotransformation of naphthalene, polycyclic aromatic hydrocarbons, and nitrotoluene (Rhee et al., 2004). Furthermore, for documentation of genes linked with detoxification and degradation of metal hydrocarbon, respectively, qPCR has been also employed as a monitoring tool (Navarro et al., 2013; Yergeau et al., 2012). Again, metagenomic methods such as 16S rRNA also applied for the identification and characterization of microbial communities involved in bioremediation of hydrocarbons, chlorinated solvents, and metals (Yergeau et al., 2012). Recently, it has been reported that the high-throughput microarrays such as PhyloChip (16S rRNA phylogenetic microarray; also monitor fluctuation in microbial communities) and GeoChip (functional gene microarray; monitor catabolic gene activity) are employed for the study of metal and organic bioremediation process (He et al., 2007; DeSantis et al., 2007; Van Nostrand et al., 2009; Conrad et al., 2010). A huge amount of data gathered either by genomics (functional genomics, metagenomics), expression-based analysis (transcriptomics, microarray, qPCR, RNA-Seq) or system biology (proteomics, metabolomics) approaches are helpful in designing of sustainable, cost-effective, and low labor-intensive model for an effective bioremediation. Former investigations on microbe-mediated bioremediation mainly relied ex situ monitoring of degradation process and which provide microbial community degradation potential (Rogers and McClure 2003).

With the advancement of various genomics tools such as DNA hybridization, polymerase chain reaction (PCR), microarray analysis, molecular connectivity, and isotope distribution analysis, metabolic footprinting analysis, and metabolic engineering are useful in the elucidation of microbe-mediated remediation processes (Villas-Boas and Bruheim, 2007). Gene transcriptional analysis, protein engineering, metagenomics, and whole-genome sequencing are the main genomic tools that have been applied in the bioremediation field of bioremediation. The microarray technology approach has been applied recent years for the microbial community assessment involved in remediation and PhyloChip allow simultaneous detection of bacterial and archaeal species involved in the process (Zhou, 2003; Wu et al., 2006; Sagaram et al., 2009; Hamady et al., 2010). Likewise, GeoChip allows functional analysis of genes involved in C, N, and P geochemical cycling; reduction of sulfate; and reduction and resistance toward metals (Gentry et al., 2006). In recent days Functional Gene Arrays (FGA) are also in use, which utilizes probes for the detection of enzyme coding genes for a specific remediation pathway (Gentry et al., 2006). The recent version of GeoChip is GeoChip 2.0, which is a wide-ranging FGA applied for the understanding of environmental remediation (He et al., 2007). DNA microarray technology–based global gene expression analyses rely on mRNA and proteins of a cell, this the major limitation of this tool, however, whole-genome sequencing of all genes presents in an organism. In this approach the contaminated samples are directly subjected to metagenomics which offer sequences, which is then subjected various bioinformatics tools to predict the gene such as ORF prediction tools. Some modern tools/software and databases for genomics studies are summarized in Table 18.3. Meta MetaRouter: bioinformatics tool is also developed for maintenance of heterogeneous information related to bioremediation (Pazos et al., 2005).

18.16 Conclusion

For a proper understanding of the system biology, both molecular and computational approaches should be considered as it can give rise to new hypothesis and discoveries through the data integration. Microbial identification and their mechanism of remediation through microarray-based technologies and bioinformatics are emerging fields employing advanced multidisciplinary approaches. Comparison of transcripts of microbes treated with pollutants versus nontreated can be addressed by analyzing their in vitro expression profiles of different genes that could be useful in the mining of pathways operate under such processes. By observing the advancement in techniques and algorithms for multiple species sequencing, data generation, analysis, and interpretation of result ensure a promising future. Microarray technologies have already been in use in various formats for different fields, as discussed in the chapter, like gene expression studies, changes in the genetic makeup, genotyping, and gene functional analysis. Recently, array-based sequencing techniques such as resequencing and minisequencing, which combines targeted hybridization with enzymatic primer extension reactions, emerged as a powerful means to scan all possible variations in DNA sequences. As with the new technologies development, a uniform and enhanced sophisticated experimental design, data management systems, statistical tools, and algorithms must be applied for data analysis. As in the current situation, microarray in combination with other techniques is leading in *omics* study with the production of a large number of accumulated datasets which are useless until they are processed. The value of current robustness free technologies depends on careful development of oligonucleotide probes, target preparation, understanding of produced data, their analysis and its integration with other biological information for efficient performance in the field of genome, transcriptome, metabolome and proteome profiling. With appropriate maturation, modification of the assay protocols, advanced statistical design, and further development in algorithms giving rise to novel routinely proper statistical tools to obtain more reasonable and accurate results, the technique may soon become the workhorses of molecular diagnostics. Bioremediation is transformation of toxic pollutants in an augmented environment to little or less toxic or other nontoxic forms for the environment. Various microbes and/or plants have been advocated out to play a vital role in bioremediation, but these microbes are not well explored to attain reliable and effective scrubbing of contaminants. For the microbiologists, it is very challenging as most of the microbes involved in the process are unculturable on synthetic media, therefore their functionality in the process not well known. Unquestionably, the microarray-based technologies in combination with bioinformatics have transformed the bioremediation field. These approaches certainly integrate numerous bioremediation tactics underneath one umbrella, which will be very valuable in understanding the complex bioremediation process.

Acknowledgment

Jyoti Singh is thankful to Council of Industrial and Scientific Research Fellowship (CSIR), New Delhi, India for Senior Research Fellowship.

References

Adams, M.D., Celniker, S.E., Holt, R.A., 2000. The genome sequence of *Drosophila melanogaster*. Science. 287 (5461), 2185–2195.

Aharoni, A., Vorst, O., 2001. DNA microarrays for functional plant genomics. Functional Genomics. Springer, Dordrecht, pp. 99–118.

Aitken, M.D., Stringfellow, W.T., Nagel, R.D., Kazunga, C., Chen, S.H., 1998. Characteristics of phenanthrene-degrading bacteria isolated from soils contaminated with polycyclic aromatic hydrocarbons. Can. J. Microbiol. 44 (8), 743–752.

Alberts, B., 2008. Molecular Biology of the Cell, 5th ed. Garland Science, New York, p. 552, ISBN 978-0-8153- 4106-2.

Allhoff, C., Sere, C., Lin, Z., 2014. Detecting differential peaks in ChIP-seq signals with ODIN. Bioinformatics 30 (24), 3467–3475.

Amin, R.P., Hamadeh, H.K., Bushel, P.R., Bennett, L., Afshari, C.A., Paules, R.S., 2002. Genomic interrogation of mechanism(s) underlying cellular responses to toxicants. Toxicology 181, 555–563.

Amin, N., Liu, A.D., Ramer, S., Aehle, W., Meijer, D., Metin, M., et al., 2004. Construction of stabilized proteins by combinatorial consensus mutagenesis. Protein Eng. Des. Sel. 17 (11), 787–793. Available from: https://doi.org/10.1093/protein/gzh091.

Arun, A., Raja, P.P., Arthi, R., Ananthi, M., Kumar, K.S., Eyini, M., 2008. Polycyclic aromatic hydrocarbons (PAHs) biodegradation by basidiomycetes fungi, *Pseudomonas* isolate, and their co-cultures: comparative in vivo and in silico approach. Appl. Biochem. Biotechnol. 151, 132–142.

Aziz, R.K., Bartels, D., Best, A.A., DeJongh, M., Disz, T., et al., 2008. The RAST Server: rapid annotations using subsystems technology. BMC Genomics. 9, 75.

Baker, K., 2015. Chromatin state analysis of the barley epigenome reveals a higher-order structure defined by H3K27me1 and H3K27me3 abundance. Plant J. 84.

Baker, W., van den Broek, A., Camon, E., Hingamp, P., Sterk, P., Stoesser, G., 2000. The EMBL nucleotide sequence database. Nucleic Acids Res. 28 (1), 19–23.

Bell, T.H., Yergeau, E., Martineau, C., Juck, D., Whyte, L.G., Greer, C.W., 2011. Identification of nitrogen incorporating bacteria in petroleum contaminated arctic soils by using [15N] DNA-based stable isotope probing and pyrosequencing. Appl. Environ. Microbiol. 77, 4163–4171.

Beller, H.R., Kane, S.R., Legler, T.C., Alvarez, P.J.J., 2002. A real-time polymerase chain reaction method for monitoring anaerobic, hydrocarbon-degrading bacteria based on a catabolic gene. Environ. Sci. Technol. 36, 3977–3984.

Benirschke, K., Kaufmann, P., 1995. Molar Pregnancies. Pathology of the Human Placenta. Springer, New York, pp. 653–685.

Benson, D.A., Karsch-Mizrachi, I., Lipman, D.J., Ostell, J., Rapp, B.A., Wheeler, D.L., 2000. GenBank. Nucleic Acids Res. 28 (1), 15–18.

Bertelsen, A.H., Velculescu, V.E., 1998. High-throughput gene expression analysis using SAGE. Drug. Discov. Today 3, 152–159.

Bharaj, B.S., Angelopoulou, K., Diamandis, E.P., 1998. Rapid sequencing of the p53gene with a new automated DNA sequencer. Clin. Chem. 44, 1397–1403.

Bishop, M.J. (Ed.), 1998. Guide to Human Genome Computing. Academic Press, San Diego, CA.

Boopathy, R., Kulpa, C.F., 1994. Biotransformation of 2,4,6-trinitrotoluene (TNT) by a *Methanococcus* sp. (strain B) isolated from a lake sediment. Can. J. Microbiol. 40 (4), 273–278.

Carrilho, E., 2000. DNA sequencing by capillary array electrophoresis and microfabricated array systems. Electrophoresis 21, 55–65.

Chakraborty, R., Wu, C.H., Hazen, T.C., 2012. Systems biology approach to bioremediation. Curr. Opin. Biotechnol. 23, 483–490.

Chalah, A., 2016. Fundamentals of Microfluidics with Applications in Biological Analysis and Discovery. Harvard Extension School.

Chen, C., Wang, J.L., 2007. Characteristics of Zn^{2+} biosorption by *Saccharomyces cerevisiae*. Biomed. Environ. Sci. 20, 478–482.

Chen, G.Y.J., Uttamchandani, M., Zhu, Q., Wang, G., Yao, S.Q., 2003. Developing a strategy for activity-based detection of enzymes in a protein microarray. ChemBioChem 4, 336–339.

Chen, X., Bhadauria, V., Ma, B., 2017. ChIP-Seq: a powerful tool for studying protein-DNA interactions in plants. Curr. Issues Mol. Biol. 27, 171–180.

Cho, R.J., Campbell, M.J., Winzeler, E.A., Steinmetz, L., Conway, A., Wodicka, L., 1998. A genome-wide transcriptional analysis of the mitotic cell cycle. Mol. Cell 2 (1), 65–73.

Chung, S.Y., Maeda, M., Song, E., Horikoshij, K., Kudo, T., 1994. A Gram-positive polychlorinated biphenyl-degrading bacterium, *Rhodococcus erythropolis* strain TA421, isolated from a termite ecosystem. Biosci. Biotechnol. Biochem. 58 (11), 2111–2113.

Conrad, M.E., Brodie, E.L., Radtke, C.W., Bill, M., Delwiche, M.E., Lee, M.H., et al., 2010. Field evidence for co-metabolism of trichloroethene stimulated by addition of electron donor to groundwater. Environ. Sci. Technol. 44, 4697–4704.

Cummings, C.A., Relman, D.A., 2000. Using DNA microarrays to study host-microbe interactions. Emerg. Infect. Dis. 6 (5), 513.

DeAngelis, M.M., Batzer, M.A., 2001. Genome sequence analysis. Encycl. Life Sci. 1–6. Available from: https://doi.org/10.1038/npg.els.0003028.

DeRisi, J.L., Iyer, V.R., Brown, P.O., 1997. Exploring the metabolic and genetic control of gene expression on a genomic scale. Science 278 (5338), 680–686.

DeSantis, T., Brodie, E., Moberg, J., Zubieta, I., Piceno, Y., Andersen, G., 2007. High-density universal 16S rRNA microarray analysis reveals broader diversity than typical clone library when sampling the environment. Microb. Ecol. 53, 371–383.

Drmanac, S., Kita, D., Labat, I., Hauser, B., Schmidt, C., Burczak, J.D., et al., 1998. Accurate sequencing by hybridization for DNA diagnostics and individual genomics. Nat. Biotechnol. 16, 54–58.

Dubus, I.G., Hollis, J.M., Brown, C.D., 2000. Pesticides in rainfall in Europe. Environ. Pollut. 10, 331–344.

Duggan, D.J., Bittner, M., Chen, Y., Meltzer, P., Trent, J.M., 1999. Expression profiling using cDNA microarrays. Nat. Genet. 21 (1 Suppl.), 10–14.

Dunham, I., 1999. The DNA sequence of human chromosome 22. Nature 402 (6761), 489–495.

Durbin, R., Eddy, S.R., Krogh, A., Mitchison, G., 1998. Biological Sequence Analysis: Probabilistic Models of Proteins and Nucleic Acids. Cambridge University Press.

Eisen, M.B., Brown, P.O., 1999. DNA arrays for analysis of gene expression. Methods Enzymol. 303, 179–205.

Fleischmann, R.D., Adams, M.D., White, O., Clayton, R.A., Kirkness, E.F., Kerlavage, A.R., 1995. Whole-genome random sequencing and assembly of *Haemophilus influenzae*. Science 269 (5223), 496–512.

Franke-Whittle, I.H., Klammer, S.H., Insam, H., 2005. Design and application of an oligonucleotide microarray for the investigation of compost microbial communities. J. Microbiol. Methods 62 (1), 37–56.

Frassinetti, S., Setti, L., Corti, A., Farrinelli, P., Montevecchi, P., Vallini, G., 1998. Biodegradation of dibenzothiophene by a nodulating isolate of *Rhizobium meliloti*. Can. J. Microbiol. 44 (3), 289–297.

Freeman, W.M., Brebner, K., Lynch, W.J., Robertson, D.J., Roberts, D.C., Vrana, K.E., 2001. Cocaine-responsive gene expression changes in rat hippocampus. Neuroscience 108, 371–380.

Friend, S.H., 1999. How DNA microarrays and expression profiling will affect clinical practice. BMJ 319 (7220), 1306–1307.

Gentry, T.J., Wickham, G.S., Schadt, C.W., He, Z., Zhou, J., 2006. Microarray applications in microbial ecology research. Microb. Ecol. 52 (2), 159–175.

Gurard-Levin, Z.A., Kilian, K.A., Kim, J., Bähr, K., Mrksich, M., 2010. Peptide arrays identify isoform-selective substrates for profiling endogenous lysine deacetylase activity. ACS Chem. Biol. 5, 863–873.

Hamady, M., Lozupone, C., Knight, R.V., 2010. Fast UniFrac: facilitating high-throughput phylogenetic analyses of microbial communities including analysis of pyrosequencing and PhyloChip data. ISME J. 4 (1), 17.

Hassan, I., Eastman, A.W., Weselowski, B., Mohamedelhassan, E., Yanful, E.K., Yuan, Z.C., 2016. Complete genome sequence of *Arthrobacter* sp. strain LS16, isolated from agricultural soils with potential for applications in bioremediation and bioproducts. Genome Announc. 4 (1), 1586–1615.

Hazen, T., Chakraborty, R., Fleming, J., Gregory, I., Bowman, J., Jimenez, L., et al., 2009. Use of gene probes to assess the impact and effectiveness of aerobic in situ bioremediation of TCE. Arch. Microbiol. 191, 221–232.

He, M., Stoevesandt, O., Palmer, E.A., Khan, F., Ericsson, O., Taussig, M.J., 2008. Printing protein arrays from DNA arrays. Nat. Methods 5, 175–177.

He, Z., Gentry, T.J., Schadt, C.W., Wu, L., Liebich, J., Chong, S.C., et al., 2007. GeoChip: a comprehensive microarray for investigating biogeochemical, ecological and environmental processes. ISME J. 1 (1), 67.

Holley, R.W., Apgar, J., Everett, G.A., Madison, J.T., Marquisee, M., Merrill, S.H., et al., 1965. Structure of a ribonucleic acid. Science 1462–1465.

Holstege, F.C.J.E., Wyrick, J.J., Lee, T.I., Hengartner, C.J., Green, M.R., Golub, T.R., et al., 1998. Dissecting the regulatory circuitry of a eukaryotic genome. Cell 95 (5), 717–728.

Hromatka, B.S., Noble, S.M., Johnson, A.D., 2005. Transcriptional response of *Candida albicans* to nitric oxide and the role of the YHB1 gene in nitrosative stress and virulence. Mol. Biol. Cell 16, 4814–4826.

Ijah, U.J.J., 1998. Studies on relative capabilities of bacterial and yeast isolates from tropical soil in degrading crude oil. Waste Manage. 18, 293–299.

Infante, J.C., De Arco, R.D., Angulo, M.E., 2014. Removal of lead, mercury and nickel using the yeast *Saccharomyces cerevisiae*. Rev. MVZ Córdoba 19, 4141–4149.

Jelinsky, S.A., Samson, L.D., 1999. Global response of *Saccharomyces cerevisiae* to an alkylating agent. Proc. Natl. Acad. Sci. U.S.A. 96 (4), 1486–1491.

Jogdand, S.N., 1995. Environ. Biotechnol., first ed. Himalaya Publishing House, Bombay, pp. 104–120.

Johnson, D.S., Mortazavi, A., 2007. Genome-wide mapping of in vivo protein–DNA interactions. Science 316, 1497–1502.

Jothi, 2008. Genome-wide identification of in vivo protein–DNA binding sites from ChIP-seq data. Nucleic Acids Res. 36 (16), 5221–5231.

Kathiresan, A., Lafitte, H.R., Chen, J., Mansueto, L., Bruskiewich, R., Bennett, J., 2006. Gene expression microarrays and their application in drought stress research. Field Crop. Res. 97 (1), 101–110.

Kaufmann, K., Muino, J.M., Østerås, M., Farinelli, L., Krajewski, P., Angenent, G.C., 2010. Chromatin immunoprecipitation (ChIP) of plant transcription factors followed by sequencing (ChIP-SEQ) or hybridization to whole genome arrays (ChIP-CHIP). Nat. Protoc. 5 (3), 457.

Koski, L.B., Gray, M.W., Lang, B.F., Burger, G., 2005. AutoFACT: an automatic functional annotation and classification tool. BMC Bioinformatics 6, 151.

Kruglyak, S., 1998. Multistage sequencing by hybridization. J. Comput. Biol. 5, 165–171.

Lal, B., Khanna, S., 1996. Degradation of crude oil by *Acinetobacter calcoaceticus* and *Alcaligenes odorans*. J. Appl. Bacteriol. 81, 355–362.

Lander, E.S., Linton, L.M., Birren, B., Nusbaum, C., Zody, M.C., Baldwin, J., 2001. Initial sequencing and analysis of the human genome. Nature 409, 860–921.

Lee, Y., Sultana, R., Pertea, G., Cho, J., Karamycheva, S., Tsai, J., 2002. Cross-referencing eukaryotic genomes: TIGR Orthologous Gene Alignments (TOGA). Genome Res. 2002 (12), 493–502.

Levitt, M., 1969. Detailed molecular model for transfer ribonucleic acid. Nature 224 (5221), 759–763.

Levitt, M., 2001. The birth of computational structural biology. Nat. Struct. Mol. Biol. 8 (5), 392.

Liang, P., Pardee, A.B., 1992. Differential display of eukaryotic messenger RNA by means of the polymerase chain reaction. Science 257, 967–971.

Liljedahl, U., Karlsson, J., Melhus, H., Kurland, L., Lindersson, M., Kahan, T., 2003. A microarray minisequencing system for pharmacogenetic profiling of antihypertensive drug response. Pharmacogenetics 13, 7–17.

Lipshutz, R.J.F.S., Gingeras, T.R., Lockhart, D.J., 1999. High density synthetic Oligonucleotide arrays. Nat. Genet. 21 (1), 20–24.

Lopez-Errasquin, E., Vazquez, C.V., 2003. Tolerance and uptake of heavy metals by *Trichoderma atroviride* isolated from sludge. Chemosphere 50, 137–143.

Lupski, J.R., 2007. Genomic rearrangements and sporadic disease. Nat. Genet. 39, S43–S47.

Malik, S., Beer, M., Megharaj, M., Naidu, R., 2008. The use of molecular techniques to characterize the microbial communities in contaminated soil and water. Environ. Int. 34 (2), 265–276.

Marchler-Bauer, A., Lu, S., Anderson, J.B., Chitsaz, F., Derbyshire, M.K., 2013. CDD: a Conserved Domain Database for the functional annotation of proteins. Nucleic acids res. 39, D225–D229.

Maxam, A.M., Gilbert, W., 1977. A new method for sequencing DNA. Proc. Natl. Acad. Sci. U.S.A. 74 (2), 560–564.

McKenzie, S.E., Mansfield, E., Rappaport, E., Surrey, S., Fortina, P., 1998. Parallel molecular genetic analysis. Eur. J. Hum. Genet. 6, 417–429.

Megharaj, M., Ramakrishnan, B., Venkateswarlu, K., Sethunathan, N., Naidu, R., 2011. Bioremediation approaches for organic pollutants: a critical perspective. Environ. Int. 37 (8), 1362–1375.

Merrifield, R.B., 1985. Solid phase synthesis (Nobel Lecture). Angew. Chem. Int. Ed. Engl. 24, 799–810.

Mirlahiji, S.G., Eisazadeh, K., 2014. Bioremediation of Uranium by *Geobacter* spp. J. Res. Dev. 1, 52–58.

Mohamed, A.T., El Hussein, A.A., El Siddig, M.A., Osman, A.G., 2011. Degradation of oxyfluorfen herbicide by soil microorganisms: biodegradation of herbicides. Biotechnology 10, 274–279.

Moriya, Y., Itoh, M., Okuda, S., Yoshizawa, A.C., Kanehisa, M., 2007. KAAS: an automatic genome annotation and pathway reconstruction server. Nucleic Acids Res. 35, 182–185.

Mohammadian Fazli, M., Soleimani, N., Mehrasbi, M., Darabian, S., Mohammadi, J., Ramazani, A., 2015. Highly cadmium tolerant fungi: their tolerance and removal potential. J. Environ. Health Sci. Eng. 13, 19. Available from: https://doi.org/10.1186/s40201-015-0176-0.

Murray-Rust, P., 1994. Bioinformatics and drug discovery. Curr. Opin. Biotechnol. 5 (6), 648–653.

Naidu, C., Suneetha, Y., 2012. Review article: Current knowledge on microarray technology—an overview. Trop. J. Pharm. Res. 11 (1).

Navarro, C.A., von Bernath, D., Jerez, C.A., 2013. Heavy metal resistance strategies of acidophilic bacteria and their acquisition: importance for biomining and bioremediation. Biol. Res. 46, 363–371.

Needleman, S.B., Wunsch, C.D., 1970. A general method applicable to the search for similarities in the amino acid sequence of two proteins. J. Mol. Biol. 48 (3), 443–453.

Newcombe, D.A., Crowley, D.E., 1999. Bioremediation of atrazine-contaminated soil by repeated applications of atrazine-degrading bacteria. Appl. Microbiol. Biotechnol. 51 (6), 877–882.

Niti, C., Sunita, S., Kamlesh, K., 2013. Bioremediation: an emerging technology for remediation of pesticides. Res. J. Chem. Environ. 17, 88–105.

Okayama, T., Tamura, T., Gojobori, T., Tateno, Y., Ikeo, K., Miyazaki, S., 1998. Formal design and implementation of an improved DDBJ DNA database with a new schema and Object-Oriented library. Bioinformatics 14 (6), 472–478.

Pai, J., Yoon, T., Kim, N.D., Lee, I.S., Yu, J., Shin, I., 2012. High-throughput profiling of peptide-RNA interactions using peptide microarrays. J. Am. Chem. Soc. 134, 19287–19296.

Pazos, F., Guijas, D., Valencia, A., De Lorenzo, V., 2005. MetaRouter: bioinformatics for bioremediation. Nucleic Acids Res. 33, 588–592. Available from: https://doi.org/10.1093/nar/gki068.

Pease, A.C., Solas, D., Sullivan, E.J., Cronin, M.T., Holmes, C.P., Fodor, S.P., 1994. Light-generated oligonucleotide arrays for rapid DNA sequence analysis. Proc. Natl. Acad. Sci. U.S.A. 91, 5022–5026.

Pedersendagger, A.G., Jensendagger, L.J., Brunak, S., Staerfeldt, H.H., Ussery, D.W., 2000. A DNA structural atlas for *Escherichia coli*. J. Mol. Biol. 299 (4), 907–930.

Pellegrini, M., Marcotte, E.M., Thompson, M.J., Eisenberg, D., Yeates, T.O., 1999. Assigning protein functions by comparative genome analysis: protein phylogenetic profiles. Proc. Natl. Acad. Sci. U.S.A. 96, 4285–4428.

Peña-Montenegro, T.D., Lozano, L., Dussán, J., 2015. Genome sequence and description of the mosquitocidal and heavy metal tolerant strain *Lysinibacillus sphaericus* CBAM5. Stand. Genomic Sci. 10, 1–10.

Pennie, W.D., 2002. Custom cDNA microarrays; technologies and applications. Toxicology 181–182, 551–554.

Pennie, W., Pettit, S.D., Lord, P.G., 2004. Toxicogenomics in risk assessment: An overview of an HESI collaborative research program. Genomics and Risk Assessment. 112, 417–419.

Perry, G.H., 2007. Diet and the evolution of human amylase gene copy number variation. Nat. Genet. 39, 1256–1260.

Perou, C., Sørlie, T., Eisen, M., et al., 2000. Molecular portraits of human breast tumors. Nature. 406, 747–752. Available from: https://doi.org/10.1038/35021093.

Pierzynski, G.M., Sims, J.T., Vance, G.F., 1994. Soils and Environmental Quality. Lewis Publishers, Ann Arbor, MI.

Rabbani, M.A., Maruyama, K., Abe, H., Khan, M.A., Katsura, K., Ito, Y., et al., 2003. Monitoring expression profiles of rice genes under cold, drought, and high-salinity stresses and abscisic acid application using cDNA microarray and RNA gel-blot analyses. Plant Physiol. 133 (4), 1755–1767.

Ravi, R.K., Pathak, B., Fulekar, M.H., 2015. Bioremediation of Persistent Pesticides in Rice field Soil Environment Using Surface Soil Treatment Reactor. Int. J. Curr. Microbiol. Appl. Sci. 4, 359–369.

Rhee, S.K., Liu, X., Wu, L., 2004. Detection of genes involved in biodegradation and biotransformation in microbial communities by using 50-mer oligonucleotide microarrays. Appl. Environ. Microbiol. 70, 4303–4317.

Rogers, S.L., McClure, N., 2003. In: Head, M., Singleton, I., Milner, M.G. (Eds.), A Critical Review in Bioremediation. Horizon Scientific Press, Wymondham, pp. 27–59.

Romualdi, A., Siddiqui, R., Glöckner, G., Lehmann, R., Sühnel, J., 2005. GenColors: acceleratedcomparative analysis and annotation of prokaryotic genomes at various stages of completeness. Bioinformatics. 21, 3669–3671.

Roth, F.P.H.J., Estep, P.W., Church, G.M., 1998. Finding DNA regulatory motifs within unaligned noncoding sequences clustered by whole-genome mRNA quantitation. Nat. Biotechnol. 16 (10), 939–945.

Safiyanu, I., Isah, A.A., Abubakar, U.S., Rita Singh, M., 2015. Review on comparative study on bioremediation for Oil spills using microbes. RJPBCS 6, 783–790.

Sagaram, U.S., DeAngelis, K.M., Trivedi, P., Andersen, G.L., Lu, S.E., Wang, N., 2009. Bacterial diversity analysis of Huanglongbing pathogen-infected citrus, using PhyloChip arrays and 16S rRNA gene clone library sequencing. Appl. Environ. Microb. 75 (6), 1566–1574.

Saleh-Lakha, S., Miller, M., Campbell, R.G., Schneider, K., Elahimanesh, P., Hart, M.M., et al., 2005. Microbial gene expression in soil: methods, applications and challenges. J. Microbiol. Methods 63 (1), 1–19.

Sanger, F., Air, G.M., Barrell, B.G., Brown, N.L., Coulson, A.R., Fiddes, J.C., et al., 1977. Nucleotide sequence of bacteriophage φX174 DNA. Nature 265 (5596), 687.

Schena, M., Shalon, D., Davis, R.W., Brown, P.O., 1995. Quantitative monitoring of gene expression patterns with a complementary DNA microarray. Science 270, 467–470.

Schena, M., Shalon, D., Heller, R., Chai, A., Brown, P.O., Davis, R.W., 1996. Parallel human genome analysis: microarray-based expression monitoring of 1000 genes. Proc. Natl. Acad. Sci. U.S.A. 93, 10614–10619.

Schuler, G.D., Epstein, J.A., Ohkawa, H., Kans, J.A., 1996. Entrez: molecular biology database and retrieval system. Methods Enzymol. 266, 141–162.

Sebat, J., 2004. Large-scale copy number polymorphism in the human genome. Science 305, 525–528.

Sebat, J., 2007. Strong association of *de novo* copy number mutations with autism. Science 316, 445–449.

Sherburne, L.A., Shrout, J.D., Alvarez, P.J., 2005. Hexahydro-1,3,5-trinitro-1,3,5-triazine (RDX) degradation by *Acetobacterium paludosum*. Biodegradation 16 (6), 539–547.

Singh, B.K., Walker, A., Morgan, J.A., Wright, D.J., 2004. Biodegradation of chlorpyrifos by Enterobacter strain B-14 and its use in bioremediation of contaminated soils. Appl. Environ. Microbiol. 70 (8), 4855–4863.

Sjoblom, T., 2006. The consensus coding sequences of human breast and colorectal cancers. Science 314, 268–274.

Sogayar, M.C., Camargo, A.A., Bettoni, F., Carraro, D.M., Pires, L.C., Parmigiani, R.B., 2004. Ludwig-FAPESP transcript finishing initiative: a transcript finishing initiative for closing gaps in the human transcriptome. Genome Res. 14, 1413–1423.

Southern, E., 1975. Detection of specific sequences among DNA fragments separated by gel electrophoresis. J. Mol. Biol. 98, 503–517.

Stears, R.L., Martinsky, T., Schena, M., 2003. Trends in microarray analysis. Nat. Med. 9 (1), 140.

Steigrad, S.J., 2003. Epidemiology of gestational trophoblastic diseases. Best. Pract. Res. Clin. Obstet. Gynaecol. 17, 837–847.

Stoughton, R.B., 2005. Applications of DNA microarrays in biology. Annu. Rev. Biochem. 74, 53–82.

Stralis-Pavese, N., Sessitsch, A., Weilharter, A., Reichenauer, T., Riesing, J., Csontos, J., et al., 2004. Optimization of diagnostic microarray for application in analysing landfill methanotroph communities under different plant covers. Environ. Microbiol. 6 (4), 347–363.

Suzuki, Y., Kelly, S.D., Kemner, K.M., Banfield, J.F., 2005. Direct microbial reduction and subsequent preservation of uranium in natural near-surface sediment. Appl. Environ. Microbiol. 71, 1790–1797.

Taniguchi, M., Miura, K., Iwao, H., Yamanaka, S., 2001. Quantitative assessments of DNA microarrays-comparison with northern blot analyses. Genomics. 71, 34–39.

Tarca, A.L., Romero, R., Draghici, S., 2006. Analysis of microarray experiments of gene expression profiling. Am. J. Obstet. Gynecol. 195 (2), 373–388. Available from: https://doi.org/10.1016/j.ajog.2006.07.001.

The C. elegans Sequencing Consortium, 1998. Genome sequence of the nematode *C. elegans*: a platform for investigating biology. Science 2012–2018.

Uttamchandani, M., Neo, J.L., Ong, B.N.Z., Moochhala, S., 2009. Applications of microarrays in pathogen detection and biodefence. Trends Biotechnol. 27 (1), 53–61.

Vallenet, D., Engelen, S., Mornico, D., Cruveiller, S., Fleury, L., Lajus, A., et al., 2009. MicroScope: a platform for microbial genome annotation and comparative genomics. Database 2009.

Van Nostrand, J.D., Wu, W.-M., Wu, L., Deng, Y., Carley, J., Carroll, S., et al., 2009. GeoChip-based analysis of functional microbial communities during the re-oxidation of a bio-reduced uranium-contaminated aquifer. Environ. Microbiol. 11, 2611–2626.

Varshney, R.K., Graner, A., Sorrells, M.E., 2005. Genomics-assisted breeding for crop improvement. Trends Plant Sci. 10 (12), 621–630.

Velculescu, V.E., Zhang, L., Vogelstein, B., Kinzler, K.W., 1995. Serial analysis of gene expression. Science. 270, 484–487.

Venter, J.C., Adams, M.D., Myers, E.W., 2001. The sequence of the human genome. Science 291 (5507), 1304–1351.

Villas-Boas, S.G., Bruheim, P., 2007. The potential of metabolomics tools in bioremediation studies. OMICS 11, 305–313.

Wagner, M., Smidt, H., Loy, A., Zhou, J., 2007. Unravelling microbial communities with DNA-microarrays: challenges and future directions. Microb. Ecol. 53 (3), 498–506.

Wang, D.G., Fan, J.B., Siao, C.J., Berno, A., Young, P., Sapolsky, R., et al., 1998. Large-scale identification, mapping, and genotyping of single-nucleotide polymorphisms in the human genome. Science 280, 1077–1082.

Wang, Z., Du, Q., Wang, F., Liu, Z., Li, B., Wang, A., 2004. Microarray analysis of gene expression on herbal glycoside recipes improving deficient ability of spatial learning memory in ischemic mice. J. Neurochem. 88, 1406–1415.
Waring, J.F., Gum, R., Morfitt, D., Jolly, R.A., Ciurlionis, R., Heindel, M., 2002. Identifying toxic mechanisms using DNA microarrays: evidence that an experimental inhibitor of cell adhesion molecule expression signals through the aryl hydrocarbon nuclear receptor. Toxicology 181, 537–550.
Waterston, R.H., Lindblad-Toh, K., Birney, E., 2002. Initial sequencing and comparative analysis of the mouse genome. Nature 420 (6915), 520–562.
Watson, J.D., Crick, F.H.C., 1953. A structure for deoxyribose nucleic acid. Nature. 171, 737–738.
Weiss, L.A., 2008. Association between microdeletion and microduplication at 16p11.2 and autism. N. Engl. J. Med. 358, 667–675.
Winzeler, E.A., Shoemaker, D.D., Astromoff, A., Liang, H., Anderson, K., Andre, B., 1999. Functional characterization of the *S. cerevisiae* genome by gene deletion and parallel analysis. Science 285 (5429), 901–906.
Wood, L.D., 2007. The genomic landscapes of human breast and colorectal cancers. Science 318.
Wood, A., Shukla, A., Schneider, J., Lee, J.S., et al., 2007. Ctk complex-mediated regulation of histone methylation by COMPASS. Mol. Cell Biol. 27 (2), 709–720.
Wu, L., Liu, X., Schadt, C.W., Zhou, J., 2006. Microarray-based analysis of sub nanogram quantities of microbial community DNAs by using whole-community genome amplification. Appl. Environ. Microbiol. 72 (7), 4931–4941.
Yamamoto, M., Wakatsuki, T., Hada, A., Ryo, A., 2001. Use of serial analysis of gene expression (SAGE) technology. J. Immunol. Methods 250, 45–66.
Yergeau, E., Sanschagrin, S., Beaumier, D., Greer, C.W., 2012. Metagenomic analysis of the bioremediation of diesel-contaminated Canadian high Arctic soils. PLoS One 7 (1), e30058. <https://doi.org/10.1371/journal.pone.0030058>.
Zhang, C., Hughes, J.B., 2003. Biodegradation pathways of hexahydro-1,3,5-trinitro-1,3,5-triazine (RDX) by *Clostridium acetobutylicum* cell-free extract. Chemosphere 50 (5), 665–671.
Zhang, J., Chiodini, R., Badr, A., Zhang, G., 2011. The impact of next-generation sequencing on genomics. J. Genet. Genomics 38, 95–109.
Zhang, Y., Liu, T., Meyer, C.A., Eeckhoute, J., Johnson, D.S., Bernstein, B.E., et al., 2008. Model-based analysis of ChIP-Seq (MACS). Genome Biol. 9 (9), R137.
Zhou, J., 2003. Microarrays for bacterial detection and microbial community analysis. Curr. Opin. Microbiol. 6 (3), 288–294.
Zhou, J., Thompson, D.K., 2002. Challenges in applying microarrays to environmental studies. Curr. Opin. Biotechnol. 13, 204–207.

Chapter 19

Application of plant–microbe systems in bioremediation of metalloid-contaminated soils

Megha Barot[1], Arpan Modi[2] and Ajay Kumar[2]

[1]*SRISTI Life Science, Grambharti, Gandhinagar, India,* [2]*Agriculture Research Organization, Ministry of Agriculture and Rural Development, Volcani Center, Rishon LeZion, Israel*

19.1 Introduction

The major cause of disturbance in any ecosystem is pollutant released through anthropogenic factors. In this era of modernization of industries, one of the most crucial parts of the ecosystem, that is, soil, is losing its physical, chemical, and biological properties. Processes such as soil erosion, pollution, and salinization play the major role in alteration of soil properties. Accumulation of heavy metals and organic pollutants in soil is the major drawback of these processes (Wu et al., 2010). Heavy metals can be defined as metallic elements with high density and high atomic weight causing toxic effects even at a concentration of less than 1 ppb. Reports from different regions of the world have been published in the form of research paper, which indicate accumulation of heavy metals in water and irrigated soil (Ahmed et al., 2019; Tchounda et al., 2019; Luo et al., 2019; Opasola et al., 2019; Islam et al., 2018; Chu et al., 2019). An exclusive survey was conducted by National Wetland Condition Assessment, more than 1000 wetland sites across the United States for human-mediated addition of heavy metals such as silver, cadmium, cobalt, chromium, copper, nickel, lead, antimony, tin, vanadium, tungsten, and zinc. Among them, it was observed that soil lead concentrations were above expectation in wetland area covering 11.3% of the total wetland area. Moreover, the data were also in a positive correlation with different landscape matrices such as road density, house unit density, population density, and percent impervious surface. Among all the zones evaluated the western zones had common heavy metals such as copper, antimony, and zinc whereas in the regions of eastern mountains and upper mid-west had more than 50 ppm of soil lead (Nahlik et al., 2019). Khalid et al. (2017) summarized the concentrations of major metals and metalloids in the soil worldwide. These elements include cadmium, lead, arsenic, zinc, nickel, copper, and chromium. Accumulation of these metals and metalloids in vegetation as well as edible parts of the plants was also discussed. Global Monitoring Program and Executive Director of the UN Environment Program had proposed the list of heavy metals such as Pb, Cd, Hg, Cu, Sn, V, Cr, Mo, Co, Ni, Sb, As, and Se. These heavy metals are considered as hazardous elements. In USSR, aerial fallout, hydrogenic contamination from industrial waste, sewage sediments, and chemicals used in agriculture were considered as sources of contamination. Such pollutants create hazardous effect in agriculture soil and thereby affect the health of plants, humans, and soil microbes (Vodyanitskii, 2013). Anthropogenic activities are not the only source for the release of heavy metals in the environment, but also several natural resources are responsible for such accumulations. Such resources are weathering of minerals, erosion and volcanic activities, forest fires and biogenic sources, and particles released by vegetation (Dixit et al., 2015). In the developing and developed countries, accumulation of heavy metals in soil is becoming a major threat to the ecosystem, and hence, the strategic planning must be employed to prevent future hazards.

Metalloids are special class of earth elements, which pose the properties different than heavy metals and nonmetals. Indeed, their properties remain in between the metals and nonmetals. By classification, there are six major metalloids, boron, silicon, germanium, arsenic, antimony, and tellurium. Apart from the listed elements, selenium, polonium, and astatine are also considered as metalloids in several literature (Vernon, 2013). Out of these nine listed, metalloids, boron, silicon, and selenium are also considered as essential elements for higher plants and humans (Bienert and

Microbe Mediated Remediation of Environmental Contaminants. DOI: https://doi.org/10.1016/B978-0-12-821199-1.00019-5

Tamás, 2018). However, elevated levels of such metalloids also pose deleterious effects on plants and other candidates of any ecosystems. According to their physical–chemical properties, these metalloids are used mostly as semiconductors except for arsenic and antimony. In general, metalloids form oxides with or without other elements and available in the oxide forms. Due to their properties between metals and nonmetals, they are conductors of heat and electricity but not as potential as metals. Metalloids group forms diagonal line through p-block elements in the periodic table. They comprise elements from the second most abundant (silicon) to the rarest earth element (astatine, being 25 g present totally on earth at given point of time). Except silicon, all the metalloids are toxic at higher concentrations. According to the US Environmental Protection Agency, ecological soil screening levels, the threshold level for antimony is 78 ppm (for vertebrates) and for arsenic is 18 ppm (for plants). After these concentrations, they impart toxicologic effects on biota of the ecosystem (Lomb and Holm, 2010). Toxic effects of these metalloids (with the concern of human health) are also very well known, which include heart, cerebrovascular, and respiratory disease; skin, lung, liver, and kidney cancer (by As); cardiotoxicity, nervous system disease, hepatic damage, and necrosis blood hemolysis (by Sb); and nausea, dizziness, hair and nail loss, and central nervous system abnormalities falling teeth (by Se). Such harmful effects caused by various heavy metals are also reviewed (Escudero et al., 2018).

19.2 Remediation strategies

Since the availability of the metalloids in soil and their easy translocation may prove as serious problem for plants, animals, and soil microflora, there should be strict regulations and operations to maintain or reduce their concentrations inside the soil. Physical and chemical types of remediation strategies are more often used to clean up the soil, these strategies are rapid, effective and can be employed for highly contaminated sites. However, these strategies on the other hand depend on type of soil, chemical, and metal types. Moreover, these strategies are also laborious and costly. Physical remediation includes soil replacement, soil isolation, vitrification, and electrokinetic remediation, whereas chemical remediation has three major means, immobilization, encapsulation, and soil washing (Khalid et al., 2017). In this section, all the strategies of physical and chemical methods are discussed.

19.2.1 Physical remediation

Physical remediation consists of physical/direct means of removal of contaminants from soil from the affected area. This is done either by soil replacement, soil isolation, vitrification, or electrophoresis. As the name suggests, soil replacement involves complete or partial removal of soil by healthy soil. Sometimes, soil spading is also carried out in which digging of soil to the deep level is done, and thereby contaminated soil is forced to move to the deeper sites of the soil layers. The rest of the area is then covered by noncontaminated healthy soil, thus dilution of heavy metal or metalloids can be expected (Yao et al., 2012). In soil isolation strategy, contaminated sites are isolated physically by means of imparting physical barrier completely or partially. This will stop the lateral movement of contaminants from affected area to nonaffected area. It will also prevent the flow of water from noncontaminated soil to contaminated soil. Walls, floors, and caps are made as vertical, bottom, and surface control barriers, respectively (Khalid et al., 2017). Vitrification is a temperature treatment applied to the contaminated soil, which makes the volatile species of heavy metals and metalloids to evaporate. The applied temperature is set up to 1850°C. This leads to the formation of vitreous material. The vitrification process is easy to apply and gives rise to efficient means of remediation (Dellisanti et al., 2009). A new and cost-effective approach under physical methods has been proposed under which electrokinetic properties of the soil are used. Contaminated soil is first isolated in a tank where electric field gradient is applied. Under the influence of electrophoretic forces, heavy metals and metalloids are separated and accumulated near electrodes. In case of lower conductivity of metalloids and metallic forms, electrolytes such as distilled water, organic acids, or synthetic chelates are used to increase their electric conductivity (Iannelli et al., 2015).

19.2.2 Chemical remediation

Chemical methods of remediation consist of chemical treatment of the contaminated soils to decrease the availability of the contaminants by chelation. These include immobilization, encapsulation, or washing. Under immobilization technique, several agents called amendments are employed to the soil on which metals and metalloids are adsorbed that makes their (bio)availability to lesser extent. Such amendments are cement, clay, zeolite, clay, phosphates, minerals, microbes, and organic amendments. Recently, the use of farmyard manure provides effective immobilization of ferrous, chromium, nickel, manganese, and lead (Khalid et al., 2017). Like immobilization, encapsulation is also used to reduce the mobility of the toxic metals and metalloids. For this purpose, various encapsulating agents are used such as polyvinyl alcohol, chitosan, alginate, agar, polyacrylamide, and polyurethanes. Several reagents and extractants are used to treat the contaminated soil in a reactor tank where precipitation, ion exchange, chelation, or adsorption of the heavy metals and metalloids take place. As a result, the contaminants are leached out in liquid phase, and then they can be separated from the leachates. Various extractants used for this method are chelating agents such as EDTA, organic acids, humic substances, surfactants, and cyclodextrins (Kulikowska et al., 2015).

However, despite these many techniques for the remediation of the contaminated soil, there are several cons due to which alternative to these strategies has to be explored. These disadvantages include waste management of the liquid slurry after treatments, relatively larger working volume that questions also on the homogeneity of the treatments, heavy engineering works are required, highly costly in terms of energy requirements, limitations with the soil types in case of electrokinetic remediation and temporary solutions with regular monitoring for recontamination in case of chemical treatments. Due to these reasons, physical and chemical methods of soil cleanup methods were not widely accepted. Therefore a need for the alternative methods was emerged. One such strategy is to employ living organisms, such as plants, microorganisms or both due to their nature of resistance. A tremendous diversity within these groups of living organisms can be seen on planet Earth, and many of them have developed a resistance against heavy metal and metalloid stress. Moreover, they also pose metabolic routes through which remediation of these metals can be made possible. The removal or conversion of toxic elements into their nontoxic counterparts from soil using living organisms may be regarded as bioremediation.

19.3 Bioremediation

It is a convenient, eco-friendly, and environment-friendly approach for the removal of toxic pollutants from the soil. The cleaning agents, here, are mostly microorganisms (either natural or genetically modified). Not only heavy metals and metalloids but also radioactive waste and oil products can also be targeted with this strategy (Mosa et al., 2016). Microorganisms can be employed for the remediation for metals and metalloids due to their resistive nature when they are in contact with metals and metalloids. They can convert the harmful form of metalloids oxides into less harmful form. They combat with these heavy metals and metalloids by four main strategies, biosorption, bioaccumulation, siderophore formation, and production of biosurfactants. In case of biosorption, both the processes of absorption and adsorption are included, and bacteria or microorganism can use any or both processes to combat the heavy metal and metalloids present in their vicinity (Escudero et al., 2018). Bacterial cell surface has many biomolecules on to which these contaminants can be adsorbed (passive and fast process). Sometimes, the organisms secrete some biomolecules that are used as adsorbents for heavy metals and metalloids. During adsorption, contaminants are adsorbed either on the surface of bacteria or the biomolecules present on the surface of bacteria, whereas absorption process consists of engulfment of the contaminants inside the cell through metallothioneins or vacuoles (Mosa et al., 2016). On the other hand, bioaccumulation process is an extension of engulfment process where the rate of absorption of contaminants is higher than the rate of their release. In such case, contaminants are concentrated at certain place that gives rise to their accumulation (Chojnacka, 2010). Metallothioneins are class of specific proteins produced by bacteria in order to increase the binding affinity toward heavy metals and metalloids and thereby increasing tolerance to the toxic effects. Both natural and engineered *Escherichia coli* were well studied for their capacity to produce these proteins, and binding capacity of these strains was increased for the heavy metals in soil (Bae et al., 2001). Molecular biological database had also been generated to find out the genes responsible for heavy metal sequestration during their high-concentration exposures. Root-associated fungus *Cadophora finlandica* showed differential response in the presence of either cadmium, lead, or zinc as revealed by transcriptomic studies. From the genomic region, Cd-regulated genes were identified. Exposure of such heavy metals showed the increased production of transporters, extracellular proteins, centaurine-type regulator proteins, B-type cyclin, and GNAT-family acetyltransferase (Gorfer et al., 2009).

In the context of remediation of heavy metals and metalloids, several bacteria, yeasts, plants, and fungi use indirect mechanism that involves the formation and secretion of secondary chemicals that act as chelating agents. They are low-molecular-weight compounds with strong binding affinity toward trivalent metal ions such as Fe^{3+}. Hydroxamate siderophores, catecholate siderophores, and carboxylate siderophores are the most common siderophores found (Neilands, 1995). Not only iron but also other heavy metals such as cadmium, lead, nickel, arsenic, aluminum, magnesium, zinc, copper, cobalt, and strontium are also subjected to remediation through siderophores production. The removal of arsenic from the soil was performed using siderophores produced by *Pseudomonas azotoformans. P. azotoformans* was able to produce two types of siderophores, hydroxamate siderophores, and catecholates type. Comparison for the efficacy of the removal of the siderophores with EDTA and citric acid was also carried out, which resulted in the removal of

92.8%, 77.3%, and 70.0% of arsenic washed (five times) by siderophores, EDTA, and citric acid, respectively (Nair et al., 2007). Arsenic resistance bacterial strain TA6 of from *Staphylococcus* sp. was also detected from the Brahmaputra Valley, India. The resistant nature of bacterium was found due to their ability to produce siderophores, which imparted tolerance of not only arsenic but also several other heavy metals such as mercury, cadmium, cobalt, nickel, and chromium. Along with siderophores, the bacterium also produced an enzyme arsenate reductase, which helped in the biotransformation of arsenate to arsenite (Das and Barooah, 2018). Binding affinity of several siderophores to boron was documented by Harris et al. (2007). They used the marine bacteria, *Marinobacter* sp. strains DG870 and DG979, to produce vibrioferrin. Apart from vibrioferrin, two other siderophores petrobactin and tiazofurin were also used for the study. Structure of borate–siderophore complex was shown and described. B–vibrioferrin complex suggested that, at the oceanic pH, this complex helps the bacteria to uptake boron from seawater for themselves or the symbiotic sea organisms. On the other hand, binding of boron to petrobactin was found to be more efficient than vibrioferrin; however, its biological significance for this affinity was not found promising. Other than siderophores, there are several high and low-molecular-weight chemicals that act as surfactants. Some microorganisms have the ability to produce such surfactants and thus they are called biosurfactants. High-molecular-weight biosurfactants bind strongly to the solid surfaces, whereas low-molecular-weight biosurfactants help to lower the surface and interface tension between two liquids (Ron and Rosenberg, 2001). Mostly, biosurfactants are used in the remediation of organic pollutants, but in some case, studies showed their efficacies in the remediation of heavy metals such as cadmium, lead, and zinc (Mosa et al., 2016). Most used biosurfactants in the nature by various microorganisms are rhamnolipids, glycolipids, lipopeptides, trehalolipids, and sophorolipids with low-molecular-weight properties and polysaccharides, proteins, lipopolysaccharides, and lipoproteins with high-molecular-weight properties (Ron and Rosenberg, 2001).

19.4 Phytoremediation

Moving further toward the biological system of higher plants also shows the properties to tolerate, accumulate, and remediate the heavy metals and metalloids from the site of contaminations. Waste materials produced from various human and natural activities are mostly received by aquatic ecosystem and wetlands. Eco-friendly remediation strategies must be employed to minimize the damage to biology of the ecosystem caused by such waste material. Often, this waste material contains heavy metal and metalloids, which needs serious attention. Apart from chemical and physical methods for the removal of heavy metals and metalloids from the soil, plant-mediated remediation strategy has also a potential to serve the same function with minimum damage to the ecosystem. Some higher plants can survive under abiotic stress condition caused by heavy metal and metalloids. They can accumulate these toxic materials inside their cells and thereby decrease the bioavailability of this material from the soil and wetlands. This process is known as phytoremediation. Phytoremediation strategy (particularly from the heavy metal– and metalloid-contaminated soils) comprises three different methods to combat with heavy metal– and metalloid-imparted stresses. These are phytoextraction, phytostabilization, and phytovolatilization. Fig. 19.1 explains how different types of phytoremediation strategies manage the toxic elements present in the soil and ecosystems (Rahman et al., 2016; Ma et al., 2016). Phytoextraction strategies are used often for not only toxic and hazardous metals but also for the recovery of precious metals such as Au, Pd, Tl, and Pt (Ali et al., 2013; Brooks et al., 1998). A novel concept of mining was explored using phytoextraction technique, which is termed as phytomining. It is ecologically and economically feasible method. In this process, plants are cultivated at the mining sites in order to extract trace elements inside them. Afterward, they are combusted for the energy production, and the remaining ash (called bio-ore) is processed for the recovery of precious metals (Rascio and Navari-Izzo, 2011). Phytoextraction method alone is not suitable for the soil polluted with multiple contaminants. The use of plant–microbe association for the remediation of such cases is discussed in detail in subsequent topics. In the phytostabilization process the roots are the only functional plant part. It does the remediation by adsorption of the pollutants on the surface of the roots, absorption and accumulation by roots, or precipitation of the pollutants near the roots or root-zones. This is the only process of remediation where the pollutants are unlikely to enter the food chain. It also prevents their leaching in the groundwater (due to reduction in mobility) and their subsequent entry in the food chain. However, the process requires continuous monitoring, and their high-concentrations may hinder the growth of the plant. Phytovolatilization process is basically used mainly for mercury-, selenium- and arsenic-contaminated soils. These toxic elements are converted into their respective volatile forms. Through the plant systems, these volatile pollutants are translocated, and finally, they are released into the environment. This method of phytoremediation only transfers the pollutants from one part of ecosystem to the other, but it does not completely remove the toxic elements from the environment. Table 19.1 shows the outcome of various plant species used for the phytoremediation of metalloids from soil. Apart from these strategies used as phytoremediation, phytodegradation, phytofiltration, and phytotransformation are

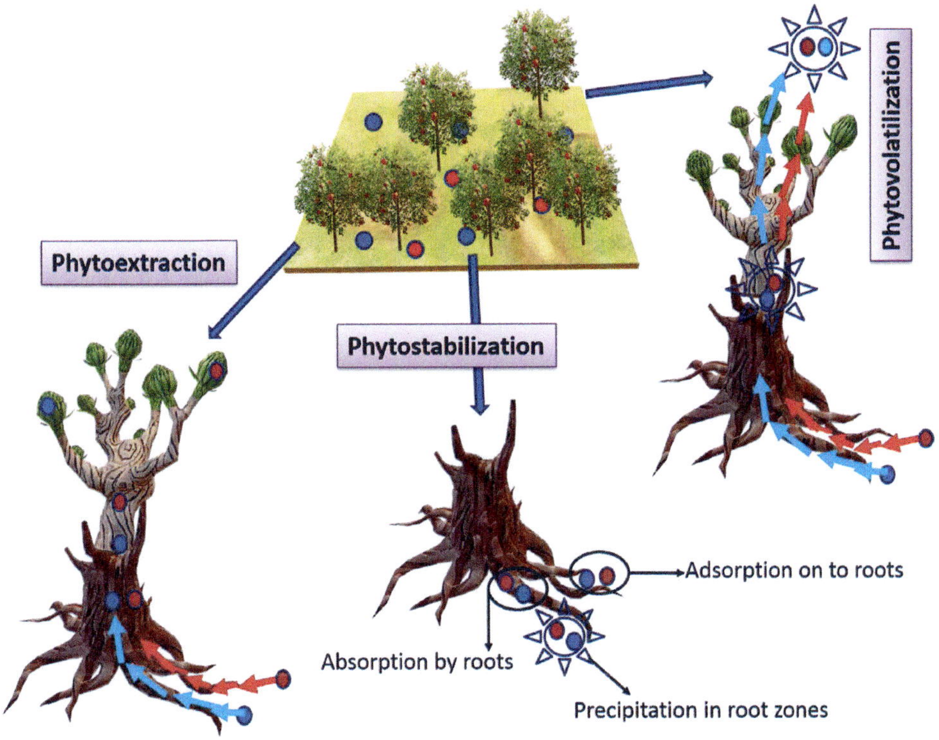

FIGURE 19.1 Various phytoremediation strategies for the removal of heavy metals and metalloids from the polluted site. It comprises three basic mechanisms, phytoextraction: phytostabilization, and phytovolatilization. Red and blue colored dots represent soil contaminants. Arrow indicates the direction of the pollutants within the plant system, and the pollutants under excited sign represent their derivatives or forms exhibiting less toxic effects in the ecosystem.

also employed for the remediation of toxic pollutants from the soil. However, they are used for the remediation of organic pollutants through the use of soil microflora (phytodegradation), or the target site is aquatic ecosystem (phytofiltration); or using soil microflora, indirect approach of remediation is governed (phytotransformation).

As mention earlier, certain bacterial and fungal strains can remove the heavy metal and metalloids from contaminated soils in the process of bioremediation. Knowing the fact that microbes can sense and metabolize these heavy metals and metalloids, certain biosensor devices have been designed. In such devices, whole cell, nucleic acid, enzymes, antibody, and peptides are used as bioreceptor entity. The reaction of recognition of pollutants by bioreceptors generates an electrical signal, which is then transduced, amplified, and processed to show the quality and quantity of the pollutants. The basic biological process involves biosorption and bioaccumulation means of bioremediations. Particularly, bacteria are preferred candidates for the remediation of heavy metals and metalloids due to their size and ability to grow in extreme conditions (Shukla et al., 2018). The combination of both the plants and microbes may result in betterment of the remediation technique due to their mutual relationship for the growth. This process is known as plant-assisted bioremediation. However, it has several limiting factors such as metal availability, metal uptake, and phytotoxicity. These factors can be overcome by genetic manipulation of microorganisms. The target for the genetic modification of microbes may be set as change in metabolic pathway for the synthesis of natural chelators such as citric acid and change in metabolic pathway to enhance the metal sequestration system. Reintroduction of the modified microbes is required for this strategy (Weyens et al., 2009). Heavy metals such as nickel, copper, chromium, cadmium, and zinc are remediated using this technique (Lodewyckx et al., 2001; Chen et al., 2005; Abou-Shanab et al., 2007; Li et al., 2007).

19.5 Mechanism of plant- and microbe-assisted remediation

In a mutual relationship between plants and endophytes, the basic function of remediation is mostly covered by endophytes either directly or indirectly. Bacteria can reduce the bioavailability of metals to the plants by producing extracellular polymeric substances through complexation or precipitation processes; producing antioxidant enzymes, which exhibit redox reaction or transformation of the pollutants; biosorption or bioaccumulation; and methylation for the degradation or reduction of Hg or Hg^{2+}. Alternatively, they may help the plant to survive in a stressful environment. They enhance the production of plant growth-promoting traits. This includes synthesis of phytohormones such as indole acetic acid (IAA) or production of enzymes such as ACC deaminase; bioavailability of nutrients to the plants (which by

TABLE 19.1 Phytoremediation of various metalloids from soil using different plant species.

Metalloids	Plant used	Outcome	Reference
Antimony (Sb)	*Chenopodium album* Linn., *Sedum emarginatum* Migo, and *Sedum lineare* Thunb	• The range of accumulated Sb concentration was from 21 to 21,148 mg/kg in studied plants. • Order of accumulation of Sb found decreased with the highest accumulation found in root followed by leaf and stem. • Among studied plants, *C. album* Linn., *S. emarginatum* Migo, and *S. lineare* Thunb showed accumulation more than 1000 mg/kg that could be considered as elevated levels.	Ning et al. (2015)
	Achillea ageratum, *Plantago lanceolata*, and *Silene vulgaris*	• Plants reported to accumulated high Sb when the extractable concentration is high (139 ± 793 mg/kg). • *A. ageratum* accumulated Sb 1367 mg/kg in basal leaves whereas *P. lanceolata* 1150 mg/kg in roots and 1164 mg/kg in *S. vulgaris* shoots.	Baroni et al. (2000)
	Ficus tikoua	• Antimony content was decreased in the tissue ordered as roots (1274.5–1580.9 mg/kg) > stems (133.5–498.9 mg/kg) > leaves (4.1–15.7 mg/kg). • The accumulation and different chemical forms of Sb were studied in tissues of the *F. tikoua*, the major accretion of Sb found in cell walls, 72.4%–87.5% followed by cytoplasmic organelles 8.2%–18.6% and cytoplasmic supernatant.	Wang et al. (2017)
	Equisetum ramosissimum	• *E. ramosissimum* showed potential for extraction by phytostabilization of the Fe, Mn, Zn, Pb, and Sb.	Ranðelović et al. (2017)
Arsenic	*Pteris vittata* L.	• Total arsenic uptake recorded for two growing cycles of *P. vittata* ranged from 8.2 to 16.9 and 5.5 to 12.0 mg, respectively. • Extraction by using *P. vittata* along with DAP resulted in better removal of arsenic and reduced Sb concentration in Rice with increased yield of rice.	Mandal et al. (2012a)
	P. vittata L.	• Phytoextraction using two successive harvests of *P. vittata* L with DAP is the best strategy for the extraction of maximum arsenic from polluted soil of West Bengal, India recorded to reduce 38% arsenic level.	Mandal et al. (2012b)
	P. Vittata	• Checked As availability in the soil by chemical extraction treatment using water, ammonium sulfate, ammonium phosphate, and Mehlich III. • As extraction was recorded as 4.6%, 7.0%, 18%, 21%, and 46% for water, ammonium sulfate, organic acids, ammonium phosphate, and Mehlich III, respectively. • Organic matter, phosphate, and magnesium concentrations in the soil positively influence As accumulation.	Gonzaga et al. (2012)
	P. vittata	• Studied species is low biomass producer that issue can be solved by adding P fertilizer or mycorrhiza fungi. • Soil contaminated with low concentration of arsenic is suitable for remediation by *P. vittata*.	Danh et al. (2014)

(*Continued*)

TABLE 19.1 (Continued)

Metalloids	Plant used	Outcome	Reference
	Brassica juncea (L.)	• At 15 and 50 ppm uptake was higher than 30 ppm whereas at 100 ppm of arsenic concentration, no plant growth was observed. • Uptake of arsenic up to 1% of the total biomass of the plant, *B. juncea* can be considered good to accumulator of arsenic.	Rahman et al. (2018)
	Leersia oryzoides	• *L. oryzoides* plants grown in an area of 180 cm^2 where the total arsenic uptake by shoots was 120, 130, and 130 g/ha at 6, 10, and 16 week, respectively. • Initial 8 weeks showed visible symptoms of arsenic stress, but then it was recovered. • Increasing growth and period of exposer the uptake was found constant in shoot. • Constant uptake of arsenic was observed by *L. oryzoides* up to 6 weeks.	Ampiah-Bonney et al. (2007)
	Canna glauca L., *Colocasia esculenta* L. Schott, *Cyperus papyrus* L., and *Typha angustifolia* L.	• *C. papyrus* L. can accumulate 130–172 mg As/kg plant, and it was high biomass producer among studied plants. • Fastest arsenic removal achieved by *C. esculenta* L with rate of 68 mg As/m^2/day, whereas removal rate of *C. glauca* L., *C. papyrus* L., and *T. angustifolia* L. was recorded as 61, 56, and 56 mg As/m^2/day, respectively.	Jomjun et al. (2010)
Arsenic and antimony (Sb)	*Pteris cretica* "Albo-Lineata" (PC), *Pteris fauriei* (PF), *Humata tyermanii* Moore (HT), and *Pteris ensiformis* Burm (PE)	• High concentration of As and Sb negatively affected the growth of PC, it was confirmed by accumulation of these metalloids in the roots. • Maximum concentration of Sb was found as 6405 mg/kg in root of *P. cretica*, while As was recorded 337 mg/kg in the rhizome of *P. fauriei*.	Feng et al. (2015)
Germanium (Ge)	*Phalaris arundinacea*	• This study revealed that citric acid and the microbial siderophore DFO-B (desferrioxamine B) may improve phytoextraction of Ge	Wiche et al. (2017)
Boron (B)	*Astragalus incanus* L., *Atriplex semibaccata* R. Br. L., *Atriplex nummularia* Lindl. L., *Brassica júncea* Czern L., *Festuca arundinacea* Schreb. L., *Brassica napus* L., *Lotus comiculatus* L., and *Hibiscus cannabinus* L.	•Among studied plants, Indian mustard grown under greenhouse accumulated maximum Se effectively in selenate-treated soil than selenite-treated soil.•Mustard grown in water culture under greenhouse treated with Se and B concentration was increased in plant tissues than in water culture.•Native soil Se and B content accumulation study was conducted where Indian mustard, canola, and tall fescue accumulated maximum concentrations when grown under greenhouse conditions.	Banuelos, 1996
	Euphorbia macroclada, *Verbascum cheiranthifolium*, and *Astragalus gummifer*	• B accumulation was recorded in roots of *E. macroclada*, *V. cheiranthifolium*, and *A. gummifer* for 25, 70, and 69 mg/kg, while in shoots it was 75, 115, and 77 mg/kg, respectively. • Soil with lower concentration of B is used to recover by the studied species.	Sasmaz (2008)
	Ricinus communis L.	• High transfer factor and transference index values achieved for boron using selected plant compare to other hyperaccumulator species.	De-Abreu et al. (2012)

(*Continued*)

TABLE 19.1 (Continued)

Metalloids	Plant used	Outcome	Reference
		• *R. communis* can tolerate high concentration of boron even in the presence of heavy metals such as Cu, Fe, Mn, and Zn. • Addition of peat along with studied plant at the boron-contaminated site may reduce 50% of time required to remove boron content in the soil.	
	Sorghum halepense L. Pers., *Cyperus rotundus* L., *Cynodon dactylon* L. Pers., *Amaranthus retroflexus* L., *Echinochloa crus-galli* L., and *C. album* L.	• Shoots accumulated high B than root. • Among studied species, species having rhizomes have high tolerance than species growing from seeds. • Studied weeds were not potential to consider for phytoextraction of boron contaminated site.	Aydın and Çakır, (2009)
	Stizolobium aterrimum	• The species found tolerant to low concentration of B and up to 240 mg/dm^3. It is moderately tolerant. • High dose effected growth of plant with symptoms of chlorosis. • Tolerance index was less than 50% so the dose 240 mg/dm^3 can be considered toxic. • Higher accumulation recorded in root tissues.	Pereira Costa et al. (2019)
	Hibiscus cannabinus, *B. juncea*, *Raphanus sativus*, and *Amaranthus cruentus*	• *H. canabinnus* was more tolerant with high dry matter and no visible toxicity symptoms were observed. • *A. cruentus* is suggested to use in field condition as it showed high indexes for decontamination of Zn and Mn along with boron. • *R. sativus* showed maximum Pb removal along with boron. • None of the studied plants found to be hyperaccumulative.	Santos et al. (2010)

DAP, Diammonium phosphate.

some mechanism inhibited due to heavy metal or metalloids stress) through the production of siderophores, biological N_2 fixation, or phosphate solubilization; or suppression the damage caused by phytopathogen (Ma et al., 2016). The major factor hindering the remediation process is the phytotoxicity imparted by heavy metal or metalloids. The plants should be provided with less toxic form of such pollutants or should be subjected to enhancement of tolerance level. This work may be efficiently governed by endophytes associated with the plants. Rajkumar et al. (2009) reported how endophytic bacteria would impart the resistance against metal toxicity through the process of acidification, or by producing iron chelators, siderophores, organic acid, or mobilizing the metal phosphate. Contradictory to this mechanism, some researchers believed that, due to the presence of endophyte, metal availability to the plants was decreased, and thus plant showed better physiological functions such as increased biomass (Madhaiyan et al., 2007). Endophytic bacteria can also impart the resistance to the host plant against metal stress. It has been demonstrated that endophytic bacteria modulated the production of certain antioxidant enzymes such as catalase, peroxidase, superoxide dismutase, glutathione peroxidase, and ascorbate peroxidase. They also modulated certain biochemical processes such as lipid peroxidation leading to increased resistance against oxidative stress imparted through heavy metal or metalloids (Li et al., 2011; Wan et al., 2012; Zhang et al., 2010). Besides reducing metal toxicity or imparting direct or indirect resistance to the host plants, endophyte also has role in metal biosorption and bioaccumulation. These processes were extensively characterized by several researchers in heavy metal–induced stresses (Guo et al., 2010; Luo et al., 2011; Shin et al., 2012). After imparting tolerance against heavy metal– and metalloid-induced oxidative stresses, the next hurdle during phytoaccumulation is their bioavailability. Again, this problem is also resolved by endophytic bacteria. As mentioned earlier, they have the ability to produce some metal chelating agents such as siderophores, low-molecular-weight organic acids and biosurfactants in order to process acidification of soil, redox activity, and phosphate solubilization. It will ultimately alter the soil pH and enhance the chelation efficiency. By this means, the bioavailability of particular pollutants will be increased and the function of the plant as hyperaccumulator may be served (Ma et al., 2011; Rajkumar et al., 2009). Nitrogen fixation under limited nitrogen conditions is also another advantage imparted by several endophytic genera such as *Burkholderia, Rahnella, Sphingomonas*, and *Acinetobacter* obtained from the stems of native black cottonwood and willow (Doty et al., 2009).

Particularly for the arsenic-contaminated soils, Zhu et al. (2014) isolated 74 endophytes from two hyperaccumulator plants *Pteris vittata* and *Pteris multifida*. All the isolated bacteria showed tolerance to arsenic at the concentration of

TABLE 19.2 Plant–microbe association for the remediation of metalloids (boron and arsenic) from contaminated soils.

Metalloids	Plant	Microorganism	Outcome	Reference
Boron	*Brassica napus*	*Bacillus megaterium*	• Accumulation of boron was concentrated in seed, and shoot parts of *B. napus* can be used as hyperaccumulators. • To decrease desired concentration for 8 mg/kg B, 4 mg/kg Pb, and 3 mg/kg Cd concentration can be decreased from the active zone of soil—approximately 2, 6, and 21 years would be necessary with only 150 mL per plot. • The study concluded inoculation with *B. megaterium* can improve plant growth and consequently increase phytoremediation efficiency.	Esringü et al. (2014)
Arsenic	*Betula celtiberica*	*Flavobacteriales, Burkholderiales, Pseudomonadales*	• The arsenic content, under in vitro conditions, triggered nonprotein thiol compound accumulation in roots, which suggested phytoextraction (due to phytochelatin complexation)-based remediation of arsenic from soil. • At the arsenic-contaminated site the siderophore and indole-3-acetic acid producing bacterial consortium improved accumulation in the leaves and roots of *B. celtiberica*. • On field studies showed that factors such as arsenic content, and pH influenced arsenic accumulation in the plant.	Mesa et al. (2017)
Arsenic	*Alnus firma*	*Bacillus thuringiensis* GDB-1	• The efficiency of the isolated organism was found to be dependent on pH and initial metal concentration. • Around 34% of total arsenic was quarantined by plant–microbe symbiosis. • Plant biomass, chlorophyll content, and nodule numbers were increased significantly after soil treatment with GDB-1. • Arsenic accumulation in *A. firma* seedlings was observed.	Babu et al. (2013)
Arsenic	*Miscanthus sinensis*	*Pseudomonas koreensis* AGB-1	• Physiological parameters such as plant biomass and chlorophyll content were increased. • Biochemical analysis showed increased protein content and increased activity of stress-responsive enzymes such as superoxide dismutase and catalase. • Metal uptake was also increased.	Babu et al. (2015)
Arsenic	*Pteris vittata*	*Glomus* spp. (AMF)	• Plant growth was enhanced and was found independent of the arsenic level of the soil. • Physiological parameters of the plant such as root surface area, chlorophyll content, and arsenic translocation factor were found increased in the AMF-treated plants.	Leung et al. (2010)

AMF, Arbuscular mycorrhizal fungi.

10 mM. These bacteria from both the plant species were able to oxidize/reduce arsenite derivatives containing different oxidation states of arsenic, namely, As(III) and As(V). Similar to this study, Lampis et al. (2015) applied selected rhizobacteria to the arsenic-contaminated soil and employed *P. vittata* to study rhizoremediation of arsenic. The consortium of different bacteria, having activities of siderophore production, arsenate reduction, and IAA production, showed 35% of arsenic removal from the contaminated soil as compared to control (15%). Boron hyperaccumulator plant *Brassica napus* showed increased remediation efficiency when inoculated or treated with *Bacillus megaterium* (Esringü et al., 2014). However, for the metalloids in particular, very limited work has been conducted using plant–microbe interaction as remediation strategy. Plant and its associated microbes (endophytes) are well characterized and optimized protocols are available for other metals and organic pollutants. Never the less, the mechanism underlying the heavy metal removal from the contaminated soils remains same for metalloids also. As mentioned earlier, some bacterial species can release siderophores (that help in the metal uptake mainly Fe^{3+}) and citric acid. One such siderophores is desferrioxamine found mostly in *Streptomyces* species and some marine bacteria such as *Streptomyces pilosus*, *Salinispora tropica*, and *Streptomyces coelicolor* (Codd et al., 2018; Mortazavi and Akbarzadeh, 2012; Chen et al., 2019; Bentley et al., 2002). Several strains of *Pseudomonas, Bacillus, Rhizobium*, and *Enterobacter* are capable of producing citric and other organic acids, which also help in sequestration of heavy metals and metalloids from the contaminated soils (Adeleke et al., 2017). Application of this siderophores desferrioxamine B (100 μM) and citric acid (100 μM) in soil containing germanium (10 μM) and other rare-earth elements showed enhancement of germanium accumulation in *Phalaris arundinacea*. However, this study was conducted in greenhouse condition, and there was no direct involvement of citric acid or siderophores producing bacteria, but more or less, the similar results may be expected with the plant–microbe association to remediate germanium from contaminated soils (Wiche et al., 2017). Table 19.2 shows the removal of boron and arsenic from the contaminated soils using plant–microbe association.

19.6 Conclusion and future prospects

As sessile organisms, plants are always exposed to certain adverse environmental conditions. These conditions may be biotic or abiotic. Under abiotic conditions, heavy metal– and metalloid-imparted stresses require serious attention. Although, not as toxic as heavy metals, some classes of elements, called metalloids (boron, silicon, germanium, arsenic, antimony, tellurium, polonium, and astatine), also need remediation strategies in order to preserve soil health for sustainable agriculture, some of them such as silicon and astatine do not impart much toxic effect in ecosystem. Metalloids such as boron, silicon, and selenium are essential elements for most of the organisms. On the other hand, arsenic, antimony, tellurium, and germanium are highly toxic and may cause abnormalities among plants and humans. They pose properties between metals and nonmetal elements. Remediation strategies comprise physical, chemical, and biological means of application. Physical and chemical means of remediation can be advantageous, but at the same time, they also require extensive treatments, large- or small-scale isolation of contaminated sites. Most of them are economically less feasible. On the other hand, biological methods of heavy metals and metalloids removal are easy to employ, and both ecologically and economically feasible. Microorganisms such as bacteria and fungi are useful for the remediation of metalloids, and they pose different strategies (involving different molecular mechanisms) for the removal of metalloids from the soil. On the other hand, several plant species are also hyperaccumulator of heavy metals and metalloids found near mining sites and polluted areas. Combining the ability of both of them, a new approach can be made using mutual relationship between plants and microorganisms. These microbes can be epiphytic or endophytic. Plants provide platform or essential nutrients for the microbes to survive and grow, and in return, the bacteria help in the remediation process in the polluted sites. By this method, arsenic, boron, germanium, and selenium have been successfully removed from the contaminated soils. However, preliminary trials may be conducted at greenhouse level before large-scale applications.

Plant-assisted bioremediation or plant–microbe association for the remediation technique is the adapted from the natural processes and relationship between several plant species and the associated microbes. However, endophytic community from one species may be isolated and employed to the metalloids affected area to enhance their uptake and removal from the soil. Apart from these methods, biotechnological interventions may be employed as a new and promising tool for the remediation of heavy metal and metalloids from the polluted area. These strategies include overexpression of genes involved in the uptake and transports of metal or metalloids, enhancing the production of metal and metalloids ligand and transformation of metals and metalloids into less toxic or volatile forms. Several genes have been isolated and characterized from microbes and higher plants that exhibit uptake and transport of heavy metals and metalloids. These include yeast cadmium factor 1 (YCF1) from yeast (Song et al., 2003), plasma membrane channel protein (*NtCBP4*) from *Nicotiana tabacum* (Sunkar et al., 2000), cyclic nucleotide–gated ion channel 1 (*AtCNGC1*) from

Arabidopsis thaliana (Zeng et al., 2015), calcium exchanger 2 (*CAX2*) from *N. tabacum* (Hirschi et al., 2000), and genes encoding aquaporin family proteins (Zhu and Rosen, 2009; Dhankher et al., 2011). Overexpression of these genes in model as well as nonmodel plants exhibits increased uptake and transport of heavy metals such as lead, cadmium, manganese, and metalloids such as arsenic, antimony, silicon, and boron. Likewise, metallothiones, phytochelatins, and glutathione coding genes are also characterized. Overexpression of these genes exhibited enhanced tolerance to cadmium (Zhu et al., 1999), arsenic (Dhankher et al., 2002; Li et al., 2004), chromium (Xie et al., 2015), and lead (Wang et al., 2015). However, gene mining for the uptake, transport, and ligand production for all the metalloids still needs to be done. In future, strategies may be designed to employ genetically modified organisms (both plants and microbes) for the effective removal of heavy metal and metalloids from soil and other candidates of ecosystems.

References

Abou-Shanab, R.A.I., Angle, J.S., Van Berkum, P., 2007. Chromate-tolerant bacteria for enhanced metal uptake by *Eichhornia crassipes* (Mart.). Int. J. Phytorem. 9 (2), 91–105.

Adeleke, R., Nwangburuka, C., Oboirien, B., 2017. Origins, roles and fate of organic acids in soils: a review. South Afr. J. Bot. 108, 393–406.

Ahmed, M., Matsumoto, M., Ozaki, A., Thinh, N.V., Kurosawa, K., 2019. Heavy metal contamination of irrigation water, soil, and vegetables and the difference between dry and wet seasons near a multi-industry zone in Bangladesh. Water 11 (3), 583.

Ali, H., Khan, E., Sajad, M.A., 2013. Phytoremediation of heavy metals—concepts and applications. Chemosphere 91 (7), 869–881.

Ampiah-Bonney, R.J., Tyson, J.F., Lanza, G.R., 2007. Phytoextraction of arsenic from soil by *Leersia oryzoides*. Int. J. Phytorem. 9 (1), 31–40.

Aydın, M., Çakır, F., 2009. Research on weed species for phytoremediation of boron polluted soil. Afr. J. Biotechnol. 8, 18.

Babu, A.G., Kim, J.D., Oh, B.T., 2013. Enhancement of heavy metal phytoremediation by *Alnus firma* with endophytic *Bacillus thuringiensis* GDB-1. J. Hazard. Mater. 250, 477–483.

Babu, A.G., Shea, P.J., Sudhakar, D., Jung, I.B., Oh, B.T., 2015. Potential use of *Pseudomonas koreensis* AGB-1 in association with *Miscanthus sinensis* to remediate heavy metal (loid)-contaminated mining site soil. J. Environ. Manage. 151, 160–166.

Bae, W., Mehra, R.K., Mulchandani, A., Chen, W., 2001. Genetic engineering of *Escherichia coli* for enhanced uptake and bioaccumulation of mercury. Appl. Environ. Microbiol. 67 (11), 5335–5338.

Banuelos, G.S., 1996. Managing high levels of boron and selenium with trace element accumulator crops. J. Environ. Sci. Health, A 31 (5), 1179–1196.

Baroni, F., Boscagli, A., Protano, G., Riccobono, F., 2000. Antimony accumulation in *Achillea ageratum*, *Plantago lanceolata* and *Silene vulgaris* growing in an old Sb-mining area. Environ. Pollut. 109 (2), 347–352.

Bentley, S.D., Chater, K.F., Cerdeño-Tárraga, A.M., Challis, G.L., Thomson, N.R., James, K.D., et al., 2002. Complete genome sequence of the model actinomycete *Streptomyces coelicolor* A3 (2). Nature 417 (6885), 141–147.

Bienert, G.P., Tamás, M.J., 2018. Molecular mechanisms of metalloid transport, toxicity and tolerance. Front. Cell Dev. Biol. 6, 99.

Brooks, R.R., Chambers, M.F., Nicks, L.J., Robinson, B.H., 1998. Phytomining. Trends Plant Sci. 3 (9), 359–362.

Chen, Y.X., Wang, Y.P., Lin, Q., Luo, Y.M., 2005. Effect of copper-tolerant rhizosphere bacteria on mobility of copper in soil and copper accumulation by *Elsholtzia splendens*. Environ. Int. 31 (6), 861–866.

Chen, J., Guo, Y., Lu, Y., Wang, B., Sun, J., Zhang, H., et al., 2019. Chemistry and biology of siderophores from marine microbes. Mar. Drugs 17 (10), 562.

Chojnacka, K., 2010. Biosorption and bioaccumulation—the prospects for practical applications. Environ. Int. 36 (3), 299–307.

Chu, H.T., Vu, T.V., Nguyen, T.K., Nguyen, H.T., 2019. Accumulation of arsenic and heavy metals in native and cultivated plant species in a lead recycling area in Vietnam. Minerals 9 (2), 132.

Codd, R., Richardson-Sanchez, T., Telfer, T.J., Gotsbacher, M.P., 2018. Advances in the chemical biology of desferrioxamine B. ACS Chem. Biol. 13 (1), 11–25.

Danh, L.T., Truong, P., Mammucari, R., Foster, N., 2014. A critical review of the arsenic uptake mechanisms and phytoremediation potential of *Pteris vittata*. Int. J. Phytorem. 16 (5), 429–453.

Das, S., Barooah, M., 2018. Characterization of siderophore producing arsenic-resistant *Staphylococcus* sp. strain TA6 isolated from contaminated groundwater of Jorhat, Assam and its possible role in arsenic geocycle. BMC Microbiol. 18 (1), 104.

De-Abreu, C.A., Coscione, A.R., Pires, A.M., Paz-Ferreiro, J., 2012. Phytoremediation of a soil contaminated by heavy metals and boron using castor oil plants and organic matter amendments. J. Geochem. Explor. 123, 3–7.

Dellisanti, F., Rossi, P.L., Valdrè, G., 2009. In-field remediation of tons of heavy metal-rich waste by Joule heating vitrification. Int. J. Miner. Process. 93 (3–4), 239–245.

Dhankher, O.P., Li, Y., Rosen, B.P., Shi, J., Salt, D., Senecoff, J.F., et al., 2002. Engineering tolerance and hyperaccumulation of arsenic in plants by combining arsenate reductase and γ-glutamylcysteine synthetase expression. Nat. Biotechnol. 20 (11), 1140.

Dixit, R., Malaviya, D., Pandiyan, K., Singh, U.B., Sahu, A., Shukla, R., et al., 2015. Bioremediation of heavy metals from soil and aquatic environment: an overview of principles and criteria of fundamental processes. Sustainability 7 (2), 2189–2212.

Doty, S.L., Oakley, B., Xin, G., Kang, J.W., Singleton, G., Khan, Z., et al., 2009. Diazotrophic endophytes of native black cottonwood and willow. Symbiosis 47 (1), 23–33.

Escudero, L.B., Quintas, P.Y., Wuilloud, R.G., Dotto, G.L., 2018. Biosorption of metals and metalloids. Green Adsorbents for Pollutant Removal. Springer, Cham, pp. 35–86.

Esringü, A., Turan, M., Güneş, A., Karaman, M.R., 2014. Roles of *Bacillus megaterium* in remediation of boron, lead, and cadmium from contaminated soil. Commun. Soil Sci. Plant Anal. 45 (13), 1741–1759.

Feng, R., Wang, X., Wei, C., Tu, S., 2015. The accumulation and subcellular distribution of arsenic and antimony in four fern plants. Int. J. Phytorem. 17 (4), 348–354.

Gonzaga, M.I.S., Ma, L.Q., Pacheco, E.P., dos Santos, W.M., 2012. Predicting arsenic bioavailability to hyperaccumulator *Pteris vittata* in arsenic-contaminated soils. Int. J. Phytorem. 14 (10), 939–949.

Gorfer, M., Persak, H., Berger, H., Brynda, S., Bandian, D., Strauss, J., 2009. Identification of heavy metal regulated genes from the root associated ascomycete *Cadophora finlandica* using a genomic microarray. Mycol. Res. 113 (12), 1377–1388.

Guo, H., Luo, S., Chen, L., Xiao, X., Xi, Q., Wei, W., et al., 2010. Bioremediation of heavy metals by growing hyperaccumulaor endophytic bacterium *Bacillus* sp. L14. Bioresour. Technol. 101 (22), 8599–8605.

Harris, W.R., Amin, S.A., Küpper, F.C., Green, D.H., Carrano, C.J., 2007. Borate binding to siderophores: structure and stability. J. Am. Chem. Soc. 129 (40), 12263–12271.

Hirschi, K.D., Korenkov, V.D., Wilganowski, N.L., Wagner, G.J., 2000. Expression of arabidopsis CAX2 in tobacco. Altered metal accumulation and increased manganese tolerance. Plant Physiol. 124 (1), 125–134.

Iannelli, R., Masi, M., Ceccarini, A., Ostuni, M.B., Lageman, R., Muntoni, A., et al., 2015. Electrokinetic remediation of metal-polluted marine sediments: experimental investigation for plant design. Electrochim. Acta 181, 146–159.

Islam, M.M., Karim, M., Zheng, X., Li, X., 2018. Heavy metal and metalloid pollution of soil, water and foods in Bangladesh: a critical review. Int. J. Environ. Res. Public Health 15 (12), 2825.

Jomjun, N., Siripen, T., Maliwan, S., Jintapat, N., Prasak, T., Somporn, C., et al., 2010. Phytoremediation of arsenic in submerged soil by wetland plants. Int. J. Phytorem. 13 (1), 35–46.

Khalid, S., Shahid, M., Niazi, N.K., Murtaza, B., Bibi, I., Dumat, C., 2017. A comparison of technologies for remediation of heavy metal contaminated soils. J. Geochem. Explor. 182, 247–268.

Kulikowska, D., Gusiatin, Z.M., Bułkowska, K., Klik, B., 2015. Feasibility of using humic substances from compost to remove heavy metals (Cd, Cu, Ni, Pb, Zn) from contaminated soil aged for different periods of time. J. Hazard. Mater. 300, 882–891.

Lampis, S., Santi, C., Ciurli, A., Andreolli, M., Vallini, G., 2015. Promotion of arsenic phytoextraction efficiency in the fern *Pteris vittata* by the inoculation of As-resistant bacteria: a soil bioremediation perspective. Front. Plant Sci. 6, 80.

Leung, H.M., Wu, F.Y., Cheung, K.C., Ye, Z.H., Wong, M.H., 2010. The effect of arbuscular mycorrhizal fungi and phosphate amendement on arsenic uptake, accumulation and growth of *Pteris vittata* in As-contaminated soil. Int. J. Phytorem. 12 (4), 384–403.

Li, Y., Dhankher, O.P., Carreira, L., Lee, D., Chen, A., Schroeder, J.I., et al., 2004. Overexpression of phytochelatin synthase in *Arabidopsis* leads to enhanced arsenic tolerance and cadmium hypersensitivity. Plant Cell Physiol. 45 (12), 1787–1797.

Li, W.C., Ye, Z.H., Wong, M.H., 2007. Effects of bacteria on enhanced metal uptake of the Cd/Zn-hyperaccumulating plant, *Sedum alfredii*. J. Exp. Bot. 58 (15–16), 4173–4182.

Li, T., Liu, M.J., Zhang, X.T., Zhang, H.B., Sha, T., Zhao, Z.W., 2011. Improved tolerance of maize (*Zea mays* L.) to heavy metals by colonization of a dark septate endophyte (DSE) Exophiala pisciphila. Sci. Total Environ. 409 (6), 1069–1074.

Lodewyckx, C., Taghavi, S., Mergeay, M., Vangronsveld, J., Clijsters, H., Lelie, D.V.D., 2001. The effect of recombinant heavy metal-resistant endophytic bacteria on heavy metal uptake by their host plant. Int. J. Phytorem. 3 (2), 173–187.

Luo, S., Wan, Y., Xiao, X., Guo, H., Chen, L., Xi, Q., et al., 2011. Isolation and characterization of endophytic bacterium LRE07 from cadmium hyperaccumulator *Solanum nigrum* L. and its potential for remediation. Appl. Microbiol. Biotechnol. 89 (5), 1637–1644.

Luo, K., Liu, H., Zhao, Z., Long, J., Li, J., Jiang, C., et al., 2019. Spatial distribution and migration of cadmium in contaminated soils associated with a geochemical anomaly: a case study in southwestern China. Pol. J. Environ. Stud. 28 (5).

Ma, Y., Rajkumar, M., Luo, Y., Freitas, H., 2011. Inoculation of endophytic bacteria on host and non-host plants—effects on plant growth and Ni uptake. J. Hazard. Mater. 195, 230–237.

Ma, Y., Rajkumar, M., Zhang, C., Freitas, H., 2016. Beneficial role of bacterial endophytes in heavy metal phytoremediation. J. Environ. Manage. 174, 14–25.

Madhaiyan, M., Poonguzhali, S., Sa, T., 2007. Metal tolerating methylotrophic bacteria reduces nickel and cadmium toxicity and promotes plant growth of tomato (*Lycopersicon esculentum* L.). Chemosphere 69 (2), 220–228.

Mandal, A., Purakayastha, T.J., Patra, A.K., Sanyal, S.K., 2012a. Phytoremediation of arsenic contaminated soil by *Pteris vittata* L. II. Effect on arsenic uptake and rice yield. Int. J. Phytorem. 14 (6), 621–628.

Mandal, A., Purakayastha, T.J., Patra, A.K., Sanyal, S.K., 2012b. Phytoremediation of arsenic contaminated soil by *Pteris vittata* LI Influence of phosphatic fertilizers and repeated harvests. Int. J. Phytorem. 14 (10), 978–995.

Mesa, V., Navazas, A., González-Gil, R., González, A., Weyens, N., Lauga, B., et al., 2017. Use of endophytic and rhizosphere bacteria to improve phytoremediation of arsenic-contaminated industrial soils by autochthonous *Betula celtiberica*. Appl. Environ. Microbiol. 83 (8), e03411–e03416.

Mortazavi, M., Akbarzadeh, A., 2012. Improvement of desferrioxamine B production of *Streptomyces pilosus* ATCC 19797 with use of protease inhibitor and minerals related to its activity. Indian J. Clin. Biochem. 27 (3), 274–277.

Mosa, K.A., Saadoun, I., Kumar, K., Helmy, M., Dhankher, O.P., 2016. Potential biotechnological strategies for the cleanup of heavy metals and metalloids. Front. Plant Sci. 7, 303.

Nahlik, A.M., Blocksom, K.A., Herlihy, A.T., Kentula, M.E., Magee, T.K., Paulsen, S.G., 2019. Use of national-scale data to examine human-mediated additions of heavy metals to wetland soils of the US. Environ. Monit. Assess. 191 (1), 336.

Nair, A., Juwarkar, A.A., Singh, S.K., 2007. Production and characterization of siderophores and its application in arsenic removal from contaminated soil. Water Air Soil Pollut. 180 (1–4), 199–212.

Neilands, J.B., 1995. Siderophores: structure and function of microbial iron transport compounds. J. Biol. Chem. 270 (45), 26723–26726.

Ning, Z., Xiao, T., Xiao, E., 2015. Antimony in the soil-plant system in an Sb mining/smelting area of southwest China. Int. J. Phytorem. 17 (11), 1081–1089.

Opasola, O.A., Adeolu, A.T., Iyanda, A.Y., Adewoye, S.O., Olawale, S.A., 2019. Bioaccumulation of heavy metals by *Clarias gariepinus* (African catfish) in Asa River, Ilorin, Kwara State. J. Health Pollut. 9 (21), 190303.

Pereira Costa, B.G., Justino, G.C., Aguiar, L.F., Souza, L.A., Camargos, L.S., 2019. Boron phytoremediation: stizolobium aterrimum is tolerant and can be used for phytomanagement of boron excess in soils. Int. J. Environ. Stud. 76 (2), 329–337.

Rahman, M.A., Reichman, S.M., De Filippis, L., Sany, S.B.T., Hasegawa, H., 2016. Phytoremediation of toxic metals in soils and wetlands: concepts and applications. Environmental Remediation Technologies for Metal-Contaminated Soils. Springer, Tokyo, pp. 161–195.

Rahman, M., Jakariya, M., Haq, N., Islam, M.A., 2018. Prospect of phytoaccumulation of arsenic by *Brassica juncea* (L.) in Bangladesh. Int. J. Phytorem. 20 (10), 1025–1032.

Rajkumar, M., Ae, N., Freitas, H., 2009. Endophytic bacteria and their potential to enhance heavy metal phytoextraction. Chemosphere 77 (2), 153–160.

Rascio, N., Navari-Izzo, F., 2011. Heavy metal hyperaccumulating plants: how and why do they do it? And what makes them so interesting? Plant Sci. 180 (2), 169–181.

Ron, E.Z., Rosenberg, E., 2001. Natural roles of biosurfactants: minireview. Environ. Microbiol. 3 (4), 229–236.

Santos, G.C.G.D., Rodella, A.A., Abreu, C.A.D., Coscione, A.R., 2010. Vegetable species for phytoextraction of boron, copper, lead, manganese and zinc from contaminated soil. Sci. Agric. 67 (6), 713–719.

Sasmaz, A., 2008. Translocation and accumulation of boron in roots and shoots of plants grown in soils of low boron concentration in Turkey's Keban Pb-Zn mining area. Int. J. Phytorem. 10 (4), 302–310.

Shin, M.N., Shim, J., You, Y., Myung, H., Bang, K.S., Cho, M., et al., 2012. Characterization of lead resistant endophytic Bacillus sp. MN3-4 and its potential for promoting lead accumulation in metal hyperaccumulator *Alnus firma*. J. Hazard. Mater. 199, 314–320.

Shukla, A., Srivastava, S., D'Souza, S.F., 2018. An integrative approach toward biosensing and bioremediation of metals and metalloids. Int. J. Environ. Sci. Technol. 15 (12), 2701–2712.

Song, W.Y., Sohn, E.J., Martinoia, E., Lee, Y.J., Yang, Y.Y., Jasinski, M., et al., 2003. Engineering tolerance and accumulation of lead and cadmium in transgenic plants. Nat. Biotechnol. 21 (8), 914–919.

Sunkar, R., Kaplan, B., Bouché, N., Arazi, T., Dolev, D., Talke, I.N., et al., 2000. Expression of a truncated tobacco NtCBP4 channel in transgenic plants and disruption of the homologous Arabidopsis CNGC1 gene confer Pb^{2+} tolerance. Plant J. 24 (4), 533–542.

Tchounda, T., Mboudou, G.M.M., Agyingi, C.M., 2019. Heavy metal concentration and distribution in stream sediments of Ndongo River, Buea, Cameroon-environmental impact. World Sci. News 131, 15–36.

Vernon, R.E., 2013. Which elements are metalloids? J. Chem. Educ. 90 (12), 1703–1707.

Vodyanitskii, Y.N., 2013. Contamination of soils with heavy metals and metalloids and its ecological hazard (analytic review). Eurasian Soil Sci. 46 (7), 793–801.

Wan, Y., Luo, S., Chen, J., Xiao, X., Chen, L., Zeng, G., et al., 2012. Effect of endophyte-infection on growth parameters and Cd-induced phytotoxicity of Cd-hyperaccumulator *Solanum nigrum* L. Chemosphere 89 (6), 743–750.

Wang, Y., Zhao, Z., Deng, M., Liu, R., Niu, S., Fan, G., 2015. Identification and functional analysis of microRNAs and their targets in *Platanus acerifolia* under lead (Pb) stress. Int. J. Mol. Sci. 16 (4), 7098–7111.

Wang, Y., Chai, L., Yang, Z., Mubarak, H., Xiao, R., Tang, C., 2017. Subcellular distribution and chemical forms of antimony in *Ficus tikoua*. Int. J. Phytorem. 19 (2), 97–103.

Weyens, N., van der Lelie, D., Taghavi, S., Vangronsveld, J., 2009. Phytoremediation: plant—endophyte partnerships take the challenge. Curr. Opin. Biotechnol. 20 (2), 248–254.

Wiche, O., Tischler, D., Fauser, C., Lodemann, J., Heilmeier, H., 2017. Effects of citric acid and the siderophore desferrioxamine B (DFO-B) on the mobility of germanium and rare earth elements in soil and uptake in *Phalaris arundinacea*. Int. J. Phytorem. 19 (8), 746–754.

Wu, G., Kang, H., Zhang, X., Shao, H., Chu, L., Ruan, C., 2010. A critical review on the bio-removal of hazardous heavy metals from contaminated soils: issues, progress, eco-environmental concerns and opportunities. J. Hazard. Mater. 174 (1–3), 1–8.

Xie, Y., Ye, S., Wang, Y., Xu, L., Zhu, X., Yang, J., et al., 2015. Transcriptome-based gene profiling provides novel insights into the characteristics of radish root response to Cr stress with next-generation sequencing. Front. Plant Sci. 6, 202.

Yao, Z., Li, J., Xie, H., Yu, C., 2012. Review on remediation technologies of soil contaminated by heavy metals. Procedia Environ. Sci. 16, 722–729.

Zeng, H., Xu, L., Singh, A., Wang, H., Du, L., Poovaiah, B.W., 2015. Involvement of calmodulin and calmodulin-like proteins in plant responses to abiotic stresses. Front. Plant Sci. 6, 600.

Zhang, X., Li, C., Nan, Z., 2010. Effects of cadmium stress on growth and anti-oxidative systems in *Achnatherum inebrians* symbiotic with *Neotyphodium gansuense*. J. Hazard. Mater. 175 (1–3), 703–709.

Zhu, Y.G., Rosen, B.P., 2009. Perspectives for genetic engineering for the phytoremediation of arsenic-contaminated environments: from imagination to reality? Curr. Opin. Biotechnol. 20 (2), 220–224.

Zhu, Y.L., Pilon-Smits, E.A., Jouanin, L., Terry, N., 1999. Overexpression of glutathione synthetase in Indian mustard enhances cadmium accumulation and tolerance. Plant Physiol. 119 (1), 73–80.

Further reading

Aguilar Pesantes, A., Peña Carpio, E., Vitvar, T., María Mahamud López, M., Menéndez-Aguado, J.M., 2019. A multi-index analysis approach to heavy metal pollution assessment in river sediments in the Ponce Enríquez area, Ecuador. Water 11 (3), 590.

Dhankher, O.P., Pilon-Smits, E.A., Meagher, R.B., Doty, S., 2012. Biotechnological approaches for phytoremediation. Plant Biotechnology and Agriculture. Academic Press, pp. 309–328.

Lombi, E., Holm, P.E., 2010. Metalloids, soil chemistry and the environment. MIPs and Their Role in the Exchange of Metalloids. Springer, New York, pp. 33–44.

Ogra, Y., Okubo, E., Takahira, M., 2010. Distinct uptake of tellurate from selenate in a selenium accumulator, Indian mustard (*Brassica juncea*). Metallomics 2 (5), 328–333.

Ranđelović, D., Mihailović, N., Jovanović, S., 2019. Potential of *Equisetum ramosissimum* Desf. for remediation of antimony flotation tailings: a case study. Int. J. Phytorem. 21 (7), 707–713.

Chapter 20

Microbe-based bioreactor system for bioremediation of organic contaminants: present and future perspective

Akhileshwar Kumar Srivastava[1], Rajesh Kumar Singh[2] and Divya Singh[3]

[1]Central Food Technological Research Institute, Mysuru, India, [2]Banaras Hindu University, Varanasi, India, [3]Central Sericultural Research and Training Institute, Mysuru, India

20.1 Introduction

From the last few decades, rapid industrialization, unsafe farming, and increasing human actions on energy pool led to several environmental issues. Pollutants such as greenhouse gases, pesticides, heavy metals, and hydrocarbons are toxic to environment and human health. In addition, contaminants such as pharmaceutical compounds, petroleum products, organic dyes, polycyclic aromatic hydrocarbons, and nitro and chlorophenols, and their derivatives become a serious problem. They can enter the environment through different ways. The major sources of organic pollutants are fuel constituents, mineral oil, and chlorinated hydrocarbons. Organic contaminants may be grouped into two categories (1) solid organic contaminants and (2) gas/liquid organic contaminants. Solid organic contaminants include foodstuffs, sludges, resins, waxes, chemicals, wood, plastic, paper, etc. Gas/liquid organic contaminants comprise volatile organic compounds (VOCs), petroleum compounds (Mandal et al., 2016), etc. About 90% of contaminated sites of the world are affected by aromatic hydrocarbons, polycyclic aromatic hydrocarbons, chlorinated hydrocarbons, and phenols (United States Environmental Protection Agency, 2003). However, chlorinated aromatics, chlorinated short-chain aliphatic, benzene–toluene–ethylbenzene–xylenes, and phenols are detected in the groundwater of the polluted sites (National Research Council, 1993, 2000; Nijenhuis et al., 2013). Pesticides can be widespread in low concentrations that may be hazardous causing chronic responses. Moreover, many pesticides belong to persistent organic pollutants that are challenging for degradation. Water treatment procedures cannot remove drugs, hormones, and pharmaceutical compounds. Thus these micro-pollutants contaminate the natural water resources. Hormones and drugs are less persistent contaminants that are transformed into different products having bioaccumulation tendency making them more toxic than the original compound. In comparison to the single compound, the mixtures of low concentrated contaminants become more toxic because of synergistic effects.

The biodegradation rate of pollutants is dependent on the physiological condition of microbes that is influenced by various environmental factors. Researchers have reported that the remediation of pollutants using microorganisms is effective and consistent. The use of microbes or their derivatives such as biomass (spent) and enzymes for cleaning of environmental pollutants is described as microbial bioremediation (Azubuike et al., 2016). Microbes are present everywhere and contaminants present in the environment come in contact with them. Then microbes may break or transform contaminants through their metabolic processes. Microorganisms can modify their metabolic pathways so that contaminants may channel in the normal process of degradation or biotransformation. Therefore applied bioremediation techniques are focusing on the use of natural catabolic potential of microbes to accumulate, transform, or degrade several synthetic compounds such as polyaromatic hydrocarbons (PAHs), polychlorinated biphenyls, hydrocarbons, metals, and radionuclides (Gargouri et al., 2011; Pino-Herrera et al., 2017). The presence of natural diversity of microorganisms increases the chances of degradation or detoxification of different varieties of contaminants (Table 20.1).

Microbial bioremediation has several advantages. Being a natural process, it has public acceptance (Chikere et al., 2012). In comparison to other techniques, it is a low-cost process for cleaning of hazardous pollutants. It can be carried

Microbe Mediated Remediation of Environmental Contaminants. DOI: https://doi.org/10.1016/B978-0-12-821199-1.00020-1

TABLE 20.1 Microorganisms involved in the bioremediation of pollutants.

Pollutants	Microbes	References
Hydroquinone, phenolic compounds, resorcinol	*Penicillium chrysogenum*	Ana Lúcia Leitão (2009), Pedro et al. (2014)
3-Chlorophenol, 4-chlorophenol, 2-flurophenol, 3-fluorophenol, 4-fluorophenol	*Penicillium simplicissimum, Penicillium frequentans* Bi 7/2	Ana Lúcia Leitão (2009)
Pyrene, benz[a]pyrene	*Penicillium janthinellum*	Ana Lúcia Leitão (2009)
Monocyclic aromatic hydrocarbon (xylene, benzene)	*Pseudomonas putida*	Safiyanu et al. (2015)
Petroleum polyaromatic hydrocarbons	*Achromobacter xylosoxidans, Aeribacillus pallidus, Mycobacterium cosmeticum, Bacillus licheniformis, Bacillus mojavensis, Sphingobium, Sphingomonas, Novosphingobium, Achromobacter, Acinetobacter, Flavobacterium, Pseudomonas alcaligenes, P. putida, Pseudomonas mendocina*	Ghosal et al. (2016), Ma et al. (2015), Mnif et al. (2014), Sani et al. (2015), Safiyanu et al. (2015)
Hydrocarbons	*Aspergillus niger, Aspergillus fumigatus, Fusarium solani, Penicillium funiculosum, Gloeophyllum trabeum, Tyromyces palustris, Trametes versicolor*	Al-Jawhari (2014), Karigar and Rao (2011)
Aromatic hydrocarbons	*Acinetobacter, Microbacterium, Ralstonia, Pseudomonas*	Simarro et al. (2013)
Poly aromatic hydrocarbons	*Gloeophyllum striatum* MTCC-1117	Yadav et al. (2011)
Benzopyrene, phenanthrene	*Candida viswanathii*	Hesham et al. (2012)
Crude oil	*A. niger, Bacillus brevis, B. licheniformis, Pseudomonas aeruginosa KH6, Bacillus sphaericus, Candida krusei, Candida glabrata, Saccharomyces cerevisiae*	Burghal et al. (2016), El-Borai et al. (2016)
Industrial dyes	*Myrothecium roridum* IM 6482, *Phanerochaete chrysosporium, Pycnoporus sanguineous, Trametes trogii, Penicillium ochrochloron*	Shedbalkar and Jadhav (2011), Yan et al. (2014), Jasińska et al. (2015)
Oxyfluorfen	*Bacillus* sp., *Pseudomonas* sp., *Arthrobacter* sp., *Aspergillus terreus, Mycobacterium* sp., *Micrococcus* sp., *Streptomyces* sp.	Mohamed et al. (2011)
Chlorpyrifos and methyl parathion	*Acinetobacter* sp., *Enterobacter* sp., *Photobacterium* sp., *Pseudomonas* sp.	Ravi et al. (2015)

out in ex situ or in situ conditions to degrade or detoxify pollutants completely, instead of transporting or transferring pollutants from one form to another. The disadvantages of the technique are that the bioremediation process takes comparatively long time to achieve the goals. It may not be effective for all the pollutants and biodegradation products that can be more toxic and persistent in comparison to original compound. Another drawback is the specificity of biological process related to microorganism population, contaminants, and environmental constraints. Also specific proficiency is required for designing and implementing the technology.

20.2 Use of microbial bioreactors

Microorganisms are able to transform complex lipophilic molecules into simple water-soluble forms. They activate their enzymatic mechanism required for the course of enrichment when exposed to organic contaminants. The enzymatic activities of microbes are affected by the presence of these contaminants mostly depending on the structural and functional groups of contaminants as well as on the selective microorganism community. Biodegradation of a pollutant is usually a consequence of the actions of several organisms. Microbial bioreactors using for bioremediation have applications in water, soil, and air environments as discussed in the following sections.

20.2.1 Treatment of wastewater and industrial effluent

The primary agents of wastewater treatment are microbes. These microbes are already present in wastewater feeding on complex materials and converting them into simple forms, thus helping in accomplishing the goal. Slurry-phase reactor, trickling filters, upflow anaerobic sludge bioreactor, and membrane bioreactor are used in the treatment of wastewater and industrial discharges.

20.2.2 Treatment of soil and land

Oily sludge, fuel oils, diesel, wood preserving substances [PAHs, pentachlorophenol (PCP) and creosote], some pesticides (Azubuike et al., 2016; Pino-Herrera et al., 2017), and coke wastes are successfully treated. The most successful remediation of petroleum hydrocarbons and other low volatile biodegradable pollutants is executed by soil bioremediation. Biofilters, stirred-tank, slurry-phase, packed, and partitioning phase microbial reactors are used in remediation of contaminated soil.

20.2.3 Control of air pollutants

Microbes are used in the remediation of inorganic or organic air contaminants in spent gases prior to release into the atmosphere (Azubuike et al., 2016). Microbes oxidize contaminants such as hydrogen sulfide, sulfur dioxide, VOCs and reduce contaminants such as nitrogen oxide thus assisting to prevent environmental health threats and nuisances. Biofilters and bioscrubbers are generally used to control air pollution.

20.2.4 Management of solid waste

Microbes are mainly accountable for the degradation of organic wastes naturally. They are involved in the decomposition processes occurring in composts and landfills. Anaerobic digesters are usually applied in the biotreatment of solid waste.

20.3 Factors affecting performance of microbial bioreactor

A number of factors play important role in all remediation technologies, including bioreactors. These factors comprise type of pollutants, community of microbes along with design of bioreactor, optimization, and monitoring of the procedure (Chikere et al., 2012; Azubuike et al., 2016; Pino-Herrera et al., 2017). Hence, the microbial bioremediation is approached by involving mainly two types of scientific approaches: (1) abiotic interactions (solubility, transport, sorption, and photolysis) and (2) biotic interactions (taxonomic diversity, physiological, genetic, and ecological interactions). Many of the abovementioned factors, including design and operation system, have to be optimized for the best performance of reactor in bioremediation (Naik and Duraphe, 2012; Srivastava et al., 2014).

Factors related to design and configuration of reactor are correlated with abiotic and biotic factors that may affect the growth of microorganisms and ultimately affect the efficiency of the reactor. Environmental conditions (moisture, pH, and temperature), contaminant concentration, contaminant mixture, and macronutrient are some factors that influence microbial growth and action (Naik and Duraphe, 2012; Srivastava et al., 2014). Reactor configuration, size, and mode of action are some factors based on reactor design.

20.3.1 Environmental factors

Environmental factors such as pH, temperature, humidity, and oxygen availability influence the growth and metabolomics of microbes. Sometimes, behavior of contaminants (solubility, volatility) is also affected by the environment (Naik and Duraphe, 2012). In the optimization process, it is essential to consider the effects of environmental conditions and optimize the process accordingly. Tekere et al. (2001) standardized the optimal growth conditions, including aeration, pH, and nutrients for deterioration of contaminants by white-rot fungi and reported that optimized process results in elevated enzyme and degradation activities.

20.3.1.1 Effect of temperature

The growth of microbes is affected by temperature. There is a temperature range indicating minimum, optimum, and maximum temperature for survivability and growth of microorganisms. Moreover, the biochemical processes accomplish at the optimum temperature to attain the required biotreatment through different microorganisms (Tekere et al., 2001). Microbial growth as well as enzyme-catalyzed reactions is affected by the extreme temperatures (too high or too low temperature). Metabolic of microbes is increased with in appropriate temperature range that enhances the bioremediation processes.

Many chemicals solubilize by increasing temperature that may lead to increase diffusion rates and fluidity also. The solubility and bioavailability of pollutants such as heavy metals and PAHs increase with temperature (Gargouri et al., 2011). Thus temperature plays a crucial role in the optimum operating performance of bioreactor for achieving the best results of remediation. The design of specialized bioreactors has temperature control system.

20.3.1.2 Effect of pH

Microbial growth and metabolic is also influenced by pH. The ionic properties of microbial cell are altered by pH that ultimately affects microbial growth. Microbes can grow in a wide pH range (minimum, optimum, and maximum). Most bacteria can grow optimally at pH 6–7.5. However, acidophiles are a group of bacteria that prefer optimal growth at acidic pH whereas alkaliphiles group of bacteria grows at high pH. Generally, fungi grow at lower pH in comparison to bacteria. The pH of reactors is set in the way that it provides optimum condition for microbial growth and enzyme activity. The performance of pollutants is also affected by pH that alters their remediation process. The redox state and solubility of metals are influenced by pH. The different metal forms and valency have unusual effects on microbes (Liu et al., 2017). However, solubility of metal enhances with decreasing medium pH whereas high pH facilitates precipitation of metal. Moreover, attachment of metal to microbial cell surface is facilitated by low pH (Gargouri et al., 2011; Liu et al., 2017). Buffers are used in the formulation of media to provide optimum pH for microbial activities. Acids and bases can be supplemented for maintaining pH during the reactor process (Tekere et al., 2001).

20.3.1.3 Effect of nutrients

Nutrients play critical role in the growth and metabolic activities of microorganisms. Various elements are involved in the development and energy production. Carbon is the fundamental element of living organisms and is required in large quantity than other elements. Other elements such as oxygen, hydrogen, nitrogen, phosphorus, sulfur, iron, magnesium, and calcium are needed for balanced nutritional medium of bioreactor depending on the type of microbe used for remediation (Tekere et al., 2001). All essential macro- and micronutrients are supplied in the nutrient media of reactor for optimal microbial growth. For degradation of pollutants, microbes utilize them as primary energy source or use nutrients as primary energy source in cometabolism of pollutants.

20.3.1.4 Effect of moisture

Water is essential to encourage microorganism growth and enzyme activity. It is maintaining osmotic pressure in the cell to support growth and chemical reactions. Water activity expresses the quantity of water available for hydration of substance. Several microbes grow at 0.98 or high water activity. On the other hand, osmotolerant microbes grow in a range of low water activity (Naik and Duraphe, 2012).

20.3.1.5 Effect of electron acceptors

The presence of electron acceptors in aerobic and anaerobic conditions also influences biodegradation process. Oxygen serves as electron acceptor in aerobic microbes whereas nitrate, sulfate, and ferric oxides are electron acceptors in anaerobic microbes capable to affect degradation process of pollutants.

20.3.2 Effect of reactor design and pollutants

The size of reactor, design and mode of operation are important factors which regulate the function of microbial bioreactors. For achieving the best bioremediation process, the reactors should be provided optimum physical, chemical and biological condition. The bioreactors should be designed with suitable physical condition for transport of solids, liquids and gases ensuring the adaptation of bioreactor for biological system. The biochemical and biophysical processes of the microbial reactor should operate at optimum level when the reactor is operated in the field. The poor design and construction of a bioreactor may affect the stability and performance of the bioremediation.

The physical and chemical properties of pollutants such as concentration, solubility, molecular complexity, volatility and toxicity also affect the microbial bioremediation. Some recalcitrant pollutants resist biodegradation i.e. they are degraded at very low rate even under optimum conditions.

20.4 Microbial bioreactors in bioremediation

Various studies of remediation have been accomplished by using microbial reactors (Tekere et al., 2005; Gros et al., 2014; Pino-Herrera et al., 2017). Pollutants present in air, water, and soil can be treated effectively through bioreactor techniques (Gargouri et al., 2011; Quintero et al., 2007). The selection of bioreactor should be based on the ease of conduct and maintenance for specific purpose and application. The flexibility in design of bioreactors for diverse remediation procedures makes their use attractive in bioremediation (Azubuike et al., 2016). The bioreactor design should contain high cell biomass, nutrient supply along with removal of waste from the system. Slurry-phase, partitioning, stirred-tank, biofilters, packed-bed, airlift, and membrane bioreactors are commonly used for remediation purpose.

20.4.1 Slurry-phase bioreactor

Slurry-phase bioreactor treats polluntants in the slurry state. Slurry-phase biological treatment and bioslurry reactor are alternative names of slurry-phase bioreactor. It is an effective technology for in situ and ex situ conditions that can be used for remediation of soils having high clay content along with organic matter affected by contaminants such as explosives, petrochemical hydrocarbons, creosotes, tars, chlorinated solvents, pesticides, and herbicides. Mostly persistent chemicals are hydrophobic so they are sorb to soil and not easily available for remediation. Various environmental parameters can be controlled in slurry reactors that lead to faster treatment of polluted soil. Batch process is the most common operation mode of slurry reactors; however, semicontinuous and continuous processes are also in use (Robles-González et al., 2008; Pino-Herrera et al., 2017). In appropriate ratio, water is mixed with soil matrix that enhances the contact of micobes, media, contaminants, and oxygen. This process enhances the availability of solubilized contaminants.

20.4.2 Partitioning bioreactor

Partitioning bioreactors are an alternative technology to slurry bioreactors. The addition of liquid solvents (water immiscible and nonbiodegradable) or solid polymer is employed to a biological process in two-phase partitioning bioreactors (TPPBs). The reactor can be designed in single or multiphase. TPPB comprises aqueous phase having microbial cells and immiscible organic phase that can segregate large quantity of toxic substrates by maintaining them at subinhibitory level in the aqueous phase. The selective partitioning of organic substrate takes place in organic phase on the basis of partition coefficient of solute that causes difference in the concentration of substrate between aqueous and organic phases. Since microbes consume substrate available in aqueous phase, additional substrate is required to maintain the thermodynamic equilibrium. Thus huge amount of harmful substrates can be supplemented to a biotreatment process without inducing cytotoxicity to microbial cells. In bioremediation of soil, partitioning phase is considered as temporary storage phase of pollutants, that is, consequently eliminated in the successive biological step thus allowing the reuse of the partitioning phase (Tomei and Daugulis, 2013). Partitioning bioreactors are used in petrochemical industry for bioremediation of toxic compounds such as VOCs and benzene. A continuous two-phase partitioning phase reactor is effective in the treatment of tannery wastewater (Angelucci et al., 2017).

20.4.3 Stirred-tank bioreactor

Stirred-tank bioreactor is one of the most conventional bioreactors. A continuously stirred-tank reactor comprises a cylindrical vessel connecting with motor-driven central shaft that holds one or many impellers (agitators). These bioreactors are mainly designed for submerged cultures. They are mechanically agitated having stirrers as main gas dispersing tools that provide excellent mixing along with high values of mass transfer rates. The advantages of stirred-tank reactor include flexible processing conditions, proper mixing of contaminants, and efficient transfer of gas to microbial cells. The major limitations of stirred-tank reactor are the mechanical agitation that requires energy and continuous stirring that may injure microbial cells. A continuous stirred-tank bioreactor was implemented in the treatment of hydrocarbon-rich industrial wastewater using acclimatized microbial consortium that accomplished successful remediation of wastewater. The residual total petroleum hydrocarbon concentration was decreased from 320 to 8 mg/L (Gargouri et al., 2011). Stirred-tank bioreactor is applied for bioremediation of benzyl alcohol, ethanol, and toluene by *Pseudomonas putida*.

20.4.4 Biofilters and bioscrubbers

Biofilters and bioscrubbers are air treatment techniques. Biofilters are one of the oldest bioremediation techniques. It is an important separation method that can be applied in the removal of organic contaminants from air and water. During biofilter operation, pollutants are passed through a media bed containing microbes that causes degradation of pollutants. Several materials (bark, composted yard waste, coarse soil, gravel, peat, or plastic shapes) are used for the preparation of bed media. Trickling filter is a characteristic example of biofilter, it is extensively used in the treatment of wastewater or liquid effluents or waste constituted into liquid. A trickling filter consists of a round vertical tank containing aggregate, plastic, or ceramic media. A vertical pipe having rotary connection with spray nozzles is placed in the middle of tank. Rotary connection is attached with a spray arm, and spray nozzles are installed along its length that helps in circulation of wastewater. Microbes growing on the packing material forms biofilm on the surface that degrades contaminants present in the wastewater. Trickling biofilter is capable to remove high concentrations of 1-butanol from the polluted air. Schmidt and Anderson (2017) evaluated the potential of biotrickling filter in the removal of excessive amount of 1-butanol from polluted air from industry. The removal efficiency of biotrickling filter was more than 80% for the butanol concentration range from 0.4 to 1.2 g/m^3 in the polluted air. The perlite-packed biotrickling filter was operated on laboratory scale for 60 days. The performance of biotrickling filter determined successful evacuation of butanol concentration varies from 0.55 to 4.65 g/m^3 along with maximum withdrawal capacity of 100 g/(m^3 h).

In the process of bioscrubber, contaminants are absorbed in aqueous phase in an absorber column and consequent biological treatment in a liquid phase bioreactor. The leaving effluent is recirculated to the top of absorber column that provides superb gas cleaning of soluble contaminants. The process of bioscrubber may be controlled in an easy way because temperature, pH, nutrient balance, and products removal can be modified in the liquid phase. However, one drawback of the reactor is that contaminants must be suspended in the liquid phase during the processing in absorber column. Bioscrubbers are suitable for the pollutabnts having high aqueous solubility that is the limiting stage in the treatment of various air odorants and contaminants (Kennes and Thalasso, 1998; Smet et al., 1998; Herrygers, 2000). Bioscrubbers are categorized into two groups based on the number of vessels in the design of reactor: single chamber and multi chamber. In the treatment of effluents having nonpolar gaseous volatile organic contaminants, the performance of bioscrubbers can be improved by using bioreactors consists of more than one chamber where mass transfer from gas phase to liquid takes place in the main chamber adding micronutrients, tensoactive compounds, etc. Then liquid having volatile organic contaminants is transferred to the next chamber for degradation (Schroeder, 2002). In single-chamber bioscrubber, pollutant mass transfer from the gas phase to liquid and degradation occurs in one chamber. Single-chamber bioscrubbers are used for the treatment of gases containing polar pollutants (alcohols and phenols) (Wei et al., 1999).

20.4.5 Trickle-bed bioreactor

Similar to biofilters, trickle-bed bioreactor consists of microbes immobilized in a biolfilm that coats surface of packing material in the column. In the operation of trickle-bed bioreactor, liquid and gas flow cocurrently in downward direction via a microbial-coated packing material. In the reactor, waste gases enter from the top in column whereas a recirculating solution of mineral salt codirected with gas in the downward direction. Aerobic oxidation through microbial populations leads to the degradation of the organic contaminants. Trickled-bed bioreactors are effective in the treatment of waste gases. Hekmat et al. (2004) applied trickle-bed bioreactor for the biodegradation of Solvesso 100. The results indicated that 67% of Solvesso 100 was removed successfully at the input rate of 120 g/(m^3 h). These trickled-bed bioreactors are suitable for the treatment of VOCs such as toluene, dichloromethane, and polyalkylated benzenes (Diks and Ottengraf, 1991; Pedersen et al., 1997; Hekmat and Vortmeyer, 1999). The procedures designed for the treatment of waste gases could be effective in the treatment of volatile contaminants present in the liquid waste.

20.4.6 Fluidized-bed bioreactor

Unlike trickle-bed bioreactor and biofilter, fluidized-bed bioreactors utilize fluidized particles as a carrier of microbial biofilm for the removal of organic contaminants under aerobic condition. Generally, activated carbon and sand are the most frequently used carriers in these reactors. In the past the processes of reactor were used for the degradation of recalcitrant contaminants containing dichloromethane and dichlorophenols. Koch et al. (1991) used an airlift-loop fluidized-bed bioreactor to compare the efficacy of activated carbon and sand biofilm carriers for the removal of *N*-heterocycles, phenolic compounds, and aromatic amines that serve as a carbon source. These compounds together provide dissolved organic carbon varying from 125 to 1000 mg/L. The results showed 59%–69% removal of dissolved organic carbon in the sand, whereas 39% dissolved organic carbon was removed by activated carbon. Using a pure oxygen aerated fluidized-bed bioreactor, Ohlen et al. (2005) demonstrated the effectiveness of volatile organic contaminants decomposition in groundwater. The reactor achieved 99% decomposition of 1,3,5-trimethylbenzene and 1,2,4-trimethylbenzene form initial concentration of 65 and 220 μg/L, respectively. In this study, fluidized-bed bioreactor was used to remove color from wastewater and transform nonbiodegradable azo dye to aromatic amines that degraded aerobically.

20.4.7 Packed-bed bioreactor

Packed-bed bioreactor provides a fixed film substratum for microbial growth in order to assure greater treatment reliability in compact reactor system. The reactor offers an advantage that pollutants present in dilute aqueous solution can be remediated at high biomass without separating treated effluent and biomass (Van Loosdrecht and Heijnen, 1993). Unlike suspension cultures, incorporation of membrane filters and centrifugation is not required for retaining the biomass during the treatment process of packed-bed biofilm. This characteristic feature of packed-bed reactor makes it useful in bioreactor system where bulky substrate flow through is required. The microbial cell concentration may be increased in a definite volume that leads to enhance productivity or efficiency of the reactor. Although high biomass can be easily retained, the rate-limiting procedure in the packed-bed bioreactor operation is the mechanism of biofilm mass transfer of substrate. There are significant variations in the microenvironment of microbes within the biofilm that depends on the distance from the biofilm surface (Van Loosdrecht and Heijnen, 1993; Bisping and Rehm, 1988; Guieysse et al., 2000).

Sources of carbon, nitrogen, and oxygen have to traverse the biofilm–liquid interface by diffusion. A diffusion gradient arises in this region. The two important processes occur in the system are (1) solute transport over biofilm and (2) combined reaction and diffusion within biofilm. These processes are important to calculate the conversion kinetics in the biofilm (Van Loosdrecht and Heijnen, 1993). In the packed-bed reactors, excess growth of microbes leads to hydraulic channeling or loss in the volume of interstitial fluid. To resolve the limitations of hydraulic hold, in the interior of bioreactor extra capillary space transverse flow has been developed.

Packing material plays important role in the remediation of pollutants through packed-bed bioreactors. Various packing materials have been used for treatment of different pollutants. For example, polyurethane foam matrices, laterite pebbles, coconut shell biochar, metal mesh, wood chip, sugarcane bagasse, calcium alginate beads are used as packing material in the treatment of malathion and gasoline contaminated water, azo dyes, Congo red dye, textile dyestuff, chlorophenols, industrial dyes, petroleum wastewater, respectively (Yum and Peirce, 1998; Kapdan et al., 2000; Senan et al., 2003; de Nardi et al., 2005; Banerjee and Ghoshal, 2017; Geed et al., 2017; Talha et al., 2018; Torres-Farradá et al., 2018). In addition, polyurethane foam and polyvinyl alcohol-alginate beads are used in packed-bed bioreactor as packing material for the removal of benzene (Kureel et al., 2017). The ideal packing material should be chemically inert, porous, and rigid to facilitate microbial growth and reusable after completion of the process.

20.4.8 Airlift bioreactor

Gaseous or volatile air pollutants can be effectively treated by airlift bioreactors. These bioreactors are widely used in wastewater treatment. The advantages of airlift bioreactors are low shearing effect, ease of scale-up, high efficiency of oxygen transfer, and regulated flow patterns. In comparison to stirred-tank bioreactors the cost of aeration and agitation is low. The limitations of these reactors include susceptibility to gas–liquid mass transfer, poor liquid phase mixing when operating at high cell density (Patel and Kumar, 2016). The performance of reactor is influenced by air pumping and circulation of liquid. The airlift bioreactor contains internal loop (draft tube) or external loop. The internal or external loop can prevent bubble coalescence by steering them in one direction. These loops provide favorable environment for cell growth by even distribution of shear stress in the reactor. These loops are helpful in increasing mass and heat transfer rates by improving cyclic movement of solution. At normal pressure and temperature, some volatile organic substances can be degenerated by microbes without producing toxic by-products (Bertollo et al., 2017). Nikakhtari and Hill (2005) evaluated an airlift bioreactor with external loops and mesh of stainless steel as packing material for biodegradation of phenol polluted air. The 100% removal of phenol was achieved at 22,160 mg/(h m^3) rate of phenol loading by *P. putida* that demonstrated the potential and novelty of the reactor for biodegradation of VOCs.

20.4.9 Membrane bioreactor

Membrane bioreactor is a combination of membrane filtration system (microporous/ultrafiltration membrane) and biological treatment. The reactor comprises a tank and a membrane component. Membrane unit can be placed either external to the reactor or in submerged state. Membrane act as a barrier for separation of liquid from solid, thus can retain suspended solids and maintain high biomass in the reactor. In comparison to other techniques the quality of the processed effluent from the membrane bioreactor is high along with reduced reactor volume and sludge production. The reactor is operating normally in aerobic conditions but can be operated anaerobically or in alteration of aerobic and anaerobic phases to increase microbial nitrification and denitrification process. Membrane bioreactor has various advantages, including requirement of smaller area than other conventional techniques. A membrane bioreactor can efficiently replace three steps (settling, activated sludge, and disinfection) in conventional treatment plant. However, the major drawback of the reactor is membrane fouling and high processing costs. The processing costs can be cut down by incorporating microbial fuel cells (Visvanathan et al., 2005; Meng et al., 2009). Membrane bioreactors are used in the processing of domestic and industrial wastewater. These bioreactors have been assessed in the bioremediation of PCP (Visvanathan et al., 2005), 2-chlorophenol, 1,2-dichlorobenzene, 1,2-dichloroethane (Carucci et al., 2007), and textile wastewater (Hossain and Ismail, 2015).

20.4.10 Moving-bed biofilm reactor

Moving-bed biofilm reactor is a continuous operating system where small carrier elements are used to grow microbial biomass that circulates in the liquid. For circulating microbial biomass, mechanical agitation is used in an anaerobic reactor whereas aeration is used in anaerobic reactor. Generally, carrier elements are prepared from plastic in different shapes (sphere, round, square, and cylindrical). They provide large surface area for the development of microbial biofilm. The carrier's density, shape, surface area, and void volume may affect the performance of the reactor. Moving-bed biofilm reactors have been applied in wastewater treatment for the removal of nitrogen, phosphorous, and organic matter. Ammonia can be removed by the operation of moving-bed biofilm reactors through traditional nitrification and denitrification processes and autotrophic bacteria (anammox). In conventional process, ammonium is converted to nitrate by nitrification process, and consequently, nitrate is converted to nitrogen gas by denitrification process. Such operations are carried out in two different reactors. Partial nitrification (nitritation) and ammonium oxidation can be accomplished in a single reactor modifying dissolved oxygen supply to biofilm. A group of autotrophic bacteria (anammox) can remove ammonium by nitritation process. In comparison to traditional process, anammox process consumes 40% energy and produces 88% less carbon dioxide. Due to slow growth rate, anammox process needs long start period at temperature 30°C–35°C. The moving-bed biofilm reactor operating system prevents slow grow microbes from washing out (Barwal and Chaudhary, 2014; Vasan, 2019). Lemaire et al. (2013) evaluated the efficiency of anammox process using moving-bed biofilm reactor for side stream reject municipal wastewater treatment and reported that up to 1.2 kg N/(m^3 d) removal rate could be achieved by this system.

20.4.11 Upflow anaerobic sludge blanket reactor

Among anaerobic treatment methods, upflow anaerobic sludge blanket (UASB) reactor is an important technique for wastewater treatment. The reactor consists of a single chamber where wastewater enters from the bottom and flows upward through activated sludge blanket filters. Generally, activated sludge remains in the form of granular aggregates that do not wash out and provide high treatment efficiency. Gases such as carbon dioxide and methane produced from wastewater under anaerobic operation lead to internal mixing that helps in the maintenance of biological granules. However, some gases produced in the process may attach to the granules. For segregation of gas, liquid, and granules, a gas–liquid–solid separator is placed on the top of the chamber. The UASB reactor is successfully applied in the treatment of industrial wastewater having high organic contaminants. Carbohydrate-rich wastewater such as starch/canning industry wastewater serves as a nutrient-rich media that is readily digested by microorganisms for anaerobic hydrogen production during the reactor operation.

UASB reactors are also effective for domestic wastewater treatment. The organic matter having high nutrient content in domestic wastewater is present in the form of blackwater. The UASB reactors are applied widely because of their facility to manage high organic loading rate, short hydraulic retention time (HRT), low nutrient and energy requirement, and less generation of biosolid wastes. The major drawbacks of this technique include long start-up period, problem of corrosion, toxicity and odor, and maintenance of a temperature range (15°C–35°C) specially in cold region and need of secondary treatment (Daud et al., 2018).

20.4.12 Rotating biological contactor

Rotating biological contactor (RBC) is also known as rotating disk bioreactor, rotating biological disks, biodisks, rotating biological filters, and rotating biological surfaces. It is a biofilm reactor alike trickling filter where microbes are attached to support material. An RBC unit consists of a series of disks mounted on a common horizontal shaft and partly or fully submerged in the wastewater. These rotating disks provide support media for biofilm. In place of rotating disks a container filled with lightweight packing material can also be used. Oxygen is supplied to the biofilm from air when the rotator disk is outside the liquid, and in submerged condition, oxygen is supplied from liquid as oxygen is mixed with the liquid by surface turbulence generated by the rotation of rotator disks. RBCs are used successfully in the secondary treatment of municipal wastewater. They have been applied in decolorization of textile dyes, colored effluents of sugar refinery, treatment of effluents from food processing, bakeries, wineries, leather tanneries, pulp and paper industry, and remediation of landfill leachates/organopollutants. The advantages of RBC are small land requirement, simple process of control and monitor, high oxygen transfer efficiency (OTE), and short HRTs. The disadvantages of the RBC system include slow start-up, complex scale-up, limited flexibility, and requirement of sufficient primary treatment and secondary clarifier (Cortez et al., 2008).

20.5 Bioremediation organisms

The active members of microbial consortium are able to degrade pollutants under different environmental conditions. These microbes include *Alcaligenes*, *Actinobacteria*, *Acinetobacter*, *Arthrobacter*, *Beijerinckia*, *Bacillus*, *Flavobacterium*, *Mycococcus*, *Mycobacterium*, *Methylosinus*, *Nocardia*, *Nitrosomonas*, *Pseudomonas*, *Phanerochaete*, *Penicillium*, *Serratio*, *Rhizoctomia*, *Xanthobacter*, and *Trametes* (Table 20.1). Individually microbes are unable to mineralize most of the hazardous compounds. Complete mineralization is a sequential degradation process of microbial consortium involving synergetic and cometabolic activities. Natural microbial communities present in different habitats have remarkable physiological adaptability. They can metabolize as well as mineralize various organic molecules. However, most of the remediation systems are operating in aerobic conditions, whereas some are operating under anaerobic conditions facilitating degradation of toxic compounds by microbes (Colberg and Young, 1995).

The isolated *P. putida* MHF 7109 from cow dung microbial consortium plays a major role in biodegradation of petroleum hydrocarbon compounds such as benzene, toluene, and *o*-xylene (BTX). Every compound has been used individually in a minimal salt medium (MSM) at different concentrations (50, 100, 250, and 500 mg/L) to analyze the degradation potential of the microbial strain. The obtained results suggest that the strain used has great capacity to degrade BTX at a concentration of 50 mg/L within a time interval of 48, 96, and 168 h, respectively, while the concentrations of 100 mg/L of benzene and toluene have been fully degraded within 120 and 168 h, respectively. 62% of *o*-xylene has degraded within 168 h at the 100 mg/L concentration level. The higher degradation rate for BTX has been observed 1.35, 1.04, and 0.51 mg/(L h), respectively. The elevated concentrations (250 and 500 mg/L) of BTX suppressed the function of microorganisms. The mass spectrometry analysis characterized the intermediates products such as catechol, 2-hydroxymuconic semialdehyde, 3-methylcatechol, *cis*-2-hydroxypenta-2,4-dienoate, 2-methylbenzyl alcohol, and 1,2-dihydroxy-6-methylcyclohexa-3,5-dienecarboxylate, for BTX, respectively. Hence, *P. putida* MHF 7109 has been explored with higher biodegradation capacity for the volatile petroleum hydrocarbons (Singh and Fulekar, 2010).

Moreover, a current study has given layout of surface soil treatment unit in which common pesticides mainly chlorpyrifos, cypermethrin, fenvalerate, and triclopyr butoxyethyl ester with different concentrations 25, 50, and 100 mg/kg have been remediated by employing cow dung microbial consortia undermaintained environmental conditions. It has been examined continuously till the parent component was transferred into intermediates or less harmful substances. After that, this would be further mineralized through part of the microbial food chain or get integrated into the humic fractions. These results expressed with highlight of potentiality of cow dung slurry consortia in bioremediation of soil contaminated with pesticides in surface soil treatment unit (Geetha and Fulekar, 2008).

The capacity of ryegrass for rhizosphere bioremediation of chlorpyrifos in mycorrhizal soil has been assessed by the greenhouse pot culture experiments. The pot cultured soil rectified at initial chlorpyrifos concentration of 10 mg/kg has been examined to be degraded totally in 7 days where the remaining modified concentrations (25–100 mg/kg) declined fast under the impact of ryegrass mycorrhizosphere as the incubation period increased to 28 days. Such bioremediation of chlorpyrifos in soil has contributed to the microorganisms occurred in the roots of ryegrass rhizosphere, so the microorganisms associated with rhizospheric soil spiked at the highest concentration (100 mg/kg) had been investigated and applied for isolation of chlorpyrifos deteriorating microorganisms. The potential degrader (*Pseudomonas nitroreducens* PS-2) had been determined by 16S rDNA analysis applying BLAST technique. Further, bioaugmentation for the increased chlorpyrifos biodegradation had been done by using PS-2 as an inoculum in the experimental setup identical to the previous one. The heterotrophic bacteria and fungi had also been isolated from the inoculated and noninoculated rhizospheric soils. In bioaugmentation experiments, it has been shown that the disappearance of chlorpyrifos was higher (100%) in the inoculated rhizospheric soil in comparison to 76.24%, 90.36%, and 90.80% in the noninoculated soil for initial concentrations of 25, 50, and 100 mg/kg at the 14th-, 21st-, and 28th-day intervals, respectively (Korade and Fulekar, 2009).

20.6 Bioremediation of pollutants using bioreactors

The potential of *Pseudomonas aeruginosa* (NCIM, 2074) has been improved with increasing its adaptability to higher level of chlorpyrifos for using at large scale during bioremediation process. *P. aeruginosa* isolate NCIM 2074 had been acclimated indifferent concentrations of chlorpyrifos (10, 20, 50, 75, and 100 mg/L) in incubator shaker at 37°C and 150 rpm. Initially, 10 mg/L concentration of chlorpyrifos has been transferred for 14 days in MSM undermaintained environmental conditions. Later, the culture has been raised up to higher content of chlorpyrifos through transferring 1 mL from the medium with 10 to 25 mg/L of the compound. This process has been consequently repeated after every 14 days, and in every repetition of process the chlorpyrifos concentration was increased in using medium. All these whole scale-up processes have been extended for 70 days. *P. aeruginosa* (NCIM 2074) have been acclimated with escalating concentration of chlorpyrifos up to 50 mg/L, although 75 and 100 mg/L were acting as inhibitory concentration for the organism. GC—MS results revealed that the complete biodegradation of chlorpyrifos at 10, 25, and 50 mg/L was observed over a period of 1, 5, and 7 days, respectively. The intermediate products 3,5,6-trichloro-2-pyridion, 2,4-bis (1,1-dimethyiethyl) phenol and 1,2-zenedicarboxylic acid were remained while bioremediation, though after long run these intermediates are converted to CO_2, biomass, and nutrients. This study concluded that the *P. aeruginosa* (NCIM 2074) is one of the potential organisms for using in bioremediation of chlorpyrifos at concentrations up to 50 mg/L, although the growth of organism is suppressed at higher concentrations, namely, higher than 50 mg/L of chlorpyrifos (Fulekar and Geetha, 2008).

The capacity of a cow dung microbial consortium has been analyzed against phenol bioremediation in a single-phase bioreactor and a TPPB. A bioreactor has been crafted and constructed so as to separate the aqueous and organic phases by providing space for aeration and stirring, a cooling system, and a sampling port. The important significances of the TPPB have been elucidated. The *P. putida* IFO 14671 had been isolated from the cow dung microbial consortium for growing in culture medium, and this organism was recognized as a high-potential phenol degrader. This method generated an advance idea to construct the bioremediation techniques for the biodegradation of organic compound such as phenol using a bioreactor. It has been also described the microorganism's capacities of cow dung as a source of biomass (Singh and Fulekar, 2009).

Bioremediation of benzene is accomplished by implicating of cow dung microflora in a bioreactor. The bioremediation of benzene has been observed 100% and 67.5%, at initial concentrations of 100 and 250 mg/L within 72 and 168 h, respectively, by impact of cow dung microflora, however, at higher concentration (500 mg/L), benzene has been observed to be inhibitory. Therefore the TPPB has been crafted and constructed to accomplish the biodegradation at elevated concentration. In this bioreactor the contaminant at 5000 mg/L concentration has been degraded up to 50.17% over a period Q1 of 168 h under influence of microorganisms. Again the *P. putida* MHF 7109 had been derived from cow dung microflora as efficient benzene degrader, and its potential to degrade benzene at different concentrations was analyzed. The obtained data have suggested that 100%, 81%, and 65% at the concentrations of 50, 100, 250 mg/L within the time period of 24, 96, and 168 h, respectively, were degraded. The data from GC—MS also reveal the presence of catechol and 2-hydroxymuconic semialdehyde that explains the developed pathway during benzene biodegradation. Current work elucidates about the potential of cow dung microflora as a huge source of biomass for benzene biodegradation in TPPB (Singh and Fulekar, 2009).

20.7 Genetic engineering

Presently, scientists are searching some novel genetically engineered microorganisms to enhance their potential to metabolize particular compounds such as hydrocarbons and pesticides. In the late 1980s the employing genetic engineering for enhancement of bioremediation process had been encouraged. It has been reported that the recombinant DNA techniques play a significant role to increase the degradation of hazardous waste under laboratory condition. The genetically engineered microorganisms have greater degradative potential for the degradation of several hazardous materials under certain conditions. Currently, genetic engineering technology is widely used as potential tool in the process of bioremediation. Bioremediation analyzes different gene diversity and metabolic diversity in microorganisms (Fulekar, 2009). The genetic construction of such microorganisms makes them beneficial in biodegradation, biotransformation, biosorption, and bioaccumulation. The essential blueprint of gene encoding for biodegradative enzymes has been found in chromosomal and extrachromosomal DNA of these microbes. Recombinant DNA techniques expedite to develop the potential of an organism to metabolize a xenobiotic by identifying these degradative genes and transforming them into particular host through appropriate vector under the tight control of suitable promoters. This depends on susceptibility to modification and exchange of genetic information. The recombinant DNA technology consists of PCR, antisense RNA technique, site-directed mutagenesis, electroporation, and particle bombardment techniques. Recently, the biotechnology associated with recombinant DNA technology has become a fine-tuning the bioremediation technology by enhancing pollutant-degrading microbes by strain alterations and genetic modification of particular regulatory and metabolic genes, which are very essential in developing effective, safe, and economical techniques for bioremediation. Although bioremediation is not efficient only for the degradation of hazardous substances, also it could be implicated to clean unwanted materials from air, soil, water, and raw materials form industrial waste.

20.8 Conclusion

A broad range of bioreactors have been applied for the remediation of different pollutants present in the air, water, and soil. These pollutants are also susceptible to microbial degradation. Diverse microorganisms have capability to degrade pollutants naturally, and the use of microbial bioreactor improves the rate of degradation.

References

Al-Jawhari, I.F.H., 2014. Ability of some soil fungi in biodegradation of petroleum hydrocarbon. J. Appl. Environ. Microbiol. 2 (2), 46–52.

Angelucci, D.M., Stazi, V., Daugulis, A.J., Tomei, M.C., 2017. Treatment of synthetic tannery wastewater in a continuous two-phase partitioning bioreactor: biodegradation of the organic fraction and chromium separation. J. Cleaner Prod. 152, 321–329.

Azubuike, C.C., Chikere, C.B., Okpokwasili, G.C., 2016. Bioremediation techniques – classification based on site of application: principles, advantages, limitations and prospects. World J. Microbiol. Biotechnol. 32 (11), 180.

Banerjee, A., Ghoshal, A.K., 2017. Biodegradation of an actual petroleum wastewater in a packed bed reactor by an immobilized biomass of *Bacillus cereus*. J. Environ. Chem. Eng. 5 (2), 1696–1702.

Barwal, A., Chaudhary, R., 2014. To study the performance of biocarriers in moving bed biofilm reactor (MBBR) technology and kinetics of biofilm for the existing aerobic treatment systems; a review. Rev. Environ. Sci. Biotechnol. 13, 285–299.

Bertollo, F.B., Lopes, G.C., Silva, E.L., 2017. Phenol biodegradation by *Pseudomonas putida* in an airlift reactor: assessment of kinetic, hydrodynamic and mass transfer parameters. Water Air Soil Pollut. 228 (10), 398.

Bisping, B., Rehm, H.J., 1988. Multistep reactions with immobilized microorganisms. Biotechnol. Appl. Biochem. 10 (2), 87–98.

Burghal, A.A., Abu-Mejdad, N.M.J.A., Tamimi, W.H.A., 2016. Mycodegradation of crude oil by fungal species isolated from petroleum contaminated soil. Int. J. Innovative Res. Sci. Eng. Technol. 5 (2), 1517–1524.

Carucci, A., Manconi, I., Manigas, L., 2007. Use of membrane bioreactors for the bioremediation of chlorinated compounds polluted groundwater. Water Sci. Technol. 55 (10), 209–216.

Chikere, C.B., Chikere, B.O., Okpokwasili, G.C., 2012. Bioreactor-based bioremediation of hydrocarbon polluted Niger Delta Marine sediment, Nigeria. 3 Biotech 2 (1), 53–66.

Colberg, P.J.S., Young, L.Y., 1995. Anaerobic degradation of no halogenated homocyclic aromatic compounds coupled with nitrate, iron or sulfate reduction. In: Young, L.Y., Cerniglia, C.E. (Eds.), Microbial Transformation and Degradation of Toxic Organic Chemicals. Wiley-Liss, New York, pp. 307–330.

Cortez, S., Teixeira, P., Oliveira, R., Mota, M., 2008. Rotating biological contactors: a review on main factors affecting performance. Rev. Environ. Sci. Biotechnol. 7, 155–172.

Daud, M.K., Rizvi, H., Akram, M.F., Ali, S., Rizwan, M., Nafees, M., et al., 2018. Review of upflow anaerobic sludge blanket reactor technology: effect of different parameters and developments for domestic wastewater treatment. J. Chem. 1–13.

de Nardi, I.R., Ribeiro, R., Zaiat, M., Foresti, E., 2005. Anaerobic packed-bed reactor for bioremediation of gasoline contaminated aquifers. Process Biochem. 40 (2), 587–592.

Diks, R.M.M., Ottengraf, S.P.P., 1991. Vertification studies of a simplified model for the removal of dichloromethane from waste gases using a biological trickling filter. Bioprocess Eng. 6, 131–140.

El-Borai, A.M., Eltayeb, K.M., Mostafa, A.R., El-Assar, S.A., 2016. Biodegradation of industrial oil-polluted wastewater in Egypt by bacterial consortium immobilized in different types of carriers. Pol. J. Environ. Stud. 25 (5), 1901–1909.

Fulekar, M.H., 2009. Bioremediation of fenvalerate by *Pseudomonas aeruginosa* in a scale up bioreactor. Rom. Biotechnol. Lett. 14 (6), 4900–4905.

Fulekar, M.H., Geetha, M., 2008. Bioremediation of chlorpyrifos by *Pseudomonas aeruginosa* using scale up technique. J. Appl. Biosci. 12, 657–660.

Gargouri, B., Karray, F., Mhiri, N., Aloui, F., Sayadi, S., 2011. Application of a continuously stirred tank bioreactor (CSTR) for bioremediation of hydrocarbon-rich industrial wastewater effluents. J. Hazard. Mater. 189 (1–2), 427–434.

Geed, S.R., Kureel, M.K., Giri, B.S., Singh, R.S., Rai, B.N., 2017. Performance evaluation of malathion biodegradation in batch and continuous packed bed bioreactor (PBBR). Bioresour. Technol. 227, 56–65.

Geetha, M., Fulekar, M.H., 2008. Bioremediation of pesticides in surface soil treatment unit using microbial consortia. Afr. J. Environ. Sci. Technol. 2 (2), 36–45.

Ghosal, D., Ghosh, S., Dutta, T.K., Ahn, Y., 2016. Current state of knowledge in microbial degradation of polycyclic aromatic hydrocarbons (PAHs): a review. Front. Microbiol. 7, 1369.

Gros, M., Cruz-Morato, C., Marco-Urrea, E., Longrée, P., Singer, H., Sarrà, M., et al., 2014. Biodegradation of X-ray contrast agent iopromide and the fluoroquinolone antibiotic ofloxacin by the white rot fungus *Trametes versicolor* in hospital wastewaters and identification of degradation products. Water Res. 60, 228–241.

Guieysse, B., Bernhoft, I., Andersson, B.E., Henrysson, T., Olsson, S., Mattiasson, B., 2000. Degradation of acenaphthrene, phenanthrene and pyrene in a packed-bed biofilm reactor. Appl. Microbiol. Biotechnol. 54 (6), 826–831.

Hekmat, D., Feuchtinger, A., Stephan, M., Vortmeyer, D., 2004. Biofilm population dynamics in a trickle-bed bioreactor used for the biodegradation of aromatic hydrocarbons from waste gas under transient conditions. Biodegradation 15 (2), 133–144.

Hekmat, D., Vortmeyer, D., 1999. Biological degradation of volatile aromatic compounds showing poor water solubility from exhaust gases. Chem. Ing. Tech. 71 (11), 1290–1294.

Herrygers, V., Langenhove, V.H., Smet, E., 2000. Biological treatment of gases polluted by volatile sulphur compounds. In: Lens, P., Hulshoff, L. (Eds.), Environmental Technology to treat Sulphur Pollution, Principles and Engineering. IWA Publishing, London, pp. 281–304.

Hesham, Ael-L., Khan, S., Tao, Y., Li, D., Zhang, Y., Yang, M., 2012. Biodegradation of high molecular weight PAHs using isolated yeast mixtures: application of meta-genomic methods for community structure analyses. Environ. Sci. Pollut. Res. 19 (8), 3568–3578.

Hossain, K., Ismail, N., 2015. Bioremediation and detoxification of pulp and paper mill effluent: a review. Res. J. Environ. Toxicol. 9 (3), 113.

Jasińska, A., Paraszkiewicz, K., Sip, A., Długoński, J., 2015. Malachite green decolorization by the filamentous fungus *Myrothecium roridum*—mechanistic study and process optimization. Bioresour. Technol. 194, 43–48.

Kapdan, I.K., Kargi, F., McMullan, G., Marchant, R., 2000. Biological decolorization of textile dyestuff by *Coriolus versicolor* in a packed column reactor. Environ. Technol. 21 (2), 231–236.

Karigar, C.S., Rao, S.S., 2011. Role of microbial enzymes in the bioremediation of pollutants: a review. Enzyme Res. 1–11.

Kennes, C., Thalasso, F., 1998. Waste gas biotreatment technology. J. Chem. Technol. Biotechnol. 72 (4), 303–319.

Koch, B., Ostermann, M., Hoke, H., Hempel, D.-C., 1991. Sand and activated carbon as biofilm carriers for microbial degradation of phenols and nitrogen containing aromatic compounds. Water Res. 25, 1–8.

Korade, D., Fulekar, M.H., 2009. Rhizosphere remediation of chlorpyrifos in mycorhizospheric soil using ryegrass. J. Hazard. Mater. 172, 1344–1350.

Kureel, M.K., Geed, S.R., Giri, B.S., Rai, B.N., Singh, R.S., 2017. Biodegradation and kinetic study of benzene in bioreactor packed with PUF and alginate beads immobilized with *Bacillus* sp. M3. Bioresour. Technol. 242, 92–100.

Leitão, A.L., 2009. Potential of *Penicillium* species in the bioremediation field. Int. J. Environ. Res. Public Health 6 (4), 1393–1417.

Lemaire, R., Thesing, G., Christensson, M., et al., 2013. Experience from start-up and operation of deammonification MBBR plants and testing of a new deammonification IFAS configuration. In: WEFTEC, The Water Environment Federation's Technical Exhibition and Conference.

Liu, S.H., Zeng, G.M., Niu, Q.Y., Liu, Y., Zhou, L., Jiang, L.H., et al., 2017. Bioremediation mechanisms of combined pollution of PAHs and heavy metals by bacteria and fungi: a mini review. Bioresour. Technol. 224, 25–33.

Ma, Y.L., Lu, W., Wan, L.L., Luo, N., 2015. Elucidation of fluoranthene degradative characteristics in a newly isolated *Achromobacter xylosoxidans* DN002. Appl. Biochem. Biotechnol. 175, 1294–1305.

Mandal, S., Kunhikrishnan, A., Bolan, N.S., Wijesekara, H., Naidu, R., 2016. Application of biochar produced from biowaste materials for environmental protection and sustainable agriculture production. Environ. Mater. Waste 73–89.

Meng, F., Chae, S.R., Drews, A., Kraume, M., Shin, H.S., Yang, F., 2009. Recent advances in membrane bioreactors (MBRs): membrane fouling and membrane material. Water Res. 43 (6), 1489–1512.

Mnif, S., Sayadi, S., Chamkha, M., 2014. Biodegradative potential and characterization of a novel aromatic-degrading bacterium isolated from a geothermal oil field under saline and thermophilic conditions. Int. Biodeterior. Biodegrad. 86, 258–264.

Mohamed, A.T., Hussein, A.A., El Siddig, M.A., Osman, A.G., 2011. Degradation of oxyfluorfen herbicide by soil microorganisms: biodegradation of herbicides. Biotechnology 10, 274–279.

Naik, M.G., Duraphe, M.D., 2012. Review paper on-parameters affecting bioremediation. Int. J. Life Sci. Pharma Res. 2 (3), L77–L80.

National Research Council, 1993. In Situ Bioremediation: When Does It Work? National Academies Press, Washington, DC.

National Research Council, 2000. Natural Attenuation for Ground Water Remediation. National Academics Press, Washington, DC.

Nijenhuis, I., Schmidt, M., Pellegatti, E., Paramatti, E., Richnow, H.H., Gargini, A., 2013. A combined stable isotope approach for source apportionment of chlorinated ethenes and related degradation processes at a complex multi-contamination event and plume field site. J. Contam. Hydrol. 153, 92–105.

Nikakhtari, H., Hill, G.A., 2005. Hydrodynamic and oxygen mass transfer in an external loop airlift bioreactor with a packed bed. Biochem. Eng. J. 27 (2), 138–145.

Ohlen, K., Chang, Y.K., Hegemann, W., Yin, C.-R., Lee, S.-T., 2005. Enhanced degradation of chlorinated ethylenes in groundwater from a paint contaminated site by two-stage fluidized-bed reactor. Chemosphere 58 (3), 373–377.

Patel, B.P., Kumar, A., 2016. Biodegradation of 4-chlorophenol in an airlift inner loop bioreactor with mixed consortium: effect of HRT, loading rate and biogenic substrate. 3 Biotech 6 (2), 117.

Pedersen, A.R., Moller, S., Molin, S., Arvin, E., 1997. Activity of toluene degrading *Pseudomonas putida* in the early growth phase of a biofilm for waste gas treatment. Biotechnol. Bioeng. 54 (2), 131–141.

Pedro, P., Francisco, J.E., João, F., Ana, L.L., 2014. DNA damage induced by hydroquinone can be prevented by fungal detoxification. Toxicol. Rep. 1, 1096–1105.

Pino-Herrera, D.O., Pechaud, Y., Huguenot, E.D., Oturan, M.A., 2017. Removal mechanisms in aerobic slurry bioreactors for remediation of soils and sediments polluted with hydrophobic organic compounds: an overview. J. Hazard. Mater. 339, 427–449.

Quintero, J.C., Lu-Chau, T.A., Moreira, M.T., Feijoo, G., Lema, J.M., 2007. Bioremediation of HCH present in soil by the white rot fungus *Bjerkandera adusta* in a slurry batch bioreactor. Int. Biodeterior. Biodegrad. 60 (4), 319–326.

Ravi, R.K., Pathak, B., Fulekar, M.H., 2015. Bioremediation of persistent pesticides in rice field soil environment using surface soil treatment reactor. Int. J. Curr. Microbiol. Appl. Sci. 4 (2), 359–369.

Robles-González, I.V., Fava, F., Poggi-Varaldo, H.M., 2008. A review on slurry bioreactors for bioremediation of soils and sediments. Microb. Cell Fact. 7 (1), 5.

Safiyanu, I., Abdulwahid, I.A., Abubakar, U.S., Rita, S.M., 2015. Review on comparative study on bioremediation for oil spills using microbes. Res. J. Pharm. Biol. Chem. Sci. 6 (6), 783–790.

Sani, I., Safiyanu, I., Rita, S.M., 2015. Review on bioremediation of oil spills using microbial approach. IJESR 3 (6), 41–45.

Schmidt, T., Anderson, W.A., 2017. Biotrickling filtration of air contaminated with 1-butanol. Environments 4 (3), 57.

Schroeder, E.D., 2002. Trends in applications of gas-phase bioreactor. Rev. Environ. Sci. Biotechnol. 1, 65–74.

Senan, R.C., Shaffiqu, T.S., Roy, J.J., Abraham, T.E., 2003. Aerobic degradation of a mixture of azo dyes in a packed bed reactor having bacteria-coated laterite pebbles. Biotechnol. Prog. 19 (2), 647–651.

Shedbalkar, U., Jadhav, J., 2011. Detoxification of malachite green and textile industrial effluent by *Penicillium ochrochloron*. Biotechnol. Bioprocess Eng. 16, 196–204.

Simarro, R., Gonzalez, N., Bautista, L.F., Molina, M.C., 2013. Assessment of the efficiency of in situ bioremediation techniques in a creosote polluted soil: change in bacterial community. J. Hazard. Mater. 262, 158–167.

Singh, D., Fulekar, M.H., 2009. Benzene bioremediation using cow dung microflora in two phase partitioning bioreactor. J. Hazard. Mater. 175, 336–343.

Singh, D., Fulekar, M.H., 2010. Biodegradation of petroleum hydrocarbons by *Pseudomonas putida* strain MHF 7109 isolated from cow dung microbial consortium. Clean Soil Air Water 38 (8), 781–786.

Smet, E., Lens, P., Van Langenhove, H., 1998. Treatment of waste gases contaminated with odorous sulphur compounds. Crit. Rev. Environ. Sci. Technol. 28 (1), 89–117.

Srivastava, J., Naraian, R., Kalra, S.J., Chandra, H., 2014. Advances in microbial bioremediation and the factors influencing the process. Int. J. Environ. Sci. Technol. 11 (6), 1787–1800.

Talha, M.A., Goswami, M., Giri, B.S., Sharma, A., Rai, B.N., Singh, R.S., 2018. Bioremediation of Congo red dye in immobilized batch and continuous packed bed bioreactor by *Brevibacillus parabrevis* using coconut shell bio-char. Bioresour. Technol. 252, 37–43.

Tekere, M., Mswaka, A.Y., Zvauya, R., Read, J.S., 2001. Growth, dye degradation and ligninolytic activity studies on Zimbabwean white rot fungi. Enzyme Microb. Technol. 28 (4–5), 420–426.

Tekere, M., Read, J.S., Mattiasson, B., 2005. Polycyclic aromatic hydrocarbon biodegradation in extracellular fluids and static batch cultures of selected sub-tropical white rot fungi. J. Biotechnol. 115 (4), 367–377.

Tomei, M.C., Daugulis, A.J., 2013. Ex situ bioremediation of contaminated soils: an overview of conventional and innovative technologies. Crit. Rev. Environ. Sci. Technol. 43 (20), 2107–2139.

Torres-Farradá, G., Manzano, A.M., Ramos-Leal, M., Domínguez, O., Sánchez, M.I., Vangronsveld, J., et al., 2018. Biodegradation and detoxification of dyes and industrial effluents by *Ganoderma weberianum* B-18 immobilized in a lab-scale packed-bed bioreactor. Biorem. J. 22 (1–2), 20–27.

United States Environmental Protection Agency, 2003. Using Bioreactors to Control Air Pollution (EPA-456/R-03-003). United States Environmental Protection Agency. Available from: <http://www.epa.gov/ttn/catc> (accessed 24.02.20.).

Van Loosdrecht, M.C., Heijnen, S.J., 1993. Biofilm bioreactors for waste-water treatment. Trends Biotechnol. 11 (4), 117–121.

Vasan, S., 2019. Nitrogen Transformation in Biofilm for Reject Water Treatment [Thesis]. University of South-Eastern Norway, Porsgrunn.

Visvanathan, C., Thu, L.N., Jegatheesan, V., Anotai, J., 2005. Biodegradation of pentachlorophenol in a membrane bioreactor. Desalination 183 (1–3), 455–464.

Wei, W.Q., Gordon, A.H., MacDonald, D.G., 1999. Bioremediation of contaminated sir using an external-loop airlift bioreactor. Can. J. Chem. Eng. 77, 955–962.

Yadav, M., Singh, S., Sharma, J., Yadav, K.D., 2011. Oxidation of polyaromatic hydrocarbons in systems containing water miscible organic solvents by the lignin peroxidase of *Gloeophyllum striatum* MTCC-1117. Environ. Technol. 32, 1287–1294.

Yan, J., Niu, J., Chen, D., Chen, Y., Irbis, C., 2014. Screening of *Trametes* strains for efficient decolorization of malachite green at high temperatures and ionic concentrations. Int. Biodeterior. Biodegrad. 87, 109–115.

Yum, K.J., Peirce, J.J., 1998. Biodegradation kinetics of chlorophenols in immobilized-cell reactors using a white-rot fungus on wood chips. Water Environ. Res. 70 (2), 205–213.

Chapter 21

Microbial biofilms for the remediation of contaminated water

Monika Singh[1], Shweta Pandey[2], Ajay Kumar[3] and K.D. Pandey[1]

[1]*Department of Botany, Centre of Advance Study in Science, Institute of Science, Banaras Hindu University, Varanasi, India,* [2]*School of Agriculture, Graphic Era Hill University, Dehradun, India,* [3]*Agriculture Research Organization, Ministry of Agriculture and Rural Development, Volcani Center, Rishon LeZion, Israel*

21.1 Introduction

In developing countries, water pollution is a major problem due to presence of high levels of inorganic and organic contaminants. For remediation, various traditional methods have been applied and these treatment technologies can be classified as physical, chemical, or biological treatment techniques. Bioremediation is basically of two types depending on location of pollutant treatment. In the case of in situ bioremediation the polluted water sample is treated in their original location, whereas in ex situ remediation, the sample is treated typically at off-site (Ateia, 2016). In situ bioremediation is a smart choice as removal of contaminants from the site is not required and as a consequence it minimizes transportation cost and site disruption. In all the biological methods of remediation, microbial biofilm mediated remediation is environment friendly and cost-effective option for cleaning up water pollutants (Manobala et al., 2019). Use of biofilms is efficient for bioremediation as biofilms absorb, immobilize, and degrade various water pollutants (Igiri et al., 2018).

Biofilms are communities composed of microbial cells including bacteria, fungi, and some cyanobacteria and mycoalgae (fungi + algae) that adhere to inert or active entities and are encased by a substrate that can be hydrated (Douterelo et al., 2014). Some microbes with strong biofilm-forming capabilities aggregate with other microorganisms to form biofilms, and these microbes can be retained in the biofilm (Rickard et al., 2004), and effectively shield the microbes from harmful toxic substances and environmental stresses to improve the adaptability of microorganisms in the given environment. The biofilm method provides an inexpensive and simple way to immobilize microbes in wastewater treatment systems (Li et al., 2013). Biofilm can provide information for pollution detection and ecological health assessment of water in the aquatic ecosystem (Xuemei et al., 2010; Dar and Bhat, 2020). Bioremediation of wastewater systems plays an important role in improving water quality for flora and fauna. So, in this chapter, we had tried to compile different biological process, especially microbial biofilm technologies to improve the water quality for human welfare (Sehar and Naz, 2016).

Wastewater is a broad term, including discharge from household wastage, agriculture, pharmaceutical, and other industries and stormwater (Deepthi et al., 2020). The organic pollutant present in wastewater includes heavy metals, detergents, pesticides, fats, and oils. In addition, different types of microorganisms, including bacteria, viruses, protozoa, and helminths, can be present in wastewater (Al-Gheethi et al., 2018). However, there is also presence of basic nutrients such as nitrogen, phosphorous, and ammonia as well as metals and inorganic materials. Several countries have regulations that involve domestic and industrial sludge treatment to reuse it, and also to fulfill water scarcity. Moreover, it is also important to detoxify contaminants from the wastewater prior to discharging into natural water habitats such as rivers and lakes. Microbial bioremediation of toxic contaminants is a cost-effective and unfailing approach for decontamination. Microbial bioremediation makes use of the metabolic activities of biofilm dwelling microorganisms which are responsible for majority of pollutant degradation in natural environments. The commonly used forms of biofilms are static biofilms, particulate biofilms, and flocs that are used in trickling filters, fluidized bed reactors, and in activated sludge processes, respectively (Rajwar et al., 2018).

Microbial biofilms comprise a number of microbes including microalgae (green algae, diatoms, and different species cyanobacteria), bacteria, fungi, and few protozoa (Callow, 2000). Wastewater bioremediation is one of the favorable

Microbe Mediated Remediation of Environmental Contaminants. DOI: https://doi.org/10.1016/B978-0-12-821199-1.00021-3

applications of microbial biofilm systems because they offer a simple energy-efficient technology for the absorption of key nutrients (Miranda et al., 2017). In natural, marine, or freshwater ecosystems, algal biofilms represent an important source of autochthonous carbon for other trophic levels and influence nutrient recycling. The same concept can be applied to wastewater treatment where algal biofilms are utilized for removal of nutrients or heavy metals through plant uptake via nutrient recycling and as a resource of carbon and oxygen for enhanced bacterial activity (Kesaano et al., 2015). Microalgae are also known to be pollutant remediator for a broad range of wastes including domestic, industrial, and agricultural sectors. Besides usual organic and inorganic contaminants such as nitrates, phosphates, ammonium, present in wastewater, microalgal cells can also assimilate or break down the highly persistent molecules such as hydrocarbons, antibiotics, and heavy metals (Delrue et al., 2016). The bacterial biomass is most directly involved in wastewater treatment. They dominate all other groups, such as fungi and algae, and cyanobacteria and also affect the process of mineralization and elimination of organic and inorganic nutrients. It is assumed that their key role in wastewater treatment is the clarification of the effluent. It has been observed that presence of protozoa increases the nitrification rate of cell, probably due to the ability of protozoa to influence bacterial growth (Madoni, 2011). Biofilms can also be formed by various protozoan groups, including naked amoebae, heterotrophic flagellates, testate amoebae, foraminiferans, heliozoans, and ciliates. Most heterotrophic flagellates are often the most abundant protozoans that form biofilms (Arndt et al., 2003). Protozoa are one of the most common components in these man-made ecosystems and play an important role in wastewater management and thought to be responsible for improving the quality of wastewater, as well as maintaining the density of dispersed bacterial populations by predation.

21.2 Biofilms

Biofilms were first described by Antonie van Leeuwenhoek in 1978. Costerton et al. (1999) coined the term biofilm and describe the significance of biofilm. The existence of biofilm started more than 3.8 billion years ago and today ubiquitous on Earth (Denkhaus et al., 2007). We now understand that biofilms are universal, occurring from our kitchen to large industries including aquatic and industrial water systems as well as a large number of environmental habitats and medical devices relevant for public health. Biofilms are microbial communities, composed mainly of some algae and microorganisms including bacteria, fungi, protozoa that accumulate at surfaces of substrata and are surrounded by extracellular polymeric substances (EPSs) matrix (Denkhaus et al., 2007; Kröpfl et al., 2006). In 1973 Characklis indicated that biofilms are not only tenacious but also showed higher resistant to the disinfectants such as chlorine. Biofilms are also known as community-level monitoring systems for detecting the effects of pollutants on aquatic systems because of capability to accumulate pollutants and their characteristic response to major changes in water quality of aquatic system (Kröpfl et al., 2006; Sabater et al., 2007; Radojevic et al., 2019).

21.3 Formation of biofilm

21.3.1 Developmental stages

Biofilm formation is a multistep process starting with the attachment of microbes to a substratum thereafter formation of microbial colony that leads to the formation of three-dimensional structure and finally ending with maturation followed by detachment. Microbial cell exhibits two types of growth mode, that is, planktonic cell or free living and sessile aggregate which are known as the biofilm. Microbial biofilms are an association of microbes in which microbial cells adhere to each other on a surface encased within matrix of EPS produced by microbe themselves (Jamal et al., 2015). Biofilm formation is a highly complex process, in which microbial cells transform from planktonic to sessile mode. The formation of microbial biofilms is dependent upon the expression of specific genes that guide establishment of biofilm. The process of biofilm formation occurs through a chain of events leading to adaptation under diverse nutritional and environmental conditions. This is a multistep process in which microorganisms undergo certain changes after attachment with a surface (Fig. 21.1).

Biofilm formation is a process that consists of a sequence of several steps. It starts with the adsorption of macromolecules and smaller molecules onto substratum. Once the surface is prepared, cells begin to attach. The first stages of biofilm formation are well documented, mostly because acquiring images of microorganisms at this stage of biofilm formation is relatively easy. The adherence of microbe to a surface is followed by the production of slimy adhesive substances, called EPSs. These are predominantly made of polysaccharides and proteins. Although the association of EPS with attached microbe has been well documented in the literature, there is little evidence suggesting that EPS

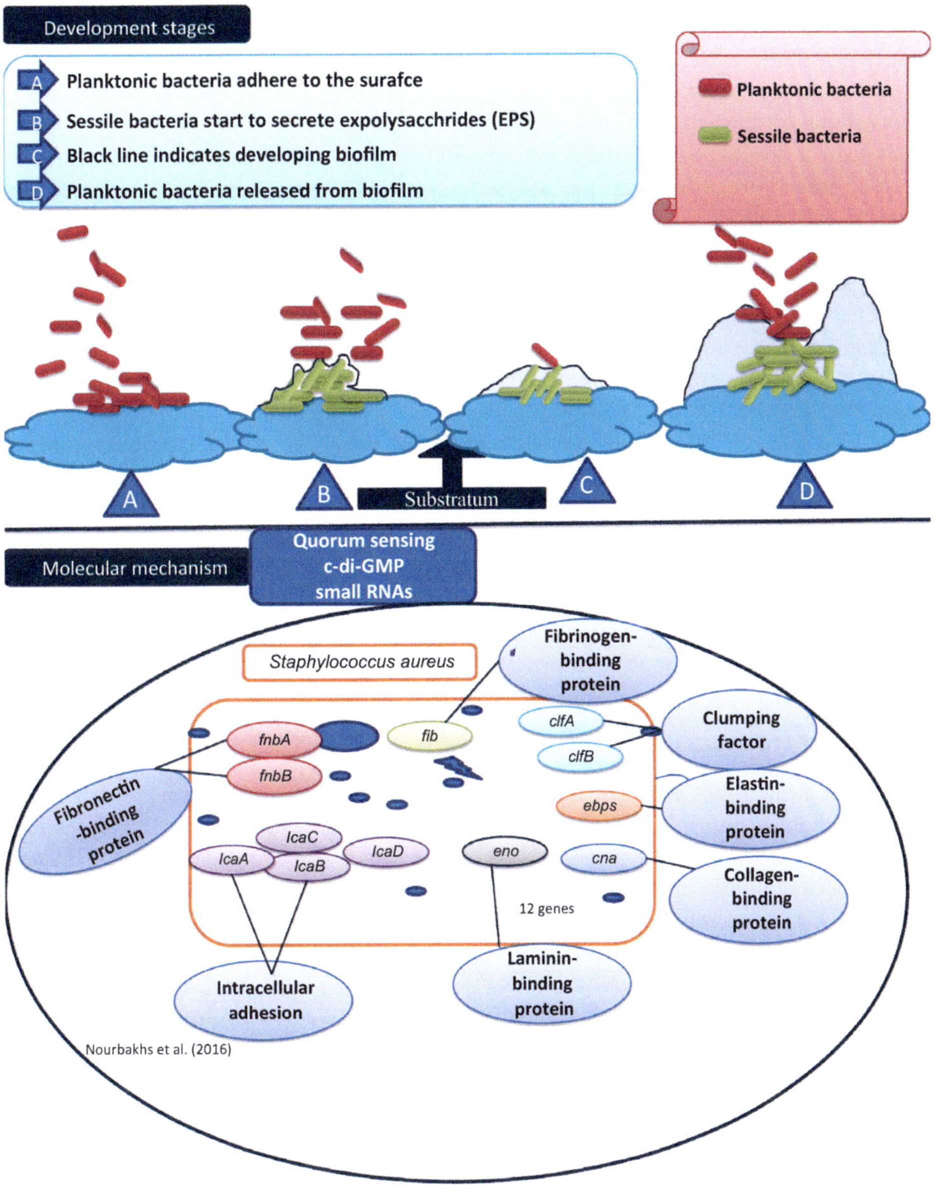

FIGURE 21.1 Schematic representation of biofilm formation by microbes. *Adapted from Nourbakhsh, F., Amirmorteza, Namvar, E., 2016. Detection of genes involved in biofilm formation in Staphylococcus aureus isolates. GMS Hyg. Infect. Control 11, 1–5.*

participates in the initial stages of adhesion. However, EPSs definitely assist the formation of mature biofilms by forming a slimy substance called the biofilm matrix (Limoli et al., 2015).

21.3.2 Genetic mechanism

Biofilm formation and dispersal are highly complex and species-specific process regulated at the genetic level of microorganisms. Quorum-sensing (QS) bis-(3′–5′)-cyclic diguanosine monophosphate (c-di-GMP) and small RNAs (sRNAs) are the main regulators of microbial biofilms. QS is considered as a special language of microbes used for intercellular communication, which is based on small, self-generated signal molecules called autoinducers (de Kievit and Iglewski, 2000; Fazli et al., 2014). This system plays a very important role during the development and dispersal of bacterial biofilms. Although these systems are not involved in the attachment and initial biofilm growth stages, they are required for further biofilm development and also are the main regulators of biofilm dispersal (Davies et al., 1998). c-di-GMP also plays an important role in the bacterial decision between planktonic and biofilm-associated lifestyle (Jenal and Malone,

2006). The factors regulated by c-di-GMP are important for three-dimensional biofilm structure development, such as synthesis of exopolysaccharides, adhesive pili and adhesins, secretion of extracellular DNA, and also for control of cell death and motility.

Small noncoding RNA molecules, sRNAs, including riboswitches, had been shown to participate in posttranscriptional gene regulation in bacteria, involving a range of metabolic processes, adaptation to stress, and microbial pathogenesis (Kang et al., 2014). The molecular mechanism of biofilm formation in methicillin-resistant *Staphylococcus aureus* strains isolated from nosocomial infections in Iran was studied. A total of 110 *S. aureus* strains were collected from three major hospitals in Isfahan. *S. aureus* is not well understood, and only a limited number of studies are available on the molecular genetic mechanism involved in biofilm production. To understand the molecular mechanism of biofilm formation, researchers defined the detection of 12 effective selected genes in biofilm formation, including the *ica ABCD* and other related genes. Because the *icaA* and *icaD* genes are regarded as necessary factors for intercellular adhesion, it could be assumed that these genes are only important for the formation of the bacterial adhesion in biofilm production. However, these 12 genes, in fact associated with both slime and biofilm formation, are *icaA*, *icaB*, *icaC*, *icaD*, *fnbA*, *fnbB*, *fib*, *clfA*, *clfB*, *ebps*, *cna*, and *eno* (Nourbakhsh et al., 2016).

Rai et al. (2019) studied the role of Histone H3 in the biofilm formation by *Candida albicans* a filamentous fungus. It is a biofilm-forming fungal pathogen both under in vitro and in vivo conditions. The transcriptional regulation during the formation of biofilms studied to be controlled by a variant histone H3, designated as $H3V_{CTG}$ because of its exclusive presence in the CTG clade of ascomycetes. A comparative transcriptome analysis of planktonic and biofilm cells revealed that the biofilm circuitry is significantly altered in $H3V_{CTG}$ null cells. $H3V_{CTG}$ binds more efficiently to the promoters of many biofilm-related genes in the planktonic cells than during biofilm growth; authors also demonstrated that cooccurrence of valine and serine at the 31st and 32nd positions in $H3V_{CTG}$, respectively, is essential for its function.

21.4 Visualization of biofilm: microscopic techniques

Microbial biofilms are formed with different microbes and can be seen with naked eye as a layer. So many techniques are developed to visualize the biofilms in high magnification. It requires the use of several types of microscopy including electron microscopy such as SEM, TEM, and some combined with various probes, such as fluorescent in situ hybridization (FISH) probes and fluorescent proteins used as reporter genes. The most popular types of microscopy among biofilm researchers are those which allow screening of living and fully hydrated biofilms. Using these electron microscopic techniques, biofilm researchers now understand that biofilms are not unstructured, homogeneous deposits of cells and accumulated slime, but rather complex communities of surface-associated cells enclosed in a polymer matrix containing open water channels (Donlan and Costerton, 2002). Using these techniques, biofilms can be analyzed for the presence of different types of microbes and their position in the biofilm, as well as their physiological states such as dead, injured, or alive. The in vitro FISH techniques, popular in medical diagnostics, require ribonucleic acid isolation from the sample and separation on a gel, and addition of fluorescent probes to the sample. The various in situ hybridization technique, which is extensively being used in biofilm research, does not require isolating ribonucleic acids prior to the use of the probes; instead, the probes are hybridized with the respective nucleotide sequences inside the cells (Delong et al., 1999; Manz et al., 1999; Biesterfeld et al., 2001; Ito et al., 2002; Jang et al., 2005).

In situ hybridization uses fluorescence-labeled complementary ribonucleic probes, often derived from fragments of DNA that have been isolated, purified, and amplified. In bacterial cells, ribosomal RNA is targeted by fluorescence-labeled oligonucleotide probes. *Leptothrix discophora* a manganese-oxidizing bacteria (MnOB) is identified with a green FISH probe and counterstained with propidium iodide (McNamara et al., 2003; GrayMerod et al., 2005; Nancharaiah et al., 2005). The use of transmission electron microscopy and specific polysaccharide stains like ruthenium red allowed researchers both to identify the nature of these extracellular fibers in biofilms and to better elucidate their association with the cells. Microbial biofilms are imaged with different types of microscopy including confocal laser scanning microscopy (CLSM) and epifluorescence microscopy (Reichhardt and Parsek, 2019). CLSM has restricted magnification, this problem is solved by the use of SEM that provides high-magnification images of single microbe that are located and interact within the biofilm. The use of CLSM and epifluorescence requires that the organisms in the biofilms be stained with fluorescent stains. These stains have characteristic to emit light at specific wavelengths and can be used to probe specific cellular functions. For example, DNA and RNA stains such as DAPI (4,6-diamidino-2-phenylindole), acridine orange, and Syto 9 will stain the ribonucleic acids of all microbial cells regardless of their viability. Other stains have been developed for probing cell viability. Propidium iodide is taken up only by cells with damaged cytoplasmic membranes, and 5-cyano-2,3-ditolyl tetrazolium chloride is taken up and reduced to 5-

cyano-2,3-ditolyl tetrazolium chloride-formazan only by cells that have a functioning cytochrome system. Using a suite of such stains allows the biofilm researcher to quantify all the cells and determine viable ones (Donlan and Costerton, 2002). However, SEM is limited by the requirement of dehydration of the microbial samples during preparation. As microbial biofilms consist mainly of water, the dehydration process of specimen might alter its shape. High magnification yet authentic images are important to understand the physiology of biofilms. However, conventional SEM, when combined with focused ion beam and CLSM techniques, does not require dehydration process. Applying all these combination methods to biofilms of Pseudomonas aeruginosa showed that the dehydration of biofilms substantially influences its appearance and that a more authentic biofilm image emerges (Alhede et al., 2012). Atomic force microscopy (AFM) is also an advanced technology that describes the nanoscale morphology of microbe within biofilm and also characterized specific morphological features of the cell wall. The AFM is a reliable technique for studying bacterial morphology and for examining microbial interactions in biofilm (Germano et al., 2013). Reichhardt and Parsek (2019) highlight microscopic approaches such as CLSM to investigate *P. aeruginosa* bacterial biofilm assembly, matrix composition, and localization. Their studies revealed novel aspects of its composition and function. They additionally described some outstanding questions and how microscopy might be used to identify the functional aspects of bacterial biofilm matrix components.

21.5 Biofilms for remediation of different water pollutants

The microbial biofilm-mediated bioremediation techniques of water pollutants are more promising alternative because of high microbial load and immobilization capacity with greater protective and adhesive proprieties (Edwards and Kjellerup, 2013; Das et al., 2017; Gupta and Diwan, 2017; Shukla et al., 2017). Bioremediation of toxic pollutants by the microbial biofilm organisms is mainly ascribed to the efficient gene transfer within the biofilm species and bacterial chemotaxis, which can induce the bioavailability of pollutants (Singh et al., 2006). Biofilms act as biomonitors and can provide information for pollution detection and ecological health assessment of water (Xuemei et al., 2010).

Anthropogenic activities have introduced heavy metals in large quantities in different water ecosystems. The heavy metals present in water are major contaminants of concern and should be essentially removed from water due to high toxicity to living beings and harmful effects on human and animal health. Remediation of these metals requires special attention to protect water quality. Chemical precipitation and membrane filtration are conventional methods for the removal of toxic metal ions; these methods are expensive and, when treating large amounts of water, generate large quantities of sludge and other toxic products and give low concentration of metal or incomplete metal removal. Biosorption and bioaccumulation are ecofriendly alternatives for metal removal. Microbial biomass, agro-wastes, and industrial byproducts suggesting as potential biosorbents for heavy metal removal due to the presence of metal-binding functional groups (Kanamarlapudi et al., 2018). The general mechanisms for metal ion removal by a microbial biofilm are very complex due to various transfers and many reactions; nowadays, it is the most popular pollutant removal method among all (Masindi and Muedi, 2018). Bioaccumulation can be developed for the removal of heavy metal ions. It is a natural biological phenomenon where microbes use proteins to uptake and sequester heavy metal ions in the intracellular space to utilize in cellular processes such as enzyme catalysis, signaling, stabilizing charges on biomolecules (Diep et al., 2018).

The biofilm is mechanically stable and homogeneous at the surface of the particle; the EPS synthesized by biofilm plays a major role in the accumulation of metal ions. It can act as a molecular sieve, sequestering cations, anions, apolar compounds, and particles from the water phase. Microbial biofilms for the remediation of toxic pollutants are well known due to their tolerance capacity towards environmental stress conditions. Heavy metals such as Zn, Cd, Ni, Cr, Cu, and dyes bind to cell walls of bacteria in activated sludge, whereas hydrophobic compounds such as benzene, toluene, and xylene are present in the matrix (Wuertz et al., 2001; Flemming and Wingender, 2003; Manobala et al., 2019). Microbial biofilms have been utilized for the last decade to neutralize, degrade, and mineralize many toxic compounds in wastewater-activated sludge (Byrns, 2001; Bertin et al., 2007; Choudhary et al., 2020).

21.5.1 Arsenic

Arsenic in drinking water is reported as the most widespread contaminant in water ecosystem worldwide and considered as one of the most toxic elements on the Earth for human beings. Groundwater contamination is a global concern and arsenic-associated human health issues have now been recognized in many developing countries. A wide variety of adverse health effects, including severe skin lesions, cardiovascular and hematological effects, and neurological disturbances, have been attributed to chronic arsenic exposure, primarily from drinking water (Palma-Laraa et al., 2020).

Furthermore, several epidemiological studies have confirmed that chronic arsenic poisoning causes skin and internal cancers. The World Health Organization (WHO) estimates that over 200 million people worldwide rely on water sources that are contaminated with arsenic above 10 ppb as the quality standard for drinking water. Mondal et al. (2008) utilized cells of *Ralstonia eutropha* immobilized on granular, activated carbon bed in a column reactor to remove arsenic from a synthetic industrial effluent. After an initial stage of adaptation and biofilm formation, the cells were able to capture As(III) and As(V). In the presence of arsenic, bacteria such as *Herminiimonas arsenicoxydans* produce large amounts of exopolymers that can be used to detoxify water pollutants with arsenic (Muller et al., 2007). Michel et al. (2007) stated that the CAsO1 bacterial consortium and pure culture of *Thiomonas arsenivorans* isolated from an arsenic-bearing waste dump, for the development of As(III)-oxidizing bioreactors (Battaglia-Brunet et al., 2002, 2005, 2006b). Biofilm development by the bacterial consortium and pure culture of *T. arsenivorans* was studied both on polystyrene microplates and on pozzolana. The As(III)-oxidase activity of *T. arsenivorans* was higher for planktonic cells than for attached ones and was induced by As(III). The results suggested that the membrane of attached bacteria is less permeable than that of planktonic cells, thus decreasing the amount of As(III) entering the cells. Biofilm structure decreases the access of As(III) inside the biofilm, and thus decreases the induction of As(III)-oxidase synthesis by As(III). Carcinogenic effects of arsenic through consumption of polluted water are an alarming threat for us. Bacterial biofilms produce polyanionic EPS that is considered an excellent absorbent material for the remediation of heavy metals and metalloids such as Arsenic. Two biofilm forming and arsenic reducer bacterial strains *Exiguobacterium profundum* PT2 and *Ochrobactrum ciceri* SW1, isolated from industrial waste, exhibited enhanced EPS production in the presence of arsenic. SEM analysis of extracted EPS revealed alterations in the structure such as the presence of closely embedded large clumps with interstitial spaces between stacked layers of the EPS of *E. profundum* PT2 strain treated with Arsenic displayed the enhanced polysaccharide content and arsenic absorption. The results revealed increased production of large number of polyanionic functional groups bacterial EPS, having capacity to sequester arsenic through electrostatic or covalent interactions. So bacterial EPS is an excellent absorbent material for arsenic bioremediation (Saba et al., 2019).

21.5.2 Cadmium

Cadmium (Cd) is a nonessential and harmful element to humans. Cadmium contamination is a serious issue for human health, especially in densely populated agroecology areas (Cai et al., 2020). Due to industrial contamination, it is present ubiquitously in the ecosystem and reported to have carcinogenic characteristics. It is a potentially toxic heavy metal and is a widespread environmental contaminant. It is known to cause renal dysfunction, hepatic toxicity, genotoxicity, and apoptotic effects depending on the dose, route, and duration of exposure. (Congiu et al., 2000; Novelli et al., 2000; WHO, 2006; Saluja et al., 2011; Nemmiche, 2017). The harmful effects of cadmium include a number of acute and chronic disorders, such as "itai-itai" disease, emphysema, hypertension, and testicular atrophy (Leyva-Ramos et al., 1997). WHO provided the drinking water guideline value as 0.005 mg Cd/L. Low concentration of cadmium (less than 5 mg/L) is difficult to treat economically using chemical precipitation methodologies. The microbial biofilm helps in remediation of cadmium from the water bodies, a biofilm-forming marine bacterium *P. aeruginosa* JP-11 was isolated from coastal region at Odisha, East Coast, India; this strain possesses *czc ABC* genes for cadmium resistance as reported in other *Pseudomonas* strains to efflux out the cadmium (Zeng et al., 2012). This gene encodes proteins expressing ion transporters by which metal ions can be pumped out of cytoplasm (Nies, 1992). The study provides information about cadmium resistance though efflux mechanism as well as removal of cadmium by binding to its biofilm (Chakraborty and Das, 2014). A bacterial strain *Enterococcus faecalis* LZ-11 isolated from the petrochemical wastewater discharge near Yellow River was able to resist and absorb cadmium (Wu et al., 2014). *Cupriavidus metallidurans* CH34 is a heavy metal-resistant bacterial strain and can tolerate millimolar concentrations of heavy metal concentrations. Biofilm formation by this bacteria is mediated by the second messenger bis-(30−50)-cyclic dimeric guanosine monophosphate (c-di-GMP). This strain forming biofilm has also been used for bioremediation, recovery, and reduction of heavy metals (Mergeay et al., 1985; Collard et al., 1994; Janssen et al., 2010). A high number of CH34 gene products confer tolerance to high concentrations of heavy metals ions, including the CzcCBA, CzcD, CzcP, and CadA proteins that are involved in cadmium resistance (Nies, 2016; Alviz-Gazitua et al., 2019).

21.5.3 Iron and manganese

The presence of high concentrations of iron and manganese ions in groundwater used as drinking water cause a negative impact causing neurotoxicity in human being. These ions cause an unpleasant metallic taste (Guzinis et al., 1998; Roth,

2006; Tekerlekopoulou and Vayenas, 2007). The concentrations of iron and manganese in drinking water should be 0.3 and 0.1 mg/L, respectively (SEPA State Environmental Protection Administration, 2002). The presence of iron (Fe) and manganese (Mn) ions in groundwater is an important concern for living beings that use it as source of drinking water. These metals may cause aesthetic and organoleptic problems that affect water quality. Biological sand filter systems are widely used to remove Fe and Mn from groundwater since they are a cost-effective technique and minimize the use of chemical oxidants. The bacterial biofilm of two biological water treatment plants from Argentina, exposed to long-term presence of manganese and with high Mn(II) removal efficiency, was characterized using 16S rRNA gene sequencing. Several different media were used to culture MnOB and a large number of known MnOB and several isolates that have never been reported before as MnOB were cultivated. These MnOB were characterized to select those with the highest Mn(II) oxidation and biofilm formation capacities and also those that can oxidize Mn(II) at different environmental growth conditions. These studies were performed to determine if the selected MnOB were able to oxidize Mn(II) present in groundwater while immobilized on sand. This work allowed the isolation of several bacterial strains adequate to develop an inoculums applicable to improve Mn(II) removal efficiency of water treatment plants (Piazza et al., 2019).

Three bacterial strains with high Fe and Mn ion oxidizing capabilities were isolated from groundwater well sludge and identified as *Acinetobacter* sp., *Bacillus megaterium*, and *Sphingobacterium* sp. The maximum removal ratios of Fe and Mn were 99.75% and 96.69%, respectively, and were obtained by an optimal combination of the bacteria at a temperature of 20.15°C, pH 7.09, and at inoculum size of 2.08%. Four biofilters (R1–R4) were simultaneously used for the removal of iron and manganese ions from groundwater. The results of these strains revealed that the Fe and Mn removal ratios of biofilter R4 were maximum, which were inoculated with iron-oxidizing bacteria and MnOB and a biofilm-forming bacterium, were approximately 95% for each metal during continuous operation, and were better than the other biofilters. This study revealed that the biofilm-forming bacterium could promote the immobilization of the iron-oxidizing bacteria and MnOB on the biofilters and may enhance the removal efficiency of Fe and Mn ions from groundwater (Li et al., 2016).

21.5.4 Copper and mercury

Copper is also an indispensable trace element in the life activities of several organisms. However, it will be harmful to the life forms and become a typical heavy metal pollutant in water ecosystem, when it reaches a certain concentration. Recently, microbial biofilm has become one of the research hotspots in dealing with heavy metal contamination because it is the composite solid phase medium that can effectively combine heavy metals (Yin et al., 2018). In recent years, the river contamination in China is a serious problem, which led to degradation of the river ecosystems and deterioration of water quality. Heavy metal contamination occupied a large proportion (Chen et al., 2013). Tien and Chen (2013) indicated the extent of contamination of water by investigating the accumulation of heavy metals in the river biofilm. Huang and Cong (2002) investigated that pH had a significant effect on the adsorption of biofilms. It indicated that the adsorption capacity of Cu by biofilms reached the maximum and biofilm also has the ability of purification. Dong et al. (2011) investigated that the cultured biofilm had obvious adsorption on heavy metals such as Copper and Lead in water.

An aquatic fungus *Mucor hiemalis* EH8 has the ability to remove toxic mercury ions from water by intracellular accumulation and reduction into elemental mercury. This strain was isolated from a microbial biofilm grown in sulfidic-reducing spring water sourced at a Marching's site located downhill from hop cultivation areas in Germany. A biodiversity survey and mercury-removal function analyses were undertaken in Bavaria (Germany) in an area of about 200 km^2 to find the role of biofilm and fungal microbe for mercury removal. After a systematic search using metal removal assays, researchers identified Marching spring's biofilm out of 18 different sulfidic springs. Fungus *M. hiemalis* EH8 formed biofilms is the only one that was capable of removing mercury ions from water. They removed approximately 99% of mercury ions from water within 10–48 hours after initial exposure to Hg(II). Scanning electron microscopic analysis revealed the occurrence of mercury in intracellular region of germinating sporangiospores exposed to mercury. Mercury was also found to be released and deposited as metallic-shiny nanospheres. Electron-dispersive X-ray analysis also confirmed nanosphere presence of mercury. *Mucor hiemalis* EH8 strain was, a first aquatic fungus has been found that is able to grow under sulfur-reducing conditions even at low temperature with promising performance in removal of Mercury ions to protect our environment, especially water system from Mercury pollution (Hoque and Fritscher, 2016).

21.5.5 Zinc and chromium

Zinc is a trace element; it serves as an enzyme activator in humans, but it can be equally toxic when present at high concentration (Volesky and Holan, 1995). Many conventional methods for chromium and zinc removal include carbon adsorption, chemical precipitation, electrochemical treatment, electro-dialysis, electrocoagulation, electro-winning, evaporation, filtration, ion exchange, membrane separation, oxidation–reduction processes, reverse osmosis, and solvent extraction (Bailey et al., 1999; Agrawal et al., 2006; Ahluwalia and Goyal, 2007). The *Escherichia coli* and *Staphylococcus epidermidis* strains biofilms are helpful in the removal of Cr(VI) and Zn(II) from aqueous solution. The adsorptive capacity of Gram-negative *E. coli* is higher than the Gram-positive *S. epidermidis* in the removal of chromium and zinc from aqueous solution (Quiton et al., 2018). The walls of Gram-positive bacteria are efficient metal chelators, and in *Bacillus* sp. the carboxyl group of glutamic acid present in peptidoglycan is the major binding site of metal deposition. Teichoic and teichuronic acids are important binding sites in *Bacillus* spp. A bacterial strain *Bacillus circulans* isolated from Anzali wetland, Iran, formed biofilm facilitating in significant removal of lead and chromium ions from aqueous solutions (Ilhan et al., 2004; Khanafari et al., 2008).

21.5.6 Organic solvents

Organic solvents as contaminants may be also found in the water ecosystem due to discharges from industrial waste. They cause severe impact on human health and it should be eliminated from water bodies with the help of microbial biofilm. A colorless organic liquid, acetonitrile is infinitely soluble in water. Because of their versatile chemical reactivity, organonitriles have many industrial uses, such as in the manufacture of plastics, synthetic fibers, and elastomers (Pozzani et al., 1959). Because of their aqueous nature, it also causes water pollution when present above standard level. Some organonitrile-degrading bacteria via the addition of biofilm-forming bacteria represent a promising technology for the treatment of organonitrile-containing wastewater. A biofilm-forming bacteria *Bacillus subtilis* N4 mixed with degrading bacteria may enhance the biodegradation efficiency. Nitrile hydratase and amidase genes, which play critical roles in organonitriles degradation, were cloned and transformed into the biofilm forming bacterial strain *B. subtilis* N4 to construct a recombinant bacterium *B. subtilis* N4/pHTnha-ami. The successful application of the immobilized recombinant bacterium offers a novel strategy for the biodegradation of organonitriles compounds (Li et al., 2018). Phenol and its derivatives are common water contaminants and include wide variety of organic chemicals. Phenol poisoning can occur by adsorption on skin inhalation, ingestion, and various different methods which can cause bad health effects. High exposures to phenol may be fatal (Pradeep et al., 2015). A bacterium *Pseudomonas stutzeri* OX1 formed biofilm isolated from the activated sludge of a wastewater treatment plant and is endowed with peculiar metabolic capabilities for the degradation of aromatic hydrocarbons such as phenols (Baggi et al., 1987; Viggiani et al., 2006) Fig. 21.2.

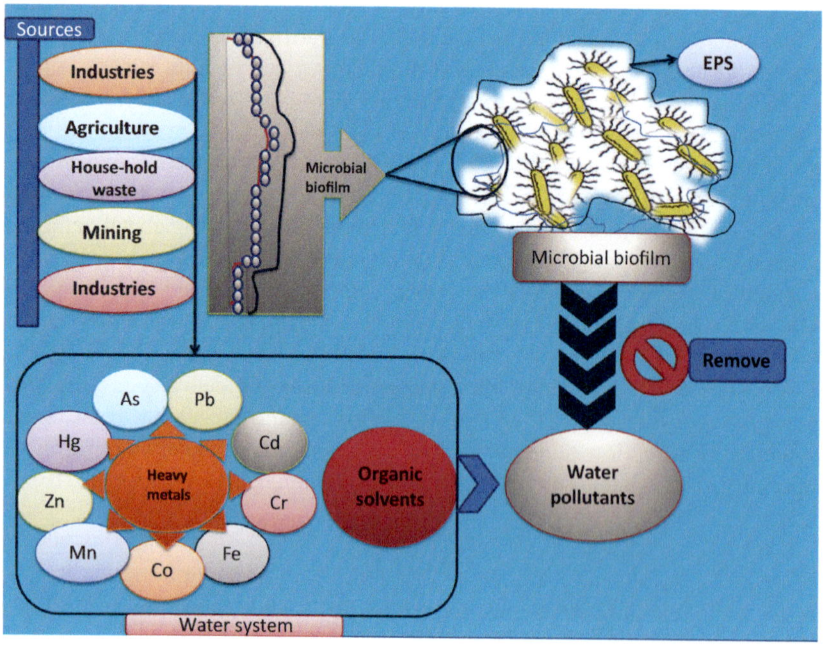

FIGURE 21.2 Graphic representation showing implications of biofilm bioremediation.

21.6 Fungal biofilm for remediation of wastewater

Fungal biofilms play an important role in water ecosystem. Biofilm formation by microbes, particularly fungi has been widely studied in last decade, and several species among filamentous, yeast, and dimorphic fungi have been described as capable of developing into microbial biofilm (Costa-Orlandi et al., 2017). Many important fungi produce biofilms, including *Candida*, *Aspergillus*, *Cryptococcus*, *Trichosporon*, *Coccidioides*, and *Pneumocystis* (Fanning and Mitchell, 2012).

Textile industries wastage is the most polluting waste as water pollutants and their treatment is greatly challenging for their safe discard. Microbial biofilm have potential ability to decolorize synthetic commercial dyes used for textile dyeing. A potential dye degrading microbial biofilms was developed from endophytic fungi and soil bacteria. Endophytic fungi were isolated from the leaves of *Eleusine indica* (Linn) plant and bacteria were isolated from soil samples obtained from textile effluent dumping site in Biyagama Industrial area, Sri Lanka. Biofilms were developed after screening the fungal and bacterial isolates with Malachite green and Nigrosin disodium dyes separately and the decolorization assay was performed for biofilms along with monocultures to evaluate their ability for dye decolorization. The highest decolorization percentages were observed by fungal and bacterial strains *Trichoderma harzianum* (F2) and *B. subtilis* (B1), respectively, for both dyes. All the biofilm combinations showed higher decolorization percentage than that of the single cultures. This study revealed that the microbial biofilms can be used as an efficient biological tool for textile effluent treatment and removal of water pollutants (Henagamage, 2019). The current challenge in intensive water ecosystem is to control the level of pollutants in the wastewater and to provide sustainable sources of proteins for feed. A synthetic lichen biofilm was developed with a fungus *Mucor indicus* and microalgae *Chlorella vulgaris*, when grown together on a polymer matrix referred to as a mycoalgae biofilm (Rajendran and Hu, 2016). When mycoalgae biofilm grows, it takes up phosphorus and nitrogen compounds and converts them to proteins and other cellular products. They clean the water from nutrient pollutants as the algae are attached to the biofilm leaving purified water at the end of the process. This process allows for easy harvesting of the mycoalgae with no energy-intensive process of separating the algae from the supernatant. The generated biofilm was composed of two organisms that have been shown to positively aid aquatic organism sheath when included as a feed supplement (Barnharst et al., 2018). A fungus *Coriolopsis* sp. (1c3) was cultured and observed to form filamentous biofilm. The biofilm of 1c3 isolate and the free-mycelium forms (control) were then used to treat two triphenylmethane dyes—Cotton Blue (CB) and Crystal Violet (CV). The results revealed that filamentous biofilm of *Coriolopsis* sp. 1c3 showed significant decolorization activities for Cotton Blue and Crystal Violet compared as compared to free-mycelium control. Biodegradation occurred in both biofilm and free-mycelium forms. The results suggested that biofilm of *Coriolopsis* sp. 1c3 was leading to enhanced decolorization activities. This technology is simple, feasible and can be adopted and further improved to obtain biofilm to enhance their dye removal efficiency (Fig. 21.3) in water habitats (Munck et al., 2018) (Tables 21.1 and 21.2).

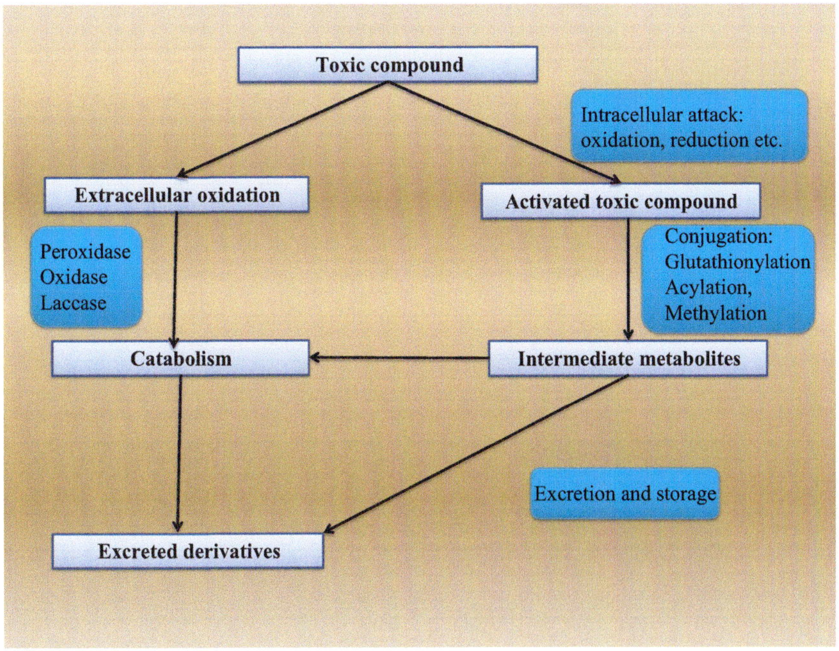

FIGURE 21.3 Bioremediation of toxic compounds by biofilm.

TABLE 21.1 Biofilms have some major characteristics that are important for bioremediation Edwards and Kjellerup, 2013.

Property	Characteristic
Tolerance toward environmental stressors (e.g., toxic chemicals, pH change, predation, and dehydration)	Extensive genetic diversification of biofilm bacteria
Communication (quorum sensing)	Critical cell density for biofilm formation
Exchange of genetic material	Horizontal transfer of genetic material between species, DNA sharing
Metabolic diversity and symbiosis	Utilizing waste products and/or accumulated products from the environment or other microorganisms
Different redox and electron acceptor diversity	Different metabolic functions with respect to electron-acceptor reduction
Porous physical structure with water channels	Allow for transfer of nutrients, electron acceptors, waste products
Surfactants	Aid in solubilizing hydrophobic or recalcitrant substrates
Varying growth rates in the biofilm	Inducing biofilm persistence due to different metabolic states in the biofilm
Microcolony and gradient formation	Redox potential and nutrient cycling because of aerobic and anaerobic processes

Source: Edwards, S.J., Kjellerup, B.V., 2013. Applications of biofilms in bioremediation and biotransformation of persistent organic pollutants, pharmaceuticals/personal care products, and heavy metals. Appl. Microbiol. Biotechnol. 97, 9909–9921.

TABLE 21.2 Microbes involved in the bioremediation of water pollutants.

S. no.	Name	Type of microbe	Pollutants	Reference
1.	*Pseudomonas*	Bacteria	Lead and Zinc	Meliani and Bensoltane (2016)
2.	*Aspergillus flavus* complex and *Bacillus cereus*	Bacteria	Hydrocarbons	Perera et al. (2019)
3.	*Pseudomonas* sp., *Sphaerotilus* sp., *Janthinobacterium* sp., *Corynebacterium aurimucosum*	Bacteria	Ammonium ions	Xu et al. (2012)
4	*Cupriavidus* sp. PDN31	Bacteria	Ammonium, nitrate, and nitrite	Srivastava and Mishra (2018)
5.	EDB-L11(*Novosphingobium* JEM-1)	Bacteria	17β-estradiol (E2)	Iasur-Kruh et al. (2011)
6.	*Pseudomonas aeruginosa*	Bacteria	Cadmium	Chellaiah (2018)
7.	*Cupriavidus metallidurans* CH34	Bacteria	Cadmium	Alviz-Gazitua et al. (2019)
8.	*Escherichia coli, Staphylococcus epidermidis*	Bacteria	Zinc(II)	Quiton et al. (2018)
9.	*S. epidermidis*	Bacteria	Chromium (VI)	Quiton et al. (2018)
10.	*Acinetobacter* sp. *Bacillus megaterium* and *Sphingobacterium* sp.	Bacteria	Iron Manganese	Li et al. (2016)
11.	*Rhodococcus rhodochrous* BX2	Bacteria	Acetonitrile	Li et al. (2013)
12.	*B. cereus* P1 (KP241859)	Bacteria	Iron Maganese	Li et al. (2016)

(*Continued*)

TABLE 21.2 (Continued)

S. no.	Name	Type of microbe	Pollutants	Reference
13.	*Bacillus subtilis* (N4-pHT01-nit)	Recombinant Bacteria	Acetonitrile	Li et al. (2016b)
14.	*B. subtilis* N4/pHTnha-ami	Recombinant Bacteria	Organonitrile	Li et al. (2018)
15.	*Aspergillus niger*	Fungi	Pb(II), Cd(II)	Tsekova et al. (2007)
16.	*Penicillium simplicissimum*	Fungi	Pb(II)	Fan et al. (2008)
17.	*Aspergillus fumigatus*	Fungi	Cr(VI)	Dhal et al. (2018)
18.	*Termitomyces clypeatus*	Fungi	Cr(VI)	Ramrakhiani and Khowala (2012)
19.	*Penicillium brevicompactum*	Fungi	Co(II)	Tsekova et al. (2007)
20.	*Saccharomyces cerevisiae*	Fungi	Cu(II)	Amirnia et al. (2015)
21.	*Trichoderma*	Fungi	Cd(II)	Bazrafshan et al. (2016)
22.	*Candida humicola*	Fungi	Arsenic	Lovely (2001)
23.	*Mucor hiemalis* EH8	Fungi	Mercury	Hoque and Fritscher (2016)

21.7 Conclusion and future perspective

In the last century, industrial processes have been creating waste all around the world, resulting in contaminated sediments and subsequent, far-reaching dispersal into water ecosystem including river, pond, groundwater, marine. The pollutants have accumulated and are still being created a major problem for us, Biofilm system is a well-developed, cost-effective, and ecofriendly technology in which different microbes such as bacteria, fungi, and microalgae form a colony. Microbial biofilm—based systems for wastewater treatment are rapidly expanding research area. Implementation of biofilm claimed beneficial thanks to the ease and simplicity of the operation compared to activated sludge processes. It has also shown high resistance to toxic pollutants and metals. This technology requires further works in large-scale wastewater researches. Microbial communities are capable of detoxifying organic pollutants and oxidizing heavy metal contaminants. Microbial biofilm provides, among other things a beneficial structure, possibility for nutrient, and genetic exchange to participating microorganisms as well as protection from the surrounding environment concerning for instance predation and chemical and shear stresses. Overall, the biofilms may be used for the bioremediation of wastewater pollutants.

Acknowledgment

The author (Monika Singh) thanks DST (Department of Science and Technology)-Inspire, New Delhi, for fellowship in the form of JRF/SRF.

References

Agrawal, A., Kumar, V., Pandey, B.D., 2006. Remediation options for the treatment of electroplating and leather tanning effluent containing chromium a review. Min. Proc. Ext. Met. Rev. 27, 99–130.

Ahluwalia, S.S., Goyal, D., 2007. Microbial and plant derived biomass for removal of heavy metals from wastewater. Bioresour. Technol. 98, 2243–2257.

Al-Gheethi, A.A., Efaq, A.N., Bala, J.D., Norli, I., Abdel-Monem, M.O., Ab Kadir, M.O., 2018. Removal of pathogenic bacteria from sewage-treated effluent and biosolids for agricultural purposes. Appl. Water Sci. 8, 1–25.

Alhede, M., Qvortrup, K., Liebrechts, R., Hoiby, N., Givskov, M., Bjarnsholt, T., 2012. Combination of microscopic techniques reveals a comprehensive visual impression of biofilm structure and composition. FEMS Immunol. Med. Microbiol. 65, 335–342.

Alviz-Gazitua, P., Fuentes-Alburquenque, S., Rojas, L.A., Turner, R.J., Guiliani, N., Seeger, M., 2019. The response of *Cupriavidus metallidurans* CH34 to cadmium involves inhibition of the initiation of biofilm formation, decrease in intracellular c-di-GMP levels, and a novel metal regulated phosphodiesterase. Front. Microbiol. 10, 1499.

Amirnia, S., Ray, M.B., Margaritis, A., 2015. Heavy metals removal from aqueous solutions using *Saccharomyces cerevisiae* in a novel continuous bioreactor–biosorption system. Chem. Eng. J. 264, 863–872.

Arndt, H., Schmidt-Denter, K., Auer, B., Weitere, M., 2003. Protozoans and biofilms. In: Krumbein, W.E., Paterson, D.M., Zavarzin, G.A. (Eds.), Fossil and Recent Biofilms. Springer, Dordrecht, pp. 161–179.

Ateia, M., 2016. In-situ biological water treatment technologies for environmental remediation: a review. J. Bioremed. Biodegrad. 07, 3–7.

Baggi, G., Barbieri, P., Galli, E., Tollari, S., 1987. Isolation of a *Pseudomonas stutzeri* strain that degrades *o*-xylene. Appl. Environ. Microbiol. 53, 2129–2132.

Bailey, S.E., Olin, T.J., Bricka, R.M., Adrian, D.D., 1999. A review of potentially low-cost sorbents for heavy metals. Water Res. 33, 2469–2479.

Barnharst, T., Rajendran, A., Hu, B., 2018. Bioremediation of synthetic intensive aquaculture wastewater by a novel feed-grade composite biofilm. Int. Biodeterior. Biodegrad. 126, 131–142.

Battaglia-Brunet, F., Dictor, M.C., Garrido, F., Crouzet, C., Morin, D., Dekeyser, K., et al., 2002. An As(III)-oxidizing bacterial population: selection, characterization, and performance in bioreactors. J. Appl. Microbiol. 93, 656–667.

Battaglia-Brunet, F., Dictor M.C., Garrido F., Michel C., Joulian C., Bourgeois F., et al., 2005. Remediation processes using biological As(III) oxidation. In: Proceedings of the 16th International Biohydrometallurgy Symposium, Cape Town, pp. 343–350.

Battaglia-Brunet, F., Itard, Y., Garrido, F., Delorme, F., Crouzet, C., Greffié, C., et al., 2006b. A simple biogeochemical process removing arsenic from a mine drainage water. Geomicrobiol. J. 23, 201–211.

Bazrafshan, E., Zarei, A.A., Mostafapour, F.K., 2016. Biosorption of cadmium from aqueous solutions by *Trichoderma* fungus: kinetic, thermodynamic, and equilibrium study. Desalin. Water Treat. 57, 14598–14608.

Bertin, L., Capodicasa, S., Occulti, F., Girotti, S., Marchetti, L., Fava, F., 2007. Microbial processes associated to the decontamination and detoxification of a polluted activated sludge during its anaerobic stabilization. Water Res. 41, 2407–2416.

Biesterfeld, S., Figueroa, L., Hernandez, M., Russell, P., 2001. Quantification of nitrifying bacterial populations in a full-scale nitrifying trickling filter using fluorescent in situ hybridization. Water Environ. Res. 73, 329–338.

Byrns, G., 2001. The fate of xenobiotic organic compounds in wastewater treatment plants. Water Res. 35, 2523–2533.

Cai, Y., Yin, Y., Li, Y., Guan, L., Zhang, P., Qin, Y., Li, Y et al., 2020. Cadmium exposure affects growth performance, energy metabolism, and neuropeptide expression in *Carassius auratus gibelio*, Fish Physiol. Biochem. 46, 187–197.

Callow, M.E., 2000. Algal biofilms. In: Evans, L.V. (Ed.), Biofilms: Recent Advances in Their Study and Control. Harwood Academic Publishers, Amsterdam, pp. 189–204.

Chakraborty, J., Das, S., 2014. Characterization and cadmium-resistant gene expression of biofilm-forming marine bacterium *Pseudomonas aeruginosa* JP-11. Environ. Sci. Pollut. Res. 21, 14188–14201.

Chellaiah, E.R., 2018. Cadmium (heavy metals) bioremediation by *Pseudomonas aeruginosa*: a mini review. Appl. Water Sci. 8, 154.

Chen, C., Zheng, B., Jiang, X., Zhao, Z., Zhan, Y., Yi, F., et al., 2013. Spatial distribution and pollution assessment of mercury in sediments of Lake Taihu, China. J. Environ. Sci. 25, 316–325.

Choudhary, P., Singh, S., Agarwal, V., 2020. Microb. Biofilms 1–14. <https://doi.org/10.5772/intechopen.90790>.

Collard, J.M., Corbisier, P., Diels, L., Dong, Q., Jeanthon, C., Mergeay, M., et al., 1994. Plasmids for heavy metal resistance in *Alcaligenes eutrophus* CH34: mechanisms and applications. FEMS Microbiol. Rev. 14, 405–414.

Congiu, L., Chicca, M., Pilastro, A., Turchetto, M., Tallandini, L., 2000. Effects of chronic dietary cadmium on hepatic glutathione levels and glutathione peroxidase activity in starlings (*Sturnus vulgaris*). Arch. Environ. Contam. Toxicol. 38, 357–361.

Costa-Orlandi, C.B., Sardi, J.C.O., Pitangui, N.S., de Oliveira, H.C., Scorzoni, L., Galeane, M.C., et al., 2017. Fungal biofilms and polymicrobial diseases. J. Fungi 3, 1–24.

Costerton, J., et al., 1999. Bacterial biofilms: a common cause of persistent infections. Science 284, 1318–1322.

Dar, S.A., Bhat, R.A., 2020. Aquatic solution stress and role of biofilms as environment cleanup technology. In: Qadri, H., et al., (Eds.), Fresh Water Pollution Dynamics and Remediation. Springer Nature Singapore Pte Ltd, pp. 293–318.

Das, N., Basak, L.V.G., Salam, J.A., 2017. Application of biofilms on remediation of pollutants—an overview. J. Microbiol. Biotechnol. Res. 2, 783–790.

Davies, D.G., Parsek, M.R., Pearson, J.P., Iglewski, B.H., Costerton, J.W., Greenberg, E.P., 1998. The involvement of cell-to-cell signals in the development of a bacterial biofilm. Science 280, 295–298.

Deepthi, D., Lakshmi, V.S., Babu, M.C., 2020. Bioremediation of wastewater using invasive bivalves. IJITEE 9, 3077–3079.

de Kievit, T.R., Iglewski, B.H., 2000. Bacterial quorum sensing in pathogenic relationships. Infect. Immun. 68, 4839–4849.

Delong, E.F., Taylor, L.T., Marsh, T.L., Preston, C.M., 1999. Visualization and enumeration of marine planktonic archaea and bacteria by using polyribonucleotide probes and fluorescent in situ hybridization. Appl. Environ. Microbiol. 65, 5554–5563.

Delrue, F., Alvarez-Diaz, P.D., Fon-Sing, S., Fleury, G., Sassi, J.F., 2016. The environmental biorefinery: Using microalgae to remediate wastewater, a win-win paradigm. Energies 9, 1–19.

Denkhaus, E., Meisen, S., Telgheder, U., Wingender, J., 2007. Chemical and physical methods for characterisation of biofilms. Microchim. Acta 158, 1–27.

Dhal, B., Abhilash, Pandey, B.D., 2018. Mechanism elucidation and adsorbent characterization for removal of Cr(VI) by native fungal adsorbent. Sustain. Environ. Res. 28, 289–297.

Diep, P., Mahadevan, R., Yakunin, A.F., 2018. Heavy metal removal by bioaccumulation using genetically engineered microorganisms. Front. Bioeng. Biotechnol. 6, 157. Available from: https://doi.org/10.3389/fbioe.2018.0015.

Dong, D., Guo, Z., Hua, X., Lan, Y., Zhou, J., Ding, X., et al., 2011. Sorption of DDTs on biofilms, suspended particles and river sediments: effects of heavy metals. Environ. Chem. Lett. 9, 361–367.

Donlan, R.M., Costerton, J.W., 2002. Biofilms: survival mechanisms of clinically relevant microorganisms. Clinical. Microbiol. Rev. 15, 167–193.

Douterelo, I., Husband, S., Boxall, J.B., 2014. The bacteriological composition of biomass recovered by flushing an operational drinking water distribution system. Water Res. 54, 100–114.

Edwards, S.J., Kjellerup, B.V., 2013. Applications of biofilms in bioremediation and biotransformation of persistent organic pollutants, pharmaceuticals/personal care products, and heavy metals. Appl. Microbiol. Biotechnol. 97, 9909–9921.

Fan, T., Liu, Y., Feng, B., Zeng, G., Yang, C., Zhou, M., et al., 2008. Biosorption of cadmium(II), zinc(II) and lead(II) by *Penicilliumsim plicissimum:* isotherms, kinetics and thermodynamics. J. Hazard. Mater. 160, 655–661.

Fanning, S., Mitchell, A.P., 2012. Fungal biofilms. PLoS Pathog. 8 (4), e1002585. Available from: https://doi.org/10.1371/journal.ppat.1002585.

Fazli, M., Almblad, H., Rybtke, M.L., Givskov, M., Eberl, L., Tolker-Nielsen, T., 2014. Regulation of biofilm formation in *Pseudomonas* and *Burkholderia* species. Environ. Microbiol. 16, 1961–1981.

Flemming, H.C., Wingender, J., 2003. The crucial role of extracellular polymeric substances in biofilms. In: Wuertz, S., Bishop, P., Wilderer, P. (Eds.), Biofilm in Wastewater Treatment. IWA Publishing, London, pp. 623–633.

Germano, F., Bramanti, E., Arcuri, C., Cecchetti, F., Marco Cicciu, M., 2013. Atomic force microscopy of bacteria from periodontal subgingival biofilm: preliminary study results. Eur. J. Dent. 7, 152–158.

GrayMerod, R., Hendrickx, L., Mueller, L.N., Xavier, J.B., Wuertz, S., 2005. Effect of nucleic acid stain Syto9 on nascent biofilm architecture of *Acinetobacter* sp BD413. Water Sci. Technol. 52, 195–202.

Gupta, P., Diwan, B., 2017. Bacterial exopolysaccharide mediated heavy metal removal: a review on biosynthesis, mechanism and remediation strategies. Biotechnol. Rep. (Amst.) 13, 58–71.

Guzinis, A., Kosmidis, N., Vayenas, D.V., Lyberatos, G., 1998. Removal of Mn and simultaneous removal of NH_3, Fe and Mn from potable water using a trickling filter. Water Res. 32, 2442–2450.

Henagamage, A.P., 2019. Bioremediation of textile dyes by fungal-bacterial biofilms. Int. J. Environ. Agric. Biotechnol. 4, 635–642.

Hoque, E., Fritscher, J., 2016. A new mercury-accumulating *Mucor hiemalis* strain from cold sulfidic spring water biofilms. Microbiol. Open 5, 763–789.

Huang, T.L., Cong, H.B., He, W.J., 2002. Determination Method of Aquatic Algae Chlorophyll. China: CN200410073542.8 (in Chinese).

Iasur-Kruh, L., Hadar, Y., Minz, D., 2011. Isolation and bioaugmentation of an estradiol-degrading bacterium and its integration into a mature biofilm. Appl. Environ. Microbiol. 77, 3734–3740.

Igiri, B.E., Okoduwa, S.I.R., Idoko, G.O., Akabuogu, E.P., Adeyi, A.O., Ejiogu, I.K., 2018. Toxicity and bioremediation of heavy metals contaminated ecosystem from tannery wastewater: a review. Hindawi J. Toxicol. 16. Article ID 2568038.

Ilhan, S., Nourbakhsh, M.N., Kilicarslan, S., Ozdag, H., 2004. Removal of chromium, lead and copper ions from industrial wastewaters by *Staphylococcus saprophyticus*. Turk. J. Biotechnol. 2, 50–57.

Ito, T., Nielsen, J.L., Okabe, S., Watanabe, Y., Nielsen, P.H., 2002. Phylogenetic identification and substrate uptake patterns of sulfate-reducing bacteria inhabiting an oxic-anoxic sewer biofilm determined by combining microautoradiography and fluorescent in situ hybridization. Appl. Environ. Microbiol. 68, 356–364.

Jamal, M., Tasnee, U., Hussain, T., Andleeb, S., 2015. Bacterial biofilm: its composition, formation and role in human infections. RRJMB 4, 1–14.

Jang, A., Okabe, S., Watanabe, Y., Kim, I.S., Bishop, P.L., 2005. Measurement of growth rate of ammonia oxidizing bacteria in partially submerged rotating biological contactor by fluorescent in situ hybridization (FISH). J. Environ. Eng. Sci. 4, 413–420.

Janssen, P.J., Van Houdt, R., Moors, H., Monsieurs, P., Morin, N., Michaux, A., et al., 2010. The complete genome sequence of *Cupriavidus metallidurans* strain CH_{34}, a master survivalist in harsh and anthropogenic environments. PLoS One 5, 10433.

Jenal, U., Malone, J., 2006. Mechanisms of cyclic-di-GMP signaling in bacteria. Annu. Rev. Genet. 40, 385–407.

Kanamarlapudi, S.L.R.K., Chintalpudi, V.K., Muddada, S., 2018. Application of biosorption for removal of heavy metals from wastewater. Biosorption . Available from: https://doi.org/10.5772/intechopen.77315.

Kesaano, M., Smith, T., Wood, J., Sims, R.C., 2015. Applications of algal biofilms for wastewater treatment and bioproduct production. In: Singh, B., Bauddh, K., Bux, F. (Eds.), Algae and Environmental Sustainability. Developments in Applied Phycology. Springer, New Delhi, pp. 7, 23–31.

Kang, Z., Zhang, C., Zhang, J., Jin, P., Zhang, J., Du, G., et al., 2014. Small RNA regulators in bacteria: powerful tools for metabolic engineering and synthetic biology. Appl. Microbiol. Biotechnol. 98, 3413–3424.

Khanafari, A., Eshghdoost, S., Mashinchian, A., 2008. Removal of lead and chromium from aqueous solution by *Bacillus circulans* biofilm. Iran. J. Environ. Health Sci. Eng. 5, 195–200.

Kröpfl, K., Vladar, P., Szabo, K., Acs, E., Borsodi, A.K., Szikora, S., et al., 2006. Chemical and biological characterisation of biofilms formed on different substrata in Tisza River (Hungary). Environ. Pollut. 144, 626–631.

Leyva-Ramos, R., et al., 1997. Adsorption of cadmium(II) from aqueous solution on to activated Carbon. Water Sci. Technol. 35, 205–211.

Li, C.Y., Li, Y., Cheng, X.S., Feng, L.P., Xi, C.W., Zhang, Y., 2013. Immobilization of *Rhodococcus rhodochrous* BX2 (an acetonitrile degrading bacterium) with biofilm-forming bacteria for wastewater treatment. Bioresour. Technol. 131, 390–396.

Li, C., Wang, S., Du, X., Cheng, X., Fu, M., Hou, N., et al., 2016. Immobilization of iron- and manganese-oxidizing bacteria with a biofilm-forming bacterium for the effective removal of iron and manganese from groundwater. Bioresour. Technol. 220, 76–84.

Li, C., Yue, Z., Feng, F., Xi, C., Zang, H., An, X., et al., 2016b. A novel strategy for acetonitrile wastewater treatment by using a recombinant bacterium with biofilm-forming and nitrile-degrading capability. Chemosphere 161, 224–232.

Li, C., Yueling, S., Zhenlei, Y., Mingyan, H., Jinming, W., Chen, X., et al., 2018. Combination of a recombinant bacterium with organonitrile-degrading and biofilm-forming capability and a positively charged carrier for organonitriles removal. J. Hazard. Mater. 353, 372–380.

Limoli, D.H., Jones, C.J., Wozniak, D.J., 2015. Bacterial extracellular polysaccharides in biofilm formation and function. Microbiol. Spectr. 3, 1–30.

Lovely, D.R., 2001. Anaerobes to the Rescue. Science 293, 1444–1446.

Madoni, P., 2011. Protozoa in wastewater treatment processes: a mini review. Ital. J. Zool. 78, 3–11.

Manobala, T., Shukla, S.K., Rao, T.S., Kumar, M.D., 2019. Uranium sequestration by biofilm-forming bacteria isolated from marine sediment collected from Southern coastal region of India. Int. Biodeterior. Biodegrad. 145, 104809. Available from: https://doi.org/10.1016/j.ibiod.2019.104809.

Manz, W., Wendt-Potthoff, K., Neu, T.R., Szewzyk, U., Lawrence, J.R., 1999. Phylogenetic composition, spatial structure, and dynamics of lotic bacterial biofilms investigated by fluorescent in situ hybridization and confocal laser scanning microscopy. Microb. Ecol. 37, 225–237.

Masindi, V., Muedi, K.L., 2018. Environmental contamination by heavy metals. Heavy Met. Available from: https://doi.org/10.5772/intechopen.76082.

McNamara, C.J., Lemke, M.J., Leff, L.G., 2003. Underestimation of bacterial numbers in starvation-survival mode using the nucleic acid stain DAPI. Arch. Hydrobiol. 157, 309–319.

Meliani, A., Bensoltane, A., 2016. Biofilm-mediated heavy metals bioremediation in PGPR *Pseudomonas*. J. Biorem. Biodegrad. 7, 370.

Mergeay, M., Nies, D., Schlegel, H.G., Gerits, J., Charles, P., Van Gijsegem, F., 1985. *Alcaligenes eutrophus* CH_{34} is a facultative chemolithotroph with plasmid-bound resistance to heavy metals. J. Bacteriol. 162, 328–334.

Michel, C., Jean, M., Coulon, S., 2007. Biofilms of As(III)-oxidising bacteria: formation and activity studies for bioremediation process development. Appl. Microbiol. Biotechnol. 77, 457–467.

Miranda, A.F., Ramkumar, N., Andriotis, C., Holtkemeier, T., Yasmin, A., Rochfort, S., et al., 2017. Applications of microalgal biofilms for wastewater treatment and bioenergy production. Biotechnol. Biofuel. 10, 1–23.

Mondal, P., Majumder, C.B., Mohanty, B., 2008. Treatment of arsenic contaminated water in a laboratory scale up-flow bio-column reactor. J. Hazard. Mater. 153, 136–145.

Muller, D., Medigue, C., Koechler, S., Barbe, V., Barakat, M., Talla, E., et al., 2007. A tale of two oxidation states: bacterial colonization of arsenic-rich environments. PLoS Genet. 3, 518–530.

Munck, C., Thierry, E., Grable, S., Chen, S.H., Ting, A.S.Y., 2018. Biofilm formation of filamentous fungi *Coriolopsis* sp. on simple muslin cloth to enhance removal of triphenylmethane dyes. J. Environ. Manage. 214, 261–266.

Nancharaiah, Y.V., Venuopalan, V.P., Wuertz, S., Wilderer, P.A., Hausner, M., 2005. Compatibility of the green fluorescent protein and a general nucleic acid stain for quantitative description of a *Pseudomonas putida* biofilm. J. Microbiol. Methods 60, 179–187.

Nemmiche, S., 2017. Oxidative signaling response to cadmium exposure. Toxicol Sci. 156, 4–10.

Nies, D.H., 1992. Resistance to cadmium, cobalt, zinc, and nickel in microbes. Plasmid 27, 17–28.

Nies, D.H., 2016. The biological chemistry of the transition metal "transportome" of *Cupriavidus metallidurans*. Metallomics 8, 481–507.

Nourbakhsh, F., Amirmorteza, Namvar, E., 2016. Detection of genes involved in biofilm formation in *Staphylococcus aureus* isolates. GMS Hyg. Infect. Control 11, 1–5.

Novelli, E.L.B., Marques, S.F.G., Almeida, J.A., Diniz, Y.S., Faine, L.A., Ribas, B.O., 2000. Toxic mechanism of cadmium exposure on cardiac tissue. Toxic Subst. Mech. 19, 207–217.

Palma-Laraa, I., Martínez-Castilloa, M., Quintana-Perezb, J.C., Arellano-Mendozab, M.G., Tamay-Cachb, F., Valenzuela-Limonc, O.L., et al., 2020. Arsenic exposure: a public health problem leading to several cancers. Regul. Toxicol. Pharmacol. 110, 104539.

Perera, M., Wijayarathna, D., Wijesundera, S., Chinthaka, M., Seneviratne, G., Jayasena, S., 2019. Biofilm mediated synergistic degradation of hexadecane by a naturally formed community comprising *Aspergillus flavus* complex and *Bacillus cereus* group. BMC Microbiol. 19, 84.

Piazza, A., Ciancio, C.L., Pacini, V.A., Sanguinetti, G., Ottado, J., Gottig, N., 2019. Environmental bacteria involved in manganese (II) oxidation and removal from groundwater. Front. Microbiol. 10, 119.

Pozzani, U.C., Carpenter, C.P., Palm, P.E., Weil, C.S., Nair, J.H., 1959. An investigation of the mammalian toxicity of acetonitrile. J. Occup. Med. 1, 634–642.

Pradeep, N.V., Anupama, S., Navya, K., Shalini, H.N., Idris, M., Hampannavar, U.S., 2015. Biological removal of phenol from wastewaters: a mini review. Appl. Water Sci. 5, 105–112.

Quiton, K.G., Doma Jr., B., Futalan, C.M., Wan, M.-W., 2018. Removal of chromium(VI) and zinc(II) from aqueous solution using kaolin-supported bacterial biofilms of Gram-negative *E. coli* and Gram-positive *Staphylococcus epidermidis*. Sustain. Environ. Res. 28, 206–213.

Radojevic, I.D., Grujic, S.M., Rankovic, B.R., Comic, L.R., Ostojic, A.M., 2019. Single-species biofilms from autochthonous microorganisms: biotechnological potential in automotive wastewater treatment. Int. J. Environ. Sci. Technol. Available from: https://doi.org/10.1007/s13762-019-02265-y.

Rai, L.S., Singha, R., Sanchez, H., Chakraborty, T., Chand, B., Bachellier-Bassi, S., et al., 2019. The *Candida albicans* biofilm gene circuit modulated at the chromatin level by a recent molecular histone innovation. PLoS Biol. 17, e300422.

Rajendran, A., Hu, B., 2016. Mycoalgae biofilm: development of a novel platform technology using algae and fungal cultures. Biotechnol. Biofuels 9, 112.

Rajwar, D., Bisht, M., Rai, J.P.N., 2018. Wastewater treatment: role of microbial biofilm and their biotechnological advances. Microbial Biotechnology in Environmental Monitoring and Cleanup. IGI Global, pp. 162–174.

Ramrakhiani, L., Khowala, S., 2012. Biosorption by biomass of *Termitomyces clypeatus* i: enhancement of hexavalent chromium removal in aqueous solution by heat pretreatment. International Journal of Integrative Sciences, Innovation and Technology (IJIIT) 1, 37–44.

Reichhardt, C., Parsek, M.R., 2019. Confocal laser scanning microscopy for analysis of *Pseudomonas aeruginosa* biofilm architecture and matrix localization. Front. Microbiol. 10, 677. Available from: https://doi.org/10.3389/fmicb.2019.00677.

Rickard, A.H., Gilbert, P., Handley, P.S., 2004. Influence of growth environment on coaggregation between freshwater biofilm bacteria. J. Appl. Microbiol. 96, 1367–1373.

Roth, J.A., 2006. Homeostatic and toxic mechanisms regulating manganese uptake, retention, and elimination. Biol. Res. 39, 45–57.

Saba, R.Y., Ahmed, M., Sabri, A.N., 2019. Potential role of bacterial extracellular polymeric substances as biosorbent material for arsenic bioremediation. Biorem. J. 1–10. Available from: https://doi.org/10.1080/10889868.2019.1602107.

Sabater, S., Guasch, H., Ricart, M., Romani, A., Vidal, G., Klunder, C., et al., 2007. Monitoring the effect of chemicals on biological communities. The biofilm as an interface. Anal. Bioanal. Chem. 387, 1425–1434.

Saluja, B., Gupta, A., Goel, R., 2011. Microbial management of cadmium and arsenic metal contaminants in soil. In: Khan, M.S., Zaidi, A., Goel, R., Musarrat, J. (Eds.), Biomanagement of Metal-Contaminated Soils. Springer-Verlag, pp. 257–275.

Sehar, S., Naz, I., 2016. Role of biofilm in wastewater treatment. In: Dhanasekaran, D. (Ed.), Microbial Biofilms—Importance and Applications. InTech Open, London, pp. 121–144.

SEPA (State Environmental Protection Administration), 2002. Monitoring and Analytical Methods for Water and Wastewater, fourth ed. China Environmental Science Press, Beijing.

Shukla, S.K., Mangwani, N., Karley, D., Rao, T.S., 2017. Bacterial biofilms and genetic regulation for metal detoxification. In: Das, S., Dash, H.R., (Eds.), Handbook of Metal-Microbe Interactions and Bioremediation, CRC Press, pp. 317-332.

Singh, R., Paul, D., Jain, R.K., 2006. Biofilms: implications in bioremediation. Trends. Microbiol. 14, 389–397.

Srivastava, M., Mishra, A.K., 2018. Nitrogen removal and metabolic profiling of a cold-adaptive and biofilm producing paddy soil bacterium *Cupriavidus* sp. PDN31. Arch. Agron. Soil Sci. 64, 1608–1621.

Tekerlekopoulou, A.G., Vayenas, D.V., 2007. Ammonia, iron and manganese removal from potable water using trickling filters. Desalination 210, 225–235.

Tien, C.J., Chen, C.S., 2013. Patterns of metal accumulation by natural river biofilms during their growth and seasonal succession. Arch. Environ. Contam. Toxicol. 64, 605–616.

Tsekova, K., Ianisa, M., Dencheva, V., Ganeva, S., 2007. Biosorption of binary mixtures of copper and cobalt by *Penicillium brevicompactum*. Z. Naturforsch. 62, 261–264.

Viggiani, A., Olivieri, G., Siani, L., Di Donato, A., Marzocchella, A., Salatino, P., et al., 2006. An airlift biofilm reactor for the biodegradation of phenol by *Pseudomonas stutzeri* OX1. J. Biotechnol. 123, 464–477.

Volesky, B., Holan, Z.R., 1995. Biosorption of heavy-metals. Biotechnol. Progr. 11, 235–250.

WHO, 2006. Guidelines for the Safe Use of Wastewater, Excreta and Greywater, Vol. I. World Health Organization, p. 95.

Wu, G., Sun, M., Liu, P., Zhang, X., Yu, Z., Zheng, Z., et al., 2014. *Enterococcus faecalis* strain LZ-11 isolated from Lanzhou reach of the Yellow River is able to resist and absorb cadmium. J. Appl. Microbiol. 16, 1172–1180.

Wuertz, S., et al., 2001. A new method for extraction of extracellular polymeric substances from biofilms and activated sludge suitable for direct quantification of sorbed metals. Water Sci. Technol. 43, 25–34.

Xu, X.Y., Feng, L.J., Zhu, L., Xu, J., Ding, W., Qi, H.Y., 2012. Biofilm formation and microbial community analysis of the simulated river bioreactor for contaminated source water remediation. Environ. Sci. Pollution. Res. 19, 1584–1593.

Xuemei, W., Jingling, L., Muyuan, M., Zhifeng, Y., 2010. Response of freshwater biofilm to pollution and ecosystem in Baiyangdian, Lake of China. Procedia Environ. Sci. 2, 1759–1769.

Yin, K., Wanga, Q., Lv, M., Chen, L., 2018. Microorganism remediation strategies towards heavy metals. Chem. Eng. J. <https://doi.org/10.1016/j.cej.2018.10.226>.

Zeng, X., Tang, J., Liu, X., Jiang, P., 2012. Response of *P. aeruginosa* E1 gene expression to cadmium stress. Curr. Microbiol. 65, 799–804.

Chapter 22

Microbial technologies for environmental remediation: potential issues, challenges, and future prospects

Ahmad Rabbani[1], Surabhi Zainith[2], Vishal Kumar Deb[1], Plaban Das[1], Priyam Bharti[1], Devendra Singh Rawat[3], Naveen Kumar[4] and Gaurav Saxena[1,2]

[1]*Laboratory of Microbiology, Department of Microbiology, School of Life and Allied Sciences, Baba Farid Institute of Technology, Dehradun, India,* [2]*Department of Environmental Microbiology (DEM), School for Environmental Sciences (SES), Babasaheb Bhimrao Ambedkar (Central) University, Lucknow, India,* [3]*Laboratory of Biotechnology, Department of Biotechnology, School of Life and Allied Sciences, Baba Farid Institute of Technology, Dehradun, India,* [4]*School of Life Sciences, The University of Warwick, Coventry, United Kingdom*

22.1 Introduction

Environmental issues have been always at the forefront of sustainable development and have become a serious matter of concern in the 21st century. Environmental sustainability with rapid industrialization is one of the major challenges of the current scenario worldwide. Industries are the key drivers in the world economy, but these are also the major polluters due to the discharge of partially treated/untreated potentially toxic and hazardous wastes containing organic and inorganic pollutants, which cause environmental (soil and water) pollution and severe toxicity in living beings (Bharagava et al., 2018a,b, 2020; Saxena et al., 2020a,b,c; Saxena and Bharagava, 2017). Among the different sources of environmental pollution, industrial waste is considered as the major source of environmental pollution because industries use cheap and poorly or nonbiodegradable chemicals to obtain the good quality of products within a short period of time and in an economic way; however, their toxicity is usually ignored. Ensuring the safety of chemicals used in many industrial processes is a major challenge for environmental safety. The governments around the globe are also strictly advocating for the mitigation of environmental pollution due to industrial wastes to promote the sustainable development of our society with low environmental impact. Being a low-cost and eco-friendly clean technology, microbial remediation technologies can be the eco-sustainable alternative to conventional technologies for the treatment and management of contaminated sites/industrial wastes to protect the public health and environment.

Remediation technologies that use microorganisms to degrade/detoxify the pollutants in a contaminated matrix for environmental remediation are known as microbial remediation technology. Microbial remediation technologies include microbial bioremediation, electrobioremediation, nanobioremediation, microbe-assisted phytoremediation, anammox, enzymatic remediation, microbial fuel cell (MFC), and MFC-constructed wetland technology. Microbial remediation technologies are mainly relied on the bioremediation phenomenon. Bioremediation is a waste management approach that utilizes microorganisms, plants, or their enzymes to degrade/detoxify the organic and inorganic pollutants from contaminated soils and wastewaters and has been approved as a sustainable remediation technology by the US Environmental Protection Agency (USEPA) (Bharagava et al., 2017, 2020; Saxena et al., 2020c; Chandra et al., 2015; Saxena and Bharagava, 2017). The biological removal of a wide range of pollutants from contaminated sites requires our increasing understanding of different degradation pathways and regulatory networks to carbon flux for their degradation and detoxification, which is utmost important for environmental safety. Therefore this chapter provides a comprehensive knowledge of the microbial remediation technologies and their applications in remediation of contaminated sites/industrial effluents with their merits and demerits as well as future directions in the field of research.

Microbe Mediated Remediation of Environmental Contaminants. DOI: https://doi.org/10.1016/B978-0-12-821199-1.00022-5

22.2 Pollution and toxicity profile of environmental contaminants

Our natural environment has been contaminated by a variety of organic and inorganic pollutants as result of environmental pollution caused by industries. A variety of highly toxic and recalcitrant pollutants are being discharged along with industrial wastewaters in the environment due to different industrial activities. Environmental pollutants are of two types: organic and inorganic. Organic pollutants mainly include phenols, nonylphenols, chlorinated phenols, azo dyes, phthalic esters, petroleum hydrocarbons, pesticides, and persistent organic pollutants (POPs), whereas inorganic pollutants comprise a variety of highly toxic nonbiodegradable heavy metals (HMs) such as arsenic (As), nickel (Ni), chromium (Cr), lead (Pb), mercury (Hg), and cadmium (Cd). A variety of organic and inorganic pollutants have been reported to cause serious soil and water pollution and severe toxic effects in living organisms (Bharagava et al., 2018a, b, 2020; Saxena et al., 2020a,b,c; Saxena and Bharagava, 2017). Hence, due to highly toxic nature, many of them have been regarded as priority pollutants by various environmental protection agencies such as the USEPA, Agency for Toxic Substances and Disease Registry (ATSDR), and World Health Organization (WHO). Table 22.1 represents the environmental hazards and toxic effects caused by various organic and inorganic pollutants in living organisms.

TABLE 22.1 Toxicity of various organic and inorganic pollutants of industrial wastewater.

Pollutants	Environmental pollution and toxicity profile
Organic pollutants	
Phenols	Most common pollutants of industrial wastewaters and associated with distilleries, pulp and paper mills, coal mines, oil refineries, wood preservation plants, pharmaceuticals, coke-oven batteries, herbicides, and pesticides as well as their wastewaters. It is also used in preparation of several chemicals such as alkylphenols, cresols, xylenols, phenolic resins, aniline, pesticides, explosives, dyes, and other compounds. Its acute exposure causes dryness of throat and mouth, nausea, vomiting, and diarrhea, while chronic exposure causes methemoglobinemia, hemolytic anemia, profuse sweating, hypotension, arrhythmia, pulmonary edema, tachycardia, and dark-colored urine excreted due to lipid peroxidation and CNS disorders leading to collapse and coma and sometimes muscular convulsions with reduction in body temperature (hypothermia). Inhalation and dermal contact causes cardiovascular diseases and skin blisters, respectively, while ingestion can cause serious gastrointestinal damage, and oral administration may result in muscle tremors and death.
Endocrine-disrupting chemicals	These are widely used as plasticizers in industries, in the manufacturing of plastic resins such as polyvinyl resins and cellulosic and polyurethane polymers. These disturb the delicate hormonal balance (endocrine system) and compromise the reproductive fitness of living beings and ultimately may lead to carcinogenic and mutagenic effects. Examples include DBP, BBP, DEHP, and NP, NP, 4-aminobiphenyl, hexachlorobenzene, and benzidine. These also cause irritation of the skin, conjunctiva, mucous membranes of oral and nasal cavities, testicular lesions, hypospadia, cryptorchidism in males and cause obesity, prolongation of the estrous cycle, and anovulation in females due to a decrease in the serum estradiol level.
Chlorinated phenols	Chlorophenols are considered as the major environmental pollutants discharged along with wastewaters from pulp and paper mills, tanneries, distilleries, dye and paint manufacturing, and pharmaceutical industries, for example, PCP. It is widely applied as herbicides and fungicides and in wood protection, tanneries, distilleries, paint manufacturing, and pulp and paper mills. It is highly carcinogenic, teratogenic, and mutagenic in nature and causes toxicity to living beings by inhibiting oxidative phosphorylation, inactivating respiratory enzymes, and damaging mitochondrial structure. However, its high concentration can also cause obstruction in the circulatory system of the lungs, heart failure, and damage to the CNS.
Azo dyes	Textile, leather, paint, acrylic, cosmetics, plastics, pharmaceutical, etc., industries use different dyes to color products. Azo dyes cause severe health hazards, such as skin irritation, digestive tract irritation, nausea, vomiting, liver and kidney damage, etc., in humans and animals.
Petroleum hydrocarbons	Refinery industry wastewaters are the major source of petroleum hydrocarbons in the environment. The most common petroleum hydrocarbons include aliphatic, branched, and cycloaliphatic alkanes, as well as monocyclic and PAHs, which include naphthalene, fluorene, phenanthrene, anthracene, fluoranthene, pyrene, benzo[a]anthracene, and benzo[a]pyrene. Inhalational of hydrocarbons in humans can lead to criminal or violent behavior, development of memory and other cognitive deficits, cerebellar dysfunction, encephalopathy, weakness, dementia, depression of the CNS, metabolic acidosis, arrhythmia, or even a fatal malignant arrhythmia termed "sudden sniffing death." In addition, the aspiration of hydrocarbons causes a potentially fatal pneumonitis characterized by cough, wheezing, respiratory distress, and hypoxia.

(Continued)

TABLE 22.1 (Continued)

Pollutants	Environmental pollution and toxicity profile
	Dermal exposure can cause dermatitis, chemical burns, and defatting injury, whereas oral exposure can cause local irritation as well as vomiting, diarrhea, and abdominal pain. Thus acute hydrocarbon exposure can result in a wide array of pathologies, such as encephalopathy, pneumonitis, arrhythmia, acidosis, and dermatitis. Intentional inhalation and accidental ingestion exposures with aspiration also lead to the greatest morbidity and mortality.
Melanoidins	Melanoidins are released as environmental pollutants by various agro-based industries, especially from cane molasses–based distilleries and fermentation industries. Discharge of melanoidins containing DWW into the environment causes several problems, such as reduction of sunlight penetration, decreased photosynthetic activity, and dissolved oxygen concentration, thereby posing deleterious effects to aquatic life. On land, it causes a reduction in soil alkalinity and inhibition of seed germination. The melanoidins containing DWW have severe toxic effects on fishes and other aquatic organisms.
Pesticides	Pesticides are used in agricultural applications to control pests and enhance crop productivity. Pesticides (such as DDT, pyrethroids, organophosphates, and carbamates); fungicides (used to kill fungi, such as hexachlorobenzene, benzothiozole, and pentachlorophenol); herbicides (used to kill weeds, such as 2,4-D, atrazine, picloram, chlorophenoxy compounds, etc.); rodenticides (used to kill rodents, such as zinc phosphide, ANTU, 4-hydroxycoumarin, and 1,3-indandiones); and fumigants (used to kill pests, such as phosphine and dibromochloropropane) cause immune suppression, hormone disruption, diminished intelligence, reproductive abnormalities, cancer upon exposure to humans and animals and also directly or indirectly affect the nontarget organisms.
POPs	POPs exhibit low water solubility but high lipid solubility and therefore bioaccumulate in fatty tissues and cause a variety of toxic effects in living beings, such as disruption of the endocrine system, neurobehavioral disorders, cancer (possibly including breast cancer), allergies, hypersensitivity, damage to the central and peripheral nervous systems, birth defects, learning disabilities, reproductive disorders, disruption of the immune system by altering the hormonal system, and mutagenicity and carcinogenicity.
Inorganic pollutants (heavy metals)	
Cadmium	It is used in rechargeable batteries, special alloys' production, coatings, pigments, platings, as a plastic stabilizer, and also present in tobacco smoke. Acute exposure of Cd causes abdominal pain, burning sensation, nausea, vomiting, salivation, muscle cramps, itai-itai disease (a combination of osteomalacia and osteoporosis), vertigo, and shock. Loss of consciousness and convulsions usually appear within 15–30 min, and gastrointestinal tract erosion; pulmonary, hepatic, or renal injury; and coma may develop depending on the route of poisoning. However, chronic exposure of Cd causes a depressive effect on levels of norepinephrine, serotonin, and acetylcholine. The mechanism of Cd toxicity is not well known but is assumed to cause cell damage through the generation of ROS that cause DNA damage.
Chromium	Industries such as metallurgical, chemical, refractory brick, leather, wood preservation, and pigments and dyes are the major consumers of chromium. It causes severe health problems such as skin irritation, nasal irritation, ulceration, eardrum perforation, and lung carcinoma.
Arsenic	Arsenic (As) causes severe disturbances of the cardiovascular and CNSs, bone marrow depression, hemolysis, hepatomegaly, melanosis, polyneuropathy, and encephalopathy, and exposure eventually leads to death. Ingestion may also cause black foot disease that is only reported in Taiwan.
Lead	The contamination of soil and water with Pb mostly occurs from anthropogenic activities, industrial wastes, mining and smelting, and the past and present use of Pb in paints, batteries, gasoline, pesticides, and explosives. Pb mainly causes injury to the CNS and also headache, poor attention span, irritability, encephalopathy (characterized by sleeplessness and restlessness), loss of memory, acute psychosis, confusion, reduced consciousness, and dullness and also adversely affects the kidneys, liver, hematopoietic system, endocrine system, and reproductive system.
Mercury	The major sources of Hg exposure include its use in dental amalgams, thermometers, sphygmomanometers, barometers, fossil fuel emissions, incandescent lights, batteries, ritualistic practices using mercury, and the incineration of medical waste. In humans and animals, it causes mental retardation, dysarthria, blindness, neurological deficits, loss of hearing, developmental defects, and abnormal muscle tone.

ANTU, α-Naphthyl thiourea; *BBP*, benzyl butyl phthalate; *CNS*, central nervous system; *DBP*, dibutyl phthalate; *DEHP*, bis(2-ethylhexyl)phthalate; *DWW*, distillery wastewater; *NP*, nonylphenol; *PAHs*, polycyclic aromatic hydrocarbons; *PCP*, pentachlorophenol; *POPs*, persistent organic pollutants; *ROS*, reactive oxygen species.
Source: Adapted from Bharagava, R.N., Saxena, G., Mulla, S.I., 2020. Introduction to industrial wastes containing organic and inorganic pollutants and bioremediation approaches for environmental management. In: Bioremediation of Industrial Waste for Environmental Safety. Springer, Singapore, pp. 1–18.

22.3 Microbial technologies for environmental remediation

Technologies that use microorganisms to degrade/detoxify the pollutants in a contaminated matrix for environmental remediation are known as microbial remediation technology. Microbial remediation technologies include microbial bioremediation, electrobioremediation, nanobioremediation, microbe-assisted phytoremediation, anammox, enzymatic remediation, MFC, and MFC-constructed wetland technology. A summarized description of each microbial remediation technology can be found in Table 22.2.

22.3.1 Microbial bioremediation technologies

Microbial bioremediation is a remediation technology wherein the inherent ability of microbes such as algae, fungi, microalgae, and bacteria is used to degrade/detoxify organic and inorganic pollutants from contaminated soil/water/wastewaters (Saxena et al., 2020c; Bharagava et al., 2017). In bioremediation process, pollutants/waste is converted into inorganic compounds such as carbon dioxide, water, and methane and, thus, leads to mineralization and detoxification (Bharagava et al., 2020). Bioremediation is chiefly depends on the metabolic capability of microbes to degrade/detoxify or transform the pollutants, which is also affected by the accessibility of pollutants and their bioavailability (Antizar-Ladislao, 2010). It can be applied both as in situ (remediation at the site) and ex situ (remediation elsewhere) remediation technology and involve bioattenuation (natural process of degradation and can be monitored by a decrease in pollutant concentration with increasing time); biostimulation (intentional stimulation of pollutants degradation by addition of water, nutrients, and electron donors or acceptors); and bioaugmentation (addition of laboratory-grown microbes with potential for degradation) (Bharagava et al., 2020; Maszenan et al., 2011). A great deal of literature can be found in the public domain on the biodegradation and bioremediation of industrial waste/pollutants (Saxena et al., 2020c; Maszenan et al., 2011; Megharaj et al., 2011).

Bioremediation technology has been successfully applied for the remediation of organic and inorganic or mixed contaminants (Zhang et al., 2013c; Lee et al., 2012). Chan et al. (2014) has successfully reported the removal of HMs from a wastewater treatment plant discharge using microalgae. According to their study, in the untreated secondary effluent trial, copper was removed up to 81.7%, reaching a lowest final concentration of 7.8 ppb after 10 days, whereas zinc was reduced by up to 94.1%, reaching 0.6 ppb after 10 days using microalgae, *Chlorella vulgaris*, *Spirulina maxima*, and a naturally growing algae sample found in the wastewater from a wastewater treatment plant [containing *Synechocystis* sp. (dominant) and *Chlorella* sp. (common) and a few cells of *Scenedesmus* sp.]. Paisio et al. (2012) reported isolated and characterized a *Rhodococcus* strain with phenol-degrading ability and use for the tannery effluent biotreatment. According to their study, *Rhodococcus* sp. CS1 strain was able to grow and completely (100%) degrade phenols from tannery effluents after 9 h of incubation. Tripathi and Garg (2013) isolated and characterized *Bacillus cereus* bacterium that efficiently remediated 57% PCP and 74% Cr^{6+} simultaneously with uptake rate of 0.65 mg Cr^{6+}/g biomass/h at initial 200 mg Cr^{6+} and 500 mg PCP/L concentration under optimized 0.4% glucose, 0.2% NH_4Cl, pH 7.0, 35°C, 1.0% inoculum during 60 h incubation. Bharagava and Mishra (2018) investigated the hexavalent chromium reduction potential of *Cellulosimicrobium* sp. isolated from common effluent treatment plant of tannery industries. According to their study, at 50, 100, 200, and 300 mg/L of Cr(VI) concentrations, the bacterium showed 99.33% and 96.98% reduction at 50 and 100 mg/L at 24 and 96 h, respectively. However, at 200 and 300 mg/L concentration of Cr (VI), only 84.62% and 62.28% reduction was achieved after 96 h, respectively. Kurade et al (2012) studied the decolorization of textile industry effluent containing disperse dye Scarlet RR by a newly developed bacterial–yeast consortium BL-GG. According to their study, an efficient decolorization of textile industry effluent containing disperse dye Scarlet RR was achieved by a developed bacterial–yeast consortium BL-GG within 48 h with 68% and 74% reduction in biochemical oxygen demand (BOD) and chemical oxygen demand (COD) values, respectively, while the same consortium decolorized the solution containing single dye Scarlet RR (50 mg/L) within 18 h (pH 9.0, 40°C, static condition) with 98% reduction in color. Sharma and Malaviya (2016) developed a consortium of four Cr-resistant ascomycetous fungi (*Cladosporium perangustum* and *Penicillium commune* isolated from tannery wastewater (TWW)–contaminated soil; *Paecilomyces lilacinus* and *Fusarium equiseti* isolated from tannery sludge) and employed for the TWW bioremediation. According to their study, the developed consortium showed good reduction in various pollution parameters [COD (82.52%), color (86.19%), Cr(VI) (100%), total Cr (99.92%), total Pb (95.91%)] of TWW after 120 h of treatment.

Bioremediation has emerged as an eco-friendly and low-cost alternative to conventional remediation technologies, which are environmentally destructive, costly, and create secondary pollution, and, thus, negatively affect the ecosystem (Bharagava et al., 2020; Daghio et al., 2017). However, it may get restricted by several factors such as (1) low or non-bioavailability of pollutants to microbes; (2) toxicity of pollutants to remediating microorganisms; (3) lack of enzymes

TABLE 22.2 Microbial technologies for environmental remediation.

Microbial technologies	Definition	Pollutants	Polluted sites	Advantages	Disadvantages
Microbial bioremediation	It is an eco-friendly waste management method that utilizes the inherent potential of microbes such as algae, fungi, and bacteria to remove organic and inorganic pollutants from industrial wastes.	Organic (phenols, chlorophenols, azo dyes, endocrine-disrupting chemicals, polyaromatic hydrocarbons, pesticides, persistent organic pollutants, polychlorinated biphenyls) and inorganic pollutants (cadmium, chromium, lead, arsenic, mercury); radionuclides industrial effluents; solid waste treatment	Contaminated soil, water, and wastewaters	Eco-friendly process, nontoxic by-products, cost-effective, highly effective treatment.	Low or nonbioavailability of pollutants to microbes, toxicity of pollutants to microbes and remediating plants, lack of enzymes responsible for the degradation and detoxification of specific environmental pollutants, recalcitrant nature of environmental pollutants, quick adaptation of the indigenous microorganisms for biodegradation of a particular contaminant, and mass transfer of electron acceptors and nutrients to microorganisms responsible for biodegradation.
Electrobioremediation	It is a popular hybrid technology that uses the combination of bioremediation and electrokinetics for the treatment of environmental pollutants. It involves the electrokinetic phenomena for the acceleration and orientation of transport of environmental pollutants and microbes for pollutant bioremediation.	Organic pollutants, inorganic pollutants, industrial effluents	Contaminated soil and wastewaters	Enhanced mobility and transport of environmental pollutants and thereby, enhanced microbial degradation.	Scale-up is required from lab-scale reactors to field-scale systems. Removal efficiency is significantly reduced if soil contains carbonates and hematite, as well as large rocks or gravel. When higher voltage is applied to the soil, the efficiency of the process decreases due to the increased temperature.
MFC	It is a bioelectrochemical device that harnesses the power of respiring microbes to convert organic substrates present in wastewater directly into electrical energy and thus wastewater treatment.	Organic pollutants and heavy metals, industrial wastewaters	Contaminated wastewaters	More efficient and maximum removal of pollutants, simultaneous removal of multiple pollutants with electricity generation.	Low- or variable-density current production.
Enzymatic bioremediation	The use of a variety of catabolic enzymes in the degradation and detoxification of various organic and inorganic pollutants from contaminated wastewater.	Organic contaminants, heavy metals, and industrial effluents	Contaminated wastewaters	Rapid, cost-effective, and highly specific remediation of contaminants, no need to use whole microbial cells for remediation, formation of nontoxic by-products.	Production cost of enzymes, and their stability and activity across a wide range of contaminants, requirement of enzyme immobilization to enhance performance of enzymes in bioremediation.

(Continued)

TABLE 22.2 (Continued)

Microbial technologies	Definition	Pollutants	Polluted sites	Advantages	Disadvantages
Nanobioremediation	It is a integration of NPs and bioremediation for sustainable remediation of environmental pollutants from contaminated matrix.	Organic contaminants, heavy metals, and industrial effluents	Contaminated soil, water and wastewaters	Increased pollutants removal due to high surface area of NPs, high reactivity, NPs can diffuse or penetrate into a contamination zone where microorganisms cannot reach, higher reactivity to redox-amenable contaminants, NPs can be used to immobilize microbial cells for further degradation process.	Sometimes, NPs are toxic to microorganisms.
Microbe assisted-phytoremediation	Microbe-assisted phytoremediation is a type of bioremediation wherein plant-associated bacteria such as rhizobacteria and endophytes are used to enhance the efficiency of remediating plants in the stressed environment via increasing the bioavailability of heavy metals in soil and plant growth promotion.	Heavy metals and organic pollutants, industrial effluents	Contaminated soil, water, and wastewaters	Enhanced phytoremediation efficiency of metal hyperaccumulating plants via increasing the bioavailability of metals in soil and plant growth promotion in the stressed environment, suppression of plant pathogens, tolerance to abiotic stress, lower metal toxicity to remediating plants through biosorption or bioaccumulation of heavy metals	The analysis of more than thousands of isolates for plant growth promoting and bioremediation traits is a tedious task, require a clear-cut understanding of plant–microbe–metal–soil interactions.
Anammox	It is a process of anaerobic ammonium oxidation, responsible for the removal of ammonia nitrogen in wastewater and eliminates the necessity of an organic carbon source for nitrification, reduces energy demand for aeration, and has a smaller production of excess sludge and lower CO_2 emissions.	Municipal wastewater treatment (sludge rejection water); organic solid waste treatment (landfills, composting, digestion); food industry effluent; manure processing industry waste; fertilizer industry waste; petrochemical industry waste; metallurgical industry waste; semiconductor industry waste; tannery wastewater	Contaminated wastewaters	Oxygen addition can be reduced (results in energy and cost savings); anammox bacteria do not require organic carbon (e.g., methanol) as they do in nitrification; production of excess sludge is reduced; reduces CO_2 emissions.	Require expense of knowledge available (skilled operation and maintenance required); high construction costs if the anammox process replaces the conventional nitrification/denitrification in treatment plants.
CW-MFC system	A MFC–coupled CW (CW-MFC) system is a latest mechanism that embeds the MFC into the CW to treat the wastewater and produce electricity.	Industrial effluents (dairy, tannery, distillery, pulp and paper mill, textile, municipal, agriculture), nutrients, dyes, heavy metals, and organic toxic compounds	Contaminated wastewaters	The produced electricity by MFC enhances the removal of pollutants from wastewater in CWs, simultaneous wastewater treatment, pollutants removal, and energy generation, production of energy without the contribution of an exogenous organic substrate, reduction of methane release.	Low or variable power densities, optimization of bioelectrochemical system.

CW, Constructed wetland; *MFC*, microbial fuel cell; *NPs*, nanoparticles.

responsible for the degradation and detoxification of specific environmental pollutants; (4) recalcitrant nature of environmental pollutants; (5) quick adaptation of the indigenous microorganisms for biodegradation of a particular contaminant; and (6) mass transfer of electron acceptors and nutrients to microorganisms responsible for biodegradation (refer to Daghio et al., 2017). Further, the use of molecular techniques may advance the bioremediation technology by developing transgenic microbes for the remediation of environmental pollutants, but the associated environmental risks such as should also be considered before field-scale application. Moreover, the strict US and Western countries' regulations on the use of genetically modified organisms also restrict their filed applications. These limitations are sufficient to discredit the applicability of bioremediation technologies and together constitute a major challenge in the way of success at field scale. Further research efforts are required to make the bioremediation technologies more efficient for environmental remediation. For the development of sustainable environmental bioremediation technologies, the future research should be focused on the (Bharagava et al., 2020) (1) search for potential microbial degraders for environmental pollutants; (2) search for catabolic enzymes or genes for the enhanced degradation/detoxification of environmental pollutants; and (3) development of transgenic/designer microbes using genetic engineering for effective bioremediation technology. However, continued efforts are required to realize the economic feasibility of bioremediation technology at the field scale.

22.3.2 Electrobioremediation technology

Electrobioremediation is becoming an increasingly popular hybrid technology that uses the combination of bioremediation and electrokinetics for the treatment of environmental pollutants (Maszenan et al., 2011). It overcomes the limitations of in situ bioremediation, which includes (Daghio et al., 2017) (1) mass transfer of electron acceptors and nutrients to microorganisms responsible for biodegradation; (2) limited bioaccessibility of contaminants (e.g., partitioning to aquifer material) for biodegradation; and (3) quick adaptation of the indigenous microorganisms for biodegradation of a particular contaminant. During electrobioremediation, direct current is applied within subsurface porous media to induce specific electrokinetic phenomena for the acceleration and orientation of transport of environmental pollutants and microbes for pollutant bioremediation (Daghio et al., 2017; Maszenan et al., 2011; Saichek and Reddy, 2005). Electrokinetic phenomenon involves the use of several phenomena such as diffusion, electrolysis, electroosmosis, electrophoresis, and electromigration and uses weak electric currents of about 0.2–2 V/cm (Daghio et al., 2017; Li et al., 2010; Maszenan et al., 2011). A number of studies are available on the use of electrobioremediation technology for contaminated soils (Wick et al., 2007; Martínez-Prado et al., 2014; Yan and Reible, 2015). In addition, the applications, potentials, and limitations of electrobioremediation technology have been reviewed by many authors (Wick et al., 2007; Maszenan et al., 2011; Gill et al., 2014).

There are many reports available in the public domain on the electrobioremediation of environmental contaminants/industrial wastes. Phenol has been bioelectrochemically degraded both by mixed cultures and a pure culture of *Cupriavidus basilensis* (Friman et al., 2013; Huang et al., 2011). The degradation of phenanthrene by *Shewanella oneidensis* MR1 14063 in a MFC was reported (Adelaja et al., 2014). *Pseudomonas aeruginosa* NCTC 10662 has been shown to degrade phenanthrene faster (54.7 μM/d) than *S. oneidensis* (25.2 μM/d) under similar operational conditions in an MFC (Adelaja et al., 2014). Morris et al. (2009) found that NO_3-reducing bacteria dominated the anodic community in a single-chamber MFC in which diesel was degraded with simultaneous current production (Morris et al., 2009). Microbial communities dominated by microorganisms belonging to the *Chloroflexi* and to the *Nitrospira* were described during degradation of polycyclic aromatic hydrocarbons (Yan et al., 2012). A recent detailed analysis performed after remediation of total petroleum hydrocarbon (TPH)-contaminated soil with a bioelectrochemical systems (BES) highlighted that Proteobacteria (especially Betaproteobacteria and Gammaproteobacteria) were the most abundant phylum after the treatment (Lu et al., 2014).

Kanagasabi et al. (2013) reported the electrobioremediation of tannery wastewaterTWW that involved an electrooxidation reactor connected to a column packed with immobilized bacterial biomass. They reported that the electrooxidation was found to obtain maximum COD reduction up to 73.1% at 1.5 A/dm^2 using raw TWW and *Bacillus* strain B proved to be superior in terms of COD reduction up to 91.5% with diluted TWW. However, the intimate coupling of electro- and biooxidation recorded good degradation for both raw and diluted TWW samples and achieved 66.2% and 76.6% COD degradation, respectively. According to Azhar et al. (2016), they investigated the effectiveness of *Pseudomonas putida* bacteria as a biodegradation agent to remediate contaminated soil. 5 kg of kaolin soil was spiked with 5 g of zinc oxide. During this process the anode reservoir was filled with *P. putida* while the cathode was filled with distilled water for 5 days at 50 V of electrical gradient. The X-ray fluorescent test indicated that there was a significant reduction of zinc concentration for the soil near the anode with 89% percentage removal. The bacteria count is high near the anode which is 1.3 × 107 cfu/gww, whereas the bacteria count at the middle and near the cathode was 5.0 × 106and 8.0 × 106 cfu/gww, respectively. The migration of ions to the opposite charge of electrodes during the electrokinetic process resulted from the reduction of zinc. The results obtained proved that the electrobioremediation reduced the level of contaminants in the soil sample. Thus the electrobioremediation technique has the potential to be used in the treatment of contaminated soil.

The principal environmental properties that influence electrobioremediation are the (Daghio et al., 2017) (1) electrolyte (i.e., groundwater chemistry or soil moisture) through which the current travels; (2) geological strata that influence electrokinetic (EK) phenomena; (3) hydrodynamics that introduce advection as an additional transport vector; (4) physical heterogeneity that can alter transport rates; (5) mixed nature of contaminants found at contaminated sites; (6) choice of an appropriate anodic material that affects the selectivity and efficiency of the hydrocarbon removal process; (7) electrode material (must be inexpensive, highly durable, physically and chemically stable, nontoxic to microbes and have electrical conductivity, catalytic activity and selectivity for the target compounds with low cost/life ratio); and (8) influence area (radius) of an electrode. These factors should be considered before applying electrobioremediation to contaminated matrix. Despite considerable scientific interest, field-scale application of electrobioremediation technology has not yet been tested and verified under fully representative conditions and currently under investigation. "Lasagna" is an electrobioremediation process developed by a consortium of industries (DuPont, General Electrics, and Monsanto) in collaboration with US Federal Agencies (EPA; DOE) that perform a large-scale demonstration of the feasibility of the technology for the in situ treatment of a variety of contaminants, including TCE (Ho et al., 1999a,b). However, the choice of appropriate electrode materials and potentials or the effect of the radius of influence and limitations during scale-up should not be overlooked during the electrobioremediation feasibility studies (Daghio et al., 2017).

Further, the future research should be focused on the (Daghio et al., 2017) (1) search for the appropriate electrode material and configuration to optimize the electrobioremediation treatment process at field scale; (2) optimization of complex physicochemical processes that can affect bioremediation performance; (3) trial of electrobioremediation on novel or mixed (organic and inorganic) contaminants; (4) feasibility of electrobioremediation coupled with other remediation technologies such as chemical oxidation/reduction and phytoremediation to achieve the better treatment efficiency; (5) development of good practice guidelines for the implementation of electrobioremediation technology at large scale without unintended adverse effects; (6) development of more robust, reliable, and advanced field equipment for electrobioremediation purpose; (7) optimization of design and operational conditions to reduce the device internal resistance and improve electrochemical processes, and (8) adverse effects of electrokinetics on microbial communities should be studied. However, more field trials are required for better cost comparison and applicability to contaminated sites and environmental impacts of electrobioremediation should also be considered before applying to the field.

22.3.3 Microbial fuel cell technology

A MFC is a bioelectrochemical device that harnesses the power of respiring microbes to convert organic substrates directly into electrical energy. MFC can be a suitable alternative to the conventional activated sludge process based–treatment systems in terms of energy-consumption and excess sludge generation. Apart from wastewater treatment and electrical energy production, MFC is a versatile research field of bioremediation, biohydrogen production, environmental sensing, desalination, and microbial carbon sequestration (Singh et al., 2019). An array of microbes (bacteria, microalgae, and fungi) are involved to catalyze the degradation of organic and inorganic waste materials (Singh et al., 2019). From different types of BES, MFCs prove to be the best solution for wastewater treatment and other applications, which rely on the electrogenic nature of bacterial communities or single strains feed on wastewater that acts as substrate for their growth (Singh et al., 2019).

MFCs offer several advantages over conventional treatment systems related to (Singh et al., 2019; Gude, 2016) (1) energy (like direct electricity generation, energy savings by anaerobic treatment due to elimination of aeration, low sludge yield); (2) environment (water reclamation, low carbon footprint, less sludge generation); (3) economic (revenue through energy and value-added products–chemicals, low operational costs); and (4) operational benefits (self-generation of microorganisms, good resistance to environmental stress, and amenable to real-time monitoring and control).

There are many reports available in the public domain on the wastewater treatment, pollutants removal, and electricity generation in MFCs. Jiang (2017) developed a combination of MFCs with microalgae cultivation for bioelectricity generation and domestic wastewater treatment. According to the study, the MFC generated a maximum power density of 268.5 mW/m^2 and achieved 67% COD removal; only 34% total phosphorus (TP) and 50% NH_4^+–N removal were obtained with MFC alone, whereas up to 97% TP and 99% NH_4^+–N removal were achieved with the combined process. Passos et al. (2016) studied the energy generation in a MFC using anaerobic sludge from a wastewater treatment plant. According to their study, at the beginning of the MFC operation, power density was low, probably due to slow microorganism growth and adhesion. The power density increased from the 15th day of operation, reaching a value of 13.5 μW/cm after c. 24 days of operation, and remained stable until the end of the process, which was very promising. Ghangrekar and Shinde (2008) studied the simultaneous sewage treatment and electricity generation in membraneless MFC. According to their study, the maximum COD removal efficiency of 91.4% and 82.7% was achieved while treating synthetic wastewater and actual sewage, respectively. Maximum current of 0.33 and 0.17 mA was produced during synthetic and actual sewage treatment, respectively. Maximum power density of 6.73 mW/m^2 (13.65 mW/m^3) and maximum current density of 70.74 mA/m^2 were obtained in this membraneless MFC with successful organic matter removal from wastewater. Dong et al. (2015) developed a combined system of MFC and intermittently aerated biological filter for energy self-sufficient wastewater treatment. According to their study, the developed two-stage combined system obtained 93.9% SCOD removal and 91.7% TCOD removal (effluent SCOD = 61 mg/L, TCOD = 82.8 mg/L). Energy analysis indicated that the MFC unit produced sufficient energy (0.27 kWh/m^3) to support the pumping system (0.014 kWh/m^3) and aeration system (0.22 kWh/m^3).

MFCs are environment-friendly technologies as they can produce clean electricity directly from organic matter in wastewater without any need for separation, purification, and conversion of the energy products and function at mild operating conditions especially at ambient temperatures (Gude, 2016). MFCs can produce up to 1.43 kWh/m^3 from a primary sludge or 1.8 kWh/m^3 from a treated effluent (Ge et al., 2013). MFCs consume only 0.024 kW or 0.076 kWh/kg COD in average (mainly for feeding and mixing in the reactor), about one order of magnitude less than activated sludge–based aerobic processes (~0.3 kW or 0.6 kWh/kg COD) (Zhang et al., 2013a,b). It means that MFCs consume only about 10% of the external energy for their operation when compared with conventional activated sludge process showing great potential for energy savings as well as possible energy recovery from wastewater treatment (Gude, 2016).

Despite several successful applications at lab scale, the application of MFC at large scale is limited due to very low– or variable-density current generation during operation. However, to expand the scope and efficacy of MFC, the future research should be focused on (Singh et al., 2019) (1) cost-effective, environment-friendly, and nontoxic electrode materials; (2) development of relevant equipment and technology to harvesting voltage (or current) potential of MFCs; (3) optimization of operational parameters for the better results; (4) coupling MFC with other remediation technologies like constructed wetlands for effective wastewater treatment, pollutants removal and power generation; (5) understanding the electrogenic mechanism of microorganisms to fully understand the functioning of MFCs for better outcomes; (6) fundamental and applied research to investigate the interactions between various microbial communities and substrates; and (7) joint effort from ecologists, environmental and wastewater treatment engineers, and microbiologists is required for the development of sustainable MFCs.

22.3.4 Enzymatic bioremediation technology

Enzyme-mediated bioremediation refers to the use of naturally occurring enzymes in microorganisms or plants to degrade or reduce harmful, undesirable, and recalcitrant environmental pollutants in order to clean contaminated sites (Karigar and Rao, 2011). The efficiency of bioremediation chiefly depends on the ability of enzymes produced by microbes to catalyze the degradation and detoxification of environmental pollutants (Saxena et al., 2020d). Enzymes are biocatalysts that lower the activation energy and facilitate quick and complete breakdown of substrates (Fan and Krishnamurthy, 1995). The smaller size of enzymes as compared to microbial cells enables them to contact contaminants easily, aiding quicker mobility, increased contact with pollutants, and facilitating more targeted rapid and effective degradation or reduction to an admissible or less harmful state (Rao et al., 2010). The advantages of using enzymes over microbes/plants in bioremediation include (Kumar and Bharadvaja, 2019) (1) applications under more harsh environmental conditions such as contaminant concentration, pH, temperature, and salinity, which are not suitable for microbial populations; (2) no need of nutrients or time for acclimatization as microbial populations do; (3) no formation of metabolic products as a result of their application; (4) reduced mass transfer limitation as compared to microbial or whole cell treatment; (5) easily controllable process as compared to microbial remediation; (6) rapid, accessible, and highly specific remediation of contaminants; and (7) reduced generation of toxic by-products. Microbial whole cells need the introduction of nutrients and air to maintain the optimum growth rate, and solvents and/or surfactants to enhance bioavailability and immobilization (Kumar and Bharadvaja, 2019). In addition, enzyme-mediated biotransformation reduces the generation of toxic by-products significantly as compared to chemical and some types of microbial remediation (Kumar and Bharadvaja, 2019).

For an enzyme to be a candidate for bioremediation, it (Sutherland et al., 2004) (1) must be able to degrade the target pollutants to substantially less toxic products; (2) should not require cofactors, such as NAD(P)H, reduced flavin or glutathione, which would be prohibitively expensive to provide in a commercial setting; (3) must function efficiently in terms of its kinetics, because it directly impact both performance and price; (4) must have a micromolar Km (substrate affinity) as well as a kcat greater than 100 turnovers each minute because most of the applications involve micromolar pollutant concentrations for the bioremediation; (5) must also be robust to the range of temperature, pH, and ionic environments; (6) must have ability to detoxify a range of substrates within a category of contaminants; and (7) must also be relatively inexpensive to produce (> 5% yield from bacterial expression), formulate, and process.

The development of enzymes for the remediation of contaminants generally involves three stages (Sutherland et al., 2004): (1) the identification of a template enzyme from natural sources; (2) isolation of the gene encoding the enzyme; and, finally, as necessary, (3) improvement of enzymatic properties in the laboratory. Although all biological sources can be mined for enzymes, the most common sources of enzymes with bioremediation potential have so far been pesticide-tolerant soil microorganisms (Sutherland et al., 2004). The source bacteria generally have adapted to growth on contaminated soils by acquiring/evolving the ability to detoxify antimicrobial pesticides or the ability to use the pesticide as a worthwhile nutrient source (Sutherland et al., 2004).

A large number of enzymes from bacteria and fungi have been reported to be involved in the biodegradation and biodetoxification of toxic organic and inorganic pollutants. Enzymes such as ligninolytic enzymes [such as lignin peroxidase (LiP), manganese peroxidase (MnP), and laccase], chrome reductase, monooxygenase and dioxygenase, and azoreductase have been reported in the degradation and detoxification of various pollutants from industrial wastewaters. For instance, Bharagava et al. (2018a,b) reported the biodegradation of crystal violet dye by a ligninolytic enzyme–producing bacterium, *Aeromonas hydrophila* isolated from textile wastewater. Paisio et al. (2012) also characterized a dioxygenase from isolated *Rhodococcus* strain with phenol-degrading ability and used in the biotreatment of tannery effluent. Kathiravan et al. (2011) also studied the ex situ bioremediation of Cr(VI)-contaminated soil by chromium reductase–producing *Bacillus* sp. Jadhav et al. (2010) also studied the decolorization and degradation of Direct Orange 39 by *P. aeruginosa* strain BCH aided by of lignin peroxidase, tyrosinase, and NADH-DCIP reductase. Kurade et al. (2012) studied the decolorization of textile industry effluent containing disperse dye Scarlet RR by a newly developed bacterial–yeast consortium BL-GG aided by veratryl alcohol oxidase, laccase, tyrosinase, and NADH-DCIP reductase.

However, the major disadvantages of using enzymes in bioremediation of contaminants include (Saxena et al., 2020d; Kumar and Bharadvaja, 2019; Sutherland et al., 2004) (1) the high cost of producing enzymes; (2) the stability and activity across a wide range of contaminants; (3) requirement of extra materials for enzyme immobilization to enhance performance of enzymes in bioremediation; and (4) typically a low yield of enzyme during production process. Further, the future research should be focused on the use of molecular tools to biocatalysis that may improve bioprospecting research, enzyme yield recovery, and enzyme specificity, thus increase the cost–benefit ratios.

22.3.5 Nanobioremediation technology

Nanobioremediation is the new concept that integrates the use of nanoparticles and bioremediation for sustainable remediation of environmental pollutants in contaminants matrix (Cecchin et al., 2017). There are several reports on the integrated use of nanoparticles and bioremediation. For instance, Le et al. (2015) conducted a study for the degradation of a solution containing Aroclor 1248 (PCB) using nZVI (1000 mg/L) and subsequent using biodegradation with bacterium, *Burkholderia xenovorans*. The researchers obtained 89% degradation of the congeners after application of nZVI. Subsequently, they observed a biodegradation of 90% in the biphenyls produced after the dechlorination of PCB by bacterial metabolism. In this study, no toxic effect toward microorganisms by the nZVI was observed. Further, Bokare et al. (2010) conducted a study into the feasibility of an integration of bioremediation process and reductive process through nanoparticles in a contaminated solution with triclosan (5 g/L). The researchers promoted a sequential degradation of the contaminant by subjecting it to an anaerobic dechlorination through the nanoparticles of Pd/Fe. Subsequently, further remediation is achieved by oxidation of the by-products through the application of the enzyme produced by *Trametes versicolor* (laccase-producing fungi). The results showed complete dechlorination of triclosan in 20 min after application, and its by-products totally oxidized by microbial enzyme.

Singh et al. (2013) succeeded (99% efficiency) in remediation of soil contaminated with lindane, in 6 days, by the combined action of CMC-Pd/Fe nanoparticles and *Sphingomonas* species. Bokare et al. (2012) conducted a study combining dechlorination processes by Pd/Fe nanoparticles and degrading action of *Sphingomonas wittichii* to remediate a contaminant solution of 2,3,7,8-tetrachlorodibenzo-p dioxin. Considering the high persistence and low biodegradability of the compound, it carried out a dechlorination process in anaerobic environment with 2,3,7,8-tetrachlorodibenzo-*p*-dioxin through the nanoparticles of Pd/Fe, yielding 100% of dechlorination in only 10 h of exposure. After dechlorination the by-product obtained (dibenzo-*p*-dioxin), not amenable to remediation by the nanoparticles, was totally degraded by microbial metabolism Thus nanobioremediation could be an excellent strategy for the remediation of contaminants in environmental matrix.

The major advantages of using nanoparticles in conjunction with bioremediation include (Koul and Taak, 2018) (1) increased pollutants removal due to high surface area of NPs; (2) highly reactive to contaminants; (3) NPs can diffuse or penetrate into a contamination zone where microorganisms cannot reach; (4) higher reactivity to redox-amenable contaminants; (5) fast degradation as compared to microbial degradation; (6) low cost of production; and (7) NPs can be used to immobilize microbial cells for further degradation process. However, the major disadvantages associated with the application of nanoparticles in conjunction with bioremediation include (Cecchin et al., 2017) (1) clogging of soil pores during nanobioremediation of contaminated soil and (2) toxicity to remediating microorganisms. Currently, the existing results of nanobioremediation to date are controversial and limited to the type of nanoparticle and tested microorganism. Further, to advance the nanobioremediation technology, the future research should be focused on the (Cecchin et al., 2017) (1) investigation of effects of the subsurface in situ conditions, such as particle size, moisture, pH, presence of nutrients, organic matter, native microorganism, type of clay mineral, among other such factors on the reaction between the microbes and the nanoparticle; (2) characterization of interactions of nanoparticles with the soil particles and the local microbiota at field scale; and (3) understanding the effects of nanoparticles on microorganisms in variable soil conditions. Overall, the fundamental research is required to engineer the sustainable nanobioremediation to optimize the degradation of soil contaminants without adversely impacting the native microorganisms.

22.3.6 Microbe-assisted phytoremediation technology

The use of plant-associated microbes with desired traits to enhance the phytoremediation efficiency of hyperaccumulating plants *via* increasing the bioavailability of metals in soil and plant growth promotion in the stressed environment is termed microbe-assisted phytoremediation (Saxena et al., 2020c). Microbes such as plant growth–promoting bacteria (PGPR), endophytic bacteria, and arbuscular mycorrhizal fungi (AMF) have been reported useful in assisting the phytoremediation of organic and inorganic pollutants (Saxena et al., 2020c). Microbe-assisted phytoremediation overcomes the limitations of phytoremediation (long time frame required for phytoremediation and physiological damage to remediating plants under toxic metal stress) and offers more sustainable way toward the remediation of HM–contaminated sites (Saxena et al., 2020c). The inoculation of plants with PGPR (root/rhizosphere colonizing bacteria) has been proved to be helpful in phytoremediation as they can suppress phytopathogens, enhance tolerance to abiotic stress, and lower the metal toxicity to remediating plants through biosorption/bioaccumulation as bacterial cells have extremely high ratio of surface area to volume as well as promote plant growth by secreting various hormones, organic acids, and antibiotics (Rajkumar et al., 2012; Ullah et al., 2015; Saxena et al., 2020c). Endophytes are also able to tolerate high metal concentration and, hence, lower phytotoxicity to remediating plants and help in growth promotion by various means and, thus, enhance phytoremediation efficiency (Ma et al., 2011; 2015). In addition, AMF (colonize plant roots) have been also reported to protect their host plants against HM toxicity through their mobilization from soil and, thus, help in the phytoremediation (Marques et al., 2009; Khan et al., 2014).

A great deal of literature can be found in the public domain on the microbe-assisted phytoremediation of HMs (Saxena et al., 2020c; Khan et al., 2014; Ma et al., 2011; 2015; Rajkumar et al., 2012; Ullah et al., 2015). For instance, Gupta et al. (2018) studied the microbe-assisted phytoremediation of TWW contaminated agricultural soils. They isolated a Cr^{6+}-resistant plant growth–promoting *Pseudomonas* sp. (strain CPSB21) from the tannery effluent–contaminated agricultural soils and evaluated for the various plant growth–promoting activities, oxidative stress tolerance, and Cr^{6+} bioremediation. Further, they applied the isolated strain for microbe-assisted phytoremediation and reported that the inoculation of strain CPSB21 alleviated the Cr^{6+} toxicity and enhanced the plant growth parameters and nutrient uptake in sunflower plant during pot experiment. Jampasri et al. (2020) studied the bacteria-assisted phytoremediation of fuel oil and lead cocontaminated soil in the salt-stressed condition by *Chromolaena odorata* and *Micrococcus luteus*. According to their study, the highest percentage of TPH degradation in the present study was found in contaminated nonsaline soil with *C. odorata* and *M. luteus* (55.1%) after 90 days of treatment. Prapagdee and Khonsue (2015) performed a field trial on the bacterial-assisted phytoremediation of cadmium by *Ocimum gratissimum* L. in polluted agricultural soil. According to their study, the highest percentage of cadmium removal was found in soil used to cultivate EDTA-treated *O. gratissimum* L., followed by an *Arthrobacter* sp.–inoculated plant. Nanekar et al. (2015) also studied the microbe-assisted phytoremediation of oil sludge and role of amendments. According to their study, 72.8% TPH degradation was observed in bulked treatment with plant, nutrients, and consortium followed by 69.6% and 65.4% in bioaugmented treatments with and without nutrients, respectively, as compared to control (33.4%).

However, microbes are sensitive to pollution, and depletion of microbial populations, both in terms of diversity and biomass, often occurs in contaminated soils (Özyiğit and Doğan, 2014). Biotic or abiotic stress through a small change in the physicochemical–biological properties of rhizosphere soils can also cause a dramatic effect on plant– microbe interaction (Özyiğit and Doğan, 2014). Moreover, the isolation and characterization of suitable plant-associated beneficial microbes is a time-consuming process (Saxena et al., 2020c). It also requires the analysis of more than thousands of isolates and thus identification of specific biomarkers may help to select the effective plant–microbe associationships for microbe-assisted phytoremediation (Rajkumar et al., 2012). Further, to ameliorate metal toxicity, plant growth promotion, and metal sequestration, extensive research efforts are also required to explore novel microbial diversity, their distribution, as well as functions in the autochthonous and allochthonous soil habitats for microbe-assisted phytoremediation of HM-contaminated sites.

22.3.7 Anammox technology

Anaerobic ammonium oxidation (anammox) is a novel way to reduce nitrogen in ammonium-rich wastewater or removal of ammonia from municipal and industrial wastewaters and it is currently emerging because of its low-cost and energy-consuming nature (Anjali and Sabumon, 2014). It involves the anoxic oxidation of ammonia with nitrite as a preferred electron acceptor and consumes 50% less oxygen and 100% less organic carbon and saves 90% of operational costs in sludge disposal as compared to the conventional nitrification/denitrification processes (Anjali and Sabumon, 2014). Therefore industries, producing wastewaters having high concentration of ammonia, are showing increased interest in anammox process. However, the long start-up time and inhibitive nature in the presence of organic carbon and NH_4–N limits its field applications (Saxena et al., 2017). Therefore it is imperative to develop the mixed consortium capable of anammox in the presence of organic compounds. Further, the development of mixed microbial consortium consisting of ammonia-oxidizing bacteria, anammox bacteria, and denitrifying bacteria is also expected to treat the wastewaters containing both ammonia and organic carbon.

During anammox, ammonium and nitrite are directly coupled under anoxic conditions to form dinitrogen gas by anammox bacteria without the need for costly aeration or an external carbon source such as methanol (Gonzalez-Martinez et al., 2018). The major advantages of anammox process over nitrification–denitrification processes are the lower sludge generation, no requirements of external carbon source, and the cost savings from aeration, due to only half of ammonia needs to be converted to nitrite (Gonzalez-Silva et al., 2017). These processes occur under two different sets of environmental conditions (Gonzalez-Martinez et al., 2018). First, a partial nitrification that requires aerobic conditions, which is followed by the anammox process under anaerobic conditions and without requirements of carbon source. The partial nitritation process involves the oxidation of half of the influent ammonium to nitrite by metabolism of autotrophic ammonium-oxidizing bacteria (AOB). In this way, Gonzalez-Martínez et al. (2014) reported that the dominant AOB in the partial nitritation processes were *Nitrosomonas eutropha*, *Nitrosomonas europaea*, and *Nitrosospira*. After the partial nitritation process, both ammonium and nitrite can be converted into dinitrogen gas due to autotrophic nitrogen removal process carried out by anammox bacteria. Some important variables to consider in the anammox process in addition to the availability of nitrite are the ammonium concentration, biomass concentration, pH, temperature, organic matter concentration, salinity, and toxic compounds (Jin et al., 2012). These parameters can inhibit partially or totally the specific anammox activity and also modify the anammox bacterial community structure (Gonzalez-Martinez et al., 2018). However, the main limitation in the way of implementation of the anammox technology is the long start-up periods, due to long duplication time of about 7–22 days for the anammox bacteria and their low yield of biomass (Gonzalez-Martinez et al., 2018; Lotti et al., 2014),. Therefore the future research should be focused to enhance the biomass retention time (Gonzalez-Martinez et al., 2018).

22.3.8 Constructed wetland-microbial fuel cell system

Integrating CWs with a MFC for wastewater treatment and electricity generation could be an innovative approach for the improved degradation of pollutants. The integration of MFC with a CW will upgrade the CW to allow it to be used for wastewater treatment and, simultaneously, electricity generation, making the CWs more sustainable and environment-friendly (Bharagava et al., 2020). The possible reasons behind the integration of MFC in CWs are (Wang et al., 2017) (1) testifying or verifying the feasibility of this integration, which includes the trails on both vertical flow and horizontal flow CWs. In addition, different types of feeding waters varied from synthetic wastewater to real polluted water (swine water or urban wastewater) were also testified; (2) utilizing the electrodes reactions for specific pollutants removal or nutrients removal enhancement; and (3) attempts to increase the power production. However, one of the main purposes of MFC integration in traditional CWs is that the produced electricity can promote pollutants removal in wastewater (Wang et al., 2017).

There are several reports available in the public domain on the integration of MFC with CWs for the wastewater treatment, pollutants removal, and power generation. According to a recent study, a maximum power density of 15.73 mW/m^2 and maximum current density of 69.75 mA/m^2 could be achieved during the treatment of synthetic wastewater containing methylene blue dye (1000 mg/L 765) with 75% COD removal in an integrated CW-MFC system planted with an ornamental plant, *Canna indica* (Yadav et al., 2012). Oon et al. (2016) designed an integrated upflow constructed wetland-MFC (UFCW-MFC) system planted with cattail for the simultaneous wastewater treatment and electricity generation. According to their study, the removal efficiencies of COD, NO^{3-}, and NH_4^+ were 100%, 40%, and 91%, respectively, with the maximum power density of 6.12 mW/m^2 and Coulombic efficiency of 8.6% at electrode spacing of anode 1 (A1) and cathode (15 cm). Fang et al. (2013) studied the performance of a MFC–coupled constructed wetland system for decolorization of azo dye and bioelectricity generation. They reported that the planted CW-MFC system achieved the highest decolorization rate of about 91.24% for azo dyes and a voltage output of about 610 mV.

22.4 Conclusion

Industrial wastes are considered as the major sources of environmental pollution and contain a variety of highly toxic organic and inorganic pollutants, which causes severe toxic effects in living beings. Physicochemical remediation technologies are environmentally destructive and require high capital investment for environmental cleanup. Microbial remediation technologies offer potential solution for environmental management and overcome the limitations of physicochemical remediation technologies. However, we can say that there is no treatment method at its best to remediate/treat the contaminated sites/industrial wastewaters and its chemicals because every technology has its own advantages and disadvantages. Moreover, it is clear that continuous efforts are required in order to develop sustainable

FIGURE 22.1 Microbial technologies for environmental remediation.

environmental microbial remediation technologies for environmental management. Further, continued efforts are required to realize the economic feasibility of microbial remediation technologies at the field scale (Fig. 22.1).

References

Adelaja, O., Keshavarz, T., Kyazze, G., 2014. Enhanced biodegradation of phenanthrene using different inoculum types in a microbial fuel cell. Eng. Life Sci. 14 (2), 218–228.

Anjali, G., Sabumon, P.C., 2014. Unprecedented development of anammox in presence of organic carbon using seed biomass from a tannery Common Effluent Treatment Plant (CETP). Bioresour. Technol. 153, 30–38.

Antizar-Ladislao, B., 2010. Bioremediation: working with bacteria. Elements 6 (6), 389–394.

Azhar, A.T.S., Nabila, A.T.A., Nurshuhaila, M.S., Shaylinda, M.Z.N., Azim, M.A.M., 2016. Electromigration of contaminated soil by electro-bioremediation technique. In: IOP Conference Series: Materials Science and Engineering, vol. 136, No. 1. IOP Publishing, p. 012023.

Bharagava, R.N., Mishra, S., 2018. Hexavalent chromium reduction potential of *Cellulosimicrobium* sp. isolated from common effluent treatment plant of tannery industries. Ecotoxicol. Environ. Saf. 147, 102–109.

Bharagava, R.N., Chowdhary, P., Saxena, G., 2017. Bioremediation: an eco-sustainable green technology: its applications and limitations. Environmental Pollutants and Their Bioremediation Approaches. CRC Press, pp. 1–22.

Bharagava, R.N., Mani, S., Mullab, S.I., Saratale, G.D., 2018a. Degradation and decolourization potential of an ligninolytic enzyme producing *Aeromonas hydrophila* for crystal violet dye and its phytotoxicity evaluation. Ecotoxicol. Environ. Saf. 156, 166–175.

Bharagava, R.N., Saxena, G., Mulla, S.I., Patel, D.K., 2018b. Characterization and identification of recalcitrant organic pollutants (ROPs) in tannery wastewater and its phytotoxicity evaluation for environmental safety. Arch. Environ. Contam. Toxicol. 75 (2), 259–272.

Bharagava, R.N., Saxena, G., Mulla, S.I., 2020. Introduction to industrial wastes containing organic and inorganic pollutants and bioremediation approaches for environmental management. Bioremediation of Industrial Waste for Environmental Safety. Springer, Singapore, pp. 1–18.

Bokare, V., Murugesan, K., Kim, Y.M., Jeon, J.R., Kim, E.J., Chang, Y.S., 2010. Degradation of triclosan by an integrated nano-bio redox process. Bioresour. Technol. 101 (16), 6354–6360.

Bokare, V., Murugesan, K., Kim, J.H., Kim, E.J., Chang, Y.S., 2012. Integrated hybrid treatment for the remediation of 2,3,7,8-tetrachlorodibenzo-*p*-dioxin. Sci. Total Environ. 435, 563–566.

Cecchin, I., Krishna, R., Thom, A., Tessaro, E.F., Schnaid, F., 2017. Nanobioremediation: Integration of nanoparticles and bioremediation for sustainable remediation of chlorinated organic contaminants in soils. Int. Biodeterior. Biodegrad. 119, 419–428.

Chan, A., Salsali, H., McBean, E., 2014. Heavy metal removal (copper and zinc) in secondary effluent from wastewater treatment plants by microalgae. ACS Sustain. Chem. Eng. 2 (2), 130–137.

Chandra, R., Saxena, G., Kumar, V., 2015. Phytoremediation of environmental pollutants: an eco-sustainable green technology to environmental management, Advances in Biodegradation and Bioremediation of Industrial Waste, first ed CRC Press, Taylor & Francis Group, Boca Raton, FL, pp. 1–30.

Daghio, M., Aulenta, F., Vaiopoulou, E., Franzetti, A., Arends, J.B., Sherry, A., et al., 2017. Electrobioremediation of oil spills. Water Res. 114, 351–370.

Dong, Y., Feng, Y., Qu, Y., Du, Y., Zhou, X., Liu, J., 2015. A combined system of microbial fuel cell and intermittently aerated biological filter for energy self-sufficient wastewater treatment. Sci. Rep. 5, 18070.

Fan, C.Y., Krishnamurthy, S., 1995. Enzymes for enhancing bioremediation of petroleum-contaminated soils: a brief review. J. Air Waste Manage. Assoc. 45 (6), 453–460.

Fang, Z., Song, H.-L., Cang, N., Li, X.-N. 2013. Performance of microbial fuel cell coupled constructed wetland system for decolorization of azo dye and bioelectricity generation. Bioresour. Technol. 144, 165–171.

Friman, H., Schechter, A., Nitzan, Y., Cahan, R., 2013. Phenol degradation in bio-electrochemical cells. Int. Biodeterior. Biodegrad. 84, 155–160.

Ge, Z., Zhang, F., Grimaud, J., Hurst, J., He, Z., 2013. Long-term investigation of microbial fuel cells treating primary sludge or digested sludge. Bioresour. Technol. 136, 509–514.

Ghangrekar, M.M., Shinde, V.B., 2008. Simultaneous sewage treatment and electricity generation in membrane-less microbial fuel cell. Water Sci. Technol. 58 (1), 37–43.

Gill, R.T., Harbottle, M.J., Smith, J.W.N., Thornton, S.F., 2014. Electrokinetic-enhanced bioremediation of organic contaminants: a review of processes and environmental applications. Chemosphere 107, 31–42.

Gonzalez-Martínez, A., Pesciaroli, C., Martínez-Toledo, M.V., et al., 2014. Study of nitrifying microbial communities in a partial-nitritation bioreactor. Ecol. Eng. 64, 443–450.

Gonzalez-Silva, B.M., Rønning, A.J., Andreassen, I.K., Bakke, I., Cervantes, F.J., Østgaard, K., et al., 2017. Changes in the microbial community of an anammox consortium during adaptation to marine conditions revealed by 454 pyrosequencing. Appl. Microbiol. Biotechnol. 101 (12), 5149–5162.

Gonzalez-Martinez, A., Muñoz-Palazon, B., Rodriguez-Sanchez, A., Gonzalez-Lopez, J., 2018. New concepts in anammox processes for wastewater nitrogen removal: recent advances and future prospects. FEMS Microbiol. Lett. 365 (6), fny031.

Gude, V.G., 2016. Wastewater treatment in microbial fuel cells—an overview. J. Clean. Prod. 122, 287–307.

Gupta, P., Rani, R., Chandra, A., Kumar, V., 2018. Potential applications of *Pseudomonas* sp. (strain CPSB21) to ameliorate Cr^{6+} stress and phytoremediation of tannery effluent contaminated agricultural soils. Sci. Rep. 8 (1), 1–10.

Ho, S.V., Athmer, C., Sheridan, P.W., Hughes, B.M., Orth, R., McKenzie, D., et al., 1999a. The Lasagna technology for in situ soil remediation. 1. Small field test. Environ. Sci. Technol. 33 (7), 1086–1091.

Ho, S.V., Athmer, C., Sheridan, P.W., Hughes, B.M., Orth, R., McKenzie, D., et al., 1999b. The Lasagna technology for in situ soil remediation. 2. Large field test. Environ. Sci. Technol. 33 (7), 1092–1099.

Huang, D.Y., Zhou, S.G., Chen, Q., Zhao, B., Yuan, Y., Zhuang, L., 2011. Enhanced anaerobic degradation of organic pollutants in a soil microbial fuel cell. Chem. Eng. J. 172 (2–3), 647–653.

Jadhav, J.P., Phugare, S.S., Dhanve, R.S., Jadhav, S.B., 2010. Rapid biodegradation and decolorization of Direct Orange 39 (Orange TGLL) by an isolated bacterium *Pseudomonas aeruginosa* strain BCH. Biodegradation 21 (3), 453–463.

Jampasri, K., Pokethitiyook, P., Poolpak, T., Kruatrachue, M., Ounjai, P., Kumsopa, A., 2020. Bacteria-assisted phytoremediation of fuel oil and lead co-contaminated soil in the salt-stressed condition by *Chromolaena odorata* and *Micrococcus luteus*. Int. J. Phytoremediat. 22 (3), 322–333.

Jiang, H.M., 2017. Combination of microbial fuel cells with microalgae cultivation for bioelectricity generation and domestic wastewater treatment. Environ. Eng. Sci. 34 (7), 489–495.

Jin, R.C., Yang, G.F., Ma, C., et al., 2012. Influence of effluent recirculation on the performance of Anammox process. Chem. Eng. J. 200–202, 176–185.

Kanagasabi, S., Kang, Y.L., Manickam, M., Ibrahim, S., Pichiah, S., 2013. Intimate coupling of electro and biooxidation of tannery wastewater. Desalination Water Treat. 51 (34–36), 6617–6623.

Karigar, C.S., Rao, S.S., 2011. Role of microbial enzymes in the bioremediation of pollutants: a review. Enzyme Res. 2011.

Kathiravan, M.N., Karthick, R., Muthukumar, K., 2011. Ex situ bioremediation of Cr(VI) contaminated soil by *Bacillus* sp.: batch and continuous studies. Chem. Eng. J. 169 (1–3), 107–115.

Khan, A., Sharif, M., Ali, A., Shah, S.N.M., Mian, I.A., Wahid, F., et al., 2014. Potential of AM fungi in phytoremediation of heavy metals and effect on yield of wheat crop. Am. J. Plant. Sci. 5, 1578–1586.

Koul, B., Taak, P., 2018. Nanobioremediation. Biotechnological Strategies for Effective Remediation of Polluted Soils. Springer, Singapore, pp. 197–220.

Kumar, L., Bharadvaja, N., 2019. Enzymatic bioremediation: a smart tool to fight environmental pollutants. Smart Bioremediation Technologies. Academic Press, pp. 99–118.

Kurade, M.B., Waghmode, T.R., Kagalkar, A.N., Govindwar, S.P., 2012. Decolorization of textile industry effluent containing disperse dye Scarlet RR by a newly developed bacterial-yeast consortium BL-GG. Chem. Eng. J. 184, 33–41.

Le, T.T., Nguyen, K.-H., Jeon, J.-R., Francis, A.J., Chang, Y.-S., 2015. Nano/bio treatment Xof polychlorinated biphenyls with evaluation of comparative toxicity. J. Hazard. Mater. 287, 335e341.

Lee, K.Y., Bosch, J., Meckenstock, R.U., 2012. Use of metal-reducing bacteria for bioremediation of soil contaminated with mixed organic and inorganic pollutants. Environ. Geochem. Health 34 (1), 135–142.

Li, T., Guo, S., Wu, B., Li, F., Niu, Z., 2010. Effect of electric intensity on the microbial degradation of petroleum pollutants in soil. J. Environ. Sci. 22 (9), 1381–1386.

Lotti, T., Kleerebezem, R., Lubello, C., Van Loosdrecht, M.C.M., 2014. Physiological and kinetic characterization of a suspended cell anammox culture. Water Res. 60, 1–14.

Lu, L., Huggins, T., Jin, S., Zuo, Y., Ren, Z.J., 2014. Microbial metabolism and community structure in response to bioelectrochemically enhanced remediation of petroleum hydrocarbon-contaminated soil. Environ. Sci. Technol. 48 (7), 4021–4029.

Ma, Y., Prasad, M.N.V., Rajkumar, M., Freitas, H., 2011. Plant growth promoting rhizobacteria and endophytes accelerate phytoremediation of metalliferous soils. Biotechnol. Adv. 29, 248–258.

Ma, Y., Oliviera, R.S., Nai, F., Rajkumar, M., Luo, Y., Rocha, I., et al., 2015. The hyperaccumulator *Sedum plumbizincicola* harbors metal-resistant endophytic bacteria that improve its phytoextraction capacity in 1206 multi-metal contaminated soil. J. Environ. Manage. 156, 62–69.

Marques, A.P.G.C., Rangel, A.O.S.S., Castro, P.M.L., 2009. Remediation of heavy metal contaminated soils: phytoremediation as a potentially promising clean-up technology. Crit. Rev. Environ. Sci. Technol. 39 (8), 622–654.

Martínez-Prado, M.A., Unzueta-Medina, J., Pérez-López, M.E., 2014. Electrobioremediation as a hybrid technology to treat soil contaminated with total petroleum hydrocarbons. Rev. Mexicana de. Ingeniería Química 13 (1), 113–127.

Maszenan, A.M., Liu, Y., Ng, W.J., 2011. Bioremediation of wastewaters with recalcitrant organic compounds and metals by aerobic granules. Biotechnol. Adv. 29 (1), 111–123.

Megharaj, M., Ramakrishnan, B., Venkateswarlu, K., Sethunathan, N., Naidu, R., 2011. Bioremediation approaches for organic pollutants: a critical perspective. Environ. Int. 37 (8), 1362–1375.

Morris, J.M., Jin, S., Crimi, B., Pruden, A., 2009. Microbial fuel cell in enhancing anaerobic biodegradation of diesel. Chem. Eng. J. 146 (2), 161–167.

Nanekar, S., Dhote, M., Kashyap, S., Singh, S.K., Juwarkar, A.A., 2015. Microbe assisted phytoremediation of oil sludge and role of amendments: a mesocosm study. Int. J. Environ. Sci. Technol. 12 (1), 193–202.

Oon, Y.L., Ong, S.A., Ho, L.N., Wong, Y.S., Dahalan, F.A., Oon, Y.S., et al., 2016. Synergistic effect of up-flow constructed wetland and microbial fuel cell for simultaneous wastewater treatment and energy recovery. Bioresour. Technol. 203, 190–197.

Özyiğit, İ.İ., Doğan, İ., 2014. Plant-Microbe Interactions in Phytoremediation.

Paisio, C.E., Talano, M.A., González, P.S., Busto, V.D., Talou, J.R., Agostini, E., 2012. Isolation and characterization of a *Rhodococcus* strain with phenol-degrading ability and its potential use for tannery effluent biotreatment. Environ. Sci. Pollut. Res. 19 (8), 3430–3439.

Passos, V.F., Aquino Neto, S., Andrade, A.R.D., Reginatto, V., 2016. Energy generation in a microbial fuel cell using anaerobic sludge from a wastewater treatment plant. Sci. Agricola 73 (5), 424–428.

Prapagdee, B., Khonsue, N., 2015. Bacterial-assisted cadmium phytoremediation by *Ocimum gratissimum* L. in polluted agricultural soil: a field trial experiment. Int. J. Environ. Sci. Technol. 12 (12), 3843–3852.

Rajkumar, M., Sandhya, S., Prasad, M.N., Freitas, H., 2012. Perspectives of plant-associated microbes in heavy metal phytoremediation. Biotechnol. Adv. 30 (6), 1562–1574.

Rao, M.A., Scelza, R., Scotti, R., Gianfreda, L., 2010. Role of enzymes in the remediation of polluted environments. J. Soil Sci. Plant. Nutr. 10 (3), 333–353.

Saichek, R.E., Reddy, K.R., 2005. Electrokinetically enhanced remediation of hydrophobic organic compounds in soils: a review. Crit. Rev. Environ. Sci. Technol. 35 (2), 115–192.

Saxena, G., Bharagava, R.N., 2017. Organic and inorganic pollutants in industrial wastes: ecotoxicological effects, health hazards, and bioremediation approaches. Environmental Pollutants and Their Bioremediation Approaches. CRC Press, pp. 23–56.

Saxena, G., Chandra, R., Bharagava, R.N., 2017. Environmental pollution, toxicity profile and treatment approaches for tannery wastewater and its chemical pollutants. Rev Environ Contam Toxicol. 240, 31–69. https://doi.org/10.1007/398_2015_5009

Saxena, G., Purchase, D., Bharagava, R.N., 2020a. Environmental hazards and toxicity profile of organic and inorganic pollutants of tannery wastewater and bioremediation approaches. Bioremediation of Industrial Waste for Environmental Safety. Springer, Singapore, pp. 381–398.

Saxena, G., Goutam, S.P., Mishra, A., Mulla, S.I., Bharagava, R.N., 2020b. Emerging and ecofriendly technologies for the removal of organic and inorganic pollutants from industrial wastewaters. Bioremediation of Industrial Waste for Environmental Safety. Springer, Singapore, pp. 113–126.

Saxena, G., Purchase, D., Mulla, S.I., Saratale, G.D., Bharagava, R.N., 2020c. Phytoremediation of heavy metal-contaminated sites: eco-environmental concerns, field studies, sustainability issues, and future prospects. Rev. Environ. Contam. Toxicol. 249, 71–131.

Saxena, G., Kishor, R., Bharagava, R.N., 2020d. Application of microbial enzymes in degradation and detoxification of organic and inorganic pollutants. Bioremediation of Industrial Waste for Environmental Safety. Springer, Singapore, pp. 41–51.

Sharma, S., Malaviya, P., 2016. Bioremediation of tannery wastewater by chromium resistant novel fungal consortium. Ecol. Eng. 91, 419–425.

Singh, R., Manickam, N., Mudiam, M.K.R., Murthy, R.C., Misra, V., 2013. An integrated (nano-bio) technique for degradation of γ-HCH contaminated soil. J. Hazard. Mater. 258, 35–41.

Singh, H.M., Pathak, A.K., Chopra, K., Tyagi, V.V., Anand, S., Kothari, R., 2019. Microbial fuel cells: a sustainable solution for bioelectricity generation and wastewater treatment. Biofuels 10 (1), 11–31.

Sutherland, T.D., Horne, I., Weir, K.M., Coppin, C.W., Williams, M.R., Selleck, M., et al., 2004. Enzymatic bioremediation: from enzyme discovery to applications. Clin. Exp. Pharmacol. Physiol. 31 (11), 817–821.

Tripathi, M., Garg, S.K., 2013. Co-remediation of pentachlorophenol and Cr^{6+} by free and immobilized cells of native *Bacillus cereus* isolate: spectrometric characterization of PCP dechlorination products, bioreactor trial and chromate reductase activity. Process. Biochem. 48 (3), 496–509.

Ullah, A., Heng, S., Munis, M., Hussain, F., Shah, F., Xiyan, Y., 2015. Phytoremediation of heavy metals assisted by plant growth promoting (PGP) bacteria: a review. Environ. Exp. Bot. 117. Available from: http://dx.doi.org/10.1016/j.envexpbot.2015.05.001.

Wang, Y., Zhao, Y., Xu, L., Wang, W., Doherty, L., Tang, C., et al., 2017. Constructed wetland integrated microbial fuel cell system: looking back, moving forward. Water Sci. Technol. 76 (2), 471–477.

Wick, L.Y., Shi, L., Harms, H., 2007. Electro-bioremediation of hydrophobic organic soil-contaminants: A review of fundamental interactions. Electrochim. Acta 52 (10), 3441–3448.

Yadav, A.K., Dash, P., Mohanty, A., Abbassi, R., Mishra, B.K., 2012. Performance assessment of innovative constructed wetland-microbial fuel cell for electricity production and dye removal. Ecol. Eng. 47, 126–131.

Yan, F., Reible, D., 2015. Electro-bioremediation of contaminated sediment by electrode enhanced capping. J. Environ. Manage. 155, 154–161.

Yan, Z., Song, N., Cai, H., Tay, J.H., Jiang, H., 2012. Enhanced degradation of phenanthrene and pyrene in freshwater sediments by combined employment of sediment microbial fuel cell and amorphous ferric hydroxide. J. Hazard. Mater. 199, 217–225.

Zhang, F., Ge, Z., Grimaud, J., Hurst, J., He, Z., 2013a. Long-term performance of literscale microbial fuel cells treating primary effluent installed in a municipal wastewater treatment facility. Environ. Sci. Technol. 47 (9), 4941–4948.

Zhang, F., Ge, Z., Grimaud, J., Hurst, J., He, Z., 2013b. In situ investigation of tubular microbial fuel cells deployed in an aeration tank at a municipal wastewater treatment plant. Bioresour. Technol. 136, 316–321.

Zhang, Y., Liu, J., Zhou, Y., Gong, T., Wang, J., Ge, Y., 2013c. Enhanced phytoremediation of mixed heavy metal (mercury)–organic pollutants (trichloroethylene) with transgenic alfalfa co-expressing glutathione S-transferase and human P450 2E1. J. Hazard. Mater. 260, 1100–1107.

Chapter 23

Microbially synthesized nanomaterials for remediation of organic contaminants

Ritu Singh[1], Prashant Kumar Jaiswal[1], Nisha Kumari[1], Monalisha Behera[1], Avimanu Sharma[1] and Sanjay Kumar Gupta[2]

[1]*Department of Environmental Science, School of Earth Sciences, Central University of Rajasthan, Ajmer, India,* [2]*Environmental Engineering, Department of Civil Engineering, Indian Institute of Technology Delhi, Delhi, India*

23.1 Introduction

In the present time of agricultural modernization and speedy development of industrial sector and urban sprawl, several organic and inorganic contaminants, such as pesticides, fertilizers, heavy metals, detergents, dyes, and pharmaceutical drugs, have become abundant in environmental compartments showing their presence in aquatic water bodies, sediments, and soils. The extensive production, uncontrolled usage, and improper management of waste materials generated from human society are responsible for the channelization of the contaminants into the environmental matrices. These waste materials either get accumulated in the environment or get biomagnified in human and animal community, in turn causing life-threatening diseases along with stern loss of ecosystem services. Hence, their removal has become an enervating task to lower their impact on environment and biota. Since decades, researchers have been working on developing different technologies, including physical, chemical, and biological methods for the removal of wide array of pollutants. Among various advanced technologies, nanotechnology has gained a lot of recognition for the distinctive features of nanoparticles (NPs), making the technology an efficient agent of pollutant removal. The nano-size of a material act differently from its bulk counter parts and possess novel characteristics such as larger surface area to volume ratio, high mobility, increased thermal stability, and can be applied both in situ and ex situ for remediation purposes. However, to achieve the nano-size of a particular material is not an easy piece of work since the NPs are often associated with various limitations such as oxidation, agglomeration, instability, and sedimentation. Hence, to remove these limitations, competent synthesis routes are adopted that are hassle free, easy to perform, prevent agglomeration, and can maintain their uniform nano-size. Microbial synthesis is one such approach that not only produces uniform and stable NPs but also is easy to handle and eco-friendly in nature.

Although there are several synthesis approaches that takes less time to prepare NPs as compared to biological synthesis, most of them have drawbacks such as demand of capping agent to stabilize the NPs, use of hazardous chemicals producing toxic by-products having cytotoxic, and carcinogenic effects (Singh et al., 2018). The basic idea during strategizing a remediation technology is the development of an agent that could remove the pollutant from contaminated matrix in a sustainable manner without producing any secondary pollutant and associated risk. Therefore biologically synthesized NP holds its significance compared to other approaches due to their nontoxic nature, plentiful existence, low cost, high efficiency, and environment friendly nature (Fig. 23.1). Some important features such as biocompatibility, natural abundance of microbial community, biodegradability, hydrophilicity, antimicrobial properties, ease of quick synthesis, and optimization of size characteristics make biosynthesized NPs an efficient remedial tool. Hence, the biological route of synthesis has emerged as a better and safer alternative to conventional methods at present and is increasingly used by researchers. This chapter briefly discusses about various synthesis approaches of NPs including physical, chemical, and biological routes. Then the chapter prominently focuses on microbial synthesis of different NPs through the application of varied biological agents followed by the discussion of prevailing environmental factors affecting microbial synthesis of NPs. Furthermore, the potential application of microbially synthesized NPs in area on environmental cleanup and future perspectives is presented.

Microbe Mediated Remediation of Environmental Contaminants. DOI: https://doi.org/10.1016/B978-0-12-821199-1.00023-7

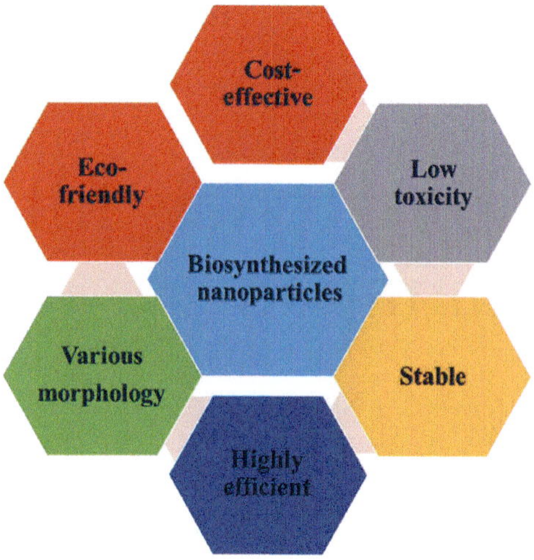

FIGURE 23.1 Benefits of biosynthesized nanoparticles.

23.2 Different synthesis approaches for nanoparticles

Numerous methods have been reported for the synthesis of NPs. Broadly these methods can be categorized by physical, chemical, and biological approaches. Physical approach includes techniques such as laser ablation, melt mixing (Pyatenko et al., 2004), and highball milling method (De Carvalho et al., 2013). Chemical approach includes chemical reduction (Khanna et al., 2005; Aguilar et al., 2019), sol–gel method, (Xu et al., 2007; Aziz et al., 2013) and so on. Synthesized nanomaterials are being used in different shapes, size, and forms such as powders, clusters, colloids, rods tubes, thin films, and wire depending on the end use application. Various studies are being conducted on already existing conventional techniques to develop novel material with better properties and functionality along with cost reduction. These techniques depend on multiple factors such as material of interest, the dimension, size, and quantity of nanomaterial. Herein, three major approaches of NP synthesis have been presented and discussed.

23.2.1 Physical approach

Physical approach mainly represents synthesis of nanomaterials through top-down approach, where synthesis is initiated with the bulk counterparts breaking down to nano-levels. It can be done through the application of mechanical pressure, high-energy radiations, thermal/electrical energy which causes abrasion, melting, evaporation, or condensation of the source material ending up with formation of NPs. The added advantage of these methods is that they do not use any solvents and generate uniform monodisperse NPs. The limitation is that it is much expensive in comparison to other routes of synthesis. Physical approach mainly includes mechanical and vapor methods which are discussed next.

23.2.1.1 Mechanical

These methods are robust and energy-efficient methods for the production of nanomaterials with varying shapes and dimensions. This mainly includes the following methods:

(a) High energy ball milling

It is a simple technique of formation of NPs of some metal and alloys in powder form. Different types of mills are used to form bulk quantities of fine particles. Majorly there are four types of milling, namely, planetary, vibratory, rod, tumbler, etc. Briefly one of the two containers is used for preparation along with hardened steel or tungsten carbide balls and powder/flakes of precursor material. The speed of rotation plays an important role in defining the size of obtained nanoforms (few nm to few tens of nm). This technique has been used for making nanocrystalline metals such as Al–Fe, Ag–Fe, Cr, Co, W, and Ni–Ti. It takes minutes to few hours to make few milligrams to kilograms of NPs (Yadav et al., 2012).

(b) Melt mixing

It is a method that trap/arrest the nanomaterial inside a glass. Glass represents an amorphous solid having no long periodic arrangement and no symmetry of the atoms/molecules. Whenever a solution is cooled below certain temperature at very high rate (10^5-10^6 K/S), it can either form crystalline or amorphous solid. Besides temperature, other deciding factors are rate of cooling and nucleation tendency that determines the formation of glass/crystalline solid. Elements such as silicon, boron, and phosphorous are used as additive, to keep the metallic glass in amorphous state. NPs can be formed within metallic glass by controlled heating and by mixing the molten stream of metal at extremely high velocity with turbulence, for example, a molten stream of Ti and Cu–B mixed to form TiB_2 NPs.

23.2.1.2 Vapor

There are many methods such as resistive heating, electron stream heating, laser heating, and sputtering, which is used to form NPs through vapor method. Through this method, NPs can be designed in the form of thin films, multilayer film, monolayer, multilayer, and clusters by evaporating the material of the interest on appropriate substance. Evaporation should be done in meticulously designed vacuum system to avoid the oxidation of synthesized materials. For evaporation, suitable filaments and crucibles are used.

(a) Laser ablation

In this method, particle size is reduced to nano level by laser radiation. Laser ablation is the process where target material (usually solid or occasionally liquid) is kept under thin layer and finally irradiated with pulsed laser radiation (Jamkhande et al., 2019; Rane et al., 2018). The irradiation of material results in fragmentation of solid material to form NPs; this remains in liquid that surrounds the target and forms colloidal solution. The material is converted to plasma at higher laser flux while at low laser flux, absorbed laser energy heat up the material, this cause material to evaporate or sublimate. When laser intensity is high enough, a continuous laser beam can ablate material instead of removing (wiki, 2017). The ablation of atoms to form particle is determined by the laser pulse energy and duration (Kruis et al., 1998).

(b) Sputtering

When target material is bombarded with energetic particles, atoms are ejected from the surface of material, this phenomenon is called sputtering, where the atoms are driven off by bombarding ions from target or cathode a momentum transfer process. Sputtered atom moves until they strike a substrate to form desired layer after deposition. Working pressure is maintained (0.05 and 0.01 Mbar) while admitting sputtering gas in evacuated vacuum chamber for sputter deposition (Jamkhande et al., 2019). The easiest source for ion sputtering is given by the commonly known phenomenon of glow discharge wherein electric field is applied between two electrodes in a gas at low pressure.

(c) Physical vapor deposition

In this method, a source of evaporation along with an inert/reactive gas such as liquid nitrogen is used for colliding precursor material vapor. Evaporated atoms/clusters get condensed on a cold finger from where they are scrapped and then compacted using piston–anvil arrangement. The particle size and distribution are largely dependent on the evaporation rate, gas pressure, and the distance of source from the cold finger inside the chamber.

(d) Ionized cluster beam deposition

This method produces high-quality single crystalline thin films of nanocrystalline material. The evaporated clusters are allowed to expand through a nozzle where they collide with electron beam and get ionized. The ionized clusters then subjected to acceleration to hit a substrate on which NPs can be deposited.

(e) Laser pyrolysis

It is also termed as laser-assisted deposition wherein reactant molecules present in gaseous form is decomposed with the help of laser beam and allowed to collide with inert gas. During this process, they interact with one another, grow and generate nanoforms which are then subjected to deposition on a cold substrate.

(f) Electric arc deposition

This method is a simplest way of producing bulk quantities of carbon nanotubes, fullerenes, graphene films, etc. The setup consists of a water-cooled vacuum chamber wherein the source material is subjected to vaporization through an arc discharge between electrodes followed by condensation, nucleation, and growth of NPs.

23.2.2 Chemical approach

As compared to the physical methods of NP synthesis, chemical methods are relatively simple and cost-efficient technique for preparation of large quantities of nanomaterials having different size and shape. Also, the possibility of synthesis at low temperatures, self-assembly, and doping with foreign atoms are added advantage of this approach. Broadly chemical synthesis is categorized in two methods: one is the reduction method that includes chemical reduction through sodium borohydride, organic acid, polyol sugar, etc., and the other is the indirect reduction via microwave irradiation, ultrasonication, gamma radiolysis, photoreduction, etc. Another method is sedimentation that includes sol−gel, coprecipitation, alkaline precipitation, and hydrothermal method. Different chemical routes of NP synthesis are presented next.

23.2.2.1 Sol−gel method

Sol−gel chemistry of transition metal oxides was reviewed for the first time by in Livage et al., 1988. Sol is basically colloidal suspension and gel is a continuous network of particles with pores filled with liquid. This method is known for the synthesis of colloidal NPs from liquid phase. Herein, sols are formed in a liquid that connects the sol particles (or other subunits capable to form a porous network) to form a network. Then the liquid is evaporated to obtain thin films/powder/monolithic solids. This method is mostly reported for preparation of metal oxides or ceramics though nitrides, borides, and sulfides could also be synthesized. The major steps in this procedure include mixing, casting, gelation, aging, drying, and densification and chief reactions are alcoholic group hydrolysis and its condensation. The parameters such as concentration and nature of salt precursor, agitation, pH, temperature, kinetics, and properties of gel influences the synthesis (Ennas et al., 1998; Ali et al., 2016). In this procedure, magnetic ordering depends on the solvent phase and its volume (Tavakoli et al., 2007).

23.2.2.2 Chemical reduction method

Chemical reduction is very prevalent and facile method for the synthesis of colloidal metal particles, due to its easy and convenient operation. In this method, metal NPs are deposited on a suitable medium by reducing ionic salt, using different reductant. Reducing agents, such as, sodium borohydride, ascorbic acid, glucose, hydroxylamine, phenyl, hydrazine, hydroquinone, pyrrole, and alkaline solutions are used (Yousefi and Cheraghizade, 2018). Among all reducing agent, sodium borohydride is used as strong and citrate as weak reductant in aqueous solution for the preparation of metal NPs (Martinez-Abad, 2011). Strong reducing agent helps in producing more smaller NPs than weak reducing agent. Sometimes reducing agents are used with stabling agent.

23.2.2.3 Hydrothermal method

Hydrothermal method works on the phenomenon where solid material is reacted with vapors of aqueous solution at elevated temperature and pressure, resulting in the deposition of small particles. Here cation is precipitated in the form of polymeric hydroxide and later these hydroxides forms metal hydroxide crystal structure, once it gets dehydrated. Another formed cation is very helpful in regulating particle formation by avoiding complex hydroxide formation while adding base to a metal salt solution (Ganachari et al., 2017). This method is helpful when chemical precursor is not easily dissolved at normal/room temperature. Hydrothermal method has been reported to be best suited synthesis approach for the synthesis of important solids such as chemical sensing oxides, complex oxide fluoride and ceramics, luminescence phosphorous, magnetic material, electronically conducting solids, supersonic conductors, and microporous crystals (Rane et al., 2018).

23.2.2.4 Sonochemical method

In this method, bubbles are generated with the help of ultrasonic waves in frequency range of ~20 kHz−2 MHz. When bubble bursts, it produces high energy increasing the temperature up to 5000°C and pressure few hundred times that of atmospheric pressure. In these conditions, reaction occurs in gas phase or the liquid phase followed by cooling down. Some of the NPs such as WO_3, ZnS, and CeO_2 have been synthesized using this method.

23.2.2.5 *Cryochemical synthesis*

It is a low temperature synthesis method. In this method, solution components having cations of target material is frozen rapidly and freeze dried under vacuum followed by thermal decomposition. Crystalline metal oxide powders can be easily obtained through this method.

The advantages and limitation of different physical and chemical methods are summarized in Table 23.1.

23.2.3 Biological approach

Biological approach for NP synthesis has several added advantages such as low toxicity, cost effectiveness, environmentally benign, and energy efficient. This approach uses bioreduction process for the fabrication of NPs by plant

TABLE 23.1 Advantages and disadvantages of physical and chemical synthesis approaches.

Synthesis approach	Advantages	Disadvantages	References
Physical 1. Ball milling	It can produce high purity NPs of superior physical properties at large scale. Like it can increase solubility of poorly water-soluble drug compound in a cost-effective manner.	Energy extensive. Prolong time of milling. Very sensitive nanostructure cannot be obtained. Steel balls may contaminate NPs.	Ullah et al. (2014), Thamizharasan and Saravanan (2017), Tavakoli et al. (2007)
2. Laser ablation	Simple and effective technique to form large number of NPs in form of suspension. Properties of NPs can be modified by choosing the laser parameter and nature of liquid accordingly. It is possible to form NPs in liquid media without adding surfactant.	Extensive laser ablation results large number of NPs in colloidal solution, the laser energy is absorbed by NPs instead of target material. This ultimately reduces the ablation rate.	Simakin et al. (2004), Ghorbani (2014)
3. Sputtering	It can be used for almost all types of material. Formed film have the same composition as of target source. Deposited film has the same composition as of the target source. The types and composition of the material to be formed can be controlled.	Sputtering targets are often expensive. Gas composition must be controlled carefully in reactive sputter deposition, to avoid poisoning of target material. Incident energy on the target mostly become heat, which must be removed.	Rane et al. (2018), Jamkhande et al. (2019), Johnson et al. (2016)
Chemical 4. Sol–gel method	Degree of porosity can be controlled. Very convenient to carry out. Particle of various size and morphology can be prepared by monitoring reaction parameter symmetrically. Synthesis can be done at low temperature. Synthesized products are of high purity. Final products are homogeneous.	Organic solvent could harm human body. Reaction time is longer.	Hasnidawani et al. (2016), Rane et al. (2018), Jamkhande et al. (2019)
b. Chemical reduction	Simple technique for the synthesis of metal NP.	There are various limiting factors associated with reducing agent such as poor reducing ability, impurities, toxicity, and expensive.	Zhang et al. (2010), Jamkhande et al. (2019)
c. Hydrothermal method	It can synthesize nanoparticle of desired shape and size. It can produce highly crystalline nanocrystals. Crystallized powder can also be produced.	This method has limited reproducibility and reliability This process is difficult to control Reaction process cannot be observed Safety issue during reaction process	Burda et al. (2005), Tavakoli et al. (2007)

NPs, Nanoparticles.

extracts, microorganisms, and biomolecules (Ramos-Ruiz et al., 2017). Every entity has its own method for the formation of NPs such as oxidation and reduction of metallic ions with sugar, protein, polyphenol, aldehyde, COO− group by microbes and plant species (Lampis et al., 2017; Kharissova et al., 2019). Plant part such as flower, leaves, fruits, and enzymes are used for synthesis of broad range of NPs. Plant extracts serves the purpose of reducing as well as stabilizing agents for NPs. For instance, leaves extract of Persimmon, Pine, Ginkgo, Platanus, and Magnolia has been successfully used as reducing agent for extracellular synthesis of metallic silver NPs (AgNPs) (Song and Kim, 2009). Other plant extracts, namely, *Salvia officinalis*, *Lippia citriodora*, *Pelargonium graveolens*, *Punica granatum*, and *Terminalia arjuna* have been utilized for the synthesis of gold NPs (AuNPs). Biosynthesized AuNPs has potential antibacterial activity against commonly found human pathogens such as *Pseudomonas aeruginosa* (NCIM 5029), *Staphylococcus aureus* (NCIM 5021), and *Salmonella typhimurium* (NCIM 2501) (Elia et al., 2014; Dudhane et al., 2019). Similarly, extracts of *Azadirachta indica* and *Chenopodium album* have also been used for synthesis of AuNPs and AgNPs. *Aspalathus linearis* extract serves as an effective chelating agent for synthesis of cobalt oxide NPs with size ~3.57 nm (Diallo et al., 2015). Another study on green synthesis was done by Ogunyemi et al. (2019), wherein zinc oxide NPs (ZnO NPs) were successfully synthesized using extracts of olive leaves (*Olea europaea*), chamomile flower (*Matricaria chamomilla* L.), and red tomato fruit (*Lycopersicon esculentum* M.). The antibacterial activity of ZnO NPs prepared from olive leaves was observed to have the highest antibacterial activity. Synthesizing metal NPs using plants has been extensively studied and has been recognized as a green and efficient way of synthesis. Several microorganisms such as fungi, bacteria, algae, yeast, and actinomycetes have also been recognized as nano-factories pertaining to their fast growth, easy cultivation, and their ability to grow at ambient condition of pH and temperature. When the microorganisms trap the target ions from the outer environment, they convert them into elemental metal through the enzymes produced during their cell metabolism. Bacterial species have special affinity toward metals and this novel metal binding property makes them useful for nano-bioremediation. Further details over microbial synthesis have been discussed in following section.

23.3 Microbially synthesized nanoparticles

Microbe mediated synthesis of NPs includes use of bacteria, yeast, and fungi, because of their easy cultivation and fast growth rate. Microorganisms have the affinity to reduce the metal ions by enzymatic activity that leads to the formation of the nanoscale material. The mechanism involved in this whole phenomenon is that metal ions are trapped inside or on the surface of microbial cell from external environment, further these captured metal ions are converted to elemental metal form through reduction process (Li et al., 2011; Fariq et al.,2017). In accordance with the location at which NPs are formed, microbial synthesis can be classified as intracellular and extracellular synthesis (Fig. 23.2). In intracellular synthesis, NPs are formed inside the cell while in extracellular synthesis NPs are formed on the surface of cell. Extracellular materials, namely, enzymes, proteins, amino acids take part in the formation of NPs. Extracellular enzymes produced by microorganisms have the potential to reduce metal ions for the formation of pure NPs. Apart from bacteria, algae and yeast have also shown the characteristic of producing large volume of proteins for the synthesis of large amount NPs (Das et al., 2018).

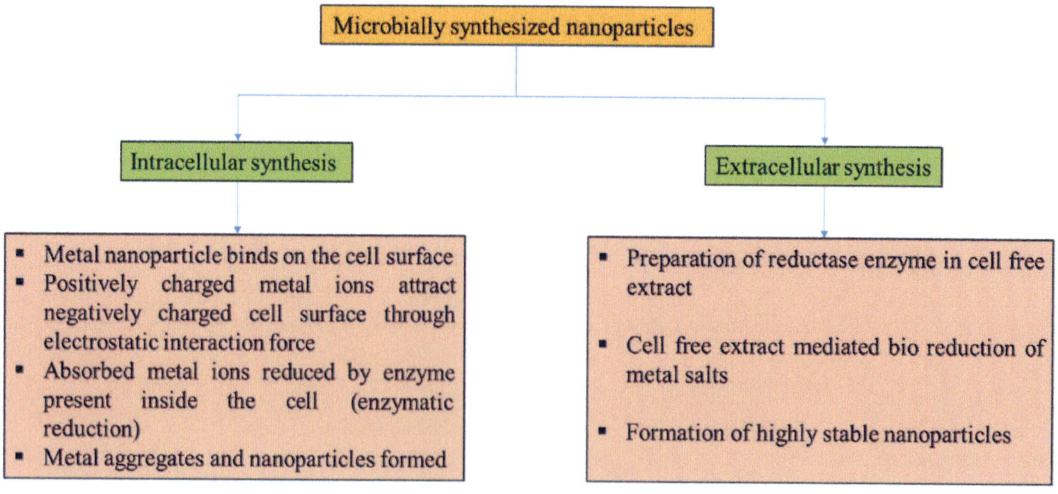

FIGURE 23.2 Mechanism of microbe-mediated synthesis of metallic nanoparticles.

23.3.1 Bacteria

Several reports are available for the production of NPs by different bacterial strain. For synthesis of metal/metal oxide NPs, actinomycetes and prokaryotic bacteria are widely used. Strain *Pseudomonas stutzeri* AG259, isolated from silver mines, showed the synthesis of variously shaped AgNPs in size range of 200 nm. Magnetic Fe_3O_4 (magnetite) NPs have been synthesized by various bacterial species including *Acinetobacter* sp., sulfate reducing bacteria, magnetotactic bacteria, namely, *Magnetospirillum magneticum*, etc. (Bharde et al., 2005; Watson et al., 1999; Mohanpuria and Rana, 2008). He et al. (2007) demonstrated the synthesis of differently shaped iron sulfide NPs at different pH with *Rhodopseudomonas capsulata*. Other species reported for synthesis of metal NPs include *Lactobacillus* sp., *Bacillus* sp., *Shewanella algae*, and *Rhodobacter sphaeroides* (Sweet et al., 2012). *Escherichia coli* is reported to produce spherical shaped AgNPs (20–30 nm) extracellularly (Parikh et al., 2008) and spherical-shaped AuNPs having 10 nm size and CdS nanocrystals through intracellular routes (Gholami-Shabani et al., 2015; Sweeney et al., 1998). *Saccharophagus degradans*, a marine bacterium, have been employed for extracellular synthesis of hexagonal- and spherical-shaped MnO_2 NPs in the size range of 35 nm (Salunke et al., 2015). In another study, *Halococcus salifodinae* BK3 was used for intracellular synthesis of tellurium NPs which showed hexagonal needle shape structure and 10 nm size (Srivastava et al. (2013).

Gurunathan et al. (2013) studied the cytotoxic effect of biologically synthesized AgNPs on MDA-MB-231 breast cancer cells and its cell death mechanism. They successfully developed green method for synthesis of stable AgNPs (20 nm) using *Bacillus funiculus* culture supernatant, which is green, eco-friendly, and cost-effective. These days, AgNPs are used as an antimicrobial agent in wound dressings and coatings in medical devices. AgNPs inhibit the growth of activated lactate dehydrogenase, reactive oxygen species generation, and activated capase-3, which is considered the most significant feature for the cellular apoptosis. Their results reveal that oxidative stress appears to be involved in NP cytotoxicity. The overall results suggest that the microbially synthesized AgNPs have antiproliferative activity through induction of apoptosis in MDA-MB-231 breast cancer cell line and might be a potential alternative agent for human breast cancer therapy. This study demonstrates that the possibility of using AgNPs to inhibit the growth of the tumor cells and their cytotoxicity for potential therapeutic treatments and offers a new method to develop molecule for cancer therapy. The biocompatibility, cost-effectiveness, and facileness in the modification of AgNPs make them viable choice in future biomedical applications. Similarly, Nadaf and Kanase (2019) adopted green approach for the extracellular synthesis of AuNPs using *Bacillus marisflavi* YCIS MN5. In their report, gold chloride solution was added into a cell free extract (acted both as reducing and stabilizing agents) of *B. marisflavi* to synthesize AuNPs at room temperature in 96 h. The synthesized NPs with an average size $\sim$14 nm was further assessed for catalytic degradation of Congo red and methylene blue dyes. Another study by Modi et al. (2015) demonstrated synthesis of AgNPs using potential silver resistant bacteria *Bacillus pumilus* and silver nitrate and silver sulfate as precursors. Synthesized Ag NPs were then assessed for dye degradation and it was observed that nano-based bioremediation is 13% more efficient than the microbial remediation.

23.3.2 Fungi

Fungi-mediated synthesis of NPs offers several benefits such as high enzymatic secretion and ease of handling. However, manipulation at genetic levels for the purpose of overexpressing enzymes of interest is relatively a difficult task. *Fusarium oxysporum* is a fungus that has been widely used for NPs synthesis. *F. oxysporum* could synthesize platinum NPs and ZnS NPs in the size range of 5–30 and 42 nm, respectively (Syed and Ahmad, 2012; Mirzadeh et al., 2013). *F. oxysporum* have also been utilized for the synthesis of AuNPs in size range of 20–40 nm (Mukherjee et al., 2002). Similarly, NPs of CdS, ZnS, MoS_2, and PbS NPs could also be synthesized with *F. oxysporum* (Ahmad et al., 2002). Other species of *Fusarium* such as *F. acuminatum* (Ingle et al., 2008) and *F. semitectum* (Basavaraja et al., 2008) are also reported for extracellular synthesis of AgNPs. *Fusarium* species could form highly stable Au–Ag alloy in size range of 8–14 nm (Senapati et al., 2005). AuNPs in different shapes (spherical, triangular, hexagonal, decahedral, and pyramidal) and sizes can be generated with the help of biomolecules secreted from *Botrytis cinereal* (Castro et al., 2014). In the case of acidophilic fungus *Verticillium luteoalbum*, pH plays an important role in defining the shape of AuNPs. Lower pH (3–5) favors formation of spherical to triangular, hexagonal, and rod-shape particles, whereas high pH (7–9) range favors the formation of irregular-shaped particles AuNPs (Gericke and Pinches, 2006). AgNPs could also be synthesized intracellularly using *Verticillium* sp. in cytoplasm (Sastry et al., 2003). Vigneshwaran et al. (2007) revealed that the proteins present on the cell wall of *Aspergillus flavus* is the reason behind the reduction and stability of AgNPs.

23.3.3 Actinomycetes

Actinomycetes possess characteristics of bacteria as well as fungi. Manivasagan et al. (2015) isolated 35 marine actinobacterial strains from marine sediments of Busan coast, South Korea and investigated their performance for AuNPs synthesis. Among them *Nocardiopsis* sp. MBRC-48 were found to extracellularly produce nonuniformly distributed, spherical shaped AuNPs with size ranges from 7 to 15 nm. The study further suggested that proteins, enzymes, and some metabolites are involved in the process of synthesis. Intracellular synthesis of AuNPs has also been reported through an alkalo-tolerant actinomycete-*Rhodococcus* sp. and *Streptomyces viridogens* strain (HM10) that could form spherical and rod-shaped NPs in size range of 5–15 and 15–20 nm, respectively (Ahmad et al., 2003; Balagurunathan et al., 2011). The strain HM10 was also observed to have greater antibacterial activity against *S. aureus* and *E. coli*. Another species of *Streptomyces* was studied by Prakasham et al. (2012) at varied pH and reaction time. The findings revealed time-dependent synthesis of AgNPs (14.5 nm) intracellularly and extracellularly at the cell surface of *Streptomyces albidoflavus*.

23.3.4 Yeast

In comparison to bacterial and fungi mediated synthesis, few studies are reported on application of yeast for NPs synthesis. Among all, *Saccharomyces cerevisiae* have been mostly researched for synthesis of metallic NPs such as Au, TiO_2, and MnO_2 (Lin et al., 2005; Jha et al., 2009; Salunke et al., 2015). Jha et al. (2009) displayed synthesis of TiO_2 NPs (8–35 nm) on cell wall of *S. cerevisiae* and revealed that during synthesis, quinones and membrane-bound oxidoreductases influences the formation of NPs. The same species also synthesized spherical- and hexagonal-shaped MnO_2 NPs with 30 nm size (Salunke et al., 2015). The synthesis of AuNPs has been reported with another fungi, namely, *Yarrowia lipolytica* (Agnihotri et al., 2009). The study reported that the reductase is involved in the synthesis of NPs, and change in pH has an important role in defining the shape of obtained NPs. Extremophilic yeast isolated from acid mine drainage have been reported to produce spherical Ag and AuNPs in size range of 20 and 30–100 nm (Mourato et al., 2011). Selvi and Das (2016) applied an integrated nano-bio hybrid system using nano zero- valent iron coated on *Candida* sp. SMNO4 for effective removal of cefdinir. Experiments reveald that nano-bio hybrid system for the degradation of cefdinir is more effective than individual system. The coated yeast showed 91% degradation of cefdinir at the concentration of 250 mg/L within 2.5 days than the noncoated NPs. Involvement of major enzyme β-lactamase along with other enzymes were also noted during cefdinir degradation. The enhanced activity of the enzymes may be possibly due to the immobilization of both intracellular and extracellular degradative enzymes on to zero-valent iron NPs.

23.3.5 Algae

Availability of documented evidence is on a lesser side supporting the use of marine algae for the synthesis of NPs. Singaravelu et al. (2007) reported the very first study on synthesis of AuNPs using brown seaweed marine alga (*Sargassum wightii*) with a size ranging 8–12 nm. An important feature of the synthesis is the formation of highly dense, extremely stable AuNPs. Another study on algae mediated synthesis of CdS NPs was reported by Mandal et al. (2016). The study reported that *Spirulina platensis* provides simple but strong prokaryotic system for biosynthesis of CdS NPs and also convert the more toxic Cd (II) species to the less toxic CdS NPs. The synthesized NPs showed efficient degradation of widely known dye pollutant Malachite green and found as potential biosorbent for Cd (II) detoxification.

Several other example of biosynthesized NPs with their application in different environmental field are mentioned in the literature and are listed in Table 23.2.

23.4 Prevailing environmental factors affecting microbial synthesis of nanoparticles

There are several factors that affect the NPs synthesis and their application. Some researchers have reported that types of adsorbate and the catalysts activity used in synthesis process changes the characteristics of synthesized NPs (Patra and Baek, 2014). Some other important factors that affect the synthesis of NPs includes pH of the solution, temperature, concentration of the extracts, size, raw materials, etc., used for the synthesis process (Baker et al., 2013) (Fig. 23.2). Some dominant factors, which affect the biosynthesis of NPs, are described in the following section.

TABLE 23.2 Biosynthesized nanoparticles and their applications.

Biological agents	Nanoparticles	Applications	Size range	References
Trichoderma viride (fungus)	AgNPs	Biosensor and bio-imaging	2–4 nm	Fayaz et al. (2010)
Dysosma pleiantha rhizome	AuNPs	Antimetastatic activity against human fibrosarcoma cancer cell line HT-1080	127 nm	Karuppaiya et al. (2013)
Brevibacterium casei SRKP2	CdS NPs	Applied in diagnostic, imaging and cancer therapies	10–30 nm	Pandian et al. (2011)
Escherichia coli (bacteria)	AgNPs	Potential to induce neuronal differentiation of SH-SY5Y cells	30 nm	Dayem et al. (2014)
Boswellia, Shorea (stem bark extracts); and *Svensonia* (leaf extract)	AgNPs	Inhibit the growth of *Pseudomonas* and *Rhizopus* species	–	Savithramma et al. (2011)
Tribulus terrestris L. (fruit bodies)	AgNPs	Antibacterial property against multidrug resistant bacteria *Streptococcus pyogenes, Pseudomonas aeruginosa, E. coli, Bacillus subtilis* and *Staphylococcus aureus*	16–28 nm	Gopinath et al. (2012)
Typha angustifolia (leaf extract)	AgNPs	Against *E. coli* and *Klebsiella pneumoniae*	8 nm	Gurunathan (2015)
Artemisia princeps (leaf extract)	AgNPs	Antibacterial and antibiofilm activity against *Helicobacter pylori* and *Helicobacter felis* and cause cytotoxic effects to human cancer cells	20 nm	Gurunathan et al. (2015)
Magnolia kobus (leaf extract)	Cu NPs	Antibacterial activity against *E. coli.*	37–110 nm	Lee et al. (2013a)
Yeast *Candida* sp. SMN04	Nanoscale zerovalent iron	Cefdinir degradation	31.7 nm	Selvi et al. (2016)
Cow milk	AgNPs	Inhibit (87.1%, 86.5%, and 83.5%) the growth of phytopathogens *Colletotrichum coccodes, Monilinia* sp., and *Pyricularia* sp., respectively	30–90 nm	Lee et al. (2013b)
Vibrio alginolyticus (bacteria)	AgNPs		50–100 nm	Rajeshkumar et al. (2013)
Streptomyces hygroscopicus (bacteria)	AgNPs	Antibacterial activity against Gram-positive bacteria (*B. subtilis* and *Enterococcus faecalis*), Gram-negative bacteria (*E. coli* and *Salmonella typhimurium*) and yeast (*Candida albicans*)	20–30 nm	Sadhasivam et al. (2010)
Padina pavonia (brown algae)	AgNPs	Medical use	49.58–86.37 nm	Abdel-Raouf et al. (2019)
Brown (*Colpomenia sinuosa*) and red (*Pterocladia capillacea*) seaweed	Fe_3O_4-NPs	Antifungal potency against *Aspergillus flavus* and *Fusarium oxysporum, P. capillacea*	1.24–33.71 and 16.85–22.47 nm, respectively	Salem et al. (2019)
Galaxaura elongate (red algae)	AuNPs	Effective against *E. coli* and *K. pneumoniae*	3.85–77.13 nm	Abdel-Raouf et al. (2017)
Gelidium amansii (marine red algae)	AgNPs	Antibacterial against pathogenic Gram-positive (*S. aureus, Bacillus pumilus*) and Gram-negative (*E. coli, P. aeruginosa, Vibrio parahaemolyticus, Aeromonas hydrophila*) pathogens		Pugazhendhi et al. (2018)
Spirulina platensis (blue green algae)	AgNPs, CrNPs, PbNPs, ZnNPs, and FeNPs	Antimicrobial activities against microorganisms includes Gram-positive bacteria (*S. aureus*), Gram-negative bacteria (*K. pneumoniae, Proteus vulgaris, Pseudomonas aeruginosa* and *E. coli*) and fungus (*Aspergillus niger*)		Gunasundari et al. (2017)

AgNPs, Silver nanoparticles; *AuNPs*, gold nanoparticles; *Fe_3O_4–NPs*, iron oxide nanoparticles.

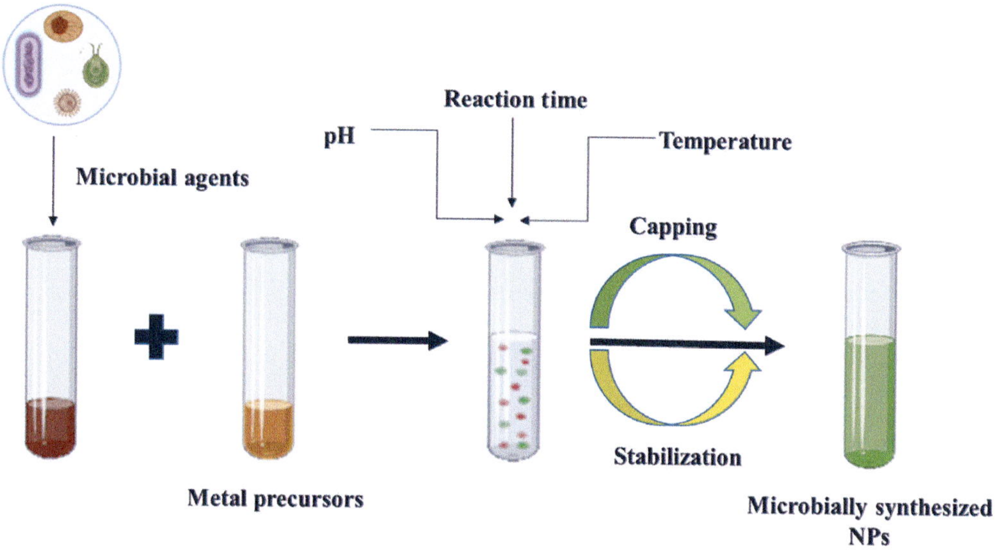

FIGURE 23.3 Prevailing factors that affect the microbial synthesis of nanoparticles.

23.4.1 pH

pH plays a very important role in the synthesis of NPs by green technology methods. Several researchers reported that pH of the solution mainly affects the size and texture of the synthesized NPs. Chitra and Annadurai (2014) extracellularly synthesized AgNPs using *Bacillus brevis* with different shape and size. Antibacterial activity of AgNPs shows better performance by changing solution: at pH 9, the smallest NPs were synthesized that showed more antibacterial activity than larger particles, which are synthesized at original pH and pH 5. Velgosová et al. (2016) successfully synthesized AgNPs by biological process using algae *Parachlorella kessleri*. It was observed at lower pH NPs showed number of aggregates. On the contrary, NPs synthesis with higher pH solution remained stable, even at very long term no change in their size were observed. Based on their experiment, they concluded that it is possible to control the size of the NPs and stability of AgNPs by changing the pH of the solution medium. Similar study was demonstrated by Kumari et al. (2016) in which AuNPs (AuNPs) were synthesized by using *Trichoderma viride* filtrate of various size and shapes. It was observed that the particle size decreased with increase in pH and at lower pH, a mixed population was synthesized. Similarly, in case of acidophilic fungus *V. luteoalbum*, pH plays an important role in defining the shape of AuNPs. Lower pH (3–5) favors the formation of spherical to triangular, hexagonal, and rod-shaped particles, whereas high pH (7–9) range favors formation of irregular-shaped particles AuNPs (Gericke and Pinches, 2006). Similar kind of effect was observed in the case of yeast (*Y. lipolytica*)-mediated synthesis of AuNPs (Agnihotri et al., 2009). With the variation in pH at 2, 7, and 9, hexagonal and triangular NPs were observed on the surface of yeast cells.

23.4.2 Temperature

The temperature of the reaction medium is a critical factor that determines the nature of the NPs. The effect of temperature on the formation of NPs has been observed in all the three approaches of synthesis, such as physical method requires the temperature >350°C, whereas chemical methods synthesis occurs at less than 350°C. In general, the biologically synthesized NPs require temperature less than 100°C or ambient temperature. Pourali et al. (2013) employed *Bacillus* sp. and *Acinetobacter schindleri* strains for the synthesis of AgNPs and evaluate the effect of temperature on the structure and antimicrobial activity of the AgNPs. It was observed that before the heat treatment, NPs had smaller sizes around 10–20 nm and were not clustered. However, heated NPs (at 100°C and 300°C) were found in clusters and the size of the particles was found around 400–800 nm. AgNPs formed a 300°C showed bigger size than those formed at 100°C. X-ray diffraction patterns of heated and nonheated AgNPs showed that the nonheated AgNPs had amorphous shapes, whereas, the heated AgNPs had crystalline structures.

23.4.3 Reaction time

Several researchers observed that every reaction needs certain time to accomplish the conversion and shape/size of the NPs varies with the change in incubation period of the reaction mixture. It has been also found that monodispersity of the synthesized NPs increases with increase of reaction time. Gericke and Pinches (2006) in their study successfully synthesized AuNPs using *V. luteoalbum* and evaluated the effect of time of microbial synthesis on the monodispersity and size of the particles. The results revealed that the particles which are exposed for 1 h to ion solution have smaller size and were monodisperse in nature, whereas the particle synthesized with 24 h of cells exposing time to the ion solution possessed greater particle size. Similarly, the synthesis of AgNPs by *S. albidoflavus* was reported to be dependent on reaction time (Prakasham et al., 2012).

23.4.4 Stirring time

Another controlling factor that facilitates the microbially mediated synthesis of NPs is stirring time. Stirring during the process of synthesis shows profound effect on the size control of NPs. The particles size sharply decreased by more than 15 times under mixing condition (3 nm) as compared to the synthesis done under mixing free condition (50 nm) in the case of *Klebsiella pneumoniae*. The monodispersity of the NPs has been better shown under mixing condition rather than mixing free condition (Zhang et al., 2011). The study very well demonstrated the remarkable contribution of synthesis-affecting parameters in the production of NPs.

23.4.5 Environment

The surrounding environment plays a vital role in determining the characteristics of synthesized NPs. Patra and Baek (2014) reported, while synthesizing ZnO NPs, change in environment from wet to dry condition instantly changed its crystalline nature.

23.5 Future prospects

Metal NPs are enormously used worldwide in multiple sectors, such as, pharmaceutics, biomedicine, optics, material science, etc. The synthesis of NPs is a very complex and dynamic process. Numerous approaches such as chemical vapor deposition, physical vapor deposition, hydrothermal method, spray pyrolysis, sol–gel method, laser ablation method, hydrothermal method, flame pyrolysis, chemical reduction, solvothermal method, electrodeposition process, supercritical fluid precipitation process, microwave technique, ultrasound technique, and green synthesis method are being used for the synthesis of NPs. Among the above-enumerated methods, each one has it is own specific advantages and disadvantages. NPs synthesis via green approach has been reported to be an eco-friendly, sustainable, usually free of chemical contaminants, and inexpensive method (Jamkhande et al., 2019). Though numerous methods have effectively emerged for the synthesis of NPs, further modification in existing technology is required to overcome the limitation and implementation at commercial level. Numerous studies revealed the application of nanotechnology in bioremediation has considerable potential for environmental contaminants. Conventional remediation technology may not always offer the most cost-effective ways for potential cleanup of various environmental contaminants present, particularly at low concentration. Nano-bioremediation may thus provide eco-friendly remediation approach especially for the of environmental contaminants without harming the nature. In contrast to the conventional technology the efficiency of nano-based remediation technique could be enhanced by modifying the size of NPs; in addition its cost could be reduced by producing it at industrial scale and considering cheaper feedstock synthesis method. Furthermore, nano-bioremediation could reduce the chances of formation of toxic intermediate, requires less space, as well as allow reuse of NPs. Microbially synthesized NPs offers the bioremediation and detoxification of environmental contaminants present in soil, water, and other environments. Further research is required to address the challenges and obstacles that NPs faces for their commercialization. In order to extend the length and quality of nano-bioremediation future research needs to be focused on development of profitable and feasible synthesis method which requires low-priced feedstock, less energy, and less time.

References

Abdel-Raouf, N., Al-Enazi, N.M., Ibraheem, I.B., 2017. Green biosynthesis of gold nanoparticles using *Galaxaura elongata* and characterization of their antibacterial activity. Arabian J. Chem. 10, S3029–S3039.

Abdel-Raouf, N., Al-Enazi, N. M., Ibraheem, I. B. M., Alharbi, R. M., Alkhulaifi, M.M., 2019. Biosynthesis of silver nanoparticles by using of the marine brown alga Padina pavonia and their characterization. Saudi J. Biol. Sci. 26 (6), 1207–1215

Agnihotri, M., Joshi, S., Kumar, A.R., et al., 2009. Biosynthesis of gold nanoparticles by the tropical marine yeast *Yarrowia lipolytica* NCIM 3589. Mater. Lett. 63, 1231–1234.

Aguilar, M.S., Esparza, R., Rosas, G., 2019. Synthesis of Cu nanoparticles by chemical reduction method. Trans. Nonferrous Met. Soc. China 29 (7), 1510–1515.

Ahmad, A., Mukherjee, P., Mandal, D., et al., 2002. Enzyme mediated extracellular synthesis of CdS nanoparticles by the fungus, *Fusarium oxysporum*. J. Am. Chem. Soc. 124 (12108), 12109.

Ahmad, A., Senapati, S., Khan, M.I., et al., 2003. Extracellular biosynthesis of monodisperse gold nanoparticles by a novel extremophilic actinomycete, *Thermomonospora* sp. Langmuir 19, 3550–3553.

Ali, A., Hira Zafar, M.Z., ul Haq, I., Phull, A.R., Ali, J.S., Hussain, A., 2016. Synthesis, characterization, applications, and challenges of iron oxide nanoparticles. Nanotechnol. Sci. Appl. 9, 49.

Aziz, M., Abbas, S.S., Baharom, W.R.W., 2013. Size-controlled synthesis of SnO_2 nanoparticles by sol–gel method. Mater. Lett. 91, 31–34.

Baker, S., Rakshith, D., Kavitha, K.S., Santosh, P., Kavitha, H.U., Rao, Y., et al., 2013. Plants: emerging as nanofactories towards facile route in synthesis of nanoparticles. BioImpacts: BI 3 (3), 111.

Balagurunathan, R., Radhakrishnan, M., Rajendran, R.B., Velmurugan, D., 2011. Biosynthesis of gold nanoparticles by actinomycete *Streptomyces viridogens* strain HM10. Indian J. Biochem. Biophys. 48, 331–335.

Basavaraja, S., Balaji, S.D., Lagashetty, A., et al., 2008. Extracellular biosynthesis of silver nanoparticles using the fungus *Fusarium semitectum*. Mater. Res. Bull. 43, 1164–1170.

Bharde, A., Wani, A., Shouche, Y., Joy, P. A., Prasad, B. L., & Sastry, M., 2005. Bacterial aerobic synthesis of nanocrystalline magnetite. J. Am. Chem. Soc. 127 (26), 9326–9327.

Burda, C., Chen, X., Narayanan, R., El-Sayed, M.A., 2005. Chemistry and properties of nanocrystals of different shapes. Chem. Rev. 105 (4), 1025–1102.

Castro, M.E., Cottet, L., Castillo, A., 2014. Biosynthesis of gold nanoparticles by extracellular molecules produced by the phytopathogenic fungus *Botrytis cinerea*. Mater. Lett. 115, 42–44.

Chitra, K., Annadurai, G., 2014. Antibacterial activity of pH-dependent biosynthesized silver nanoparticles against clinical pathogen. Biomed. Res. Int. 2014.

Das, S., Chakraborty, J., Chatterjee, S., Kumar, H., 2018. Prospects of biosynthesized nanomaterials for the remediation of organic and inorganic environmental contaminants. Environ. Sci.: Nano 5 (12), 2784–2808.

Dayem, A.A., Kim, B., Gurunathan, S., Choi, H.Y., Yang, G., Saha, S.K., et al., 2014. Biologically synthesized silver nanoparticles induce neuronal differentiation of SH-SY5Y cells via modulation of reactive oxygen species, phosphatases, and kinase signaling pathways. Biotechnol. J. 9 (7), 934–943.

De Carvalho, J.F., De Medeiros, S.N., Morales, M.A., Dantas, A., Carriço, A.S., 2013. Synthesis of magnetite nanoparticles by high-energy ball milling. Appl. Surf. Sci. 275, 84–87.

Diallo, A., Beye, A.C., Doyle, T.B., Park, E., Maaza, M., 2015. Green synthesis of Co_3O_4 nanoparticles via *Aspalathus linearis*: physical properties. Green Chem. Lett. Rev. 8 (3–4), 30–36.

Dudhane, A.A., Waghmode, S.R., Dama, L.B., Mhaindarkar, V.P., Sonawane, A., Katariya, S., 2019. Synthesis and characterization of gold nanoparticles using plant extract of *Terminalia arjuna* with antibacterial activity. Int. J. Nanosci. Nanotechnol. 15 (2), 75–82.

Elia, P., Zach, R., Hazan, S., Kolusheva, S., Porat, Z.E., Zeiri, Y., 2014. Green synthesis of gold nanoparticles using plant extracts as reducing agents. Int. J. Nanomed. 9, 4007.

Ennas, G., Musinu, A.N.N.A., Piccaluga, G., Zedda, D., Gatteschi, D., Sangregorio, C., et al., 1998. Characterization of iron oxide nanoparticles in an $Fe_2O_3 - SiO_2$ composite prepared by a sol – gel method. Chem. Mater. 10 (2), 495–502.

Fariq, A., Khan, T., Yasmin, A., 2017. Microbial synthesis of nanoparticles and their potential applications in biomedicine. J. Appl. Biomed. 15 (4), 241–248.

Fayaz, M., Tiwary, C.S., Kalaichelvan, P.T., Venkatesan, R., 2010. Blue orange light emission from biogenic synthesized silver nanoparticles using *Trichoderma viride*. Colloids Surf. B: Biointerfaces 75 (1), 175–178.

Ganachari, S.V., Banapurmath, N.R., Salimath, B., Yaradoddi, J.S., Shettar, A.S., Hunashyal, A.M., et al., 2017. Synthesis techniques for preparation of nanomaterials. Handbook of Ecomaterials. Springer, Cham, https://doi.org/10.1007/978-3-319-48281-1_149-1.

Gericke, M., Pinches, A., 2006. Biological synthesis of metal nanoparticles. Hydrometallurgy 83 (1–4), 132–140.

Gholami-Shabani, M., Shams-Ghahfarokhi, M., Gholami-Shabani, Z., et al., 2015. Enzymatic synthesis of gold nanoparticles using sulfite reductase purified from *Escherichia coli*: a green eco-friendly approach. Process Biochem. 50, 1076–1085.

Ghorbani, H.R., 2014. A review of methods for synthesis of Al nanoparticles. Orient. J. Chem. 30 (4), 1941–1949.

Gopinath, V., MubarakAli, D., Priyadarshini, S., Priyadharsshini, N.M., Thajuddin, N., Velusamy, P., 2012. Biosynthesis of silver nanoparticles from *Tribulus terrestris* and its antimicrobial activity: a novel biological approach. Colloids Surf. B: Biointerfaces 96, 69–74.

Gunasundari, E., Kumar, P.S., Christopher, F.C., Arumugam, T., Saravanan, A., 2017. Green synthesis of metal nanoparticles loaded ultrasonic-assisted *Spirulina platensis* using algal extract and their antimicrobial activity. IET Nanobiotechnol. 11 (6), 754–758.

Gurunathan, S., 2015. Biologically synthesized silver nanoparticles enhances antibiotic activity against Gram-negative bacteria. J. Ind. Eng. Chem. 29, 217–226.

Gurunathan, S., Han, J.W., Eppakayala, V., Jeyaraj, M., Kim, J.H., 2013. Cytotoxicity of biologically synthesized silver nanoparticles in MDA-MB-231 human breast cancer cells. Biomed. Res. Int. 2013.

Gurunathan, S., Jeong, J.K., Han, J.W., Zhang, X.F., Park, J.H., Kim, J.H., 2015. Multidimensional effects of biologically synthesized silver nanoparticles in *Helicobacter pylori*, *Helicobacter felis*, and human lung (L132) and lung carcinoma A549 cells. Nanoscale Res. Lett. 10 (1), 1–17.

Hasnidawani, J.N., Azlina, H.N., Norita, H., Bonnia, N.N., Ratim, S., Ali, E.S., 2016. Synthesis of ZnO nanostructures using sol-gel method. Procedia Chem. 19, 211–216.

He, S., Guo, Z., Zhang, Y., et al., 2007. Biosynthesis of gold nanoparticles using the bacteria *Rhodopseudomonas capsulata*. Mater. Lett. 61, 3984–3987.

Ingle, A., Gade, A., Pierrat, S., et al., 2008. Mycosynthesis of silver nanoparticles using the fungus *Fusarium acuminatum* and its activity against some human pathogenic bacteria. Curr. Nanosci. 4, 141–144.

Jamkhande, P.G., Ghule, N.W., Bamer, A.H., Kalaskar, M.G., 2019. Metal nanoparticles synthesis: an overview on methods of preparation, advantages and disadvantages, and applications. J. Drug. Deliv. Sci. Technol. 101–174.

Jha, A.K., Prasad, K., Kulkarni, A.R., 2009. Synthesis of TiO_2 nanoparticles using microorganisms. Colloids Surf. B Biointerfaces 71, 226–229.

Johnson, G.E., Moser, T., Engelhard, M., Browning, N.D., Laskin, J., 2016. Fabrication of electrocatalytic Ta nanoparticles by reactive sputtering and ion soft landing. J. Chem. Phys. 145 (17), 174701.

Karuppaiya, P., Satheeshkumar, E., Chao, W.T., Kao, L.Y., Chen, E.C.F., Tsay, H.S., 2013. Anti-metastatic activity of biologically synthesized gold nanoparticles on human fibrosarcoma cell line HT-1080. Colloids Surf. B: Biointerfaces 110, 163–170.

Khanna, P.K., Singh, N., Charan, S., Subbarao, V.V.V.S., Gokhale, R., Mulik, U.P., 2005. Synthesis and characterization of Ag/PVA nanocomposite by chemical reduction method. Mater. Chem. Phys. 93 (1), 117–121.

Kharissova, O.V., Kharisov, B.I., Oliva González, C.M., Méndez, Y.P., López, I., 2019. Greener synthesis of chemical compounds and materials. R. Soc. Open Sci. 6 (11), 191378.

Kruis, F.E., Fissan, H., Peled, A., 1998. Synthesis of nanoparticles in the gas phase for electronic, optical and magnetic applications—a review. J. Aerosol Sci. 29 (5–6), 511–535.

Kumari, M., Mishra, A., Pandey, S., Singh, S.P., Chaudhry, V., Mudiam, M.K.R., et al., 2016. Physico-chemical condition optimization during biosynthesis lead to development of improved and catalytically efficient gold nano particles. Sci. Rep. 6 (1), 1–14.

Lee, H.J., Song, J.Y., Kim, B.S., 2013a. Biological synthesis of copper nanoparticles using *Magnolia kobus* leaf extract and their antibacterial activity. J. Chem. Technol. Biotechnol. 88 (11), 1971–1977.

Lampis, S., Zonaro, E., Bertolini, C., Cecconi, D., Monti, F., Micaroni, M., et al. 2017. Selenite biotransformation and detoxification by Stenotrophomonas maltophilia SeITE02: novel clues on the route to bacterial biogenesis of selenium nanoparticles. J Hazard Mater. 324, 3–14.

Lee, K.J., Park, S.H., Govarthanan, M., Hwang, P.H., Seo, Y.S., Cho, M., et al., 2013b. Synthesis of silver nanoparticles using cow milk and their antifungal activity against phytopathogens. Mater. Lett. 105, 128–131.

Li, X., Xu, H., Chen, Z.S., Chen, G., 2011. Biosynthesis of nanoparticles by microorganisms and their applications. J. Nanomater. 2011.

Lin, Z., Wu, J., Xue, R., Yang, Y., 2005. Spectroscopic characterization of Au^{3+} biosorption by waste biomass of *Saccharomyces cerevisiae*. Spectrochim. Acta, A: Mol. Biomol. Spectrosc. 61, 761–765.

Livage, J., Henry, M., Sanchez, C., 1988. Sol-gel chemistry of transition metal oxides. Prog Solid State Ch. 18 (4), 259-341

Mandal, R.P., Sekh, S., Sarkar, N.S., Chattopadhyay, D., De, S., 2016. Algae mediated synthesis of cadmium sulphide nanoparticles and their application in bioremediation. Mater. Res. Express 3 (5), 55007.

Manivasagan, P., Alam, M.S., Kang, K.-H., et al., 2015. Extracellular synthesis of gold bionanoparticles by *Nocardiopsis* sp. and evaluation of its antimicrobial, antioxidant and cytotoxic activities. Bioprocess Biosyst. Eng. Available from: https://doi.org/10.1007/s00449-015-1358-y.

Martinez-Abad, A., 2011. Silver-based antimicrobial polymers for food packaging. Multifunctional and Nanoreinforced Polymers for Food Packaging. Woodhead Publishing., pp. 347–367.

Mirzadeh, S., Darezereshki, E., Bakhtiari, F., et al., 2013. Characterization of zinc sulfide (ZnS) nanoparticles biosynthesized by *Fusarium oxysporum*. Mater. Sci. Semicond. Process. 16, 374–378.

Modi, S., Pathak, B., Fulekar, M.H., 2015. Microbial synthesized silver nanoparticles for decolorization and biodegradation of azo dye compound. J. Environ. Nanotechnol. 4 (2), 37–46.

Mohanpuria, P., Rana, N.K., 2008. Biosynthesis of nanoparticles: technological concepts and future applications. J. Nanopart. Res. 10, 507–517.

Mourato, A., Gadanho, M., Lino, A.R., Tenreiro, R., 2011. Biosynthesis of crystalline silver and gold nanoparticles by extremophilic yeasts. Bioinorg. Chem. Appl. 2011, 546074.

Mukherjee, P., Senapati, S., Mandal, D., et al., 2002. Extracellular synthesis of gold nanoparticles by the fungus *Fusarium oxysporum*. ChemBioChem 3, 461–463.

Nadaf, N.Y., Kanase, S.S., 2019. Biosynthesis of gold nanoparticles by *Bacillus marisflavi* and its potential in catalytic dye degradation. Arabian J. Chem. 12 (8), 4806–4814.

Ogunyemi, S.O., Abdallah, Y., Zhang, M., Fouad, H., Hong, X., Ibrahim, E., et al., 2019. Green synthesis of zinc oxide nanoparticles using different plant extracts and their antibacterial activity against *Xanthomonas oryzae* pv. oryzae. Artif. Cells Nanomed. Biotechnol. 47 (1), 341–352.

Pandian, S.R.K., Deepak, V., Kalishwaralal, K., Gurunathan, S., 2011. Biologically synthesized fluorescent CdS NPs encapsulated by PHB. Enzyme Microb. Technol. 48 (4–5), 319–325.

Parikh, R.Y., Singh, S., Prasad, B.L.V., et al., 2008. Extracellular synthesis of crystalline silver nanoparticles and molecular evidence of silver resistance from *Morganella* sp.: towards understanding biochemical synthesis mechanism. ChemBioChem 9, 1415–1422.

Patra, J.K., Baek, K.H., 2014. Green nanobiotechnology: factors affecting synthesis and characterization techniques. J. Nanomater. 2014.

Pourali, P., Baserisalehi, M., Afsharnezhad, S., Behravan, J., Ganjali, R., Bahador, N., et al., 2013. The effect of temperature on antibacterial activity of biosynthesized silver nanoparticles. BioMetals 26 (1), 189–196.

Prakasham, R.S., Buddana, S.K., Yannam, S.K., Guntuku, G.S., 2012. Characterization of silver nanoparticles synthesized by using marine isolate *Streptomyces albidoflavus*. J. Microbiol. Biotechnol. 22, 614–621.

Pugazhendhi, A., Prabakar, D., Jacob, J.M., Karuppusamy, I., Saratale, R.G., 2018. Synthesis and characterization of silver nanoparticles using *Gelidium amansii* and its antimicrobial property against various pathogenic bacteria. Microb. Pathog. 114, 41–45.

Pyatenko, A., Shimokawa, K., Yamaguchi, M., Nishimura, O., Suzuki, M., 2004. Synthesis of silver nanoparticles by laser ablation in pure water. Appl. Phys. A 79 (4–6), 803–806.

Rajeshkumar, S., Malarkodi, C., Paulkumar, K., Vanaja, M., Gnanajobitha, G., Annadurai, G., 2013. Intracellular and extracellular biosynthesis of silver nanoparticles by using marine bacteria *Vibrio alginolyticus*. Nanosci. Nanotechnol. 3 (1), 21–25.

Ramos-Ruiz, A., Sesma-Martin, J., Sierra-Alvarez, R., Field, J.A., 2017. Continuous reduction of tellurite to recoverable tellurium nanoparticles using an upflow anaerobic sludge bed (UASB) reactor. Water Res. 108, 189–196.

Rane, A.V., Kanny, K., Abitha, V.K., Thomas, S., 2018. Methods for synthesis of nanoparticles and fabrication of nanocomposites. Synthesis of Inorganic Nanomaterials. Woodhead Publishing, pp. 121–139.

Sadhasivam, S., Shanmugam, P., Yun, K., 2010. Biosynthesis of silver nanoparticles by *Streptomyces hygroscopicus* and antimicrobial activity against medically important pathogenic microorganisms. Colloids Surf. B: Biointerfaces 81 (1), 358–362.

Salem, D.M., Ismail, M.M., Aly-Eldeen, M.A., 2019. Biogenic synthesis and antimicrobial potency of iron oxide (Fe_3O_4) nanoparticles using algae harvested from the Mediterranean Sea, Egypt. Egypt. J. Aquat. Res. 45 (3), 197–204.

Salunke, B.K., Sawant, S.S., Lee, S.-I., Kim, B.S., 2015. Comparative study of MnO_2 nanoparticle synthesis by marine bacterium *Saccharophagus degradans* and yeast *Saccharomyces cerevisiae*. Appl. Microbiol. Biotechnol. 99, 5419–5427.

Sastry, M., Ahmad, A., Islam Khan, M., Kumar, R., 2003. Biosynthesis of metal nanoparticles using fungi and actinomycete. Curr. Sci. 85, 162–170.

Savithramma, N., Rao, M.L., Rukmini, K., Devi, P.S., 2011. Antimicrobial activity of silver nanoparticles synthesized by using medicinal plants. Int. J. ChemTech Res. 3 (3), 1394–1402.

Selvi, A., Das, N., 2016. Nano-bio hybrid system for enhanced degradation of cefdinir using *Candida* sp. SMN04 coated with zero-valent iron nanoparticles. J. Appl. Pharm. Sci. 6 (09), 009–017.

Selvi, B. C. G., Madhavan, J., & Santhanam, A., 2016. Cytotoxic effect of silver nanoparticles synthesized from Padina tetrastromatica on breast cancer cell line. Adv Nat Sci-Nanosci. 7 (3), 035015.

Senapati, S., Ahmad, A., Khan, M.I., et al., 2005. Extracellular biosynthesis of bimetallic Au-Ag alloy nanoparticles. Small 1, 517–520.

Simakin, A.V., Voronov, V.V., Kirichenko, N.A., Shafeev, G.A., 2004. Nanoparticles produced by laser ablation of solids in liquid environment. Appl. Phys. A 79 (4–6), 1127–1132.

Singaravelu, G., Arockiamary, J.S., Kumar, V.G., Govindaraju, K., 2007. A novel extracellular synthesis of monodisperse gold nanoparticles using marine alga, *Sargassum wightii* Greville. Colloids Surf. B: Biointerfaces 57 (1), 97–101.

Singh, J., Dutta, T., Kim, K.H., Rawat, M., Samddar, P., Kumar, P., 2018. 'Green' synthesis of metals and their oxide nanoparticles: applications for environmental remediation. J. Nanobiotechnol. 16 (1), 84.

Song, J.Y., Kim, B.S., 2009. Rapid biological synthesis of silver nanoparticles using plant leaf extracts. Bioprocess. Biosyst. Eng. 32 (1), 79.

Srivastava, P., Bragança, J., Ramanan, S.R., Kowshik, M., 2013. Synthesis of silver nanoparticles using haloarchaeal isolate *Halococcus salifodinae* BK3. Extremophiles 17, 821–831.

Sweeney, R.Y., Mao, Chuanbin, X.G., Justin, L., Burt, A.M.B., 1998. Bacterial biosynthesis of cadmium sulfide nanocrystals. Development 125, 2457–2467.

Sweet, M.J., Chesser, A., Singleton, I., 2012. Review: metal-based nanoparticles; size, function, and areas for advancement in applied microbiology. Adv. Appl. Microbiol. 80, 113.

Syed, A., Ahmad, A., 2012. Extracellular biosynthesis of platinum nanoparticles using the fungus *Fusarium oxysporum*. Colloids Surf. B Biointerfaces 97, 27–31.

Tavakoli, A., Sohrabi, M., Kargari, A., 2007. A review of methods for synthesis of nanostructured metals with emphasis on iron compounds. Chem. Pap. 61 (3), 151–170.

Thamizharasan, S., Saravanan, N.A., 2017. Nanosization of drug biomaterials and its solubility enhancement by high energy ball milling. J. Nanosci. Technol. 3, 237–239.

Ullah, M., Ali, M., Hamid, S.B.A., 2014. Surfactant-assisted ball milling: a novel route to novel materials with controlled nanostructure—a review. Rev. Adv. Mater. Sci. 37.

Velgosová, O., Mražíková, A., Marcinčáková, R., 2016. Influence of pH on green synthesis of Ag nanoparticles. Mater. Lett. 180, 336–339.

Vigneshwaran, N., Ashtaputre, N.M., Varadarajan, P.V., et al., 2007. Biological synthesis of silver nanoparticles using the fungus *Aspergillus flavus*. Mater. Lett. 61, 1413–1418.

Watson, J.H.P., Ellwood, D.C., Soper, A.K., Charnock, J., 1999. Nanosized strongly-magnetic bacterially-produced iron sulfide materials. J. Magn. Magn. Mater. 203, 69–72.

Weblink: <https://en.wikipedia.org/wiki/Laser_AblationIST12.10Pm>, (18.04.17).

Xu, J., Yang, H., Fu, W., Du, K., Sui, Y., Chen, J., et al., 2007. Preparation and magnetic properties of magnetite nanoparticles by sol–gel method. J. Magn. Magn. Mater. 309 (2), 307–311.

Yadav, T.P., Yadav, R.M., Singh, D.P., 2012. Mechanical milling: a top down approach for the synthesis of nanomaterials and nanocomposites. Nanosci. Nanotechnol. 2 (3), 22–48.

Yousefi, R., Cheraghizade, M., 2018. Semiconductor/graphene nanocomposites: synthesis, characterization, and applications. Applications of Nanomaterials. Woodhead Publishing, pp. 23–43.

Zhang, Q.L., Yang, Z.M., Ding, B.J., Lan, X.Z., Guo, Y.J., 2010. Preparation of copper nanoparticles by chemical reduction method using potassium borohydride. Trans. Nonferrous Met. Soc. China 20, s240–s244.

Zhang, X., Yan, S., Tyagi, R.D., Surampalli, R.Y., 2011. Synthesis of nanoparticles by microorganisms and their application in enhancing microbiological reaction rates. Chemosphere 82 (4), 489–494.

Chapter 24

Bioremediation of radionuclides by plant-microbe system: current progress and challenges

Bhupendra Koul and Khyati Adlakha
Department of Biotechnology, School of Bioengineering and Biosciences, Lovely Professional University, Phagwara, India

24.1 Introduction

Radionuclides (radioactive nuclide) are isotopes (same atomic number but difference in atomic mass due to number of neutrons) of elements that have high nuclear energy, which makes them unstable. Any chemical element can be found as radionuclide, even the lightest of all, hydrogen can exist as tritium in the environment. Radionuclides are persistently being added to the environment by natural cosmic radiation, bombardment of stable nucleus or through anthroprogenic activities like military operations, disposal of radioactive waste, fuel-processing plants, nuclear testing, fuel fabrication units, research laboratories, and radiological incidents such as Fukushima natural disaster and Chernobyl incident. Fly ash containing radionuclides are even added through coal-fired power plants. Radionuclides decay according to their half-life. The naturally occurring radionuclides tend to have long half-lives and are more stable. Radium is an example of a highly radioactive natural element and exists as four isotopes ^{223}Ra, ^{224}Ra, ^{226}Ra, and ^{228}Ra (Sharma et al., 2014). There are three ways by which radionuclides are formed: (1) *natural*: these are present in trace amounts in nature; (2) *nuclear fission*: they are produced accidently through nuclear fission and thermonuclear explosions; and (3) *synthetic*: they are deliberately produced using nuclear reactors.

Radionuclides can be produced either naturally or artificially through (1) *particle accelerators*—protons are accelerated along linear path with the help of voltage to attain high energies; (2) *fission*—it results in splitting of large atom/nucleus producing gamma rays along with high energy neutrons, like uranium-235 (235U) or plutonium-239 (239Pu). The neutrons that are produced as a byproduct of fission can be slowed down in nuclear reactors by using the rods of cadmium or boron; (3) *nuclear activation*—in nuclear reactors, neutrons are bombarded with stable neutrons to produce radionuclides. This method has poor specific activity so other production methods are generally preferred; (4) *cyclotron*—they accelerate high energy charged particles such as protons, deuterons, triton, and α-particles to high velocities to penetrate orbital electrons of the host atom. The radionuclides thus produced are neutron deficient and decay by electron capture or β^+ emission (Qaim, 2012). Small cyclotrons (9–11 MeV) are deployed to produce fluorine-18 for positron emission tomography (PET), while the medium-sized cyclotrons (15 MeV) are customized to produce large array of radionuclides such as ^{13}N, ^{11}C, and ^{15}O. On the contrary, larger cyclotrons (30 MeV) produce ^{67}Ga, ^{201}Tl, and ^{123}I, and (5) *radionuclide generators*—used to separate parent and daughter nuclei that are present in the transient state where parent is longer lived (e.g., ^{90}Mo) than daughter nuclei (e.g., ^{99m}Tc). The production of carrier free radionuclide provides greater radiopharmaceutical stability (Currie et al., 2011).

24.1.1 Uses of radionuclides

In the world, there are 447 nuclear reactors where United States is the most nuclear generating country. There are around 22 nuclear reactors in India, where 7 are operational with installed capacity of 6780 MW. The nuclear energy is helpful in producing electricity. About three quarters of electricity is produced by nuclear energy in France (71.7%), followed by Slovakia (55%), Ukraine (53%), Hungary (50.6%), and Sweden (40.3%) (IEAE, 2018). Radionuclides have

Microbe Mediated Remediation of Environmental Contaminants. DOI: https://doi.org/10.1016/B978-0-12-821199-1.00024-9

importance in both diagnostics and therapeutic applications (Qaim, 2012). They should release sufficient destructive radiation and stay in the target site. Several β-emitting radionuclides, for example, ^{64}Cu, ^{89}Sr, ^{90Y}, ^{109}Pd, ^{153}Sm, ^{165}Dy, ^{166}Ho, ^{169}Er, ^{177}Lu, ^{186}Re, and ^{198}Au have been reported for radionuclide therapy for bone pain palliation, for use as radiopharmaceuticals, radiation synovectomy, and tumor therapy (Neves et al., 2002). The radiation dose for a patient is adjusted such that it is safe and facilitates the imaging and diagnostics. The γ-emitters such as ^{99m}Tc and ^{123}I have found application in single photon emission computed tomography (SPECT) and positron emitters in PET (Qaim, 2012). Some of the popular radionuclides and their applications are as follows: Cobalt-60: it is used in seizing the development of cancer cells; Iodine-131: it is effective in the treatment of hyperthyroidism; Strontium-89: it relieves pain in bone metastasis; and Carbon-14: it is used in breath-test to detect ulcer causing bacteria *Heliobacter pylori* (Yeong et al., 2015). The most suitable radionuclide for therapy should be chosen according to its half-life constituting both biological half-life (T_b) and physical half-life (T_p). Some examples of radionuclides along with their half-lives are as follows: (1) Carbon-14: 5000 years, (2) Cobalt-58: 71 days, (3) Cobalt-60: 5.2 years, (4) Copper-64: 12.8 hours, (5) Caesium-135: 2,300,000 years, (6) Caesium-137: 30 years, (7) Hafnium-181: 40 days, (8) Iodine-129: 15,800,000 years, (9) Iodine-131: 8 days, (10) Iron-55: 2.9 years, (11) Manganese-54: 300 days, (12) Neptunium-237: 2,100,000 years, (13) Neptunium-239: 2.35 days, (14) Plutonium-239: 24,400 years, (15) Plutonium-240: 6540 years, (16) Plutonium-244: 380,000 years, (17) Polonium-210: 138.4 years, (18) Radon-222: 3.83 days, (19) Sodium-22: 2.6 years, (20) Sodium-24: 16 hours, (21) Strontium-90: 27 years, (22) Tantalum-182: 111 days, (23) Thorium-230: 77,000 years, (24) Thorium-232: 14,000,000,000 years, (25) Tritium-15 hours, (26) Uranium-233:1,62,000 years, (27) Uranium-235: 710,000 years, (28) Uranium-238: 4,500,000 years, (29) Uranium-239: 23.5 minutes, and (30) Zirconium-95: 65 days.

Apart from the application of radionuclides in medical field, they are the pioneers in solving Earth science problems. They help to determine the chronology of sediments and other sediments from glacial and fossil lakes (Krishnaswami and Lal, 1978). Man has been taking advantage of radiation in agriculture, biology, and electric power generation (Masok et al., 2016). Food can also be preserved through radiations as it stops the sprouting of root crops after harvesting. In industries, welds and leaks are detected through the same. Radionuclides can also be used to trace the environmental pollutants and water runoffs from rain and snow.

24.1.2 Effects of radionuclides on health and environment

Despite various applications of radionuclides in different fields, the exposure of radiation has various harmful effects on living organisms and the environment. The Fukushima nuclear accident in 2011 at Fukushima Daiichi, Japan is one such example where site got contaminated with Iodine-131. Isotopes of Cs (^{134}Cs, ^{136}Cs, and ^{137}Cs), Te (^{129m}Te, ^{129}Te, and ^{132}Te), I (^{131}I and ^{132}I), Ba (^{140}Ba), and La (^{140}La) were released (Endo et al., 2012). Unfortunately, the population of butterflies, cicadas, and other animals got reduced at locations where the contamination was highest. But, much damage to marine organisms has not been reported. It has been reported that mammals and birds are most radionuclide sensitive, while viruses and molluscs are most radioresistant. The resistance of marine organisms to radioactive particles leads to accumulation of these particles that is then transferred to the food chain and proves to be harmful to higher organisms. The frequent high consumption of fish-based products has adverse effects on health (Milenkovic et al., 2019). The most radioactive sensitive ecosystem is coniferous forests, while the most radioresistant are meadow plant community (Geras'kin, 2016).

The radionuclides can be originated from biotic as well as abiotic sources. Soil cannot be alone the source of radiation exposure to human beings (Doyi et al., 2017). The radionuclides can be leached into soil and transported through groundwater, drainage, and dust, which can then be incorporated into the food chain. The disposal of radionuclides in the water bodies negatively impacts agricultural activities and threaten food security. According to a report from World Health Organization, the continuous exposure to radon leads thousands of death, due to lung cancer annually all over the globe. Radionuclide exposure to humans can be both internal and external. The gamma radiation and cosmic rays constitute the external exposure while the inhalation and digestion through food and drinking is internal exposure (Masok et al., 2016). The internal exposure is same as that of chemicals and results from accumulation in the alimentary canal. The external exposure is due to close contact with contaminated sediments, soil, or water. The released radionuclides increase the temperature in tissues, excite and ionize the atoms, break the chemical bonds, and hence the biological effects. The biological exposure to internally present radionuclides depends on the pattern of ionizing radiation dose to sensitive cells, tissues, or organs. It may cause death or other competing risks (Raabe, 2010). As the DNA repair mechanisms are activated, they are prone to errors resulting in cell death, chromosome aberrations, and mutations. The mutation depends on the cell type being affected. If it is within a somatic cell, it can lead to cell death. If there has been misrepair, it can transform into cancerous state. If the mutation is in the germ line cells, it can reduce the number of gametes, increase embryo mortality, or can be inherited by the offspring. The level of lethality varies from organism to organism. However, biological effects of radiation depend mostly on factors such as: (1) type of radiation, (2) individual biological susceptibility, (3) amount of exposure, (4) rate of exposure, and (5) part of body exposed to radiation. The unit of radiation absorbed after exposure is expressed as "Gray" (1 gray is equal to 1 joule of radiation energy/kg of matter) in SI unit. The ill effect of radiation exposure can be in the following manner: (1) if the dose is 0.25 Gy, no discernible effect is observed; (2) if the dose is 1.0 Gy, blood changes would occur, although no illness is observed; (3) if dose is 2.0 Gy, radiation sickness may be induced, although death may not take place; (4) if dose is 4.5 Gy, approximate 50% irradiated people will die; and (5) if dose is 10 Gy, 100% of irradiated people may die (Ojovan and Lee, 2014).

These damages caused by radiation in the cell can be summed up as follows: (1) damages at molecular level: macromolecules such as enzymes, RNA, and DNA get damaged through ionization and cross-linkages; (2) damages at subcellular level: damages to cell membranes, nuclei, chromosomes such as fragmentation, mitochondria, lysozymes, etc.; (3) damages at cellular level: inhibition of cell division, death, decay, and transformation to malignant state, etc.; (4) damages to tissues and organs: loss of sight, disruption of central nervous system, inactivation of bone marrow activity resulting into malignancy, blood cancer, and ulceration of intestinal tract, etc.; and (5) damages to an individual and population: death or shortening of life, change in characteristics due to mutation, etc.

The coal combustion has led to the production of coal-derived radionuclides which have serious health effects on human and ecosystem. These include ^{235}U, ^{238}U, and ^{232}Th. Mining, atmospheric emissions of respirable particulates and volatile elements, and cleaning and storage operations may culminate in the accumulation of radionuclides in the environment (Van Hook, 1979). In a report after study of radionuclides in Richards Bay precinct, it was found that the nuclides after the natural decay of ^{238}U rich soil proved to be less contaminated when used for fertilizer production (Masok et al., 2016).

24.1.3 Treatment of radionuclide pollution

Since the radionuclides are hazardous and detrimental for all living organisms, it is very crucial to treat them. The main objective for treatment of radioactive wastes is that it should be harmless when released into environment. There are several types of radioactive wastes that can be classified as (1) low-level waste (LLW): the radioactive content does not exceed alpha activity up to 4 GBq/t or beta-gamma activity up to 12 GBq/t. It is usually generated from hospitals and industry, contains low amount of shorter lived radioactivity, thus it is suitable for disposal; (2) intermediate-level waste (ILW): it is more radioactive than LLW, so it is not suitable for disposal and storage. However, nonsolids and smaller sized particles may be condensed in concrete or bitumen for discardation; (3) high-level waste (HLW): it is highly radioactive, where on decay increase the temperature of surroundings. It has both short-lived and long-lived components, which depends on the half-life of radionuclides (Radioactive Waste Management, 2018).

The natural decay of radionuclides takes thousands or even millions of years. The primary practice involved for treatment is disposal of radionuclides in space outside earth's gravitational field or deep underground surface, burial under sea bed or specially constructed buildings under human care, reprocessing of radionuclides or transmutation to isotopes with shorter half-lives. The major challenge that lies with these methods is that they are not very effective and when disposed underground; they may leach through the soil and accumulate in the groundwater, thus causing the contamination of water table. The radionuclide contaminated soils are also treated with chemicals and chelating agents. The agriculture-based methods like the use of ammonium-ferric-hexacyano-ferrate(II) (AFCF) to reduce the amount of radiocesium contaminants from soil is one such example. Various potassium (K)- and calcium (Ca)-based fertilizers can also be used to remediate radiocesium and radiostrontium contaminated soil, respectively. But it has low applicability as it can be applied to small sites (Zhu and Shaw, 2000). The drawbacks of these methods are their low feasibility and very high operation costs. The negative impacts of radioactive exposure on human and animal health and high treatment/disposal cost makes it crucial to find an alternative method for treatment of radioactive wastes. The use of living organisms to decontaminate the polluted sites is an attractive alternative and has the potential to treat the aforementioned contaminants.

24.2 Bioremediation

Bioremediation is the use of biological agents like bacteria, fungi, and plants (lower and higher plants) to degrade and detoxify the environmental contaminants into least toxic forms. It can be beneficial where environmental conditions enable the growth of living organisms (Gupta et al., 2020; Singh et al., 2020). Through their metabolic processes,

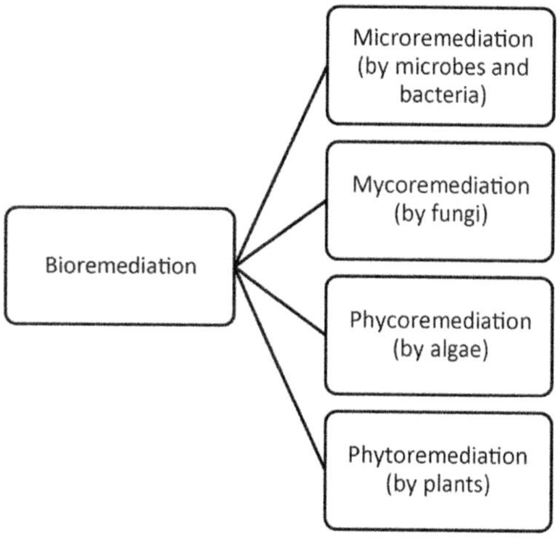

FIGURE 24.1 Types of bioremediation.

organisms metabolize radionuclides into simpler forms. Though radionuclides cannot be completely destroyed, they can either be transformed or contained.

There are various types of in situ and ex situ bioremediation techniques currently available for treatment of radioactive pollutants. In situ methods are applicable for areas with contaminants dispersed over a large area, present in dilute concentration or are inaccessible. Here the remediation process involves immobilization of contaminants to decrease the solubility. Ex situ methods require pumping and excavations above ground level located on site or off-site. They are easier to control and monitor. This includes extraction, separation, recovery of radionuclides and treatment and disposal of secondary waste. The lab tests have proved that microorganisms can change the oxidation state of some radionuclides by using them as electron acceptors. For remediation of radionuclide contaminants in soil, ex situ is the most common method which involves solidification/stabilization (Francis and Nancharaiah, 2015). However, the in situ method is cost-effective and involves minimal radiation exposure to site workers (Hazen and Tabak, 2005). There are four broad types of bioremediation techniques: (1) microremediation, (2) mycoremediation, (3) phycoremediation, and (4) phytoremediation (Fig. 24.1).

24.2.1 Characteristics of microorganisms

The adaptability, growth, and ecofriendly activities of microorganisms make them suitable to be deployed for remediation procedures. They are reliable for biogeochemical cycling of metals and remediation of environmental contaminants (solid wastes, pesticides, heavy metals, radionuclides, etc.). The property of electron transfer is important for cycling various elements in the environment (Francis and Nancharaiah, 2015). The most commonly known resistant genus for remediation of U, Ni, Cu, Co, and Cd are *Bacillus*, *Arthrobacter*, *Serratia*, *Rhodococcus*, and *Pseudomonas* (Cox and Battista, 2005). The gamma rays and other ionizing radiations have known to be lethal as they disrupt the DNA of these organisms. This can lead to formation of lesions. It is interesting to note that *Deinococcus radiodurans* is a bacterium that survives under gamma radiations. Some species of the genus *Rubrobacter*, *Chroococcidiopsis*, *Micrococcus rhodanobacter*, and *Desulfuromusa ferrireducens* are also known to exhibit similar resistant characteristics (Prakash et al., 2013; Shukla et al., 2017).

24.2.2 Techniques of microremediation

There are various mechanisms involved for the remediation of radioactive waste, which includes biotransformation, bioaccumulation, biosorption, bioprecipitation, biosolubilization and biomineralization, bioleaching, biovolatilization, and siderophore-mediated bioremediation. Fig. 24.2 shows all the abovementioned remediation mechanisms.

24.2.2.1 Biotransformation

Several radionuclides after interacting with microorganisms transform them from one oxidation state to the other. The transformation process either increases the mobility of the contaminant or precipitates it out of the solution so that it becomes immobilized (Hazen and Tabak, 2005). The solubility of oxidized form of radionuclides in aqueous medium is high, whereas the reduced species are mostly insoluble, which makes them precipitate in solution. Radionuclides

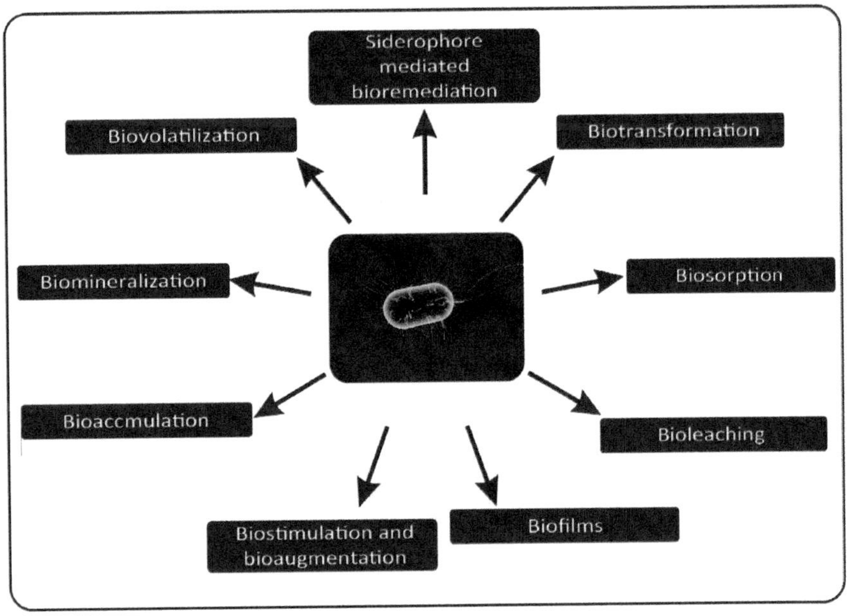

FIGURE 24.2 Techniques of microremediation.

including uranium (U), chromium (Cr), and technetium (Tc) are present in oxidized forms which are highly miscible in aqueous media and contaminate ground waters. There are two ways of enzymatic microremediation of radionuclides: (1) direct and (2) indirect.

Shewanella putrefaciens directly converts a soluble species into insoluble one through its surface. The c-type cytochrome present on the periplasm of this bacterium is required for reduction of U(VI). In vitro reduction of U(VI) is known to be carried out with the help of *Desulfovibrio vulgaris* cytochrome c3 protein using U(VI) reductase along with hydrogenase (Elias et al., 2004). Trihaem periplasmic cytochrome *c*7 (PpcA), a homologous cytochrome of Fe (III) reducing bacteria *Geobacter sulfurreducens* and homologous cytochrome c3 of *Desulfovibrio desulfuricans* can also perform hydrogen dependent U(VI) reduction (Lloyd et al., 2003; Prakash et al., 2013). Another periplasmic cytochrome, MacA, efficiently reduces U(VI) (Newsome et al., 2013). Although, technetium Tc(VII) has longer half-life which is hazardous for the environment and human health, but it can be degraded (reduced and precipitated) to low-valency oxide Tc(IV), using *S. putrefaciens* and *Geobacter metallireducens*. The anaerobic strain of *Escherichia coli* has been studied for biochemical pathway of Tc(VII) reduction. The formate hydrogenlyase catalyzes the transfer of electrons from dihydrogen to Tc(VII) (Lloyd et al., 1997). *Geobacter* sp. and *Rhodoferax ferrireducens* are evaluated as iron reducing bacteria which reduce radionuclides enzymatically due to its metabolic potential (Kim et al., 2012). *S. putrefaciens* can reduce the highly soluble Np(V) to insoluble Np(IV) and the insoluble phosphate biomineral can be removed by phosphate liberating *Citrobacter* sp. (Lloyd et al., 2000a; Prakash et al., 2013).

In indirect enzymatic microremediation of radionuclides, contaminants in subsurface and sedimentary environments are reduced by sulfate-reducing or metal-reducing microorganisms. The oxidation of organic compounds or hydrogen and reduction of iron Fe(III) to Fe(II) or sulfur S(VI) to S(II) can be simultaneous. The microbes may also release metabolites such as citric acid and ethylenediaminetetraacetic acid (EDTA) which can reduce Pu(VI) to Pu(IV) and Th(IV) to Th(IV) (Panak and Nitsche, 2001). *Arthrobacter aurescens*, *Pantoea agglomerans*, *D. vulgaris*, and *Pseudomonas aeruginosa* have been isolated from chromate adulterated waters, sediments, and oil to reduce highly soluble Cr(III) to $Cr(OH)_3$ (Kumar et al., 2007). The reduction of Pu(IV), Np(V) and Tc(VI) by *Shewanella oneidensis* is enhanced by adding Riboflavin (Cherkouk et al., 2016; Shukla et al., 2017).

The major challenge that stands up for the biotransformation is regarding the extent of its stability. The presence oxygen or change in pH may reverse the process of reduction. Moreover, uranium usually remains insoluble at very narrow range (Hazen and Tabak, 2005).

24.2.2.2 Bioaccumulation and biosorption

Bioaccumulation and biosorption of radionuclides is one of the strategies of microremediation wherein the microorganism concentrate (enrichment of contaminants) the radionuclides thousand times higher than the ambient concentrations.

It involves formation of radionuclide complexes which are insoluble or least bioavailable. The electrostatic attraction force exists between metal (radionuclide) cations and negatively charged cell surface. For example, uranium has no biological activity, so it may be taken up by the increased permeability of cell membrane (Newsome et al., 2013). Both bioconcentration and biomagnification contribute to bioaccumulation of radionuclides (Kumar et al., 2007). The process of bioaccumulation can either be active or passive. Active accumulation is a slow and more energy-consuming process. It can either be dependent or independent of microbial metabolism. On the other hand, passive accumulation is a fast process, requires low energy, and is suitable in areas of nutrient limitation. Most commonly used microorganisms for bioremediation of radioactive waste are *Arthrobacter nicotianae, Bacillus megaterium, Micrococcus luteus, Citrobacter* sp., *Pseudomonas* sp., *Arthrobacter ilicis*, and *Bacillus vallismortis* are reported to be useful for bioaccumulation of uranium (Newsome et al., 2013). *Bacillus subtilis* enhance the accumulation of iodide in the presence of glucose (Amachi et al., 2005).

To give effect to biosorption the ionizable groups such as phosphate and carboxylate in the lipopolysaccharides in gram negative bacteria and teichoic acids, peptidoglycan, and teichuronic acids in gram positive bacteria associate with the soluble substances/contaminants. To facilitate and enhance the bacterial activity the bacterial cell may be overlayered by several surface structures that are composed of carbohydrate polymers (capsule) or proteinaceous surface layers (S-layers). The S-layer of *Bacillus sphaericus* is reported to accumulate uranium. The bacterial cell surface is crucial for adsorption and tolerance to radionuclides (Douglas and Beveridge, 1998; Kumar et al., 2007). The binding of nuclides to cell wall is a passive process that is mediated due to the presence of cell wall ligands such as phosphate, amine, carboxyl, and hydroxyl (Shukla et al., 2017). *Citrobacter freudii* and *Firmicutes* exhibit efficient radionuclide biosorbent activity (Haferburg et al., 2007; N'Guessan et al., 2008; Xie et al., 2008). The sorption of Cd(II), Co(II), Ni (II), Pb(II), and Cr(IV) has been effectively done by using extracellular polymeric sequences (EPS) of *Rhodococcus opacus* and *Rhodococcus rhodochrous* (Shukla et al., 2017). The process of biosorption is influenced by aeration, pH, temperature, growth of cells, presence of organic or inorganic content and metabolites, secretion or production of exopolymers, electrostatic attraction, chemical interaction of functional groups extracellular biopolymers, and metal ions. The presence of electronegative charge on the bacterial cell envelope is able to attract cations from radionuclides that further sorb and immobilize on the bacterial surface (Newsome et al., 2013). The challenge lies that due to the competition of cations for the binding site on bacterial cell surface, the sorption may not be stable. Moreover, the in-sufficiency of bacterial biomass can be a limiting factor in sorption of heavy radionuclide contamination. Biosorption is best suitable for effluents with a range of low to medium radionuclide concentration, because binding of contaminants to cell wall is easier than its uptake by the cell wall.

24.2.2.3 Bioprecipitation and biomineralization

This is one of the recent techniques in the field of radionuclide-microbe interaction. Precipitation converts the soluble form of nuclides into insoluble (reduced) form (carbonates, hydroxides, phosphate, and sulfide) (Barkay and Schaefer, 2001; Shukla et al., 2017). The concentration of ligand produced by the cell decides the site of precipitation which is called "nucleation site" (Jabbar and Wallner, 2015). The organic matter is oxidized and sulfate is reduced to sulfide with the help of sulfate-reducing bacteria in anaerobic environments (Francis and Nancharaiah, 2015). *Acinetobacter* spp. bioprecipitates the metal phosphates via the hydrolysis of polyphosphates (Lloyd and Macaskie, 2000b). *Citrobacter* sp., *Rahnella* sp., *Bacillus* sp., *and Aeromonas* sp. are reported for uranium biomineralization (Macaskie et al., 1992; Newsome et al., 2013). The *Citrobacter* accumulates and sorbs polycrystalline $NaUO_2PO_4$ in and around the cell wall lipopolysaccharide. This mineral deposition is done by a gradient of incoming UO_4 and an outgoing PO_4 which results in complete removal of uranium. There are several chelating agents and organic acids [tartaric acid, oxalic acid, hydroxyl acetic acid, citric acid, EDTA, hydroxyethyl ethylenediamine triacetic acid (HEDTA), nitrilotriacetic acid (NTA), and diethylenetriamine pentaacetic acid (DTPA)] present in the wastes that form a complex with the radionuclides. After the aerobic or anaerobic degradation of these metal chelates, precipitates are formed as water soluble hydroxide or oxide. Citrate is one of the promising chelating agent which forms stable complex with radionuclides (metal) which are susceptible to microbe- mediated degradation. It forms bidentate, tridentate, and polynuclear stable complexes with uranium. It was observed that bidentate complex was biodegradable as compared to tridentate (Francis, 1998). Moreover, the addition of inorganic phosphate allows uranium to precipitate directly because phosphate rapidly precipitates with aqueous metals (metal recovery) to restrict their movement in the environment (Newsome et al., 2013). In *P. aeruginosa* a gene was cloned which encodes polyphosphate kinase (pkk). This gene shifted the bacterial ability to grow from high PO_4 to low PO_4 and resulted in precipitation of complex which contained both uranium and phosphorus on the bacterial cell surface (Barkay and Schaefer, 2001).

24.2.2.4 Biofilms

These are large population of microbial cells which can attach to aerobic or anaerobic surfaces with the help of their extracellular polymeric substances (EPS). They not only support large biomass which results in high microbial activities but also enhance the mineralization of radionuclides in the vicinity of cell. The biofilms of *D. desulfuricans* (sulfate-reducing bacteria) has been reported to mineralize U(VI) (Beyenal et al., 2004). Biofilms of *P. aeruginosa* have been reported to bind (biosorption) Fe, Au, and La [with oxidation state (III)] more efficiently to the cell surface (Barkay and Schaefer, 2001).

24.2.2.5 Bioleaching/biosolubilization

The autotrophic metabolism of microorganisms leaches out the radionuclides from their solid matrices. The energy is obtained by the reduced Fe or S with simultaneous solubilization of nuclides. The acidophilic or mesophilic bacteria perform bioleaching/biosolubilization in acidic pH, moisture, and oxygen to oxidize Fe and S and leach out radionuclides. Some of the solubilizers reported are *Leptospirillum ferrooxidans*, *Sulfolobus*, *Acidithiobacillus thiooxidans*, and *Acidithiobacillus ferrooxidans* (Ewart and Hughes, 1991; Munoz et al., 1995; Bosecker, 1997). Citric acid efficiently enhances the process of biosolubilization (Francis et al., 1992). In Germany a pilot-scale study on bioremediation successfully removed almost 62% of Cd, Ni, Co, and Zn were leached from the dredged sediments (oxygen rich conditions with high permeability) (Seide et al. 2004; Hazen and Tabak, 2005).

24.2.2.6 Biostimulation and bioaugmentation

Biostimulation is the addition of nutrient sources to increase the number and activity of microflora which enhances the bioremediation process. Specific strains of microorganisms are used to bioremediate radionuclides. To prevent the spread and bioavailability of radionuclides the in situ method was developed by promoting the activity of dissimilatory sulfate and iron reducing bacteria. In nitric acid cocontaminated sediments the nitrate serves as the preferred electron acceptor for metal-reducing bacteria. Uranium, being one of the most important radionuclide found at contaminated sites, can fairly be reduced to insoluble U(IV) from U(VI) by biostimulation. The presence of heavy metals inhibits the reduction of nitrate in bioremediation of U(VI). As a remedy, an ex situ method needs to be followed prior to in situ process. The bioreduction of nitrate could be enhanced by the addition of ethanol followed by successful reduction and immobilization U(VI). The addition of carbon source (ethanol) and raising the pH can accelerate the reduction of uranium at areas with high contamination and low pH by stimulating the growth of denitrifying and radionuclide reducing bacteria (Istok et al., 2004; Nyman et al., 2006). Members of Geobacteraceae family can accelerate in situ immobilization of uranium after acetate treatment to uranium-contaminated aquifer (Ortiz-Bernad et al., 2004). *Geobacter* sp., *Shewanella* sp., and *Desulfovibrio* sp. are reported to carry out reduction of radionuclides that forms as a precipitate. These organisms are resistant to noninducible heavy metals that could limit their in situ activity.

At the site of contamination, certain new species or strains of mixed species of microorganisms are added to facilitate and hasten the process of transformation or degradation of contaminants. The ex situ treatment can also be applied along with the aforementioned in situ treatment for bioaugmentation. However, bioaugmentation has not proved to be much useful in bioremediation of chlorinated solvents and only one bacterium *Dehalococcoides ethenogenes* has shown promise for dehalorespiration of chlorine-containing solvents (Hazen and Tabak, 2005).

24.2.2.7 Biovolatilization

Microorganisms can remediate the heavy metals (Hg and Se) by methylation. In a pilot-scale experiment in Florida, 50% success (polonium volatilization) was achieved when groundwater contaminated with organopolonium was treated with nitrogen. However, the results were not reproducible because of the variability in the organic polonium complexation and the presence of organic ligands that regulates differently polonium adsorption in soil and water (Hussain et al., 1995). Biovolatilization of radionuclides is safe for contaminated lakes and wetlands (Hazen and Tabak, 2005). As with heavy metals, the microorganisms volatilize the radionuclides by methylation. This mechanism was further confirmed when polonium was lost, after addition of methylcobalamin in polonium-contaminated water (Momoshima et al., 2001).

24.2.2.8 Siderophore mediated bioremediation

Siderophores are chelating agents with low molecular weight that are produced by the bacterial cell to bind iron (Fe) and transport it into the cell through energy dependent process. Some of the products such as siderophores, extracellular metabolites, and organic acids are produced by *Microbacterium flavescens* which can bind with radionuclides such as uranium(U), thorium(Th), and plutonium(Pu) and dissolve and mobilize them into the soil in the presence of siderophore desferrioxamine (John et al., 2001). Siderophores are highly specific for iron [Fe (III)] and bind to actinides of radionuclides. The Pu-siderophore complexes are degraded slower than the Fe-siderophore complex, whereas U-siderophore complex does not degrade at all. The structure of complexes Pu(IV)−DF and Fe(III)−DF are similar and can be identified by same bacterial system, so they compete for the binding site and transport mechanisms (Kumar et al., 2007).

24.3 Mycoremediation

Soil is the major source of nutrients for the growth and development of plant along with the microflora present in it. Fungi are one of the main components in soil, especially in the acidic soils. The fruiting bodies of fungi are known to accumulate high quantity of radiocesium from the time of Chernobyl incident in 1986. Fungi were the potential pool for radiocesium immobilization. The fungi locks up the radionuclides and inhibit soil pollution and transfer to the environment (Zhu and Shaw, 2000).

The symbiotic relationship between fungi and terrestrial plant is called mycorrhiza. The plant provides carbon in form of sugars to fungus, while fungus enhances the ability to uptake water and minerals from roots. The hyphae broaden the surface area to accumulate radionuclides along with the roots. The root zone or rhizosphere serves as the area of interface between soil, roots, and mycorrhiza. Radionuclides that are found in low concentration in the soil, that is, ^{137}Cs and ^{90}Sr are readily absorbed by mycorrhizal plants. In ectomycorrhizae the sheath or Hartig net play the role in absorbing radionuclides. Plants do vary in their potential to establish mycorrhiza because plant physiological and environmental conditions play a major role to establish relationship with fungi (Entry et al., 1997). This relationship alters the nutrient acquisition patterns. Legume *Sesbania rostrata* forms nitrogen fixing nodules with *Azorhizobium caulinodans* on both roots and stem, which tolerates heavy metals such as Pb. Arbuscular mycorrhiza (AM) fungi (AMF) are known to exist with every terrestrial plant. Soil contaminated with uranium reduces the root and shoot biomass, leghaemoglobin content, and photosynthetic activity of *S. rostrata*. It can also lead to oxidative stress and the production of reactive oxygen species (ROS). This may culminate to metabolic imbalance and death of local tissue. The association of rhizobia, *Sesbania rostrata*, and AMF were tested for the remediation of uranium-contaminated soil. The triple symbiosis was enhanced by ORS571 and GE which would increase the nodule content. The uranium concentrations were found in the roots and shoots of *S. rostrata*. AMF hyphae could take up uranium and transfer them to the roots (Ren et al., 2018). The process includes precipitation of radionuclides in phosphate granules, chelation of pollutants by secretion of ligand molecules inside the fungus, adsorption of granules on fungus cell wall, etc. (Gohre and Paszkowski, 2006). Better management and biotechnological strategies need to be developed for land contaminated with radionuclides.

24.4 Phycoremediation

The degradation of contaminants with the help of algae is called phycoremediation. Freshwater or terrestrial strains were selected. Some of the algae to eliminate highly radioactive cesium (Cs) were freshwater eustigmatophycean algae *Vacuoliviride crystalliferum* and freshwater florideophycean algae *Batrachospermum virgatodecaisneanum*. To remediate strontium (Sr), strains of freshwater cyanobacterium *Stigonema ocellatum*, the freshwater chlorophycean alga *Oedogonium* sp., and the freshwater Magnoliopsida *Egeria densa* are reported. The remediation of radioactive iodine was actively done by all strains except the green alga *Stigeoclonium aestivale*. Isolation and identification of *V. crystalliferum* as a new genus bioaccumulates radiocesium. The cells of this alga are crystalline and have a large reddish globule, which can be grown in harsh environmental conditions such as high osmotic pressure, diffused light and high and low temperatures. The culture of this strain needs minimal effort, so it proves to be the best for the remediation of radiocesium. Along with *V. crystalliferum*, unicellular red alga *Galdieria sulphuraria* eliminates radiocesium in mixotrophic conditions. It prefers acidic pH below 4.0, so it may be useful for remediation of acidic radionuclide contaminated waste water (Iwamoto and Minoda, 2018).

24.5 Phytoremediation

The use of green plants to remove contaminants (radionuclides) from the environment is called phytoremediation. The efficiency of this method depends upon the ability of the plant to grow and survive at the contaminated site. The rate of phytoremediation depends upon soil microbe interactions, addition of chelators and agronomic practices. Chelators enhance the phytoremediation process by accumulating, stabilizing, and degrading the contaminants. Natural chelators released by plants such as phytosiderophores (PS) in the rhizosphere, organic acid anions such as oxalate, succinate, phthalate, acetate, citrate, malate, tartrate, and salicylate help in formation of complex and enhance radionuclide degradation activity of plants. Synthetic chelators such as organic acid molecules such as HEDTA, ethylenediaminetetraacetic acid (EDTA), ethylenediamine-N,N-bis(2 hydroxyphenyl)acetic acid (EDDHA), -1,2-cyclohexylenedinitrilotetraacetic acid (CDTA), NTA, and DTPA trans increase the bioavailability of radionuclides to the roots. The plants recruited for phytoremediation should have certain characteristics; they should (1) be resistant to plants and diseases, (2) be fast growing, (3) have radionuclide tolerance, and (4) be less attractive so that the contaminants are not consumed by organisms and passed on to higher trophic levels. Plants are beneficial to remediate the aquatic and terrestrial ecosystems. Table 24.1 summarizes some the

TABLE 24.1 Some of the plants used for phytoremediation of strontium, cesium, and uranium.

Radionuclide	Plant species	Reference
Strontium	*Calotropis gigantea* R.Br. (giant milky weed)	Eapen et al. (2006)
	Broussonetia papyrifera L. *Vent.* (paper mulberry)	Li et al. (2011)
	Parthenocissus quinquefolia L. *Planch.* (Virginia creeper)	Li et al. (2011)
	Cannabis sativa L. (hemp)	Hoseini et al. (2012)
	Euphorbia macroclada Bioss. (*spurge*)	Sasmaz and Sasmaz (2009)
	Verbascum cheiranthifolium Bioss. (Mullein)	Saśmaz and Sasmaz (2009)
	Astragalus gummifer L. (tragant)	Sasmaz and Sasmaz (2009)
	Vetiveria zizanioides L. Roberty. (khus)	Singh et al. (2008)
	Funaria hygrometrica Hedw. (moss)	Balarama Krishna et al. (2004)
	Amaranthus retroflexus L. (redroot *pigweed*)	Eapen et al. (2006)
	Brassica juncea L. Czern. (Indian mustard)	Fuhrmann et al. (2002)
	Phaseolus acutifolius A. Gray (tepary bean)	Fuhrmann et al. (2002)
		Fuhrmann et al. (2002)
Cesium	*Lycopersicon esculentum* Mill. (tomato)	Brambilla et al. (2002)
	Beta vulgaris var. cicla L. (beet)	Schuller et al. (2004)
	Catharanthus roseus L. (Madagascar periwinkle)	Fulekar et al. (2010)
	Amaranthus chlorostachys Willd. (redroot pigweed)	Moogouei et al. (2011)
	Calendula alata L. (marigold)	Moogouei et al. (2011)
	Chenopodium album L. (lamb's quarters)	Moogouei et al. (2011)
	Brassica campestris L. pekinensis (cabbage)	Chiang et al. (2005)
	F. hygrometrica Hedw. (moss)	Balarama Krishna et al. (2004)
	Hordeum vulgare L. (barley)	Bange and Overstreet (1960)
	A. retroflexus L. (redroot pigweed)	Fuhrmann et al. (2002)
	B. juncea L. Czern. (Indian mustard)	Fuhrmann et al. (2002)
	P. acutifolius A. Gray (tepary bean)	Fuhrmann et al. (2002)
	C. gigantea R.Br. (giant milky weed)	Eapen et al. (2006)
	Chromolaena odorata L. (Christmas bush)	Singh et al. (2009)
	C. sativa L. (hemp)	Hoseini et al. (2012)
	Agropyron spicatum (pursh) Scribn and Smith (blue bunch wheatgrass)	Cook et al. (2009)
	Leymus cinereus Scribn and Merr. (Great Basin wild rye)	Cook et al. (2009)
	Bromus tectorum L. (cheat grass)	Cook et al. (2009)
	Agropyron cristatum L. Gaertn. (crested wheatgrass)	Cook et al. (2009)
	V. zizanioides L. Roberty. (khus)	Singh et al. (2008)
Uranium	*B. juncea* (oriental mustard)	Huang et al. (1998)
	Brassica chinensis (cabbage)	Li et al. (2011)
	Phragmites australis	Mkandawire et al. (2004)
	Lemna gibba L.	Stojanović et al. (2012)
	Nicotiana tabacum L. (*tobacco*)	Huhle et al. (2008)
	Helianthus annuus (*sunflower*)	Favas et al. (2014)
	Aquatic plants—*Fontinalis antipyretica, Callitriche stagnalis, Callitriche hamulata, Ranunculus peltatus* subsp. *saniculifolius, Callitriche lusitanica, Ranunculus trichophyllus*	

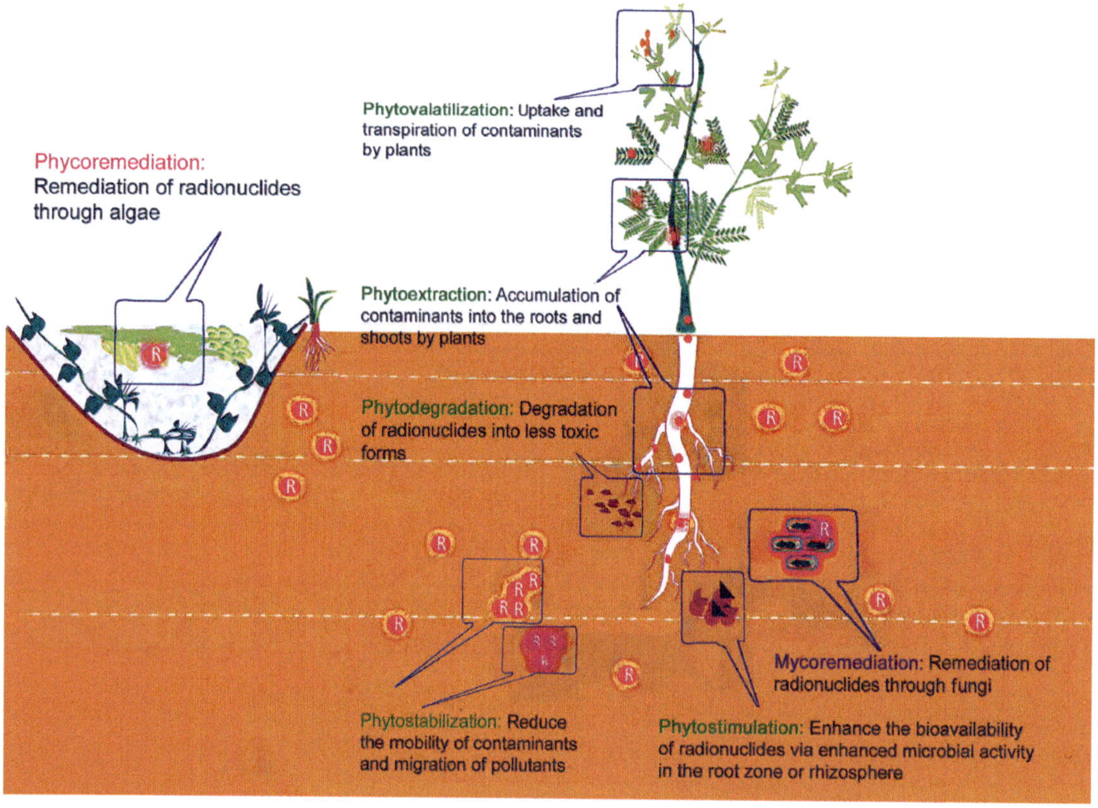

FIGURE 24.3 Bioremediation by plants, algae, and fungi.

plants used in phytoremediation of commonly present radionuclides, that is, strontium, cesium, and uranium. The plants recruited for phytoremediation should exhibit high transfer factor (i.e., ratio of contaminant concentration in plant to that in the soil) and the occurrence of AM colonies because abundant arbuscules help in catalyzing radionuclides. There can be two types of phytoremediation: (1) direct phytoremediation: it involves the uptake of pollutants through roots and translocate them to shoot. (2) Explanta phytoremediation: it is also called rhizoremediation, where roots and rhizosphere play the key role in degradation and accumulation of roots (Alkorta and Garbisu, 2001). Fig. 24.3 depicts different modes of phytoremediation applied for decontamination of radionuclides, namely, phytoextraction, rhizofiltration, phytostabilization, phytovolatalization, and phytodegradation along with phycoremediation and mycoremediation.

24.5.1 Phytoextraction

Plant materials are known to accumulate radionuclides from soil. The natural ability of vascular plants is utilized to take up the radionuclides from the soil and translocate them to above-ground biomass (Eapen et al., 2007). This biomass gets accumulated with radionuclides is harvested, processed, and disposed off as radioactive waste. It reduces the transfer of radionuclides from soil to the food chain. The soil structure and soil fertility are not much affected. The large areas with low contamination can easily be treated through this method. The bioavailability of radionuclides, uptake of radionuclides by roots, and translocation to shoot affect the remediation process. The laboratory test for various elements has been done; however, only ^{137}Cs, ^{90}Sr, and $^{235,238}U$ have found their way in the field. Organic soil is reported to seize the radionuclides, for which sandy soil is preferred for extraction. Microbes in rhizosphere increase the root surface area which enhances the ability to accumulate radionuclides (Ramaswami et al., 2001). Radiostrontium is one of the few radionuclides that have the ability to translocate itself easily. The major challenge lies when radionuclide co-precipitates with oxides and adsorbs to clay minerals, which may impact the bioavailability of radionuclides. Manipulating soil pH, addition of chelators (citrate, in the case of U remediation) and chemicals stimulate radionuclide desorption, interaction with microorganisms and induces hyperaccumulation in normal plants. Synthetic chelators such as EDTA keep the bioaccumulation of radionuclides higher in stems and leaves (Barren, 2012). In the case of uranium, if stable U-phosphate complexes are formed, they are retained in roots, inhibiting their translocation to the shoots (Dushenkov, 2003). As compared to bushes and grasses, trees are known to accumulate higher amount of contaminants due to higher biomass. Members of family chenopodiaceae, Fabaceae, and related families have been reported to accumulate radioactive waste (Zhu and Shaw, 2000; Ribeiro de Souza et al. 2012; Sharma et al., 2014). The leguminous woody plants the concentration is maximum in the root zone; however, under stressed conditions toxicity can be observed in chlorophyll and carotenoid contents. *Amaranthus retroflexus* accumulates ^{137}Cs with high shoot biomass production. Imran et al. (2019) reported accumulation of U by *Bidens pilosa* L. in the roots and translocating them to higher parts of plant, that is, leaves and stem. Brassicaceae family can accumulate U efficiently than the grasses. High concentration of U, an increase up to 1000 folds with addition of citric acid was observed in shoots of *Brassica juncea*, *Brassica chinensis*, *Amaranthus cruentus*, *Beta vulgaris*, and *Helianthus annuus* (McGrath et al., 2002). Radionuclides such as ^{244}Cm, ^{137}Cs, ^{238}pu, ^{226}Ra, and ^{90}Sr are reported to accumulate within the trees such as *Liriodendron tulipifera*, *Liquidambar styraciflua*, and *Acer rubrum*. Addition of phosphorus and potassium inhibit the process of taking up ^{137}Cs by the trees. Herbaceous plants such as *Festuca arundinacea*, *Festuca rubra*, *Lolium perenne*, *Trifolium repens*, and *Cerastium fontanum* accumulate ^{137}Cs (Dahlman et al., 1969; Salt et al., 1992). Aquatic plants and macrophytes (plants growing in or near water which are categorized as merged, submerged and free-floating plants, e.g., *Hydrocotyle* spp.) have been well demonstrated for remediation of ^{137}Cs and ^{90Sr} (Entry et al., 1997). NH_4NO_3 and citric acid (ideal chelating agent) enhance the rate of phytoextraction (Ebbs et al., 1998; Lasat et al., 1998).

24.5.2 Rhizofiltration

The use of plant roots to absorb, concentrate, precipitate harmful pollutants from aqueous streams is called rhizofiltration. The pollutants either get adsorbed on the roots or absorbed in the roots. The plants with dense root system are generally opted for the higher concentration of pollutants, which can grow well hydroponically. It employs hydrophytes to contain environmental contamination. The first pilot-scale system for cleaning radionuclide contaminated water brought up the idea of using sequence of ponds (from reservoir to first tank, followed by second tank, followed by third tank and then the collecting tank) with a slow water flow to clean contaminated water. It was tested for contamination of uranium in the wastewater. Samples of water, sludge, and plants were taken regularly. The radioactivity was measured as 97.2% in first tank, out of which 94.9% was absorbed by sludge and water plants and 2.3% retained in water. The second and third tank contained 1.8% and 0.7% radioactivity, respectively. Currently, terrestrial plant roots have been employed for the removal of contaminants from aqueous streams. Sunflower (*H. annuus* L.) has proved to be more effective than Indian mustard by removing greater quantity of Uranium, strontium, and cesium contaminated water in greenhouse (Dushenkov, 2003). Rhizofiltration has proved to be one of the possible approach for removal of radionuclides from aqueous streams with little optimization and economic evaluation as compared to conventional methods (Eapen et al., 2007).

24.5.3 Phytostabilization

It is the most appropriate remediation method for radionuclide contaminated sites where contaminants are held up at one place so that their bioavailability, migration, and exposure can be prevented at further places (Dushenkov, 2003). There is a formation of static complex where metal cations bind and form strong complex with -H groups on surface of minerals, due to which there is precipitation in the root zone. Chelating substances (organic acids) are released by the microorganisms present in the root zone that immobilize the pollutants by forming the complex. Pollutants are prevented for leaching in the groundwater (percolation in the soil) and air when absorption, adsorption, and accumulation take place by vacuole sequestration or cell wall binding in the roots. There is no further translocation to the shoots. Without involving plant tissues for accumulation, phytostabilization mainly focuses on desolating pollutants in the rhizosphere (Lee, 2013). Supplementing soil with organic wastes (manure, straw, digested sewage sludge, etc.) to inorganic waste site assist in enhanced binding of radionuclides to the plants. Addition of lime and limestone neutralizes acid soils to help in binding of cations from metals to inorganic waste. The plants selected for phytostabilization should poorly translocate contaminants to above ground tissue. They should have the potential of tolerating high toxicity of contaminants with sufficient growth of dense root system. The main challenge for removal of radionuclides lies in this method as phytostabilization does not remove radioactivity which may affect wildlife and humans due to continuous exposure (Eapen et al., 2007).

24.5.4 Phytovolatilization

Phytovolatilization is a procedure for contaminant remediation that transfers the contaminants from original or contaminated site to the atmosphere. The plants are used to remove contaminants that are dispersed off into the atmosphere by evaporation/volatilization from leaf surfaces (transpiration). The most common radioactive element decontaminated through this method is tritium (^3H) with half-life of around 12 years. Tritiated water is the common form for existence of this element in environment. The irrigation of forest surfaces can be done by tritiated water that enters the hydrological cycle. This water evaporates from soil surface, absorbed by plants or percolates through soil to the water table. The absorption of ^3H by plant roots persists in the plant tissue in form of largely exchangeable hydroxyl ions or is incorporated into organic molecules through photosynthesis (Dushenkov, 2003). Terry et al. (2000) reported that selenium (Se) with half-life of 327,000 years can be remediated through phytovolatilization. Se is majorly present in five oxidation states (−2, 0, +2, +4, and +6), of which +6 is most common in terrestrial ecosystems. The inorganic Se is volatilized by the activity of plant enzymes. The volatilized forms of Se are named as dimethyl selenide (DMSe), dimethyl selenone, dimethyl selenylsulfide, dimethyl diselenide (DMDSe), and methaneselenol. These volatilized molecules are further degraded in the atmosphere or remain as air pollutants in the environment. Phytovolatilization has reduced the risk of spills, exposure to workers, and avoided secondary waste treatment (De Filippis, 2015). This is a union of hydraulic control and evapotranspiration (Eapen et al., 2007). More research and assessment are necessary so that the method can be accepted by regulators and public (Lee, 2013).

24.5.5 Phytodegradation

The use of plants to uptake, metabolize, and degrade contaminants mainly with the above ground parts of plant lead to a decline in toxicities produced by radionuclides. Plants release various enzymes such as peroxidase, dehalogenase, nitrilase, nitroreductase, and laccase which assist in degrading contaminants within the plant, soil, surface water, or ground water (Lee, 2013). Without utilizing the activity of microorganisms, this process utilizes plants which can take up contaminants and translocate them to stem or leaves. Phytodegradation has mainly focused on break down of organic compounds, such as herbicides, pesticides, 2,4,6-trinitrotoluene (TNT), polychlorinated biphenyls (PCB), polycyclic aromatic hydrocarbons (PAH), and inorganic nutrients. There no such reports of research using phytodegradation with radionuclides (De Filippis, 2015).

24.6 The challenges

The advantages that lie with the bioremediation of radionuclides are that it is eco-friendly, specific, easy to adapt, and the bioproducts can be easily recycled. The molecular composition of organisms, specially the enzymes and metabolites, can easily be modified through the use of genetic engineered tools that can be used for detoxification. However, bioremediation is a slow process and is difficult to monitor, which may lead to the accumulation of pollutants in the environment that can be hazardous. The feasibility of microbial remediation is a major challenge. For the bioremediation of radionuclides, outer membrane proteins of bacteria needs to be modified to improve the binding capacity of element to microbes (Jabbar and Wallner, 2015). In in situ bioremediation process the growth conditions of microbial populations, that is, pH, temperature, moisture content, contaminant level, and suitable levels of nutrients, are required to be closely operated. The implementations of in situ bioremediation with genetically engineered microbes require additional steps for safe environmental cleanup.

The major challenge that lies with the application of phytoremediation is that the necessity of soil remediation destroys many physical and chemical properties of soil that may have impact on microbiological process which effects the plant growth. The addition of chelators does not necessarily remove radionuclides but also soil nutrients. The growth of plants on physically, chemically, and biologically contaminated soil is a formidable challenge (Entry et al., 1997). The contaminants can reach up to phytotoxic level, which can make the plant ineffective. Bioremediation is still fairly new and underdeveloped area of research. It is inconvenient to implement this method from bench and pilot-scale studies to full-scale field operations. The contaminants are emerging, so there is need for the organisms to be versatile enough so that they can target a larger group of radionuclides by making it faster, recyclable, and monitored.

24.7 Genetic engineering and bioremediation

The recombinant DNA technology is used to develop organisms with alteration in genetic material to enhance bioremediation. In plants, specific genes are expressed from organisms such as bacteria and mammals to increase tolerance for radionuclides and metabolism of such contaminants. The bacterial cell surfaces have been equipped with metal binding polypeptides, polyhistidines, metallothioneins, and synthetic phytochelatines (Schembri et al., 1999; Valls et al., 2000; Bae et al., 2000). These protein constructs enhance the metal binding. For example, the gene denoted as phoK is responsible for alkaline phosphatase activity which was isolated from *Sphingomonas* sp. strain BSAR-1 (Misra et al., 2012). It was then cloned and expressed in *E. coli* strain BL21 (DE3) and bacterium *D. radiodurans*, respectively. The precipitation by recombinant strain was evaluated to be faster in the case of uranium (Jabbar and Wallner, 2015). The transporter genes *nixA* from *Helicobacter pylori* and *merTP* from *Serratia marcescens* were altered and transferred to *E. coli* which increased the sorption ability of U(VI) up to five times (Beckwith et al., 2001). *D. radiodurans* has been reported to detoxify Cr(VI), U(VI), and Tc(VII) from soil (Fredrickson et al., 2000). Recombinant *D. radiodurans* has been developed by transferring *E. coli* gene (*merA*) that utilizes carbon and energy from catabolism of toluene and mercury (Brim et al., 2000). The genetic engineering of *G. metallireducens* was successful after harboring gene *dcuB* from *G. sulfurreducens* which encodes for fumarate transporter, where fumarate acts as a terminal electron acceptor with extended respiratory abilities (Butler et al., 2006). Transfer of Gene NiCoT extracted from *Rhodopseudomonas palustris* strain CGA0009 and *Novosphingobium aromaticivorans* strain F-199 in *E. coli* removed cobalt up to 85%. (Raghu et al., 2008). The reduction of uranium(VI) to uranium(IV) was done by genetically engineered *D. vulgaris* by gene *ctyc3* where hydrogen acts as electron donor (Payne et al., 2002). The incorporation of *czc* and *fccA* from *Ralstonia eutropha* CH34 and *S. oneidensis* helps in the remediation of cadmium, zinc, cobalt, nitrate, fumarate, trimethylamine *N*-oxide (TMAO), dimethyl sulfoxide, insoluble iron, and manganese oxides (Saffarini et al., 2003). The development of synthetic biofilm is another approach for the remediation of radionuclides. Biofilm is the assemblage of microbial cells where microbes are gripped together by EPS. It can vary from single-cell monolayer to thick mucous microcolonies. The bEPS from *Shewanella* sp. and *D. desulfuricans* G20 have been successfully used for immobilization of uranium. In plants, *Arabidopsis thaliana* has been reported for 150 different cation transporters. The merA and merB genes in *Arabidopsis thaliana* had the ability to volatilize mercury (Lee, 2013). Yeast CUP1 gene transferred in cauliflower showed 16-fold higher accumulation of cadmium. Genetically engineered canola and tobacco lowered down the toxication of mercury by volatilizing it. The detoxifying enzymes, nitroreductase, peroxidases, phosphatases, catalases (monooxygenases), dehalogenases, and cytochrome (oxidase or reductase) can be increased in activity in genetically modified plants. The transgenics should be fast growing, able to survive in toxic environments and should accumulate large amount of radionuclides. The sequencing of genetic material of plants and microbes, protein coding and noncoding transcripts, and molecular markers in the recent times is providing new and novel information to generate modified organisms (De Filippis, 2015). Bioinformatics can certainly contribute to the research being pursued by the scientists to designing of efficient bioremediation methods which can surpass the public and regulatory concerns.

24.7.1 "-omics" integrated approach

Microorganisms have always proved with a considerable range of radionuclide detoxification abilities. Different catabolic pathways need to be explored to reduce certain nuclides. The entire set of hereditary information that is converted to mRNA for protein translation is called genome. Advanced genomics, transcriptomics (DNA microarray analysis), and proteomics have enhanced the study of potential enzymes and genes in radionuclide-resistant organisms. Genome-wide transcriptome analysis provides better understanding of metabolic activities of microorganisms. The proteome of an organism is the complete set of proteins with all the enzymes that are being expressed in an organism. It is crucial to study the structure and function of various interactions between proteins and other metabolites to identify useful genes, proteins, and enzymes involved for bioremediation of radionuclides. The proteins are different in each cell with the variability of biochemical interactions taking place in the environment. Tian et al. (2010) reported 552 genes that are involved in the resistance of organism to radionuclides present on the cell membrane of *Deinococcus geothermalis*. *S. oneidensis* has the ability to grow under aerobic and anaerobic conditions utilize a variety of electron donors, including nitrite, iron and manganese (Marshall et al., 2006). The c-type cytochromes of *S. oneidensis* are important for the reduction of U(VI) and formation of extracellular uranium oxide (UO_2). The enzymatic proteins on outer cell membrane of *S. oneidensis* has been reported in reduction of radionuclides, namely, chromium, technetium, and uranium by transforming them into precipitates which makes them insoluble. The reduction genes, such as *mtrA*, *mtrB*, and *mtrC*, present in *S. oneidensis* encode for protein of unknown function from outer membrane, decaheme *c*-type cytochrome from periplasm and decaheme *c*-type cytochrome from outer membrane, respectively (Beliaev et al., 2001; Beliaev and Saffarini, 1998). The genes well known for DNA replication and repair in radioresistant bacterium, *D. radiodurans*, *RecA*, and *PprA* were significantly upregulated (Lu et al., 2009; Prakash et al., 2013). X-ray crystallography has been used to determine the three dimensional (3-D) protein structure and function in respect to transformation of radionuclides. The recognization and isolation of special genes have paved a pathway for remediation of radionuclides.

The linking of genome to the proteome and phenotypic characteristics of an organism is called transcriptome. *G. sulfurreducens* has been evaluated with more than 100 genes encoding c-type cytochrome in its own genome and various cytochromes were included in radionuclide reduction pathways of U(VI) and Tc(VII). DNA microarray analysis (transcriptomics) of *S. oneidensis* reported upregulation of about 121 genes during reduction of U(VI) (Methe et al., 2003) Phylogenetic characteristics and localization of microbes have been provided by the study of 16S rRNA involved in the bioremediation of radionuclides. Stimulation of U(VI) reduction has been reported by supplementing acetate in contaminated groundwater for removal of uranium. Promotion of microbial reduction of uranium from U(VI) to U(IV) in uranium-contained ground waters was done by sequence analysis of 16S rRNA which increased the number of *Geobacter* species at the contaminated site (Chandler et al., 2006). Among *Archaea, Thermococcus gammatolerans* has been found to be radioresistant to cope with the stress of radionuclides which expresses genes, such as glutaredoxin-like protein (tg1302) and thioredoxin reductase (tgo180) along with two peroxiredoxins (tg1253 and tg1220) (Zivanovic et al., 2009). NiCoT gene was highly expressed in *N. aromaticivorans* F-199 and *R. palustris* CGA009 when they are allowed to grow where radioactive cobalt exists (Raghu et al., 2008).

The integration of transcriptomics and proteomics is required for the comprehensive view of complete gene expression which can be applied for an effective radionuclide bioremediation. The combine efforts of "-omics"-based technologies would facilitate to accomplish the goal for association of microbe to radionuclide that would decontaminate them efficiently from the environment.

24.8 Conclusion

Radionuclides occur in the environment when unstable atoms seek stability by emitting energy in the form of radiation. Their application in biological, physiological, medical, and geological processes promotes the production of radionuclides. Little amount of radioactivity can be present in every biotic and abiotic environment. This presence of radionuclides can interfere in the functioning of environment, as they may enter the food chain and become the part of geological cycles. This interference is inhibited by removing the radionuclides through bioremediation. Microremediation process may transform radionuclides and can have beneficial effects to the humans. As mentioned before, various microbes such as *D. radiodurans*, *Rhodanobacter*, *D. ferrireducens*, *Rubrobacter*, *Chroococcidiopsis, Bacillus, Arthrobacter, Serratia, Rhodococcus*, and *Pseudomonas* have been extensively used for remediation of radionuclides. The application of algae and fungi including mycorrhizae are the recent advancements in the field of bioremediation. Phytoremediation is another technique of removing radionuclides with the help of plants. Terrestrial and aquatic plants (e.g., *A. retroflexus, B. pilosa, H. annuus, A. cruentus, Fontinalis antipyretica*, *Callitriche stagnalis*, and *Callitriche hamulata*) both can remediate radionuclide contaminated sites. It is an effective, eco-friendly, economic, and versatile way of cleaning environment over conventional method. To meet the major challenge proposed by the bioremediation of availability of soils to organisms and their efficiency to control the levels of radionuclides, the microorganisms are genetically engineered by integrating genes of interest (*nixA*, *merTP*, *merA*, *dcuB*, *NiCoT*, etc.) into them so that the remediation potential of both plants and microbes can be increased. Microbial transformation using "-omics" integrated technologies would prove to be effective for the removal of radionuclides from the environment for stable bioremediation.

Acknowledgment

We are thankful to Lovely Professional University, Phagwara, Punjab, India for providing the infrastructural support.

References

Alkorta, I., Garbisu, C., 2001. Phytoremediation of organic contaminants in soil. Bioresour. Technol. 79, 273–276.

Amachi, S., Fujii, T., Shinoyama, H., Muramatsu, Y., 2005. Microbial influences on the mobility and transformation of radioactive iodine in the environment. J. Nucl. Radiochem. Sci. 6 (1), 21–24.

Bae, W., Chen, W., Mulchandani, A., Mehra, R.K., 2000. Enhanced bioaccumulation of heavy metals by bacterial cells displaying synthetic phytochelatins. Biotechnol. Bioeng. 70, 518–524.

Balarama Krishna, M.V., Arunachalam, J., Murali, M.S., Kumar, S., Manchanda, V.K., 2004. Performance of immobilized moss in the removal of 137Cs and 90Sr from actual low level waste solution. J. Radioanal. Nucl. Chem. 261, 551–557.

Bange, G.G.J., Overstreet, R., 1960. Some observation on absorption of cesium by excised barley roots. Plant. Physiol. 35, 605–608.

Barkay, T., Schaefer, J., 2001. Metal and radionuclide bioremediation, issues, considerations and potentials. Curr. Opin. Microbiol. 4 (3), 318–323.

Barren, F., 2012. Chelate assisted phytoextraction using oilseed *Brassicas*. Plant. Fam. Brassicaceae 21, 289–311.

Beckwith, C.S., McGee, D.J., Mobley, H.L., Riley, L.K., 2001. Cloning, expression, and catalytic activity of *Helicobacter hepaticus* urease. Infect. Immun. 69, 5914–5920.

Beliaev, A.S., Saffarini, D.A., 1998. *Shewanella putrefaciens* mtrB encodes an outer membrane protein required for Fe(III) and Mn(IV) reduction. J. Bacteriol. 180, 6292–6297.

Beliaev, A.S., Saffarini, D.A., Mclaughlin, J.L., Hunnicutt, D., 2001. MtrC, an outer membrane decahaem c cytochrome required for metal reduction in *Shewanella putrefaciens* MR-1. Mol. Microbiol. 39, 722–730.

Beyenal, H., Sani, R.K., Peyton, B.M., Dohnalkva, A.C., Amonette, J.E., Lewandowski, Z., 2004. Uranium immobilization by sulfate-reducing biofilms. Environ. Sci. Technol. 38, 2067–2074.

Bosecker, K., 1997. Bioleaching: metal solubilization by microorganisms. FEMS Microbiol. Rev. 20 (3–4), 591–604.

Brambilla, M., Fortunati, P., Carini, F., 2002. Foliar and root uptake of 134Cs, 85Sr and 65Zn in processing tomato plants (*Lycopersicon esculentum* Mill.). J. Environ. Radioact. 60, 351–363.

Brim, H., McFarlan, S.C., Fredrickson, J.K., Minton, K.W., Zhai, M., Wackett, L.P., et al., 2000. Engineering *Deinococcus radiodurans* for metal remediation in radioactive mixed waste environments. Nat. Biotechnol. 18, 85–90.

Butler, J.E., Glaven, R.H., Esteve-Nunez, A., Nunez, C., Shelobolina, E.S., Bond, D.R., et al., 2006. Genetic characterization of a single bifunctional enzyme for fumarate reduction and succinate oxidation in *Geobacter sulfurreducens* and engineering of fumarate reduction in *Geobacter metallireducens*. J. Bacteriol. 188, 450–455.

Chandler, D.P., Jarrell, A.E., Roden, E.R., Golova, J., Chernov, B., Schipma, M.J., et al., 2006. Suspension array analysis of 16S rRNA from Fe and $SO(4)^{2-}$ reducing bacteria in uranium-contaminated sediments undergoing bioremediation. Appl. Environ. Microbiol. 72, 4672–4687.

Cherkouk, A., Law, G.T., Rizoulis, A., Law, K., Renshaw, J.C., Morris, K., 2016. Influence of riboflavin on the reduction of radionuclides by *Shewanella oneidensis* MR-1. Dalton Trans. 45 (12), 5030–5037.

Chiang, P.N., Wand, M.K., Wang, J.J., Chih, C., 2005. Low-molecular-weight organic acids exudation of rape (*Brassica campestris*) roots in cesium-contaminated soils. Soil Sci. 170, 726–733.

Cook, L.L., Inouye, R.S., McGonigle, T.P., 2009. Evaluation of four grasses for use in phytoremediation of cs-contaminated arid land soil. Plant Soil 324, 169–184.

Cox, M.M., Battista, J.R., 2005. *Deinococcus radiodurans*- the consummate survivor. Nat. Rev. Microbiol. 3 (11), 882–892.

Currie, G.M., Wheat, J.M., Davidson, R., Kiat, H., 2011. Radionuclide production. Radiographer 58 (3), 46–52.

Dahlman, R.C., Auerbach, S.I., Dunaway, P.B., 1969. Behaviour of 137Cs-tagged particles in a fescue meadow. In: Contribution IAEA-SM-117; 72 to Proc. seminar Environmental Contamination by Radioactive Materials, Vienna, 24–28 March. I.A.E.A. Vienna. A slightly updated version is to be found in Auerbach and Dunaway (1970) op. cit.

De Filippis, L.F., 2015. Role of phytoremediation in radioactive waste treatment. Soil Remediation Plants: Prospects and Challenges, pp. 207–254.

Douglas, S., Beveridge, T.J., 1998. Mineral formation by bacteria in natural microbial communities. FEMS Microbiol. Rev. 26, 79–88.

Doyi, I.N., Essumang, D.K., Dampare, S.B., Duah, D., Ahwireng, A.F., 2017. Evaluation of radionuclides and decay simulation in a terrestrial environment for health risk assessment. Sci. Rep. 7 (1), 1–11.

Dushenkov, S., 2003. Trends in phytoremediation of radionuclides. Plant. Soil. 249, 167–175.

Eapen, S., Singh, S., D'Souza, S.F. 2007. Phytoremediation of metals and radionuclides. In: Singh, S.N., Tripathi, R.D. (Eds.), Environmental Bioremediation Technologies. Springer, Berlin, Heidelberg.

Eapen, S., Singh, S., Thorat, V., Kaushik, C.P., Raj, K., D'Souza, S.F., 2006. Phytoremediation of radiostrontium (90Sr) and radiocesium (137Cs) using giant milky weed (*Calotropis gigantea* R.Br.) plants. Chemosphere 65, 2071–2073.

Ebbs, S.D., Brady, D.J., Kochian, L.V., 1998. Role of uranium speciation in the uptake and translocation of uranium by plants. J. Exp. Bot. 49 (324), 1183–1190.

Elias, D.A., Suflita, J.M., McInerney, M.J., Krumholz, L.R., 2004. Periplasmic cytochrome c3 of *Desulfovibrio vulgaris* is directly involved in H-2-mediated metal but not sulfate reduction. Appl. Environ. Microbiol. 70, 413–420.

Endo, S., Kimura, S., Takatsuji, T., Nanasawa, K., Imanaka, T., Shizuma, K., 2012. Measurement of soil contamination by radionuclides due to the Fukushima Dai-ichi nuclear power plant accident and associated estimated cumulative external dose estimation. J. Environ. Radioact. 111, 18–27.

Entry, J.A., Vance, N.C., Hamilton, M.A., Zabowski, D., Watrud, L.S., Adriano, D.C., 1997. Phytoremediation of soil contaminated with low concentrations of radionuclides. Water Air Soil Poll. 88, 167–176.

Ewart, D.K., Hughes, M.N., 1991. The extraction of metals from ores using bacteria. Adv. Inorg. Chem. 36, 103–135.

Favas, P.J.C., Pratas, J., Varun, M., D'Souza, R., Paul, M.S., 2014. Accumulation of uranium by aquatic plants in field conditions, prospects for phytoremediation. Sci. Total. Environ. 470–471, 993–1002.

Francis, A., Dodge, C., Gillow, J., 1992. Biodegradation of metal citrate complexes and implications for toxic-metal mobility. Nature 356 (6365), 140–142.

Francis, A.J., 1998. Biotransformation of uranium and other actinides in radioactive wastes. J. Alloy. Comp. 271–273, 78–84.

Francis, A.J., Nancharaiah, Y.V., 2015. In situ and ex situ bioremediation of radionuclide-contaminated soils at nuclear and norm sites. In: Velzen, L. V. (Eds.), Environmental Remediation and Restoration of Contaminated Nuclear and Norm Sites. Woodhead Publishing, pp. 185–236.

Fredrickson, J.K., Kostandarithes, H.M., Li, S.W., Plymale, A.E., Daly, M.J., 2000. Reduction of Fe(III), Cr(VI), U(VI), and Tc(VII) by *Deinococcus radiodurans* R1. Appl. Environ. Microbiol. 66, 2006–2011.

Fuhrmann, M., Lasat, M.M., Ebbs, S.D., Kochian, L.V., Cornish, J., 2002. Uptake of cesium-137 and strontium-90 from contaminated soil by three plant species; application to phytoremediation. J. Environ. Qual. 31, 904–909.

Fulekar, M.H., Singh, A., Thorat, V., Kaushik, C.P., Eapen, S., 2010. Phytoremediation of 137Cs from low level nuclear waste using *Catharanthus roseus*. Indian J. Pure App. Phys. 48, 516–519.

Geras'kin, S.A., 2016. Ecological effects of exposure to enhanced levels of ionizing radiation. J. Environ. Radioact. 162–163, 347–357.

Gohre, V., Paszkowski, U., 2006. Contribution of the arbuscular mycorrhizal symbiosis ton heavy metal phytoremediation. Planta 223, 1115–1122.

Gupta, A., Singh, S.K., Singh, V.K., Singh, M.K., Modi,A., Zhimo, V.Y., et al., 2020. Endophytic microbe approaches in bioremediation of organic pollutants. In: Kumar, A., Singh, V.K. (Eds.), Microbial Endophytes: Prospects of Sustainable Agriculture. Elsevier, pp. 157–174.

Haferburg, G., Merten, D., Buchel, G., Kothe, E., 2007. Biosorption of metal and salt tolerant microbial isolates from a former uranium mining area. Their impact changes rare earth Elem patterns acid mine drainage. J. Basic Microbiol. 47, 474–484.

Hazen, T.C., Tabak, H.H., 2005. Developments in bioremediation of soils and sediments polluted with metals and radionuclides: 2. Field research on bioremediation of metals and radionuclides. Rev. Environ. Sci. Biol. 4, 157–183.

Hoseini, P.S., Poursafa, P., Moattar, F., Amin, M.M., Rezaei, A.H., 2012. Ability of phytoremediation for absorption of strontium and cesium from soils using *Cannabis sativa*. Int. J. Environ. Health Eng. 1, 1–5.

Huang, J.W., Blaylock, M.J., Kapulnik, Y., Ensley, B.D., 1998. Phytoremediation of uranium-contaminated soils: role of organic acids in triggering uranium hyperaccumulation in plants. Environ. Sci. Technol. 32 (13), 2004–2008.

Huhle, B., Heilmeier, H., Merkel, B., 2008. Potential of *Brassica juncea* and *Helianthus annuus* in phytoremediation for uranium. In: Merkel, B.J., Hasche-Berger, A. (Eds.), Uranium, Mining and Hydrogeology. Springer-Verlag, Berlin, Heidenberg.

Hussain, N., Ferdelman, T.G., Church, T.M., Luther, G.W., 1995. Bio-volatilization of polonium: results from laboratory analyses. Aquat. Geochem. 1, 175–188.

IEAE. 2018.Power Reactor Information System (PRIS) Database-U.S. Energy Information retrieved from, <https://www.world-nuclear.org/information-library/current-and-future-generation/nuclear-power-in-the-world-today.aspx>.

Imran, M., Hu, S., Luo, X., Cao, Y., Samo, N., 2019. Phytoremediation through *Bidens pilosa* L., a nonhazardous approach for uranium remediation of contaminated water. Int. J. Phytoremediat. 21, 1–8.

Istok, J.D., Senko, J.M., Krumholz, L.R., Watson, D., Bogle, M.A., Peacock, A., et al., 2004. *In situ* bioreduction of technetium and uranium in a nitrate-contaminated aquifer. Environ. Sci. Technol. 38, 468–475.

Iwamoto, K., Minoda, A., 2018. Bioremediation of Biophilic Radionuclides by Algae. In: Wong, Y.K. (Ed.), Algae. From: <https://www.intechopen.com/books/algae/bioremediation-of-biophilic-radionuclides-by-algae> (retrieved 05.11.18.).

Jabbar, T., Wallner, G., 2015. Biotransformation of radionuclides, trends and challenges. In: Walther, C., Gupta, D.K. (Eds.), Radionuclides in the Environment, Springer Nature, Switzerland, pp.169–184.

John, S.G., Ruggiero, C.E., Hersman, L.E., Tung, C.S., Neu, M.P., 2001. Siderophore mediated plutonium accumulation by *Microbacterium flavescens* (JG-9). Environ. Sci. Technol. 35, 2942–2948.

Kim, S.J., Koh, D.C., Park, S.J., Cha, I.T., Park, J.W., Na, J.H., 2012. Molecular analysis of spatial variation of iron-reducing bacteria in riverine alluvial aquifers of the Mankyeong River. J. Microbiol. 50, 207–217.

Krishnaswami, S., Lal, D., 1978. Radionuclide limnochronology. In: Lerman, A., (Eds.), Lakes, Springer-Verlag, Berlin, Heidelberg, pp. 153–177.

Kumar, R., Singh, S., Singh, O.V., 2007. Bioremediation of radionuclides, emerging technologies. Omics: J. Integr. Biol. 11 (3), 295–304.

Lasat, M., Fuhrmann, M., Ebbs, S.D., Cornish, J.E., Kochian, L.V., 1998. Phytoremediation of a radiocesium-contaminated soil, evaluation of cesium-137 bioaccumulation in the shoots of three plant species. J. Environ. Qual. 27, 165–169.

Lee, J.H., 2013. An overview of phytoremediation as a potentially promising technology for environmental pollution control. Biotechnol. Bioproc. E 18, 431–439.

Li, G., Hu, N., Ding, D., Zheng, J., Liu, Y., Wang, Y., et al., 2011. Screening of plant species for phytoremediation of uranium, thorium, barium, nickel, strontium and lead contaminated soils from a uranium mill tailings repository in south China. B Environ. Contam. Tox. 86 (6), 646–652.

Lloyd, J., Macaskie, L., 2000b. Bioremediation of radionuclide-containing wastewaters. In: Lovley, D. (Ed.), Environmental Microbe-Metal Interactions. ASM Press, Washington, DC, pp. 277–327.

Lloyd, J.R., Cole, J.A., Macaskie, L.E., 1997. Reduction and removal of heptavalent technetium from solution by *Escherichia coli*. J. Bacteriol. 179, 2014–2021.

Lloyd, J.R., Leang, C., Myerson, A.L.H., Coppi, M.V., Cuifo, S., Methe, B., et al., 2003. Biochemical and genetic characterization of PpcA, a periplasmic c-type cytochrome in *Geobacter sulfurreducens*. Biochem. J. 369 (1), 153–161.

Lloyd, J.R., Sole, V.A., Van Praagh, C.V., Lovley, D.R., 2000a. Direct and Fe(II)-mediated reduction of technetium by Fe(III)-reducing bacteria. Appl. Environ. Microbiol. 66, 3743–3749.

Lu, H., Gao, G., Xu, G., Fan, L., Yin, L., Shen, B., et al., 2009. *Deinococcus radiodurans* PprI switches on DNA damage response and cellular survival networks after radiation damage. Mol. Cell Proteom. 8, 481–494.

Macaskie, L.E., Empson, R.M., Cheetham, A.K., Grey, C.P., Skarnulis, A.J., 1992. Uraniumbioaccumulationby a Citrobacter sp. as a result of enzymicallymediated growth of polycrystallineHUO2PO4. Science 257, 782–784.

Marshall, M.J., Beliaev, A.S., Dohnalkova, A.C., Kennedy, D.W., Shi, L., Wang, Z., et al., 2006. c-Type cytochrome-dependent formation of U(IV) nanoparticles by *Shewanella oneidensis*. PLoS Biol. 4 (e268), 1324–1333.

Masok, F.B., Masiteng, P.L., Mavunda, R.D., Maleka, P.P., 2016. Health effects due to radionuclides content of solid minerals within port of Richards Bay, South Africa. Int. J. Environ. Res. Public Health 13 (12), 1180.

McGrath, S.P., Zhao, F.J., Lombi, E., 2002. Phytoremediation of metals, metalloids, and radionuclides. Adv. Agron. 75, 1–56.

Methe, B.A., Nelson, K.E., Eisen, J.A., Paulsen, I.T., Nelson, W., Heidelberg, J.F., et al., 2003. Genome of *Geobacter sulfurreducens*: metal reduction in subsurface environments. Science 302, 1967–1969.

Milenkovic, B., Stajic, J.M., Stojic, N., Pucarevic, M., Strbac, S., 2019. Evaluation of heavy metals and radionuclides in fish and seafood products. Chemosphere 229, 324–331.

Misra, C.S., Appukuttan, D., Kantamreddi, V.S.S., Rao, A.S., Apte, S.K., 2012. Recombinant *D. radiodurans* cells for bioremediation of heavy metals from acidic/neutral aqueous wastes. Bioengineered 3 (1), 44–48.

Mkandawire, M., Taubert, B., Dudel, E.G., 2004. Capacity of *Lemna gibba* L. (duckweed) for uranium and arsenic phytoremediation in mine tailing waters. Int. J. Phytoremediat. 6 (4), 347–362.

Momoshima, N., Song, L.-X., Osaki, S., Maeda, Y., 2001. Formation and emission of volatile polonium compound by microbial activity and polonium methylation with methylcobalamin. Environ. Sci. Tech. 35, 2956–2960.

Moogouei, R., Borghei, M., Arjmandi, R., 2011. Phytoremediation of stable Cs from solutions by *Calendula alata*, *Amaranthus chlorostachys* and *Chenopodium album*. Ecotoxicol. Environ. Saf. 74, 2036–2039.

Munoz, J., Gonzalez, F., Blazquez, M., Ballester, A., 1995. A study of the bioleaching of a Spanish uranium ore. Part. I: a Rev. bacterial leaching Treat. uranium ores. Hydrometallurgy 38 (1), 39–57.

N'Guessan, A.L., Vrionis, H.A., Resch, C.T., Long, P.E., Lovley, D.R., 2008. Sustained removal of uranium from contaminated groundwater following stimulation of dissimilatory metal reduction. Environ. Sci. Technol. 42, 2999–3004.

Neves, M., Kling, A., Lambrechst, R.M., 2002. Radionuclide production for therapeutic radiopharmaceuticals. Appl. Radiat. Isot. 57 (5), 657–664.

Newsome, L., Morris, K., Lloyd, J.R., 2013. The biogeochemistry and bioremediation of uranium and other priority radionuclides. Chem. Geol. 363, 164–184.

Nyman, J.L., Marsh, T.L., Ginder-Vogel, M.A., Gentile, M., Fendorf, S., Criddle, C., 2006. Heterogeneous response to biostimulation for U(VI) reduction in replicated sediment microcosms. Biodegradation 17, 303–316.

Ojovan, M.I., Lee, W.E., 2014. Background radiation. In: Ojovan, M.I., Lee, W.E. (Eds)., An Introduction to Nuclear Waste Immobilisation. Elsevier, pp. 41–47.

Ortiz-Bernad, I., Anderson, R.T., Vrionis, H.A., Lovley, D.R., 2004. Resistance of solid-phase U(VI) to microbial reduction during *in situ* bioremediation of uranium-contaminated groundwater. Appl. Environ. Microbiol. 70, 7558–7560.

Panak, P., Nitsche, H., 2001. Interaction of aerobic soil bacteria and plutonium(VI). Radiochim. Acta 89, 499–504.

Payne, R.B., Gentry, D.M., Rapp-Giles, B.J., Casalot, L., Wall, J.D., 2002. Uranium reduction by *Desulfovibrio desulfuricans* strain G20 and a cytochrome c3 mutant. Appl. Environ. Microbiol. 68 (6), 3129–3132.

Prakash, D., Gabani, P., Chandel, A.K., Ronen, Z., Singh, O.V., 2013. Bioremediation, a genuine technology to remediate radionuclides from the environment. Microb. Biotechnol. 6 (4), 349–360.

Qaim, S.M., 2012. The present and future of medical radionuclide production. Radiochim. Acta 100 (8–9), 635–651.

Raabe, O.G., 2010. Concerning the health effects of internally deposited radionuclides. Health Phys. 98 (3), 515–536.

Radioactive Waste Management, 2018. Retrieved from <https://www.world-nuclear.org/information-library/nuclear-fuel-cycle/nuclear-wastes/radioactive-waste-management.aspx>

Raghu, G., Balaji, V., Venkateswaran, G., Rodrigue, A., Maruthi Mohan, P., 2008. Bioremediation of trace cobalt from simulated spent decontamination solutions of nuclear power reactors using *E. coli* expressing NiCoT genes. Appl. Microbiol. Biotechnol. 81 (3), 571–578.

Ramaswami, A., Carr, P., Burkhardt, M., 2001. Plant-uptake of uranium: hydroponic and soil system studies. Int. J. Phytoremediat. 3, 189–201.

Ren, C.G., Kong, C.C., Wang, S.X., Xie, Z.H., 2018. Enhanced phytoremediation of uranium-contaminated soils by arbuscular mycorrhiza and rhizobium. Chemosphere 217, 773–779.

Ribeiro de Souza, S.C., López, A., de Andrade, S., Anjos de Souza, L., Schiavinato, M.A., 2012. Lead tolerance and phytoremediation potential of Brazilian leguminous tree species at the seedling stage. J. Environ. Manage. 110, 299–307.

Saffarini, D.A., Schultz, R., Beliaev, A., 2003. Involvement of cyclic AMP (cAMP) and cAMP receptor protein in anaerobic respiration of *Shewanella oneidensis*. J. Bacteriol. Res. 185 (12), 3668–3671.

Salt, C.A., Mayes, R.W., Elston, D.A., 1992. Effects of season, grazing intensity and diet composition on the radiocaesium intake by sheep on reseeded hill pasture. J. Appl. Ecol. 29, 378–387.

Sasmaz, A., Sasmaz, M., 2009. The phytoremediation potential for strontium of indigenous plants growing in a mining area. Environ. Exp. Bot. 67, 139–144.

Schembri, M.A., Kjaergaard, K., Klemm, P., 1999. Bioaccumulation of heavy metals by fimbrial designer adhesins. FEMS Microbiol. Lett. 170, 363–371.

Schuller, P., Bunzi, K., Voigt, G., Krarup, A., Castillo, A., 2004. Seasonal variation of the radiocaesium transfer soil-to-swiss chard (*Beta vulgaris* var. cicla L.) in allophonic soils from the lake region, Chile. J. Envion. Radioact. 78, 21–23.

Seide, 1, H., Loser, C., Zehnsdorf, A., Hoffmann, P., Schmerold, R., 2004. Bioremediation process for sediments contaminated by heavy metals: Feasibility study on a pilot scale. Environ. Sci. Technol. 38 (5), 1582–1588.

Sharma, S., Singh, B., Manchanda, V.K., 2014. Phytoremediation, role of terrestrial plants and aquatic macrophytes in the remediation of radionuclides and heavy metal contaminated soil and water. Environ. Sci. Pollut. Res. Int. 22 (2), 946–962.

Shukla, A., Parmar, P., Saraf, M., 2017. Radiation, radionuclides and bacteria, an in-perspective review. J. Environ. Radioact. 180, 27–35.

Singh, P., Singh, V.K., Singh, R., Borthakur, A., Madhav, S., Ahamad, A., et al., 2020. Bioremediation: a sustainable approach for management of environmental contaminants. In: Singh, P., Kumar, A., Borthakur, A. (Eds.) Abatement of Environmental Pollutants. Elsevier, pp. 1–23.

Singh, S., Eapen, S., Thorat, V., Kaushik, C.P., Raj, K., D'Souza, S.F., 2008. Phytoremediation of cesium and strontium from solutions and lowlevel nuclear waste by *Vetiveria zizanioides*. Ecotoxicol. Environ. Saf. 69, 306–331.

Singh, S., Thorat, V., Kaushik, C.P., Raj, K., Eapen, S., D'Souza, S.F., 2009. Potential of *Chromolaena odorata* for phytoremediation of 137Cs from solution and low level nuclear waste. J. Hazard. Mater. 162, 743–745.

Stojanović, M.D., Mihajlović, M.L., Milojković, J.V., Lopičić, Z.R., Adamović, M., Stanković, S., 2012. Efficient phytoremediation of uranium mine tailings by tobacco. Environ. Chem. Lett. 10, 377.

Terry, N., Zayed, A.M., De Souza, M.P., Tarun, A.S. 2000. Selenium in higher plants. Annu. Rev. Plant Bio. 51 (1), 401–32.

Tian, B., Wang, H., Ma, X., Hu, Y., Sun, Z., Shen, S., et al., 2010. Proteomic analysis of membrane proteins from a radioresistant and moderate thermophilic bacterium *Deinococcus geothermalis*. Mol. Biosyst. 6, 2068–2077.

Valls, M., Atrian, S., de Lorenzo, V., Fernandez, L.A., 2000. Engineering a mouse metallothionein on the cell surface of *Ralstonia eutropha* CH34 for immobilization of heavy metals in soil. Nat. Biotechnol. 18, 661–665.

Van Hook, I.R., 1979. Potential health and environmental effects of trace elements and radionuclides from increased coal utilization. Environ. Health Perspect. 33, 227–247.

Xie, S., Yang, J., Chen, C., Zhang, X., Wang, Q., Zhang, C., 2008. Study on biosorption kinetics and thermodynamics of uranium by *Citrobacter freudii*. J. Environ. Radioact. 99, 126–133.

Yeong, C.H., Cheng, M.-H., Ng, K.-H., 2015. Therapeutic radionuclides in nuclear medicine, current and future prospects. J. Zhejiang Univ. Sci. A 15 (10), 845–863.

Zhu, Y.G., Shaw, G., 2000. Soil contamination with radionuclides and potential remediation. Chemosphere 41 (1–2), 121–128.

Zivanovic, Y., Armengaud, J., Lagorce, A., Leplat, C., Guerin, P., Dutertre, M., et al., 2009. Genome analysis and genome-wide proteomics of *Thermococcus gammatolerans*, the most radioresistant organism known amongst the Archaea. Genome Biol. 10, R70.

Chapter 25

Environmental pollution and their socioeconomic impacts

Fidelis O. Ajibade[1,2,3], Bashir Adelodun[4,5], Kayode H. Lasisi[1,3,6], Oluniyi O. Fadare[3,7], Temitope F. Ajibade[1,3,6], Nathaniel A. Nwogwu[8], Ishaq D. Sulaymon[3,9], Adamu Y. Ugya[10,11], Hong Cheng Wang[2] and Aijie Wang[2]

[1]*Department of Civil and Environmental Engineering, Federal University of Technology, Akure, Nigeria,* [2]*Key Laboratory of Environmental Biotechnology, Research Centre for Eco-Environmental Sciences, Chinese Academy of Sciences, Beijing, P.R. China,* [3]*University of Chinese Academy of Sciences, Beijing, P.R. China,* [4]*Department of Agricultural and Biosystems Engineering, University of Ilorin, Ilorin, Nigeria,* [5]*Department of Agricultural Civil Engineering, Kyungpook National University, Daegu, Republic of Korea,* [6]*Institute of Urban Environment, Chinese Academy of Sciences, Xiamen, P.R. China,* [7]*State Key Laboratory of Environmental Chemistry and Ecotoxicology, Research Centre for Eco-Environmental Sciences, Chinese Academy of Sciences, Beijing, P.R. China,* [8]*Department of Agricultural and Bioresources Engineering, Federal University of Technology, Owerri, Nigeria,* [9]*Huairou Eco-Environmental Observatory, Chinese Academy of Sciences, Beijing, P.R. China,* [10]*Department of Environmental Management, Kaduna State University, Kaduna, Nigeria,* [11]*College of New Energy and Environment, Jilin University, Changchun, P.R. China*

25.1 Background

Environmental pollution is the world's largest cause of untimely death and disease (Landrigan et al., 2019). Globally, the term "environmental pollution" has long existed and its impacts have greatly been felt on the whole of humanity and its environment whether knowingly or otherwise ever since the first congregation and groups of people lived for some long time in one particular location of the Earth (Easterbrook, 1995; Botkin, 1995). Archaeologists have discovered several remains of tools, shell mounds, clothing, and rubble heaps of such people (Keller, 1988). During their existence on Earth, pollution was not seen as a serious problem because of the large availability of space for individuals and groups of people to live in. Although some of their activities (if not most) were not pollution-free but human beings were not affected by the menace or nuisance been caused then (Miller, 1997, 2004). However, as the population increased rapidly, resulting in the formation of more permanent colonies by numerous numbers of people, the subject of pollution resulted in becoming a menace to both humanity and its environment, and it has remained a major one ever since (Jackson and Jackson, 1996; Spellman, 2017).

History has it that, in ancient time, both cities and major towns where major civilization emerged from were baneful areas, soiled by debris and the wastes from human. At about 1000 CE the utilization of coal as a source of fuel resulted in considerable air pollution. From the inception of the 17th century all through the commencement of the 19th century, conversion of coal to coke, which was used for iron smelting, further aggravated the problem. Through the 19th century, there was an extension of this menace to water bodies also. Largely, solid waste accretion of congested urban areas was the problem during the 19th century but, due to human population's rapid industrialization and geometrical growth most especially in urban cities, pollution has become a universal problem and a major threat to human's existence (Alina, 2018; Sulaymon et al., 2020). Landrigan et al. (2017) documented that pollution is accountable for the 16% of all deaths worldwide and 9 million deaths yearly three times more deaths than malaria, tuberculosis, and AIDS altogether. Environmental pollution is responsible for greater than one death in four (Landrigan et al., 2019). Moreover, Suk et al. (2016) asserted that children are impeccably responsive to pollution.

25.2 Concept of environmental pollution

The Earth is faced today by numerous negative occurrences and challenges (caused by human beings, animals, organisms, and many more) and each was triggered to the dangerous level as consequence of Industrial Revolution

Microbe Mediated Remediation of Environmental Contaminants. DOI: https://doi.org/10.1016/B978-0-12-821199-1.00025-0

(Spellman, 1999). The impacts of this menace increase day by day, causing irreparable damage to the Earth through increasing various anthropogenic activities such as incessant combustion of fossil fuels, which emit both particulate and gaseous pollutants, which are toxic in nature; disposal of toxic effluents by factories into water bodies, rendering them unusable and destroying aquatic life; sprawling urbanization resulting in deforestation and thus causing the quality of air to be poor; damaging the productivity of soil via plastic waste littering, just to mention a few. Despite all these accompanying impacts of environmental pollution, there have been little to none efforts to intentionally and pragmatically address its environmental effects and human health risks.

Defining the term "environmental pollution" can be a bit difficult, as various definitions rendered by various authors in the past have been mainly given, based on their various perspectives of the subject matter. One of such definitions by Spellman and Whiting (2006), which seems to capture this concept entirely is stated as "the contamination of air, water, or soil (food) in such a manner as to cause real or potential harm to human health or well-being, or to damage or harm nonhuman nature without justification." As encompassing as this definition is, a layperson may find it challenging to comprehend, thus resulting in sidelining and neglecting some classes of humanity who also have contributed massively to this long-aged world menace. To avoid the repetitions and catalogs of definitions as used in the past and also to make it all-inclusive, we define environmental pollution in its simplest form as "the pollution of the various aspect of the environment making up our natural resources such as soil, water, air, forests and et cetera." All that surround us make up the environment we inhabit and its pollution is hence termed "environmental pollution."

25.3 Pollution: what does it mean?

Over the years, the state of clearly distinguishing succinctly between the terms "environmental pollution" and "pollution" had often resulted in some mild arguments in the environmental world. Some had termed both concepts to be the same in meaning, while others think otherwise. Even in various colleges and higher institutions of learning around the world, students find it tricky to glaringly differentiate between the two terms, especially as regards the definition of terms. Most of their responses are identical and lie in describing the outright effects of the terms, having witnessed some forms of this scenario firsthand (Spellman, 1999, 2017). As pollution is ubiquitous, seeming to take after one of the characteristics of divinity, giving it a definite meaning is complicated (just like the case of environmental pollution mentioned in Section 25.2). It cannot be easily defined. What pollution clearly represents and vice versa is judgmental. We were even taught by nature, the intimate interconnection of most minute elements with other one, which are ultimately the consequential effects of pollution (Pepper et al., 2006). Let us quickly look at some definitions given by some specialists in matters relating to the environment. They are stated as follows:

Pollution is defined as "the man-made or man-induced alteration of the physical, biological, and radioactive integrity of water" (USEPA, 1989 under the Clean Water Act).

Pollution is explained as the casting off of any substance in the category of solid, liquid, gas or any form of energy (in form of heat, sound, or radioactivity) to the environment at a rapid rate greater than those substances can be dispersed, diluted, decomposed, recycled, or stored in some harmless form. (Jerry A. Nathanson – An environmentalist writer under the auspices of Encyclopedia Britannica Inc.)

Pollution is the state of making every part of the environment including air, land and water dirty and not safe or suitable to use. This is often accomplished via the instauration of contaminants into the natural environment, which may not need to be tangible. Light, temperature and sound as simple as they may look can largely be termed as pollutants when introduced intentionally and artificially into an environment. (Alina Bradford – Live Science Contributor)

Pollution is described as one of the several serious problems facing humanity which occurs at the release of substances into the environment in manners or quantities that forbid the environment from effectively handling it, and thus leading to detrimental effects on the ecosystem. (Harrison, 1996)

Any undesirable change of the environment which can be characterized physically, chemically or biologically, and is harmful to human either directly or indirectly through animals, plants, and other living or non-living organisms or entities is termed pollution. (Titiksha, 2017)

Pollution is the "introduction of harmful materials into the environment" (National Geographic, 2018).

It can be vividly noted from the previous definitions, the various perspectives of authors on pollution with each of their definitions displaying their judgmental manner of approach. This cycle of judgmental inclusion will continue even in thousands of definitions that can be expressed on the subject matter, as it has several manifestations in nature. Fig. 25.1 clearly gives these manifestations. To therefore save ourselves from the stress of giving a universally acceptable and concise definition of pollution, we can state based on its two major classifications, point source and

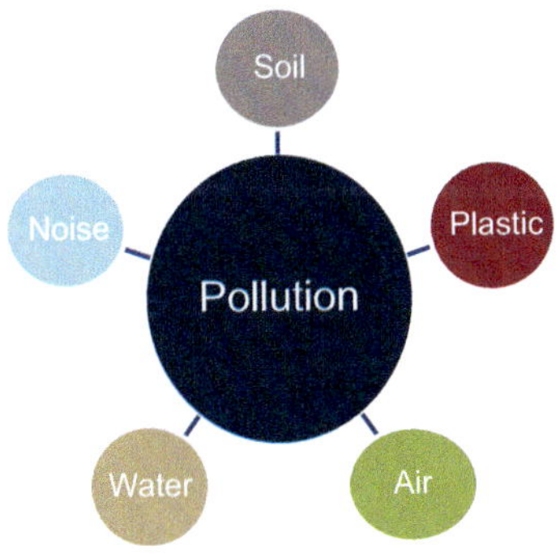

FIGURE 25.1 Environmental pollution and its major types.

nonpoint source, which largely reduce the complexity of defining the term. More discussion will be provided on both in subsequent sections of this chapter.

25.4 Pollutants and their various categories

It is of great importance to look at what pollutants are and how they really affect the environment. The word "pollutant" is closely related to "contaminant" and this may raise some questions if they are not properly defined. Most people use the terms "pollution" and "contamination" interchangeably since they both deal with unwanted elements. According to Mondal (2018), it is simplest to say that the substances that cause pollution are termed "pollutants" while those that lead to contamination are called "contaminants." Citing as an example, a waste material in a body of water may be termed a contaminant or a pollutant. Also, the cleanup processes that occur in remediation, the ones for pollution are similar to that of contamination but they are linked in a way that the introduction of harmful contaminants leads to pollution (Mondal, 2018). There is more negativity attached to pollution due to its more parlous impacts as compared to contamination. Truly, these occurrences are interconnected and comparable, but still have notable differences.

A pollutant is any substance which can be chemical or physical in nature, introduced into air, soil or water in a quantity that threatens humanity, its enjoyment and its proper functioning in any aspect of the environment.

A contaminant is the availability of grand concentrations of substances which may be chemical, biological, or radiological, present in an environment which is beyond the normal datum level for both the area and the organism in the environment.

25.4.1 Categories of pollutants

Pollutants can generally be categorized into two major parts. Although, most often, there exist other categories depending on their existence in nature, but when these are critically analyzed, they will fall under one of these two categories. They are *natural* and *artificial* pollutants. They can also be termed *quantitative* and *qualitative pollutants*, respectively.

25.4.1.1 Natural pollutants

These are materials or substances occurring naturally but later attain the position of a pollutant when their amount or concentration in a given sphere of environment escalates as a result of the mindless actions and activities of man. That, they are natural does not mean man is not involved, but his unawareness of contribution is unknown to him. For example, when some gas such as CO_2 (caused by automobiles and industries) is present in greater quantity in the atmosphere than expected, it results in immeasurable effects on animals, plants, and humans.

25.4.1.2 Artificial (man-made) pollutants

These are materials or substances that do not occur naturally in nature but are introduced into the environment via humans and their activities. Depending on the configuration they assume after their discharge into the environment, they can be classified as *primary* and *secondary* pollutants. Furthermore, based on their natural disposal as related to ecosystem, pollutants can be further classified as *biodegradable* and *nondegradable* pollutants. They are all briefly explained as follows:

1. *Primary pollutants*

These are pollutants that are directly emitted into the environment and they persist in the form in which they are released from their various sources. Their emission could be through natural ways or human actions. Gas, ash, smoke, dust, nitric oxide, sulfur(iv) oxide, and hydrocarbons are few examples under this category. The emission of gases and ash from a volcanic reaction are primary pollutants released in a natural way but carbon monoxide (CO) gas emitted from vehicles or generator machines is also a primary pollutant released via human activities. The emission of primary pollutants into the environment or the air is very harmful and can cause serious health and environmental problems such as respiratory diseases, global warming, and acid rains. The major sources of primary pollutants are motor vehicles and industries and, most often, they are precursors for secondary pollutants.

2. *Secondary pollutants*

   These are pollutants not passed off directly into the environment but are formed from primary pollutants. This can be via a chemical interaction between two or more primary pollutants or by reaction with some constituents present in the atmosphere. Examples of these pollutants include ozone, aldehydes, ketones, sulfur(vi) oxide, and nitrogen dioxide.

3. *Biodegradable pollutants*

   These are pollutants broken down by the various actions of microorganism naturally into smaller, nontoxic, substances in due course of time. Domestic garbage, wood, agricultural remains, fecal matter, cattle dung, vegetable remains, and sewage are typical examples of these pollutants.

4. *Nondegradable pollutants*

These are pollutants that cannot be broken down (or they degrade at a very slow rate) in the environment. Polythene bags, plastics, aluminum plates and cans, synthetic fibers, silver foils, mercury, and lead, arsenic are examples of non-biodegradable pollutants. Some of these nondegradable pollutants conglomerate and are biologically enlarged as they transit along the food chain and the biogeochemical cycle of the ecosystem. Dichlorodiphenyltrichloroethane (DDT), as a distinctive example, moves to the streams and it is taken in by the phytoplanktons that are eaten by the fishes. Therefore the initial ingested dose of DDT by the phytoplankton, which was not harmful, will accumulate over time in the fish and becomes very harmful.

25.4.2 Pollution sources

As mentioned earlier in Section 25.4.1, the two major categories of pollution, point source and nonpoint source will be discussed next.

25.4.2.1 Point source pollution

Point sources represent single and discrete locations or facilities that emit pollution. Examples of these are automobile engine, factory, ditch, tunnel, smokestack, and pipe. This category of pollution makes it easy to monitor and control the discharge of pollutants. Regulatory bodies such as the United States Environmental Protection Agency (EPA) have established standards for emission of some chemicals and compounds in order to control the outflow from any point source and to check and monitor that discharge levels abide by the set regulations.

25.4.2.2 Nonpoint source pollution

Nonpoint sources have no definite origin of concentration and are therefore termed being diffuse and widespread. Discharge of pollutants and contaminants are dispersed into the environment uncontrolled. Wind blows some of these pollutants into the air while others are swept into waterways by rainfall. Some pop up from unrecognized sources, such as oil dripping on roads from faulty vehicles, pesticides used on lawns and fields, livestock and pets droppings, or disturbed soil during excavation. It is difficult to regulate emission of nonpoint sources as the discharge of pollutants cannot be measured at the source but at destination. Samples are collected from any of the sources of pollution in polluted areas and measured. The contribution of various nonpoint sources can only then be estimated. It is difficult for EPA to channel their regulations toward specific individuals or businesses in this case, instead they directed it at municipalities. The federal standards for permissible levels of chemicals in drinking water have been set so that communities are responsible for treating their water until it meets those standards.

25.5 Types of pollution

25.5.1 Water pollution

Water occupies 71% of the Earth out of which 95.6% is held in the oceans, which is not directly available for human consumption such as drinking without arduous desalination process (Jeevanantham et al., 2019). Water is a necessity for both human and industrial growth as it remains one of the vital and defenseless natural resources, which is impossible to survive without its availability (Ajibade et al., 2020; Akosile et al., 2020; Opafola et al., 2020; Ilori et al., 2019; Afroza et al., 2015; Ajibade et al., 2015). The two common types of water are surface and groundwater. Surface water exists in various forms such as ponds, wetlands, streams, rivers, oceans, and lakes. Out of these forms, oceans hold the highest volume of the total water yet are not commonly used directly (unlike the other forms) as a result of high natural salt content (Jeevanantham et al., 2019). The remaining small fraction represents the other forms of water, which we refer to as freshwater on which man mainly depends on. Due to population explosion as well as some anthropogenic activities, which alter the cleanliness of water, the demand for clean water is also increasing highly (Ashraf et al., 2013). Water-related problems are common in developing countries usually due to poor regulatory management, poor design, and inadequate financing (Afroza et al., 2015); however, Juahir et al. (2011) suggested that a joint decision-making process is a sure scheme of action for overcoming these water resource problems.

Water pollution is one of the global environmental problems that threaten both human and industrial development. Though people at present subconsciously associate dirt to disease, the transmission of disease by pathogenic organisms was recognized in the middle of the 19th century (Peirce et al., 1998). Water quality in rivers and other reservoirs is presently threatened by rapid urbanization and industrialization, mining activities, higher level of usage of chemicals in agricultural sectors, poor environmental management (Afroza et al., 2015; Jeevanantham et al., 2019), and indiscriminate disposal of waste (Ajibade et al., 2019a; Adewumi et al., 2019). Pollution of groundwater is commonly by chemical compounds that are harmful to health. Water pollution emanates from two major sources, which are point sources and nonpoint sources. The point sources are recognized as pollution occurring from sources having well-defined point of discharge. Examples of these come from municipal and industrial discharges where the industrial facilities and municipal wastewater treatment plants (WWTPs) are the main sources of water pollution because it contains some toxic heavy chemicals and other pollutants (Adeniran et al., 2016). However, some of these heavy chemicals such as Cu, Cr, Mn, Co, Zn, and Mo, which are heavy metals, are needed in small quantities for plant and animal growth and development but become harmful when supplied in large quantities (Wijayawardena et al., 2016). Nonpoint sources are pollution having their origin of discharge diffused, making it difficult to relate their discharge to a well-defined location. These sources are recognized as dry weather pollutants that enter the waterways via pipes and channels. However, watershed runoff and storm drainages (even though water may be channeled to waterways by pipes) are considered as a nonpoint source (Peirce et al., 1998). Other nonpoint sources include agriculture, silviculture, abandoned mines drainage, solid waste, and leaching from land disposal.

25.5.1.1 Sources of water pollution

When water is polluted, it poses serious threat to the entirety of the environment and, most importantly, the lives of humans (Inyinbor et al., 2018). There are several sources through which water can be polluted, most of which are largely from group of substances that are organic and inorganic in nature such as oxygen-demanding wastes, pathogens, nutrients, suspended solids, salts, heavy metals, pesticides, volatile organic compounds (VOCs), and persistent organic pollutants (POPs). Heavy metals (inorganic) and organic pollutants formed a large part of these sources and they are briefly expatiated next.

1. *Heavy metal*

   Heavy metals find their routes into the environment via various activities of industries, agricultural, mining, and so on (Ugya et al., 2018; Cao et al., 2017) and humans and animals are exposed to the toxicity of this metals through inhalation, food chain, and direct consumption of water having metal (Popa and Petrus, 2017). The effects of these metals' toxicity mostly on human include severe headache (which result in mild eye, nose, and sometimes skin irritations); dizziness to organ dysfunction; vomiting; stomach ache; hematemesis; and diarrhea (Dada et al., 2015). Even some heavy metals considered as essential elements (cobalt, copper, iron, manganese, vanadium, and zinc) are required in small quantity in the body for various biochemical processes while metals such as arsenic, lead, mercury, and cadmium are greatly considered threat or danger in the body (Inyinbor et al., 2018). Water polluted with arsenic when ingested can stimulate lungs, liver, and bladder cancer while the one with cadmium can result in kidney and lung damage (Rana et al., 2017). Brain damage, memory loss, and hampering of learning can result from exposure to lead (Verma and Schneider, 2017; Sun et al., 2017) while mercury (although unique amidst other heavy metals) can travel a wide range of distance in the body of humans and animals, it goes into the bloodstream and can only be flushed through urine or feces. It has the potential of existing in the urinal system for 2 months or less causing renal dysfunction. It is thus classified a global pollutant (Li et al., 2015).

2. *Organic pollutants*

Organic pollutants have several varieties accompanied with huge range of toxicity. Aquatic organisms, humans, and plants have greatly been affected by some of these pollutants such as personal care products wastes, petroleum and its waste, dyes, plant and animal pharmaceuticals, a member of endocrine disruptive chemicals (EDCs) (Kar and Roy, 2010; Aguirre-Martínez et al., 2016). As pharmaceuticals are used universally and unavoidably, they have polluted the environment by the indiscriminate disposal of hospital and household wastes via landfill leaching, drainage water, and sewage (Archer et al., 2017). Dyes are water-soluble giant chemical that is greatly used in many industries to impact color on products but its presence in water has greatly affected aquatic organisms, plants, and humans (Adelodun et al., 2020). Humans are exposed to dye toxicity via consumption of vegetables and fish, which bioaccumulate dyes (Oplatowska et al., 2011). Dyes are considered as carcinogenic and mutagenic; thus their removal from wastewater before disposal is ultimately important (Inyinbor et al., 2018). In human, EDCs increase cancer risk via abnormal endocrine activities while on aquatic lives, their effects range from endocrine system disruption through the reduction in eggs and sperm cells production to feminization of female aquatics (Jung et al., 2015; Akanyeti et al., 2017; Ng et al., 2017).

25.5.2 Air pollution

Air pollution is increasingly becoming an issue globally with a large number of concerns arising from developing cities. It occurs when the atmospheric air is contaminated in such a mode or way that poses potential or real threat to human health, his natural surrounding, and his well-being (Sonibare and Jimoda, 2009). "Unless we clean up the air, by the middle of the century one person will die prematurely every 5 seconds from outdoor air pollution" (OECD, 2016). The use of various modes of sophisticated motorized transports, machineries, and industrial works is expected to increase, thus potentially deteriorating air quality. Also, increase in the concentration of harmful gases and particles being released into the atmosphere causes human and environmental health problems (Sulaymon et al., 2020). The effects of air pollution from natural sources especially volcanic eruption are more severe and significantly last longer than those from anthropogenic sources. For instance, pollutants from volcanic eruption can get to high altitude of the atmosphere and reflect sunlight with consequent decrease in atmospheric temperature for few years or more (Spellman, 2017). Generally, air pollutants can be classified in several ways as follows:

1. *Primary air pollutants*

   These pollutants are released right from the source contributor; an example of primary air pollutant is factory chimney. Moreover, primary air pollutants can also be classified into sources of stationary air pollutants (e.g., power plants) or mobile air pollutants (e.g., automobiles).
2. *Secondary air pollutants*

   These are pollutants that originate from their source contributors and subsequently combine chemically (usually involving sunlight) with other substances to form toxic compounds. An example of secondary air pollutant is the conversion of nonmethane VOCs to ozone (O_3).
3. *Indoor and outdoor (or ambient) air pollutants*

Other classification of air pollutants is indoor and outdoor pollutants, which can be further classified into criteria and hazardous air pollutants. The National Ambient Air Quality Standards (NAAQS) declared under the Clean Air Act (CAA) that the following are the six criteria of air pollutants: carbon(ii) oxide (CO), VOCs or ozone (O_3) particulates, nitrogen oxides (NO_x), sulfur(iv) oxide (SO_2), and particulate lead (Pb) (Spellman, 2017). Table 25.1 depicts several kinds of standards depending on the receptors and severity of the impacts.

25.5.2.1 Sources of air pollution

These are referred to as the various locations, activities, or factors that are responsible for the release of air pollutants into the atmosphere. There are two classes, the natural and artificial sources.

TABLE 25.1 Air quality guidelines for WHO and Europe Union (EU).

Pollutant	Air quality guidelines	EU Air Quality Standards EU Directive 2008/50/EC
PM2.5	10 μg/m³ annual mean 25 μg/m³ 24-h mean	25 μg/m³ annual mean (labeled target value as of 2010, and as limit value as of 2015)
PM10	20 μg/m³ annual mean 50 μg/m³ 24-h mean	50 μg/m³ 24-h mean (not to be exceeded more than 35 times a calendar year) 40 μg/m³ annual mean
O_3	100 μg/m³ 8-h mean	240 μg/m³ 1-h mean
NO_2	40 μg/m³ annual mean 200 μg/m³ 1-h mean	40 μg/m³ annual mean 200 μg/m³ 1-h mean (not to be exceeded more than 18 times a calendar year)
SO_2	20 μg/m³ 24-h mean 500 μg/m³ 10-min mean	125 μg/m³ 24-h mean (not to be exceeded more than 3 times a calendar year) 350 μg/m³ 1-h mean (not to be exceeded more than 24 times a calendar year)

Source: Data from WHO (World Health Organisation), 2006. Air Quality Guidelines. Global Update 2005. Particulate Matter, Ozone, Nitrogen Dioxide and Sulfur Dioxide. Available from: <http://www.euro.who.int/en/health-topics/environment and-health/Housing-and-health/publications/pre-2009/air-quality-guidelines.-global-update-2005.-particulate-matter,-ozone,-nitrogendioxide-and-sulfur-dioxide>; EEA (European Environmental Agency), 2017. Air quality in Europe—2017 report. In: EEA Report No 13-2017. Available from: <https://www.eea.europa.eu/publications/air-quality-ineurope-2017>.

1. *Natural sources*

   These occur due to numerous natural phenomena that include dust and noble gas from radioactive decay (EIA, 2008). Smoke and carbon (ii) oxide from wildfires react with anthropogenic pollutants (such as NO_x, SO_2, and organic carbon compounds) to produce a seasonal haze of secondary pollutants (Goldstein et al., 2009).
2. *Artificial sources*

   These are due to various human activities and are predominantly related to the burning of different kinds of fuel and materials, which release significant high concentration of pollutants into the atmosphere. The various forms of artificial sources are discussed next.

   a. Mobile sources: Several motorized vehicles numbered up to millions ply the Nigerian highways on daily basis as it is found in many other countries of the world and this leads to vehicles consuming millions of liters of petrol (Ndoke and Jimoh, 2000). Hence, vehicular emission remains a threat to environmental health problem, which is expected to increase reasonably as vehicle ownership increases in the world (Abam and Unachukwu, 2009). Over 600 million people globally are exposed to hazardous level of traffic generated pollutants [WHO (World Health Organisation), 2006]. These pollutants are as a result of incomplete combustion of fuels such as natural gas, coal, or wood. Though it should be realized that O_2 used in internal combustion process is a component of air, which contains 78.08% nitrogen and as combustion takes place, oxides of nitrogen are equally formed. Hence, traffic emissions contribute about 50%–80 % of NO_2 and CO concentration in developing countries (Fu, 2001). Furthermore, the increase of this traffic-related pollution is not only based on the aforementioned factors only, but also on low-quality fuel, poor traffic regulation, and lack of air quality implementation force (Abam and Unachukwu, 2009).

   b. Stationary sources: The most significant stationary combustion sources include energy production facilities such as municipal power plants, waste incineration, and residential combustion. Several industrial processes, such as iron and steel production, also involve combustion of fossil fuels or biomass for generating power and heat needed for the process (Sonibare, 2010). Most of these sources are considered point sources, although smaller and more widespread sources such as residential combustion could also be considered as an area source. Physical and chemical characteristics of the particles emitted from these source categories depend on the combustion process itself, and the type of fuel burnt (solid, liquid, or gas). Combustion processes and properties of particulate matter emitted from these sources have been comprehensively reviewed by Holmes and Morawska (2006). These include smokestacks of power plants (Sonibare, 2010), manufacturing facilities, and waste incinerators, as well as furnaces and other types of fuel-burning heating devices (Anf and Emad, 2014). The emissions from these facilities are majorly products of combustion processes. The industries emit various gases of different magnitude into the atmosphere on daily basis (Anf and Emad, 2014).

25.5.3 Soil or land pollution

Soil is a universal sink originating from living organisms and materials of inorganic substances, which bears the greatest burden of environmental pollution (Ajibade et al., 2019b). Spellman (2017) defines soil as the upmost stratum of weathered rocks and decomposed organic matters, which can support life as it primarily comprises air, moisture, nutrients. Soil is also vital as it supports biogeochemical nutrients cycling. From the global land mass of 13.07 × 109 ha, arable land takes 11.3% only; pasture land takes 24.6%; timberland takes 34.1%; and residences, municipals, industries, and roads take the remaining fraction of 31% (Alloway, 2001). Soil pollution among other degrading phenomena such as erosion, flood, and urbanization is not just a great threat to the sustainability of soil resources only but also to human health as potential toxic substances move through the food chain and groundwater, which are in turn used and consumed by man (Akinbile et al., 2016a,b; Adewumi and Ajibade, 2015). Okrent (1999) defines soil pollution as the presence of toxic compounds, chemicals, salts, radioactive materials, or disease-causing agents built up in soil, which have adverse effects on plant, human, and animal health. Some sources of soil and subsurface contaminants include graveyards, landfills, septic tanks, sump and dry wells, underground storage tanks, and pipelines. Moreover, soil contamination could also occur below the water table as a result of mining operations, test holes, agricultural drainage pits, and canals (Spellman, 2017). Most technologically advanced regions of the world experience more land pollution than undeveloped areas (Alloway, 2001). Through biogeochemical cycles, nutrients contaminate the soil and these contaminants contribute to air pollution. Some of these pollutants include sulfur(iv) oxide, nitrogen gases (NO, NO_2, NH_3), hydrogen sulfide, hydrocarbons, carbon(ii) oxide, and ozone. However, due to the negligible changes that happen in the soil, their effects toward air pollution are usually neglected when analyzing air pollution. Two typical aerial particulate soil pollutions include the accumulation of heavy metals within smelters, and soils within municipalities where exhaust fumes connected with vehicle and industrial machinery emissions (Spellman, 2017).

25.5.3.1 Sources of soil/land pollution

The sources of soil pollution can be categorized under two major types, (1) soil pollution from agricultural sources and (2) soil pollution from nonagricultural sources, for example, industrial effluents, solid wastes, and other urban activities. All these affect the surface soil, underground soil, and soil profile (Ashraf et al., 2014).

1. *Agricultural sources*

   This type of pollution emanates from different agricultural practices such as use of fertilizers, pesticides, insecticides, and herbicides. Animal wastes in large quantities also result in soil pollution.
2. *Nonagricultural sources*

This type usually emanates from urbanization caused by rapidly increasing population, which in turn caused an increase in the per capita output of waste. These wastes are dumped indiscriminately on the soil and with time find their entry into the soil system (Oluwatuyi et al., 2020). Their long persistence and accumulation result in toxic concentration and thus become sources of pollution. Some of those most important soil pollutants are inorganic toxic compounds (Ashraf et al., 2014).

25.5.4 The interface of air, water, and land pollution

There exists a relationship between air, water, and land pollution, as the pollutants contaminating these three resources are interrelated (Nathanson, 2007). The phenomenon of studying these three major types of pollution—air, water, and land and their combined interactions with the environment is termed *pollution Interfacial system.* Fig. 25.2 gives a representation of the interrelationship of environmental pollution problems pertaining to air, water, and soil quality. This interrelationship was duly explained as follows by Spellman (2017).

Number "1" represents *acid rain* caused by air pollution but destroyed both aquatic and terrestrial ecosystem. Number "2" represents *incineration* that indicates burning of waste products to preserve land resources but causes air

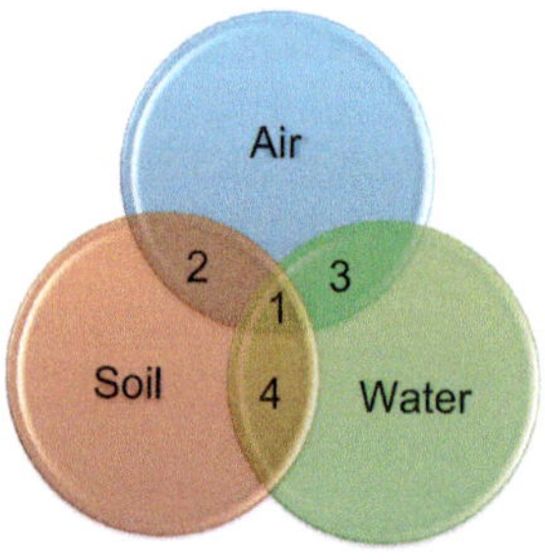

FIGURE 25.2 Environmental pollution issues relating to water, soil, and air quality interrelationship. *Source: Adapted from Nathanson, J.A., 2007. Basic Environmental Technology, fifth ed. Prentice Hall, Upper Saddle River, NJ.*

pollution problems. Number "3" represents *air control system pollution* connoting some certain air pollution control systems such as "scrubbers" but produces a cycle of dirty water, which can cause water pollution, and Number "4" represents *leachate production* indicating that leachate seeping through landfills has high tendency of contaminating both the surface water and groundwater.

25.5.5 Noise pollution

Detection and making of sound are important attributes, which help humans in communicating with one another and in receiving valuable information such as sound signal from a whistling tea kettle, warning as from a fire alarms, and entertainment as in music. Besides these beneficial and enjoyable sounds, there is noise that is an unwanted and irregular sound that commonly comes from such products of development as trucks, industrial machinery, airplanes, trains, air conditioners, and the likes. This phenomenon is termed noise pollution. Noise pollution has been recognized as an urban environmental issue in the world specifically in developing nations. It is well established that noise pollution has been affecting the global population increasingly, mainly in large cities (de Souza et al., 2020) and the intensification of noise pollution has been aggravated since the Industrial Revolution in the 18th and 19th centuries, as a result of the growing use of machinery (Murgel, 2007). Recently, it was reported by the World Health Organization (WHO) (2018) that about 100 million people are affected by road noise in the European Union (EU) and no less than 1.6 million years of healthy life are lost annually. Rinkesh (2019) defines noise pollution as a phenomenon that takes place when there is either an excessive amount of noise or an unpleasant sound that causes a temporary disruption in the natural balance. Psychological health problems such as anxiety, depression, stress, and fatigue can be generated when exposed to high levels of sound pressure (Brazilian Association of Technical Standards, 2019), as well as cardiac, auditory, or cognitive problems, affecting from fetuses to the adults [World Health Organization (WHO), 2018]. In overall, teaching and learning, mental activities, verbal communication, and sleep are frequently the most affected (Brazilian Association of Technical Standards, 2017). Also, it has developed to cognitive impairment in children and tinnitus and can be a risk factor for diabetes (Dzhambov, 2015). Environmental noise is an ancient sensation that has been of great concern. For instance, owing to distraction during sleep, it is said that Julius Caesar prohibited the driving of chariots at night on the cobblestone streets of Rome (Peirce et al., 1998). Urban noise is a gradual and harmful pollutant that has some negative physiological and psychological impacts (sometimes leading to permanent damages) on man. Besides these effects on humans, noise pollution also contributes negatively to the distribution of terrestrial wildlife (Sarikavak and Boxall, 2019). Table 25.2 gives a detail summary of all the types of pollution enumerated in Sections 25.5.1–25.5.5.

25.5.5.1 Sources of noise pollution

The main significant sources of noise pollution are road traffic, industry, railways, and airports. According to European Commission (2017), 69,000 hospital admissions and 15,900 instances of early death are traceable to noise pollution in the EU every year and nearly 6 million suffered highly disrupted sleep as an aftereffect.

25.5.6 Plastic and microplastic pollution

The proliferation and accumulation of plastic materials in the natural environment causing undesirable impacts on living organisms and their habitats are referred to as plastic pollution. Plastic pollution occurs in different forms in the environment, such as plastic litter, marine debris, and plastic particles. For many years, plastic pollution has been recognized as a threat to the ecosystem and has become an issue of global concern (Wagner and Reemtsma, 2019). A large percentage of plastics produced are single-use and are released into the natural environment due to poor management, lack of effective recovery methods, as well as limited success in recycling technology. Since the majority of these materials are resistant to biodegradation, they can remain in the environment for decades, resulting in the massive presence of mismanaged plastics in the environment (Fadare et al., 2019). In effect, about 79% of plastic wastes ended up in landfills (Nature Communications, 2018) while the oceans have become a reservoir for mismanaged plastics through different water channels such as rivers, lakes, and freshwater. It is estimated that by 2050, with the present unabated rate of plastic production, the quantity (in weight) of plastics in the oceans will be more than that of fish (World Economic Forum, 2016). So far, about 700 marine species have been reported to interact with marine debris (Gall and Thompson, 2015), due to the accumulation of plastic waste in oceans, sandy beaches, and seabed. As a result, mismanaged plastics have become a major challenge raising concerns about sanitation, environmental, and human health issues.

Recently, Thompson observed the degradation of mismanaged plastics via natural processes in the environment, which he termed "microplastics" (Thompson et al., 2004). Since then, awareness about the impacts of plastic particles

TABLE 25.2 Environmental pollution.

	Environmental pollution			
Types of pollution	**Water pollution**	**Air pollution**	**Soil/land pollution**	**Noise pollution**
Source contributors	Industries, solid and liquid waste disposal, hydrocarbons, USTs, septic tanks, eutrophication, acidification, biosolids, agriculture, landfills, combustion, oil and gas brine pits, abandoned well sites, saltwater intrusion, surface impoundments	Industries, transportation, solid waste disposal, combustion, miscellaneous, nuclear operations, geothermal operations	Industrial practices, solid and liquid waste disposal, stockpiles, tailing and spoils, salt spreading on roads, animal feedlots, compost of organic wastes, oil spillage, seepage from landfills and illegal dumps, miscellaneous, mining, geothermal operations, agrochemicals, agriculture, hydrocarbons, radionuclides	Railway, airport, industries
Common pollutants	Industrial wastes, WWTPs, breweries, milk processing plants, paper mills, nutrients (e.g., O_2, N, and P), suspended solids, BOD, pH, pathogenic microorganisms, FOG, PPCPs, POPs, toxic (e.g., Zn, Pb, Cd, Ni, and Cu) and nontoxic substances (e.g., Cl, Fe, NH_3, and phenols), pathogenic parasites, Bioconcentrated metals, agricultural wastes, acid rain and acid washed in from watershed and vegetation, pH, fertilizers, fungicides, herbicides, and pesticides, PPCPs, plastics, shipwrecks, heavy metals	Industrial wastes, methane, CFCs, particulate matter, sulfur dioxide, VOCs, CO_2, hydrogen oxides, solid wastes, nitrogen oxides, hydrocarbons, miscellaneous	Heavy metals, leachates, FOG, hydrocarbons, fertilizers, fungicides, herbicides, and pesticides, industrial chemicals, animal feeds, pharmaceutical wastes, biosolids, livestock manures, radioactive substances, pathogenic organisms, plastics	Trucks, industrial machinery, airplanes, trains, air conditioners, generators, dynamites
Effects	Cholera, typhoid, heart and kidney problems, poor blood circulation, vomiting, skin lesions, nervous system problems, damage to aquatic ecosystem, death, diarrhea, loss of occupation	Damage to vegetation, building materials, respiratory system, eyes, nervous system; visibility reduction, death, climate change, low temperature, chronic obstructive pulmonary disease	Underground water pollution, toxicological problems in ecosystems and humans, damage to structures and services	Death, diseases (insomnia, cardiovascular), blood pressure imbalance (hypertension), blood clotting issues, stress, deafness, annoyance, heart attack, stroke, dementia, discomfort
Control/ remediation	Prevention at source, control of nutrients, disinfection, water/ wastewater treatment, thermal treatment, bioremediation, phytoremediation, proper	Pollution prevention, air emission reduction, clearing the air, selection and use of sustainable air pollution control methods, equipment and systems, removal of dry	Adoption of land treatment, thermal treatment, asphalt incorporation, solidification/ stabilization, chemical extraction, excavation,	Modification of the sound producers (e.g., muffling), interrupting the path of the sound, protecting the recipient (e.g., by well, vegetation), effective implementation of

(*Continued*)

TABLE 25.2 (Continued)

Types of pollution	Environmental pollution			
	Water pollution	Air pollution	Soil/land pollution	Noise pollution
	siting, designing, construction and management of potential point source, use of sewerage and proper septic tank designs, proper siting and design of landfills, effective implementation of government policies on sustainable water quality and environment	particulate matters using effective equipment (e.g., gravity settler, cyclone collectors, electrostatic precipitators, wet or venturi scrubbers, baghouse, or fabric filters); absorption, adsorption, condensation, effective combustion, control of crankcase and evaporative emissions, use of catalytic converters, effective implementation of government policies on sustainable environment and land development	manure treatment, waste control technology, waste minimization, input substitutions, process modification, good operating practices, recycling of waste materials, biological treatment, effective implementation of government policies on sustainable air quality and environment	government policies on sustainable environment

BOD, Biochemical oxygen demand; *CFCs*, chlorofluorocarbons; *FOG*, fat, oil, and grease; *pH*, potential of hydrogen; *POPs*, persistent organic pollutants; *PPCPs*, pharmaceuticals and personal care products; *USTs*, underground storage tanks; *VOCs*, volatile organic compounds; *WWTPs*, wastewater treatment plants.

has significantly increased. Micro-/nanoplastics (NPs) are products of mismanaged plastics via degradation or intentionally produced plastic particles that are used in consumer care products or for industrial applications, released into the environment. Plastic debris breaks down over time into smaller particles of various sizes and shapes to form macroplastics (>25 mm), mesoplastics (5–25 mm), microplastics MPs (0.1–5 mm), and NPs (<100 nm) (Alimi et al., 2018), under natural processes such as UV radiation and mechanical abrasion (Fig. 25.3). These are known as secondary microplastics (Mattsson et al., 2015). On the other hand, plastic particles that are produced intentionally for industrial applications (plastic pellets), pharmaceuticals or microbeads used to enhance consumer care products are called primary microplastics. Nevertheless, other sources are yet to be classified, for instance, plastic debris from newly manufactured plastic products, which are not as a result of natural degradation processes or plastic particles produced intentionally.

25.5.6.1 Sources of plastic and microplastic pollution

Plastics and microplastics can cause pollution in the environment at any stage of product manufacturing, consumption, and disposal. However, the primary source of plastic pollution is mismanaged plastic materials from anthropogenic activities in the environment with terrestrial and freshwater ecosystems as emission pathways (Wagner and Reemtsma,

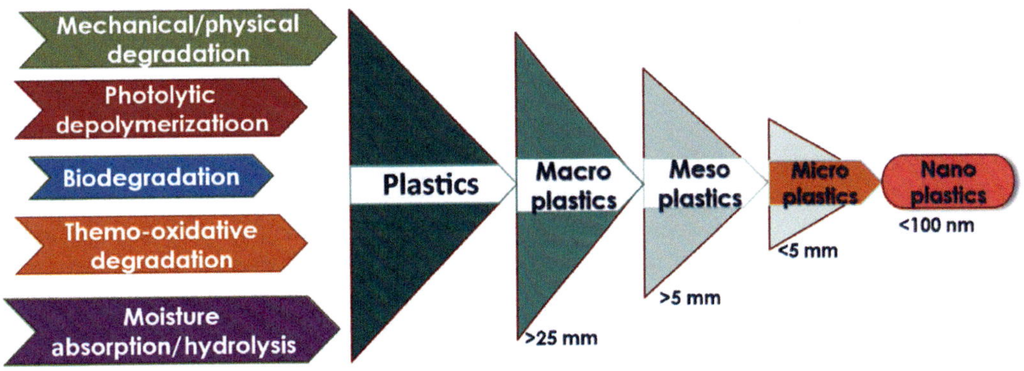

FIGURE 25.3 Plastics degradation processes in the natural environment.

2019). For instance, in the terrestrial system, single-use plastics are found in landfills, and along water channels due to indiscriminate disposal of used plastics into the environment as well as weak recycling system, causing environmental nuisance and unpleasant effects. Also, accidental loss/discharge of nurdles used in plastic production during transportation and illegal discharge of plastic wastes are some of the activities contributing to plastic pollution in the environment (Clunies-ross, 2019).

On the other hand, the presence of mismanaged plastics in the aquatic medium is mainly from land- and ocean-based sources (Li et al., 2016). Plastics that are improperly disposed of are transported from their sources through various water channels and WWTPs into the aquatic ecosystem (Browne, 2015). Land-based sources of plastic litter contribute about 80% of the mismanaged plastics found in the aquatic environment. This usually comes from the industrialized locality with high human presence and activities, which include recreation, tourism, WWTPs effluents, surface water drains, and refuse dumpsites (Lattin et al., 2004; van Wezel et al., 2016). For instance, a recent study identified coastal recreational activities as a significant source of mismanaged plastic wastes in the beach (Clunies-ross, 2019). Ocean-based plastic pollution comes from direct human activities in the oceans; this includes abandoned/lost/discarded fishing nets (Andrady, 2011), shipwreck/accidental discharge, recreational/tourist activities, as well as the illegal discharge of plastic waste in the oceans.

Furthermore, a more significant percentage of plastic particles in the environment emanated from degrading mismanaged plastic materials from landfills. Other sources include wash off microbeads in consumer care products, such as exfoliants, hand cleaners, facial cleaners, toothpaste, airblast cleaning media, and microplastics introduced into fertilizer/herbicides formulation. Also, synthetic clothing is believed to be the primary source of airborne plastic particles (Dris et al., 2016). Other sources are plastic particles from house furniture, building materials, erosion of car tires, industrial emissions, synthetic particles used in garden soils, and plastic incineration (Dris et al., 2016; Dris et al., 2017; Liebezeit and Liebezeit, 2015).

25.6 Socioeconomic impacts of pollution

To better understand the socioeconomic impacts of pollution generally, it is pertinent to know what the term "socioeconomic" means. In brief, it is the study on how both social and economic activities affect the totality of the environment. The socioeconomic standing of a given society is commonly evaluated by defining the state of education, income, health, occupation, and sometimes a combination of these aspects within the populace and their economic growth. There are numerous ways this can be explained when relating it to pollution. It makes no sense to further emphasize that environmental pollution has demean and stretched its poisonous grasping hands into the very basic human needs, which include water, air, and soil via numerous activities. All these activities in the long run tend to have several negative impacts on the environment in which humans are most important. Several studies reveal that pollution has enormous negative impacts on the total environment and people living therein. That is to say that, environmental pollution does not pose threat to human alone but also to nonhuman nature. This section explains the socioeconomic impacts of pollution on:

1. human health,
2. the environment, and
3. economic growth and development.

It could be seen that the previous three categories are a composition of the determining factors (education, income, health, occupation, and economic growth) of the socioeconomic status of a society. These three categories are elaborated one after the other next.

25.6.1 On human health

Environment has a great influence on health, hence, the need to encourage good and healthy environment as the basis of reliable health security. Health is wealth is a common adage because health is very vital for the development and viable growth. The World Health Organization (WHO) estimates that 8.9 million deaths are caused by pollution (Landrigan and Fuller, 2016). Environmental impacts of pollution on health are dependent upon concentration of pollutants, exposure response coefficients, and other related attributes (Feng et al., 2019). It is evident that the severity of most human respiratory diseases such as asthma, bronchitis, sputum, lung cancer, and coughing due to cold increases with prolonged exposure to air pollutants at some degrees of concentrations and nearness (Srinivasan, 2013; Feng et al., 2019); however, the health effects of air pollution vary significantly among individuals with highly susceptible and high-risk groups as adults, infants, and pregnant women with children at the highest risk level because they are prominently active outdoors and their lungs are still developing (Srinivasan, 2013). The WHO appraises that the maximum amount of suspended particulate matter (SPM), which cannot upset human health is 90 $\mu g/m^3$ (Srinivasan, 2013). It is estimated that about 2 million premature deaths in developing countries result from indoor air pollution (Srinivasan, 2013), and from drinking contaminated water mainly in developing countries (Afroza et al., 2015) where almost half of these deaths are due to pneumonia in children below 5 years of age, whereas about 1.3 million deaths globally result from outdoor air pollution; and people living in middle-income countries unevenly suffer this problem [World Health Organization (WHO), 2011]. USEPA (US Environmental Protection Agency) (2013) equally affirms that air pollution is also responsible for some acute and chronic lung diseases and other airborne diseases. It is important to note that air pollution also affect crop plants and animals and, hence, indirectly impair human health when consumed. Air pollution affects not only local environment but also the adjacent areas (Ma et al., 2014; Feng et al., 2019).

Besides air pollution, water pollution is also another vital aspect that greatly affects human health (Ashraf et al., 2013; Afroza et al., 2015) because water resources are very germane for socioeconomic development (Ugya et al., 2018). Increase in the level of water pollution could be attributed to higher development of many fields and anthropogenic activities (Jeevanantham et al., 2019). Agriculture is known to consume water at high level globally and in some cases, polluted water is used. However, polluted water has both advantages and disadvantages and partially treated polluted water is used to irrigate about 20 million hectares of agricultural land in over 50 countries (Ashraf et al., 2010). Polluted water contains toxic and nontoxic metals and microorganisms and when animals (including avian and aquatic lives) drink and/or inhale it or plants absorb it, it gets into them with the pollutant and these plants and animals are in turn consumed by man; hence, health is endangered (Afroza et al., 2015; Jeevanantham et al., 2019). However, assimilating contaminants from water through the consumption of crops by man is less pronounced owing to the fact that plants have power of detoxifying heavy metals and this is achieved through different mechanisms such as mycorrhizae, organic acids, phytochelatins, metallothioneins, root reductases, and heat shock proteins (Kumar et al., 2016; Jeevanantham et al., 2019), though traces of contaminants can be received through this consumption. These heavy metals commonly affect areas like heart (carcinogenic and teratogenic impacts), kidneys, nervous system, liver, pancreas, skin, and reproductive system of man (Jeevanantham et al., 2019). Diseases that can spread as a result of consuming polluted water include typhoid, cholera, poor blood circulation, vomiting, skin leissons, and nervous system problems (Afroza et al., 2015). Ghafoor et al. (1994) affirm that water pollution is the major cause of human death (Ghafoor et al., 1994). About 10% of diseases worldwide are caused by water pollution, hygiene, and sanitation (Pruss-Ustun et al., 2008). It was estimated that half a million population die of malaria as well as 1.4 million deaths recorded for diarrhea among children (Pruss-Ustun et al., 2008; Afroza et al., 2015). UNICEF (2008) concluded that about 900 million people lack access to safe water, whereas 2.5 million people lack proper sanitation systems (Afroza et al., 2015). Water stress is experienced by one-third of the population globally (IWMI, 2007). Health impacts resulting from water pollution have been highly negative that in 2002 alone; more than 3.5 million people died as a result of water pollution, sanitation, and hygiene (Pruss-Ustun et al., 2008); and approximate cases of diarrhea of 4 million reported yearly (Afroza et al., 2015).

Noise pollution is also another subject of interest due to its negative impacts. Studies showed that noise instigates displeasure at levels above 55 and 85 dB in office and in industry, respectively; 55–116 dB causes hypertension, 75 dB causes loss in hearing; and a noise level above 81 dB leads to permanent deafness [World Health Organization (WHO), 2004; Sarikavak and Boxall, 2019]. All these burdens boil down to high medical costs and death in most cases.

25.6.2 On the environment

The frequent penetration of pollutants into the soil affects the organisms living therein. Nevertheless, the effect caused by the presence of pollutants in soil or the lithosphere (on both terrestrial animals and ecosystems) is much more appreciable as these substances cumulate in food chains. The roving component of the soil environment such as water or, more precisely, the underground water can diffuse the contaminants through the soil and soil-forming rocks somewhat rapidly. Ipingbemi (2009) reports that about 5%–10% of the Nigerian mangrove ecosystems have been eradicated by oil spillage and other adverse impacts of pollution on the environment are death and loss of aquatic resources, eutrophication of water bodies, and ecological destruction. Landrigan et al. (2010) revealed that at every level of income in most countries, polluting industries and hazardous waste sites are disproportionately located in poor, minority, and marginalized communities, a phenomenon termed "environmental injustice" (Landrigan and Fuller, 2016). Similarly, some US-owned companies in developing nations make use openly, industrial practices not allowed in the United States, thus creating some significant levels of pollution in these countries due to their lack of knowledge on the impending dangers, the economic ability, or the political structure to protect their environment (Spellman, 2017). A problem long unnoticed, soil pollution generated by industrial contamination, management of superfund sites, exploration and production, mining, and nuclear industrial practices (among others) has affected soil quality at levels we have only recently begun to understand. It is difficult to determine the degree of soil damage in many areas, because assessment itself is debatable; however, based on strong evidences, staggeringly large amounts of polluted soils have been generated (Testa, 1997). Spellman (2017) reported that in Oklahoma, for example, contaminated soil accounts for about 90% of the waste generated as a one-time occurrence.

25.6.3 On the economic growth and development

When it comes to economic growth and development, environmental pollution has tremendous adverse effects on the livelihoods, businesses, education, and occupation of the populace in general. Moreover, other hostile effects of pollution on the environment and economic growth include loss of biodiversity and diminished food and agricultural production levels. In some cases, for instance, the Niger Delta region (where the livelihoods of people come mainly from land and water) of Nigeria that are commonly faced with oil spillage, many residents lose their occupation such as fishing, canoe carving, and forestry, which account for about 70% of the total employment in the region; this often leads to dropping out of school by children of most families who can no longer pay their wards' school fees (Ipingbemi, 2009). It is evident that this menace leads to reduced family income level and, hence, perpetuating poverty. Therefore pollution could be said to have a close relationship with poverty. Some disease like arsenicosis is very dangerous as its patients most likely do not feel or look ill at the early stage but still are treated as "dangerous people" and stripped of their social status (Rahmana et al., 2018), and, hence, face many economic difficulties such as reduction in work efficiency, financial losses, sequestration, and failure to get suitable jobs as a result of judgment (Ahmad et al., 2007; Rahmana et al., 2018). This problem of ostracism of arsenicosis patients has been reported in Bangladesh (Ahmad et al., 2007). These patients of school-age are most times discouraged from appearing in public, attending schools and social event, as well as have been abandoned by their friends and classmates (Alam et al., 2002; Rahmana et al., 2018). In some cases, arsenicosis has led to marriage failure where either husband or wife left each other or remarried in the fear of contracting the disease (Chowdhury et al., 2006). Generally, the diseases initiated by pollution inflict pronounced economic costs globally as well as direct medical costs, opportunity costs resulting in reduced productivity of populace impaired by pollution and cost to health-care systems (Landrigan and Fuller, 2015). Specifically, Trasande and Liu (2011) reported that the United States spends about US$ 76.6 billion yearly on diseases in children as a result of environmental pollution, while the cost of occupational diseases and injuries is US$ 250 billion (Leigh, 2011; Landrigan and Fuller, 2016).

25.6.4 Environmental effects of plastic/microplastic pollution

Plastic wastes can cause adverse environmental impacts, which can affect every organ of the ecosystem (Fig. 25.4). The environmental effects of plastic debris include physical damage to animals, disruption of different habitats, alteration of biodiversity, propagation and transmission of disease-causing organisms, and reduction in aesthetic and recreational worth. Also, the presence of plastics in the environment has been reported as contributing significantly to climate change due to carbon emission and a threat to the global food chain (Reid et al., 2019; Shen et al., 2019). Plastics in landfills have been identified as causing significant changes in the geochemistry and biophysical properties of soils, in addition to its deleterious effects on microbial and plant community, invertebrate, and vertebrate animals (Abel et al., 2018). More so, plastic litters can serve as habitats for disease-causing/transmitting organisms in the environment, such as mosquitoes or reptiles.

In the aquatic environment, plastic debris has been reported to cause physical damage in aquatic animals. For instance, fishes are entangled in plastic materials such as abandoned/lost fishing net that often affects animal's mobility and ability to escape predators or procreate (Clunies-ross, 2019). Ingestion of nylon bags and suffocation (Vince and Hardesty, 2016) are also some of the detrimental effects of plastic litter. The presence of plastics on beaches has resulted in damage to the intrinsic beauty and recreational value of the environment (Andrady, 2011). In some cases, the beaches no longer enjoy human patronage, which can affect social behavior and societal well-being.

Furthermore, plastic pollution can aid or significantly increase the process of rafting, which is the wide mobility of invasive species and the spreading of diseases across diverse aquatic space due to colonization of new favorable environments. Plastics can also enter the food chain when organisms ingest plastic wastes, which can lead to biomagnification and a potential threat to higher organisms, including humans. Plastic-associated chemicals have been detected in wildlife and marine animals as well as in the human blood, urine, and serum (Talsness et al., 2009).

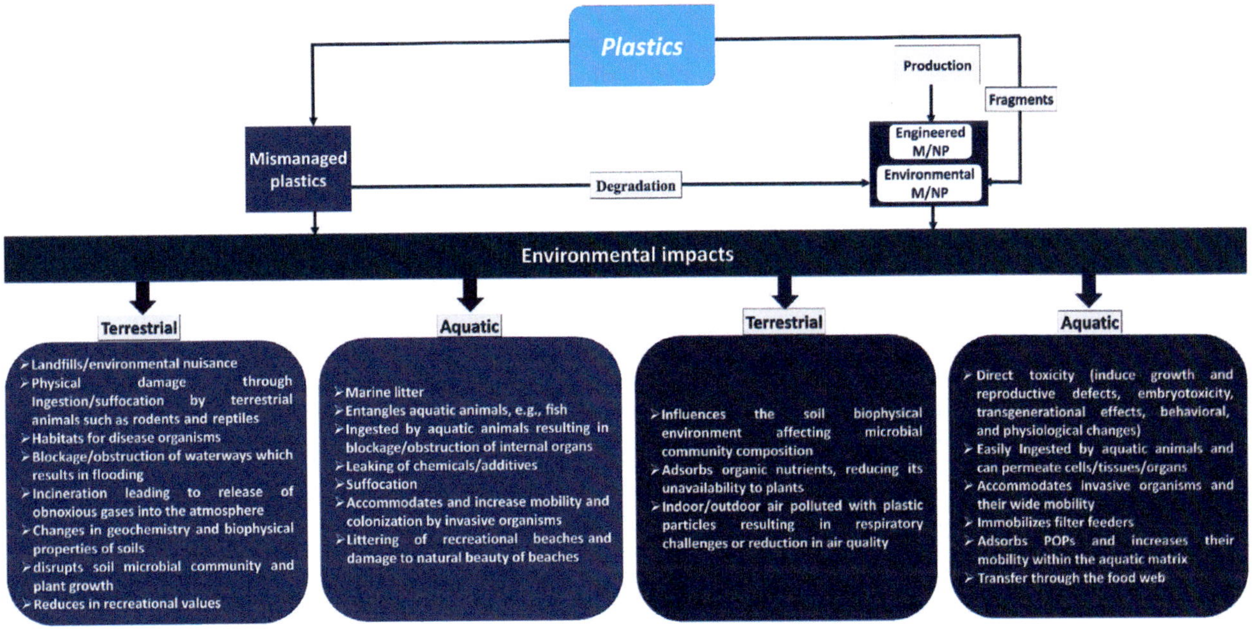

FIGURE 25.4 Plastic and microplastic pollution in the environment with their environmental effects.

In the same vein, several studies have reiterated that microplastic particles have polluted the world's oceans, and various sizes of plastics have been isolated from both freshwaters and marine systems (Dris et al., 2015; Willis et al., 2017; Halle et al., 2017). The presence of plastic particles in bottled water (Eerkes-Medrano et al., 2018) and seafoods (Li et al., 2015; Santillo et al., 2017; Karami et al., 2018) have also been reported. Likewise, microplastics have been reported in urban indoor and outdoor air (Liu et al., 2019). Pets' and human feces have also been demonstrated to contain microplastics (Schwabl et al., 2019; Zhang et al., 2007).

To date, little is known about the detrimental effects of microplastics on living organisms and the environment. However, laboratory studies have demonstrated various possible impacts of plastic particles both in the terrestrial and aquatic environments. Inhalation of polluted air with microplastics (plastic fibers and particles) can induce respiratory discomfort caused by airway and interstitial inflammatory responses in exposed workers (Prata, 2018). Plastic particles can act as vectors for the transport of xenobiotic contaminants, such as POPs and heavy metal. Similarly, the accumulation of plastic particles in soils has been observed to result in a change in soil structure and an adverse effect on soil microbes and plant growth (Awet et al., 2018; Qi et al., 2018; Boots et al., 2019). It has also been shown that microplastics can move through the food chain when, for instance, humans consumed seafood such as bivalves, fish, and shrimps that have previously ingested plastic particles (Li et al., 2015). Besides, several studies have isolated plastic particles from different sea products supporting the transfer of microplastics through the food web (Van Cauwenberghe and Janssen, 2014; Santillo et al., 2017). Plastic particles can penetrate and be retained in the gut wall for a long time and are likely to induce abrasion and internal tissue/organ damage to animals (Wright et al., 2013; Rochman et al., 2013).

Moreover, in smaller aquatic invertebrates, microplastics have been reported to induce growth and reproductive defects, immobility, embryotoxicity, transgenerational effects, and behavioral and physiological changes (Klaper, 2007; Besseling et al., 2013; Browne et al., 2013; Della Torre et al., 2014).

25.7 Remedial measures to control environmental pollution

To promote a sustainable environment, we need to embrace the "green engineering." Green professionals are those environment-friendly and concerned experts such as engineers, scientists, planners, or decision makers who try to incorporate the significance of sustainability while delivering their services. Green engineering, therefore, is a systematic engineering for effectiveness, efficiency, and sustainability. To control or prevent most environmental problems faced today, we need to think and embrace the green engineering by adopting better means of using materials and energy. Pollution as a great threat to the environment and public health has gained enough concern globally. Waste products or pollutants enter the environment in various forms and means and endanger the quality of water, air, and land, which in turn, endangers the public health through the flow of these pollutants within the food chain and, in some cases, direct consumption. The need to prevent pollution has led to the adoption of four types of control, which include legal, social, economic, and technological measures. These measures however help to control environmental pollution by various techniques or methods of operations. These techniques are further applied according to the environment and characteristics of the pollutants and their source contributors. The control actions available for preventing water, air, noise, and land pollution are enumerated as follows.

25.7.1 Corrective actions to control water pollution

Water pollution is a great threat to health as the accumulation and concentration of the pollutants in food chain by absorption is quickly enhanced by biochemical processes, which often lead to high level of toxicity. Hence, it is greatly important to study the methods of treating waste products and eliminating them from aqueous environment. In water pollution control, it would be helpful to consider the various sources of water pollution as whether point sources or nonpoint sources. To control water pollution from nonpoint sources, the following can be used:

1. Judicious use of agrochemicals such as pesticides, and fertilizers, which will reduce their surface runoff and leaching. However, this option is not effective on sloping lands.
2. Use of nitrogen-fixing plants to supplement the use of fertilizers.
3. Adopting integrated pest management to reduce greater use of pesticides.
4. Diversion of manure runoff to basin for settlement instead of flow of manure to water bodies. The manure in the basin can later be used as fertilizer in the fields.
5. Planting trees does not only reduce soil erosion but can also help in trapping sediments, which would have polluted water bodies.
6. Separate drainage of sewage and rain water should be provided to prevent overflow of sewage with rain water.

It should be recalled that point sources of water pollution are municipal and industrial WWTPs. In consequence of the associated health, discomfort problems and environmental degradation from wastewater pollution, the wastewater should be properly treated before being discharged to the environment. While the treatment of wastewater can be costly especially when chemicals and advanced processes such as electrodialysis and ion exchange are used, the natural biomass materials can provide cost-effective, efficient, and sustainable alternative wastewater treatment within the permissible environmental limit (Adelodun et al., 2019). For the control of water pollution from point sources, the following treatment technologies are usually employed.

1. *Adsorption*

   This is a technique that involves the accumulation of the pollutants on the surfaces of specially made chemical, physical, and biological materials known as adsorbents for the removal of pollutants. Adsorption is an economical and effective technique. Examples of commonly used adsorbents include metal organic frameworks, carbon materials, aluminosilicates, zeolites, and mesoporous materials. Biosorbents are biomass-based natural or engineered adsorbents. Other materials used as adsorbents are organic soils, organic carbon framework, activated carbon, activated hydrochars, biochars, carbon nanotubes, graphenes, and clay (Jeevanantham et al., 2019).
2. *Nanotechnology*

   This is a special technology that uses good adsorbent materials called nano-adsorbents for removal of pollutants from water environments. Removal of pollutants is done by of nanotubes, carbon- and carbon-based nanomaterials, nanofiltration, nanoparticles, nanofibers, nanoclusters, and nanocomposites. Common nano-adsorbents used for removal of nitrate from water environment include Fe_3O_4 nanoparticles, magnetic nanomaterials (those with magnetic properties), chitosan and aluminum oxide nanofibers, oxidized carbon nanotubes, polystyrene and zinc, nanocomposites of carbon, silicon, chitosan, polyvinyl acetate, and polyethylene glycol.
3. *Electrochemical treatment*

   This technology removes heavy metals from water environment by the application of electric current. Electrochemical treatments include (1) electrochemical precipitation method that removes metals by forming precipitation as a result of direct current supply, (2) electrodialysis method that removes pollutants by the use of membranes, (3) electrochemical oxidation method, and (4) membrane electrolysis.
4. *Biological treatments*

   This involves removal of heavy metals and other pollutants by using biological substances such as bacteria and microorganisms. Biological treatment methods include (1) biosorption, (2) using bacteria and microorganisms, (3) activated sludge process, (4) biofilter, (5) anaerobic digestion, (6) stabilization ponds, and (7) biofilms.

5. *Phytoremediation*

This is a comparatively contemporary method in pollution control especially in constructed wetlands (Omotade et al., 2019; Ajibade and Adewumi, 2017; Akinbile et al., 2015; Vymazal, 2007). It involves the use of plants that have great ability to store metals or diminish accessible metals in the polluted environment. As these plants grow, they take up the heavy metals and nutrients present in the environment (soil or water) and by some mechanisms convert the toxic contaminants to nontoxic forms. Phytoremediation treatments include phytostabilization, rhizodegradation, rhizofiltration, phytoextraction, phytoaccumulation, and phytovolatilization. According to Jeevanantham et al. (2019), phytoremediation procedure taken by phytoremediators in the removal of contaminants can be shown in the following order:

Upgraded take-up by transporters → Bioactivation of metals → Detoxification by dispersion → Vacuole sequestration.

25.7.2 Corrective actions to control air pollution

Emissions into the air are basically random and indefinitely numerous. Precipitation is the commonly known air cleaner though, not quite efficient and its seasonal occurrence. Sustainable air quality is dependent upon effective prevention of pollution and possible emission of pollutants. It is technically difficult and costly to prevent or control air pollution. However, some methods or techniques have been employed and confirmed efficient for the control of air pollution. These air pollution control techniques are discussed as follows:

1. *Source correction (including raw materials substitution, process/equipment modification)*: Prevention is generally affirmed to be better than cure. Hence, it is better to change or eliminate the practice that causes emission of pollutants into air than striving to confine the pollutants when already released. For instance, the further refining of gasoline for proper removal of lead and, hence, eliminating lead emission to air. In the same way the removal of sulfur from coal, and oil also helps in the elimination of SO_2 during fuel combustion. Some processes such as the control of odors from municipal incinerators by incinerating at high temperature capable of completely oxidizing the odorants. The 1990 CAA in an effort to improve air quality by reducing CO emissions from automobiles decrees that oxygenated fuels should be used in urban areas (Peirce et al., 1998). On the other hand, the substitution of raw materials, and modification of process and/or equipment are all resourceful in the control of air pollution.
2. *Pollutants collection*: There is a great challenge often experienced in the collection of pollutants, which often makes it often almost impossible especially for preventing mobile air pollutants sources, for example, that from automobile exhaust. Recycling of exhaust gases is confirmed a good control process. Though automobiles cannot meet 1990 exhaust emission standards through recycling exhaust and blowby gases, the method is still proven to be a promising control technique for controlling emission from automobiles (Peirce et al., 1998). However, recycling better employed in stationary industrial machines and satisfactory results are obtained. Carbon monoxide (CO) and VOCs are the usually recycled. CO serves as fuel as it releases heat when burnt to CO_2 (Peirce et al., 1998). Channeled exhaust gases through one or more stacks are easier to be collected, but emissions through windows, doors, some cracks in walls, and dust particles raised during on-site transportation of materials offer great hitches to collection; hence, there is need for industries to renovate the entire airflow arrangement for efficient control.
3. *Cooling*: By cooling the temperatures of some pollutants drop below their condensation points hence, making it easier to collect them as liquids. Some common techniques often employed in carrying out this are dilution, quenching, and heat exchange among which quenching relatively higher advantage of cleaning of some gases and particulates, nonetheless, its drawback is that it yields dirty, hot liquid that equally needs to disposed of. For heat conservation, use of cooling coils method is preferred.
4. *Treatment:* What is important in this regard is the selection of the most efficient device for each or a particular set of pollutants since pollutants even though they may come from the same stack vary in size, characteristics, and magnitude. Considering the basic nature of air pollutants, the control or treatment devices are divided into:
   a. Particulate matter control devices: They are cyclones, fabric (baghouse) filters, wet (venture) collectors, electrostatic precipitators, and gravity settlers.
   b. Gaseous pollutants control devices: These can be classified according to classification of the mobility of the source contributors. Hence, treatment processes for controlling pollutants (and other gaseous emissions) from stationary sources include absorption, adsorption, condensation, and incineration (which could be any of: direct flame combustion or flaring, thermal combustion or afterburners, catalytic combustion). While treatment processes for controlling pollutants (and other gaseous emissions) from mobile sources often involve the modification of the devices or their components. Common technologies used include (1) positive crankcase ventilation;

(2) catalytic converters (types include oxidizing catalytic converters, reducing catalytic converters, and three-way catalytic converters); (3) hydrocarbon adsorbers; (4) ceramic wall-flow particulate filters or traps; (5) De-NO_x technologies (selective catalytic reduction, and NO_x adsorber); and (6) installation of a canister (under fuel tank) filled with activated charcoal to adsorb hydrocarbon emissions, hence, controlling evaporative emissions.

25.7.3 Corrective actions to control soil/land pollution

It is necessary to perform some sorts of remediation or cleanup a contaminated soil where the degree of contamination is high enough beyond the recommended standard. Common options employed include excavating the polluted soil and either disposing of it safely in an approved landfill (also known as "dig and dump") or taking it out of site (also called ex situ remediation) (Alloway, 2001). Nevertheless, due to high cost associated with haulage and landfilling, the adoption of in situ remediation techniques has been commonly used. Concise definitions of common soil/land pollution control methods are as follows:

1. *Containment*
   This is the restriction of contaminants by making of physical barriers such as trenches and reinforcing them with impermeable clay, for example, bentonite; or use of hydraulic barriers in order to maintain fluid pressure differentials by the removal or introduction of water for the confinement of the contaminant column in aquifers.
2. *Pump and treat*
   This is a system employed in treating polluted groundwater. In this technique the groundwater is pumped to the surface from custom-built drilled wells and subsequently remedial treatment is employed to remove the contaminants and afterward, the cleaned water is returned to the aquifer.
3. *Air sparging*
   This technique is used when dealing with VOCs, for example, chlorinated solvents. It involves the aeration of extracted water, hence, enhancing the volatilization of the organic pollutants in the water.
4. *In situ soil washing*
   This technique involves the injection of water through contaminated soil for the dissolution and extraction of the contaminants. Alloway (2001) affirms that this technique can be applied in extracting several contaminants even those that are not water-soluble, and including sulfate salts.
5. *Soil vapor extraction*
   This is an enhanced in situ recovery technique that is carried out by aerating the soil of enough macropores to provide enough air penetrability and then collected from special extraction wells where the volatile contaminating compounds can either be burnt in situ or adsorbed into activated carbon. It should be noted, however, that soil pollutants that are adsorbed into the soil solids are highly difficult to treat (Alloway, 2001).
6. *Bioremediation*
   This is typically used in treating soils contaminated with organic compounds. It involves improvement of encouraging conditions for microbial decomposition of the contaminating chemicals.
7. *Phytoremediation*
   Ajibade et al. (2013) defines phytoremediation as a promising cleanup technology for contaminated soils, groundwater, and wastewater that requires both low tech and low cost. Although numerous decontaminating plants with great capacities have been proven, most of them do not yield adequate biomass to excellently extract substantial quantity of metals out of polluted soil. Hence, it is better used to treat soils of low contamination level and in treating severely contaminated soil, further treatment may be recommended.
8. *Natural attenuation*

This is a natural decomposition and adsorption of organic pollutants. Common soil/land pollution control techniques can be summarized as shown in Table 25.3.

25.7.4 Corrective actions to control noise pollution

Noise polluted has been identified as another problem of concern when dealing with environmental pollution as it has various significant psychological and health hazards. Transmission of sound involves three stages, production, transmission pathway (or medium), and receiver. To control noise pollution (in industries, communities, and homes) effectively, various techniques have to be employed across these three stages of its transmission. Generally, it should

TABLE 25.3 Common soil/land pollution control methods.

S/N	method	Pollution problem
Civil engineering–based techniques		
1	Excavation and landfill	"Trouble spot" pollution at concentrations too high for safe in situ treatment, especially inorganic chemicals such as heavy metals
2	Physical containment	Organic NAPLs, landfill leachates
3	Hydraulic control	Supporting containment and/or for treatment of polluted surface or groundwater
Process-based techniques		
1	Thermal treatment	Incineration of soils polluted with POPs, such as polycyclic aromatic hydrocarbons (PAHs), PCBs, PCDDs
2	Physical treatment to separate contaminants from soils or different fractions of contaminated media (commonly ex situ)	Soil washing to remove sulfates, PAHs metals, and other contaminants adsorbed in silt and clay particles
3	Chemical treatment (may be in situ or ex situ): this could be for removal, destruction, or modification of pollutants	Dechlorination of PCB-polluted soils, acid leaching of metal-polluted soils
4	Biological treatment (may be in situ or ex situ): using microorganisms for removal, destruction, or modification of pollutants	Organic pollutants, hydrocarbons, pesticides, solvents, tars, PAHs
5	Stabilization/solidification (may be in situ or ex situ)	Heavy metals, POPs

NAPLs, Non-aqueous phase liquids; PCBs, Polychlorinated biphenyl; *PCDDs,* polychlorinated dibenzodioxins; *POPs,* persistent organic pollutants.
Source: Courtesy Harris, M., Herbert, S. 1994. 'Contaminated Land', Institution of Civil Engineers, Thomas Telford Publishers, London (Harris and Herbert, 1994).

be noted that noise pollution is commonly associated with industry, community, and home. Noise pollution is generally controlled by:

1. reducing the sound production,
2. interposing the sound path, and
3. defending the sound receiver.

It should also be noted that noise pollution control involves money and private enterprises tend to avoid incurring such expenditure; hence, government or the public comes into noise pollution by making and enforcing some sustainable policies that would help regulate noise in the environment.

1. *Industrial noise control*

   In general, industrial noise control primarily has to do with the substitution of noisy machinery with quieter substitutes. Examples of such technique is the reduction of noise from air fan by increasing its number of blades and decreasing its rotational speed, while gaining the same airflow efficiency. By interposing the sound path, industrial noise can be controlled, for instance, the use of insulating material for covering some necessary parts of a noisy motor. By trying to defend the receivers (workers) in an industry, protective hearing devices should be provided to the workers however; such devices should be effective enough not to cause impairment on workers from hearing normal human communication and cautionary signals in the workstation.
2. *Community noise control*

   Community noise comes mainly from aircraft, highway traffic, construction (Peirce et al., 1998), and railway. Peirce et al. (1998) suggests that noise from construction sites should be regulated by local ordinances (unless federal funds are involved). The control techniques usually involve use of mufflers in air compressors, jack hammers, hand compactors, and other similar equipment. In aircraft noise regulation the US Federal Aviation Administration provides major laws on aircraft-associated noise. These are as follows:

   a. A certain limit of aircraft engine noise is allowed in airports, hence, compelling industrialists to develop a quieter engine that would still yield desired thrust.

b. Diversion of flight pathways away from populated areas, and, whenever necessary, mandates pilots to use less than maximum power when the takeoff transports them above a noise-restricted area.
c. Another regulation is the banning of the use of supersonic commercial aircraft. This is as a result of great noise produced by the engines and damage to property due to high sonic boom.

Traffic is another major source of community noise. The components of cars, trucks, and trains such as the exhaust system, tires, engine, gears, wheels, and transmission are the major contributors of traffic noise. Moreover, this noise is often increased when vehicles move over raised highways and bridges owing to the resonation of the bridge and the raised highway with the motion. Generally, the worst contributors of traffic noise are trucks and high-speed trains.

Nevertheless, there are three basic useful choices that can be adopted in the control of traffic noise. They are basically noise from the source, which can be regulated by producing quieter vehicles; second, highways should be distantly directed from populous zones; and third, noise can also be controlled by walling or planting vegetation (as in boulevards) along the highways. However, these options have their imperfections: there is possibility of noise bouncing off the walls and creating little or no noise reduction, walling also creates poor ventilation along the highway, hence, increasing the concentration of CO- and traffic-related pollutants. Design noise levels for various land uses as set up by the US Department of Transportation are as given in Table 25.4.

Similarly, in an effort to regulate noise problem across Europe, the EU adopted some regulations, which are based on interoperability directives specified in the technical specification for interoperability (TSI). They are classified into two, conventional railways and high-speed railways. These EU directives and TSIs are listed in Table 25.5.

Furthermore, the Ministry of Environment and Urban Planning of the Turkish Republic stated their regulations on environmental noise criteria and pollution (mainly in urban centers) in their Official Gazette published June 4, 2004 in issue number 27601. The threshold values for railway systems are stated in Section 4, Article 19, thus in the daytime (7 a.m.−7 p.m.), it should be less than 65 dB (A); in the evening (7 p.m.−11 p.m.), less than 60 dB (A); and at night (11 p.m.−7 a.m.), less than 50 dB (A) (Sarikavak and Boxall, 2019).

3. *Noise in the home*

Individual dwellings are becoming noisier due to internally produced sound and also sound from the external environment. This is also becoming more pronounced as modern homes buy and use more noise-making gadgets (e.g., the New Year's Eve noisemakers). To curb noise from simple appliances in the homes, it is imperative to ask during

TABLE 25.4 Design noise levels for various land uses set up by the Federal Highway Administration.

Land category	Design noise levels (L_{10})	Description of land use
A	60 dB (A) exterior	Activities requiring special qualities of serenity and quietness, such as amphitheaters
B	70 dB (A) exterior 55 dB (A) exterior	Residences, motels, hospitals, schools, parks, libraries Residences, motels, hospitals, schools, parks, libraries
C	75 dB (A) exterior	Developed lands not included in categories A and B
D	No limit	Undeveloped land

Source: Courtesy Peirce, J.J., Weiner, R.F., Vesilind, P.A., 1998. Environmental pollution and control, fourth ed. Butterworth-Heinemann.

TABLE 25.5 Interoperability and European Union (EU) derivatives in terms of noise.

Type	EU directive	Technical specifications for interoperability
High-speed traffic	Directive 96/48/EC; Interoperability of the trans-European high-speed rail system	Commission Decision 2002/735/EC—TSI relating to high-speed rolling stock Commission Decision 2002/732/EC—TSI relating to high-speed railway infrastructures Commission Decision 2008/232/EC—TSI relating to the rolling stock subsystem of the trans-European high-speed rail system

TSI, Technical specification for interoperability.
Source: Courtesy International Union of Railways (UIC), 2010. UIC high speed rail. www. uic.org/download.php/publication/521E.pdf. (accessed 18.06.15); European Commission, 2015. Existing directives relating to noise sources: http://ec.europa.eu/environment/noise/sources.htm (accessed 25.06.15) (International Union of Railways, 2010; European Commission, 2015).

shopping for how much noise an appliance makes just as one invariably asks for how much an appliance cost. Also, use of dissipative silencers as effective modifications of appliances should be used.

25.7.5 Corrective actions to control plastic/microplastic pollution

The reports on the environmental implications of plastic/microplastic pollution have provoked ideas on how to address this global problem. Plastic pollution is a problem caused as a result of human activities in the environment; therefore human behavioral change is key to solving the problem. From the individual user of plastic materials to plastic products' manufacturers, nongovernmental organizations, government regulatory agencies, government at various levels, and international environmental monitoring organizations (Vince and Hardesty, 2016) must come together to address the plastic pollution and its associated problems. Behavioral change is a major panacea in addressing global plastic waste proliferation and pollution. Individuals must cultivate the habit of proper waste disposal to avoid mismanaged plastics in the environment. The principle of three Rs (reducing, reusing, and recycling) of plastic materials must be adopted and fully incorporated at all levels as one of the control measures in combating the impacts of the accumulated plastics in the natural environment. Plastic product manufacturers must take responsibility for their products, even after use. Likewise, more resources must be invested into the development of low cost, effective and efficient plastic waste evacuation, and recovery and recycling techniques. Researchers must come up with alternative uses for plastic wastes that will help to minimize or eradicate plastic pollution in the environment. For instance, some countries such as India (Laskar and Kumar, 2019) have utilized some of their plastic wastes in road construction. Nongovernmental organizations must step up their campaign, awareness, and sensitization of the public on the impacts of plastic pollution on the environment and human health.

Moreover, advocacy and education on proper usage and disposal of plastic waste must be incorporated to the elementary schools' curriculum. Stringent regulation on production and consumption of single-use plastics and plastic additives, for instance, more tax may be introduced to discourage unnecessary use of disposable plastics. Regulatory agencies must adopt strict procedures in enforcing compliance both by the users and the producers of plastics. New laws must be promulgated to punish any violation of plastic pollution control laws. Global sanction/regulations on plastic wastes must be adopted; illegal discharge of plastic waste in another nation's territory must be met with a strict sanction. Global constant monitoring and assessment of various efforts in combating plastic pollution will also contribute immensely to reducing the effect of plastic waste on the environment and human health.

25.7.6 Bioremediation technique of controlling environmental pollution

The presence of chemical pollutants in freshwater bodies such as lakes, rivers, stream, and pond is referred to as freshwater chemical pollution. Their occurrences alter the freshwater characteristics thereby making freshwater unsafe for drinking, recreational activities, irrigation, and sometimes habitat due to the negative effects that chemical pollutants exert on the organisms inhabiting in freshwater bodies (Malaj et al., 2014; Choudri et al., 2017).

Anthropogenic generated chemical pollutants that contaminate freshwater habitat can be classified in two major groups, organic and inorganic pollutants (Ugya et al., 2019). Table 25.6 shows various organic and inorganic pollutants that have been shown to be responsible for the alteration of physicochemical and nutrient status of different freshwater bodies. Organic pollutants are chemical pollutants containing carbon with the exception of carbonates, cyanides, and oxalates (Schweitzer and Noblet, 2018; Ugya, 2015). These pollutants tend to contain other elements such as oxygen, chlorine, bromine, sulfur, nitrogen, phosphorus, and hydrogen, in addition to carbon atom that is the principal element. These pollutants are produced as a result of the usage and production of synthetic organic compounds such as pesticides, herbicides, fertilizers, and petrochemicals (Carpenter, 2011; Ugya and Imam, 2017). Organic pollutants can be classified into two classes based on polarity, hydrophilic and hydrophobic organic pollutants (Schweitzer and Noblet, 2018). Hydrophilic organic pollutants are polar and soluble in water while hydrophobic organic pollutants are nonpolar and insoluble in water (Nanny and Ratasuk, 2002). Hydrophilic organic pollutants include alcohol and carboxylic acid with relatively small alkyl group (CH_3, C_2H_5, C_3H_7, and C_4H_5). Another vital example of hydrophilic organic pollutants is methyl tertiary butyl ether (MTBE). These pollutants do not exert pressure on freshwater due to their nonpersistent nature and their ability to mix well with water (Qin et al., 2010; Gabrielska et al., 1997). Hydrophobic organic pollutants, on the other hand, include organic pollutants, including DDT; polychlorinated biphenyls (PCBs); polybrominated diphenyl ethers (PBDEs); polychlorinated dibenzodioxins (PCDDs); polychlorinated dibenzofurans (PCDFs); and benzene, toluene, ethylbenzene, and xylene (BTEX) (Jaffe, 1991). These pollutants are problematic due to their persistent

TABLE 25.6 The ecological fate of chemical pollutants in freshwater habitat.

S/N	Pollutant types	Case study examples	Environmental fate of chemical pollutant	References
1	DDT	Amazon River, Brazil, Sayong River Watershed, Malaysia	Accumulate in sediment, water, and organisms of freshwater habitat, is reported to cause reproductive failure in fishes and aquatic birds that prey on fishes.	Torres et al. (2002) and Ghani et al. (2017)
2	PCB	Willamette River Basin, Arkansas River Basin	Accumulate in sediment, water, and organisms of freshwater habitat, is reported to cause reproductive failure in fishes and aquatic birds that prey on fishes.	Hope (2008) and Belden et al. (2000)
3	PBDE	Chaohu Lake China, Asunle Stream, Nigeria	Accumulate in sediment, water, and organisms of freshwater habitat, is reported to cause reproductive failure in fishes and aquatic birds that prey on fishes.	Liu et al. (2018) and Olutona et al. (2017)
4	PCDD	Detroit and Rounge, Saginaw River	Accumulate in sediment, water, and organisms of freshwater habitat, is reported to cause reproductive failure in fishes and aquatic birds that prey on fishes.	Kannan et al. (2001, 2008)
5	PCDF	Jukskei and Klip/Vaal South Africa, Xiangjiang River	Accumulate in sediment, water, and organisms of freshwater habitat, is reported to cause reproductive failure in fishes and aquatic birds that prey on fishes.	Rimayi et al. (2016) and Chen et al. (2012)
6	PAH	Columbia Cauca River, Lower Fox River	PAH is utilized by aquatic organisms, including mollusks, crustaceans, polychaetes, and fish, but at higher concentration PAH tends to be toxic to aquatic organisms leading reproductive and endocrine failure in fish.	Sarria-Villa et al. (2016), Brewster et al. (2018), and Collier et al. (2013)
7	BTEX	Tegas River and Marrecas Stream, Buriganga River	BTEX are moderately toxic to aquatic flora and fauna at higher concentration, which are resulted from spillage but are unlikely to cause any toxicity at normal environmental concentration because they degraded easily.	Fernandes et al. (2014) and Mottaleb et al. (2003)
8	Ammonia	Yellow River, Grand River	Causes eutrophication and algae blooms.	Zhang et al. (2007) and Sonthiphand et al. (2013)
9	Nitrate	Lowa's River	Causes eutrophication and algae blooms.	Li et al. (2013)
10	Phosphate	Poyang Lake	Causes eutrophication and algae blooms.	Wang and Liang (2015)
11	Heavy metals	River Pra	Effect depends on the type of heavy metal but toxicity of all heavy metal in freshwater organisms results when the uptake rate exceeds the combined rates of detoxification.	Duncan et al. (2018) and May et al. (2008)

BTEX, Benzene, toluene, ethylbenzene, and xylene; *DDT*, dichlorodiphenyltrichloroethane; *PBDE*, polybrominated diphenyl ether; *PCB*, polychlorinated biphenyl; *PCDD*, polychlorinated dibenzodioxin; *PCDF*, polychlorinated dibenzofuran.

nature, leading to bioaccumulation in freshwater habitat until they reach an elevated concentration, which is injurious to freshwater flora and fauna (Zhang and Kelly, 2018).

Inorganic pollutants are pollutants, including sulfur and its compounds, ammonia, nitrate, phosphate, and heavy metals, that are released into freshwater bodies from industrial and power plants, from agricultural and urban runoff, and acid mine drainage. These pollutants tend to facilitate the depleting of oxygen in freshwater habitat and also lead to the change of the physical chemistry of freshwater bodies thereby exerting negative impact on the flora and fauna organisms inhabiting freshwater habitat (Camargo and Alonso, 2006; Zhou et al., 2014).

The ecological fates of chemical pollutants in freshwater habitat for some particular chemical pollutants have been summarized in Table 25.6. The most similar effect attributing to almost all the pollutants is eutrophication. Eutrophication has been defined by Walker et al. (2019) to mean the incessant accumulation of organic and inorganic material in the aquatic environment. The continuous accumulation of these nutrients stimulates the increase of algae biomass, which is as a result of the utilization of the accumulated nutrients by the algae species. The increase in algae biomass leads to toxic algae blooms that are associated with the release of toxins, affecting both the flora and fauna of freshwater habitat (Heisler et al., 2008). The blooms also encourage the growth of bacteria which used up the available dissolve oxygen present in freshwater habitat, thereby contributing to the negative effect on the flora and fauna of freshwater habitat (Ramanan et al., 2016; Yarimizu et al., 2018). Persistence organic chemical pollutants such as DDT, PCB, PBDE, PCDF, PAH, and BTEX tend to accumulate and magnify in sediments and across aquatic food chain and web, although scanty literature exist showing acute toxicity to freshwater fauna resulting from these organic pollutants. However, the reproductive failure has been reported to occur in fishes and some aquatic birds (Opopol, 2008). The presence of these persistence organic compounds in freshwater habitat can stimulate the growth of bacteria, which used up the available dissolve oxygen in the freshwater environment, thereby affecting the biodiversity of the habitat in enormous ways (Doolotkeldieva et al., 2018). The environmental fate of heavy metals tends to vary between heavy metals. Mercury, for instance, tend to accumulate and magnify in bottom sediment of freshwater habitat leading to the formation of methylmercury, which is a highly toxic chemical to the flora and fauna of freshwater (Rice et al., 2014), and across the food chain and food web (Tchounwou et al., 2012). The concentration of mercury accumulated in small fishes tends to double as compared to in the planktons that were directly exposed to the heavy metal (Rajeshkumar and Li, 2018). Similarly, the concentration of mercury accumulated in larger fishes was higher than the concentration accumulated in the small fish, with the accumulated concentration getting higher across the food chain and web (Kim et al., 2006). Arsenic and cadmium tend to accumulate in water, sediment, and tissues of freshwater organisms, thereby resulting in deleterious effects on aquatic fauna (Perera et al., 2016). Selenium is another heavy metal that accumulates but does not magnify like mercury. Selenium affects the survival and growth rate of juvenile fishes and has been linked to ill development of offspring of fishes (May et al., 2008).

25.7.6.1 Microbial world and wastewater treatment

Microorganisms are unicellular organisms that are either prokaryotic or eukaryotic microscopic organisms that are classified into major groups such as algae, bacteria, fungi, and protist. They are crucial in wastewater treatment plants for water decontamination to safeguard public and environmental health (Wu et al., 2019). These organisms have shown great importance to man and other organisms in many ways, including fermentation of food, clinical importance, veterinary importance, treatment of sewages, production of biofuel, production of enzymes, production of bioactive compounds, and wastewater treatment. The process by which microorganisms are used for biological treatment of wastewater includes bioremediation (bacteria), mycoremediation (fungi), and phycoremediation (microalgae). Recently, an emerging biological process of treatment of wastewater is the use of algae biofilm, which is a complex assemblage of algae, bacteria, and fungi that multiply by forming thick submerged mats in an aquatic environment and terrestrial surfaces under suitable condition.

25.7.6.2 Bioremediation of wastewater

The potentiality of bacteria in the wastewater remediation is on the rise with different bacteria consortia employed by either converting the pollutants into nontoxic substances or by acting as a catalyst in the degradation of pollutants into nonharmful substances. The mechanism involved in the bioremediation of organic and inorganic pollutants by bacteria is a redox reaction in which an electron must be donated by one substance and accepted by a different substance for degradation to occur. The ability of bacteria to bioremediate inorganic substances such as heavy metals has been proved to be due to the ability of the bacteria to tolerate heavy metals due to their high adsorption potential. The techniques in bioremediation in the treatment of wastewater are categorized into aerobic and anaerobic methods.

1. *Aerobic method*
   Aerobic method of wastewater treatment, which is commonly called activated sludge, involves the degradation of organic and some inorganic compounds present in wastewater in the presence of oxygen. The roles of the activated sludge in the wastewater treatment include oxidation and flocculation. A convectional activated sludge process consists of an aeration tank and a sedimentation tank. The aeration tank is an enclosure where aerobic oxidation of organic pollutants occurs while the sedimentation tank is use for the sedimentation of microbial sludge produce during the process of aerobic oxidation in the aeration tank.
2. *Anaerobic method*
   The anaerobic method of wastewater treatment involves the breakdown of inorganic and organic compound by microbe in the absence of oxygen. The method of wastewater treatment using anaerobic biological method is

employed in septic tank, upflow anaerobic sludge blanket, anaerobic filter, anaerobic attached-film expanded-bed and fluidized-bed reactor and rotating biological contactors (RBCs).

a. *Septic tank*

A septic tank is a system consisting of an underground chamber that could be in the form of a tank, concrete, or fiberglass via which wastewater is stored for the biological digestion of organic pollutant in a wastewater under anaerobic condition. The process of sedimentation also occurs in the chamber and the septic tank also contains an absorption field, which is surrounded by some gravel or crushed stones and functions for the treatment and disposal of wastewater after the removal of contaminant and impurities.

b. *Upflow anaerobic sludge blanket*

This is a system that is used for anaerobic wastewater treatment; this system consists of packed sludge forming a bottom layer, a liquid upper layer, and a sludge blanket in between. During wastewater treatment, wastewater tends to flow upward via the sludge bed and the organic pollutants in it are degraded. The organic pollutants are degraded because the sludge bed is covered with flocs of active bacteria in the form of floating blanket. The treated water is separated from the sludge flocs by the settler screen while the methane gas produce is collected at the top of the reactor.

c. *Anaerobic filter*

This system is in the form of bioreactor consisting of many filtration chambers connected in series forming a fixed bed of biological reactors. These filters tend to correlate positively with the anaerobic trickling filter, the support media present in anaerobic filters are in the form of either a rock, gravel, or plastic with empty spaces. The filter media serve as attachment medium for the microorganism although some microorganisms tend to form flocs that are trapped within the filter medium. This system is advantageous due to low sludge production and high efficiency in solids and biological oxygen demand (BOD) reduction.

d. *Anaerobic* attached-*film expanded-bed and fluidized-bed reactor*

This efficiency of this system in the treatment of wastewater depends on the upflow of wastewater, which is greater in expanded beds than the fluidized beds. Wastewater tends to flow upward via the sand bed of diameter of less than 1 mm, which serves as surface for the growth of the biofilm. The high flow rate causes the recirculation of the wastewater via the bed leading to the treatment of organic pollutants present in wastewater.

e. *Anaerobic RBC*

RBC is a system of fixed bed reactor that contains stacks of rotating disk placed on a horizontal shaft. These rotating disks are partially submerged and rotated as wastewater flow via the system. These systems are used in the convectional WWTP as a secondary treatment method after the use of sedimentation for the treatment of domestic gray or brackish water. The microbial community is alternately exposed to the atmosphere and the wastewater allowing both aeration and assimilation of dissolved organic pollutants and nutrients for degradation.

25.7.6.3 Fungi (mycoremediation)

Mycoremediation of wastewater is the form of landfill remediation that involves the remediation of pollutants present in wastewater by fungi. This method is very cheap and has been proved to have high efficiency in the removal of both organic and inorganic pollutants. The ability of fungi to mycoremediate inorganic substances such as heavy metals has been proved to be due to the high hyperaccumulating potentials of these organisms and their the ability to tolerate heavy metals due to their high adsorption potential. Mycoremediation of compounds such as dye (Congo dye, azo dye, malachite dye, nigrosin dye, and trypan blue dye); pesticides (insecticide endosulfan, imazalil, thiophanate methyl, orthophenylphenol, diphenylamine, and chlorpyrifos); organic compounds (BTEX); PAH (polycyclic aromatic hydrocarbon); and MTBE (methyl tert-buthyl ether) has been proved to be due to the ligninolytic enzyme such as laccase, and manganese peroxidase, released by fungi (Chen et al., 2019; Colborne et al., 2019). These enzymes bring about the degradation of these compounds.

25.7.6.4 Microalgae (phycoremediation)

Phycoremediation of wastewater is the process by which both macro- and microalgae bring about the transformation of pollutant present in landfill leachates. The mechanism involved in the phycoremediation of wastewater is based on the fact that algae are autotrophic organism so are able to perform photosynthesis, during the process of photosynthesis, algae produce reactive oxygen species (ROS) such as hydrogen peroxide (H_2O_2), super oxide (O_2^-), and hydroxide (OH^-) radicals. Many researchers have shown the ability of the ROS produced by algae to be the reason why algae bring about the degradation of organic compounds and also reduction of metal oxide, which is in turn used by the algae for biomass production since the process of phycoremediation is also based on the ability of microalgae to use pollutants for biomass production.

25.7.6.5 Biotechnology and wastewater treatment

The biotechnological technologies used in wastewater treatment include bioaugumentation technology, immobilized cell technology, membrane technology, and bioelectrochemical technology.

1. *Bioaugumentation technology in wastewater treatment*
   The advancement in the field of genetic engineering has led to the isolation and production of commercially available microorganism for the degradation of pollutants such as pesticides and other chemicals in wastewater. These process of addition of microbial world and enzymes for rapid degradation of contaminant in wastewater is referred to as bioaugmentation. Wastewater contains microbial world of origin that has the potential of breaking the contaminant present in wastewater at a slow rate with low-reduction efficiency but with bioaugumentation the efficiency is increased. The strain of microbial world used for bioaugumentation is isolated from substrate such as biosolids, compost, soil, and even the wastewater using the convectional enrichment techniques, among the samples isolated from the substrate is enzyme that includes catalase, phosphatase, aminopeptidase, and esterases, which are also added to wastewater to enhance the degradation of pollutants in the wastewater.
2. *Immobilized cell technology*
   Immobilize cell technology is a technology used in the treatment processes such as activated sludge, trickling filter, or RBCs. This technology is of advantageous because it protects the cell from biotic and abiotic stresses, toxic chemicals and increases survival rate and catabolic and anabolic activities. This technology is used in the removal of brown lignin compound from paper mill effluent, biodegradation of phenolic compounds, treatment of cyanide wastewater, dehalogenation of chloroaromatics, methane production, and enhancement of nitrification, denitrification, and the efficiency of activated sludge
3. *Membrane technology*
   Membrane technology is the combination of biological and physical processes to treat wastewater, it is commonly known as membrane bioreactors (MBRs) because it involves microfiltration or ultrafiltration combined with biological wastewater treatment process. MBRs are efficient in the removal of chemical oxygen demand (COD), biochemical oxygen demand (BOD), trace organic compounds, and heavy metals. MBRs can be operated with high concentration of biomass and settling does not depend on the formation of biological flocs in the aeration tank thereby do not lead to bulking problems. MBRs tend to retain planktonic microorganism and particles and are widely used in industrial treatment of wastewater and gray water. The disadvantage of MBRs is the formation of inorganic and organic pore clogging and sludge cake formation known as membrane fouling. Inorganic membrane fouling is resulted due to chemical precipitation mainly as calcium carbonate and to biological precipitation caused as a result of interaction between biopolymer and metal ion. Organic membrane fouling is as a result of microbial biomass, bound extracellular polymeric substance (EPS) colloidal organic matter, adsorbed dissolved organic matter and biofilm formation on the membrane which increases with time.
4. *Bioelectrochemical technology*

This is a technology that is based on the use of electrochemically active microbes as catalysts through the electron transfer inside of a cell to an electrode to interconvert both chemical and electrical energy. The categories of bioelectrochemical systems include microbial fuel cell, microbial electrolysis cell, and phototrophic microbial fuel cells. In a bioelectrochemical system, electromotive force (emf) of the overall reaction can be represented as follows:

$$\text{emf} = \frac{\Delta G}{nF} \tag{25.1}$$

where ΔG is Gibbs free energy (J/mol), n is amount of electron (mol), and F is Faraday's constant (96, 485.3 C/mol). The three categories of bioelectrochemical system are briefly described as follows:

a. Microbial fuel cells: negative ΔG and positive emf leading to electricity production.
b. Microbial electrolysis cell: ΔG is positive and emf is negative, requires energy but produces H_2.
c. Phototrophic microbial fuel cell: involves photoautotrophs such as algae (Croese et al., 2011).

25.8 Ensuring environmental quality and sustainability: preventing pollution

One of the major goals of this chapter is to bring awareness to the general public (including government, social and political groups, environmental, industry, as well as a several body of ordinary people who are marginally concerned), the urgent need of action in combating the threats and menace resulting from environmental pollution. We have discussed about pollution and some of its types, sources, and effects in previous sections of this chapter. It is not enough to point out both potential and current issues on pollution effects on the present environment, it is also imperative to stress the need for pollution reduction via pragmatic approaches thereby providing a sustainable environment for future generations to live in. Ensuring proper environmental quality and sustainability especially in reducing and preventing pollution is the key to securing safe future habitation on the planet Earth.

25.8.1 Environmental quality and sustainability

There exists a strong interaction between quality of life and the environment (Diener and Suh, 1997; UNECE, 2009; Streimikiene, 2015). It is the truth that one of the key factors in people's well-being is *environmental quality* as their lives are greatly affected by the health status of their physical environment (Holman and Coan, 2008; Kahn, 2002; Streimikiene, 2015). Environmental quality is essential because most people value and care about the depletion of its environment natural resources and the health of the environment they live (Brajša-Žganec et al., 2011). Preserving environmental and natural resources is therefore one of the most important factors in ensuring the preservation of well-being overtime (VanLiere and Dunlap, 1980).

Elleuch et al. (2018) defines *environmental sustainability* as the process of meeting our current needs without affecting or jeopardizing future generation's right and ability to meet theirs. The focus of sustainability (and sustainable development) is hinged on balancing the line between our competing need to advance both economically and technologically and also the needs to protect the environments in which we live with others. Kates et al. (2005) established the fact that sustainability is not limited to the environment but also extends to our health as a society in order to ensure that no areas of our life suffer. Sustainability also extends to the state of examining human's actions and their long-term effects and how this can be prevented or improved (United Nations General Assembly, 1987). A key issue for sustainable environment is environmental pollution with its health impacts (United Nations General Assembly, 1987).

25.8.2 Pollution prevention

The rising in both human and animal population has made pollution become a great concern and this scenario will continue because various activities of human will increase more (meaning this is a problem that is inevitable). Therefore, with an all-encompassing program of pollution prevention, most pollution can be majorly be reduced, reused, or prevented (Thiel et al., 2015).

The term *pollution prevention* has generally not been in use as pollution control, but the basic idea of preventing pollution instead of fixing problems is highly pertinent for efficient and economically processes and it also helps to address many environmental problems (Harland, 2012). It is therefore this way; pollution prevention will make an essential component of sustainability. The amount of pollution generated via processes and activities in industries, agriculture, and so on is reduced through the concept of pollution prevention but pollution control strategies generally seek to manage and reduce a pollutant and its impact upon the environment after its emission, which most often is more strenuous and costly especially for nonpoint source pollutants (Elleuch et al., 2018). Pollution prevention approach, however, increases the efficiency of pollution generated processes by reducing the amount of pollution generated at its source (Sherman et al., 2016). Another best approach to this is to prevent the pollution from being created in the first place by finding new efficiencies and ways of doing things astutely.

Jorgenson and Wilcoxen (1990) emphasized the need for people to possess primal ideas of preventing pollution in every sphere with new materials and new approaches, rather than just fixing problems. It holds both economic and environmental benefits to reduce the input of materials and energy, minimize the amount of waste for treatment and disposal, and make sure that the wastes that are produced are reusable or can be disposed of easily without harm to the environment (Harland, 2012). To tidy it up the first watchword within any environmental management hierarchy is prevention. Pollution prevention should therefore be everyone's goal. Individuals across cultures are expected to utilize their knowledge to take actions that are protective of human health and the environment.

25.9 Conclusion

Environment is the primary niche of all living organism and it describes the overall condition of one's immediate surroundings. The degradation of the environment via several activities, caused majorly by man over the years, has made it hard for his convenient living. This chapter has clearly explained the concept of environmental pollution, the widely recognized types of pollution (namely, land/soil, water, air, noise, and plastic/microplastic) and their respective socioeconomic impacts on health, environment, and economic growth and development. Some remedial measures to controlling environmental pollution were provided alongside pragmatic approach toward environmental quality and sustainability, which will largely help to enhance public health and well-being of man and his immediate environment.

References

Abam, F.I., Unachukwu, G.O., 2009. Vehicular emissions and air quality standards in Nigeria. Eur. J. Sci. Res. 34 (4), 550–560.

Abel, A., Machado, D.S., Hempel, S., Rillig, M.C., Kloas, W., Zarfl, C., 2018. Microplastics as an emerging threat to terrestrial ecosystems. Global Change Biol. 24, 1405–1416. Available from: https://doi.org/10.1111/gcb.14020.

Adelodun, B., Ajibade, F.O., Abdulkadir, T.S., Bakare, H.O., Choi, K.S., 2020. SWOT analysis of agro-waste based adsorbents for persistent dye pollutants removal in water and wastewater treatment. In: Kumar, V., Singh, J., Kumar, P. (Eds.), Environmental Degradation Causes and Remediation Strategies, Volume 1. Agro Environ Media, Haridwar, pp. 88–103. Available from: https://doi.org/10.26832/aesa-2020-edcrs-07.

Adelodun, B., Ajibade, F.O., Ogunshina, M.S., Choi, K.S., 2019. Dosage and settling time course optimization of *Moringa oleifera* in municipal wastewater treatment using response surface methodology. Desalin. Water Treat. 167, 45–56. Available from: https://doi.org/10.5004/dwt.2019.24616.

Adeniran, K.A., Adelodun, B., Ogunshina, M., 2016. Artificial neural network modelling of biochemical oxygen demand and dissolved oxygen of rivers: case study of Asa River. Environ. Res. Eng. Manage. 72 (3), 59–74. Available from: https://doi.org/10.5755/j01.erem.72.3.14120.

Adewumi, J.R., Ajibade, F.O., 2015. The pollution effects of indiscriminate disposal of wastewater on soil in Semi-Urban Area. J. Appl. Sci. Environ. Manage. 19, 412–419. Available from: https://doi.org/10.4314/jasem/v19i3.10.

Adewumi, J.R., Ejeh, O.J., Lasisi, K.H., Ajibade, F.O., 2019. A GIS-AHP based approach in sitting MSW landfills in a Nigerian Confluence State. SN Appl. Sci. 11528. Available from: https://doi.org/10.1007/s42452-019-1500-6.

Afroza, R., Banna, H., Masudc, M.M., Akhtar, R., Yahaya, S.R., 2015. Household's perception of water pollution and its economic impact on human health in Malaysia. Desalin. Water Treat. 1–9. Available from: https://doi.org/10.1080/19443994.2015.1006822.

Aguirre-Martínez, G.V., Okello, C., Salamanca, M.J., Garrido, C., Del Valls, T.A., Martín-Díaz, M.L., 2016. Is the step-wise tiered approach for ERA of pharmaceuticals useful for the assessment of cancer therapeutic drugs present in marine environment? Environ. Res. 144, 43–59.

Ahmad, S.A., Sayed, M.H., Khan, M.H., Karim, M.N., Haque, M.A., Bhuiyan, M.S., et al., 2007. Sociocultural aspects of arsenicosis in Bangladesh: community perspective. J. Environ. Sci. Health, A: Environ. Sci. Eng.Toxic Hazard. Subst. Control 42 (12), 1945–1958. Available from: https://doi.org/10.1080/10934520701567247.

Ajibade, F.O., Adewumi, J.R., 2017. Performance evaluation of aquatic macrophytes as a constructed wetland for municipal wastewater treatment. FUTA J. Eng. Eng Tech. (FUTAJEET) 11, 1–11.

Ajibade, F.O., Olajire, O.O., Ajibade, T.F., Nwogwu, N.A., Lasisi, K.H., Alo, A.B., et al., 2019a. Combining multicriteria decision analysis with GIS for suitably siting landfills in a Nigerian State. Environ. Sustainability Indic. 3–4, 100010. Available from: https://doi.org/10.1016/j.indic.2019.100010.

Ajibade, F.O., Adelodun, B., Ajibade, T.F., Lasisi, K.H., Abiola, C., Adewumi, J.R., et al., 2019b. The threatening effects of open dumping on soil at waste disposal sites of Akure city, Nigeria. Int. J. Environ. Waste Manage. (Manuscript accepted). Available from: https://www.inderscience.com/info/ingeneral/forthcoming.php?jcode = ijewm.

Ajibade, F.O., Adeniran, K.A., Egbuna, C.K., 2013. Phytoremediation efficiencies of water hyacinth in removing heavy metals in domestic sewage (a case study of University of Ilorin, Nigeria). Int. J. Eng. Sci. 2 (12), 16–27.

Ajibade, F.O., Adewumi, J.R., Ojo, O.M., Babatola, J.O., Oguntuase, A.M., 2015. Issues, challenges and management of water supply and sanitation in Nigeria: an overview. In: 13th International Civil Engineering Conference and Annual General Meeting held at Kwara Hotel, Ilorin, Nigeria, October 28–30, 2015. Nigerian Institution of Civil Engineers, pp. 19–34.

Ajibade, T.F., Nwogwu, N.A., Ajibade, F.O., Adelodun, B., Idowu, T.E., Ojo, A.O., et al., 2020. Potential dam sites selection using integrated techniques of remote sensing and GIS in Imo State, Southeastern, Nigeria. Sustain. Water Resour. Manag 6, 57. Available from: https://doi.org/10.1007/s40899-020-00416-5.

Akanyeti, I., Kraft, A., Ferrari, M., 2017. Hybrid polystyrene nanoparticle-ultrafiltration system for hormone removal from water. J. Water Process. Eng. 17, 102–109.

Akinbile, C.O., Ajibade, F.O., Ofuafo, O., 2016a. Soil quality analysis for dumpsite environment in a University Community in Nigeria. FUTA J. Eng. Eng. Technol. 10, 68–73.

Akinbile, C.O., Erazua, A.E., Babalola, T.E., Ajibade, F.O., 2016b. Environmental implications of animal wastes pollution on agricultural soil and water quality. Soil. Water Res. 11, 172–180. Available from: https://doi.org/10.17221/29/2015-SWR.

Akinbile, C.O., Ogunrinde, T.A., Che bt Man, H., Aziz, H.A., 2015. Phytoremediation of domestic wastewaters in free water surface constructed wetlands using Azolla pinnata. Int. J. Phytoremediation 18 (1), 54–61. Available from: https://doi.org/10.1080/15226514.2015.1058330.

Akosile, S.I., Ajibade, F.O., Lasisi, K.H., Ajibade, T.F., Adewumi, J.R., Babatola, J.O., et al., 2020. Performance evaluation of locally produced ceramic filters for household water treatment in Nigeria. Sci. Afr. 7, e00218. Available from: https://doi.org/10.1016/j.sciaf.2019.e00218.

Alam, M.G., Allinson, G., Stagnitti, F., Tanaka, A., Westbrooke, M., 2002. Arsenic contamination in Bangladesh groundwater: a major environmental and social disaster. Int. J. Environ. Health Res. 12 (3), 235–253. Available from: https://doi.org/10.1080/0960312021000000998.

Alimi, O.S., Farner Budarz, J., Hernandez, L.M., Tufenkji, N., 2018. Microplastics and nanoplastics in aquatic environments: aggregation, deposition, and enhanced contaminant transport. Environ. Sci. Technol. 52, 1704–1724. Available from: https://doi.org/10.1021/acs.est.7b05559.

Alina, B., 2018. Pollution Facts and Types of Pollution. Available from: <https://www.livescience.com/22728-pollution-facts.html> (accessed 25.11.19).

Alloway, B.J., 2001. Soil pollution and land contamination. In: Alloway, B.J., Arthur, D., Ayres, J., Chester, R., Clift, R., Crathorne, B., et al., Pollution: Causes, Effects and Control, fourth ed. The Royal Society of Chemistry, Cambridge, pp. 352–377.

Andrady, A.L., 2011. Microplastics in the marine environment. Mar. Pollut. Bull. 62, 1596–1605. Available from: https://doi.org/10.1016/j.marpolbul.2011.05.030.

Anf, H.Z., Emad, S., 2014. An environmental impact assessment of the open burning of scrap tyres. J. Appl. Sci. 14 (21), 2695–2703.

Archer, E., Petrie, B., Kasprzyk-Hordern, B., Wolfaardt, G.M., 2017. The fate of pharmaceuticals and personal care products (PPCPs), endocrine disrupting contaminants (EDCs), metabolites and illicit drugs in a WWTW and environmental waters. Chemosphere 174, 437–446.

Ashraf, M.A., Maah, M.J., Yusoff, I., 2014. Soil Contamination, Risk Assessment and Remediation. Intechopen (online publisher). Available from: http://dx.doi.org/10.5772/57287.

Ashraf, M.A., Maah, M.J., Yusoff, I., Mehmood, K., 2010. Effects of polluted water irrigation on environment and health of people in Jamber, District Kasur, Pakistan. Int. J. Basic Appl. Sci. 10, 37–57.

Ashraf, M.A., Rehman, M., Alias, Y., Yusoff, I., 2013. Removal of Cd(II) onto *Raphanus sativus* peels biomass: equilibrium, kinetics, and thermodynamics. Desalin. Water Treat. 51, 4402–4412.

Awet, T.T., Kohl, Y., Meier, F., Straskraba, S., Grün, A.L., Ruf, T., et al., 2018. Effects of polystyrene nanoparticles on the microbiota and functional diversity of enzymes in soil. Environ. Sci. Eur. 30 (11), 1–10. Available from: https://doi.org/10.1186/s12302-018-0140-6.

Belden, J.B., Eaton, H.J., Lydy, M.J., 2000. Occurrence of polychlorinated biphenyls in fish and sediment of the Arkansas River Basin, Kansas. Trans. Kans. Acad. Sci. 103 (1903), 177–184.

Besseling, E., Wegner, A., Foekema, E.M., Van Den Heuvel-Greve, M.J., Koelmans, A.A., 2013.). Effects of microplastic on fitness and PCB bioaccumulation by the lugworm *Arenicola marina* (L.). Environ. Sci. Technol. 47, 593–600. Available from: https://doi.org/10.1021/es302763x.

Brewster, C.S., Sharma, V.K., Cizmas, L., McDonald, T.J., 2018. Occurrence, distribution and composition of aliphatic and polycyclic aromatic hydrocarbons in sediment cores from the Lower Fox River, Wisconsin, US. Environ. Sci. Pollut. Res. Int. 25, 4974–4988.

Boots, B., Russell, C.W., Green, D.S., 2019. Effects of microplastics in soil ecosystems: above and below ground. Environ. Sci. Technol. 53, 11496–11506. Available from: https://doi.org/10.1021/acs.est.9b03304.

Botkin, D.B., 1995. Environmental Science: Earth as a Living Planet. John Wiley & Sons, New York.

Brajša-Žganec, A., Merkaš, M., Šverko, I., 2011. Quality of life and leisure activities: how do leisure activities contribute to subjective well-being. Soc. Indic. Res. 102 (1), 80–97.

Brazilian Association of Technical Standards, 2017. ABNT, NBR 10.152 – Acoustics – Sound Pressure Levels of Indoor Environments. Rio de Janeiro.

Brazilian Association of Technical Standards, 2019. ABNT, NBR 10151 Acoustics—Measurement and Evaluation of Sound Pressure Levels in Inhabited Environments—Application for General Use. Rio de Janeiro.

Browne, M.A., 2015. Sources and pathways of microplastics to habitats. In: Bergmann, M., Gutow, L., Klages, M. (Eds.), Marine Anthropogenic Litter. Springer International Publishing, Cham, pp. 229–244. Available from: https://doi.org/10.1007/978-3-319-16510-3_9.

Browne, M.A., Niven, S.J., Galloway, T.S., Rowland, S.J., Thompson, R.C., 2013. Microplastic moves pollutants and additives to worms, reducing functions linked to health and biodiversity. Curr. Biol. 23, 2388–2392. Available from: https://doi.org/10.1016/j.cub.2013.10.012.

Camargo, J.A., Alonso, A., 2006. Ecological and toxicological effects of inorganic nitrogen pollution in aquatic ecosystems: a global assessment. Environ. Int. 32, 831–849.

Carpenter, D.O., 2011. Health effects of persistent organic pollutants: the challenge for the Pacific Basin and for the world. Rev. Environ. Health 26, 61–69.

Cao, Y., Zhang, S., Wang, G., Li, T., Xu, X., Deng, O., et al., 2017. Enhancing the soil heavy metals removal efficiency by adding HPMA and PBTCA along with plant washing agents. J. Hazard. Mater. 339, 33–42.

Chen, H., Zhou, Y., Hu, X., Tian, K., Zhang, J., 2019. Effects of chlortetracycline on biological nutrient removal from wastewater. Sci. Total Environ. 647, 268–274.

Chen, Z., Yang, B., Mengoni, A., Dong, J., Peng, X., 2012. Distribution patterns of polychlorinated dibenzo-*p*-dioxins and polychlorinated dibenzofurans in sediments of the Xiangjiang River, China. Environ. Monit. Assess. 184, 7083–7092.

Choudri, B.S., Charabi, Y., Baawain, M., Ahmed, M., 2017. Effects of pollution on freshwater organisms. Water Environ. Res. 89, 1676–1703.

Chowdhury, M.A.I., Uddin, M.T., Ahmed, M.F., Ali, M.A., Rasul, S.M.A., Hoque, M.A., et al., 2006. Collapse of socio-economic base of Bangladesh by arsenic contamination in groundwater. Pak. J. Biol. Sci. 9, 1617–1627. Available from: https://doi.org/10.3923/pjbs.2006.1617.1627.

Colborne, S.F., Maguire, T.J., Mayer, B., Nightingale, M., Enns, G.E., Fisk, A.T., et al., 2019. Water and sediment as sources of phosphate in aquatic ecosystems: the Detroit River and its role in the Laurentian Great Lakes. Sci. Total Environ. 647, 1594–1603.

Collier, T.K., Anulacion, B.F., Arkoosh, M.R., Dietrich, J.P., Incardona, J.P., Johnson, L.L., et al., 2013. 4 – Effects on fish of polycyclic aromatic hydrocarbons (PAHS) and naphthenic acid exposures. In: Tierney, K.B., Farrell, A.P., Brauner, C.J. (Eds.), Fish Physiology. Academic Press.

Croese, E., Pereira, M.A., Euverink, G.J.W., Stams, A.J.M., Geelhoed, J.S., 2011. Analysis of the microbial community of the biocathode of a hydrogen-producing microbial electrolysis cell. Appl. Microbiol. Biotechnol. 92, 1083–1093.

Clunies-ross, P., 2019. Plastics in the Environment Te Ao Hurihuri – The Changing World. Royal Society Te Apārangi (NZ). 49 p.

Dada, A.O., Adekola, F.A., Odebunmi, E.O., 2015. Kinetics and equilibrium models for sorption of cu (II) onto a novel manganese nano-adsorbent. J. Dispers. Sci. Technol. 37 (1), 119–133.

de Souza, T.B., Alberto, K.C., Barbosa, S.A., 2020. Evaluation of noise pollution related to human perception in a university campus in Brazil. Appl. Acoust. 157, 107023.

Della Torre, C., Bergami, E., Salvati, A., Faleri, C., Cirino, P., Dawson, K.A., et al., 2014. Accumulation and embryotoxicity of polystyrene nanoparticles at early stage of development of sea urchin embryos *Paracentrotus lividus*. Environ. Sci. Technol. 48, 12302–12311. Available from: https://doi.org/10.1021/es502569w.

Diener, E., Suh, E., 1997. Measuring quality of life: economic, social, and subjective indicators. Soc. Indic. Res. 40 (1–2), 189–216.

Doolotkeldieva, T., Konurbaeva, M., Bobusheva, S., 2018. Microbial communities in pesticide-contaminated soils in Kyrgyzstan and bioremediation possibilities. Environ. Sci. Pollut. Res. Int. 25, 31848–31862.

Dris, R., Gasperi, J., Saad, M., Mirande, C., Tassin, B., 2016. Synthetic fibers in atmospheric fallout: a source of microplastics in the environment? Mar. Pollut. Bull. 104, 290–293. Available from: https://doi.org/10.1016/j.marpolbul.2016.01.006.

Dris, R., Imhof, H., Sanchez, W., Gasperi, J., Galgani, F., Tassin, B., et al., 2015. Beyond the ocean: contamination of freshwater ecosystems with (micro-) plastic particles. Environ. Chem. 12, 539–550. Available from: https://doi.org/10.1071/EN14172.

Duncan, A.E., De Vries, N., Nyarko, K.B., 2018. Assessment of heavy metal pollution in the sediments of the River Pra and its tributaries. Water Air Soil Pollut. 229, 272.

Dzhambov, A.M., 2015. Long-term noise exposure and the risk for type 2 diabetes: a metaanalysis. Noise Health 17, 23–33.

Easterbrook, G., 1995. A Moment on the Earth: The Coming Age of Environmental Optimism. Viking, Bergenfield, NJ.

EEA (European Environmental Agency), 2017. Air quality in Europe—2017 report. In: EEA Report No 13-2017. Available from: <https://www.eea.europa.eu/publications/air-quality-ineurope-2017>.

Eerkes-Medrano, D., Leslie, H.A., Quinn, B., 2018. Microplastics in drinking water: a review and assessment. Curr. Opin. Environ. Sci. Heal. 7, 69–75. Available from: https://doi.org/10.1016/j.coesh.2018.12.001.

EIA, 2008. Environmental Impact Assessment Report. Industrial Hygiene Services. Environmental Division, Vadodara, for Madhuvan Cement Industries, Junagadh.

Elleuch, B., Bouhamed, F., Elloussaief, M., Jaghbir, M., 2018. Environmental sustainability and pollution prevention. Environ. Sci. Pollut. Res. 25 (19), 18223–18225. Available from: https://doi.org/10.1007/s11356-017-0619-5.

European Commission, 2015. Existing directives relating to noise sources: http://ec.europa.eu/environment/noise/sources.htm (accessed 25.06.15).

European Commission, 2017. Second Report from the Commission to the European Parliament and the Council on the implementation of the Environmental Noise Directive in accordance with Article 11 of the Directive. <http://eur-lex.europa.eu/legal-content/EN/TXT/?qid = 1490941871748&uri = COM:2017:151:FIN>.

Fadare, O.O., Wan, B., Guo, L.H., Xin, Y., Qin, W., Yang, Y., 2019. Humic acid alleviates the toxicity of polystyrene nanoplastic particles to *Daphnia magna*. Environ. Sci. Nano 6 (6), 1466–1477. Available from: https://doi.org/10.1039/c8en01457d.

Feng, Y., Cheng, J., Shen, J., Sun, H., 2019. Spatial effects of air pollution on public health in China. Environ. Resour. Econ. 73, 229–250. Available from: https://doi.org/10.1007/s10640-018-0258-4.

Fernandes, A.N., Gouveia, C.D., Grassi, M.T., Da Silva Crespo, J., Giovanela, M., 2014. Determination of monoaromatic hydrocarbons (BTEX) in surface waters from a Brazilian subtropical hydrographic basin. Bull. Environ. Contam. Toxicol. 92, 455–459.

Fu, L., 2001. Assessment of vehicle pollution in China. J. Air Waste Manage. 51 (5), 658–668.

Gabrielska, J., Sarapuk, J., Przestalski, S., 1997. Role of hydrophobic and hydrophilic interactions of organotin and organolead compounds with model lipid membranes. Z. Naturforsch. C 52, 209–216.

Gall, S.C., Thompson, R.C., 2015. The impact of debris on marine life. Mar. Pollut. Bull. 92, 170–179. Available from: https://doi.org/10.1016/j.marpolbul.2014.12.041.

Ghafoor, A., Rauf, A., Arif, M., Muzaffar, W., 1994. Chemical composition of effluents from different industries of the Faisalabad City, Pakistan. J. Agric. Sci. 31, 367–370.

Ghani, Z.A., Anuar, A.N., Majid, Z.A., Yoneda, M., 2017. Multimedia environmental fate and transport model of dichlorodiphenyltrichloroethane (DDT): case study Sayong River Watershed, Johor, Malaysia. Sains Malays. 41, 2383–2392.

Goldstein, A.H., Charles, D.K., Colette, L.H., Inez, Y.F., 2009. Biogenic carbon and anthropogenic pollutants combine to form a cooling haze over the Southeastern United States. Proc. Natl Acad. Sci. U.S.A. 4 (5), 66–72.

Halle, A.T., Jeanneau, L., Martignac, M., Jarde, E., Pedrono, B., Brach, L., et al., 2017. Nanoplastic in the North Atlantic Subtropical Gyre. Environ. Sci. Technol. 51, 13689–13697. Available from: https://doi.org/10.1021/acs.est.7b03667.

Harland, J., 2012. Why the Road to Sustainability Starts With Pollution Prevention? P2 Impact Article. Available from: <https://www.greenbiz.com/blog/2012/03/06/why-road-sustainability-starts-pollution-prevention> (accessed 15.12.19).

Harris, M., Herbert, S., 1994. Contaminated Land', Institution of Civil Engineers. Thomas Telford Publishers, London.

Harrison, R.M. (Ed.), 1996. Pollution: Causes, Effects and Control. third ed. Royal Society of Chemistry, Cambridge.

Heisler, J., Glibert, P., Burkholder, J., Anderson, D., Cochlan, W., Dennison, W., et al., 2008. Eutrophication and harmful algal blooms: a scientific consensus. Harmful Algae 8, 3–13.

Holman, M.R., Coan, T.G., 2008. Voting Green. Soc. Sci. Q. 89, 1121–1135.

Holmes, N.S., Morawska, L., 2006. A review of dispersion modelling and its application to the dispersion of particles: an overview of different dispersion models available. Atmos. Environ. 40 (30), 5902–5928.

Hope, B.K., 2008. A model for the presence of polychlorinated biphenyls (PCBs) in the Willamette River Basin (Oregon). Environ. Sci. Technol. 42, 5998–6006.

Ilori, B.A., Adewumi, J.R., Lasisi, K.H., Ajibade, F.O., 2019. Qualitative assessment of some available water resources in Efon-Alaaye, Nigeria. J. Appl. Sci. Environ. Manage. 23 (1), 29–34.

Inyinbor, A.A., Adebesin, B.O., Oluyori, A.P., Adelani-Akande, T.A., Dada, A.O., Oreofe Toyin, A., 2018. Water Pollution: Effects, Prevention, and Climatic Impact. Intechopen (online publisher). Available from: http://dx.doi.org/10.5772/intechopen.72018.

International Union of Railways (UIC), 2010. UIC high speed rail. www.uic.org/download.php/publication/521E.pdf. (accessed 18.06.15).

Ipingbemi, O., 2009. Socio-economic implications and environmental effects of oil spillage in some communities in the Niger delta. J. Integr. Environ. Sci. 6 (1), 7–23. Available from: https://doi.org/10.1080/15693430802650449.

IWMI, 2007. *Comprehensive Assessment of Water Management in Agriculture, London.* International Water Management Institute, Geneva, Earthscan, and Colombo.

Jackson, A.R., Jackson, J.M., 1996. Environmental Science: The Natural Environment and Human Impact. Pearson Longman, New York.

Jaffe, R., 1991. Fate of hydrophobic organic pollutants in the aquatic environment: a review. Environ. Pollut. 69, 237–257.

Jeevanantham, S., Saravanan, A., Hemavathy, R.V., Senthil Kumar, P., Yaashikaa, P.R., Yuvaraj, D., 2019. Removal of toxic pollutants from water environment by phytoremediation: a survey on application and future prospects. Environ. Technol. Innov. 13, 264–276. Available from: https://doi.org/10.1016/j.eti.2018.12.007.

Jorgenson, D.W., Wilcoxen, P., 1990. Environmental regulation and U.S. economic growth. RAND J. Econ. 21 (2), 314–340.

Juahir, H., Zain, S.M., Yusoff, M.K., Hanidza, T.T., Armi, A.M., Toriman, M.E., et al., 2011. Spatial water quality assessment of Langat River Basin (Malaysia) using environmetric techniques. Environ. Monit. Assess. 173, 625–641.

Jung, C., Son, A., Her, N., Zoh, K., Cho, J., Yoon, Y., 2015. Removal of endocrine disrupting compounds, pharmaceuticals, and personal care products in water using carbon nanotubes: a review. J. Ind. Eng. Chem. 27, 1–11.

Kahn, M.E., 2002. Demographic change and the demand for environmental regulation. J. Policy Anal. Manage. 21 (1), 45–62.

Kannan, K., Kober, J.L., Kang, Y.S., Masunaga, S., Nakanishi, J., Ostaszewski, A., et al., 2001. Polychlorinated naphthalenes, biphenyls, dibenzo-*p*-dioxins, and dibenzofurans as well as polycyclic aromatic hydrocarbons and alkylphenols in sediment from the Detroit and Rouge Rivers, Michigan, USA. Environ. Toxicol. Chem. 20, 1878–1889.

Kannan, K., Yun, S.H., Ostaszewski, A., Mccabe, J.M., Mackenzie-Taylor, D., Taylor, A.B., 2008. Dioxin-like toxicity in the Saginaw River Watershed: polychlorinated dibenzo-*p*-dioxins, dibenzofurans, and biphenyls in sediments and floodplain soils from the Saginaw and Shiawassee Rivers and Saginaw Bay, Michigan, USA. Arch. Environ. Contam. Toxicol. 54, 9–19.

Kar, S., Roy, K., 2010. First report on interspecies quantitative correlation of ecotoxicity of pharmaceuticals. Chemosphere 81, 738–747.

Karami, A., Golieskardi, A., Choo, C.K., Larat, V., Karbalaei, S., Salamatinia, B., 2018. Microplastic and mesoplastic contamination in canned sardines and sprats. Sci. Total Environ. 612, 1380–1386. Available from: https://doi.org/10.1016/j.scitotenv.2017.09.005.

Kates, R.W., Parris, T.M., Leiserowitz, A.A., 2005. What is sustainable development? Goals, indicators, values, and practice. Environ. Sci. Policy Sustain. Dev. 47 (3), 8–21. Available from: https://doi.org/10.1080/00139157.2005.10524444.

Keller, E.A., 1988. Environmental Geology. Merrill, Columbus, OH.

Kim, E.H., Kim, I.K., Kwon, J.Y., Kim, S.W., Park, Y.W., 2006. The effect of fish consumption on blood mercury levels of pregnant women. Yonsei Med. J. 47, 626–633.

Klaper, R., 2007. Behavioral and physiological changes in *Daphnia magna* when exposed to nanoparticle suspensions (titanium dioxide, nano-C_{60}, and $C_{60}H_xC_{70}H_x$). Environ. Sci. Technol. 41 (12), 4465–4470.

Kumar, R., Mishra, R.K., Mishra, V., Qidwai, A., Pandey, A., Shukla, S.K., et al., 2016. Detoxification and tolerance of heavy metals in plants. Plant. Met. Interact. 335–359. Available from: https://doi.org/10.1016/B978-0-12-803158-2.00013-8.

Landrigan, P.J., Fuller, R., Acosta, N.J.R., Adeyi, O., Arnold, R., Basu, N.N., et al., 2017. The Lancet Commission on pollution and health. Lancet 391 (10119), 462–512. Available from: https://doi.org/10.1016/S0140-6736(17)32345-0.

Landrigan, P., Fuller, R., 2015. Environmental pollution: an enormous and invisible burden on health systems in low- and middle-income countries. World Hosp. Health Serv. 50 (4), 35–41.

Landrigan, P.J., Fuller, R., 2016. Pollution, health and development: the need for a new paradigm. Rev. Environ. Health 31 (1), 121–124. Available from: https://doi.org/10.1515/reveh-2015-0070.

Landrigan, P.J., Fuller, R., Fisher, S., Suk, W.A., Sly, P., Chiles, T.C., et al., 2019. Pollution and children's health. Sci. Total Environ. 650 (part 2), 2389–2394. Available from: https://doi.org/10.1016/j.scitotenv.2018.09.375.

Landrigan, P., Rauh, V., Galvez, M., 2010. Environmental justice and the health of children. Mt. Sinai J. Med. 77, 178–187.

Laskar, N., Kumar, U., 2019. Plastics and microplastics: a threat to environment. Environ. Technol. Innov. 14, 100352. Available from: https://doi.org/10.1016/j.eti.2019.100352.

Lattin, G.L., Moore, C.J., Zellers, A.F., Moore, S.L., Weisberg, S.B., 2004. A comparison of neustonic plastic and zooplankton at different depths near the southern California shore. Mar. Pollut. Bull. 49, 291–294. Available from: https://doi.org/10.1016/j.marpolbul.2004.01.020.

Leigh, J., 2011. Economic burden of occupational injury and illness in the United States. Milbank Q. 89 (4), 728–772.
Li, D., Chan, K.S., Schilling, K.E., 2013. Nitrate concentration trends in Iowa's Rivers, 1998 to 2012: what challenges await nutrient reduction initiatives? J. Environ. Qual. 42, 1822–1828.
Li, P., Du, B., Chan, M.C., Feng, X., 2015. Human inorganic mercury exposure, renal effects and possible pathways in Wanshan mercury mining area, China. Environ. Res. 140, 198–204.
Li, W.C., Tse, H.F., Fok, L., 2016. Plastic waste in the marine environment: a review of sources, occurrence and effects. Sci. Total Environ. 567, 333–349. Available from: https://doi.org/10.1016/j.scitotenv.2016.05.084.
Liebezeit, G., Liebezeit, E., 2015. Origin of synthetic particles in honeys. Pol. J. Food Nutr. Sci. 65, 143–147. Available from: https://doi.org/10.1515/pjfns-2015-0025.
Liu, C., Li, J., Zhang, Y., Wang, L., Deng, J., Gao, Y., et al., 2019. Widespread distribution of PET and PC microplastics in dust in urban China and their estimated human exposure. Environ. Int. 128, 116–124. Available from: https://doi.org/10.1016/j.envint.2019.04.024.
Liu, J., Lu, G., Zhang, F., Nkoom, M., Yan, Z., Wu, D., 2018. Polybrominated diphenyl ethers (PBDEs) in a large, highly polluted freshwater lake, China: occurrence, fate, and risk assessment. Int. J. Environ. Res. Public Health 15, 1529.
Ma, L., Zhang, X., Cass, G., 2014. The spatial effect of China's Haze pollution and the impact from economic change and energy structure. China Ind. Econ. 31 (4), 19–31.
Malaj, E., Von Der Ohe, P.C., Grote, M., Kühne, R., Mondy, C.P., Usseglio-Polatera, P., et al., 2014. Organic chemicals jeopardize the health of freshwater ecosystems on the continental scale. Proc. Natl. Acad. Sci. U.S.A. 111, 9549–9554.
Mattsson, K., Hansson, L.A., Cedervall, T., 2015. Nano-plastics in the aquatic environment. Environ. Sci. Process. Impacts 17, 1712–1721. Available from: https://doi.org/10.1039/C5EM00227C.
May, T.W., Fairchild, J.F., Petty, J.D., Walther, M.J., Lucero, J., Delvaux, M., et al., 2008. An evaluation of selenium concentrations in water, sediment, invertebrates, and fish from the Solomon River Basin. Environ. Monit. Assess. 137, 213–232.
Miller, G.T., 1997. Environmental Science: Working with the Earth, fifth ed. Wadsworth, Belmont, CA.
Miller, G.T., 2004. Environmental Science, 10th ed. Brooks Cole, Pacific Grove, CA.
Mondal, P., 2018. Pollutants: Definition and Classification of Pollutants. Available from: <http://www.yourarticlelibrary.com/essay/pollutants-definition-and-classification-of-pollutants/28277> (accessed 15.12.19).
Mottaleb, M.A., Abedin, M.Z., Islam, M.S., 2003. Determination of benzene, toluene, ethylbenzene and xylene in river water by solid-phase extraction and gas chromatography. Anal. Sci. 19, 1365–1369.
Murgel, E., 2007. Fundamentos de Acústica Ambiental. Senac, São Paulo, Brazil, 2007.
Nanny, M.A., Ratasuk, N., 2002. Characterization and comparison of hydrophobic neutral and hydrophobic acid dissolved organic carbon isolated from three municipal landfill leachates. Water Res. 36, 1572–1584.
Nathanson, J.A., 2007. Basic Environmental Technology: Water Supply, Waste Management and Pollution Control, fifth ed. Prentice Hall, Upper Saddle River, NJ.
National Geographic, 2018. Pollution. National Geographic Periodicals October, 2018.
Nature Communications, 2018. The future of plastic. Nat. Commun. 9, 1–3. Available from: https://doi.org/10.1038/s41467-018-04565-2.
Ndoke, P.N., Jimoh, D.O., 2000. Impact of traffic emission on air quality in developing city of Nigeria. Dep. Civ. Engineering, Fed. Univ. Technology, Minna Niger. vol. 2(2), pp. 14–23.
Ng, C.K., Bope, C.D., Nalaparaju, A., Cheng, Y., Lu, L.B., Cao, W.R., 2017. Concentrating synthetic estrogen 17a-ethinylestradiol using microporous polyethersulfone hollow fiber membranes: experimental exploration and molecular simulation. Chem. Eng. J. 314, 80–87.
OECD, 2016. The Economic Consequences of Outdoor Air Pollution. OECD Publishing, Paris.
Okrent, D., 1999. On intergenerational equity and its clash with intragenerational equity and on the need for policies to guide the regulation of disposal of wastes and other activities posing very long-time risks. Risk Anal. 19, 877–901.
Olutona, G.O., Oyekunle, J.A.O., Ogunfowokan, A.O., Fatoki, O.S., 2017. Concentrations of polybrominated diphenyl ethers (PBDEs) in water from Asunle Stream, Ile-Ife, Nigeria. Toxics 5, 13.
Oluwatuyi, O.E., Ajibade, F.O., Ajibade, T.F., Adelodun, B., Olowoselu, A.S., Adewumi, J.R., et al., 2020. Total concentration, contamination status and distribution of elements in a Nigerian State dumpsites soil. Environ. Sustain. Indic. 100021. Available from: https://doi.org/10.1016/j.indic.2020.100021.
Omotade, I.F., Alatise, M.O., Olanrewaju, O.O., 2019. Recycling of aquaculture wastewater using charcoal based constructed wetlands. Int. J. Phytoremediation 21, 399–404. Available from: https://doi.org/10.1080/15226514.2018.1537247.
Opafola, O.T., Oladepo, K.T., Ajibade, F.O., David, A.O., 2020. Potability assessment of packaged sachet water sold within a tertiary institution in southwestern Nigeria. J. **King Saud Univ**. Sci 32 (3), 1999–2004. Available from: https://doi.org/10.1016/j.jksus.2020.02.004.
Oplatowska, M., Donnelly, R.F., Majithiya, R.J., Kennedy, D.G., Elliott, C.T., 2011. The potential for human exposure, direct and indirect, to the suspected carcinogenic triphenylmethane dye brilliant green from green paper towels. Food Chem. Toxicol. 49, 1870–1876.
Opopol, N., 2008. Impact of pops on the Republic of Moldova environment and public health. In: Mehmetli, E., Koumanova, B. (Eds.), The Fate of Persistent Organic Pollutants in the Environment. Springer Netherlands, Dordrecht, pp. 405–424.
Peirce, J.J., Weiner, R.F., Vesilind, P.A., 1998. Environmental Pollution and Control, fourth ed. Butterworth-Heinemann.
Pepper, I.L., Gerba, C.P., Brusseau, M.L., 2006. Environmental and Pollution Science. Burlington, MA.
Perera, P.A.C.T., Sundarabarathy, T.V., Sivananthawer, T., Kodithuwakku, S.P., Edirisinghe, U., 2016. Arsenic and cadmium contamination in water, sediments and fish is a consequence of paddy cultivation: evidence of river pollution in Sri Lanka. Achiev. Life Sci. 10, 144–160.

Popa, C., Petrus, M., 2017. Heavy metals impact at plants using photoacoustic spectroscopy technology with tunable CO_2 laser in the quantification of gaseous molecules. Microchem. J. 134, 390–399.

Prata, J.C., 2018. Airborne microplastics: consequences to human health? Environ. Pollut. 234, 115–126. Available from: https://doi.org/10.1016/j.envpol.2017.11.043.

Pruss-Ustun, A., Bos, R., Gore, F., Safar, J.B., 2008. Water, Better Health: Costs, Benefits and Sustainability of Interventions to Protect and Promote Health. World Health Organization, Geneva.

Qi, Y., Yang, X., Mejia, A., Huerta, E., Beriot, N., Gertsen, H., et al., 2018. Macro- and micro- plastics in soil-plant system: effects of plastic mulch film residues on wheat (*Triticum aestivum*) growth. Sci. Total Environ. 645, 1048–1056. Available from: https://doi.org/10.1016/j.scitotenv.2018.07.229.

Qin, W.C., Su, L.M., Zhang, X.J., Qin, H.W., Wen, Y., Guo, Z., et al., 2010. Toxicity of organic pollutants to seven aquatic organisms: effect of polarity and ionization. SAR QSAR Environ. Res. 21, 389–401.

Rahmana, M.A., Rahman, A., Khanc, M.Z., Renzaho, A.M., 2018. Human health risks and socio-economic perspectives of arsenic exposure in Bangladesh: a scoping review. Ecotoxicol. Environ. Saf. 150, 335–343. Available from: https://doi.org/10.1016/j.ecoenv.2017.12.032.

Rajeshkumar, S., Li, X., 2018. Bioaccumulation of heavy metals in fish species from the Meiliang Bay, Taihu Lake, China, Toxicology Report, 5. pp. 288–295. Available from: https://doi.org/10.1016/j.toxrep.2018.01.007.

Ramanan, R., Kim, B.H., Cho, D.H., Oh, H.M., Kim, H.S., 2016. Algae–bacteria interactions: evolution, ecology and emerging applications. Biotechnol. Adv. 34, 14–29.

Rana, R.S., Singh, P., Kandari, V., Singh, R., Dobhal, R., Gupta, S., 2017. A review on characterization and bioremediation of pharmaceutical industries' wastewater: an Indian perspective. Appl. Water Sci. 7, 1–12. Available from: https://doi.org/10.1007/s13201-014-0225-3.

Reid, A.J., Carlson, A.K., Creed, I.F., Eliason, E.J., Gell, P.A., Johnson, P.T.J., et al., 2019. Emerging threats and persistent conservation challenges for freshwater biodiversity. Biol. Rev. 94, 849–873. Available from: https://doi.org/10.1111/brv.12480.

Rice, K.M., Walker, E.M.J.R., Wu, M., Gillette, C., Blough, E.R., 2014. Environmental mercury and its toxic effects. J. Prev. Med. Public Health Yebang Uihakhoe Chi 47, 74–83.

Rimayi, C., Chimuka, L., Odusanya, D., De Boer, J., Weiss, J., 2016. Distribution of 2,3,7,8-substituted polychlorinated dibenzo-p-dioxin and polychlorinated dibenzofurans in the Jukskei and Klip/Vaal catchment areas in South Africa. Chemosphere 145, 314–321.

Rinkesh, B., 2019. What is Noise Pollution? Conserve Energy Future. Available from: <https://www.conserve-energy-future.com/causes-and-effects-of-noise-pollution.php> (assessed 26.12.19).

Rochman, C.M., Hoh, E., Kurobe, T., Teh, S.J., 2013. Ingested plastic transfers hazardous chemicals to fish and induces hepatic stress. Sci. Rep. 3, 1–7. Available from: https://doi.org/10.1038/srep03263.

Santillo, D., Miller, K., Johnston, P., 2017. Microplastics as contaminants in commercially important seafood species. Integr. Environ. Assess. Manage. 13, 516–521. Available from: https://doi.org/10.1002/ieam.1909.

Sarikavak, Y., Boxall, A., 2019. The impacts of pollution for new high-speed railways: the case of noise in Turkey. Acoust. Aust. Available from: https://doi.org/10.1007/s40857-019-00154-5.

Sarria-Villa, R., Ocampo-Duque, W., Paez, M., Schuhmacher, M., 2016. Presence of PAHs in water and sediments of the Colombian Cauca River during heavy rain episodes, and implications for risk assessment. Sci. Total Environ. 540, 455–465.

Schwabl, P., Koppel, S., Konigshofer, P., Bucsics, T., Trauner, M., Reiberger, T., et al., 2019. Detection of various microplastics in human stool: a prospective case series. Ann. Intern. Med. 171, 453–457. Available from: https://doi.org/10.7326/M19-0618.

Schweitzer, L., Noblet, J., 2018. Chapter 3.6 – Water contamination and pollution. In: Török, B., Dransfield, T. (Eds.), Green Chemistry. Elsevier.

Shen, M., Ye, S., Zeng, G., Zhang, Y., Xing, L., Tang, W., et al., 2019. Can microplastics pose a threat to ocean carbon sequestration? Mar. Pollut. Bull. 150, 110712. Available from: https://doi.org/10.1016/j.marpolbul.2019.110712.

Sherman, J., McGain, F., Fanzca, F.C., 2016. Environmental sustainability in anesthesia. Pollut. Prev. Patient Saf. 4, 47–61.

Sonibare, J.A., 2010. Air pollution implications of Nigeria's present strategy on improved electricity generation. Energy Policy 38, 5738–5789.

Sonibare, J.A., Jimoda, L.A., 2009. Criteria air pollutants from some anthropogenic combustion processes in Lagos, Nigeria. Energy Res., A: Recovery Util. Environ. Eff. 31 (11), 923–935.

Sonthiphand, P., Cejudo, E., Schiff, S.L., Neufeld, J.D., 2013. Wastewater effluent impacts ammonia-oxidizing prokaryotes of the Grand River, Canada. Appl. Environ. Microbiol. 79, 7454–7465.

Spellman, F.R., 1999. The Science of Environmental Pollution. CRC Press, Boca Raton, FL.

Spellman, F.R., 2017. Soil pollution. In: Spellman, F.R. (Ed.), The Science of Environmental Pollution, third ed. NW: CRC Press, Taylor & Francis Group, Boca Raton, FL, London, New York, pp. 335–361.

Spellman, F.R., Whiting, N.E., 2006. Environmental Science and Technology Concepts and Applications, second ed. Government Institutes, Rockville, MD.

Srinivasan, S., 2013. Socio-economic impact of air pollution on dwellers of and around the industrial area in Tamil Nadu, India. EJBE 3 (1), 58–94.

Streimikiene, D., 2015. Environmental indicators for the assessment of quality of life. Intellect. Econ. 9, 67–79.

Suk, W.A., Ahanchian, H., Asante, K.A., Carpenter, D.O., Diaz-Barriga, F., Ha, E.H., et al., 2016. Environmental pollution: an underrecognized threat to children's health, especially in low- and middle-income countries. Environ. Health Perspect. 124 (3), A41–A45. Available from: https://doi.org/10.1289/ehp.1510517.

Sulaymon, I.D., Mei, X., Yang, S., Chen, S., Zhang, Y., Hopke, P.K., et al., 2020. PM 2.5 in Abuja, Nigeria: chemical characterization, source apportionment, temporal variations, transport pathways and the health risks assessment. Atmos. Res. 237, 104833. Available from: https://doi.org/10.1016/j.atmosres.2019.104833.

Sun, B., Zhang, X., Yin, Y., Sun, H., Ge, H., Li, W., 2017. Effects of sulforaphane and vitamin E on cognitive disorder and oxidative damage in lead-exposed mice hippocampus at lactation. J. Trace Elem. Med. Biol. 44, 88–92.

Talsness, C.E., Andrade, A.J.M., Kuriyama, S.N., Taylor, J.A., vom Saal, F.S., 2009. Components of plastic: Experimental studies in animals and relevance for human healt. Philosophical Transactions of the Royal Society B 364, 2079–2096.

Tchounwou, P.B., Yedjou, C.G., Patlolla, A.K., Sutton, D.J., 2012. Heavy metal toxicity and the environment. Exper. Suppl. 101, 133–164.

Testa, S.M., 1997. The Reuse and Recycling of Contaminated Soil. CRC Press, Boca Raton, FL.

Thiel, C.L., Eckelman, M., Guido, R., 2015. Environmental impacts of surgical procedures: life cycle assessment of hysterectomy in the United States. Environ. Sci. Technol. 49, 1779–1786.

Thompson, R.C., Olsen, Y., Mitchell, R.P., Davis, A., Rowland, S.J., John, A.W.G., et al., 2004. Lost at Sea: where is all the plastic? Brevia (Rome) 304, 838. Available from: https://doi.org/10.1111/sum.12058.

Titiksha, T., 2017. Biology Discussion: Biodegradable pollutants and Non-biodegradable pollutants. Available from: https://www.biologydiscussion.com/articles/biodegradable-pollutants-and-non-biodegradable-pollutants/2382. Accessed on 28th January, 2020.

Torres, J.P., Pfeiffer, W.C., Markowitz, S., Pause, R., Malm, O., Japenga, J., 2002. Dichlorodiphenyltrichloroethane in soil, river sediment, and fish in the Amazon in Brazil. Environ. Res. 88, 134–139.

Trasande, L., Liu, Y., 2011. Reducing the staggering costs of environmental disease in children, estimated at $76.6 billion in 2008. Health Affairs 30 (5), 863–870. Available from: https://doi.org/10.1377/hlthaff.2010.1239.

Ugya, A.Y., Ajibade, F.O., Ajibade, T.F., 2018. Water pollution resulting from mining activity: an overview. In: Proceedings of the 2018 Annual Conference of the School of Engineering & Engineering Technology (SEET). The Federal University of Technology, Akure, Nigeria, 17–19 July, pp. 703–718.

Ugya, A.Y., Hua, X., Agamuthu, P., Ma, J., 2019. Molecular approach to uncover the function of bacteria in petrochemical refining wastewater: a mini review. Appl. Ecol. Environ. Res. 17 (2), 3645–3665. Available from: https://doi.org/10.15666/aeer/1702_36453665.

Ugya, A.Y., 2015. The efficiency of *Lemna minor* L. in the phytoremediation of Romi stream: a case study of Kaduna refinery and petrochemical company polluted stream. J. Appl. Biol. Biotechnol. 3, 11–14.

Ugya, A.Y., Imam, T.S., 2017. Temporal heavy metals variation in vegetables sampled at Kasuwan Mata, Kaduna Metropolis, Nigeria. Malays. J. Sci. 36, 63–73.

UNECE, 2009. Measuring Sustainable Development, United Nation Economic Commission for Europe. United Nations Publication, New York/Geneva.

United Nations General Assembly, 1987. Report of the world commission on environment and development: our common future. In: Transmitted to the General Assembly as an Annex to Document A/42/427 – Development and International Co-Operation: Environment. Our Common Future, Chapter 2: Towards Sustainable Development; Paragraph 1. United Nations General Assembly World Commission on Environment and Development (1987) Our common future. Oxford University Press, Oxford, p. 27.

UNICEF, 2008. Progress on Drinking Water and Sanitation: Special Focus on Sanitation, in Progress on Drinking Water and Sanitation: Special Focus on Sanitation. WHO/JMP, Geneva.

USEPA, 1989. Glossary of Environmental Terms and Acronyms List. U.S. Environmental Protection Agency, Washington, DC.

USEPA (US Environmental Protection Agency), 2013. Effects of Air Pollutants – Health Effects. <http://www.epa.gov/eogaptil/course422/ap7a.html>.

Van Cauwenberghe, L., Janssen, C.R., 2014. Microplastics in bivalves cultured for human consumption. Environ. Pollut. 193, 65–70. Available from: https://doi.org/10.1016/j.envpol.2014.06.010.

VanLiere, K.D., Dunlap, R.E., 1980. The social bases of environmental concern: a review of hypotheses, explanations and empirical evidence. Public Opin. Q. 44 (2), 181–197.

van Wezel, A., Caris, I., Kools, S.A.E., 2016. Release of primary microplastics from consumer products to wastewater in the Netherlands. Environ. Toxicol. Chem. 35, 1627–1631. Available from: https://doi.org/10.1002/etc.3316.

Verma, M., Schneider, J.S., 2017. Strain specific effects of low-level lead exposure on associative learning and memory in rats. Neurotoxicology 62, 186–191.

Vince, J., Hardesty, B.D., 2016. Plastic pollution challenges in marine and coastal environments: from local to global governance. Restor. Ecol. 25, 123–128. Available from: https://doi.org/10.1111/rec.12388.

Vymazal, J., 2007. Removal of nutrients in various types of constructed wetlands. Sci Total Environ 380, 48–65.

Wagner, S., Reemtsma, T., 2019. Things we know and don't know about nanoplastic in the environment. Nat. Nanotechnol. 14, 300–301. Available from: https://doi.org/10.1038/s41565-019-0424-z.

Walker, D.B., Baumgartner, D.J., Gerba, C.P., Fitzsimmons, K., 2019. Chapter 16 – Surface water pollution. In: Brusseau, M.L., Pepper, I.L., Gerba, C.P. (Eds.), Environmental and Pollution Science, third ed. Academic Press.

Wang, L., Liang, T., 2015. Distribution characteristics of phosphorus in the sediments and overlying water of Poyang Lake. PLoS One 10, e0125859.

Wijayawardena, M.A., Megharaj, M., Naidu, R., 2016. Exposure, toxicity, health impacts, and bioavailability of heavy metal mixtures. Adv. Agron. 138, 175–234. Available from: https://doi.org/10.1021/jf035501t.

Willis, K.A., Eriksen, R., Wilcox, C., Hardesty, B.D., 2017. Microplastic distribution at different sediment depths in an urban estuary. Front. Mar. Sci. 4, 419. Available from: https://doi.org/10.3389/fmars.2017.00419.

World Economic Forum, 2016. The New Plastics Economy: Rethinking the Future of Plastics. Ellen MacArthur Foundation, and McKinsey & Company. <newplasticseconomy.org/about/publications>.

World Health Organization (WHO), 2018. Environmental Noise Guidelines for the European Region. p. 160.

World Health Organization (WHO), 2011. Air Quality and Health. <http://www.who.int/mediacentre/factsheets/fs313/en/index.html>.

WHO (World Health Organisation), 2006. Air Quality Guidelines. Global Update 2005. Particulate Matter, Ozone, Nitrogen Dioxide and Sulfur Dioxide. Available from: <http://www.euro.who.int/en/health-topics/environmentand-health/Housing-and-health/publications/pre-2009/air-quality-guidelines.-global-update-2005.-particulate-matter,-ozone,-nitrogendioxide-and-sulfur-dioxide>.

World Health Organization (WHO), 2004. Protection of the human environment occupational noise. In: Environmental Burden of Disease Series. No. 9.

Wright, S.L., Thompson, R.C., Galloway, T.S., 2013. The physical impacts of microplastics on marine organisms: a review. Environ. Pollut. 178, 483–492. Available from: https://doi.org/10.1016/j.envpol.2013.02.031.

Wu, L., Ning, D., Zhang, B., Li, Y., Zhang, P., Shan, X., et al., 2019. Global diversity and biogeography of bacterial communities in wastewater treatment plants. Nat Microbiol. 4 (7), 1183–1195. Available from: https://doi.org/10.1038/s41564-019-0426-5.

Yarimizu, K., Cruz-López, R., Carrano, C.J., 2018. Iron and harmful algae blooms: potential algal-bacterial mutualism between *Lingulodinium polyedrum* and *Marinobacter algicola*. Front. Mar. Sci. 5, 180. Available from: https://doi.org/10.3389/fmars.2018.00180.

Zhang, H., Kelly, B.C., 2018. Sorption and bioaccumulation behavior of multi-class hydrophobic organic contaminants in a tropical marine food web. Chemosphere 199, 44–53.

Zhang, X.Q., Xia, X.H., Yang, Z.F., 2007. Reasons of high concentration ammonium in Yellow River, China. Huan Jing Ke Xue 28, 1435–1441.

Zhou, G.J., Ying, G.G., Liu, S., Zhou, L.J., Chen, Z.F., Peng, F.Q., 2014. Simultaneous removal of inorganic and organic compounds in wastewater by freshwater green microalgae. Environ. Sci. Process. Impacts 16, 2018–2027.

Further reading

Andersen, Z.J., Kristiansen, L.C., Andersen, K.K., Olsen, T.S., Hvidberg, M., Jensen, S.S., et al., 2011. Stroke and long-term exposure to outdoor air pollution from nitrogen dioxide: a cohort study. Stroke 2 (6), 23–32.

Fadare, O.O., Wan, B., Guo, L., Zhao, L., 2020. Microplastics from consumer plastic food containers: Are we consuming it? Chemosphere 253, 126787. Available from: https://doi.org/10.1016/j.toxrep.2018.01.007.

Nathanson, J.A. (2019). An Article on Pollution – Environment. Available from: <https://www.britannica.com/science/light-pollution> (accessed on 29.11.19).

Neidell, M.J., 2004. Air pollution, health, and socio-economic status: the effect of outdoor air quality on childhood asthma. J. Health Econ. 23, 1209–1236. Available from: https://doi.org/10.1016/j.jhealeco.2004.05.002.

Pepper, I.L., Gerba, C.P., Brusseau, M.L., 1996. Pollution Science. Academic Press, San Diego, CA.

Saha, B., Karounou, E., Streat, M., 2010. Removal of 17b-oestradiol and 17a-ethinyl oestradiol from water by activated carbons and hypercrosslinked polymeric phases. React. Funct. Polym. 70, 531–544.

Chapter 26

Remediation action on persistent organic pollutants by wonder weeds and associated microbiomes

Megha Barot[1] and Nirmal Kumar[2]

[1]*SRISTI Life Science, Grambharti, Gandhinagar, India,* [2]*PG Department of Environment Science and Technology, ISTAR, Anand, India*

26.1 Introduction

Persistent organic pollutants (POPs), a broad term, which includes pesticides, herbicides, and chemicals produced from industries, have drawn global attention because of their chemical properties and persistency in environment and named "forever compounds"/"POPs"/"xenobiotic compounds." These compounds are persistent in environment and, being persistent, enter the food chain of ecosystem imparting toxic effects on all the associated biological agents. POPs are synthetic organic chemicals and have drawn everyone's attention (as environment disturbing agents) globally due to their potential to transport across the continent mainly via wind and oceanic currents. By means of food, air, outdoor, indoor, and at workplace humans are easily exposed to these chemicals. Lifestyle of human being comprises many products that contain POPs, including dichlorodiphenyltrichloroethane (DDT), polychlorinated biphenyls (PCBs), and some by-products from industries, polychlorinated dibenzo-*p*-dioxins and dibenzofurans (PCDD/Fs), results of which show their existence on Earth in measurable concentrations. POPs are lipophilic in nature and accumulate in adipose tissues and magnify so in food chain maximum concentration is found in higher level of the food chain. Being at topmost position on food chain, human beings are also expected to accumulate these chemicals in substantial amount within their body. These toxic compounds can lead to many health issues and a risk of life-threatening diseases such as cancer, immune system dysfunction, reproductive problems, neurological disorder, endocrine disruption, and genetic toxicity (Jones and De Voogt, 1999). However, these persistent compounds occupy very small fractions of total known chemicals and pollutants in the marine ecosystem (Dachs and Méjanelle, 2010) and are the contributors of total potential pollutants up in the environment (Muir and Howard, 2006). Some of the persistent compounds found everywhere in the world in biotic and abiotic systems (Gioia et al., 2006, 2008; Gilman et al., 1997; Jaward et al., 2004). Now, even mostly, we focus on limited chemical compounds such as PCBs, hexachlorocyclohexane (HCHs), DDT, and polycyclic aromatic hydrocarbons (PAHs). These chemical groups should be considered as "indicators" or "surrogates" of the other pollutants in marine ecosystem with identical physicochemical properties (Gioia et al., 2011). Besides perfluorinated acids such as perfluorooctanesulfonate (PFOS) and perfluorooctanoate (PFOA) are pollutants with completely different properties such as greater water solubility and low volatility could reach to the sea and open oceans via tributaries/rivers and exposed to different environmental conditions. Yamashita et al. (2008) reported that because of the high water solubility of these compounds the open-ocean water column is suggested to be the final sink for PFOS and PFOA, which are considered as global environmental contaminants. Even though, some newly emerging organic compounds have been reported to behave as passive tracers in waters (Yamashita et al., 2008), their behavior was found relatively different from that shown by principal POPs such as PCBs or HCHs. These pollutants change their properties because of their ionic characteristic (Gioia et al., 2011). In aquatic system, due to their hydrophobicity and lipophilic properties, they remain inside the aquatic biota and thereby, aquatic food chain. Under environmental temperatures, most of them are semivolatiles and thus long-range transport from one part of ecosystem to other is often recorded. With these two important reasons, POPs are well documented as pollutants with long-range transporter and accumulator in food chain for longer period (Jones and De Voogt, 1999).

Microbe Mediated Remediation of Environmental Contaminants. DOI: https://doi.org/10.1016/B978-0-12-821199-1.00026-2

26.2 Characteristics of persistent organic pollutants

First, persistence organic pollutants are synthetic compounds produced from industries and utilized intentionally or unintentionally in daily life. They remain persistent in environment for longer time (due to their long half-life) and may take decades or centuries to degrade. Due to this reason, these POPs are extensively studied from water, soil, air, and food chain.

Manmade organic chemicals: These are synthetic compounds produced from industries and released intentionally/unintentionally as by-products by different process. (1) They remain persistent in environment for long time and it may take decades or centuries to get degrade. (2) Long-range /global transport: There are some POPs that will mostly found in different parts of the world while studying from water, air, food, and even soil that indicates their long-range transport. (3) Lipophilic: Their deposition in fat tissues shows their lipophilic nature. (4) Accumulate in food chain: These organic compounds entered food chain from ecosystem and accumulate in higher organisms as they eat from smaller one. Aquatic mammals are accumulating maximum POPs. (5) Toxicity: High-level POPs exposure shows acute toxic effects for some pesticides, for example, aldrin, dieldrin, and toxaphene. Polychlorinated biphenyls caused mass poisoning and it was well-recorded episodes called "Yusho" and "Yu Cheng," which happened in coastal countries China, Taiwan, and Japan.

26.3 The "dirty dozen"

According to the Stockholm Convection on POPs, 12 POPs were identified as initial "dirty dozen." However, in the convection, nine new POPs were also introduced (Fiedler, 2008; Thorne and Stryker, 2018). They are also known as "legacy POPs" and are banned globally under Stockholm Convention for their adverse effect on environment and human health. Information about these categorized (dirty dozen) POPs is presented in Table 26.1. Initially, 12 POPs were identified in Stockholm Convention, later at the Fourth Conference of the Parties-4 (COP-4) meeting held in May 2009 included 9 new chemicals, and therefore presently there are 21 POPs in the convention list. These nine new pollutants were α-HCH, β-HCH, chlordecone, γ-HCH, hexabromobiphenyl, hexabromodiphenyl ether and heptabromodiphenyl ether (PBDE), pentachlorobenzene, perfluorooctane sulfonic acid (PFOS), and tetrabromodiphenyl ether and pentabromodiphenyl ether (PBDE). These 21 pollutants are either intentionally or unintentionally produced. Industrial chemicals and organochlorine pesticides are intentionally produced POPs, whereas PAHs, dioxins, and furans are produced as a by-product of the industrial preparations.

26.4 Global environmental threats of persistent organic pollutants

Persistent organic compounds, at very low level, transport long range as they are semivolatile mostly travel by water and by air with particulate matter, which result wide range global distribution even reported where they have never been used (Buccini, 2003). These compounds surprisingly reported in Arctic regions; however, they have been banned in United States and Canada since many years. POPs persist in colder regions as they travel toward colder regions and get accumulate; due to temperature here they are resistant to degradation and there they accumulate in plants and subsequently in the higher animals. In the late 1970s POPs came in light; in Arctic regions pesticides detected in bear fat tissue these regions (Ashraf, 2017).

26.5 Sources of persistent organic pollutants in aquatic system and their reservoirs

Chemicals in sediments come from various sources and can be divided into various categories, namely, industrial chemicals (anthropogenic); incomplete combustion of fuels; petroleum oils; natural disasters such as forest and grass fires; degraded waste through organisms; diffusion from the sediment layers—mantle and petroleum source rocks or reservoirs (Readman et al., 2002; Kim et al., 1999) (Fig. 26.1). PCBs are most dangerous, toxic, and carcinogenic among POPs as they are persistent, highly resistant to biodegradation and can accumulate in fat tissues (Meagher, 2000). These chemicals (PCBs) released during production spills and disposal from industries making hydraulic fluids, lubricants, plasticizers, adhesives, transformer fluids (Scragg, 2005). The POPs such as PCDD/Fs are persist in sediments since decades, so these chemicals are still found in deep sea and oceans as they are found recalcitrant to abiotic and biotic degradation processes (Uchimiya and Masunaga, 2007). Many chlorinated POPs are more toxic and carcinogenic for humans (Kaiser, 2000). Moreover, other groups of chlorinated compounds produced as solvent and used in wood treatment such as carbon tetrachloride, trichloroethene, and pentachlorophenol are found commonly in environment (Scragg, 2005). Plastic debris is the main factor of transport of POPs and hydrophobic contaminants in marine environment (Mato et al., 2001; Teuten et al., 2007). POPs are found on plastic debris because of the hydrophobic nature of the

TABLE 26.1 List of dirty dozen persistent organic pollutants (POPs) categorized in Stockholm Convention.

POPs	Category	Use	Adverse effects	Half-life/ persistency
Aldrin	Pesticides	Kills termites, grasshoppers, corn rootworms, and other insects and pests	Poisonous for birds, fish, and humans. Fatal dose for adult male is 5 g	–
Chlordane	Pesticides	Termite control and broad-spectrum insecticides	Lethal for mallard duck, bobwhite quail, pink shrimps. Known to reduce immunity and induce cancer in humans	1 year (half-life)
DDT	Pesticides	Controls malaria, typhus, and other insect-mediated diseases	Thinning of eggshell in birds. Induce chronic diseases in humans	10–15 years (persistency)
Dieldrin	Pesticides	Controls termite, insect-borne diseases in agricultural crops and textile pests	Highly toxic to fish and other aquatic animals especially frogs	5 years (half-life)
Endrin	Pesticides	Controls insects and rodents in agricultural crops such as cotton and grains	Highly toxic to fish, early hatching, reproductive malfunction	12 years (persistency)
Heptachlor	Pesticides	Control of soil insects and termites, cotton insects, grasshoppers, and malaria-carrying mosquitoes	Responsible for the decreased population of wild birds especially Canadian Geese and American Kestrels	–
Hexachlorobenzene	Industrial chemicals (fungicides) and by-products	Control of fungi in food crops mainly wheat bunt	Imparts photosensitive skin lesion, colic, and debilitation; metabolic disorder such as porphyria turcica	–
Mirex	Pesticides	Control of ants, fire ants, and termites. Fire retardants in plastic, rubber, and electrical goods	Carcinogen for humans. Toxic to several plant species, fishes, and crustaceans	10 years (half-life)
PCBs	Industrial chemicals	As heat exchanger fluids, electric transformers and capacitors, and as an additive in paints	Lethal to fish at higher doses and can cause spawning failure at lower doses. Causes reproductive failure and suppression of immune system in various wild animals. Causes developmental delays and behavioral problems, poor short-term memory function, suppression of human immune system and cancer	10 days to 1.5 years (half-life)
PCDFs	By-products	Used as dioxins and in the production of PCB	Carcinogen	–
PCDDs	By-products	Produced due to incomplete combustion	Immune system disorders, enzyme disorders, and chloracne. Increases the risk of birth defects in laboratory animals	10–12 years (persistency)
Toxaphene	Pesticides	Control of insects of cotton, cereal grains, fruits, nuts, and vegetables		12 years (persistency)

PCBs, Polychlorinated biphenyls; *PCDDs*, polychlorinated dibenzo-*p*-dioxins; *PCDFs*, polychlorinated dibenzofurans.

compounds, PCB and other pollutants absorbed on plastic surface. The organic pollutants have different affinity toward the plastic. For example, PCBs have high affinity for polyethylene than polypropylene (Endo et al., 2005; Ogata et al., 2009) which was confirmed by in situ and laboratory experiments where plastic debris were ingested by seabirds detected with POPs (Mato et al., 2001). Moreover, plastic resins were also found in coastal areas (Karapanagioti and Klontza, 2007, 2008). There are many reports on the POPs' accumulation and translocation in food chain. Organochlorine residue pattern was reported in human breast milk from South India (Tanabe et al., 1990). POPs were also found in migratory birds (Tanabe et al., 1998) and zooplanktons and fish from Arabian sea (Shailaja and Nair, 1997). Persistent, long-chained perfluorinated carboxylates reported in seabirds from the Canadian Arctic study

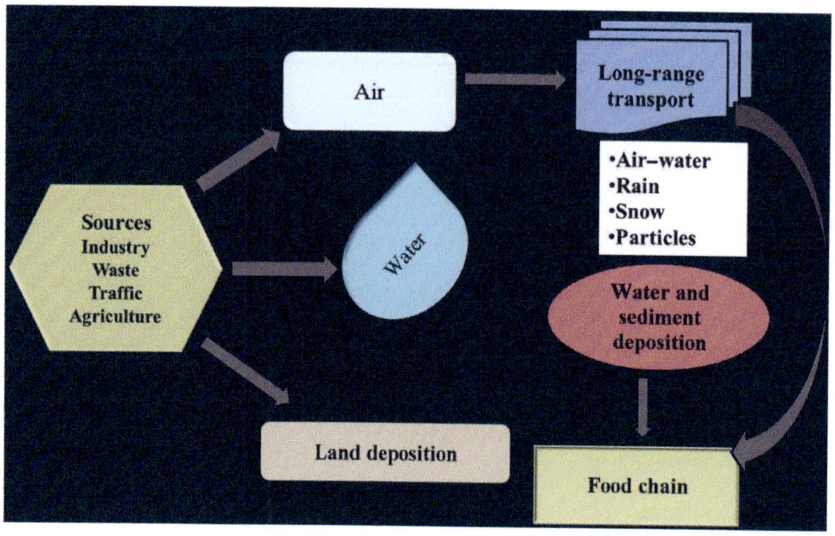

FIGURE 26.1 Schematic diagram explaining distribution of various POPs in an environment. *POPs*, Persistent organic pollutants.

represented data made between 1975 and 2004 (Butt et al., 2007). Organochlorine insecticides and PCBs were observed in the Mediterranean Sea (El-Dib and Badawy, 1985). Degradation of DDT, dieldrin, aldrin, and endrin and their effect on marine microorganisms were also found casual (Patil et al., 1972; Lal and Saxena, 1982). Accumulation of DDT and benzene hexachloride (BHC) is found in adipose tissue of Indian males by Siddiqui and Saxena (1985). Siddiqui et al. (1981) reported concentration of PCBs and organochlorine pesticides in human milk and blood from the year 1972 to 1977. Chlorinated pesticides are detected in Indian people (Dale et al., 1965). Reports on POPs in edible fish indicate human and environment health issues (Corsolini et al., 2005); PCBs and organochlorine pesticides are encountered in seabirds ingested plastics (Colabuono et al., 2010).

26.6 Conventional method of remediation of organic pollutants

After the recognition of hazardous consequences of these organic pollutants, researchers around the globe developed various protocols to combat with these POPs. Using physical and chemical methods (conventional methods), many of these pollutants have been remediated successfully until date. However, these methods also have some disadvantages and due to which, an alternative protocol for the remediation of these pollutants needed to be developed. Widely used conventional methods are summarized in Table 26.2. Considering their application for the remediation of POPs, one must consider several drawbacks associated with them. For example, in an electrochemical oxidation low selectivity and poor reaction rates are the major two disadvantages. Moreover, adsorption method has a limitation of functionality when the pollutants are distributed in two different medium, solid and liquid. In addition to that, it is very difficult to regenerate the adsorbent when these pollutants are strongly attached. On the other hand, brighter side of these methods also exists with some advantages such as supportive chemicals, high temperature or pressure is not required (in the case of electrochemical oxidation), and cost-effective (in case of adsorption). However, cost-effectiveness is also not applicable to other conventional methods, namely, solvent extraction and precipitation.

Apart from electrochemical oxidation method, there are several other oxidation processes, generally summarized as advanced oxidation processes (AOPs) as described by Machulek et al. (2012). There are seven different AOPs employed for the remediation of these pollutants. These AOPs include ozonation, UV mediated, UV/H_2O_2 mediated, O_3/UV/H_2O_2 mediated, Fenton reaction, photo-Fenton reaction, and heterogenous photocatalysis. Ozone is a strong oxidizing agent and thus efficient primary oxidation of pollutants may be achieved in ozonation treatment. As compared to other oxidizing agents, ozone is required in very small quantity. However, ozonization alone is not enough to remediate organic pollutants and therefore it should be followed by other oxidation methods. With this method, degradation of organic and inorganic materials can be achieved efficiently with the disadvantage of high electricity consumption as well as high operating and maintenance cost (Ikehata et al., 2008). Numbers of reports have been published using UVA-, UVB-, and UVC-based photocatalytic oxidation and subsequent sequestration of organic pollutants in marine ecosystem (Matafonova and Batoev, 2018). UV-based oxidation is generally suitable for the remediation of colored water and would not leave any harmful byproducts with effective degradation of organic pollutants. However, high particulate matters are more resistant to such treatments as they absorb UV light. Using the intrinsic properties of hydrogen

TABLE 26.2 Conventional methods used to remove organic pollutants from water sources.

Methods	Description	References
Electrochemical oxidation	• Does not require any supportive chemicals and other conditions such as high temperature or pressure while low selectivity and reaction rates are disadvantages of this technique.	Dominguez-Ramos and Irabien (2013) and Martinez-Huitle and Ferro (2006)
Membrane process	• Compact • High effluent quality • High volumetric load possible • High rate of degradation • Aeration limitations • Stress on sludge in external MBR • Membrane pollution • Cost price.	Prodanović and Vasić (2013)
Coagulation	• It reduces the time of settling suspended solid. • Very effective in removing even fine particles. • Can be effective in removing many protozoa, bacteria, and viruses. • It is costly. • It requires accurate dosing. Frequent monitoring is needed.	Papić et al. (2004) and Lee et al. (2006)
Adsorption	• Method is cost-effective, easy to operate, and most profitable than the other conventional methods such as precipitation and solvent extraction but there are some disadvantages over benefits such as it is difficult to remove pollutants from two different medium, solid and liquid while expensive to regenerate mainly when pollutants are strongly attached with adsorbent.	Ilisz et al. (2002), Messina and Schulz (2006), Jain et al. (2007), and Rafatullah et al. (2010)
Precipitation	• Permanent: can be effective for years • Immediate results: rapid in removal of contaminants • Efficient • Easily implemented • Easy to monitor • High cost • Not applicable for all cases • Requires O&M • Requires power • May generate a waste product.	Zhu et al. (2007)
Sonochemical degradation	Require simple handling, can achieve rapid degradation and wide applicable pH range.	Vecitis et al. (2010), Stock et al. (2000)

O&M, Operation and maintenance; *MBR*, Membrane bioreactor.

peroxide as oxidizing and reducing agent, it is used as an extension of UV-based oxidation. Hydrogen peroxides can oxidize varieties of pollutants containing sulfides, hydrogen sulfides, thiosulfates, cyanides, sulfites, chlorine, and nitrogenous compounds and do not produce any harmful secondary products. It can be used with wide range of temperature and pH with high water solubility and stability. UV treatment along with the hydrogen peroxide could serve as effective strategy for the remediation of organic pollutants (Simonenko et al., 2015). Operational cost may become a limitation while using UV$-H_2O_2$ method. Another method comprises ozone, UV, and H_2O_2 showed highly efficient degradation of micropollutants such as pesticides, herbicides, pharmaceuticals, and endocrine-disrupting compounds (Liu et al., 2019). Most widely used Fenton oxidation process is acidic Fenton reaction, which is carried out in the pH range of 2–4. Fenton reagent is a solution of hydrogen peroxide and ferrous ion (Fe^{2+}), which is highly potential to oxidize POPs (mostly pesticides) present in the soil (Gan and Ng, 2012). On the other hand, in photo-Fenton oxidation reaction, instead of hydrogen peroxides as oxidizing agent, UV light is used to induce photooxidation. Photo-Fenton reaction can be applied on variety of organic pollutants. Maintenance of acidic pH in photo-Fenton oxidation is the only limitation of this strategy (Machulek et al., 2012). Lastly, under heterogenous photocatalysis method, photoexcitation of semiconductor oxides creates a removal of electron in pair and produces electron deficiency. In an aqueous solution, such reaction would result in generation of superoxide ions (O_2^-), which is a strong oxidizing agent. The process ends up with the complete mineralization of pollutants (Augugliaro et al., 2006).

26.7 Biodegradation and bioremediation

As the name suggests, bioremediation process comprises biological agents for the removal and control of organic and inorganic pollutants from ecosystems (Singh et al., 2020; Gupta et al., 2020). Before they can be employed as regulators of organic pollutants, they are monitored as biomarker of adverse aquatic environment. These include change in temperature, pH, salinity, and the level of both organic and inorganic pollutants. Indicator organisms used in biomonitoring are zooplankton, algae, bivalve mollusks, insect, macrophyte, fish, amphibian, and gastropods. Their precision in the process of monitoring is well established and thus through the estimation of their population density, concentration of the inorganic pollutants (mostly metal content) can be estimated. Particularly seaweeds are model for such monitoring due to their size, biodiversity, longevity, presence at polluted sites, bioaccumulation of metals, and ease of identification (Phang et al., 2015).

The conventional techniques such as incineration, land filings, even dredging and related techniques when applied to the cleanup of system cost very high (Kuiper et al., 2004); therefore economical and more cost-effective techniques are required. Marine ecosystems are sinking of wide verities of the pollutants coming from different sources such as water runoff, rivers, and from anthropogenic activities. Things coming from different sources get settled down and contaminate the natural sediments of this aquatic system with organic and inorganic pollutants (Zoumis et al., 2001). Chemical pollutant may be accumulated or mobilized in the aquatic bed subjects to processes of transformation and activation (Martínez-Jerónimo et al., 2008). Aquatic beds are hub of anaerobic microorganisms due to high carbon source and lack of oxygen (Himmelheber et al., 2007). These microbes act on settled chemical pollutants as carbon source and they potential to degrade the organic pollutants. As discussed before, pollutants may enter aquatic system and become persistent. Their accumulation and toxicity depend on their physicochemical properties. Among them, compounds such as PAHs and PCBs are carcinogenic [IARC (International Agency for Research on Cancer), 1983]. Organic pollutants may interact with particulate and suspended matters in the water bottom and so they can be available to the organisms by the process of sorption and desorption, which determines the bioavailability and toxicity (Viganò, 2000). Scientist suggested that release of chemical compounds from the sediments through desorption process can be the main route of exposure to the organisms. For the toxicity study, both sediment and overlying water characteristics must be considered (Zoumis et al., 2001). The chemical diffusion in water through changes in sediment chemistry, seasonal variation, degradation by microbes, and gets diffuse in water by concentration gradient leads to the generation of more suspended forms of compounds (Booij et al., 1992). Conventional techniques have several disadvantages and made the researchers to realize the need of eco-friendly robust technology. After that, bioremediation methods were (re) discovered and found highly competent and sustainable. Both the targets of environmental protection and cost reduction at the same time may be achieved. Bioremediation can be defined as the process where biological organisms, plants, and their association involve in degradation of the inorganic or organic pollutants (Curtis et al., 2002). Microbes transform themselves to survive even in critical condition and their ubiquitous nature, high catalytic activity, and their genetics make them competent enough to degrade the available material (Chen et al., 1999; Mishra et al., 2001; Watanabe, 2001; Paul et al., 2005). Various organic pollutants, remediated by biodegradation or bioremediation processes, are listed in Table 26.3.

The bioremediation approach can be the following: (1) monitored natural recovery, under which, natural self-healing properties of the indigenous organisms with natural physical and chemical process. (2) Biostimulation that can enrich the site with indigenous microbial population by influencing positive factors such as nutrients and air, which enhance the growth (Singh et al., 2011). (3) Bioaugmentation, the process comprises addition of species from outside for degradation of the contaminants on-site (Stroo et al., 2012). (4) Phytoremediation, in which, remediation or removal of pollutants is carried out using plants and algae from the system (Azubuike et al., 2016; Singh, 2004). The chemical composition found in sediments and associated water column describes the effects of eutrophication on transformation of compounds. Few researchers had studied biogeochemical cycle of eutrophication and associated microbes' interaction in lake (Sanz-Lázaro et al., 2015; Spivak and Ossolinski, 2016). Some researchers also reported that there is a significant change in microbial community with increasing load of organic pollution (Sinkko et al., 2013; Xiong et al., 2014). Study also suggested that area rich with seaweeds diversity has found changed microbial diversity with effect of nutrient load may be due to carbon supply from different seaweeds (Bowen et al., 2009; Lee et al., 2015). Seaweed blooms are associated with nutrient load, which may affect benthic microbial community this study needs to explore yet (Aires et al., 2019). Realizing macroalgal bloom and high concentration of POPs study has been conducted by Cheney et al. (2019) on green tide occur every year at site of New Bedford Harbor, MA, United States; *Ulva rigida*, the green seaweeds were reported for accumulation of PCB with 95−99 mg/kg. Seaweeds involve in degradation of organic pollutants were discussed in many studies (Table 26.4).

TABLE 26.3 Microorganisms used for the bioremediation of organic pollutants from aquatic ecosystems.

Pollutants	Organisms	References
Petroleum hydrocarbons	*Methylococcus, Methylosinus, Methylocystis, Methylomonas, Methylocella, Aspergillus niger, Aspergillus fumigatus, Fusarium solani,* and *Penicillium funiculosum*	Das and Chandran (2011), Prince (1993), Swannell et al. (1996), Head and Swannell (1999), Head et al. (2006), Atlas (1995), Al-Jawhari (2014), and McDonald et al. (2006)
Organic matter	*Alcanivorax* spp., *Cycloclasticus* spp., *Oleiphilus* spp., *Oleispira* spp., *Chloroflexi/Firmicutes, Alphaproteobacteria, Gammaproteobacteria, Delta-proteobacteria, Actinobacteria,* and *Verrucomicrobia*	Dell'Anno et al. (2012), Beolchini et al. (2010), Kasai et al. (2002), and Danovaro et al. (2001)
PAHs	*Mycobacterium flavescens, Rhodococcus* species, and *Aeromonas* sp.	Dean-Ross et al. (2002), Wilson and Jones (1993), Cerniglia (1993), Bauer and Capone (1988), Tiehm and Fritzsche (1995), Shuttleworth and Cerniglia (1995), Hesham et al. (2012), and Kiyohara et al. (1976)
Aromatic hydrocarbons	*Rhodopseudomonas* spp.	Smith (1990)
Pyrene	*M. flavescens; Rhodococcus* sp.	Dean-Ross and Cerniglia (1996) and Walter et al. (1991)
Fluoranthene	*Mycobacterium* sp.	Mueller et al. (1989) and Kelley and Cerniglia (1991)
Anthracene	*Rhodococcus* species	Dean-Ross et al. (2001)
Anthracene and phenanthrene	*Pseudomonas fluorescens* 5R	Menn et al. (1993)
Aliphatic hydrocarbon	*Fusarium* sp. F 092	Hidayat and Tachibana (2012)
Textile azo dyes and textile dye (Remazol Black B)	*Micrococcus luteus, Listeria denitrificans, Nocardia atlantica,* and *Bacillus* spp.	Hassan et al. (2013), Shah et al. (2013), and Das et al. (2016)
Sulfonated di-azo dye Reactive Red HE8B	*Pseudomonas aeruginosa*	Patel and Gupte (2016)
Persistent pesticides	*Pseudomonas putida* sp. strain B1, *Acinetobacter* sp. strain B2, *Arthrobacter* sp. strain B3. *Acinetobacter radioresistens, Pseudomonas frederiksbergensis, Bacillus pumilus, Serratia liquefaciens, Serratia marcescens,* and *Burkholderia gladioli*	Ravi et al. (2015), Mónica et al. (2016), and Hussaini et al. (2013)

PAHs, Polycyclic aromatic hydrocarbons.

26.8 The seaweed and associated microflora

It has been already established that terrestrial and aquatic plants can cope up with organic pollutants. As reported previously, seaweeds are excellent bioindicators; they are also good source of two basic processes of remediation, bioaccumulations and biodegradation. Apart from these two processes, they also work with other associated microbes to provide the platform for aerobic and anaerobic oxidation (degradation) of organic pollutants. Several characteristics of seaweed such as tolerance to high nutrient levels; high tolerance to physical factors such as temperature, pH, light, and salinity; and diversity and high productivity support this function (Phang et al., 2015). Naturally, they are capable of combating with organic and inorganic pollutants but in some cases (like transgenic plants and acquired immunity); they are modified to show bioaccumulation and subsequent remediation of these pollutants. In general, such type of remediation is known as phytoremediation. It covers all the remediation techniques such as rhizoremediation, phyto-stabilization, phytoextraction, rhizofiltration, phyto-volatilization, phyto-degradation, and hydraulic control exhibited by terrestrial and aquatic plants (Singh et al., 2018). A number of reports have been published describing methods for the remediation of organic pollutants from contaminated soils, rivers, and lakes. Use of persistent pesticides and other chemicals will ultimately result in their accumulation in aquatic ecosystem. Therefore it is very important to employ green algae for the bioaccumulation/biodegradation of such POPs. The seaweed is probably the best candidate for the remediation of heavy metals as well as xenobiotic degradation (Chekroun et al., 2014). Before their recognition as POPs, a study was conducted with one such organic pollutants, naphthalene in which cyanobacteria and microalgae were employed to degrade the naphthalene under photoautotrophic conditions. Degradation of naphthalene was achieved through oxidation process governed by these

TABLE 26.4 Seaweeds studied for bioremediation of persistent organic pollutants.

Pollutants	Species	Function	References
Alkylphenol and alkylphenolpolyethoxylates	–	Degradation	Priac et al. (2017)
Polychlorinated biphenyls	*Ascophyllum nodosum, Fucus* sp., *Caepidium antarcticum*, and *Desmarestia* sp.	Bioremediation	Lara et al. (1989), Chekroun et al. (2014)
Organic pollutant	Green seaweeds	Bioremediation	Sivakumar et al. (2012)
PCBs, PAHs, and Me-PAHs	*Ulva lactuca*	Bioaccumulation	Net et al. (2015)
PAHs, PCBs, and pesticides	*Ulva* sp., *Gracilaria* sp., *Porphyra* sp., *Grateloupia* sp., *Undaria* sp., *Fucus* sp., and *Cystoseira* sp.	Bioaccumulation	Pavoni et al. (2003)
Phenol	*Lessoniani grescens* Bory and *Macrocystisin tegrifolia* Bory	Biosorption	Navarro et al. (2008)
Aromatic compounds	Microalgae	Biodegradation	Semple et al. (1999)
Lindane, naphthalene, phenol, and mirex	*Chlamydomonas* spp.	Bioaccumulation and biotransformation	Kobayashi and Rittmann (1982)
DDT, benzene, and chlorobenzene, phenanthrene	*Selenastrum capricornutum, Euglena gracilis, Scenedesmus obliquus*, and *Cylindrotheca* sp.	Bioaccumulation and biotransformation	Kobayashi and Rittmann (1982)
Phenylalanine	Planktonic algae	Biodegradation	Vose et al. (1971)
BaP	*S. capricornutum*	Biodegradation	Lindquist and Warshawsky (1985)
BaP PAHs	*S. capricornutum, Scenedesmus acutus*, and *Ankistrodesmus braunii*	Biotransformation	Warshawsky et al. (1995)
Bisphenol A	*Monoraphidium braunii*—fresh water algae	Remediation	Gattullo et al. (2012)
Phenanthrene and fluoranthene	*Skeletonema costatum* and *Nitzschia* sp.	Accumulation and biodegradation	Hong et al. (2008)
Prometryne	*Chlamydomonas reinhardtii*	Bioaccumulation and degradation	Jin et al. (2012)
Phenolics, hydrocarbons, pesticides, and biphenyls	Algae and cyanobacteria	Remediation	Subashchandrabose et al. (2013)
Fenamiphos	Green algae and cyanobacteria	Biodegradation	Cáceres et al. (2008)
Gamma-hexachlorocyclohexane and 4-chlorobenzoate	Filamentous cyanobacteria	Biodegradation	Kuritz and Wolk (1995)

BaP, Benzo(a)pyrene; *DDT*, dichlorodiphenyltrichloroethane; *PAHs*, polycyclic aromatic hydrocarbons; *PCBs*, polychlorinated biphenyls.

organisms (Cerniglia et al., 1980). Thus it is obvious that, like chemical and physical methods of remediation, biological method also comprises similar principle of oxidation, where the pollutants are being oxidized through enzymes or oxidizing agents generated by employed organisms. To treat the polluted area with conventional methods, the affected ecosystem generally does not favor for the remediation processes, on the other hand, biological agents such as seaweed and associated microbiomes can be easily employed for the remediation. The biochemical basis of algal–bacterial consortium uses the same such principle. Algae, upon illumination, conduct photosynthesis for the cheaper, safer, and reliable source of mechanical aeration. Associated aerobic bacteria then utilize this molecular oxygen and in return, they produce carbon dioxide (because of respiration), which is a readily available source algal photosynthesis (Munoz et al., 2003). Acquired immunity, on the other hand, can also be seen in some green algae. The most obvious effect of these pollutants on aquatic plants is the generation of reactive oxygen species (ROS) results in peroxide accumulation of DNA degradation. However, Zhang et al. (2011) studied the effect of fluroxypyr (herbicide) on the green alga *Chlamydomonas reinhardtii* and observed that the plant became adaptive to the exposure of fluroxypyr. In the range of 0.05–1.00 ppm treatment of pollutant, the plant accumulated and degraded this organic xenobiotic. Hong et al. (2008) reported the accumulation and biodegradation of two PAHs, phenanthrene and fluoranthene by two algae from mangrove aquatic ecosystem, *Skeletonema coastatum* and *Nitzschia* sp. They observed that *S. coastatum* was more resistant to PAHs exposure; however, fluoranthene was more recalcitrant than phenanthrene. Degradation of PAHs by microbial community may be classified through three different routes. First is the bacterial degradation in which hydroxylation of ring via dioxygenase enzymes takes place.

After that, through the action of bacterial dehydrogenase, catechol derivatives are produced, which undergo fission reactions to produce less toxic acid and aldehyde open-chain derivatives along with the release of carbon dioxide. Second pathway comprises ligninolytic fungal degradation by the action of lignin and manganese peroxidase enzymes to form quinone derivatives. Quinone rings are then subjected to fission reaction and same pathway as previous is followed. Lastly, there are several cytochrome P450 monooxygenase enzymes, which can degrade such pollutants. Action of these cytochrome P450s will lead to the production of certain oxides, which are then further oxidized to less harmful compounds. Under anaerobic condition, with the series of carboxylation, hydrogenation, and decarboxylation reactions, PAHs can be degraded. Despite this, there are several factors affecting degradation of PAHs such as temperature, pH, oxygen availability, nutrient availability, bioavailability, and toxicity of end products (Bamforth and Singleton, 2005). In some cases, oxidative stress imparted by POPs causes the production of oxidative stress–related enzymes such as arginine decarboxylase (ADC), *S*-adenosylmethionine decarboxylase (SAMDC) due to enzymatic and nonenzymatic production of antioxidants, polyamines. Subsequently, the plant gets protection against oxidative damage toward important cellular components such as plasma membrane and chloroplast. The role of polyamines is very well studied under oxidative damage. Burritt (2008) studied the effect of application of polyamines to PAHs-exposed aquatic plant *Riccia fluitans* L. After prolonged exposure of phenanthrene and simultaneous application of polyamines, the plant recovered from the oxidative damage, whereas untreated plants were died due to inhibition of synthesis of ADC and SAMDC. Several seaweed species used for the remediation of POPs are listed in Table 26.4.

26.9 Conclusion

Microorganisms and higher plants have been extensively used for the remediation of organic and inorganic contaminants (including heavy metals and metalloids). Among them, the seaweeds have higher biomass and cover large part of sea and coastal area. They have ample environment for the accumulation and degradation of toxic and xenobiotic compounds. As an individual, they produce certain enzymes, which help in oxidation of xenobiotic compounds, whereas, with the mutualistic relation with associated microbes, they supply oxygen to its symbioant for the aerobic degradation of pollutants and in return get carbon dioxide for the completion of photosynthesis process. Role of natural agents such as seaweeds and associated microbes as well as their utilization in wastewater treatment was found highly significant for the reduction of contaminant dumps in marine ecosystem. The overall environmental health can be achieved by improving the system to utilize the natural remedies. Being stable and persistent in many situations, POPs became a global issue because their production, transport, long-range travel, and accumulation cause serious hazards. Although there are plenty of techniques available for their removal, the task is still challenging especially for the aquatic sites. We must look at the naturally happening bioremediation process and must adapt it at certain level to mitigate and remediate these toxic substances. A detailed study on seaweeds and associated microbes mainly benthic organisms is at most needed so that we can develop it as model to remediate them not only in marine but also in other ecosystem. Bioremediation method can be improved by biotechnical and molecular approach to identify potential organisms, which withstand against POPs so that it can become more robust techniques than other invasive existing methods. Plenty of species of seaweed and many marine microorganisms associated with them are available to be employed for the remediation of POPs. These include *Methylococcus* sp., *Aspergillus* sp., *Fusarium solani*, *Penicillium funiculosum*, *Mycobacterium flavescens*, *Rhodococcus* sp., *Aeromonas* sp., as microbes and *Ascophyllum nodosum*, *Fucus* sp., *Caepidium antarcticum*, *Desmarestia* sp., *Ulva* sp., *Gracilaria* sp., *Porphyra* sp., *Grateloupia* sp., and *Undaria* sp. as seaweeds.

Remediation methods used so far with seaweeds and microbiomes may be improved. Knowing the fact that seaweeds are also good source of nutraceuticals and their use in food industries is increasing day by day, one has to think about the darker side of the coin that most of them are bioaccumulation and biodegradation agents and, thus, they are a source of organic pollutants at the time of harvest from natural resources. Thus strict regulatory protocol should be optimized before and after the use of seaweed in the food industries. Apart from that, for the remediation purpose, genetically modified microbes can be designed and applied on/off the site, depending on the regulatory norms for their usage and disposal.

References

Al-Jawhari, I.F.H., 2014. Ability of some soil fungi in biodegradation of petroleum hydrocarbon. J. Appl. Environ. Microbiol. 2 (2), 46–52.

Aires, T., Muyzer, G., Serrão, E.A., Engelen, A.H., 2019. Seaweed loads cause stronger bacterial community shifts in coastal lagoon sediments than nutrient loads. Front. Microbiol. 9, 3283.

Ashraf, M.A., 2017. Persistent organic pollutants (POPs): a global issue, a global challenge. Environ. Sci. Pollut. Res. 24, 4223–4227. doi: 10.1007/s11356-015-5225-9.

Atlas, R.M., 1995. Bioremediation of petroleum pollutants. Int. Biodeterior. Biodegrad. 35 (1–3), 317–327.

Augugliaro, V., Litter, M., Palmisano, L., Soria, J., 2006. The combination of heterogeneous photocatalysis with chemical and physical operations: a tool for improving the photoprocess performance. J. Photochem. Photobiol., C: Photochem. Rev. 7 (4), 127–144.

Azubuike, C.C., Chikere, C.B., Okpokwasili, G.C., 2016. Bioremediation techniques—classification based on site of application: principles, advantages, limitations and prospects. World J. Microbiol. Biotechnol. 32 (11), 180.

Bamforth, S.M., Singleton, I., 2005. Bioremediation of polycyclic aromatic hydrocarbons: current knowledge and future directions. J. Chem. Technol. Biotechnol. 80 (7), 723–736.

Bauer, J.E., Capone, D.G., 1988. Effects of co-occurring aromatic hydrocarbons on degradation of individual polycyclic aromatic hydrocarbons in marine sediment slurries. Appl. Environ. Microbiol. 54 (7), 1649–1655.

Beolchini, F., Rocchetti, L., Regoli, F., Dell'Anno, A., 2010. Bioremediation of marine sediments contaminated by hydrocarbons: experimental analysis and kinetic modeling. J. Hazard. Mater. 182 (1–3), 403–407.

Booij, K., Achterberg, E.P., Sundby, B., 1992. Release rates of chlorinated hydrocarbons from contaminated sediments. Neth. J. Sea Res. 29 (4), 297–310.

Bowen, J.L., Crump, B.C., Deegan, L.A., Hobbie, J.E., 2009. Salt marsh sediment bacteria: their distribution and response to external nutrient inputs. ISME J. 3 (8), 924–934.

Buccini, J., 2003. The development of a global treaty on persistent organic pollutants (POPs). Persistent Organic Pollutants. Springer, Berlin, Heidelberg, pp. 13–30.

Burritt, D.J., 2008. The polycyclic aromatic hydrocarbon phenanthrene causes oxidative stress and alters polyamine metabolism in the aquatic liverwort *Riccia fluitans* L. Plant, Cell Environ. 31 (10), 1416–1431.

Butt, C.M., Mabury, S.A., Muir, D.C., Braune, B.M., 2007. Prevalence of long-chained perfluorinated carboxylates in seabirds from the Canadian Arctic between 1975 and 2004. Environ. Sci. Technol. 41 (10), 3521–3528.

Cáceres, T.P., Megharaj, M., Naidu, R., 2008. Biodegradation of the pesticide fenamiphos by ten different species of green algae and cyanobacteria. Curr. Microbiol. 57 (6), 643–646.

Cerniglia, C.E., 1993. Biodegradation of polycyclic aromatic hydrocarbons. Curr. Opin. Biotechnol. 4 (3), 331–338.

Cerniglia, C.E., Gibson, D.T., Van Baalen, C., 1980. Oxidation of naphthalene by cyanobacteria and microalgae. Microbiology 116 (2), 495–500.

Chekroun, K.B., Sánchez, E., Baghour, M., 2014. The role of algae in bioremediation of organic pollutants. J. Issues ISSN 2360, 8803.

Chen, W., Brühlmann, F., Richins, R.D., Mulchandani, A., 1999. Engineering of improved microbes and enzymes for bioremediation. Curr. Opin. Biotechnol. 10 (2), 137–141.

Cheney, D., Logan, J.M., Gardner, K., Sly, E., Wysor, B., Greenwood, S., 2019. Bioaccumulation of PCBs by a seaweed bloom (*Ulva rigida*) and transfer to higher trophic levels in an estuarine food web. Mar. Ecol. Prog. Ser. 611, 75–93.

Colabuono, F.I., Taniguchi, S., Montone, R.C., 2010. Polychlorinated biphenyls and organochlorine pesticides in plastics ingested by seabirds. Mar. Pollut. Bull. 60 (4), 630–634.

Corsolini, S., Ademollo, N., Romeo, T., Greco, S., Focardi, S., 2005. Persistent organic pollutants in edible fish: a human and environmental health problem. Microchem. J. 79 (1–2), 115–123.

Curtis, T.P., Sloan, W.T., Scannell, J.W., 2002. Estimating prokaryotic diversity and its limits. Proc. Natl. Acad. Sci. U.S.A. 99 (16), 10494–10499.

Dachs, J., Méjanelle, L., 2010. Organic pollutants in coastal waters, sediments, and biota: a relevant driver for ecosystems during the Anthropocene? Estuaries Coasts 33 (1), 1–14.

Dale, W.E., Copeland, M.F., Hayes Jr, W.J., 1965. Chlorinated insecticides in the body fat of people in India. Bull. World Health Organ. 33 (4), 471.

Danovaro, R., Armeni, M., Dell'Anno, A., Fabiano, M., Manini, E., Marrale, D., et al., 2001. Small-scale distribution of bacteria, enzymatic activities, and organic matter in coastal sediments. Microb. Ecol. 177–185.

Das, A., Mishra, S., Verma, V.K., 2016. Enhanced biodecolorization of textile dye remazol navy blue using an isolated bacterial strain *Bacillus pumilus* HKG212 under improved culture conditions. J. Biochem. Technol. 6 (3), 962–969.

Das, N., Chandran, P., 2011. Microbial degradation of petroleum hydrocarbon contaminants: an overview. Biotechnol. Res. Int. 2011.

Dean-Ross, D., Cerniglia, C.E., 1996. Degradation of pyrene by *Mycobacterium flavescens*. Appl. Microbiol. Biotechnol. 46 (3), 307–312.

Dean-Ross, D., Moody, J., Cerniglia, C.E., 2002. Utilization of mixtures of polycyclic aromatic hydrocarbons by bacteria isolated from contaminated sediment. FEMS Microbiol. Ecol. 41 (1), 1–7.

Dean-Ross, D., Moody, J.D., Freeman, J.P., Doerge, D.R., Cerniglia, C.E., 2001. Metabolism of anthracene by a *Rhodococcus* species. FEMS Microbiol. Lett. 204 (1), 205–211.

Dell'Anno, A., Beolchini, F., Rocchetti, L., Luna, G.M., Danovaro, R., 2012. High bacterial biodiversity increases degradation performance of hydrocarbons during bioremediation of contaminated harbor marine sediments. Environ. Pollut. 167, 85–92.

Dominguez-Ramos, A., Irabien, A., 2013. Analysis and modeling of the continuous electro-oxidation process for organic matter removal in urban wastewater treatment. Ind. Eng. Chem. Res. 52 (22), 7534–7540.

El-Dib, M.A., Badawy, M.I., 1985. Organochlorine insecticides and PCBs in water, sediment, and fish from the Mediterranean Sea. Bull. Environ. Contam. Toxicol. (US) 34 (2).

Endo, S., Takizawa, R., Okuda, K., Takada, H., Chiba, K., Kanehiro, H., et al., 2005. Concentration of polychlorinated biphenyls (PCBs) in beached resin pellets: variability among individual particles and regional differences. Mar. Pollut. Bull. 50 (10), 1103–1114.

Fiedler, H., 2008. Stockholm convention on POPs: obligations and implementation. The Fate of Persistent Organic Pollutants in the Environment. Springer, Dordrecht, pp. 3–12.

Gan, S., Ng, H.K., 2012. Current status and prospects of Fenton oxidation for the decontamination of persistent organic pollutants (POPs) in soils. Chem. Eng. J. 213, 295–317.

Gattullo, C.E., Bährs, H., Steinberg, C.E., Loffredo, E., 2012. Removal of bisphenol A by the freshwater green alga *Monoraphidium braunii* and the role of natural organic matter. Sci. Total Environ. 416, 501–506.

Gilman, A., Dewailly, E., Feeley, M., Jerome, V., Kuhnlein, H., Kwavnick, B., et al., 1997. Chapter 4: Human health. In: Jensen, J., Adhare, K., Shearer, R. (Eds.), Canadian Arctic Contaminants Assessment Report. Department of Indian and Northern Affairs, Ottawa, ON, pp. 295–377.

Gioia, R., Dachs, J., Nizzetto, L., Berrojalbiz, N., Galbán, C., Del Vento, S., et al., 2011. Sources, transport and fate of organic pollutants in the oceanic environment. Persistent Pollution–Past, Present and Future. Springer, Berlin, Heidelberg, pp. 111–139.

Gioia, R., Nizzetto, L., Lohmann, R., Dachs, J., Temme, C., Jones, K.C., 2008. Polychlorinated biphenyls (PCBs) in air and seawater of the Atlantic Ocean: sources, trends and processes. Environ. Sci. Technol. 42 (5), 1416–1422.

Gioia, R., Steinnes, E., Thomas, G.O., Mejier, S.N., Jones, K.C., 2006. Persistent organic pollutants in European background air: derivation of temporal and latitudinal trends. J. Environ. Monit. 8 (7), 700–710.

Gupta, A., Singh, S.K., Singh, V.K., Singh, M.K., Modi, A., Zhimo, V.Y., et al., 2020. Endophytic microbe approaches in bioremediation of organic pollutants. In: Kumar, A., Singh, V.K. (Eds.), Microbial Endophytes: Prospects of Sustainable Agriculture. Elsevier, pp. 157–174. , Pp.

Hassan, M.M., Alam, M.Z., Anwar, M.N., 2013. Biodegradation of textile azo dyes by bacteria isolated from dyeing industry effluent. Int. Res. J. Biol. Sci. 2 (8), 27–31.

Head, I.M., Swannell, R.P., 1999. Bioremediation of petroleum hydrocarbon contaminants in marine habitats. Curr. Opin. Biotechnol. 10 (3), 234–239.

Head, I.M., Jones, D.M., Röling, W.F., 2006. Marine microorganisms make a meal of oil. Nat. Rev. Microbiol. 4 (3), 173–182.

Hesham, A.E.L., Khan, S., Tao, Y., Li, D., Zhang, Y., Yang, M., 2012. Biodegradation of high molecular weight PAHs using isolated yeast mixtures: application of meta-genomic methods for community structure analyses. Environ. Sci. Pollut. Res. 19 (8), 3568–3578.

Hidayat, A., Tachibana, S., 2012. Biodegradation of aliphatic hydrocarbon in three types of crude oil by *Fusarium* sp. F092 under stress with artificial sea water. J. Environ. Sci. Technol. 5 (1), 64–73.

Himmelheber, D.W., Pennell, K.D., Hughes, J.B., 2007. Natural attenuation processes during in situ capping. Environ. Sci. Technol. 41 (15), 5306–5313.

Hong, Y.W., Yuan, D.X., Lin, Q.M., Yang, T.L., 2008. Accumulation and biodegradation of phenanthrene and fluoranthene by the algae enriched from a mangrove aquatic ecosystem. Mar. Pollut. Bull. 56 (8), 1400–1405.

Hussaini, S.Z., Shaker, M., Iqbal, M.A., 2013. Isolation of bacterial for degradation of selected pesticides. Adv. Biores. 4 (3), 82–85.

IARC (International Agency for Research on Cancer), 1983. IARC Monographs on the Evaluation of the Carcinogenic Risk of Chemicals to Humans: Polynuclear Aromatic Compounds Part I. IARC Press, Lyon.

Ikehata, K., Gamal El-Din, M., Snyder, S.A., 2008. Ozonation and advanced oxidation treatment of emerging organic pollutants in water and wastewater. Ozone: Sci. Eng. 30 (1), 21–26.

Ilisz, I., Dombi, A., Mogyorósi, K., Farkas, A., Dékány, I., 2002. Removal of 2-chlorophenol from water by adsorption combined with TiO_2 photocatalysis. Appl. Catal., B: Environ. 39 (3), 247–256.

Jain, R., Mathur, M., Sikarwar, S., Mittal, A., 2007. Removal of the hazardous dye rhodamine B through photocatalytic and adsorption treatments. J. Environ. Manage. 85 (4), 956–964.

Jaward, F.M., Meijer, S.N., Steinnes, E., Thomas, G.O., Jones, K.C., 2004. Further studies on the latitudinal and temporal trends of persistent organic pollutants in Norwegian and UK background air. Environ. Sci. Technol. 38 (9), 2523–2530.

Jin, Z.P., Luo, K., Zhang, S., Zheng, Q., Yang, H., 2012. Bioaccumulation and catabolism of prometryne in green algae. Chemosphere 87 (3), 278–284.

Jones, K.C., De Voogt, P., 1999. Persistent organic pollutants (POPs): state of the science. Environ. Pollut. 100 (1–3), 209–221.

Kaiser, J., 2000. Just how bad is dioxin? Science 288 (5473), 1941–1944.

Karapanagioti, H.K., Klontza, I., 2007. Investigating the properties of plastic resin pellets found in the coastal areas of Lesvos Island. Global NEST Int. J. 9 (1), 71–76.

Karapanagioti, H.K., Klontza, I., 2008. Testing phenanthrene distribution properties of virgin plastic pellets and plastic eroded pellets found on Lesvos island beaches (Greece). Mar. Environ. Res. 65 (4), 283–290.

Kasai, Y., Kishira, H., Harayama, S., 2002. Bacteria belonging to the genus *Cycloclasticus* play a primary role in the degradation of aromatic hydrocarbons released in a marine environment. Appl. Environ. Microbiol. 68 (11), 5625–5633.

Kelley, I., Cerniglia, C.E., 1991. The metabolism of fluoranthene by a species ofMycobacterium. J. Ind. Microbiol. 7 (1), 19–26.

Kim, G.B., Maruya, K.A., Lee, R.F., Lee, J.H., Koh, C.H., Tanabe, S., 1999. Distribution and sources of polycyclic aromatic hydrocarbons in sediments from Kyeonggi Bay, Korea. Mar. Pollut. Bull. 38 (1), 7–15.

Kiyohara, H., Nagao, K., Nomi, R., 1976. Degradation of phenanthrene through *o*-phthalate by an *Aeromonas* sp. Agric. Biol. Chem. 40 (6), 1075–1082.

Kobayashi, H., Rittmann, B.E., 1982. Microbial removal of hazardous organic compounds. Environ. Sci. Technol. 16 (3), 170A–183A.

Kuiper, I., Lagendijk, E.L., Bloemberg, G.V., Lugtenberg, B.J., 2004. Rhizoremediation: a beneficial plant-microbe interaction. Mol. Plant-Microbe Interact. 17 (1), 6–15.

Kuritz, T., Wolk, C.P., 1995. Use of filamentous cyanobacteria for biodegradation of organic pollutants. Appl. Environ. Microbiol. 61 (1), 234–238.

Lal, R., Saxena, D.M., 1982. Accumulation, metabolism, and effects of organochlorine insecticides on microorganisms. Microbiol. Rev. 46 (1), 95.

Lara, R.J., Wiencke, C., Ernst, W., 1989. Association between exudates of brown algae and polychlorinated biphenyls. J. Appl. Phycol. 1 (3), 267.

Lee, J.W., Choi, S.P., Thiruvenkatachari, R., Shim, W.G., Moon, H., 2006. Evaluation of the performance of adsorption and coagulation processes for the maximum removal of reactive dyes. Dye Pigment. 69 (3), 196–203.

Lee, P.O., McLellan, S.L., Graham, L.E., Young, E.B., 2015. Invasive dreissenid mussels and benthic algae in Lake Michigan: characterizing effects on sediment bacterial communities. FEMS Microbiol. Ecol. 91 (1), 1–12.

Lindquist, B., Warshawsky, D., 1985. Identification of the 11, 12-dihydro-11, 12-dihydroxybenzo(*a*)pyrene as a major metabolite produced by the green alga, *Selenastrum capricornutum*. Biochem. Biophys. Res. Commun. 130 (1), 71–75.

Liu, Z., Hosseinzadeh, S., Wardenier, N., Verheust, Y., Chys, M., Hulle, S.V., 2019. Combining ozone with UV and H_2O_2 for the degradation of micropollutants from different origins: lab-scale analysis and optimization. Environ. Technol. 40 (28), 3773–3782.

Machulek Jr, A., Quina, F.H., Gozzi, F., Silva, V.O., Friedrich, L.C., Moraes, J.E.F., 2012. Fundamental mechanistic studies of the photo-Fenton reaction for the degradation of organic pollutants. Organic Pollutants Ten Years After the Stockholm Convention – Environmental and Analytical Update. IntechOpen, pp. 271–292.

Martinez-Huitle, C.A., Ferro, S., 2006. Electrochemical oxidation of organic pollutants for the wastewater treatment: direct and indirect processes. Chem. Soc. Rev. 35 (12), 1324–1340.

Martínez-Jerónimo, F., Cruz-Cisneros, J.L., García-Hernández, L., 2008. A comparison of the response of *Simocephalus mixtus* (Cladocera) and *Daphnia magna* to contaminated freshwater sediments. Ecotoxicol. Environ. Saf. 71 (1), 26–31.

Matafonova, G., Batoev, V., 2018. Recent advances in application of UV light-emitting diodes for degrading organic pollutants in water through advanced oxidation processes: a review. Water Res. 132, 177–189.

Mato, Y., Isobe, T., Takada, H., Kanehiro, H., Ohtake, C., Kaminuma, T., 2001. Plastic resin pellets as a transport medium for toxic chemicals in the marine environment. Environ. Sci. Technol. 35 (2), 318–324.

McDonald, I.R., Miguez, C.B., Rogge, G., Bourque, D., Wendlandt, K.D., Groleau, D., et al., 2006. Diversity of soluble methane monooxygenase-containing methanotrophs isolated from polluted environments. FEMS Microbiol. Lett. 255 (2), 225–232.

Meagher, R.B., 2000. Phytoremediation of toxic elemental and organic pollutants. Curr. Opin. Plant Biol. 3 (2), 153–162.

Menn, F.M., Applegate, B.M., Sayler, G.S., 1993. NAH plasmid-mediated catabolism of anthracene and phenanthrene to naphthoic acids. Appl. Environ. Microbiol. 59 (6), 1938–1942.

Messina, P.V., Schulz, P.C., 2006. Adsorption of reactive dyes on titania–silica mesoporous materials. J. Colloid Interface Sci. 299 (1), 305–320.

Mishra, V., Lal, R., Srinivasan, 2001. Enzymes and operons mediating xenobiotic degradation in bacteria. Crit. Rev. Microbiol. 27 (2), 133–166.

Mónica, P., Darwin, R.O., Manjunatha, B., Zúñiga, J.J., Diego, R., Bryan, R.B., et al., 2016. Evaluation of various pesticides-degrading pure bacterial cultures isolated from pesticide-contaminated soils in Ecuador. Afr. J. Biotechnol. 15 (40), 2224–2233.

Mueller, J.G., Chapman, P.J., Pritchard, P.H., 1989. Action of a fluoranthene-utilizing bacterial community on polycyclic aromatic hydrocarbon components of creosote. Appl. Environ. Microbiol. 55 (12), 3085–3090.

Muir, D.C., Howard, P.H., 2006. Are there other persistent organic pollutants? A challenge for environmental chemists. Environ. Sci. Technol. 40 (23), 7157–7166.

Munoz, R., Guieysse, B., Mattiasson, B., 2003. Phenanthrene biodegradation by an algal-bacterial consortium in two-phase partitioning bioreactors. Appl. Microbiol. Biotechnol. 61 (3), 261–267.

Navarro, A.E., Portales, R.F., Sun-Kou, M.R., Llanos, B.P., 2008. Effect of pH on phenol biosorption by marine seaweeds. J. Hazard. Mater. 156 (1–3), 405–411.

Net, S., Henry, F., Rabodonirina, S., Diop, M., Merhaby, D., Mahfouz, C., et al., 2015. Accumulation of PAHs, Me-PAHs, PCBs and total mercury in sediments and marine species in coastal areas of Dakar, Senegal: contamination level and impact. Int. J. Environ. Res. 9 (2).

Ogata, Y., Takada, H., Mizukawa, K., Hirai, H., Iwasa, S., Endo, S., et al., 2009. International pellet watch: global monitoring of persistent organic pollutants (POPs) in coastal waters. 1. Initial phase data on PCBs, DDTs, and HCHs. Mar. Pollut. Bull. 58 (10), 1437–1446.

Papić, S., Koprivanac, N., Božić, A.L., Meteš, A., 2004. Removal of some reactive dyes from synthetic wastewater by combined Al(III) coagulation/carbon adsorption process. Dyes Pigm. 62 (3), 291–298.

Patel, Y., Gupte, A., 2016. Evaluation of bioremediation potential of isolated bacterial culture YPAG-9 (*Pseudomonas aeruginosa*) for decolorization of sulfonated di-azodye Reactive red HE8B under optimized culture conditions. Int. J. Curr. Microbiol. Appl. Sci. 5 (8), 258–272.

Patil, K.C., Matsumura, F., Boush, G.M., 1972. Metabolic transformation of DDT, dieldrin, aldrin, and endrin by marine microorganisms. Environ. Sci. Technol. 6 (7), 629–632.

Paul, D., Pandey, G., Pandey, J., Jain, R.K., 2005. Accessing microbial diversity for bioremediation and environmental restoration. TRENDS Biotechnol. 23 (3), 135–142.

Pavoni, B., Caliceti, M., Sperni, L., Sfriso, A., 2003. Organic micropollutants (PAHs, PCBs, pesticides) in seaweeds of the lagoon of Venice. Oceanol. Acta 26 (5–6), 585–596.

Phang, S.M., Chu, W.L., Rabiei, R., 2015. Phycoremediation. In: Sahoo, D., Seckbach, J. (Eds.), The Algae World. Cellular Origin, Life in Extreme Habitats and Astrobiology, vol. 26. Springer, Dordrecht.

Priac, A., Morin-Crini, N., Druart, C., Gavoille, S., Bradu, C., Lagarrigue, C., et al., 2017. Alkylphenol and alkylphenolpolyethoxylates in water and wastewater: a review of options for their elimination. Arab. J. Chem. 10, S3749-S3773.

Prince, R.C., 1993. Petroleum spill bioremediation in marine environments. Crit. Rev. Microbiol. 19 (4), 217–240.

Prodanović, J.M., Vasić, V.M., 2013. Application of membrane processes for distillery wastewater purification—a review. Desalin. Water Treat. 51 (16–18), 3325–3334.

Rafatullah, M., Sulaiman, O., Hashim, R., Ahmad, A., 2010. Adsorption of methylene blue on low-cost adsorbents: a review. J. Hazard. Mater. 177 (1–3), 70–80.

Ravi, R.K., Pathak, B., Fulekar, M.H., 2015. Bioremediation of persistent pesticides in rice field soil environment using surface soil treatment reactor. Int. J. Curr. Microbiol. Appl. Sci. 4 (2), 359–369.

Readman, J.W., Fillmann, G., Tolosa, I., Bartocci, J., Villeneuve, J.P., Catinni, C., et al., 2002. Petroleum and PAH contamination of the Black Sea. Mar. Pollut. Bull. 44 (1), 48–62.

Sanz-Lázaro, C., Valdemarsen, T., Holmer, M., 2015. Effects of temperature and organic pollution on nutrient cycling in marine sediments. Biogeosciences 12 (15), 4565–4575.

Scragg, A.H., 2005. Environmental Biotechnology. Oxford University Press, New York.

Semple, K.T., Cain, R.B., Schmidt, S., 1999. Biodegradation of aromatic compounds by microalgae. FEMS Microbiol. Lett. 170 (2), 291–300.

Shah, M.P., Patel, K.A., Nair, S.S., Darji, A.M., 2013. Microbial degradation of textile dye (Remazol Black B) by *Bacillus* spp. ETL-2012. J Bioremed. Biodeg. 4 (2), 1–5.

Shailaja, M.S., Nair, M., 1997. Seasonal differences in organochlorine pesticide concentrations of zooplankton and fish in the Arabian Sea. Mar. Environ. Res. 44 (3), 263–274.

Shuttleworth, K.L., Cerniglia, E., 1995. Environmental aspects of PAH biodegradation. Appl. Biochem. Biotechnol. 54 (1–3), 291–302.

Siddiqui, M.K.J., Saxena, M.C., 1985. Placenta and milk as excretory routes of lipophilic pesticides in women. Hum. Toxicol. 4 (3), 249–254.

Siddiqui, M.K.J., Saxena, M.C., Krishna Murti, C.R., 1981. Storage of DDT and BHC in adipose tissue of Indian males. Int. J. Environ. Anal. Chem. 10 (3–4), 197–204.

Simonenko, E., Gomonov, A., Rolle, N., Molodkina, L., 2015. Modeling of H_2O_2 and UV oxidation of organic pollutants at wastewater post-treatment. Procedia Eng. 117, 337–344.

Singh, A., 2004. Biodegradation and Bioremediation, vol. 2. Springer Science & Business Media.

Singh, A., Parmar, N., Kuhad, R.C., Ward, O.P., 2011. Bioaugmentation, biostimulation, and biocontrol in soil biology. Bioaugmentation, Biostimulation and Biocontrol. Springer, Berlin, Heidelberg, pp. 1–23.

Singh, P., Singh, V.K., Singh, R., Borthakur, A., Madhav, S., Ahamad, A., et al., 2020. Bioremediation: a sustainable approach for management of environmental contaminants. In: Singh, P., Kumar, A., Borthakur, A. (Eds.), Abatement of Environmental Pollutants. Elsevier, pp. 1–23.

Singh, V.K., Singh, A.L., Singh, R., Kumar, A., 2018. Iron oxidizing bacteria: insights on diversity, mechanism of iron oxidation and role in management of metal pollution. Environ. Sustain. 1 (3), 221–231.

Sinkko, H., Lukkari, K., Sihvonen, L.M., Sivonen, K., Leivuori, M., Rantanen, M., et al., 2013. Bacteria contribute to sediment nutrient release and reflect progressed eutrophication-driven hypoxia in an organic-rich continental sea. PLoS One 8 (6).

Sivakumar, G., Xu, J., Thompson, R.W., Yang, Y., Randol-Smith, P., Weathers, P.J., 2012. Integrated green algal technology for bioremediation and biofuel. Bioresour. Technol. 107, 1–9.

Smith, M.R., 1990. The biodegradation of aromatic hydrocarbons by bacteria. Biodegradation 1 (2–3), 191–206.

Spivak, A.C., Ossolinski, J., 2016. Limited effects of nutrient enrichment on bacterial carbon sources in salt marsh tidal creek sediments. Mar. Ecol. Prog. Ser. 544, 107–130.

Stock, N.L., Peller, J., Vinodgopal, K., Kamat, P.V., 2000. Combinative sonolysis and photocatalysis for textile dye degradation. Environ. Sci. Technol. 34 (9), 1747–1750.

Stroo, H.F., Leeson, A., Ward, C.H. (Eds.), 2012. Bioaugmentation for Groundwater Remediation, vol. 5. Springer Science & Business Media.

Subashchandrabose, S.R., Ramakrishnan, B., Megharaj, M., Venkateswarlu, K., Naidu, R., 2013. Mixotrophic cyanobacteria and microalgae as distinctive biological agents for organic pollutant degradation. Environ. Int. 51, 59–72.

Swannell, R.P., Lee, K., McDonagh, M., 1996. Field evaluations of marine oil spill bioremediation. Microbiol. Mol. Biol. Rev. 60 (2), 342–365.

Tanabe, S., Gondaira, F., Subramanian, A., Ramesh, A., Mohan, D., Kumaran, P., et al., 1990. Specific pattern of persistent organochlorine residues in human breast milk from South India. J. Agric. Food Chem. 38 (3), 899–903.

Tanabe, S., Senthilkumar, K., Kannan, K., Subramanian, A.N., 1998. Accumulation features of polychlorinated biphenyls and organochlorine pesticides in resident and migratory birds from South India. Arch. Environ. Contam. Toxicol. 34 (4), 387–397.

Teuten, E.L., Rowland, S.J., Galloway, T.S., Thompson, R.C., 2007. Potential for plastics to transport hydrophobic contaminants. Environ. Sci. Technol. 41 (22), 7759–7764.

Thorne, B.M., Stryker, J.P. 2018. The "Dirty Dozen" tax scams plus 1. Shippensburg University, ABD Journal, 7 (1), pp.1–22.

Tiehm, A., Fritzsche, C., 1995. Utilization of solubilized and crystalline mixtures of polycyclic aromatic hydrocarbons by a *Mycobacterium* sp. Appl. Microbiol. Biotechnol. 42 (6), 964–968.

Uchimiya, M., Masunaga, S., 2007. Time trends in sources and dechlorination pathways of dioxins in agrochemically contaminated sediments. Environ. Sci. Technol. 41 (8), 2703–2710.

Vecitis, C.D., Wang, Y., Cheng, J., Park, H., Mader, B.T., Hoffmann, M.R., 2010. Sonochemical degradation of perfluorooctanesulfonate in aqueous film-forming foams. Environ. Sci. Technol. 44 (1), 432–438.

Viganò, L., 2000. Assessment of the toxicity of River Po sediments with *Ceriodaphnia dubia*. Aquat. Toxicol. 47 (3–4), 191–202.

Vose, J.R., Cheng, J.Y., Antia, N.J., Towers, G.H.N., 1971. The catabolic fission of the aromatic ring of phenylalanine by marine planktonic algae. Can. J. Bot. 49 (2), 259–261.

Walter, U., Beyer, M., Klein, J., Rehm, H.J., 1991. Degradation of pyrene by *Rhodococcus* sp. UW1. Appl. Microbiol. Biotechnol. 34 (5), 671–676.

Warshawsky, D., Cody, T., Radike, M., Reilman, R., Schumann, B., LaDow, K., et al., 1995. Biotransformation of benzo[*a*]pyrene and other polycyclic aromatic hydrocarbons and heterocyclic analogs by several green algae and other algal species under gold and white light. Chem. Biol. Interact. 97 (2), 131–148.

Watanabe, K., 2001. Microorganisms relevant to bioremediation. Curr. Opin. Biotechnol. 12 (3), 237–241.

Wilson, S.C., Jones, K.C., 1993. Bioremediation of soil contaminated with polynuclear aromatic hydrocarbons (PAHs): a review. Environ. Pollut. 81 (3), 229–249.

Xiong, J., Ye, X., Wang, K., Chen, H., Hu, C., Zhu, J., et al., 2014. Biogeography of the sediment bacterial community responds to a nitrogen pollution gradient in the East China Sea. Appl. Environ. Microbiol. 80 (6), 1919–1925.

Yamashita, N., Taniyasu, S., Petrick, G., Wei, S., Gamo, T., Lam, P.K., et al., 2008. Perfluorinated acids as novel chemical tracers of global circulation of ocean waters. Chemosphere 70 (7), 1247–1255.

Zhang, S., Qiu, C.B., Zhou, Y., Jin, Z.P., Yang, H., 2011. Bioaccumulation and degradation of pesticide fluroxypyr are associated with toxic tolerance in green alga *Chlamydomonas reinhardtii*. Ecotoxicology 20 (2), 337–347.

Zhu, M.X., Lee, L., Wang, H.H., Wang, Z., 2007. Removal of an anionic dye by adsorption/precipitation processes using alkaline white mud. J. Hazard. Mater. 149 (3), 735–741.

Zoumis, T., Schmidt, A., Grigorova, L., Calmano, W., 2001. Contaminants in sediments: remobilisation and demobilisation. Sci. Total Environ. 266 (1–3), 195–202.

Further reading

Baghour, M., 2019. Algal degradation of organic pollutants. In: Martínez, L., Kharissova, O., Kharisov, B. (Eds.), Handbook of Ecomaterials. Springer, Cham.

Chapter 27

Sources, fate, and treatment of polycyclic aromatic hydrocarbons from the polluted environment

Akanksha Gupta, Anubhuti Singh and Virendra Kumar Mishra
Institute of Environment and Sustainable Development, Banaras Hindu University, Varanasi, India

27.1 Introduction—what are polycyclic aromatic hydrocarbons

The entry of any matter in our environment which can hamper the functioning of the ecosystem and its elements, for example, animals and soil, is called pollutants. These pollutants can be in the form of chemicals or energy; these can be already present in the atmosphere from an early time and due to anthropogenic or natural activities their percentage can rise up (Scullion, 2006). In the category of chemicals there are many organic (polyaromatic hydrocarbons, explosives) and inorganic (Hg, Pb, and Cd), natural (fly ash) or human-made (plastic waste), biodegradable (insecticides, chemical fertilizers), or nonbiodegradable ones (metallic compounds) (Singh et al., 2020; Gupta et al., 2018).

27.1.1 Organic pollutants

The term "polycyclic aromatic hydrocarbon" (PAH) is derived from the structure of these compounds, which contains aromatic rings of carbon and hydrogen (Di Toro et al., 2000; Arey and Atkinson, 2003a,b). PAHs are pervasive, generally colorless, noxious, and are found in mixtures. Compounds that contains up to six benzene rings are called small PAHs, whereas compounds that consist more than six polycyclic rings are tabulated in large PAHs category (Abdel-Shafy and Mansour, 2016). Anthracene and phenanthrene are made up of only three benzene rings, so they are the smallest of all PAHs (Fernández-Luqueño et al., 2011). A total of 17 PAHs (acenaphthene, anthracene, acenaphthylene, benz[*a*]anthracene, benzo[*b*]fluoranthene, benzo[*k*]fluoranthene, benzo[*g,h,i*]perylene, benzo[*a*]pyrene, chrysene, dibenz [*a,h*]anthracene, fluorene, fluoranthene, indeno[1,2,3-*c,d*]pyrene, naphthalene, naphthylamine, pyrene, and phenanthrene) are suggested by the environmental protection agency (EPA) to be removed from the environment immediately; around 15 PAHs are well known for their carcinogenic nature in that list (Liu et al., 2001).

27.2 Sources, levels, and fate of polycyclic aromatic hydrocarbons in the environment

PAHs are released from three chief sources to the environment which are biological, pyrogenic, and petrogenic. Biological sources are very less known. PAHs can be produced by some plants/microbes or during some biotic activities like the breakdown of litter. Pyrogenic as the name suggests are formed due to incomplete combustion of organic products in less O_2 concentration. Mainly pyrogenic products are formed at high heat (350°C–1200°C), but there is an exception, crude oil possesses PAHs that are formed at a low temperature around 100°C–150°C, over a very long era. When this crude oil is gone through maturation, it produces petrogenic PAHs. They are found very easily due to the global use of crude oil and its products. Transportation leakage and oil spills are part of petrogenic sources (Fig. 27.1). PAHs can be produced naturally (volcanic activities, corrosion of sedimentary rocks enclosing petroleum hydrocarbons, forest fire) as well as by human activities such as incineration, vehicular exhaust, and cigarette smoking (Fig. 27.2) (Abdel-Shafy and Mansour, 2016).

Microbe Mediated Remediation of Environmental Contaminants. DOI: https://doi.org/10.1016/B978-0-12-821199-1.00027-4

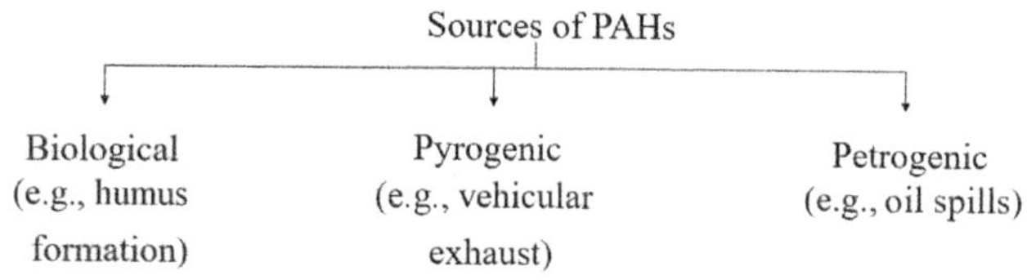

FIGURE 27.1 Classification of the sources of PAHs in the environment (on the basis of origin). *PAH*, Polycyclic aromatic hydrocarbon.

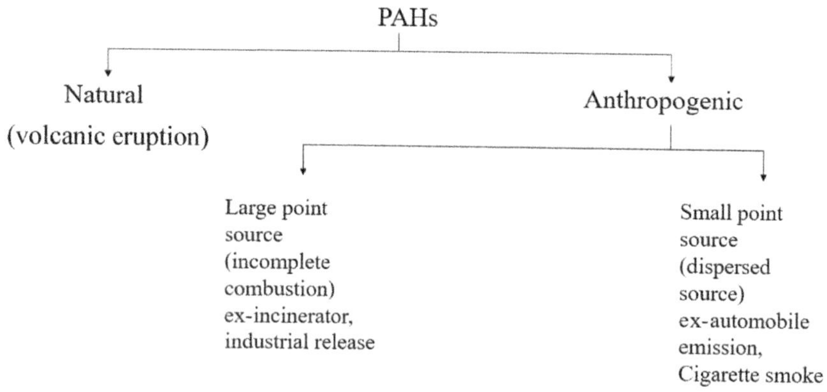

FIGURE 27.2 Origin of PAHs based on the release points in the ecosystem (on the basis of release). *PAH*, Polycyclic aromatic hydrocarbon.

According to Johnsen et al. (2005), PAHs present in the atmosphere can be introduced to soil with rainfall, and from water bodies, this pollutant can settle down to the sediments. So here, soil acts as a bed for PAHs. Nearby sources of PAHs can be of emission from automobiles or air can carry PAHs from more distant places which can be wet or dry deposited; dumping of waste is also a big cause of the release of polyaromatic hydrocarbons in soil (Masih and Taneja, 2006). Cachada et al. (2012) have revealed that when these polyaromatic hydrocarbons are leaked from the dumped waste, they are in mobile form. To transform them into an immobile form it has to bind with soil particles. The size of soil particles and space among them are the factors influencing this phenomenon (Riccardi et al., 2013). After sticking to the soil particles PAHs cannot move freely, their movement will be depends on soil particle movement; the affinity of sticking to the soil particles are reliant on both PAHs as well as characters of soil such as organic carbon, conductivity, and K_{ow} (EPRI, 2000). An increase in octanol–water partitioning coefficient (K_{ow}) is proportional to the lowering in solubility of PAHs in water (Schwarzenbach et al., 2003). Table 27.1 contains K_{ow} of 17 PAHs listed by EPA (Table 27.1).

From time to time, several studies have reported different levels of PAH, that is, Gupta and Kumar (2020) have monitored levels of diverse PAH compounds in soils at a number of places in Delhi, the capital of India where they recorded average values of 1714 μg/kg during before winter seasons and 1909 μg/kg in after winter seasons. Table 27.2 shows levels of PAHs in the soil at different places.

27.2.1 Fate of polycyclic aromatic hydrocarbons in environment

Most of the PAHs are soluble in fat substances, so they have the tendency of biomagnification (Van der Oost et al., 2003). Due to this property, proper management should be taken care of for their releases; otherwise, fortuitous discharge of PAHs can create a problem for hundreds of years to dissolve (Chaineau et al., 2005). After entering the environment PAHs persist for a long duration (Abdel-Shafy and Mansour, 2016). The half-life of PAH molecule varies from around minimum of 13 h (naphthylamine) to several years (pyrene and benzo[*b*]fluoranthene) (Coover and Sims, 1987; NCBI) (Table 27.3).

After aging for their half-life span, these compounds are subjected to degradation in the matrix. Some go under photo-degradation, some are degraded by UV induced oxidation, and most of the compounds become a substrate for microbial degradation and eventually get mineralized (Samanta et al., 2002) (Table 27.4).

Fig. 27.3 shows the environmental cycle of PAH starting from its release point to its fate in the environment.

TABLE 27.1 Octanol–water partition coefficient of polycyclic aromatic hydrocarbons (PAHs).

PAHs	Octanol–water partition coefficient (K_{ow})	References
Anthracene	4.63	Ma et al. (2009)
Acenaphthene	3.97	
Acenaphthylene	3.80	
Benz[a]anthracene	5.83	
Benzo[*b*]fluoranthene	5.83	
Benzo[*k*]fluoranthene	5.85	
Benzo[g,h,i]perylene	6.60	NCBI
Benzo[a]pyrene	5.99	Ma et al. (2009)
Chrysene	5.67	
Dibenz[*a,h*]anthracene	6.50	
Fluorene	4.14	
Fluoranthene	4.98	
Indeno[l,2,3-*c,d*]pyrene	6.53	
Naphthalene	3.39	
Naphthylamine	3.39	NCBI
Pyrene	5.06	Ma et al. (2009)
Phenanthrene	4.49	

TABLE 27.2 Containing levels of polycyclic aromatic hydrocarbons (PAHs) recorded at various places.

PAH	Level (ng/g)	References
Soil across lake Son-Kul of Tian Shan Mountain (Kyrgyztan, Central Asia)	54.5–428	Li et al. (2020)
Surface soils from Dajiuhu Sub-alpine Wetland, central China	42.4 ± 38.5	Xing et al. (2020)
Urban soil of Delhi, India Pre winter Post winter	956–2348 1230–2714	Gupta and Kumar (2020)
Northern Bavaria forest soil, Germany	24–15,056	Marx et al. (2019)
Ahvaz metropolis urban soil	1732.89	Mohit et al. (2019)
Istanbul, Turkey	684	Cetin et al. (2017a)
Kocaeli, Turkey	1323	Cetin et al. (2017b)
Aliaga, Turkey	4628	Bozlaker et al. (2008)
Greece	1475	Chrysikou et al. (2008)
Bayreuth, Germany	1475	Krauss and Wilcke (2003)
Hanpu, China	233,678	Zhang et al. (2013)
Dilovası, Turkey	≤6726	Cetin (2016)
Scotland	12560	Rhind et al. (2013)
Beijing, China	13,141	Peng et al. (2011)
Glasgow, Europe	1487–51,822	Morillo et al. (2007)
New Orleans, USA	2927	Mielke et al. (2004)
Tallinn, Estonian rural soil	100	Trapido (1999)
Bangkok, Thailand	380	Wilcke et al. (1999)
Shanghai, China	3780	Jiang et al. (2009)

TABLE 27.3 The half-life of 17 polycyclic aromatic hydrocarbon (PAH) listed by the EPA.

PAH	Half-life	References
Anthracene	139 days	Sims and Overcash (1983)
Acenaphthene	121 days	NCBI
Acenaphthylene	121 days	Coover and Sims (1987)
Benz[*a*]anthracene	261 days	Park et al. (1990)
Benzo[*b*]fluoranthene	Several months to years	Coover and Sims (1987)
Benzo[*k*]fluoranthene	65–1400 days	NCBI
Benzo[*g,h,i*]perylene	173–863 days	
Benzo[*a*]pyrene	309 days	
Chrysene	371 days	Park et al. (1990)
Dibenz[*a,h*]anthracene	361 to 940 days	NCBI
Fluorene	5.01 days	
Fluoranthene	377 days	
Indeno[l,2,3-*c,d*]pyrene	120 days	Mackay et al. (2006)
Naphthalene	1.1 days	NCBI
Naphthylamine	0.58 days	
Pyrene	Several weeks to years	
Phenanthrene	26 days	NCBI

27.3 Need for treatment

If the soil is polluted with high levels of polyaromatic hydrocarbons, it can employ toxic effects including malfunctioning of the reproductive system, tumor development, developmental errors on the invertebrates of that area (Dong et al., 2012). Mammals absorb the PAHs from their surroundings via inhaling, through the skin or via the gastrointestinal route (Beyer et al., 2010; Veltman et al., 2011). Instead, plants can uptake these pollutants through their roots and distribute them to the aerial parts, though the toxicity caused by PAHs in plants are infrequent and less documented. So there is a need for more studies on this. Soil polluted with PAHs must be treated to remove these pollutants from it, these treatments can be physical, chemical, biological, or can be a mixture of two or more to achieve a significant result (Abdel-Shafy and Mansour, 2016).

27.3.1 Light-induced degradation

When a compound starts reactions by absorbing sunlight that leads to the destruction of that complex, this process is called photodegradation (Manahan, 2017). When light falls upon the chemical compound, it excites the outer orbital electrons which unstabilize the structure. This wobbly complex is a suitable substrate for many reactions (Schwarzenbach et al., 1993). Finlayson-Pitts and Pitts (1997) have exposed that PAHs compounds are more vulnerable to degradation when they are in a liquid or gaseous state. Photodegradation is dependent on various factors, that is, the surface area of the particle, structure of pollutant, color of soil, and bioavailability of the compound. Linear/ringed/clustered PAHs are more susceptible than angular ones like phenanthrene because they are comparatively more stable (Korfmacher et al., 1980). Silica or alumina base gives a much higher degradation rate than carbon black; half-life of anthracene on alumina is half an hour but it takes more than 12 days of half-life on carbon black (Abdel-Shafy and Mansour, 2016). Low molecular weight compounds are more readily available for degeneration due to their higher bioavailability and extended exposure time (Korfmacher et al., 1980). Niu et al. (2007) have discovered that when PAHs are bound to particles of fly ash they used to safeguard themselves from sunlight degradation.

TABLE 27.4 Some polycyclic aromatic hydrocarbon (PAH) degrading microbial strains and their break down products.

PAHs	Biodegradative strain	Degradation products	References
Anthracene	*Mycobacterium* sp. PYR-1	9,10-Anthraquinone and 1-methoxy-2-hydroxyanthracene	Van Herwijnen et al. (2003)
Acenaphthene	*Acinetobacter* sp. strain AGAT-W	Catechol	Ghosal et al. (2013)
Acenaphthylene	*Rhizobium* sp. CU-A1, *Stenotrophomonas* sp. RMSK and *Beijerinckia* sp.	1-Naphthoic acid	Selifonov et al. (1996)
Benz[a] anthracene	*Irpex lacteus*	2-Hydroxymethyl benzoic acid or monomethyl and dimethylesters of phthalic acid and 1-tetralone	Cajthaml et al. (2006)
Benzo[*k*] fluoranthene	*Sphingobium* sp. strain KK22	1,8-Naphthalic anhydride and 1-naphthoic acid	Maeda et al. (2014)
Benzo[*g,h,i*] perylene	Yeast consortium, YC04	Oxalic acid, 6-ethyloct-3-yl isohexyl ester	Mandal and Das (2018)
Benzo[a]pyrene	*Beijerinckia* sp. mutant	*cis*-1,2-Dihydroxy-1,2-dihydrobenz[*a*]anthracene and *cis*-8,9- and *cis*-10,11-dihydrodiols	Gibson et al. (1975)
Chrysene	*Sphingomonas* sp. CHY-1	*cis*-Dihydrodiol	Lal and Khanna (1996)
Fluorene	*Sphingomonas* sp. LB126, *Brevibacterium* sp. DPO1361, and *Pseudomonas* sp. F274	Phthalic acid and 2-hydroxypenta-2,4-dienoate	Wattiau et al. (2001)
Fluoranthene	*Alcaligenes denitrificans* strain WW1	7-Hydroxyacenaphthylene, 7-acenaphthenone, and 3-hydroxymethyl-4,5-benzocoumarine	Weissenfels et al. (1991)
Naphthalene	*Pseudomonas* sp. N7	Salicylic acid, catechol, and 2-hydroxymuconic semi aldehyde	Jia et al. (2008)
Pyrene	*Mycobacterium* spp. strain KR2	*cis*-4,5-Pyrene dihydrodiol, 4–5-phenanthrene dicarboxylic acid, 1-hydroxy-2-naphthoic acid, 2-carboxybenzaldehyde, phthalic acid, and protocatechuic acid	
Phenanthrene	*Pleurotus ostreatus*	2,2′-Diphenic acid and phenanthrene *trans*-9,10-dihydrodiol	Bezalel et al. (1996)

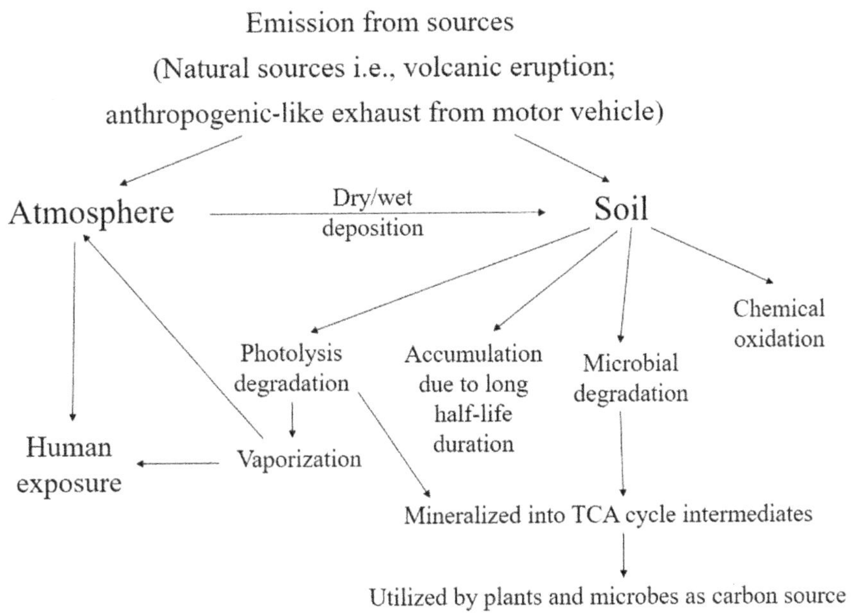

FIGURE 27.3 Environmental fate of PAH. *PAH*, Polycyclic aromatic hydrocarbon.

27.3.2 Degradation through oxidation

Oxidation plays a negligible role in the degradation process of PAHs compared to other mechanisms. It can be natural or a part of other procedures (Abdel-Shafy and Mansour, 2013). The oxidation rate is a subject of the physical form of the pollutant, molecular weight of PAH, power of the oxidizing substance, and structure of PAH. Moursy and Abdel-Shafy (1983) showed that PAHs can be eliminated from surface water using an oxidation process. Some experiments have suggested that fluoranthene is very rigid for oxidation using O_3 as an oxidizing agent (Alebic-Juretic et al., 1990); this study also justifies the higher reporting of fluoranthene in most soil samples analysis. Some studies have shown that when ozone exposure is combined with a catalyst like ZnO or TiO_2, it is capable of degrading many PAH in polluted soil sites (Zhang et al., 2008). Wang et al. (2009b) have experienced that oxalic acid and Fe-oxides can enhance the degradation rate of pyrene under UV-sunlight. Some experiments were also done on the breakdown of PAHs using high-frequency ultrasound waves (Manariotis et al., 2011).

27.4 Bioremediation approach

Since physical and chemical treatments of soil contaminated with PAHs are costly, Habe and Omori (2003) suggested adapting for biological ones that are sustainable and cost-friendly. Bioremediation is a process in which pollutants are removed (mineralized or degraded to less harmful forms) from the matrix by using biological entities (Garbisu and Alkorta, 1999; Paul et al., 2005). This process is affected by physiochemical characteristics of the targeted chemical (pH, molecular weight, octanol/water partition coefficient, etc.) and matrix (water content, pH, other substrates present in it, minerals, ions, pore size, etc.) (Kim et al., 2005; Aichberger et al., 2006; Stokes et al., 2005; Zhang et al., 2006; Núnez et al., 2009). Bioremediation can be of mainly two types—one is biostimulation, in which the matrix is supported by adding nutrients, surfactants, organic matter, etc., to upsurge the speed of removal; and second is bioaugmentation in which different kind of microorganisms with the capacity of degrading PAHs are mixed in the soil. These microbes can be the native flora or can be introduced from other place or they can be engineered (Fernández-Luqueño et al., 2011).

27.4.1 Mechanism of degradation by bacteria

Bacteria-mediated degradation is very popular for PAHs removal. There is two kinds of biodegradation process—one is under aerobic conditions and another is under anaerobic conditions. Although anaerobic degradation is very rare, recently this process is reported under denitrifying conditions by Peng et al. (2008) as well as by Haritash and Kaushik, 2009. To utilize PAHs as their energy source the bioavailability of these compounds to bacteria is very important (Cerniglia, 2003; Dandie et al., 2004; Fredslund et al., 2008); bioavailability depends on many factors like the physical state of PAHs (Kim et al., 2007). Adsorption of pollutant on soil particles decreases their bioavailability because the enzymes that take parts in degradation process cannot work on them unless these compounds are again desorbed by the soil particles and became free which depends on the aging of PAH molecule (Rappert et al., 2006; Uyttebroek et al., 2007). If simple carbon sources are present in the surrounding of PAH degrading bacterial community, they will be preferred over PAH molecules as an energy source. Uptake of PAH is also dependent on the nature of enzymes involved in the degradation; if the enzymes are nonspecific to the substrate, competitive inhibition can occur which will result in the persistence of PAH in the matrix (Wang et al., 2009a; Abdel-Shafy and Mansour, 2016). A study conducted by a group of researchers unveiled that anthracene can breakdown to basic elements by *Rhodococcus*, *Beijerinckia*, *Sphingomonas*, *Paracoccus*, and *Nocardia*, in this process dihydriol is produced as a first intermediate product of oxygenation reaction (Teng et al., 2010). Samanta et al. (2002) have described a general pathway of PAHs degeneration by microbial activity in the presence of oxygen. Below is a diagrammatic scheme that shows the general pathway of PAHs utilization by microbial enzymes (Fig. 27.4).

Under aerobic conditions, first, dioxygenase enzyme converts PAH molecule in *cis*-dihydrodiol which undergoes rearomatization under the influence of dehydrogenase and produce dihydroxylated intermediates; at the end, these intermediate compounds go through ring cleavage and ultimately generate tricarboxylic acid (TCA) cycle intermediates. These TCA intermediates will be harvested by the microbes in energy production (Samanta et al., 2002). In the absence of oxygen, Meckenstock et al. (2004) have found that PAHs which contain up to three aromatic rings can be break down by microbial activity. So far no evidence is found which can claim that PAHs of more than three rings can be biodegradable anaerobically (Zhang et al., 2006). Different PAHs are degradable under different anaerobic environment, that is, naphthalene needs nitrate and sulfate reducing conditions (Bregnard et al., 1996; Coates et al., 1996).

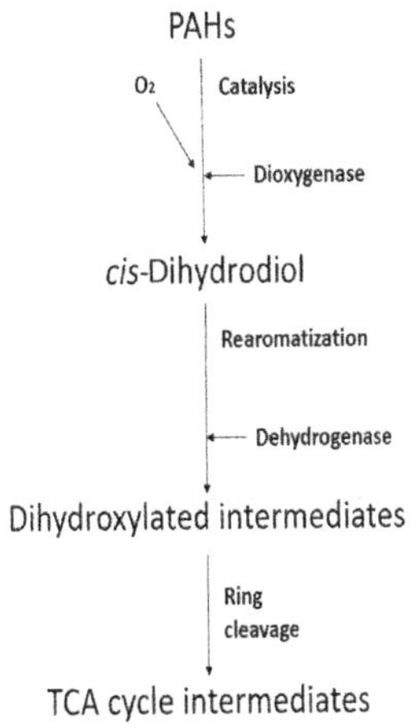

FIGURE 27.4 Flow chart of PAHs degradation by microbes under aerobic conditions. *PAH*, Polycyclic aromatic hydrocarbon.

TABLE 27.5 Listing of some genetically engineered microbes (GEMs) reported for degradation of polycyclic aromatic hydrocarbons (PAHs)-.

GEMs	PAHs	Comments	References
Streptomyces lividans	Phenanthrene and 1-Methoxynaphthalene	phdABCD gene cluster in a marine bacterium *Nocardioides* sp. strain KP7 codes for the multicomponent enzyme phenanthrene dioxygenase	Chun et al. (2001)
Pseudomonas fluorescens HK44	Naphthalene	This stain harbors *lux* gene, hence produces bioluminescence	Ripp et al. (2000)
Comamonas testosterone	Naphthalene	Classical gene (*nah*) from *Pseudomonas* strain NCIB 9816-4 incorporates to *C. testosteroni*	Goyal and Zylstra (1996)
P. fluorescens	Naphthalene	*lux* gene cassette from *Vibrio fischeri* into *P. fluorescens*	King et al. (1990)
Escherichia coli	Phenanthrene, dibenzothiophene, and pyrene	The genes from *Mycobacterium* sp. strain PYR-1 were subcloned and overexpressed in *E. coli*	Khan et al. (2001)
Sphingomonas sp. GY2B	Phenanthrene	*Pseudomonas* sp. GP3A, (Protoplast fusion)	Lu et al. (2013)
Pseudomonas sp. CGMCC2953-pk	Phenanthrene	*Pseudomonas* sp. *CGMCC2953* and catechol 2,3-dioxygenase (C23O) was the key enzyme in the phenanthrene biodegradation pathway	Zhou et al. (2013)
E. coli BL21 (DE3)	Naphthalene, phenanthrene, or salicylate	*Sphingomonas yanoikuyae* B1, transformation of *bphA3-bphA2c-bphA1c* genes	Cho et al. (2005)
Acinetobacter sp. BS3-C23O	PAHs	*Pseudomonas putida* BNF1	Xie et al. (2014)

27.4.2 Genetically engineered microorganisms to speed up the remediation

Though microbes possess natural power to degrade PAH but this supremacy is limited by some environmental factors that reduce the efficacy of bioremediation (Shukla et al., 2013). This slow down reminds us for the advancement of the existing technique. The effectiveness of the prevailing process can be heightened by developing genetically engineered microorganisms. Human activities exposed the natural microbial flora to these chemicals which trigger resistant microorganisms to develop metabolic pathways for survival. By means of horizontal gene transfer, microbes acquire new genes to get adapted for the new stress in the environment (Wang et al., 2010). Various studies conducted on endophytic organisms that showed hydrocarbon-degrading abilities have revealed that they contain alkB and CYT153 genes that are known for producing alkane monooxygenase and cytochrome P450 hydrolase, respectively (Khan et al., 2013; Afzal et al., 2014). Table 27.5 which contains some studies dealing with application of genetically engineered microorganisms.

27.5 Future perspective

Researches regarding bioremediation are evolving with passing time and now engulfed biotechnology and recombinant technology to fight the major restrictions. However, most of the studies are confined in vitro, because it is too difficult to conduct these experiments on field-scale due to variable environmental factors. Thus upcoming studies should be dedicated to factors that can enhance the pace of the bioremediation process. Application of genetically engineered microbes (GEMs) has its own pros and cons; low public acceptance and rules regarding the release of GEMs are the major hurdles in applying the more enhanced version of bioremediation. Perhaps the selection of natural flora for the development of genetically engineered one gives a ray of hope from the banning. Bioremediation is a cost-friendly technique but the development of genetically engineered microorganisms cost us a fortune, so there is a need for the cost-cutting in this potential tool that will result in treasure.

Acknowledgment

The author is thankful to CSIR for providing fellowship in the form of JRF.

References

Abdel-Shafy, H.I., Mansour, M.S., 2013. Removal of selected pharmaceuticals from urine via fenton reaction for agriculture reuse. J. SSP 17, 20–29.

Abdel-Shafy, H.I., Mansour, M.S., 2016. A review on polycyclic aromatic hydrocarbons: source, environmental impact, effect on human health and remediation. Egypt. J. Pet. 25 (1), 107–123.

Afzal, M., Shabir, G., Tahseen, R., Islam, E.U., Iqbal, S., Khan, Q.M., et al., 2014. Endophytic *Burkholderia* sp. strain Ps JN improves plant growth and phytoremediation of soil irrigated with textile effluent. CLEAN—Soil Air Water 42 (9), 1304–1310.

Aichberger, H., Loibner, A.P., Celis, R., Braun, R., Ottner, F., Rost, H., 2006. Assessment of factors governing biodegradability of PAHs in three soils aged under field conditions. Soil. Sediment. Contam. 15 (1), 73–85.

Alebic-Juretic, A., Cvitas, T., Klasinc, L., 1990. Heterogeneous polycyclic aromatic hydrocarbon degradation with ozone on silica gel carrier. Environ. Sci. Technol. 24 (1), 62–66.

Arey, J., Atkinson, R., 2003a. Photochemical reactions of PAHs in the atmosphere. PAHs: Ecotoxicol. Perspect. 47–63.

Arey, J., Atkinson, R., 2003b. Photochemical reactions of PAH in the atmosphere. In: Douben, P.E.T. (Ed.), PAHs: An Ecotoxicological Perspective. John Wiley and Sons Ltd, New York, pp. 47–63.

Beyer, J., Jonsson, G., Porte, C., Krahn, M.M., Ariese, F., 2010. Analytical methods for determining metabolites of polycyclic aromatic hydrocarbon (PAH) pollutants in fish bile: a review. Environ. Toxicol. Pharmacol. 30 (3), 224–244.

Bezalel, L.E.A., Hadar, Y., Fu, P.P., Freeman, J.P., Cerniglia, C.E., 1996. Metabolism of phenanthrene by the white rot fungus *Pleurotus ostreatus*. Appl. Environ. Microbiol. 62 (7), 2547–2553.

Bozlaker, A., Muezzinoglu, A., Odabasi, M., 2008. Atmospheric concentrations, dry deposition and air–soil exchange of polycyclic aromatic hydrocarbons (PAHs) in an industrial region in Turkey. J. Hazard. Mater. 153 (3), 1093–1102.

Bregnard, T.P.A., Höhener, P., Häner, A., Zeyer, J., 1996. Degradation of weathered diesel fuel by microorganisms from a contaminated aquifer in aerobic and anaerobic microcosms. Environ. Toxicol. Chem. 15 (3), 299–307.

Cachada, A., Pato, P., Rocha-Santos, T., da Silva, E.F., Duarte, A.C., 2012. Levels, sources and potential human health risks of organic pollutants in urban soils. Sci. Total Environ. 430, 184–192.

Cajthaml, T., Erbanová, P., Šašek, V., Moeder, M., 2006. Breakdown products on metabolic pathway of degradation of benz[*a*]anthracene by a ligninolytic fungus. Chemosphere 64 (4), 560–564.

Cerniglia, C.E., 2003. Recent advances in the biodegradation of polycyclic aromatic hydrocarbons by *Mycobacterium* species. The Utilization of Bioremediation to Reduce Soil Contamination: Problems and Solutions. Springer, Dordrecht, pp. 51–73.

Cetin, B., 2016. Investigation of PAHs, PCBs and PCNs in soils around a Heavily Industrialized Area in Kocaeli, Turkey: concentrations, distributions, sources and toxicological effects. Sci. Total Environ. 560, 160–169.

Cetin, B., Ozturk, F., Keles, M., Yurdakul, S., 2017a. PAHs and PCBs in an Eastern Mediterranean megacity, Istanbul: their spatial and temporal distributions, air-soil exchange and toxicological effects. Environ. Pollut. 220, 1322–1332.

Cetin, B., Yurdakul, S., Keles, M., Celik, I., Ozturk, F., Dogan, C., 2017b. Atmospheric concentrations, distributions and air-soil exchange tendencies of PAHs and PCBs in a heavily industrialized area in Kocaeli, Turkey. Chemosphere 183, 69–79.

Chaineau, C.H., Rougeux, G., Yepremian, C., Oudot, J., 2005. Effects of nutrient concentration on the biodegradation of crude oil and associated microbial populations in the soil. Soil Biol. Biochem. 37 (8), 1490–1497.

Cho, O., Choi, K.Y., Zylstra, G.J., Kim, Y.S., Kim, S.K., Lee, J.H., et al., 2005. Catabolic role of a three-component salicylate oxygenase from *Sphingomonas yanoikuyae* B1 in polycyclic aromatic hydrocarbon degradation. Biochem. Biophys. Res. Commun. 327 (3), 656–662.

Chrysikou, L., Gemenetzis, P., Kouras, A., Manoli, E., Terzi, E., Samara, C., 2008. Distribution of persistent organic pollutants, polycyclic aromatic hydrocarbons and trace elements in soil and vegetation following a large scale landfill fire in northern Greece. Environ. Int. 34 (2), 210–225.

Chun, H.K., Ohnishi, Y., Misawa, N., Shindo, K., Hayashi, M., Harayama, S., et al., 2001. Biotransformation of phenanthrene and 1-methoxynaphthalene with *Streptomyces lividans* cells expressing a marine bacterial phenanthrene dioxygenase gene cluster. Biosci. Biotechnol. Biochem. 65 (8), 1774–1781.

Coates, J.D., Anderson, R.T., Lovley, D.R., 1996. Oxidation of polycyclic aromatic hydrocarbons under sulfate-reducing conditions. Appl. Environ. Microbiol. 62 (3), 1099–1101.

Coover, M.P., Sims, R.C., 1987. The effect of temperature on polycyclic aromatic hydrocarbon persistence in an unacclimated agricultural soil. Hazard. Waste Hazard. Mater. 4 (1), 69–82.

Dandie, C.E., Thomas, S.M., Bentham, R.H., McClure, N.C., 2004. Physiological characterization of *Mycobacterium* sp. strain 1B isolated from a bacterial culture able to degrade high-molecular-weight polycyclic aromatic hydrocarbons. J. Appl. Microbiol. 97 (2), 246–255.

Di Toro, D.M., McGrath, J.A., Hansen, D.J., 2000. Technical basis for narcotic chemicals and polycyclic aromatic hydrocarbon criteria. I. Water and tissue. Environ. Toxicol. Chem. 19 (8), 1951–1970.

Dong, C.D., Chen, C.F., Chen, C.W., 2012. Determination of polycyclic aromatic hydrocarbons in industrial harbor sediments by GC-MS. Int. J. Environ. Res. Public Health 9, 2175–2188.

EPRI (Electric Power Research Institute). Literature review of background polycyclic aromatic hydrocarbons. In: Final Report; March 2000.

Fernández-Luqueño, F., Valenzuela-Encinas, C., Marsch, R., Martínez-Suárez, C., Vázquez-Núñez, E., Dendooven, L., 2011. Microbial communities to mitigate contamination of PAHs in soil—possibilities and challenges: a review. Environ. Sci. Pollut. Res. 18 (1), 12–30.

Finlayson-Pitts, B.J., Pitts, J.N., 1997. Tropospheric air pollution: ozone, airborne toxics, polycyclic aromatic hydrocarbons, and particles. Science 276 (5315), 1045–1051.

Fredslund, L., Sniegowski, K., Wick, L.Y., Jacobsen, C.S., De Mot, R., Springael, D., 2008. Surface motility of polycyclic aromatic hydrocarbon (PAH)-degrading mycobacteria. Res. Microbiol. 159 (4), 255–262.

Garbisu, C., Alkorta, I., 1999. Utilization of genetically engineered microorganisms (GEMs) for bioremediation. J. Chem. Technol. Biotechnol. 74 (7), 599–606.

Ghosal, D., Dutta, A., Chakraborty, J., Basu, S., Dutta, T.K., 2013. Characterization of the metabolic pathway involved in assimilation of acenaphthene in *Acinetobacter* sp. strain AGAT-W. Res. Microbiol. 164 (2), 155–163.

Gibson, D.T., Mahadevan, V., Jerina, D.M., Yogi, H., Yeh, H.J., 1975. Oxidation of the carcinogens benzo[*a*]pyrene and benzo[*a*]anthracene to dihydrodiols by a bacterium. Science 189 (4199), 295–297.

Goyal, A.K., Zylstra, G.J., 1996. Molecular cloning of novel genes for polycyclic aromatic hydrocarbon degradation from *Comamonas testosteroni* GZ39. Appl. Environ. Microbiol. 62 (1), 230–236.

Gupta, A., Singh, M., Pandey, K.D., Kumar, A., 2018. Biodegradation of xenobiotics using cyanobacteria. J. Sci. Res. 62, 35–40.

Gupta, H., Kumar, R., 2020. Distribution of selected polycyclic aromatic hydrocarbons in urban soils of Delhi, India. Environ. Technol. Innov 17, 100500.

Habe, H., Omori, T., 2003. Genetics of polycyclic aromatic hydrocarbon metabolism in diverse aerobic bacteria. Biosci. Biotechnol. Biochem. 67 (2), 225–243.

Haritash, A.K., Kaushik, C.P., 2009. Biodegradation aspects of polycyclic aromatic hydrocarbons (PAHs): a review. J. Hazard. Mater. 169 (1–3), 1–15.

Jia, Y., Yin, H., Ye, J.S., Peng, H., He, B.Y., Qin, H.M., et al., 2008. Characteristics and pathway of naphthalene degradation by *Pseudomonas* sp. N7. Huan jing ke xue = Huanjing kexue 29 (3), 756–762.

Jiang, Y.F., Wang, X.T., Wang, F., Jia, Y., Wu, M.H., Sheng, G.Y., et al., 2009. Levels, composition profiles and sources of polycyclic aromatic hydrocarbons in urban soil of Shanghai, China. Chemosphere 75 (8), 1112–1118.

Johnsen, A.R., Wick, L.Y., Harms, H., 2005. Principles of microbial PAH-degradation in soil. Environ. Pollut. 133 (1), 71–84.

Khan, A.A., Wang, R.F., Cao, W.W., Doerge, D.R., Wennerstrom, D., Cerniglia, C.E., 2001. Molecular cloning, nucleotide sequence, and expression of genes encoding a polycyclic aromatic ring dioxygenase from *Mycobacterium* sp. strain PYR-1. Appl. Environ. Microbiol. 67 (8), 3577–3585.

Khan, S., Afzal, M., Iqbal, S., Khan, Q.M., 2013. Plant–bacteria partnerships for the remediation of hydrocarbon contaminated soils. Chemosphere 90 (4), 1317–1332.

Kim, Y.H., Freeman, J.P., Moody, J.D., Engesser, K.H., Cerniglia, C.E., 2005. Effects of pH on the degradation of phenanthrene and pyrene by *Mycobacterium vanbaalenii* PYR-1. Appl. Microbiol. Biotechnol. 67 (2), 275–285.

Kim, S.J., Kweon, O., Jones, R.C., Freeman, J.P., Edmondson, R.D., Cerniglia, C.E., 2007. Complete and integrated pyrene degradation pathway in *Mycobacterium vanbaalenii* PYR-1 based on systems biology. J. Bacteriol. 189 (2), 464–472.

King, J.M.H., DiGrazia, P.M., Applegate, B., Burlage, R., Sanseverino, J., Dunbar, P., et al., 1990. Rapid, sensitive bioluminescent reporter technology for naphthalene exposure and biodegradation. Science 249 (4970), 778–781.

Korfmacher, W.A., Wehry, E.L., Mamantov, G., Natusch, D.F.S., 1980. Resistance to photochemical decomposition of polycyclic aromatic hydrocarbons vapor-adsorbed on coal fly ash. Environ. Sci. Technol. 14 (9), 1094–1099.

Krauss, M., Wilcke, W., 2003. Polychlorinated naphthalenes in urban soils: analysis, concentrations, and relation to other persistent organic pollutants. Environ. Pollut. 122 (1), 75–89.

Lal, B., Khanna, S., 1996. Degradation of crude oil by *Acinetobacter calcoaceticus* and *Alcaligenes odorans*. J. Appl. Bacteriol. 81 (4), 355–362.

Li, Q., Wu, J., Zhou, J., Sakiev, K., Hofmann, D., 2020. Occurrence of polycyclic aromatic hydrocarbon (PAH) in soils around two typical lakes in the western Tian Shan Mountains (Kyrgyzstan, Central Asia): Local burden or global distillation? Ecol. Indic. 108, 105749.

Liu, K., Han, W., Pan, W.P., Riley, J.T., 2001. Polycyclic aromatic hydrocarbon (PAH) emissions from a coal-fired pilot FBC system. J. Hazard. Mater. 84 (2-3), 175–188.

Lu, J., Guo, C., Li, J., Zhang, H., Lu, G., Dang, Z., et al., 2013. A fusant of *Sphingomonas* sp. GY2B and *Pseudomonas* sp. GP3A with high capacity of degrading phenanthrene. World J. Microbiol. Biotechnol. 29 (9), 1685–1694.

Ma, Y.G., Lei, Y.D., Xiao, H., Wania, F., Wang, W.H., 2009. Critical review and recommended values for the physical-chemical property data of 15 polycyclic aromatic hydrocarbons at 25 C. J. Chem. Eng. Data 55 (2), 819–825.

Mackay, D., Shiu, W.Y., Lee, S.C., 2006. Handbook of Physical-Chemical Properties and Environmental Fate for Organic Chemicals. CRC Press.

Maeda, A.H., Nishi, S., Hatada, Y., Ozeki, Y., Kanaly, R.A., 2014. Biotransformation of the high-molecular weight polycyclic aromatic hydrocarbon (PAH) benzo[*k*]fluoranthene by *Sphingobium* sp. strain KK 22 and identification of new products of non-alternant PAH biodegradation by liquid chromatography electrospray ionization tandem mass spectrometry. Microb. Biotechnol. 7 (2), 114–129.

Manahan, S., 2017. Environmental Chemistry. CRC Press.

Manariotis, I.D., Karapanagioti, H.K., Chrysikopoulos, C.V., 2011. Degradation of PAHs by high frequency ultrasound. Water Res. 45 (8), 2587–2594.

Mandal, S.K., Das, N., 2018. Biodegradation of perylene and benzo[*g,h,i*]perylene (5-6 rings) using yeast consortium: kinetic study, enzyme analysis and degradation pathway. J. Environ. Biol. 39 (1), 5–15.

Marx, M., Ackermann, J., Schmidt, S., Utermann, J., Bussian, B.M., 2019. Occurrence and Spatial Distribution of Selected Organic Substances in Germany's Forest Soils. Status and Dynamics of Forests in Germany. Springer, Cham, pp. 231–259.

Masih, A., Taneja, A., 2006. Polycyclic aromatic hydrocarbons (PAHs) concentrations and related carcinogenic potencies in soil at a semi-arid region of India. Chemosphere 65 (3), 449–456.

Meckenstock, R.U., Safinowski, M., Griebler, C., 2004. Anaerobic degradation of polycyclic aromatic hydrocarbons. FEMS Microbiol. Ecol. 49 (1), 27–36.

Mielke, H.W., Wang, G., Gonzales, C.R., Powell, E.T., Le, B., Quach, V.N., 2004. PAHs and metals in the soils of inner-city and suburban New Orleans, Louisiana, USA. Environ. Toxicol. Pharmacol. 18 (3), 243–247.

Mohit, A., Keshavarzi, B., Moore, F., 2019. Polycyclic aromatic hydrocarbons (PAHs) in urban soils of Ahvaz metropolis; contamination, composition, distribution, potential sources, and cancer risk. Hum. Ecol. Risk Assess. 25 (4), 935–948.

Morillo, E., Romero, A.S., Maqueda, C., Madrid, L., Ajmone-Marsan, F., Grcman, H., et al., 2007. Soil pollution by PAHs in urban soils: a comparison of three European cities. J. Environ. Monit. 9 (9), 1001–1008.

Moursy, A.S., Abdel-Shafy, H.I., 1983. Removal of hydrocarbons from Nile water. Environ. Int. 9 (3), 165–171.

Niu, J., Sun, P., Schramm, K.W., 2007. Photolysis of polycyclic aromatic hydrocarbons associated with fly ash particles under simulated sunlight irradiation. J. Photochem. Photobiol., A 186 (1), 93–98.

Núnez, E.V., Gaytán, A.G., Luna-Guido, M., Marsch, R., Dendooven, L., 2009. Impact of moisture dynamic and sun light on anthracene removal from soil. Biodegradation 20 (2), 191–198.

Park, K.S., Sims, R.C., Dupont, R.R., Doucette, W.J., Matthews, J.E., 1990. Fate of PAH compounds in two soil types: influence of volatilization, abiotic loss and biological activity. Environ. Toxicol. Chem. 9 (2), 187–195.

Paul, D., Pandey, G., Pandey, J., Jain, R.K., 2005. Accessing microbial diversity for bioremediation and environmental restoration. Trends Biotechnol. 23 (3), 135–142.

Peng, R.H., Xiong, A.S., Xue, Y., Fu, X.Y., Gao, F., Zhao, W., et al., 2008. Microbial biodegradation of polyaromatic hydrocarbons. FEMS Microbiol. Rev. 32 (6), 927–955.

Peng, C., Chen, W., Liao, X., Wang, M., Ouyang, Z., Jiao, W., et al., 2011. Polycyclic aromatic hydrocarbons in urban soils of Beijing: status, sources, distribution and potential risk. Environ. Pollut. 159 (3), 802–808.

Rappert, S., Botsch, K.C., Nagorny, S., Francke, W., Müller, R., 2006. Degradation of 2, 3-diethyl-5-methylpyrazine by a newly discovered bacterium, *Mycobacterium* sp. strain DM-11. Appl. Environ. Microbiol. 72 (2), 1437–1444.

Rhind, S.M., Kyle, C.E., Kerr, C., Osprey, M., Zhang, Z.L., Duff, E.I., et al., 2013. Concentrations and geographic distribution of selected organic pollutants in Scottish surface soils. Environ. Pollut. 182, 15–27.

Riccardi, C., Di Filippo, P., Pomata, D., Di Basilio, M., Spicaglia, S., Buiarelli, F., 2013. Identification of hydrocarbon sources in contaminated soils of three industrial areas. Sci. Total Environ. 450, 13–21.

Ripp, S., Nivens, D.E., Ahn, Y., Werner, C., Jarrell, J., Easter, J.P., et al., 2000. Controlled field release of a bioluminescent genetically engineered microorganism for bioremediation process monitoring and control. Environ. Sci. Technol. 34 (5), 846–853.

Samanta, S.K., Singh, O.V., Jain, R.K., 2002. Polycyclic aromatic hydrocarbons: environmental pollution and bioremediation. Trends Biotechnol. 20 (6), 243–248.

Schwarzenbach, R.P., Gschwend, P.M., Imboden, D.M., 1993. Sorption of neutral organic chemicals to soils and sediments. Environmental Organic Chemistry, Illustrative Examples, Problems and Case Studies. John Wiley and Sons, Inc, New York, pp. 265–284.

Schwarzenbach, R.P., Gschwend, P.M., Imboden, D.M., 2003. Organic liquid–water partitioning. Environmental Organic Chemistry. Wiley, Hoboken, NJ, pp. 213–244.

Scullion, J., 2006. Remediating polluted soils. Naturwissenschaften 93 (2), 51–65.

Selifonov, S.A., Grifoll, M., Eaton, R.W., Chapman, P.J., 1996. Oxidation of naphthenoaromatic and methyl-substituted aromatic compounds by naphthalene 1,2-dioxygenase. Appl. Environ. Microbiol. 62 (2), 507–514.

Shukla, K.P., Sharma, S., Singh, N.K., Singh, V., Bisht, S., Kumar, V., 2013. Rhizoremediation: a promising rhizosphere technology. In: Applied Bioremediation: Active and Passive Approaches, 333-352.

Sims, R.C., Overcash, M.R., 1983. Fate of polynuclear aromatic compounds (PNAs) in soil-plant systems. Residue Reviews. Springer, New York, NY, pp. 1–68.

Singh, P., Singh, V.K., Singh, R., Borthakur, A., Madhav, S., Ahamad, A., et al., 2020. Bioremediation: a sustainable approach for management of environmental contaminants. Abatement of Environmental Pollutants. Elsevier, pp. 1–23.

Stokes, J.D., Paton, G.I., Semple, K.T., 2005. Behaviour and assessment of bioavailability of organic contaminants in soil: relevance for risk assessment and remediation. Soil Use Manage. 21, 475–486.

Teng, Y., Luo, Y., Sun, M., Liu, Z., Li, Z., Christie, P., 2010. Effect of bioaugmentation by *Paracoccus* sp. strain HPD-2 on the soil microbial community and removal of polycyclic aromatic hydrocarbons from an aged contaminated soil. Bioresour. Technol. 101 (10), 3437–3443.

Trapido, M., 1999. Polycyclic aromatic hydrocarbons in Estonian soil: contamination and profiles. Environ. Pollut. 105 (1), 67–74.

Uyttebroek, M., Vermeir, S., Wattiau, P., Ryngaert, A., Springael, D., 2007. Characterization of cultures enriched from acidic polycyclic aromatic hydrocarbon-contaminated soil for growth on pyrene at low pH. Appl. Environ. Microbiol. 73 (10), 3159–3164.

Van der Oost, R., Beyer, J., Vermeulen, N.P., 2003. Fish bioaccumulation and biomarkers in environmental risk assessment: a review. Environ. Toxicol. Pharmacol. 13 (2), 57–149.

Van Herwijnen, R., Springael, D., Slot, P., Govers, H.A., Parsons, J.R., 2003. Degradation of anthracene by *Mycobacterium* sp. strain LB501T proceeds via a novel pathway, through o-phthalic acid. Appl. Environ. Microbiol. 69 (1), 186–190.

Veltman, K., Huijbregts, M.A., Rye, H., Hertwich, E.G., 2011. Including impacts of particulate emissions on marine ecosystems in life cycle assessment: The case of offshore oil and gas production. Integr. Environ. Assess. Manage. 7 (4), 678–686.

Wang, C., Sun, H., Li, J., Li, Y., Zhang, Q., 2009a. Enzyme activities during degradation of polycyclic aromatic hydrocarbons by white rot fungus *Phanerochaete chrysosporium* in soils. Chemosphere 77 (6), 733–738.

Wang, Y., Liu, C.S., Li, F.B., Liu, C.P., Liang, J.B., 2009b. Photodegradation of polycyclic aromatic hydrocarbon pyrene by iron oxide in solid phase. J. Hazard. Mater. 162 (2–3), 716–723.

Wang, Y., Li, H., Zhao, W., He, X., Chen, J., Geng, X., et al., 2010. Induction of toluene degradation and growth promotion in corn and wheat by horizontal gene transfer within endophytic bacteria. Soil Biol. Biochem. 42 (7), 1051–1057.

Wattiau, P., Bastiaens, L., van Herwijnen, R., Daal, L., Parsons, J.R., Renard, M.E., et al., 2001. Fluorene degradation by *Sphingomonas* sp. LB126 proceeds through protocatechuic acid: a genetic analysis. Res. Microbiol. 152 (10), 861–872.

Weissenfels, W.D., Beyer, M., Klein, J., Rehm, H.J., 1991. Microbial metabolism of fluoranthene: isolation and identification of ring fission products. Appl. Microbiol. Biotechnol. 34 (4), 528–535.

Wilcke, W., Müller, S., Kanchanakool, N., Niamskul, C., Zech, W., 1999. Polycyclic aromatic hydrocarbons in hydromorphic soils of the tropical metropolis Bangkok. Geoderma 91 (3-4), 297–309.

Xie, Y., Yu, F., Wang, Q., Gu, X., Chen, W., 2014. Cloning of catechol 2,3-dioxygenase gene and construction of a stable genetically engineered strain for degrading crude oil. Indian J. Microbiol. 54 (1), 59–64.

Xing, X., Mao, Y., Hu, T., Tian, Q., Chen, Z., Liao, T., et al., 2020. Spatial distribution, possible sources and health risks of PAHs and OCPs in surface soils from Dajiuhu Sub-alpine Wetland, central China. J. Geochem. Explor. 208, 106393.

Zhang, X.X., Cheng, S.P., Cheng-Jun, Z.H.U., Shi-Lei, S.U.N., 2006. Microbial PAH-degradation in soil: degradation pathways and contributing factors. Pedosphere 16 (5), 555–565.

Zhang, L., Li, P., Gong, Z., Li, X., 2008. Photocatalytic degradation of polycyclic aromatic hydrocarbons on soil surfaces using TiO2 under UV light. J. Hazard. Mater. 158 (2–3), 478–484.

Zhang, J., Yang, J.C., Wang, R.Q., Hou, H., Du, X.M., Fan, S.K., et al., 2013. Effects of pollution sources and soil properties on distribution of polycyclic aromatic hydrocarbons and risk assessment. Sci. Total Environ. 463, 1–10.

Zhou, Y., Wei, J., Shao, N., Wei, D., 2013. Construction of a genetically engineered microorganism for phenanthrene biodegradation. J. Basic. Microbiol. 53 (2), 188–194.

Chapter 28

Electrochemical technologies for wastewater treatment and resource reclamation

Ramya Thangamani[1], Lakshmanaperumal Vidhya[2] and Sunita Varjani[3]

[1]*Department of Chemical Engineering, King Mongkut's University of Technology North Bangkok, Bangkok, Thailand,* [2]*Department of Chemical Engineering, Sethu Institute of Technology, Virudhunagar, India,* [3]*Gujarat Pollution Control Board, Gandhinagar, India*

28.1 Introduction

The analytical methods of electrochemistry and electrochemical processes have a great influence on our routine activities in several ways. The significant examples are batteries, corrosion, commodity, metallurgy, and electroplating (Beranek, 2011).

In the year 1889, in the United Kingdom, electricity was first recommended to treat wastewater. Since the end of the last century, electrochemical methods are known yet it has not attained a wide dissemination owing to its higher energy intake, nonavailability of inexpensive electrode materials, and defectiveness in the manufacturing of the equipment (Shanthi et al., 2016).

But there has been increasing interest in environmental electrochemistry for the past two decades. Electrochemical techniques were successfully adopted in the purification of industrial effluents (Deng and Zhao, 2015). The main drawback of advanced oxidation techniques is the handling and mixing of hydrogen peroxide and ozone. Since the same could be generated in situ by electrochemical techniques, the problems of handling and mixing could be avoided (Verschueren and De Borggraeve, 2019).

The cost of operation is reliant on the consumption of power, it is expressed as:

$$P = U \times I \tag{28.1}$$

P implies power consumption (W), voltage is represented as (V), and current as (I). By ohm's law ($U = R \times I$), R is resistance in ohms. Eq. (28.1) can be denoted as:

$$P = R \times I^2 \tag{28.2}$$

$$P = \frac{U^2}{R} \tag{28.3}$$

The correlation between electrical current and electrical power consumption is shown in Eq. (28.1), and a variation in electrical current will alter the electrical power consumption in the second power. The quantity of gas evolved in the electrodes is reliable on the current flowing across the electrodes (Philippe et al., 2005). In order to cut down the electrical power intake without varying the current and the degree of separation, you can reduce the resistance in the electrolyte (Liu et al., 2016). The reduction of the distance between the electrodes or an increase in conductivity of the electrolyte will reduce power consumption without varying the degree of separation (Mazloomi et al., 2012). Since the current is not changed.

According to Ohm's law, the power consumption will reduce as the distance between the two electrodes is reduced and as the conductivity of the electrolyte increases (Opu, 2015). The power of consumption will be disturbed by the

Microbe Mediated Remediation of Environmental Contaminants. DOI: https://doi.org/10.1016/B978-0-12-821199-1.00028-6

electrolytic conductivity in the same way as the electrodes. In accordance with the ohm's law, a higher conductivity in the electrolyte and a smaller distance between the will provide power consumption.

28.2 Physical methods for wastewater treatment

28.2.1 Electrodialysis

Membrane treatment process is otherwise termed electrodialysis. During electrodialysis the ions are passed through semipermeable membrane under electric potential condition. In general, the membranes are cation or anion selective. The polyelectrolytes are cation membranes and are negatively charged in nature due to which they allow the positive ions to pass through while they reject the negative ions. During electrodialysis using multiple membranes in an order, the negatively or positively charged ions are allowed to flow through the ions that can be removed from wastewater. Electrodialysis is mainly applied to the removal of ions from wastewater. Those particles that do not possess any electrical charge are not removed. Generally, sulfonated polystyrene is often used as cation membrane, whereas the polystyrene with quaternary ammonia membranes is used as anion-selective membranes. Some wastewater contains suspended solids sized in the range of 10 μm that needs to be pretreated. On the other hand, a set of chosen effective membranes is disturbed through large-sized organic anions, iron oxide, colloids, and manganese oxide in order to neutralize the membrane. Before electrodialysis the pretreatment methods such as carbon filtration, flocculation, and filtration methods are carried out.

28.2.2 Membrane filtrations

Nanofiltration and reverse osmosis (RO) are mainly applied in wastewater treatment processes either as the main treatment or posttreatment process in separation, purification, and reuse of salts in addition to large-sized molecules that include dyes used in dye bath effluents and the wastewater generated by textile-processing units (Crossley, 1995, Vant Hul et al., 1997; Sójka-Ledakowicz et al., 1998; Koyuncu et al., 2004; Kim et al., 2005). In RO process, moderate pressure is applied onto the effluent across a semipermeable membrane acceptable to create a concentrate as well as purified permeate (Southern, 1995).

Close to 98% of the water impurities, with more than 100 relative molecular mass, can be removed when this process is used. However, in the case of nanofiltration, the membrane used in that process filters the wastewater and retains the material with more than 200 relative molecular mass (Southern, 1995).

Though there are so many advantages associated with these two processes, still there exist some disadvantages (Cooper, 1993; Vandevivere et al., 1998; Hao et al., 2000). Regular cleaning of the membranes should be performed to retain its surface characteristics since these membranes are prone to attack by dye materials or other effluent constituents. Further, it is a high-investment and cost-consuming process. The filtration process that has the biggest pores (i.e., ultrafiltration and microfiltration) is generally useless because the membrane has a big pore size to restrain the dye molecules from moving through. So this step is considered as the preliminary treatment (Rozzi et al., 1999; Marcucci et al., 2001; Koyuncu, 2003; Ciardelli et al., 2003).

28.2.3 Sorption

Whenever deploying any kind of adsorbents such as high-surface-area inorganic material, activated carbon, or ion exchanger in the process of removal of species present in a liquid stream, the entire process is primarily dependent on the equilibrium that exists between the free and adsorbed species (Southern, 1995). Being a multicomponent mixture, the dye effluents possess a wide range of concentrations and absorption degrees. On the other hand, sometimes weaker bonds are created with the adsorbent due to which the materials get released back into the stream itself. In the studies conducted earlier, various kinds of adsorbents are discussed, for instance, chitin- and cellulose-based adsorbents (poly-*N*-acetylglucosamine), high-surface-area inorganic materials, activated carbons, synthetic ion-exchange resins, fiber-based bioadsorbents, and synthetic cellulose.

In general, the standard ion-exchange systems are not preferred to treat the dye effluents, thanks to heavy monetary thing involved in organic solvents for ion-exchanger regeneration and enormous amount of inorganic load of the effluent (Southern, 1995; Slokar and Le Marechal, 1998; Robinson et al., 2001). Comparatively, the activated carbon seems to be quite efficient in the removal of different dyes present in the aqueous streams (Slokar and Le Marechal, 1998; Robinson et al., 2001). But the regeneration process that demands high temperature, high replacement cost, and waste sludge yield makes it unviable for the SMEs. So, a wide range of cheap adsorbents were tested instead of activated carbon. For the treatment of dyes, researchers already investigated using inorganic adsorbents, for instance, clays, cinder ash, and high-surface-area silica. The output or the success of these adsorbents is primarily dependent on the relative charge of the dye molecule or based on the dye type present in the effluent stream. Being cost-effective and naturally occurring biodegradable polymers (or their synthetic derivatives), the bioadsorbents possess extreme dye-binding capacity that can also perform the function of ion exchangers. Various biomaterials can be deployed as bioadsorbents such as yeast biomass, cotton waste, fungal biomass, wheat, bagasse pith, cellulose, sawdust, wood chips, rice husks, corn, bacterial biomass, and bark. The study that was conducted in the recent times suggest to make use of purified chitin or chitosan since it pH is extremely acidic and it also possesses excellent binding capacity with reactive dye adsorption (Pereira et al., 2003; Faria et al., 2004; Forgacs et al., 2004; Golob and Ojstrsek, 2005).

28.3 Chemical methods for wastewater treatment

28.3.1 Coagulation and flocculation

For a known period of time the inorganic coagulants such as iron salts, magnesium, aluminum, and lime are in use among the textile-processing units to treat the wastewater by removing the total suspended solids (TSS), COD, and BOD (Semerjian and Ayoub, 2003; Aguilar et al., 2002; Golob and Ojstrsek, 2005). The working principle in this process is that when a coagulant is added, there is a quick association occurs between the pollutants and the coagulant. These coagulates formed, or otherwise the flocks, get precipitated over a period of time and are removed through various processes such as filtration, settling, flotation, or simply any other physical technique (Aguilar et al., 2002; Semerjian and Ayoub, 2003; Papić et al., 2004; Golob and Ojstrsek, 2005; Mishra and Bajpai, 2005). The result of this process is sludge formation that is retreated in order to mitigate its toxicity and water content. In the recent years, nonionic coagulant polymers, cationic and organic anionic ones are also being used to remove the color. These are preferred nowadays since the advantages in these surpass the inorganics in terms of enhanced color removal, reduced toxicity, and low amount of sludge production (Al-Mutairi et al., 2004; Zouboulis et al., 2004).

28.3.2 Advanced oxidation processes

Several researchers have done experiments for the last two decades on an exceptional group of oxidation techniques, that is, advanced oxidation process (AOP) indicating its distinctive role in the purification of wastewater (Ollis and Ekabi, 1993). It has been verified that AOP has the possibility to successfully get over the biorecalcitrant pollutant issues in water that will arise at or near the ambient temperature and pressure (Bolton and Cater, 1994).

When discussing about redox reaction, it remains useful to know about the oxidation number or otherwise the oxidation state. This can otherwise be explained as the charge possessed by the bonded atom when the electrons in every bond are offered to highly electronegative atom. For instance, in H_2O, being the highly electronegative element in the group, the oxygen is offered with bond electrons. Due to this scenario the oxidation number of "oxygen" becomes -2 whereas in the case of hydrogen, it becomes 1 which when added, becomes 0 for the neutral molecule. It is to be noted that the oxidation number or state of a given atom is showcased as sign that precedes the number unlike the corresponding charge designation for an ion that has completely different scenario, that is, sign next to number (e.g., O_2^- for oxide anions whereas H^+ in case of protons). The same chemical feature is applicable in whole AOPs, that is, generation of OH• hydroxyl radicals. Being highly reactive species, these radicals predominantly attack each and every organic molecule that is with rate constants in the order of 10^6-10^9 mol/L/s (Hoigne, 1997). OH• has another important characteristic, that is, low-selectivity attack that remains helpful when an oxidant is deployed in the treatment of wastewater and in solving the pollution-related problems. The AOP's versatility is further increased, thanks to different OH• production methods that allow enhanced alignment with precise treatment needs. In Table 28.1, it has been shown that the OH• possesses the highest thermodynamic oxidation potential. This might be the reason behind OH•-based oxidation processes gained momentum among the oxidation technology developers. Further, a number of contaminants of environment react 1 M to 1 B times rapidly with OH• compared to O_3 which is traditionally used oxidant. Photochemical advanced oxidation (APO) actions are followed under an interesting class of AOP since the basis of these processes is nothing, but a free-radical mechanism that is kindled by optimum energy–photon interaction with the chemical molecules represents in the solvent or catalyst. The following sections deal with four broad categories of APO technologies

TABLE 28.1 Oxidation potential of selected oxidizing agents.

Oxidizing agent	EOP (V)
Fluorine	3.06
Hydroxyl radical	2.80
Oxygen (atomic)	2.42
Ozone	2.08
Hydrogen peroxide	1.78
Hypochlorite	1.49
Chlorine	1.36
Chlorine dioxide	1.27
Oxygen (molecular)	1.23

EOP, Electrochemical oxidation potential.

present such as photochemical advanced oxidation, heterogeneous photocatalysis in addition to homogeneous photocatalysis as well as the photo-Fenton process (Mantzavinos et al., 1997; Luck, 1999).

$$\text{abstraction of hydrogen: } \bullet OH + RH \rightarrow R\bullet + H_2O \tag{28.i}$$

$$\text{addition reactions: } \bullet OH + PhX \rightarrow HOPhX\bullet \tag{28.ii}$$

$$\text{electron transfer: } \bullet OH + RH \rightarrow [R-H]^{+\bullet} + HO- \tag{28.iii}$$

28.3.3 Ozone

Being the most powerful and quickly reacting oxidation agent, ozone is able to interact with most of the elements that possess multiple bonds, for instance, C = C, C = N, and N = N. In some instances, it can also bond with simple oxidizable ions, for example, S^{2-}, resulting in the creation of oxyanions such as SO_3^{2-} and SO_4^{2-} (Gogate and Pandit, 2004). There is a quick decolorization occurs when the ozone is applied in wastewater containing water-soluble dyes, but it processes slow reaction with nonsoluble dyes such as disseminate dyes and indigo dyes. In addition, the wastewater produced by textile units generally contains refractory constituents that enter into reaction with ozone resulting in its increased demand (Özbelge et al., 2003; Pera-Titus et al., 2004, Muthukumar et al., 2005). The ozone needs extremely basic pH value in the range of pH >10 for it to get decomposed (Aplin and Wait, 2000; Chu and Ma Chi, 2000). In alkaline solutions, there is a rapid reaction occurs between ozone and compounds of any type present in the reacting medium. This results in the conversion of organic compounds into biodegradable and small-sized molecules (Peralta-Zamora et al., 1999). As a consequence, once the ozone treatment is completed, the biological methods are used so as to achieve holistic mineralization. One of the major disadvantages of ozonation is that its relatively high-cost production of ozone in addition to its very short half-life (Gogate and Pandit, 2004).

28.3.4 Photochemical advanced oxidation technologies

In the event of introducing photochemical alterations in a molecule, the light must be irradiated in the ultraviolet (UV)–visible layout within the arrangement itself. The visible spectrum lies in the wavelength range of 400–800 nm. Some applications tape the sunlight irradiation though one must consider that a mere 3%–5% UV light is only present in the solar spectrum. Generally, the UV light is sourced from pulsed-UV (P-UV) xenon arc lamps, continuous-wave medium-pressure mercury vapor lamps (MP-UV), and continuous-wave low-pressure mercury vapor lamps (LP-UV). In this case, MP-UV as well as LP-UV mercury vapor lamps create a series of line outputs. But there is a continuous output spectrum produced by xenon arc lamp.

LP-UV and MP-UV are predominantly used in the conventional applications of UV irradiation with H_2O_2 or O_3. But the attention is being focused on MP-UV nowadays since it has wider capability for direct photolysis. Further, the MP-UV lamps are able to radiate in a wider range of wavelengths (200–400 nm) in comparison with LP-UV lamps. This enables the easier production of hydroxyl radicals in the presence of hydrogen peroxide. The absorption rate of hydrogen peroxide is high in greater wavelengths such as 250–300 nm. In addition, the electrical efficiency of LP-UV lamp seems to be higher than the MP-UV lamp, whereas the MP-UV lamp generates high amount of UV irradiation per lamp. So, the MP-UV systems can be predicted to utilize a few lamps, consume less space, and need minimal maintenance.

In general, the AOTs' reaction rate is increased by light when compared to similar technology when there is no illumination. So, high-pressure mercury or xenon arc lamps can be used as good light source with excellent emission characteristics in near-UV range. There are a few applications that may need short-UV irradiation, which are discussed in the later section, during which cost-effective germicide lamps are used. There is a reduction in the operating costs, thanks for consumption of less power in the production of HO• than the cost-consuming AOPs, for instance, supercritical technologies or radiolysis. When solar light can be used, one can experience the production of electrical power with safe industrial installations. Since the light is entirely pointed toward the system, the photochemical industrial equipment utilized is compact due to which one can employ small tanks. The system flexibility is generally increased when using light thus allowing the utilization of different oxidations and operational conditions. Further, the photochemical technical has another advantage, that is, there is no need of drastic pH changes in effluents as in case of alkaline ozonation. It is critical to note that the light-mediated AOPs, especially homogeneous processes, to be specific, are not sufficient to treat the mixture of elements that has high absorbing capability or that possesses high amounts of suspended solids. This might be attributed to the decreased quantum efficiency because of loss of light, dispersion, and/or by competitive light absorption.

28.3.5 Heterogeneous photocatalysis

Fujishima and Honda (1972) published their research work on photocatalytic water decomposition using TiO_2 electrodes. After their research, photocatalysis gained enormous attention across the globe in degrading a prism of contaminants such as alkenes, alkanes, phenols, dyes, carboxylic acids, aromatic hydrocarbons, PCBs, halogenated alkanes, surfactants, and pesticides. Based on the catalyst irradiation, the heterogeneous photocatalysis technology works. In this technology a semiconductor is used, which undergoes photoexcitation resulting in the formation of electron-donor sites (or otherwise called reducing sties) and electron-acceptor sites (oxidizing sites) thus providing it with a vast scope as redox reagents. As two active phases such as solid and liquid are involved, this process is termed heterogeneous.

In the presence of various semiconductors, the titanium dioxide seemed to the most viable and optimum used one in environmental applications due to its cost-effectiveness and stability from photo- and chemical corrosion. Titanium dioxide possesses an optimum range of energetic separation from its valence and conduction bands such as VB and CB, and this can be outperformed by solar photon energy. The titanium deoxidation has an estimated VB energy of +3.1 volts and CB energy of −0.1 volt denoting the fact that it has a bandgap energy of 3.2 eV and absorbs the UV light (wavelength $\lambda < 387$ nm). As per the research studies conducted earlier (Legrini et al., 1993; Valavanidis et al., 2006), the hydroxyl radical (•OH) remains the predominant oxidizing species that is held accountable for photooxidation of organic compounds. The electron/hole pairs (Eq. 28.iv) are created, through the separation between CB and VB remains the primary event after the near-UV radiations got absorbed at $\lambda < 387$ nm.

$$TiO_2 + h\nu \rightarrow TiO_2(e^- + h^+) \tag{28.iv}$$

Fig. 28.1 presents various actions that take place after the UV light is absorbed by titanium dioxide. The concurrent production and separation of electrons (e^-_{CB}) as well as holes (h^+_{VB}) are also illustrated in the figure. To be precise, there are three oxidation reactions observed experimentally such as electron transfer from R, H_2O, and adsorption of ^-OH on the catalyst surface. Studies proved that the photocatalytic degradation of water contaminants can be improved when hydrogen peroxide is added since it is a comparatively better electron accepted than the oxygen. This reaction is deduced by the generation of hydroxyl radicals at the time of reaction with CB electrons as per Eq. (28.iv). It confines the recombination rate of electron–hole and enhances the concentration of hydroxyl radicals at the surface of TiO_2.

28.3.6 Homogeneous photodegradation

The earlier stage usages of homogeneous photodegradation (single-phase system) predominantly made use of UV/H_2O_2 and UV/O_3 in the treatment of contaminated water. There are two principal areas to be focused when using the UV light in the photodegradation of pollutants:

- Direct photodegradation in which the pollutant is directly excited by UV light.

$$M^{n+} + H_2O_2 \rightarrow M^{(n+1)+} + OH^- + \bullet OH \tag{28.v}$$

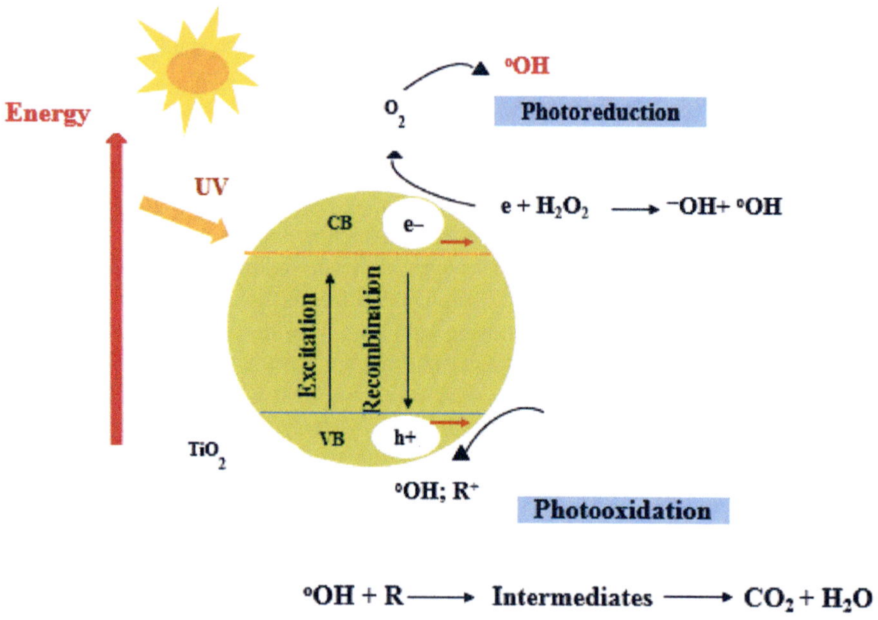

FIGURE 28.1 Actions occur after the UV light absorbed by titanium dioxide. *UV*, Ultraviolet.

- Photooxidation in which the oxidative processes are led by light after being initiated by hydroxyl radicals. In this process the oxidant is used to produce radicals that attack the organic pollutants so as to start the oxidation process. Three primary oxidants used are hydrogen peroxide (H_2O_2), ozone, and photo-Fenton system (Fe^{3+}/H_2O_2).

$$Fe^{2+} + H_2O_2 \rightarrow Fe^{3+} + OH^- + \bullet OH \quad K = 63\ L/mol/s \tag{28.vi}$$

$$Fe^{2+} + H_2O_2 \rightarrow Fe^{2+} + HO_2\bullet + H^+ \quad K = 3.1 \times 10^{-3}\ L/mol/s \tag{28.vii}$$

$$Fe^{2+} + \bullet OH \rightarrow Fe^{3+} + OH^- \quad K = 3 \times 10^8\ L/mol/s \tag{28.viii}$$

$$\bullet OH + H_2O_2 \rightarrow HO_2\bullet + H_2O \quad K = 3.3 \times 10^7\ L/mol/s \tag{28.ix}$$

$$Fe^{3+} + HO_2\bullet \rightarrow Fe^{2+} + H^+ + O_2 \quad K = 2 \times 10^3\ L/mol/s \tag{28.x}$$

$$Fe^{3+} + O_2\bullet^- \rightarrow Fe^{2+} + O_2 \quad K = 5 \times 10^7\ L/mol/s \tag{28.xi}$$

$$Fe^{2+} + HO_2\bullet \rightarrow Fe^{3+} + HO_2^- \quad K = 1.2 \times 10^6\ L/mol/s \tag{28.xii}$$

Fenton reagents. The Fenton's reagent-based oxidation system, that is, hydrogen peroxide (i.e., with the availability of ferrous salt), is used to treat both organic as well as inorganic substances. This process is on the basis of production of ROS that can effectively degrade the wastewater pollutants though these elements' nature is still discussed for further usage. The primary reactive radical species involve the availability of hydroxyl radicals, whereas the higher oxidized iron species may be generated. There has been an acceptance that hydroxyl in addition to ferryl complexes presents in Fenton's mechanism in which one is able to predominate based on the operating conditions such as the addition of scavengers, metal–peroxide ratio, and nature of substrate. The oxidation system can be better deployed to degrade the nonbiodegradable toxic waste effluents and make it optimal so that it undergoes a secondary biological treatment. A wide range of dyes can be decolorized through Fenton oxidation process and compared to ozonation, this process is cost-effective, and high amount of COD reduction is possible in this process. But the Fenton oxidation has a disadvantage, that is, it demands low pH, whereas the pH level in textile processed wastewaters is high. At high pH values, this process leads to the generation of huge volume of sludge due to the precipitation of ferric iron salts. This eventually results in the loss of effectiveness of the process.

28.3.7 Fenton- and photo-Fenton-assisted reactions

In the year 1894 the Fenton reaction was discovered by H.J.H. Fenton, whereas in 1934 two researchers Haber and Waiser found that the Fenton reaction has an efficient oxidative agent, that is, hydroxyl radical (•OH). After this major finding, various researchers attempted to decode the entire mechanism (Walling, 1975; Sychev and Isak, 1995).

When there are no light or complexing ligands, for instance, water, the hydrogen peroxide decomposition mechanism that occurs in acid homogeneous aqueous solution involves the production of hydroxyperoxyl ($HO\bullet_2/O_2^-$) and hydroxyl radicals (•OH) (DeLaat and Gallard, 1999; Gallard and De Laat, 2000). The •OH radical present in the solution starts attacking every organic compound. There are various pathways followed in the metal regeneration process (Sychev and Isak, 1995). There is a strong influence upon Fenton reaction rates with irradiation using UV/visible light (Rupert et al., 1993). This photo-assisted reaction is otherwise called "photo-Fenton reaction." The positive impact created by the irradiation upon the rate of degradation might be attributed to the photochemical regeneration of Fe^{2+} through the ferric ion (Fe^{3+}) photoreduction. The newly produced ferrous ion enters into reaction with hydrogen peroxide thus resulting in strong •OH radical and ferric ion, whereas the cycle continues to process. During these scenarios the iron can be completely accounted as a real catalyst.

The photo-Fenton process has a major advantage, that is, light sensitivity up to 600 nm (35% of the solar radiation) wavelength. The high light penetration and close contact between the oxidizing agent and pollutant are present when using homogeneous solution. Both photo-Fenton and Fenton processes are applied in a successful manner in the treatment of numerous contaminants, for instance, halogenated alkanes, alkenes, phenols, dyes, and pesticides. Various parameters that authorize the photo-Fenton system kinetics were reported such as availability of radical scavengers, H_2O_2 concentration, initial pollutant concentration, iron concentration, temperature, pH, irradiation source, iron species, and concentration of hydrogen peroxide (Chen and Pignatello, 1997; Kiwi et al., 2000).

28.4 Challenges and perspectives

In recent years, electrooxidation is found to be the effectual methods for unspecific and total mineralization of dissolved organic pollutants. For the past two decades, more focus was on the research of efficiently oxidizing different pollutants on various electrodes, upgrading the electrocatalytic activity and electrochemical stability of electrode materials, exploration of factors upsetting the process performance, and investigation of the mechanisms and kinetics of pollutant deprivation. The stable electrode metals such as Ti, Pt, and Pb were used with noble metal oxide coated, and concurrent oxidation–reduction reaction of organic pollutants will take place on electrode surfaces without adding any chemicals. Here in this process, no sludge will be formed in this process. Due to its very low concentrated solutions, its efficiency is restricted.

Till now, the electro-Fenton processes, and specifically the methods based on the electrochemical production of hydrogen peroxide, were efficaciously employed for the degradation of several pollutants from wastewater, sludge, and water. The pollutants that let in PAHs, aromatic compounds, refractory wastes, and chlorinated solvents were the target.

Oxidizing solution has active oxidant agents in aqueous systems that were confirmed by electrochemical processes. Let us take an example of reactive chlorine species, hypochlorous acid (HClO), chlorine (Cl_2), and hypochlorite ions (ClO^-) can be electrochemically generated from a solution containing chloride, for example, NaCl used as electrolyte solution, due to the NaCl electrolysis:

$$\text{Anode: } 2Cl^- \rightarrow Cl_2 + 2e^- \tag{28.xiii}$$

$$\text{Cathode: } 2H_2O + 2e^- \rightarrow H_2 + 2OH^- \tag{28.xiv}$$

$$\text{Bulk solution: } Cl_2 + H_2O \rightarrow HClO + H^+ + Cl^- \tag{28.xv}$$

$$HOCl \rightarrow H^+ + OCl^- \tag{28.xvi}$$

Increase in the redox potential the reactive chloride species will form which will oxidize the organic pollutants during the electrolysis process furthermore without adding oxidizing solutions.

28.5 Conclusion

In recent years, electrochemical technology is found to be one of the most efficient methods for undefined and total mineralization of dissolved organic pollutants. For the past 20 years, research work has been concentrated on the effectiveness in oxidating several pollutants on various electrodes. While using the stable electrode metals such as Ti, Pt, and Pb, with noble metal oxide coated, simultaneous oxidation–reduction reaction of organic pollutants will take place on electrode surfaces without the addition of any chemicals. No sludge will be formed in this process. But its efficiency is restricted to very low concentrated solutions.

To date, the electro-Fenton processes, and in specific the methods based on the production of hydrogen peroxide, were efficaciously employed for the degradation of various pollutants from wastewater, sludge, and water. The pollutants targeted were PAHs, aromatic compounds, refractory wastes, and chlorinated solvents. Further, in wastewater treatments, isolation of good COD showed to be achieved within one or a few hours with current densities of 100–200 mA/cm^2. In electrochemical processes for the aqueous systems, active oxidant agents are found to be generated in the presence of an oxidizing solution.

References

Aguilar, M.I., Sáez, J., Lloréns, M., Soler, A., Ortuño, J.F., 2002. Nutrient removal and sludge production in the coagulation–flocculation process. Water Res. 36 (11), 2910–2919.

Al-Mutairi, N.Z., Hamoda, M.F., Al-Ghusain, I., 2004. Coagulant selection and sludge conditioning in a slaughterhouse wastewater treatment plant. Bioresour. Technol. 95 (2), 115–119.

Aplin, R., Wait, T.D., 2000. Comparison of three advanced oxidation processes for degradation of textile dyes. Water Sci. Technol. 42 (5), 345–354.

Beranek, R., 2011. Photo Electrochemical Methods for the Determination of the Band Edge Positions of TiO_2 Based Nanomaterials. Hindawi Publishing Corporation. Available from: http://doi.org/10.1155/2011/786759.

Bolton, J.R., Cater, S.R., 1994. Homogeneous photodegradation of pollutants in contaminated water: an introduction. Aquatic and Surface Photochemistry. Lewis Publishers, Boca Raton, FL, pp. 467–490.

Chen, R., Pignatello, J.J., 1997. Role of quinone intermediates as electron shuttles in Fenton and photo-assisted Fenton oxidations of aromatic compounds. Environ. Sci. Technol. 31 (8), 2399–2406.

Chu, W., Ma Chi, W., 2000. Quantitative prediction of direct and indirect dye ozonation kinetics. Water Res. 34 (4), 3153–3160.

Ciardelli, G., Ciabatti, I., Ranieri, L., Capannelli, G., Bottino, A., 2003. Membrane contactors for textile wastewater ozonation. Ann. N. Y. Acad. Sci. 984 (1), 29–38.

Cooper, P., 1993. Removing colour from dyehouse waste waters – a critical review of technology available. J. Soc. Dyers Colour. 109 (53), 97–100.

Crossley, C., 1995. Membrane technology for the separation of dyehouse effluent. In: Cooper, P. (Ed.), Colour in Dyehouse Effluent. Society of Dyers and Colourists, Bradford, England, pp. 155–170.

De Laat, J., Gallard, H., 1999. Catalytic decomposition of hydrogen peroxide by Fe(III) in homogeneous aqueous solution: mechanism and kinetic modeling. Environ. Sci. Technol. 33 (16), 2726–2732.

Deng, Y., Zhao, R., 2015. Advanced oxidation processes (AOPs) in wastewater treatment. Curr. Pollut. Rep. 1, 167–176.

Faria, P.C.C., Órfão, J.J.M., Pereira, M.F.R., 2004. Adsorption of anionic and cationic dyes on activated carbons with different surface chemistries. Water Res. 38 (8), 2043–2052.

Forgacs, E., Cserhati, T., Oros, G., 2004. Removal of synthetic dyes from wastewaters: a review. Environ. Int. 30 (7), 953–971.

Fujishima, A., Honda, K., 1972. Electrochemical photolysis of water at a semiconductor electrode, Nature 238, 37–38.

Gallard, H., De Laat, J., 2000. Kinetic modelling of Fe(III)/H_2O_2 oxidation reactions in dilute aqueous solution using atrazine as a model organic compound. Water Res. 34 (12), 3107–3116.

Gogate, P.R., Pandit, A.B., 2004. A review of imperative technologies for wastewater treatment II: hybrid methods. Adv. Environ. Res. 8 (3), 553–597.

Golob, V., Ojstrsek, A., 2005. Removal of vat and disperse dyes from residual pad liquors. Dye Pigment. 64 (1), 57–61.

Hao, O.J., Kim, H., Chiang, P.C., 2000. Decolorization of wastewater. Crit. Rev. Environ. Sci. 30 (4), 449–505.

Hoigne, J., 1997. Inter-calibration of OH radical sources and water quality parameters. Water Sci. Technol. 35 (4), 351–358.

Kim, T.H., Park, C., Kim, S., 2005. Water recycling from desalination and purification process of reactive dye manufacturing industry by combined membrane filtration. J. Clean. Prod. 13 (8), 779–786.

Kiwi, J., Lopez, A., Nadtochenko, V., 2000. Mechanism and kinetics of the OH-radical intervention during Fenton oxidation in the presence of significant amount of radical scavenger (Cl –). Environ. Sci. Technol. 34 (11), 2162–2168.

Koyuncu, I., 2003. Influence of dyes, salts and auxiliary chemicals on the nanofiltration of reactive dye baths: experimental observations and model verification. Desalination 154 (1), 79–88.

Koyuncu, I., Topacik, D., Yuksel, E., 2004. Reuse of reactive dyehouse wastewater by nanofiltration: process water quality and economical implications. Sep. Pur. Technol. 36 (1), 77–85.

Legrini, O., Oliveros, E., Braun, A.M., 1993. Photochemical processes for water-treatment. Chem. Rev. 93 (2), 671–698.

Liu, C., Neale, Z.G., Cao, G., 2016. Understanding electrochemical potentials cathode materials in rechargeable batteries. Mater. Today 19 (2), 109–123.

Luck, F., 1999. Wet air oxidation: past, present and future. Catal. Today 53, 81–91.

Mantzavinos, D., Lauer, E., Hellenbrand, R., Livingston, A.G., Metcalfe, I.S., 1997. Wet oxidation as a pretreatment method for wastewaters contaminated by bioresistant organics. Water Sci. Technol. 36 (2), 109–116.

Marcucci, M., Nosenzo, G., Capannelli, G., Ciabatti, I., Corrieri, D., Ciardelli, G., 2001. Treatment and reuse of textile effluents based on new ultrafiltration and other membrane technologies. Desalination 138, 75–82.

Mazloomi, K., Sulaiman, N. b, Moayedi, H., 2012. Electrical efficiency of electrolytic hydrogen production. Int. J. Electrochem. Sci. 7 (4), 3314–3326.

Mishra, A., Bajpai, M., 2005. Flocculation behaviour of model textile wastewater treated with a food grade polysaccharide. J. Hazard. Mater. 118 (1), 213–217.

Muthukumar, M., Sargunamani, D., Senthilkumar, M., Selvakumar, N., 2005. Studies on decolouration, toxicity and the possibility for recycling of acid dye effluents using ozone treatment. Dye Pigment. 64 (1), 39–44.
Ollis D.F., Ekabi H.A. (1993) Photocatalytic Purification and Treatment of Water and Air. Elsevier Pvt. Ltd, Amsterdam.
Opu, M.S., 2015. Effect of operating parameters on performance of alkaline water electrolysis. Int. J. Therm. Environ. Eng. 9 (2), 53–60.
Özbelge, T.A., Erol, F., Ozbelge, H.O., 2003. A kinetic study on the decolorization of aqueous solutions of Acid Red-151 by ozonation. J. Environ. Sci. Health A: Tox. Hazard. Subst. Environ. Eng. 38 (8), 1597–1614.
Papić, S., Koprivanac, N., Loncaric Bozic, A., Metes, A., 2004. Removal of some reactive dyes from synthetic wastewater by combined Al(III) coagulation/carbon adsorption process. Dye Pigment. 62 (3), 291–298.
Pera-Titus, M., García-Molina, V., Baños, M.A., Giménez, J., Esplugas, S., 2004. Degradation of chlorophenols by means of advanced oxidation processes: a general review. Appl. Cat. B: Environ. 47 (4), 219–256.
Peralta-Zamora, P., Kunz, A., de Moraes, S.G., Pelegrini, R., Moleiro, P.D., Reyes, J., et al., 1999. Degradation of reactive dyes – I. A comparative study of ozonation, enzymatic and photochemical processes. Chemosphere 38 (4), 835–852.
Pereira, M.F.R., Soares, S.F., Órfão, J.J.M., Figueiredo, J.L., 2003. Adsorption of dyes on activated carbons: influence of surface chemical groups. Carbon 41, 811–821.
Philippe, M., Jerome, H., Sebastien, B., Gerard, P., 2005. Modelling and calculation of the current density distribution evolution at vertical gas-evolving electrodes. Electrochim. Acta 51 (6), 1140–1156.
Robinson, T., McMullan, G., Marchant, R., Nigam, P., 2001. Remediation of dyes in textile effluent: a critical review on current treatment technologies with a proposed alternative. BioRes. Technol. 77 (3), 247–255.
Rozzi, A., Malpei, F., Bonomo, L., Bianchi, R., 1999. Textile wastewater reuse in Northern Italy (COMO). Water Sci. Technol. 39 (5), 122–128.
Rupert, G., Bauer, R., Heisler, G., 1993. The photo-Fenton reaction—an effective photochemical wastewater treatment process. J. Photochem. Photobiol. A 73 (1), 75–78.
Semerjian, L., Ayoub, G.M., 2003. High-pH-magnesium coagulation flocculation in wastewater treatment. Adv. Environ. Res. 7, 389–403.
Shanthi, V., Ramanathan, K., Ahmed Basha, C., 2016. Domestic sewage treatment using batch stirred tank electrochemical reactor. Int. J. Chem. Tech. Res. 3 (3), 1711–1721.
Slokar, Y.M., Le Marechal, A.M., 1998. Methods of decoloration of textile wastewaters. Dye Pigment. 37 (4), 335–356.
Sójka-Ledakowicz, J., Koprowski, T., Machnowski, W., Knudsen, H.H., 1998. Membrane filtration of textile dyehouse wastewater for technological water reuse. Desalination 119 (1), 1–9.
Southern, T.G., 1995. Technical solutions to the colour problem: a critical review. In: Cooper, P. (Ed.), Colour in Dyehouse Effluent. Society of Dyers and Colourists, Bradford, England, pp. 73–91.
Sychev, A.Y., Isak, V.G., 1995. Iron compounds and the mechanisms of the homogeneous catalysis of the activation of O_2 and H_2O_2 and of the activation of organic substrates. Russ. Chem. Rev. 64 (12), 1105–1129.
Valavanidis, A., Vlahogianni, T., Dassenakis, M., Scoullos, M., 2006. Molecular biomarkers of oxidative stress in aquatic organisms in relation to toxic environmental pollutants. Ecotoxic. Environ. Saf. 64 (2), 178–189.
Vandevivere, P.C., Bianchi, R., Verstraete, W., 1998. Treatment and reuse of wastewater from the textile wet-processing industry: review of emerging technologies. J. Chem. Technol. Biotechnol. 72 (4), 289–302.
Vant Hul, K.P., Racz, I.G., Reith, T., 1997. The application of membrane technology for reuse of process water and minimization of wastewater in textile washing range. J. Soc. Dyers Colour. 113, 287–295.
Verschueren, R.H., De Borggraeve, W.M., 2019. Electrochemistry and photoredox catalysis a comparative evaluation in organic synthesis. Molecules 24 (11), 338.
Walling, C., 1975. Fenton's reagent revisited. Acc. Chem. Res. 8 (19), 125–131.
Zouboulis, A.I., Chai, X.L., Katsoyiannis, I.A., 2004. The application of bioflocculant for the removal of humic acids from stabilized landfill leachates. J. Environ. Manage. 70, 35–41.

Chapter 29

Sources, fate, and impact of pharmaceutical and personal care products in the environment and their different treatment technologies

Gurudatta Singh, Anubhuti Singh, Priyanka Singh, Akanksha Gupta, Reetika Shukla and Virendra Kumar Mishra

Institute of Environment and Sustainable Development, Banaras Hindu University, Varanasi, India

29.1 Pharmaceuticals and personal care products

The occurrence of PPCPs in the environment has emerged as a serious health concern worldwide. Pharmaceuticals and personal care products (PPCPs) are the groups of compounds that are frequently termed emerging contaminants. Pharmaceuticals are defined as prescription, over the counter, human and veterinary curative drugs cast off to treat human and animal diseases. In contrast, personal care products (PCPs) are generally used to improve the quality of daily life (Boxall et al., 2012). Currently, about 4000–5000 pharmaceutical compounds are in use, which may be released into the environment, and under the present scenario, there is no legally permissible concentration of pharmaceutical compounds in the environment (Hernandez Leal et al., 2010).

The consumption of pharmaceuticals is rising rapidly, and about 8–9 million kg of antibiotics were used in the year 2013 (Pennington et al., 2017). Pharmaceutical contaminants are continuously released into the aquatic environments through multiple pathways, including wastewater treatment plants, domestic wastewater, hospital discharge, improper disposal, and runoff of veterinary medicine (Daughton, 2003; Leung et al., 2012; Liu and Wong, 2013); therefore these compounds may be increased continuously into the environment. Since the last few decades, there have been increasing reports of the presence of PPCPs in the water bodies (water, sediments, and biota) at concentrations capable of causing harmful effects to the aquatic organisms. This becomes a foremost concern because PPCPs are widely used in human and veterinary medicine, resulting in their continuous release to the environment (Nikolaou et al., 2007). The existence of emerging pollutants, such as pharmaceuticals and PCPs, demands more attention toward this matter (Daughton and Ternes, 1999; Diaz-Cruz et al., 2003).

Pharmaceutical contaminants and PCPs entering water and soils may either remain in their original form or may be metabolized into different forms (Kepp et al., 1997). These compounds were found to be in the range of ng/L to moderately high concentrations of μg/L (Ternes, 2007; Schwarzenbach et al., 2006). The higher levels of these compounds are reported in wastewater and hospital waste, from where they can reach the surrounding aquatic ecosystems. Pharmaceutical compounds from the industrial and domestic sectors are expelled out in their original form into sewage treatment plants (STPs) (Yadav et al., 2019), from these STPs they might reach into the aquatic environment. The various studies identified its impact on ecology, and the qualitative function of water even at a very low concentration. The toxicity of these pharmaceutical compounds may affect the ecological functionality of any ecosystems and ultimately some serious impacts on human life might be resulted (Hernandez Leal et al., 2010).

The different PPCPs (Fig. 29.1) reported from the water and wastewater in different concentrations are discussed (Table 29.1). Most of the pharmaceutical compounds have a molecular mass less than 500 Da (Lipinski et al., 1997) and are composed of multifaceted chemical, which differs in structure, function, molecular weight, and form. These are

Microbe Mediated Remediation of Environmental Contaminants. DOI: https://doi.org/10.1016/B978-0-12-821199-1.00029-8

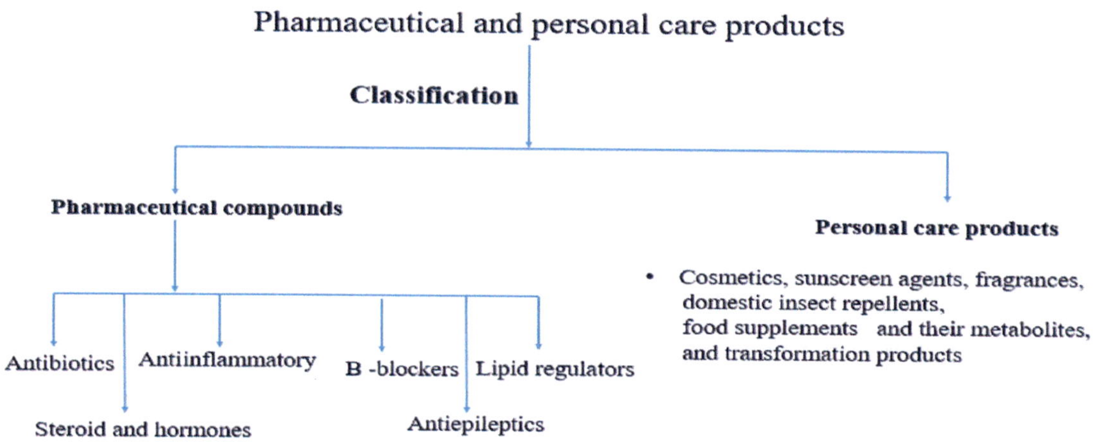

FIGURE 29.1 Classification of pharmaceutical and personal care products.

TABLE 29.1 Pharmaceuticals and personal care products (PPCPs) in the wastewater.

Name of PPCPs	Compound classification	Concentration (ng/L)	References
Triclosan	Antiseptic	800; 13–82	Jim et al. (2006) and Kasprzyk-Hordern et al. (2009)
Phenytoin	Anticonvulsant	450; 90–373	Jim et al. (2006) and Reungoat et al. (2011)
Ketoprofen	NSAID	1200; 22–1081; 330	Jim et al. (2006), Rabiet et al. (2006), and Bendz et al. (2005)
Ibuprofen	NSAID	1900; 65–491; 420	Jim et al. (2006) and Kasprzyk-Hordern et al. (2009)
Chlorophene	Antiseptic	750	Jim et al. (2006)
Diclofenac	NSAID	6–496; 271–486; 599; 187–855	Kasprzyk-Hordern et al. (2009), Rabiet et al. (2006), and Teijon et al. (2010)
Biphenylol	Antiseptic	900	Jim et al. (2006)
p-Chloro-*m*-cresol	Antiseptic	400	Jim et al. (2006)

NSAID, Nonsteroidal antiinflammatory drug.

polar molecules with 1–2 functional groups. These compounds have a degree of ionization, may be lipid-soluble or water-soluble (Rivera-Utrilla et al., 2013). The nature and behavior of pharmaceutical and PCPs are discussed in Table 29.2.

29.1.1 Personal care products

PCPs are a group of compounds used in soaps, toothpaste, fragrances, lotions, and sunscreens. The primary classes of PCPs include insect repellants (DEET), preservatives (parabens), disinfectants [triclosan (TCS)], fragrances (musks), and UV filters (methyl benzylidene camphor). Unlike pharmaceuticals, these are intended for external on the human body and thus are not subjected to metabolic alterations. Therefore large quantities of PCPs enter the environment unaltered through regular usage (Ternes et al., 2004). Some PCP is used in large amounts, and recent studies have indicated that many of them are environmentally persistent, bioactive, and have the potential for bioaccumulation (Peck, 2006; Mackay and Barnthouse, 2010). PCPs are among the most commonly detected compounds in surface water throughout the world (Peck, 2006). However, in comparison to pharmaceuticals, relatively little is known about the toxicity of PCPs (Daughton and Ternes, 1999).

It is frequently found at high concentrations in various environments such as surface water, wastewater, and even in mother's milk. Different PCPs have been found associated with adverse impacts on ecosystems, for example, the

TABLE 29.2 List of pharmaceutical compounds and their nature with the chemical structure of some compounds.

Name of compounds	Nature of pharmaceutical contaminants	Examples	Chemical structure of some compounds
Antibiotics	Hydrophilic	Ciprofloxacin Clarithromycin Erythromycin Norfloxacin Ampicillin Azithromycin	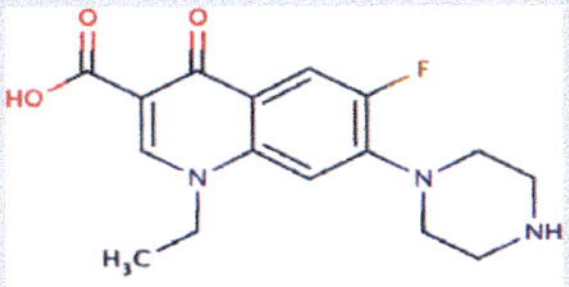 Norfloxacin
Antiinflammatory	Hydrophilic Varies Moderate hydrophobic Moderate hydrophobic Varies Moderate hydrophobic	Aspirin Diclofenac Ibuprofen Irgasan Clotrimazole Econazole	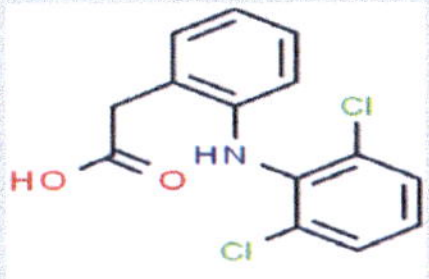 Diclofenac
Antiepileptics	Moderate hydrophobic	Carbamazepine Phenobarbital Phenytoin	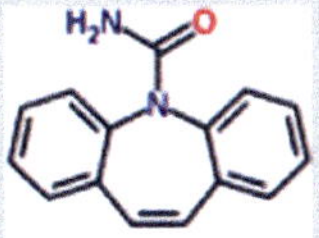 Carbamazepine
Lipid regulators	Moderate hydrophobic Hydrophobic Hydrophobic	Clofibric acid Bezafibrate Fenofibric acid	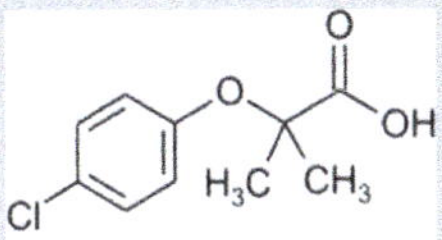 Clofibric acid
Artificial sweeteners	Hydrophilic Varies Moderate hydrophobic	Cyclamate Saccharin Sucralose Acesulfame K	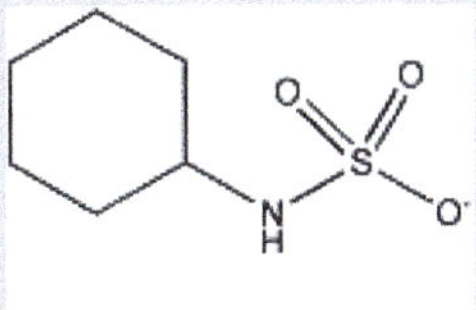 Cyclamate
Beta-blockers	Hydrophilic Hydrophilic Hydrophilic	Metoprolol Acebutolol Atenolol	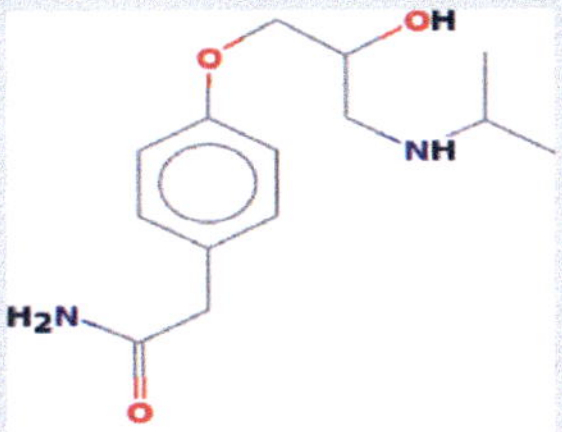 Atenolol
Personal care products	Hydrophobic Hydrophilic	DEET—insect repellent Parabens—Bacteriostatic and fungistatic, cosmetic Triclosan—biocide Sunscreen agents Synthetic musk	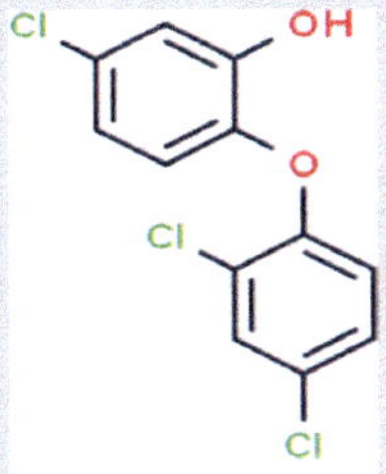 Triclosan

toxicity evaluation of TCS exposed it as one of the important reasons behind abortion in rats and humans (Cullinan et al., 2012; Wang et al., 2015). Studies also revealed that the high TCS level causes the assortment of pathogenic bacteria into multidrug-resistant bacteria. TCS makes target site modification, membrane resistance, and efflux pumps in pure cultures, leading toward cross-resistance against antibiotics (Carey and McNamara, 2015).

29.2 Source of pharmaceutical and personal care products in the aquatic environment

Increasing needs of the human population are the main driving force responsible for the production and discharge of pharmaceutical and PCPs into the different environmental matrices (water, wastewater, soil, and sediments) (Daughton, 2007). Human pharmaceutical compounds are consumed, metabolized, and excreted into the domestic sewage as a mixture of the parent compound and its metabolites, comprising mostly of transformation products and conjugated glucuronides (Heberer, 2002). Conjugated compounds of pharmaceuticals have previously been shown to be easily cleaved during wastewater treatment, releasing the compound into the treated wastewater, and subsequently into the environment (Ternes, 1998; Jelic et al., 2011).

The pharmaceutical and PCPs are released into the environment mainly by two means, that is, the point sources and the diffuse sources. The point sources of contamination are the distinct sites discharging the pollutants from a specific point and the number of discharges through them can be measured easily via mathematical modeling (Lapworth et al., 2012). For example, effluents generate from industries, hospitals, and STPs, as well as the septic tanks, are the point sources for the soil zone and the water resources. On the other hand, in the case of diffused sources, pollutants are released from large area and combined into several points and hardly recognized as the distinct site (Lapworth et al., 2012). When comparing to the point source of pollutants, diffused pollution is advantageous as it usually has less environment loading as it is attenuated naturally in the subsurface and shallow soil (Murray et al., 2010). There are different sources and pathways through which drugs are discharged transported to various matrices (Fig. 29.2). Majorly there exist several sources, namely, landfill, animal waste, freshwater aquaculture waste, hospital waste, industrial waste, and municipal waste from where these compounds may reach to either in soil or aquatic ecosystems (Heberer, 2002; Eggen et al., 2010; Pal et al., 2010; Lapworth et al., 2012).

29.2.1 Point source

29.2.1.1 Municipal landfills

Municipal landfills are the foundation of varied ranges of hazardous complexes concerning ecological, wildlife, and human health (Eggen et al., 2010). Wastes dumped in the landfills may comprise additives that possibly will unfavorably affect the health of the environment. Besides, municipal landfilling generate leachate, which includes remarkable

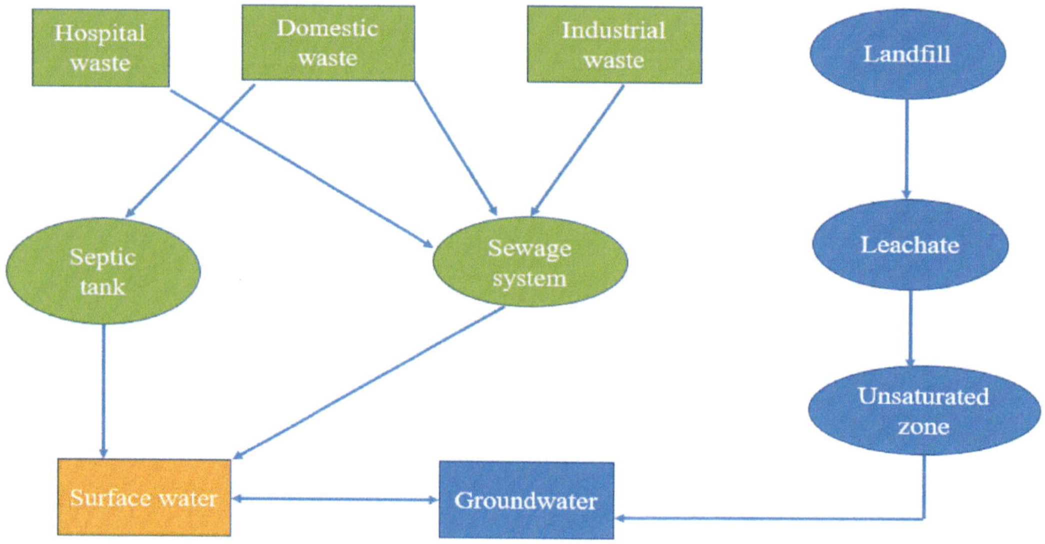

FIGURE 29.2 Pathway of pharmaceutical compounds in the aquatic environment.

dissolved organic matters, such as pesticides, pharmaceuticals, and various other contaminants such as heavy metals, nutrients (Li et al., 2009). It is wastewater, which is the infiltrating water cooperating with fluids, solids, and gases within the buried waste (Buszka et al., 2009). According to the study conducted by Buszka et al. (2009), it was found that there are numerous types of pollutants such as pharmaceuticals, PCPs, hormones, and fire retardants in the wells that locate at its nearest from the landfill. Another research by Barnes et al. (2004) in the United States has the same discovery that those wells that are closest to the landfills have the highest concentration of pharmaceuticals. It is persistent in the groundwater as the landfills generate the carbon-based wastewater and the leachates that it is flowing, penetrate the underground water, and get discharged to the surface water.

Landfill sites also have the existence of pharmaceutical compounds, but their presence and fate are mostly neglected. Once disposed of in municipal solid waste or domestic sewage, the pharmaceutical compounds path toward the landfill, which may undergo adsorption, degradation at the site or enter the leachate, and lastly, disposed of in the landfill (Musson and Townsend, 2009). Many studies and experiments conducted to provide evidence relevant to the life cycle analysis of pharmaceuticals and when joined with landfill leachate data aids the valuation of the efficiency of landfills disposal of medicines. In addition, if destruction did along with domestic waste, complexes ended on landfills site, and from there they reach to the effluent (Metzger, 2004). Holm et al. (1995) found the distribution of pharmaceuticals originate from waste from the pharmaceutical industries in the downgradient of a landfill.

29.2.1.2 Wastewater (domestic and industrial)

Wastewater is considered as the most important point source out of the many other sources of the pollutants to the aquatic environments (Glassmeyer et al., 2005). Numerous researches conducted in various regions such as England, Australia, the United States, and Spain reported the huge number of drugs and medicines from the wastewater (Petrovic et al., 2006; Gros et al., 2006; Kasprzyk-Hordern et al., 2008; Batt et al., 2008). The abovementioned studies suggest that the wastewaters from hospitals, aquaculture, trading, and municipals are the major sources of PPCPs in the environment (Figs. 29.3 and 29.4). Most of the treated wastewater is discharged from the STPs into the surrounding environment, with this discharge several PPCPs may enter the environment (Daughton and Ternes, 1999). Recently it has been studied that the septic system in the United States holds comparable types of antibiotics, which include ibuprofen, paracetamol, salicylic acid, and TCS. The presence of these compounds may be hazardous to the ground and surface water resources (Carrara et al., 2008; Godfrey et al., 2007; Conn et al., 2010; Lapworth et al., 2012).

29.2.1.3 Sewage sludge

Carbon-based pollutants such as pharmaceuticals and antibiotics enter the lands as well as freshwater resources through various means, and sewage sludge is one of the utmost significant sources (Lapworth et al., 2012). Sewage sludge is

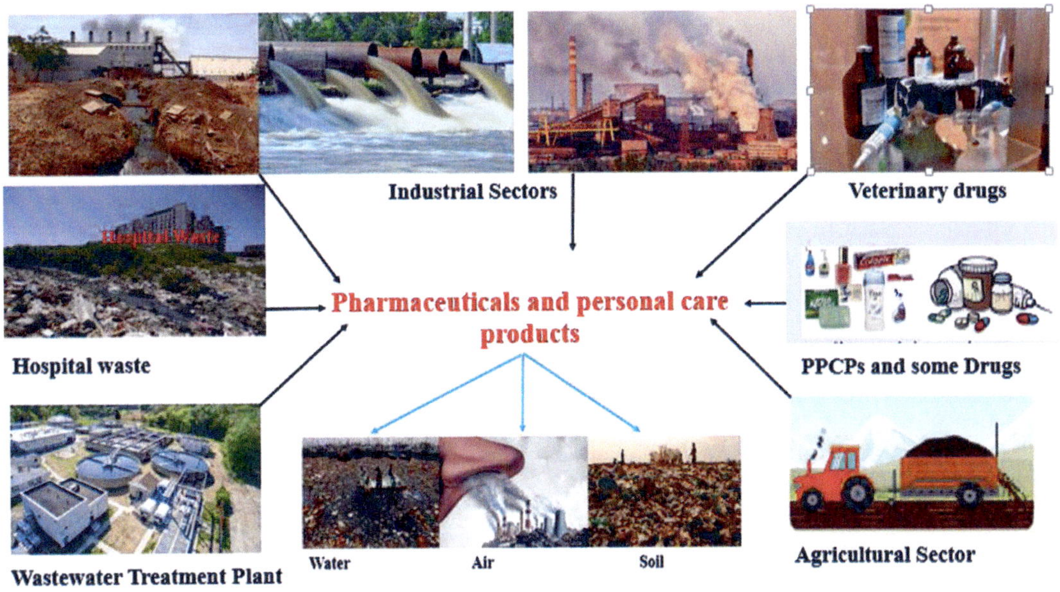

FIGURE 29.3 Source of pharmaceutical compounds in the environment (Water, Soil, and Air).

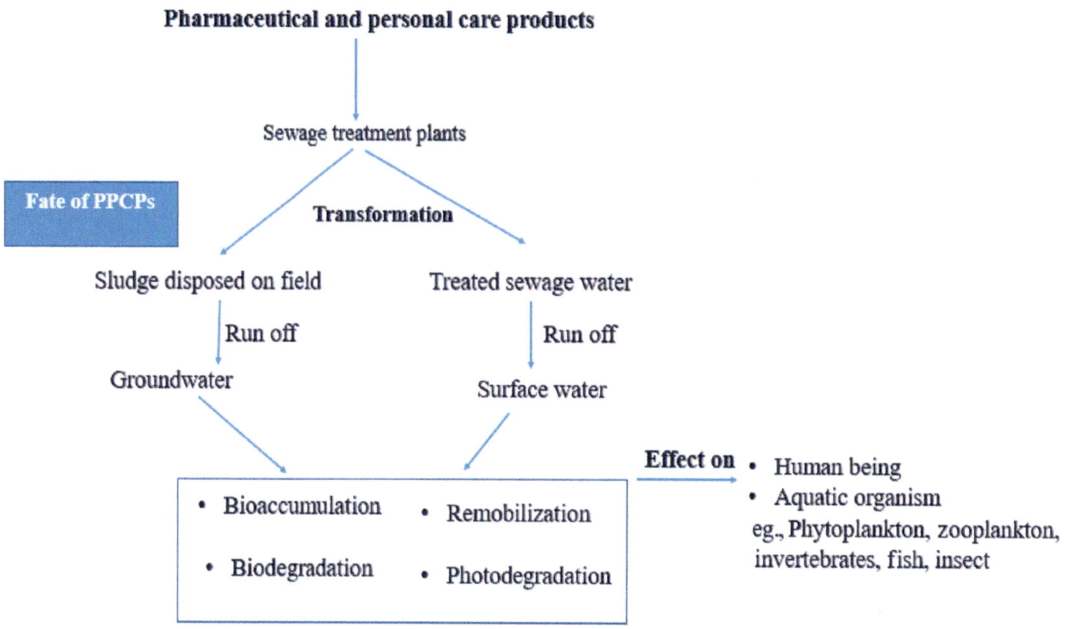

FIGURE 29.4 Fate of the pharmaceutical compounds in the environment.

also known as biosolid; it is the filtrate of the treatment of wastewater. In recent times in America, it is generally used as soil modification, for example, it was estimated that 8106 dry tons of sludge emerged; 50% of which were used for landfilling. Since it consists of nutrients as well as a carbon-based substance, which can improve the soil qualities (Kinney et al., 2008), therefore it can be used additive in various land applications. According to a study conducted by Kinney et al. (2008), the maximum amount of antibiotics that originate in the biosolid was thiabendazole (about 5000 mg/kg). Besides this, other kinds of medicines, such as caffeine and carbamazepine, were also found. Likewise, because of the increasing amount of antibiotics as well as high solubility of two kinds of halogenated hydrocarbon in biosolid, the use of biosolids may lead to possible freshwater pollution (Lapworth et al., 2012).

29.2.1.4 Hospitals

Pharmaceutical residues are the major constituent of hospital wastewater (Schuster et al., 2008). The concentrations of pharmaceutical compounds in hospital wastewater are much higher than municipal sewage. However, substance flow in the wastewater of the hospital is much lower than the municipal wastewater because of the much smaller share of effluent in developed countries. The study of Kümmerer and Helmers (2000) observed that hospital wastewater is 100 times more dilute than the municipal sewage.

29.2.1.5 Private households

Therapeutic compounds that crossed their date of expiry and those that are unused used to get disposed of in domestic premises and ultimately reach to sewage. Following European Union (EU) legislation, the disposal of medicines that are not in use through household waste has been permitted since 1994. According to the reports of Rönnefahrt, 2005, approximately one-third of the whole volume of pharmaceutical compounds traded in Germany and around 25% of that used in Austria (Sattelberger, 1999) are disposed of into the domestic sewage. In a public poll, it has been found that 17.7% of those surveyed dispose of their unused and expired pills by draining them into the toilet, and about 20%–25% do the same with unused liquid pharmaceutical compounds (Götz and Keil, 2007). Some models demonstrated that the elimination of unused pharmaceutical pollutants is done either at the household level or through the sewers. This main rout of disposal needs great attention (Bound and Voulvoulis, 2005).

In research, conducted in the United States, more than half of the population disposed of the unused medicines in the toilets, only 22.9% of reported return to the pharmacy for disposal. Nearly 20% were advised for proper discarding by a health-care provider (Seehusen and Edwards, 2006). In a study completed in Kuwait (Abahussain et al., 2006), almost half of the patients (45.4%) found drugs prescribed more than three times a year, and nearly all had an undesirable prescription of medicines in their homes. The most conventional method used for disposing of the medicine was a throw of unwanted medicines in the trashes (76.5%) or flushes them in the drainage system (11.2%). These studies suggest that there is a need of awareness related to the proper disposal of the not used and expired medicines is critical in all countries (Niquille and Bugnon, 2008).

29.3 The fate of pharmaceutical and personal care products in the environment

PPCPs released into the aqueous environment have no definite future. Most of the pharmaceuticals end in municipal STPs. Most of the STPs are actually not meant for the treatment of these compounds; however, the pharmaceutical compounds are transformed during the treatment process in STPs. At the stage of this transformation, there is a possibility of the conversion of these compounds into the more toxic and higher persistent form (Celiz et al., 2009). Moreover, the number of altered products in the water may be found higher than the parental component, as verified for carbamazepine, diclofenac, and atorvastatin (Langford and Thomas, 2011). The stage under which the quantity of the parent component is underneath the state of detection, the formed product after also needs to be detected (Jakimska et al., 2014). The removal of pharmaceuticals and its changing aspects in aqueous setting depends upon numerous features that are:

- the intrinsic physicochemical properties of the compounds
- sunlight accessibility
- temperature
- pH
- oxygen availability
- microorganisms

Removal of PPCPs in aqueous setting involves the conversion by the process of either photo or biodegradation. The method of photodegradation is more feasible as PPCPs consist of heteroatoms or aromatic chain or structural complexity, which is more susceptible to be degraded by light called direct photolysis or photogenerated transient species (indirect photolysis) (Khetan and Collins, 2007). Microbes involved in the degradation of pharmaceuticals are not defined abundantly, but it is found that there is more defined microbial transformation than photodegradation in surface water contamination (Khetan and Collins, 2007). There is a need to identify the microbial strains that are involved in the process of biodegradation as there is valuable potential to utilize these identified strains into the biological removal of PPCPs in wastewater treatment plants. The persistent nature of PPCPs is not confirmed yet (Bu et al., 2016). According to the United Nations Environmental Programme (UNEP), the half-life of any compound to persist in the aquatic environment is 60 days. Understandably, the persistence of pharmaceuticals is depended upon the inherent physicochemical properties and several environmental conditions. As an example, the half-life for carbamazepine in lab and in-field researches has been described to be varied from 3.5 to 233 days (Yamamoto et al., 2009) and from 63 to 1200 days, respectively (Tixier et al., 2003; Zou et al., 2015). The occurrence of PPCPs in shallow water leads to its contacts to the aquatic biota. The effect of these pharmaceutical entities has been studied chiefly for the fishes including altered reproductive behavior, aggression, boldness, activity, sociality, and feeding rate (Brodin et al., 2014). The frequent entry of antibiotics prompts the development of antibiotic resistance genes and indorses the rise in antibiotic-resistant microorganisms not individually in hospital wastewaters and animal production wastewaters, but also in domestic sewage, surface water, and groundwater (Zhang et al., 2009).

29.4 Impact and toxicity of pharmaceutical and personal care products

The harmful effects of pharmaceuticals are much more traced in aquatic biota nowadays than on human beings, which is thought to be negligible, it is believed that aqueous life is at much more risk. Very limited researches related to the harmful impact on human beings are available. The study indicates that the effect of the exposure of pharmaceutical compounds is nearly analogous for any population. Therefore there are not much differences in the consumption for a fetus, babies, kids, to the aged (Kümmerer, 2009). Long-term occurrence of the pharmaceutical compounds in the aquatic environment could raise their resistance to its control via medicines (Cooper et al., 2008). As medications are intended to have biological possessions in minor quantities and most of the pharmaceutical compounds are water solvable. They are less prone to biological magnification (Sherer, 2006), that is, as many compounds build up in the fat of organisms. The most noticeable risk related to the discoveries in the study of Fick et al. (2009) was the high levels of broad-spectrum antibiotics could encourage the expansion of antibiotic-resistant microbes. The rise in the existence of multiresistant pathogens is a severe worldwide risk to human well-being. It is supported by the extensive usage of antibiotics in medications of humans and veterinaries. One would choose the industrial wastewater containing antibiotics for the selection of resistant strains of bacteria in the environment (Guardabassi et al., 1998).

29.4.1 Microorganisms

PPCPs have been used to prevent from large number of microbes; however, the excessive use of these medicines is resulting in the production of resistant microbes. For example, the oral intake of pain-relieving, antiinflammatory, and antipyretic medicines such as ibuprofen (Sanyal et al., 1993; Reynolds, 1989), for the treatment of mild-to-moderate rheumatic pain and other muscular–skeletal ailments, has been quite prevalent. The same authors noted that *Staphylococcus aureus* is also susceptible to ibuprofen. Growth of *S. aureus* was inhibited by ibuprofen concentrations higher than 150 mg/L at initial pH 7, while at pH 6, such levels prevent growth. Furthermore, Lee and Bird described that the calanoid copepods *Temora turbinata*, if cultured in waste containing pharmaceutical compounds having a concentration about 1 ppm, results in small size of an adult, reduction in the production of egg, and an abnormal growth pattern.

29.4.2 Phytoplankton

Phytoplanktons are very sensitive toward a change in physicochemical quality of the water body. Pharmaceuticals such as antibiotic named streptomycin prohibited the typical growth pattern of some cyanobacterial species in an examination performed by Harrass et al. (1985), at concentrations 0.09–0.86 mg/L. The phytoplanktons such as *Chlorella vulgaris*, *Scenedesmus obliquus*, and *Ulothrix* sp. cultivated in streptomycin concentration less than 21 mg/L, resulted in prohibited growth of *Chlamydomonas reinhardtii* at a level of 0.66 mg/L. Growing algae at a sublethal concentration of streptomycin was decelerated or delayed, and the maximum density achieved by numerous species was reduced.

29.4.3 Plants

A large number of studies related to impact of pharmaceuticals on plants are available. Different classes of pharmaceuticals have their own impact on plants. According to a study conducted by Batchelder (1981, 1982), the effect of antibiotics such as chlortetracycline and oxytetracycline on plants varied from species to species. The recorded delicate plant species in the Batchelder's study was pinto beans growing on the sandy loam soil. The severe toxic effect of furazolidone, which is mainly used in fish feed, has been examined by Macri et al. (1988), and the researchers also have some findings of significant toxicity of the compounds on *Daphnia magna*, while *Artemia salina* proved to be the less sensitive. Migliore et al. (1997) presented the toxic effect of numerous agricultural antibiotics on *Artemia*. Acute toxicity effects of four of the antibiotics, aminosidine, bacitracin, erythromycin, and lincomycin, all of them are utilized as a feed additive or bulk remedy in intensive farming. The impacts of pharmaceuticals are more important in the case of edible plants where toxicity can directly affect human being.

29.4.4 Fish

The effect of antibiotics and pharmaceuticals on fishes is not extensively described in literature till now. ISO laboratory examines with ibuprofen showed that the compounds were not quite toxic on bluegill sunfish and *Sheepshead minnow*. Up-to-date studies detects the effect of some pharmaceuticals (Lagesson et al., 2016) that assess to what extent five medicines (diphenhydramine, oxazepam, trimethoprim, diclofenac, and hydroxyzine) are uptaken by the fishes (European perch), by outlining their bioconcentrations over several months in a seminatural large-scale (pond) system. The results described the difference between drugs in their capacity to bioaccumulation and species in uptake.

29.5 Different techniques for removal of pharmaceutical compounds

Removing pharmaceuticals by sustainable means from different matrices will reduce the ecological effect on natural water bodies and soil (Hernandez Leal et al., 2010). There are several techniques to remove pharmaceutical contaminants from water and wastewater, that is, conventional and nontraditional. These techniques have been evolved to remove and recover various pollutants such as emerging contaminants (pesticides, pharmaceuticals, antibiotics, etc.) (Ratola et al., 2012) from water and sewage effluents. In addition, an optimization of these techniques may restrict these contaminants to reach the freshwater, which is a growing concern worldwide (Ratola et al., 2012). Adding to this, the use of bio-based techniques also proved to be potential, affordable, and useful for removing PPCPs compared to other technologies such as adsorption, oxidation, ozonation, and activated carbon (Langenhoff, 2009; Metcalf, 2003). The microbial process can be beneficial for degrading these pharmaceutical compounds to less toxic form. These microbes, including bacteria, algae, and fungi, have been used for remediation of pharmaceutical compounds (Tiehm et al., 2011) and soil. Some of the previous studies on pharmaceutical compounds and their concentration are given in Table 29.1. Bacteria degrade pharmaceutical and initiate its biodegradation, for example, there is only one bacterial strain discovered yet that reduces ibuprofen and utilizes ibuprofen as the basis of its carbon and energy (Murdoch and Hay, 2005; Rodarte-Morales et al., 2012).

Among various techniques natural reduction of these composites into the environmental setting is a modest, cost-effective approach that is applied physically, chemically, and biologically. Some of the important techniques used for the treatment of pharmaceuticals are dispersion, dilution, volatilization, photolysis, sorption, biodegradation, biotransformation, membrane bioreactor, ozonation, and phytoremediation, etc. (Khan et al., 2004; Kuppusamy et al., 2016), nevertheless all of them has its own advantages and drawbacks. Traditional STPs, which usually include a necessary arrangement of physicochemical treatment adding secondary treatment by activated sludge bioreactor, were predictable technology for efficient removal of trace pharmaceuticals. There are two categories of techniques to remove pharmaceutical compounds, that is, conventional and nonconventional.

29.5.1 Conventional treatment

Traditional wastewater treatment comprises preliminary, primary, secondary, and tertiary procedures that are based on biological, physical, and chemical processes. The most common biological process of sewage water treatment is a treatment with activated sludge. Traditional wastewater treatment facility conducts activated sludge processes or additional forms of biological treatments such as biofiltration. These processes have been able to remove various pharmaceuticals ranged from 20% to more than 90%. The competence of these processes for the remediation of medicines diverges within and among various studies and is dependent on the effective construction of the wastewater treatment facility. Features influence elimination including sludge age, activate sludge tank temperature, and hydraulic retention time. Relatively, progressive wastewater treatment processes, such as reverse osmosis, ozonation, and advanced oxidation technologies, can attain higher removal rates for pharmaceuticals.

29.5.2 Nonconventional treatment

Nonconventional methods to treat medicines are easy to handle, cost-effective, and are less complex to operate and design. Nonconventional treatment processing, such as ozonation, advanced oxidation, activated carbon, and membrane filtration (e.g., nanofiltration, reverse osmosis), was able to achieve high removal rate (above 99%) for a target pharmaceuticals in numerous studies. Following are some important methods to treat pharmaceuticals from wastewater.

1. Nanofiltration

   Nanofiltration utilizes membranes for removal of PPCPs impurities from the effluents. Under this technique the water and wastewater are passed through filters under high pressure for the removal of particles and then through the membrane with the pore size of 0.1−1.0 nm, to separate the influx into the permeate (treated effluent) and concentrate (contaminants), which were supposed to be recycled again for the treatment. The initial elimination of these pollutants is done by physical filtering, by sieves following adsorption and electrostatic repulsion, which achieve higher removal efficacy and avoid the formation of by-products (Bellona et al., 2004; Bolong et al., 2009; Fu and Wang, 2011).
2. Solar photo-Fenton (SPF)

   The SPF technique is a progressive oxidation procedure known for the simplest process and highly efficient in the treatment of the most organic pollutants. This procedure contains the addition of catalyst and H_2O_2 into the influent, which is then passed through reactors irradiated by solar light to generate OH• radicals and oxidize PPCPs. To increase the treatment efficiency, the influent should be kept at low pH, followed by its neutralization before the discharge into the environment (Muñoz et al., 2007).
3. Ozonation

This treatment works through direct and indirect reactions of PPCPs with hydroxyl radicals generated by ozone decomposition in the contactors. This method can achieve a proper removal but can also produce harmful by-products

(Margot et al., 2013; Tripathi and Tripathi, 2011; Westerhoff et al., 2005). The treatment efficiency is directly dependent on the pH, contact time, and organic matter content, that is, pharmaceutical contaminants. After the treatment the effluent is balanced by the addition of sodium hydroxide. The ozone can be generated either from liquid oxygen or air, with the latter mainly used at large-scale facilities; both methods have significant electricity requirements (Kim and Tanaka, 2011; Wang et al., 2005).

29.5.3 Membrane bioreactor

It is another technique used to treat industrial, domestic, and municipal wastewaters (Yang et al., 2006) containing pharmaceuticals. This technique involves the addition of biological methods of remediation and then removal of biomass physically from the aqueous phase by the porous membrane. There are several advantages of membrane bioreactors such as full elimination of suspended solids, compacted size, and fewer resource demands, high degradation rate, less skill full about its operation, less generation of sludges, cleansing and decontamination is easy, odor controller, long microbial retention, and treating recalcitrant and lethal organics and inorganics contaminant. Experiment by Cheng et al. (2018) accomplished the treatment of sewage (wastewater) containing pharmaceutical compounds. Some technicians used ZeeWeed membrane bioreactor for treating pharmaceutical wastewater (Noble, 2006).

29.5.4 Phytoremediation

It is an emerging cost-effective, environment-friendly technique experimenting across the world for the treatment of various contaminants such as organic and inorganic pollutants. It uses the natural capacity of degradation and accumulation of green plants as an efficient tool to act as an agent for the treatment (Amin et al., 2013). Researches showed that plants have exceptional ability in treating the pharmaceuticals from wastewater and soil (Billore et al., 2001). It was found by various workers that pharmaceuticals can be treated by the aid of plants especially *Typha latifolia* and *Phragmites karka*. There is also reported that phenol generated from industrial wastewater is phytoremediation by the aid of peroxidases of tomato hairy root cultures (Gonzalez et al., 2006).

PPCPs are an emerging group of organic pollutants present at trace concentrations in surface waters. These compounds can suppress. Bioactive PPCPs in surface waters also prompt concerns about human health effects, especially when considering lifelong exposure (Pomati et al., 2006). These compounds are difficult to remove by traditional STPs and have been reported in secondary treated discharges. Phytoremediation can be an option for the removal of these compounds from various environmental matrices. Active plant processes, including uptake and metabolism, can remove these compounds in wetland systems. Duckweeds (*Lemna minor*) have been found efficient in uptake and metabolization of chlorinated and fluorinated phenols. Recently, few researchers have reported the removal and uptake of veterinary pharmaceuticals from fortified soil media and manure.

29.5.5 Constructed wetlands

Constructed wetlands (CWs) are land-based wastewater treatment systems that are consisted of the shallow ponds, beds, or trenches that contain floating or emergent rooted wetland vegetation (Cole, 1998). In subsurface-horizontal flow CWs, wastewater infiltrates through vegetated gravel beds confined by a liner. The wastewater treatment relies on biological, chemical, and physical processes occurring in a natural environment (Cole, 1998). CWs are cost-effective and natural wastewater treatment technology that were extensively investigated for the removal of organic matter and nutrients (nitrogen and phosphorus) (Kadlec and Wallace, 2009; Vymazal, 2013; Ilyas and Masih, 2017, 2018), as well as PPCPs from wastewater (Imfeld et al., 2009; Xi et al., 2015; Verlicchi and Zambello, 2014; Ekperusi et al., 2019). The role of design parameters, including physical configuration, hydraulic mode, and vegetation species, are considered as very important in the removal of PPCPs (Li et al., 2017). The removal processes and the influence of design and operation parameters on the removal of four pharmaceutical contaminants by CWs (azithromycin, clarithromycin, diclofenac, and erythromycin), which are on the priority list of the EU have been performed successfully by various workers (Gorito et al., 2017).

29.6 Removal of pharmaceutical contaminants by bacterial treatment (bioremediation)

Bioremediation, that is, the use of microbes especially bacteria for the treatment of contaminants is one of the favored methods for the treatment of pharmaceuticals. It is only possible with natural resistant microbial strains that are capable of transforming the hazardous form to its less toxic form (Saratale et al., 2011; Singh et al., 2020). Earth is provided with the most biologically rich, diverse habitat to various life forms reaching toward 1 trillion species out of which the most abundant and the oldest one are bacteria (Locey and Lennon, 2016). They are not marking their presence in nature only, but even they also exist in the number of millions of cells inside the human body (Kallmeyer et al., 2012). Various cycles of metabolism performed by them are necessary for sustainable development and the establishment of life on this planet Earth. Bacteria impact all systematic disciplines including arenas of biology, cell biology, ecology, immunology, geology, chemistry, physics, climatology, alternative energy production, and bioremediation (Ramírez-Durán et al., 2017). Due to the vast potential of these life forms toward other biological phenomena, they have limitless influences on different scientific areas. Technology utilizing sludges (activated, granular, aerobic, etc.) has been fruitful in the handling of household sewage (Adav et al., 2008) is possible mainly due to the existence of bacteria.

Though many researchers also recommend that removing pharmaceuticals by the aid of bacteria is generally less efficient in comparison to other techniques and there is a risk of development of antibiotic-resistant bacteria also (Xiong et al., 2018; Aydin, 2016). But in spite of the fact, bioremediation is an excellent substitute for various chemically and physically treatment methods, since microbes naturally have the capacity to degrade pollutants. This procedure can be stimulated by the addition of nutrients for biodegrading microbes, known as biostimulation, or sometimes actively furnishing microbial areas. It is added as enhancer toward remediation this technique identified as bioaugmentation (Amer et al., 2015). The biodegradation of pharmaceuticals for removing pharmaceuticals from wastewater and surface-water is getting more attention worldwide (Andersen et al., 2005; Bagnall et al., 2012). This technique generates plentiful degradation and transformation products, the impact of which on the environment must be a matter of concern (Petrie et al., 2015). Some notable examples of microbes associated with the degradation of pharmaceuticals are fungus *Trametes versicolor* that incorporates to eliminate naproxen and ibuprofen (Marco-Urrea et al., 2009). *T. versicolor* associates to the recalcitrant analgesics, antiinflammatories, antibiotics, and psychotropic drugs (Mir-Tutusaus et al., 2017). *Candida tropicalis*, *Debaryomyces polymorphus*, and *Candida zeylanoides* utilized for eliminating synthetic dyes (Yang et al., 2003; Rai and Mallick, 1992). *Aspergillus oryzae*, *Aspergillus variabilis*, and *Tolypothrix ceytonica* removed organic matter, copper, and zinc (El-Bestawy, 2008).

Studies estimating the dynamics of the bacteria in the elimination of traces of pharmaceuticals such as alprenolol, bisoprolol, metoprolol, propranolol, venlafaxine, salbutamol, fluoxetine, and norfluoxetine in an aerobic granular sludge membrane reactor reported a diverse bacterial community. The most abundant phyla were Proteobacteria, Bacteroidetes, and Actinobacteria as well as the bacterial groups Acidobacteria, Chlamydiae, Chloroflexi, Cyanobacteria, Epsilonproteobacteria, Firmicutes, Gemmatimonadetes, Tenericutes, and Verrucomicrobia (Amorim et al., 2018.). Other researches (Muter et al., 2017) incorporate activated sludge in lab-scale treatment and evaluate the elimination of pharmaceuticals such as atenolol, atorvastatin, azithromycin, caffeine, carbamazepine, ciprofloxacin, clarithromycin, diclofenac, erythromycin, fluoxetine, ibuprofen, ketoprofen, losartan, metoprolol, naproxen, paracetamol, simvastatin, sulfamethoxazole, trimethoprim, valsartan, and xylazine at concentrations of 13.2 ng/L–51.8 μg/L in sewage water in Latvia describing that biostimulation, that is, adding nutrients to the sludges can enhance the rate of biodegradation of many of the pharmaceuticals by rising the activity of phyla Verrucomicrobia, Gemmatimonadetes, Nitrospirae, Chlorobi, Acidobacteria, Chloroflexi, Actinobacteria, Firmicutes, Bacteroidetes, Proteobacteria, and others. In addition, there are many genera and species defined to remove pharmaceuticals such as alprenolol, bisoprolol, metoprolol, propranolol, venlafaxine, salbutamol, fluoxetine, norfluoxetine, 17β-estradiol, and gemfibrozil have been distinguished having 90% or higher removal competency. Table 29.3 describes the removal efficiency of various genera toward the removal of multiple pharmaceuticals contaminants.

Bacterial metabolism and its physiological progressions showed its reliable capacity toward removing trace pharmaceuticals. On the other hand, there is still a high demand for exploring the ability of these genera toward the removal of these contaminants of emerging concern. The composite relations and exchanges of this community in removing the pharmaceuticals were previously being considered a substitute for remediation of these emerging contaminants biologically. Despite the fact, there is a vast range of experiments conducted relative to removing pharmaceuticals as a contaminant, but the process of biotransformation has not usually been appraised nor has the subsequent product of these transformations (Verlicchi et al., 2012; Majewsky et al., 2011). As favorable responses of bacteria toward its high metabolic rate, tolerance nature, stress response, and diversity, these can be utilized as a model organism for removing these pharmaceuticals of emerging concern.

TABLE 29.3 Removal of pharmaceutical contaminants through bacterial remediation.

Pharmaceutical contaminants	Source of contaminants	Species/genus	Removal efficiency
Paracetamol	Bioreactor	*Delftia tsuruhatensis* *Pseudomonas aeruginosa*	99.9%
Cefalexin Sulfamethoxazole Caffeine Salicylic acid Chloramphenicol	Activated sludge	*Pseudomonas* sp. *CE21*	46.7% 87.5%–95% 35.9%–67.9% 36.5%–100% 55.2%–85.4%
Cefalexin Sulfamethoxazole Caffeine Salicylic acid Chloramphenicol	Activated sludge	*Pseudomonas* sp. *CE22*	< 90% 88.1%–100% 74.1%–66.3% 64.5%–96.9% 77.8%–87.1%
Triclosan Bisphenol Ibuprofen 17β-Estradiol Gemfibrozil	Sewage treatment plant	*Novosphingobium* sp. *Sphingomonas* sp. *Sphingopyxis* sp. *Sphingobium* sp. *Isoptericola* sp. *Nubsella* sp. *Rhodococcus* sp. *Bacillus* sp.	0.2–8340 ng/L
17α-Ethynylestradiol	Wastewater	*Acinetobacter* sp. *Phyllobacterium myrsinacearum* *(AY785315)* *Ralstonia pickettii (AY268180)* *P. aeruginosa (AY738263)* *Pseudomonas* sp. *(DQ303435)* *Acinetobacter* sp. *(CR543861)* *Acinetobacter* sp. *(CR543861)*	–
Salicylic acid	In vitro	*Pseudomonas fluorescens* *HK44*	80%–90%
17α-Ethynylestradiol	Bioreactor	*Nitrosomonas europaea*	96%
Methylamine Trimethylamine Dimethylamine	Biofilter	*Paracoccus* sp. *CP2* *Arthrobacter* sp.	96%

29.6.1 Removal of pharmaceutical compounds by fungal treatment

Fungi can also be utilized as an agent for remediating the various contaminants, including pharmaceuticals, from the contaminated environment. There is a vast range of fungal genera, which can be proved to be an efficient tool for remediating wastes generated from industries or households, but the only limitation is the long day's growth cycle and formation of filaments and spores. The research by Spina et al. (2012) linked the fungal treatment with the activated sludge for treating the pharmaceuticals present in industrial wastewater. In the study, they utilize fungal strain, *Bjerkandera adusta* MUT 2295, over which they attained removed various pharmaceuticals. Various fungal strains such as *Aspergillus niger*, *Aspergillus fumigatus*, and *Aspergillus niveus* showed reduction of pharmaceuticals from diluted industrial wastes at diverse concentrations. A group of fungi mainly belongs to *Ascomycetes* also plays a significant role in treating industrial sewage containing pharmaceuticals, for example, *Penicillium decumbens* and *Penicillium lignorum* strain of fungi showed a substantial reduction in COD, phenol, and color (Mohammad et al., 2006; Angayarkanni et al., 2003).

29.7 Conclusion

An array of pharmaceutical compounds and PCPs ranging from analgesics, antibiotic to stimulants are investigated in the different dimensions of the environment from the ground and surface water resources to soil as well. These residues entered the environment from different point and diffuse sources. Relating the concentration of PPPCs in a comparative manner in the ground, freshwater resource, groundwater, and the effluent generated from STPs/WTTPs, it is concluded that the level of it in the natural environment is somewhat lesser than that of STPs. This might be due to the natural process, for instance, biotransformation, photolysis, and sorption, is always working in nature to diminish the pharmaceuticals. These contaminants of ecological concern demand for new approaches to remediate these compounds. As none of the existing methods is able to treat these compounds in a feasible manner, there is a need for improved STPs/WWTPs in order to reduce the concentrations of pharmaceuticals discharged to the environment and thus eliminate its potential effects on natural resources. Bioremediation or biological treatment proved as an effective technique for removing contaminants organic in nature, such as pharmaceuticals, antibiotics, and aromatics generated in effluents. The benefits of this method are abundant: it includes efficiency and easiness in abolishing the pollutants and not demanding unnecessary technology, skills, and equipment. In addition, there is a need for comprehensive studies related to bioremediation of pharmaceuticals, specifically the ones that are nonbiodegradable and persevere in nature for the long term. It is predicted that this method results in a sustainable remediation technique while impacting a reduction in environmental footprint.

References

Abahussain, E.A., Ball, D.E., Matowe, W.C., 2006. Practice and opinion towards disposal of unused medication in Kuwait. Med. Princ. Pract. 15 (5), 352–357.

Adav, S.S., Lee, D.J., Show, K.Y., Tay, J.H., 2008. Aerobic granular sludge: recent advances. Biotechnol. Adv. 26 (5), 411–423.

Amer, R.A., Mapelli, F., El Gendi, H.M., Barbato, M., Goda, D.A., Corsini, A., et al., 2015. Bacterial diversity and bioremediation potential of the highly contaminated marine sediments at El-Max District (Egypt, Mediterranean Sea). BioMed. Res. Int. 2015.

Amin, A., Naik, A.R., Azhar, M., Nayak, H., 2013. Bioremediation of different waste waters—a review. Cont. J. Fish. Aquat. Sci. 7 (2), 7.

Amorim, C.L., Alves, M., Castro, P.M., Henriques, I., 2018. Bacterial community dynamics within an aerobic granular sludge reactor treating wastewater loaded with pharmaceuticals. Ecotoxicol. Environ. Saf. 147, 905–912.

Andersen, H.R., Hansen, M., Kjølholt, J., Stuer-Lauridsen, F., Ternes, T., Halling-Sørensen, B., 2005. Assessment of the importance of sorption for steroid estrogens removal during activated sludge treatment. Chemosphere 61 (1), 139–146.

Angayarkanni, J., Palaniswamy, M., Swaminathan, K., 2003. Biotreatment of distillery effluent using *Aspergillus niveus*. Bull. Environ. Contam. Toxicol. 70 (2), 0268–0277.

Aydin, S., 2016. Enhanced biodegradation of antibiotic combinations via the sequential treatment of the sludge resulting from pharmaceutical wastewater treatment using white-rot fungi *Trametes versicolor* and *Bjerkandera adusta*. Appl. Microbiol. Biotechnol. 100 (14), 6491–6499.

Bagnall, J.P., Evans, S.E., Wort, M.T., Lubben, A.T., Kasprzyk-Hordern, B., 2012. Using chiral liquid chromatography quadrupole time-of-flight mass spectrometry for the analysis of pharmaceuticals and illicit drugs in surface and wastewater at the enantiomeric level. J. Chromatogr. A 1249, 115–129.

Barnes, K.K., Christenson, S.C., Kolpin, D.W., Focazio, M.J., Furlong, E.T., Zaugg, S.D., et al., 2004. Pharmaceuticals and other organic waste water contaminants within a leachate plume downgradient of a municipal landfill. Ground Water Monit. Rem. 24 (2), 119–126.

Batchelder, A.R., 1981. Chlortetracycline and oxytetracycline effects on plant growth and development in liquid cultures. J. Environ. Qual. 10 (4), 515–518.

Batchelder, A.R., 1982. Chlortetracycline and oxytetracycline effects on plant growth and development in soil systems. J. Environ. Qual. 11 (4), 675–678.

Batt, A.L., Kostich, M.S., Lazorchak, J.M., 2008. Analysis of ecologically relevant pharmaceuticals in wastewater and surface water using selective solid-phase extraction and UPLC–MS/MS. Anal. Chem. 80 (13), 5021–5030.

Bellona, C., Drewes, J.E., Xu, P., Amy, G., 2004. Factors affecting the rejection of organic solutes during NF/RO treatment—a literature review. Water Res. 38 (12), 2795–2809.

Bendz, D., Paxeus, N.A., Ginn, T.R., Loge, F.J., 2005. Occurrence and fate of pharmaceutically active compounds in the environment, a case study: Höje River in Sweden. J. Hazard. Mater. 122 (3), 195–204.

Billore, S.K., Singh, N., Ram, H.K., Sharma, J.K., Singh, V.P., Nelson, R.M., et al., 2001. Treatment of a molasses based distillery effluent in a constructed wetland in central India. Water Sci. Technol. 44 (11–12), 441–448.

Bolong, N., Ismail, A.F., Salim, M.R., Matsuura, T., 2009. A review of the effects of emerging contaminants in wastewater and options for their removal. Desalination 239 (1–3), 229–246.

Bound, J.P., Voulvoulis, N., 2005. Household disposal of pharmaceuticals as a pathway for aquatic contamination in the United Kingdom. Environ. Health Perspect. 113 (12), 1705–1711.

Boxall, A.B., Rudd, M.A., Brooks, B.W., Caldwell, D.J., Choi, K., Hickmann, S., et al., 2012. Pharmaceuticals and personal care products in the environment: what are the big questions? Environ. Health Perspect. 120 (9), 1221–1229.

Brodin, T., Piovano, S., Fick, J., Klaminder, J., Heynen, M., Jonsson, M., 2014. Ecological effects of pharmaceuticals in aquatic systems—impacts through behavioural alterations. Philos. Trans. R. Soc. B: Biol. Sci. 369 (1656), 20130580.

Bu, Q., Shi, X., Yu, G., Huang, J., Wang, B., Wang, J., 2016. Pay attention to non-wastewater emission pathways of pharmaceuticals into environments. Chemosphere 165, 515–518.

Buszka, P.M., Yeskis, D.J., Kolpin, D.W., Furlong, E.T., Zaugg, S.D., Meyer, M.T., 2009. Waste-indicator and pharmaceutical compounds in landfill-leachate-affected ground water near Elkhart, Indiana, 2000–2002. Bull. Environ. Contam. Toxicol. 82 (6), 653–659.

Carey, D.E., McNamara, P.J., 2015. The impact of triclosan on the spread of antibiotic resistance in the environment. Front. Microbiol. 5, 780.

Carrara, C., Ptacek, C.J., Robertson, W.D., Blowes, D.W., Moncur, M.C., Sverko, E.D., et al., 2008. Fate of pharmaceutical and trace organic compounds in three septic system plumes, Ontario, Canada. Environ. Sci. Technol. 42 (8), 2805–2811.

Celiz, M.D., Tso, J., Aga, D.S., 2009. Pharmaceutical metabolites in the environment: analytical challenges and ecological risks. Environ. Toxicol. Chem. 28 (12), 2473–2484.

Cheng, H., Ngo, H., Guo, W., Soon, Y.L., et al., 2018. Anaerobic membrane bioreactors for antibiotic wastewater treatment: Performance and membrane fouling issues. Biores. Technol. 267, 714–724.

Cole, S., 1998. The emergence of treatment wetlands. Environ. Sci. Technol. 32 (9), 218A–223A.

Conn, K.E., Lowe, K.S., Drewes, J.E., Hoppe-Jones, C., Tucholke, M.B., 2010. Occurrence of pharmaceuticals and consumer product chemicals in raw wastewater and septic tank effluent from single-family homes. Environ. Eng. Sci. 27 (4), 347–356.

Cooper, E.R., Siewicki, T.C., Phillips, K., 2008. Preliminary risk assessment database and risk ranking of pharmaceuticals in the environment. Sci. Total. Environ. 398 (1–3), 26–33.

Cullinan, M.P., Palmer, J.E., Carle, A.D., West, M.J., Seymour, G.J., 2012. Long term use of triclosan toothpaste and thyroid function. Sci. Total Environ. 416, 75–79.

Daughton, C.G., 2003. Cradle-to-cradle stewardship of drugs for minimizing their environmental disposition while promoting human health. I. Rationale for and avenues toward a green pharmacy. Environ. Health Perspect. 111 (5), 757–774.

Daughton, C.G., 2007. Pharmaceuticals in the environment: sources and their management. Compr. Anal. Chem. 50, 1–58.

Daughton, C.G., Ternes, T.A., 1999. Pharmaceuticals and personal care products in the environment: agents of subtle change? Environ. Health Perspect. 107 (Suppl. 6), 907–938.

Diaz-Cruz, M.S., de Alda, M.J.L., Barcelo, D., 2003. Environmental behavior and analysis of veterinary and human drugs in soils, sediments and sludge. TrAC Trends Anal. Chem. 22 (6), 340–351.

Eggen, T., Moeder, M., Arukwe, A., 2010. Municipal landfill leachates: a significant source for new and emerging pollutants. Sci. Total Environ. 408 (21), 5147–5157.

Ekperusi, A.O., Sikoki, F.D., Nwachukwu, E.O., 2019. Application of common duckweed (*Lemna minor*) in phytoremediation of chemicals in the environment: state and future perspective. Chemosphere 223.

El-Bestawy, E., 2008. Treatment of mixed domestic–industrial wastewater using cyanobacteria. J. Ind. Microbiol. Biotechnol. 35 (11), 1503–1516.

Fick, J., Söderström, H., Lindberg, R.H., Phan, C., Tysklind, M., Larsson, D.J., 2009. Contamination of surface, ground, and drinking water from pharmaceutical production. Environ. Toxicol. Chem. 28 (12), 2522–2527.

Fu, F., Wang, Q., 2011. Removal of heavy metal ions from wastewaters: a review. J. Environ. Manage. 92 (3), 407–418.

Glassmeyer, S.T., Furlong, E.T., Kolpin, D.W., Cahill, J.D., Zaugg, S.D., Werner, S.L., et al., 2005. Transport of chemical and microbial compounds from known wastewater discharges: potential for use as indicators of human fecal contamination. Environ. Sci. Technol. 39 (14), 5157–5169.

Godfrey, E., Woessner, W.W., Benotti, M.J., 2007. Pharmaceuticals in on-site sewage effluent and ground water, western Montana. Groundwater 45 (3), 263–271.

Gonzalez, P.S., Capozucca, C.E., Tigier, H.A., Milrad, S.R., Agostini, E., 2006. Phytoremediation of phenol from wastewater, by peroxidases of tomato hairy root cultures. Enzyme Microb. Technol. 39 (4), 647–653.

Gorito, A.M., Ribeiro, A.R., Almeida, C.M.R., Silva, A.M., 2017. A review on the application of constructed wetlands for the removal of priority substances and contaminants of emerging concern listed in recently launched EU legislation. Environ. Pollut. 227, 428–443.

Götz, K., Keil, F., 2007. Medikamentenentsorgung in privaten Haushalten: Ein Faktor bei der Gewässerbelastung mit Arzneimittelwirkstoffen? Umweltwiss. Schadst.-Forsch. 19 (3), 180–188.

Gros, M., Petrovic, M., Barceló, D., 2006. Development of a multi-residue analytical methodology based on liquid chromatographyetandem mass spectrometry (LCeMS/MS) for screening and trace level determination of pharmaceuticals in surface and wastewaters. Talanta 70, 678e690.

Guardabassi, L., Petersen, A., Olsen, J.E., Dalsgaard, A., 1998. Antibiotic resistance in *Acinetobacter* spp. isolated from sewers receiving waste effluent from a hospital and a pharmaceutical plant. Appl. Environ. Microbiol. 64 (9), 3499–3502.

Harrass, M.C., Kindig, A.C., Taub, F.B., 1985. Responses of blue-green and green algae to streptomycin in unialgal and paired culture. Aquat. Toxicol. 6 (1), 1–11.

Heberer, T., 2002. Tracking persistent pharmaceutical residues from municipal sewage to drinking water. J. Hydrol. 266 (3–4), 175–189.

Hernandez Leal, L., Vieno, N., Temmink, H., Zeeman, G., Buisman, C.J., 2010. Occurrence of xenobiotics in gray water and removal in three biological treatment systems. Environ. Sci. Technol. 44 (17), 6835–6842.

Holm, J.V., Rügge, K., Bjerg, P.L., Christensen, T.H., 1995. Occurrence and distribution of pharmaceutical organic compounds in the groundwater downgradient of a landfill (Grindsted, Denmark). Environ. Sci. Technol. 29 (5), 1415–1420.

Ilyas, H., Masih, I., 2017. The performance of the intensified constructed wetlands for organic matter and nitrogen removal: a review. J. Environ. Manage. 198, 372–383.

Ilyas, H., Masih, I., 2018. The effects of different aeration strategies on the performance of constructed wetlands for phosphorus removal. Environ. Sci. Pollut. Res. 25 (6), 5318–5335.

Imfeld, G., Braeckevelt, M., Kuschk, P., Richnow, H.H., 2009. Monitoring and assessing processes of organic chemicals removal in constructed wetlands. Chemosphere 74 (3), 349–362.

Jakimska, A., Śliwka-Kaszyńska, M., Reszczyńska, J., Namieśnik, J., Kot-Wasik, A., 2014. Elucidation of transformation pathway of ketoprofen, ibuprofen, and furosemide in surface water and their occurrence in the aqueous environment using UHPLC-QTOF-MS. Anal. Bioanal. Chem. 406 (15), 3667–3680.

Jelic, A., Gros, M., Ginebreda, A., Cespedes-Sánchez, R., Ventura, F., Petrovic, M., et al., 2011. Occurrence, partition and removal of pharmaceuticals in sewage water and sludge during wastewater treatment. Water Res. 45 (3), 1165–1176.

Jim, T.Y., Bouwer, E.J., Coelhan, M., 2006. Occurrence and biodegradability studies of selected pharmaceuticals and personal care products in sewage effluent. Agric. Water Manage. 86 (1–2), 72–80.

Kadlec, R.H., Wallace, S.D., 2009. Treatment Wetlands, second ed. CRC Press, Boca Raton, FL.

Kallmeyer, J., Pockalny, R., Adhikari, R.R., Smith, D.C., D'Hondt, S., 2012. Global distribution of microbial abundance and biomass in subseafloor sediment. Proc. Natl Acad. Sci. U.S.A. 109 (40), 16213–16216.

Kasprzyk-Hordern, B., Dinsdale, R.M., Guwy, A.J., 2008. The occurrence of pharmaceuticals, personal care products, endocrine disruptors and illicit drugs in surface water in South Wales, UK. Water Res. 42 (13), 3498–3518.

Kasprzyk-Hordern, B., Dinsdale, R.M., Guwy, A.J., 2009. The removal of pharmaceuticals, personal care products, endocrine disruptors and illicit drugs during wastewater treatment and its impact on the quality of receiving waters. Water Res. 43 (2), 363–380.

Kepp, D.R., Sidelmann, U.G., Hansen, S.H., 1997. Isolation and characterization of major phase I and II metabolites of ibuprofen. Pharm. Res. 14 (5), 676–680.

Khan, F.I., Husain, T., Hejazi, R., 2004. An overview and analysis of site remediation technologies. J. Environ. Manage. 71 (2), 95–122.

Khetan, S.K., Collins, T.J., 2007. Human pharmaceuticals in the aquatic environment: a challenge to green chemistry. Chem. Rev. 107 (6), 2319–2364.

Kim, I., Tanaka, H., 2011. Energy consumption for PPCPs removal by O_3 and O_3/UV. Ozone: Sci. Eng. 33 (2), 150–157.

Kinney, C.A., Furlong, E.T., Kolpin, D.W., Burkhardt, M.R., Zaugg, S.D., Werner, S.L., et al., 2008. Bioaccumulation of pharmaceuticals and other anthropogenic waste indicators in earthworms from agricultural soil amended with biosolid or swine manure. Environ. Sci. Technol. 42 (6), 1863–1870.

Kümmerer, K., 2009. The presence of pharmaceuticals in the environment due to human use—present knowledge and future challenges. J. Environ. Manage. 90 (8), 2354–2366.

Kümmerer, K., Helmers, E., 2000. Hospital effluents as a source of gadolinium in the aquatic environment. Environ. Sci. Technol. 34 (4), 573–577.

Kuppusamy, S., Palanisami, T., Megharaj, M., Venkateswarlu, K., Naidu, R., 2016. In-situ remediation approaches for the management of contaminated sites: a comprehensive overview, Reviews of Environmental Contamination and Toxicology, vol. 236. Springer, Cham, pp. 1–115.

Lagesson, A., Fahlman, J., Brodin, T., Fick, J., Jonsson, M., Byström, P., et al., 2016. Bioaccumulation of five pharmaceuticals at multiple trophic levels in an aquatic food web—insights from a field experiment. Sci. Total Environ. 568, 208–215.

Langenhoff, A.A.M., 2009. Bioremediation of areas polluted with chlorinated and non-chlorinated hydrocarbons. Land Contam. Reclam. 17 (3/4), 619–625.

Langford, K., Thomas, K.V., 2011. Input of selected human pharmaceutical metabolites into the Norwegian aquatic environment. J. Environ. Monit. 13 (2), 416–421.

Lapworth, D.J., Baran, N., Stuart, M.E., Ward, R.S., 2012. Emerging organic contaminants in groundwater: a review of sources, fate and occurrence. Environ. Pollut. 163, 287–303.

Leung, H.W., Minh, T.B., Murphy, M.B., Lam, J.C., So, M.K., Martin, M., et al., 2012. Distribution, fate and risk assessment of antibiotics in sewage treatment plants in Hong Kong, South China. Environ. Int. 42, 1–9.

Li, H.S., Zhou, S.Q., Sun, Y.B., Feng, P., 2009. Advanced treatment of landfill leachate by a new combination process in a full-scale plant. J. Hazard. Mater. 172 (1), 408–415.

Li, J., Zhou, Q., Campos, L.C., 2017. Removal of selected emerging PPCP compounds using greater duckweed (*Spirodela polyrhiza*) based lab-scale free water constructed wetland. Water Res. 126, 252–261.

Lipinski, C.A., Lombardo, F., Dominy, B.W., Feeney, P.J., 1997. Experimental and computational approaches to estimate solubility and permeability in drug discovery and development settings. Adv. Drug Deliv. Rev. 23 (1–3), 3–25.

Liu, J.L., Wong, M.H., 2013. Pharmaceuticals and personal care products (PPCPs): a review on environmental contamination in China. Environ. Int. 59, 208–224.

Locey, K.J., Lennon, J.T., 2016. Scaling laws predict global microbial diversity. Proc. Natl Acad. Sci. U.S.A. 113 (21), 5970–5975.

Mackay, D., Barnthouse, L., 2010. Integrated risk assessment of household chemicals and consumer products: addressing concerns about triclosan. Integr. Environ. Assess. Manage. 6 (3), 390.

Macri, A., Stazi, A.V., Di Delupis, G.D., 1988. Acute toxicity of furazolidone on *Artemia salina*, *Daphnia magna*, and *Culex pipiens* molestus larvae. Ecotoxicol. Environ. Saf. 16 (2), 90–94.

Majewsky, M., Gallé, T., Bayerle, M., Goel, R., Fischer, K., Vanrolleghem, P.A., 2011. Xenobiotic removal efficiencies in wastewater treatment plants: residence time distributions as a guiding principle for sampling strategies. Water Res. 45 (18), 6152–6162.

Marco-Urrea, E., Pérez-Trujillo, M., Vicent, T., Caminal, G., 2009. Ability of white-rot fungi to remove selected pharmaceuticals and identification of degradation products of ibuprofen by *Trametes versicolor*. Chemosphere 74 (6), 765–772.

Margot, J., Kienle, C., Magnet, A., Weil, M., Rossi, L., De Alencastro, L.F., et al., 2013. Treatment of micropollutants in municipal wastewater: ozone or powdered activated carbon? Sci. Total Environ. 461, 480–498.

Metcalf, W., 2003. Metcalf and Eddy wastewater engineering: treatment and reuse. Wastewater Engineering: Treatment and Reuse. McGraw Hill, New York.

Metzger, J.W., 2004. Drugs in municipal landfills and landfill leachates. Pharmaceuticals in the Environment. Springer, Berlin, Heidelberg, pp. 133–137.

Migliore, L., Civitareale, C., Brambilla, G., Di Delupis, G.D., 1997. Toxicity of several important agricultural antibiotics to Artemia. Water Res. 31 (7), 1801–1806.

Mir-Tutusaus, J.A., Parladé, E., Llorca, M., Villagrasa, M., Barceló, D., Rodriguez-Mozaz, S., et al., 2017. Pharmaceuticals removal and microbial community assessment in a continuous fungal treatment of non-sterile real hospital wastewater after a coagulation-flocculation pretreatment. Water Res. 116, 65–75.

Mohammad, P., Azarmidokht, H., Fatollah, M., Mahboubeh, B., 2006. Application of response surface methodology for optimization of important parameters in decolorizing treated distillery wastewater using *Aspergillus fumigatus* UB2 60. Int. Biodeterior. Biodegrad. 57 (4), 195–199.

Muñoz, I., Peral, J., Antonio Ayllón, J., Malato, S., José Martin, M., Yves Perrot, J., et al., 2007. Life-cycle assessment of a coupled advanced oxidation-biological process for wastewater treatment: comparison with granular activated carbon adsorption. Environ. Eng. Sci. 24 (5), 638–651.

Murdoch, R.W., Hay, A.G., 2005. Formation of catechols via removal of acid side chains from ibuprofen and related aromatic acids. Appl. Environ. Microbiol. 71 (10), 6121–6125.

Murray, K.E., Thomas, S.M., Bodour, A.A., 2010. Prioritizing research for trace pollutants and emerging contaminants in the freshwater environment. Environ. Pollut. 158 (12), 3462–3471.

Musson, S.E., Townsend, T.G., 2009. Pharmaceutical compound content of municipal solid waste. J. Hazard. Mater. 162 (2–3), 730–735.

Muter, O., Ingus, P., Selga, T., Berzins, A., Gudra, D., Radovica-Spalvina, I., et al., 2017. Removal of pharmaceuticals from municipal wastewaters at laboratory scale by treatment with activated sludge and biostimulation. Sci. Total Environ. 584, 402–413.

Nikolaou, A., Meric, S., Fatta, D., 2007. Occurrence patterns of pharmaceuticals in water and wastewater environments. Anal. Bioanal. Chem. 387 (4), 1225–1234.

Niquille, A., Bugnon, O., 2008. Pharmaceuticals and environment: role of community pharmacies. Pharmaceuticals in the Environment. Springer, Berlin, Heidelberg, pp. 467–473.

Noble, J., 2006. GE ZeeWeed MBR technology for pharmaceutical wastewater treatment. Membr. Technol. 2006 (9), 7–9.

Pal, A., Gin, K.Y.H., Lin, A.Y.C., Reinhard, M., 2010. Impacts of emerging organic contaminants on freshwater resources: review of recent occurrences, sources, fate and effects. Sci. Total Environ. 408 (24), 6062–6069.

Peck, A.M., 2006. Analytical methods for the determination of persistent ingredients of personal care products in environmental matrices. Anal. Bioanal. Chem. 386 (4), 907–939.

Pennington, M.J., Rothman, J.A., Jones, M.B., McFrederick, Q.S., Gan, J., Trumble, J.T., 2017. Effects of contaminants of emerging concern on *Megaselia scalaris* (Lowe, Diptera: Phoridae) and its microbial community. Sci. Rep. 7 (1), 8165.

Petrie, B., Barden, R., Kasprzyk-Hordern, B., 2015. A review on emerging contaminants in wastewaters and the environment: current knowledge, understudied areas and recommendations for future monitoring. Water Res. 72, 3–27.

Petrovic, M., Gros, M., Barcelo, D., 2006. Multi-residue analysis of pharmaceuticals in wastewater by ultra-performance liquid chromatography–quadrupole–time-of-flight mass spectrometry. J. Chromatogr. A 1124 (1–2), 68–81.

Pomati, F., Castiglioni, S., Zuccato, E., Fanelli, R., Vigetti, D., Rossetti, C., Calamari, D., 2006. Effects of a Complex Mixture of Therapeutic Drugs at Environmental Levels on Human Embryonic Cells. Environ. Sci. Technol. 40 (7), 2442–2447.

Rabiet, M., Togola, A., Brissaud, F., Seidel, J.L., Budzinski, H., Elbaz-Poulichet, F., 2006. Consequences of treated water recycling as regards pharmaceuticals and drugs in surface and ground waters of a medium-sized Mediterranean catchment. Environ. Sci. Technol. 40 (17), 5282–5288.

Rai, L.C., Mallick, N., 1992. Removal and assessment of toxicity of Cu and Fe to *Anabaena doliolum* and *Chlorella vulgaris* using free and immobilized cells. World J. Microbiol. Biotechnol. 8 (2), 110–114.

Ramírez-Durán, N., Moreno-Pérez, P.A., Sandoval-Trujillo, A.H., 2017. Bacterial treatment of pharmaceutical industry effluents. Ecopharmacovigilance. Springer, Cham, pp. 175–187.

Ratola, N., Cincinelli, A., Alves, A., Katsoyiannis, A., 2012. Occurrence of organic microcontaminants in the wastewater treatment process. A mini review. J. Hazard. Mater. 239, 1–18.

Reungoat, J., Escher, B.I., Macova, M., Keller, J., 2011. Biofiltration of wastewater treatment plant effluent: effective removal of pharmaceuticals and personal care products and reduction of toxicity. Water Res. 45 (9), 2751–2762.

Reynolds, J., 1989. Sodium Cromoglycate and Related Anti-Allergic Agents.

Rivera-Utrilla, J., Sánchez-Polo, M., Ferro-García, M.Á., Prados-Joya, G., Ocampo-Pérez, R., 2013. Pharmaceuticals as emerging contaminants and their removal from water. A review. Chemosphere 93 (7), 1268–1287.

Rodarte-Morales, A.I., Feijoo, G., Moreira, M.T., Lema, J.M., 2012. Biotransformation of three pharmaceutical active compounds by the fungus *Phanerochaete chrysosporium* in a fed batch stirred reactor under air and oxygen supply. Biodegradation 23 (1), 145–156.

Rönnefahrt, I., 2005. Verbrauchsmengen in der Bewertung des Umweltrisikos von Humanarzneimitteln. Arzneimitttel in der Umwelt–Zu Risiken und Nebenwirkungen fragen Sie das Umweltbundesamt. Dessau, UBA-Texte, 29(05).

Sanyal, A.K., Roy, D., Chowdhury, B., Banerjee, A.B., 1993. Ibuprofen, a unique anti-inflammatory compound with antifungal activity against dermatophytes. Lett. Appl. Microbiol. 17 (3), 109–111.

Saratale, R.G., Saratale, G.D., Chang, J.S., Govindwar, S.P., 2011. Bacterial decolorization and degradation of azo dyes: a review. J. Taiwan Inst. Chem. Eng. 42 (1), 138–157.

Sattelberger, R., 1999. Arzneimittelrückstände in der Umwelt. Bestandsaufnahme und Problemdarstellung. Report R-162. Federal Environmental Agency, Vienna.

Schuster, A., Hädrich, C., Kümmerer, K., 2008. Flows of active pharmaceutical ingredients originating from health care practices on a local, regional, and nationwide level in Germany—is hospital effluent treatment an effective approach for risk reduction? Water Air Soil Pollut. Focus. 8 (5–6), 457–471.

Schwarzenbach, R.P., Escher, B.I., Fenner, K., Hofstetter, T.B., Johnson, C.A., Von Gunten, U., et al., 2006. The challenge of micropollutants in aquatic systems. Science 313 (5790), 1072–1077.

Seehusen, D.A., Edwards, J., 2006. Patient practices and beliefs concerning disposal of medications. J. Am. Board Family Med. 19 (6), 542–547.

Sherer, J.T., 2006. Pharmaceuticals in the environment. Am. J. Health-System Pharm. 63 (2), 174–178.

Singh, P., Singh, V.K., Singh, R., Borthakur, A., Madhav, S., Ahamad, A., et al., 2020. Bioremediation: a sustainable approach for management of environmental contaminants. In: Singh, P., Kumar, A., Borthakur, A. (Eds.), Abatement of Environmental Pollutants. Elsevier, pp. 1–23.

Spina, F., Anastasi, A.E., Prigione, V.P., Tigini, V., Varese, G., 2012. Biological treatment of industrial wastewaters: a fungal approach.

Teijon, G., Candela, L., Tamoh, K., Molina-Díaz, A., Fernández-Alba, A.R., 2010. Occurrence of emerging contaminants, priority substances (2008/105/CE) and heavy metals in treated wastewater and groundwater at Depurbaix facility (Barcelona, Spain). Sci. Total Environ. 408 (17), 3584–3595.

Ternes, T., 2007. The occurrence of micopollutants in the aquatic environment: a new challenge for water management. Water Sci. Technol. 55 (12), 327–332.

Ternes, T.A., 1998. Occurrence of drugs in German sewage treatment plants and rivers. Water Res. 32 (11), 3245–3260.

Ternes, T.A., Joss, A., Siegrist, H., 2004. Peer reviewed: scrutinizing pharmaceuticals and personal care products in wastewater treatment.

Tiehm, A., Schmidt, N., Stieber, M., Sacher, F., Wolf, L., Hoetzl, H., 2011. Biodegradation of pharmaceutical compounds and their occurrence in the Jordan Valley. Water Resour. Manage. 25 (4), 1195–1203.

Tixier, C., Singer, H.P., Oellers, S., Müller, S.R., 2003. Occurrence and fate of carbamazepine, clofibric acid, diclofenac, ibuprofen, ketoprofen, and naproxen in surface waters. Environ. Sci. Technol. 37 (6), 1061–1068.

Tripathi, S., Tripathi, B.D., 2011. Efficiency of combined process of ozone and bio-filtration in the treatment of secondary effluent. Bioresour. Technol. 102 (13), 6850–6856.

Verlicchi, P., Zambello, E., 2014. How efficient are constructed wetlands in removing pharmaceuticals from untreated and treated urban wastewaters? A review. Sci. Total Environ. 470, 1281–1306.

Verlicchi, P., Al Aukidy, M., Zambello, E., 2012. Occurrence of pharmaceutical compounds in urban wastewater: removal, mass load and environmental risk after a secondary treatment—a review. Sci. Total Environ. 429, 123–155.

Vymazal, J., 2013. The use of hybrid constructed wetlands for wastewater treatment with special attention to nitrogen removal: a review of a recent development. Water Res. 47 (14), 4795–4811.

Wang, L.K., Hung, Y.T., Shammas, N.K. (Eds.), 2005. Physicochemical Treatment Processes. Humana Press.

Wang, J., Mao, D., Mu, Q., Luo, Y., 2015. Fate and proliferation of typical antibiotic resistance genes in five full-scale pharmaceutical wastewater treatment plants. Sci. Total Environ. 526, 366–373.

Westerhoff, P., Yoon, Y., Snyder, S., Wert, E., 2005. Fate of endocrine-disruptor, pharmaceutical, and personal care product chemicals during simulated drinking water treatment processes. Environ. Sci. Technol. 39 (17), 6649–6663.

Xi, J.C., Kong, Q.Q., Wang, X.G., 2015. Spatial polarization of villages in tourist destinations: a case study from Yesanpo, China. J. Mt. Sci. 12 (4), 1038–1050.

Xiong, J.Q., Kurade, M.B., Jeon, B.H., 2018. Can microalgae remove pharmaceutical contaminants from water? Trends Biotechnol. 36 (1), 30–44.

Yadav, S., Jadeja, N.B., Dafale, N.A., Purohit, H.J., Kapley, A., 2019. Pharmaceuticals and personal care products mediated antimicrobial resistance: future challenges. Pharmaceuticals and Personal Care Products: Waste Management and Treatment Technology. Butterworth-Heinemann, pp. 409–428.

Yamamoto, H., Nakamura, Y., Moriguchi, S., Nakamura, Y., Honda, Y., Tamura, I., et al., 2009. Persistence and partitioning of eight selected pharmaceuticals in the aquatic environment: laboratory photolysis, biodegradation, and sorption experiments. Water Res. 43 (2), 351–362.

Yang, Q., Yang, M., Pritsch, K., Yediler, A., Hagn, A., Schloter, M., et al., 2003. Decolorization of synthetic dyes and production of manganese-dependent peroxidase by new fungal isolates. Biotechnol. Lett. 25 (9), 709–713.

Yang, W., Cicek, N., Ilg, J., 2006. State-of-the-art of membrane bioreactors: worldwide research and commercial applications in North America. J. Membr. Sci. 270 (1–2), 201–211.

Zhang, X.X., Zhang, T., Fang, H.H., 2009. Antibiotic resistance genes in water environment. Appl. Microbiol. Biotechnol. 82 (3), 397–414.

Zou, H., Radke, M., Kierkegaard, A., MacLeod, M., McLachlan, M.S., 2015. Using chemical benchmarking to determine the persistence of chemicals in a Swedish lake. Environ. Sci. Technol. 49 (3), 1646–1653.

Chapter 30

Mycoremediation: A novel approach for sustainable development

Bhupendra Koul[1], Waseem Ahmad[2] and Joginder Singh[3]

[1]*Department of Biotechnology, School of Bioengineering and Biosciences, Lovely Professional University, Phagwara, India,* [2]*Government Higher Secondary School Nehalpora, Srinagar, India,* [3]*Department of Microbiology, School of Bioengineering and Biosciences, Lovely Professional University, Jalandhar, India*

30.1 Introduction

Although the world is administratively partitioned into several blocks, as a matter of fact, the environment is what we have in common globally. Therefore, any inconsistency in the smooth maneuver of the vital environmental components spreads a wave of disturbance that affects the places far away from the point of the origination. So, it is evident that the global ethics and maneuvers be united to combat such issues of common interest. The advancement of medical and agricultural technology recently has paralleled the upsurge in the population of human beings. The increasing population has created a demand for more food hence, more cultivable land is required, globally. Besides creating a lousy demand of the cultivable land resources, it has led to the demand for clean water throughout the global human settlements (Pimentel et al., 2007). The anthropogenic activities have posed a tremendous strain on the resources globally, mainly in the form of contamination of the land and water, the major contaminants being organic, inorganic, or simply biological (Koul and Taak, 2018). The contamination of land and water has created severe concerns for the farmers in general and policymakers in particular. The farmers all over the globe, especially in the arid zone, are claiming that the rising levels of the salinity of soils is due to unchecked irrigation, and reckless use of synthetic fertilizers This in turn has drastically decreased the yields of the most important crops since the last few decades. Another concern of utmost importance here is the accumulation of heavy metals and dyes which have rendered the land and water resources unfit for cultivation and daily use. The heavy metals and dyes are generated from innumerable industries comprising textile, leather, pharmaceutical, and various other synthetically active sectors (Ismail et al., 2019). Heavy metal ions are the elements from the fourth period of the periodic table, typically chromium (Cr), cobalt (Co), nickel (Ni), copper (Cu), zinc (Zn), arsenic (As), lead (Pb), and mercury (Hg) all being the foremost pollutants present in water (Burakov et al., 2018; Koul and Taak, 2018). The dyes being the most detrimental elements of human aesthetics are used in bulk in the textile industry. The present scenario of growing population, food crisis, land and water contamination intensified by the novel emerging agents of contaminations are the factors that notify us to initiate the intentional maneuvers for the remediation of the contaminated land and water so that the mounting global demands of cultivable land and clean drinking water be ensured at the earliest.

The present scenario of food and health insecurity has created a havoc world over—on one side the farmers continuously experience the crop failures due to unchecked water contamination and on the other side, the global population is severely hit by the food crisis due to population explosion and partly accelerated due to the degradation of the arable land. This dire situation has alarmed the researchers globally to look for cost-effective and efficient strategies to tackle the contaminants. The most frequently engaged conventional remediation techniques such as membrane separation, coagulation, precipitation, and flocculation against the heavy metal contamination have now been replaced by the bioremediation strategies recently due to the desired features of the later such as high affinity, capacity, and selectivity.

30.2 Bioremediation

Bioremediation is the novel remediation methodology that uses the biological agents to remediate the polluted land and waters. It encompassses eco-friendly, cost-effective, efficient, target-specific, and versatile strategies. Microscopic, as

Microbe Mediated Remediation of Environmental Contaminants. DOI: https://doi.org/10.1016/B978-0-12-821199-1.00030-4

well as the macroscopic life forms, have been deployed in bioremediation, but the microscopic life forms have proven to be more reliable. Mycoremediation (*fungi-mediated bioremediation*) is emerging as a promising non-conventional and cost-effective bioremediation approach to tackle the emerging crisis of the resource contaminations worldwide (Fig. 30.1). It ensures the biosorption of the contaminants selectively with desired efficiency and targeted capacity. Any bioremediation technique aims at biosorption of the contaminants using biological agents with prime emphasis on sustainable environment and rejuvenation of the contaminated resources. The biosorption mediated by the biological agents is actually the surface phenomenon involving adsorption of the contaminants by the components of the biological agent. Many microorganisms have been frequently used as bio-sorbents since decades and have proved to be a promising alternative over the conventional strategies in many respects. The advantages of the bioremediation over the conventional methods have been mentioned in Table 30.1.

The biosorption neutralizes the land and water resources in innumerable ways (Fig. 30.2). Several living or dead microorganisms such as bacteria, fungus, and algae are comprehensively used for heavy metal abstraction for the intention that of high adsorption capacity, low cost, and its readiness in large extents (Rastegari et al., 2019). The use of algae in the bioremediation has certain drawbacks as they have been reported to convert the contaminating dyes into potent toxins. Another well-documented drawback associated with the use of algae in the decontamination against dyes is the decolorization of the dyes rather than their removal. The emergence of the fast-growing textile, leather, and pharmaceutical industry very recently has four folded the discharge of several new dyes that too in escalated volumes. The

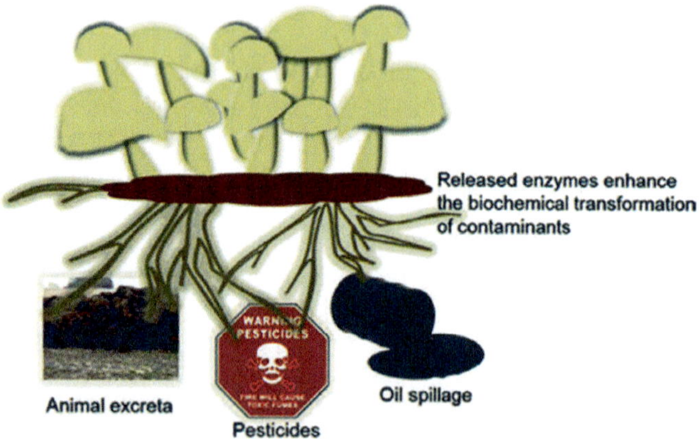

FIGURE 30.1 Bioremediation potential of fungi.

TABLE 30.1 Comparison of the conventional and nonconventional techniques of remediation.

Conventional remediation	Nonconventional remediation
Expensive	Cost-effective
Eco hazardous factors	Ecofriendly
Limited domains	Versatility
Nonselective and nonspecific	Selectivity and specificity
Low efficiency	High efficiency

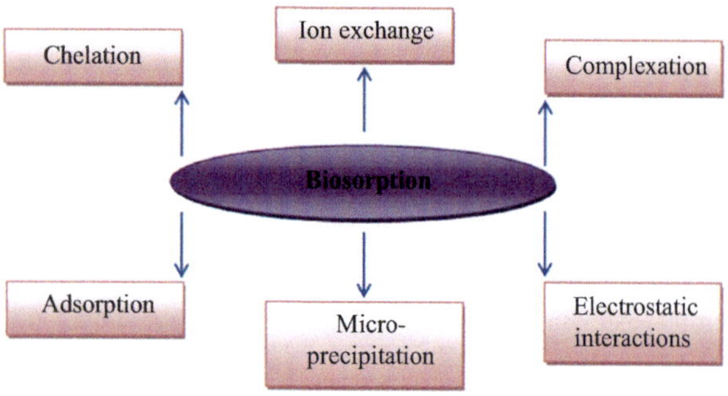

FIGURE 30.2 Biosorption and its mechanisms.

increase in the dye concentrations in the water reservoirs has led to severe threats to flora and fauna of the aquatic ecosystems. The dyes being severely carcinogenic and toxic to a broad spectrum of the biological world have now been rated as the main cause of the deterioration of water bodies. Use of fungi for wastewater management is one of the promising alternatives to physical and chemical processes (Couto, 2009; Yadav et al., 2016, 2019b).

30.2.1 Mycosorbents: fungi in biosorption

The fungi fits well into the kingdom of heterotrophic eukaryotes with operative osmotrophic properties that can be used to challenge the contemporary crunch of contamination. They are achlorophyllous and filamentous organisms with spore formation as the foremost mode of reproduction. The mycoremediation approach involves the use of fungi as biological tools to degrade, transform, or immobilize the contaminants. The fungal cell wall components such as chitin and chitosan have been extensively tested to have biosorption properties making them operative ploys to slurp heavy metals, dyes, and phenolic compounds. The biochemical conformation of the chitin and chitosan makes them chemically well-suited to bind with the heavy metal ions. Their chemical structure is given in Fig. 30.3. Most frequently used fungi for the remediation or sequestration of dyes from wastewater are the diverse forms of fungal sorbents that are used as fungal pellets, mycelium, or dead fungus by many investigators (Yagub et al., 2014). Numerous molds and filamentous microorganisms such as the *Aspergillus niger*, *Penicillium simplicissimum*, *Aspergillus fumigatus*, *Termitomyces clypeatus*, *Penicillium brevicompactum*, *Saccharomyces cerevisiae*, and *Trichoderma* are used for remediation of heavy metals and dyes (Rana et al., 2019; Yadav et al., 2019a,b).

The most reliable feature of the fungi which makes them compatible to be used as bio-sorbents in heavy metal contamination is their resistance to high metal ion concentrations. Most of the fungal taxon has been stated to withstand 20–25 mM of Pb followed by 15–20 mM of Cu and Zn and 10–15 mM of Cr. Heavy metal adsorption by fungi through ion exchange and coordination is due to the existence of chitin–chitosan, glucuronic acid, phosphate, and polysaccharides present in/on the cells of fungi (Kumar et al., 2019). Several functional groups such as amine, carboxyl, hydroxyl, phosphate, and sulfhydryl play a dynamic role in the adsorption of heavy metals and dyes by fungal strains (Yin et al., 2018). Fungal enzymes (lactase, lignin peroxidase, and Mn peroxidase) present in white-rot fungi can bioremediate numerous xenobiotic compounds present in wastewaters (Yadav et al., 2017a,b; 2018). Table 30.2 enlists the fungal species involved in mycoremediation of heavy metal and hydrocarbon contaminants from polluted soils and waters.

30.2.2 Culturing fungi for mycoremediation

The culture of fungi for mycoremediation can be done in vitro at a pilot level first, then subjected to proliferation or can directly be screened from the industrial wastes such as *A. niger* (screened from the waste of Citric acid industry) and *S. cerevisiae* (wastes from Brewing industry). The isolation of fungi is further enhanced by the formation of high-density pellets in suspension cultures. The linear increase in the pellet size and mass with time makes the isolation and use of the fungi for the remediation more reliable. The characteristic features that make fungi reliable for remediation are large surface to volume ratios along with hyphae that makes penetration into the substrates more easy and efficient. The fungi are easy to culture, isolate, and manage, these properties makes mycoremediation a promising strategy for environment clean-up. They are efficient in the accumulation of substances into the thalli, which makes them more promising to be put into use. One of the main advantages associated with the mycoremediation is the fact that the fungi have been reported to promote the growth of various plant taxon, so the inoculation of an experimental field with the fungi always parallels with the rapid promotion of the various plant taxa. There is evidence that they act as the nutrient

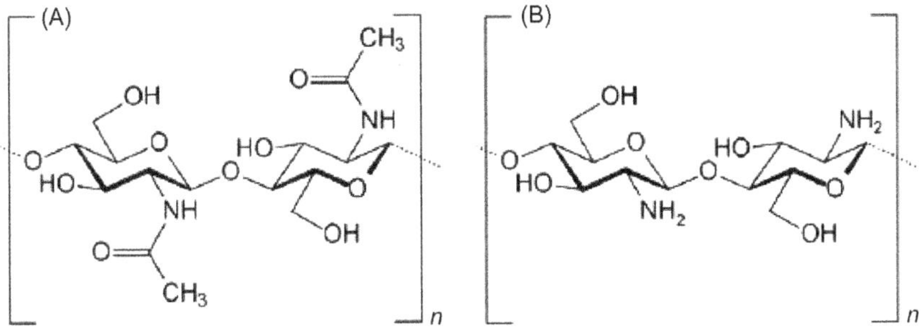

FIGURE 30.3 The biochemical structure of (A) chitin and (B) chitosan.

TABLE 30.2 Mycoremediation of heavy metals and hydrocarbons.

Fungi	Heavy metal	Reference(s)
Absidia cylindrospora	Fluorene	Garon et al. (2004)
Acremonium strictum	Cr	Ahluwalia and Goyal (2010)
Agaricus macrospores	Cd, Cu, Hg	Garcia et al. (2005)
Agaricus bisporus, *Lactarius piperatus*	Cd(II) ions	Nagy et al. (2013)
Amanita rubescens	Pb	Sarı and Tuzen (2009)
Aspergillus fumigatus	Cr, Pb	Iram et al. (2013)
Aspergillus versicolor	Pb	Cabuk et al. (2005)
Aspergillus flavus	Cr, Hg, Pb	Iram et al. (2013), Kurniati et al. (2014)
Aspergillus foetidus	Cr	Ahluwalia and Goyal (2010)
Aspergillus lentulus	Pb, Cu, Cr, Ni	Mishra and Malik (2014)
Aspergillus niger	Cd, Cu, Cr, Ni, Pb, Zn	Kapoor et al. (1999), Ahluwalia and Goyal (2007), Thakur and Srivastava (2011), Ali (2013), Iram et al. (2013), Sallau et al. (2014)
Aspergillus oryzae	Cr	Sepehr et al. (2005)
Aspergillus terreus	Fe	Dias et al. (2002)
Aspergillus terricola	Cr	Ahluwalia and Goyal (2010)
Aureobasidium pullulans	Cr	Ahluwalia and Goyal (2010
Auricularia polytricha	Cd, Cu, Pb	Zhang et al. (2011)
Candida spp.	Cd, Cr, Cu, Fe, Hg, Ni, Pb, Zn	Anaemene (2012), Ali (2013)
Circinella sp.	Ni	Alpat et al. (2010)
Cladosporium resinae	Cr	Ahluwalia and Goyal (2010)
Claviceps paspali	Zn	Luef et al. (1991)
Coniothyrium sp., *Fusarium* sp.	PAHs	Jacques et al. (2008)
Coriolus versicolor MKACC 52492	PAH	Jang et al. (2009)
Cunninghamella sp.	Pb	El-Morsy (2004)
Cyberlindnera fabianii	Cr	Bahafid et al. (2013)
Flammulina velutipes	Cu	Luo et al. (2013)
Fomes fasciatus	Cu(II)	Sutherland and Venkobachar (2013)
Ganoderma lucidum	Cu	Muraleedharan and Venkobachar (1990)
Hirsutella sp.	Cr	Srivastava and Thakur (2006)
Hymenoscyphus ericae	Zn, Cd, and Fe	Binsadiq (2015)
Lentinula edodes	2,4-DCP	Tsujiyama et al. (2013)
Metarhizium anisopliae	Pb	Cabuk et al. (2005)
Mucor rouxii	Cd, Ni, Pb, Zn	Yan and Viraraghavan (2003)
Mucor sp.	Pb, Cd, Ni, Zn, Cu	Zhang et al. (2017)
Paecilomyces lilacinus	Cd, Cr	Sharma and Adholeya (2011), Zeng et al. (2013)

(*Continued*)

TABLE 30.2 (Continued)

Fungi	Heavy metal	Reference(s)
Paecilomyces sp.	Cr	Cárdenas-González and Acosta-Rodríguez (2010)
Paecilomyces variotii	Cr	Ahluwalia and Goyal (2010)
Paxillus involutus	Cd	Blaudez et al. (2000)
Penicillium lanosumcoeruleum	Cu, Ni, Pb	İlhan et al. (2004)
Penicillium chrysogenum	Cd, Cr, Cu, Ni, Pb, Zn	Ahluwalia and Goyal (2007)
Penicillium citrinum	Cu	Verma et al. (2013)
Penicillium sp.	Zn, Ni, Pb	Mohsenzadeh and Shahrokhi (2014)
Penicillium sp.	Cr	Arévalo-Rangel et al. (2013)
Penicillium verrucosum	Pb	Cabuk et al. (2005)
Phanerochaete chrysosporium	Cd, Cr, Cu, Ni, Pb, Zn	Say et al. (2001), Gopal et al. (2002), Çeribasi and Yetis (2004), Li et al. (2004), Marandi et al. (2010), Murugavelh and Mohanty (2012), Xu et al. (2013)
P. chrysosporium	Pb, Cd	Xu et al. (2014)
Pisolithus sp.	Pb, Cu, Cr, Ni, Zn	Tam (1995)
Pleurotus ostreatus	Pb	Kapahi and Sachdeva (2017)
	Cd	Kapahi and Sachdeva (2017)
	Cd	Favero et al. (1990a,b)
	Hg	Mandal et al. (1998)
	Cd, Hg	Yayçinkaya et al. (2002)
	Pb	Xiangliang et al. (2005)
	Cd	Das et al. (2007)
	Cr	Javaid and Bajwa (2007)
	Cu, Cr, Ni, Zn	Javaid and Bajwa (2008)
	Co	Xiangliang et al. (2009)
	Pb	Tay et al. (2009)
	Crude oil	Olusola and Anslem (2010)
	Pb	Liew et al. (2010)
	Cd	Tay et al. (2011)
	Zn	Jibran and Milsee Mol (2011)
	PAH	Okparanma et al. (2011)
	Cd	Tay et al. (2011)
	Cu, Cr, Ni, Zn	Javaid et al. (2011)
	Radioactive waste	Eskander et al. (2012)
	Cu, Cr, Fe, Zn	Arbanah et al. (2012)
	Heavy metals	Oyetayo et al. (2012)
	Cr	Puentes-Cárdenas et al. (2012)
	Pb	Prasad et al. (2013)
	Cu, Cd, Fe, Ni, Pb, Zn	Lamrood and Ralegankar (2013)

(Continued)

TABLE 30.2 (Continued)

Fungi	Heavy metal	Reference(s)
	Cu, Pb	de Almeida and Burgess (2013)
	Cr	Arbanah et al. (2013)
	Oxo-biodegradable plastic	Da Luz et al. (2013)
	Pb	Dulay et al. (2015)
	Mn	Wu et al. (2016)
Pleurotus platypus, A. bisporus, Calocybe indica	Cu, Zn, Fe, Cd, Pb, Ni	Lamrood and Ralegankar (2013)
Rhizopus arrhizus	Zn	Preetha and Viruthagiri (2005)
Rhizopus nigricans	Cd, Cr, Cu, Ni, Pb, Zn	Kogez and Pavko (2001), Ahluwalia and Goyal (2007)
Rhizopus sp. Pb	Pb, Cd, Cr, Cu, Ni, Zn	Pal et al. (2010), Luo and Xiao (2010), Shoaib et al. (2012), Fourest and Roux (1992)
Saccharomyces cerevisiae	Al, Cd, Co, Cr, Cu, Fe, Hg, Ni, Pb, Sr, Zn	Avery and Tobin (1992), Mapolelo and Torto (2004), Benazir et al. (2010), Kumar and Rao (2011), Gohari et al. (2013)
Schizophyllum commune, Polyporous sp.	Malachite green	Rajput et al. (2011)
Trichoderma sp.	Cu, Pb, Cd, As, Zn	Fazli et al. (2015)
Trichosporon cutaneum	Cr	Bajgai et al. (2012)
Volvariella volvacea	Cd, Co, Cu, Pb	Purkayastha and Mitra (1992)

DCP, Dichlorophenol; *PAHs*, polycyclic aromatic hydrocarbons.

sources for various organisms, most important to mention here is that various arthropods dwell as refugee in the fungal colonies. So the fungi inevitably facilitates other organisms, hence intensify the colonization of the area under experimentation.

30.2.3 Methods of isolation

There are several methods to isolate the terrestrial as well as aquatic fungi. Once isolated, they can be assayed for their remediation potential. Given below are the important methods of isolation of the fungi.

30.2.3.1 Isolation of the terrestrial fungi

Numerous immersion methods have been established and put into use to isolate the terrestrial fungi for various scientific studies. The pioneering technique was developed by Chesters (1949) since then modified by a number of scientists in order to make the technique more efficient, reliable, and cheap. All the modified versions of the basic technique developed by Chesters (1949) have the same basic principle "the isolating medium is positioned in the soil in such a way that it is partitioned from the soil by an air space." The medium is left undisturbed for a period of 5–7 days after which the small samples are plated on the selection medium. The fungi are then isolated, purified, and cultured for further use. Since the establishment of this isolation method, more and more abridged methods have been proposed time to time including the simple immersion method described by Mueller and Durrell (1957), hyphal isolation by Warcup (1995), and detachable fungal propagules isolation by Harley and Waid (1955) (Table 30.2).

30.2.3.2 Isolation of the aquatic fungi

The isolation methods of the fungi from the aquatic samples have been developed by Jones (1971) and the media for culture formulated by Jones (1971) and Booth (1971). The important routinely adopted for the isolation of the aquatic fungi are Baiting method, Dilution method (Parkes, 1982), Particle method, and Concentration methods. The aquatic fungi are easy to isolate as compared to the terrestrial fungi.

30.3 Biochemical aspects of the mycoremediation

The fungi in general and the mushrooms, in particular, predominantly produce extracellular peroxidases, ligninases, cellulases, pectinases, xylanases, and oxidases. These enzymes are able to oxidize the recalcitrant contaminants in the soil. They have the potential to degrade the nonpolymeric, recalcitrant pollutants such as nitrotoluenes, polycyclic aromatic hydrocarbons (PAHs), organic and synthetic dyes, and pentachlorophenol.

Fig. 30.4 describes the biochemical mechanics of most of the bioremediation pathways adopted by the various fungi. Recently, it has been reported that many taxa of mushrooms degrade the polymers such as plastics (Da Luz et al., 2013). The rhizosphere of the plant is a thin layer of soil, adhering to the root surface. The fungi have proved to be mediators in the alterations of the rhizosphere in promoting the plant growth, community-onset and finally the rejuvenation of the land.

Plant related microbes that are directly or indirectly involved in growth promotion and yield improvement are used for plant growth promotion. These microorganisms that are associated with plant growth are called plant growth-promoting organisms (Kloepper et al., 2004). Similarly, the fungal taxa have been confirmed to be the mediators of environmental rejuvenation and remediation. The soils, water, sediments, and air are enormously polluted with the heavy metals. The recurrently mounting heavy metal contaminations are posing a threat to the life and livelihood of many sections of the society. Various tests demonstrate that the copper (Cu) is highly toxic to most of the microbial life forms even in minute concentrations, but the fungi such as *Aspergillus niger, Penicillium chrysogenum*, and *Rhizopus stolonifer* are able to tolerate as high as 500 μg/mL of Cu concentration. It has been documented that the fungal bodies can accumulate Cu in the rage of 3000–6000 μg/g. The fungal taxa such as *Hymenoscyphus ericae, Pisolithus tinctorius*, and *Vaccinium macrocarpon* were able to predominate in the liquid media having 400 to 500 μg/mL of zinc (Zn), cadmium (Cd), and iron (Fe). The fungi are not able only to withstand high concentrations of eavy metals, but also they are able to accumulate, tolerate and sequester them for longer durations. This makes the fungi comparatively more reliable in remediation.

The fungi such as *A. niger* and *Phanerochaete chrysosporium* have been used repeatedly for the biodegradation and detoxification of various dyes such as malachite green, nigrosin, and basic fuchsin. Another fungi, *Perenniporia tephropora*, has been used in the discoloration and biodegradation of the reactive blue 4 and methyl orange. The existing chapter focuses on the most important contaminants of the world including heavy metals, mineral oils, PAHs, nonaromatic hydrocarbons (NAHs), phenols chlorinated hydrocarbons (CHCs), endocrine-disrupting chemicals (EDCs) etc.

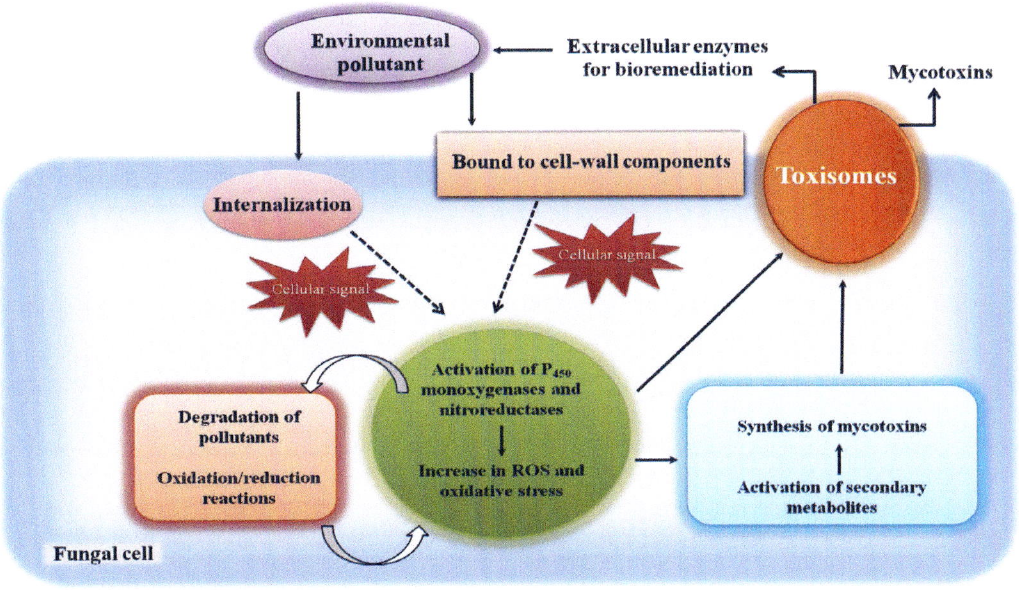

FIGURE 30.4 The mechanistic illustration of the mycoremediation.

Researches all over the globe are deploying both micro as well as macrofungi to remediate the resources. The *Basidiomycetes* in general and the White rot fungi (WRF) in particular are used frequently in the bioremediation processes for land reclamation. The WRF has been reported to efficiently degrade the lignin and cellulosic polymers, thus play an important role in the mineralization and nutrient recycling of immobilized resources. The fungi including *P. chrysosporium*, *Pleurotus ostreatus*, *Trametes versicolor*, *Bjerkandera adusta*, *Lentinula edodes*, *Irpex lacteus*, *Agaricus bisporus*, *Pleurotus tuber-regium*, and *Pleurotus pulmonarius* have been documented to release the extracellular enzymes such as lignin peroxidases, manganese dependent peroxidases and versatile peroxidase to degrade environmental pollutants. Among these fungi, *P. chrysosporium* has been explored the most for its capability to destroy toxic or insoluble compounds to CO_2 and H_2O, more competently than other fungi. In 1985, for the first time, Bumpus et al. proposed the application of *P. chrysosporium* in bioremediation.

30.4 Parameters affecting mycoremediation

The physiochemical aspects of the surrounding environment (pH, temperature, hydration level, redox potential) in general and that of the immediate microenvironment, in particular, affect the rate and efficacy of the remediation process. The availability of the micro- and macronutrients, the contaminant type, the concentration level, the toxicity, and the mobility of the contaminants are the important determining factors of the bioremediation. It has been observed that if the appropriate of the necessary nutrients and particular electron donor species are not available, their careful addition improves the rate and effectiveness of the remediation process (biostimulation). In another regulating step, that is, bioaugmentation, allochthonous, or supplemented autochthonous microorganisms, capable of metabolizing a definite contaminant are introduced in the soil. In both cases the homogeneous distribution of the added biomass and its propagation, in competition with innate microorganisms, is a crucial challenge. Furthermore, bioaugmentation and biostimulation could also be coupled in order to further excite the introduced biomass. At higher concentrations of the contaminants, most of the bioremediation agents have been testified to flop, but the fungal mediators have been stated to tolerate and work efficiently even under the higher concentration of the contaminants. Filamentous fungi which create mycotoxins (e.g. *Aspergillus* and *Penicillium* spp.) have the potential to degrade a wide variety of pharmaceutical compounds among them are the emergent pollutants EDCs which are ineffectively degraded by bacteria. An aspect of utmost importance in the mycoremediation programs is to manage/balance the factors that enhance the establishment of the fungal inoculum successfully.

30.5 Conclusions and future prospects

The use of biological agents in the remediation of the contaminants is a promising strategy for environment clean-up. The technique of mycoremediation can be deployed for the biodegradation of a variety of environmental contaminants such as heavy metals, mineral oils, PAHs, NAHs, phenols, and CHCs, together with the emerging contaminants and pharmaceutical-personal care products. Many fungal taxa have been reported to be reliable mycoremediation agents. The important parameters in mycoremediation include the biochemical and physiochemical aspects that inhibit or stimulate the process of remediation. Research is being done to use mycoremediation in national defense against chemical and biological warfare. Another approach to the use of fungal-derived microbial systems is to synthesize nanoparticles (clean and green nanotechnology) that can provide an additional advantage in environmental clean-up.

References

Ahluwalia, S.S., Goyal, D., 2007. Microbial and plant derived biomass for removal of heavy metals from wastewater. Biores. Technol. 98, 2243–2257.

Ahluwalia, S.S., Goyal, D., 2010. Removal of Cr6 + from aqueous solution by fungal biomass. Eng. Life Sci. 10 (5), 480–485.

Ali, S.A.A., 2013. Removal of heavy metals from synthesis industrial waste water using local isolated (*Candida utilis* and *Aspergillus niger*). Int. J. Biotechnol. 2 (5), 83–90.

Alpat, S., Alpat, S.K., Çadirci, B.H., Özbayrak, Ö., Yasa, İ., 2010. Effects of biosorption parameter: kinetics, isotherm and thermodynamics for Ni(II) biosorption from aqueous solution by *Circinella* sp. Electron. J. Biotechnol. 13, 4–5.

Anaemene, I.A., 2012. The use of *Candida* sp. in the biosorption of heavy metals from industrial effluent. Eur. J. Exp. Biol. 2, 484–488.

Arbanah, M., Miradatul Najwa, M.R., Ku Halim, K.H., 2012. Biosorption of Cr(III), Fe(II), Cu(II), Zn(II) ions from liquid laboratory chemical waste by *Pleurotus ostreatus*. Int. J. Biotechnol. Wellness Ind. 1, 152–162.

Arbanah, M., Miradatul Najwa, M.R., Ku Halim, K.H., 2013. Utilization of *Pleurotus ostreatus* in the removal of Cr(VI) from chemical laboratory waste. Int. Refreed J. Eng. Sci. 2 (4), 29–39.

Arévalo-Rangel, D.L., Cárdenas-González, J.F., Martínez-Juárez, V.M., Acosta-Rodríguez, I., 2013. Hexavalent chromate reductase activity in cell free extracts of *Penicillium* sp. Bioinorg. Chem. Appl. 2013, 1–6.

Avery, S.V., Tobin, J.M., 1992. Mechanisms of strontium uptake by laboratory and brewing strains of *Saccharomyces cerevisiae*. Appl. Environ. Microbiol. 58 (12), 3883–3889.

Bahafid, W., Joutey, N.T., Sayel, H., Ghachtouli, N.E.L., 2013. Mechanism of hexavalent chromium detoxification using *Cyberlindnera fabianii* yeast isolated from contaminated site in Fez (Morocco). J. Mater. Environ. Sci. 4 (6), 840–847.

Bajgai, R.C., Georgieva, N., Lazarova, N., 2012. Bioremediation of chromium ions with filamentous yeast *Trichosporon cutaneum* R57. J. Biol. Earth Sci. 2 (2), 70–R75.

Benazir, J.F., Suganthi, R., Rajvel, D., Pooja, M.P., Mathithumilan, B., 2010. Bioremediation of chromium in tannery effluent by microbial consortia. Afr. J. Biotech. 9 (21), 3140–3143.

Binsadiq, A.R.H., 2015. Fungal absorption and tolerance of heavy metals. Ind. Wastewater J. Agric. Sci. Technol. 5, 77–80.

Blaudez, D., Jacob, C., Turnau, K., Colpaert, J.V., Ahonen-Jonnarth, U., Finlay, R., et al., 2000. Differential responses of ectomycorrhizal fungi to heavy metals in vitro. Myco Res. 104, 1366–1371.

Booth, C., 1971. Methods in microbiology. Academic Press.

Burakov, A.E., Galunin, E.V., Burakova, I.V., Kucherova, A.E., Agarwal, S., Tkachev, A.G., et al., 2018. Adsorption of heavy metals on conventional and nanostructured materials for wastewater treatment purposes: a review. Ecotoxicol. Environ. Saf. 148, 702–712.

Cabuk, A., Ilhan, S., Filik, C., Çaliskan, F., 2005. Pb^{2+} biosorption by pretreated fungal biomass. Turk. J. Biol. 29, 23–28.

Cárdenas-Gonźalez, J.F., Acosta-Rodríguez, I., 2010. Hexavalent chromium removal by a *Paecilomyces* sp. fungal strain isolated from environment. Bioinorg. Chem. Appl. 2010, 1–6.

Çeribasi, I.H., Yetis, U., 2004. Biosorption of Ni^{2+} and Pb^{2+} by *Phanerochaete chrysosporium* from a binary metal system-kinetics. Water Soil. Air Res. 27 (1), 15–20.

Chesters, G.C., 1949. Presidential address concerning fungi inhabiting soil. Trans. Br. Mycol. Soc. 32(3-4), IN1-216.

Couto, S.R., 2009. Dye removal by immobilised fungi. Biotechnol. Adv. 27(3), 227–235.

Da Luz, J.M.R., Paes, S.A., Nunes, M.D., da Silva, M.C.S., Kasuya, M.C.M., 2013. Degradation of oxo-biodegradable plastic by *Pleurotus ostreatus*. PLoS One 8 (8), 69386.

Das, N., Charumathi, D., Vimala, R., 2007. Effect of pretreatment on Cd^{2+} biosorption by mycelia biomass of *Pleurotus florida*. Afr. J. Biotechnol. 6, 2555–2558.

de Almeida, L.K., Burgess, J.E., 2013. Biosorption and Bioaccumulation of Copper and Lead by *Phanerochaete* and *Pleurotus ostreatus*.

Dias, M.A., Lacerda, I.C.A., Pimentel, P.F., De Castro, H.F., Rosa, C.A., 2002. Removal of heavy metals by an *Aspergillus terreus* strain immobilized in a polyurethane matrix. Lett. Appl. Microbiol. 34 (1), 46–50.

Dulay, R.M.R., De Castro, M.A.E.G., Coloma, N.B., Bernardo, A.P., Cruz, A.G.D., Tiniola, R.C., et al., 2015. Effects and myco-remediation of lead (Pb) in five Pleurotus mushrooms. Int. J. Biol. Pharm. Allied Sci. 4 (3), 1664–1677.

El-Morsy, E.S.M., 2004. *Cunninghamella echinulata* a new biosorbent of metal ions from polluted water in Egypt. Mycologia 96, 1183–1189.

Eskander, S.B., Abd El-Aziz, S.M., El-Sayaad, H., Saleh, H.M., 2012. Cementation of Bioproducts Generated From Biodegradation of Radioactive Cellulosic-Based Waste Simulates by Mushroom.

Favero, N., Bressa, G., Costa, P., 1990a. Response of *Pleurotus ostreatus* to cadmium exposure. Ecotoxicol. Environ. Safe 20 (1), 1–6.

Favero, N., Costa, P., Paolo Rocco, G., 1990b. Role of copper in cadmium metabolism in the basidiomycetes *Pleurotus ostreatus*. Comp. Biochem. Physiol., C: Comp Pharmacol. 97 (2), 297–303.

Fazli, M.M., Soleimani, N., Mehrasbi, M., Darabian, S., Mohammadi, J., Ramazani, A., 2015. Highly cadmium tolerant fungi: their tolerance and removal potential. J. Environ. Health Sci. Eng. 13, 19.

Fourest, E., Roux, J.C., 1992. Heavy-metal biosorption by fungal mycelial byproducts-mechanisms and influence of pH. Appl. Microbiol. Biotechnol. 37, 399–403.

Garcia, M.A., Alonso, J., Melgar, M.J., 2005. Agaricus macrosporus as a potential bioremediation agent for substrates contaminated with heavy metals. J. Chem. Technol. Biotechnol. 80, 325–330.

Garon, D., Sage, L., Wouessidjewe, D., Seigle-Murandi, F., 2004. Enhanced degradation of fluorene in soil slurry by *Absidia cylindrospora* and maltosyl-cyclodextrin. Chemosphere 56 (2), 159–166.

Gohari, M., Hosseini, S.N., Sharifnia, S., Khatami, M., 2013. Enhancement of metal ion adsorption capacity of *Saccharomyces cerevisiae*'s cells by using disruption method. J. Taiwan. Inst. Chem. Eng. 44 (4), 637–645.

Gopal, M., Pakshirajan, K., Swaminathan, T., 2002. Heavy metal removal by biosorption using *Phanerochaete chrysosporium*. Appl. Biochem. Biotechnol. 102 (1–6), 227–237.

Harley, J.L., Waid, J.S., 1955. A method of studying active mycelia on living roots and other surfaces in the soil. Trans. Brit. Mycol. Soc. 38, 104–118.

İlhan, S., Çabuk, A., Filik, C., Çalişkan, F., 2004. Effect of pretreatment on biosorption of heavy metals by fungal biomass. Trak. Univ. J. Sci. 5 (1), 11–17.

Iram, S., Perveen, K., Shuja, N., Waqar, K., Akhtar, I., Ahmad, I., 2013. Tolerance potential of different species of *Aspergillus* as bioremediation tool-comparative analysis. J. Microbiol. Res. 1 (1), 001–008.

Ismail, M., Akhtar, K., Khan, M.I., Kamal, T., Khan, M.A., et al., 2019. Pollution, toxicity and carcinogenicity of organic dyes and their catalytic bio-remediation. Curr. Pharm. Des. 25 (34), 3645–3663.

Jacques, R.J.S., Okeke, B.C., Bento, F.M., Teixeira, A.S., Peralba, M.C.R., Camargo, F.A.O., 2008. Microbial consortium bioaugmentation of a polycyclic aromatic hydrocarbons contaminated soil. Bioresour. Technol. 99 (7), 2637–2643.

Jang, K.Y., Cho, S.M., Seok, S.J., Kong, W.S., Kim, G.H., Sung, J.M., 2009. Screening of biodegradable function of indigenous ligno-degrading mushroom using dyes. Mycobiology 37, 53–61.

Javaid, A., Bajwa, R., 2007. Biosorption of Cr(III) ions from tannery wastewater by *Pleurotus ostreatus*. Mycopathologia 5, 71–79.

Javaid, A., Bajwa, R., 2008. Biosorption of electroplating heavy metals by some basiodiomycetes. Mycopathologia 6, 1–6.

Javaid, A., Bajwa, R., Shafique, U., Anwar, J., 2011. Removal of heavy metals by adsorption on *Pleurotus ostreatus*. Biomass Bioenergy 35, 1675–1682.

Jibran, A.K., Milsee Mol, J.P., 2011. *Pleurotus sajor-caju* protein: a potential biosorptive agent. Adv. Biol. Tech. 11, 25–27.

Jones, E.G., 1971. Chapter XII Aquatic Fungi, Methods in microbiology, 4. Academic Press, pp. 335–365.

Kapahi, M., Sachdeva, S., 2017. Mycoremediation potential of *Pleurotus* species for heavy metals: a review. Bioresour. Bioprocess. 4 (1), 32.

Kapoor, A., Viraraghavan, T., Cullimore, D.R., 1999. Removal of heavy metals using the fungus *Aspergillus niger*. Bioresour. Technol. 70 (1), 95–104.

Kloepper, J.W., Ryu, C.-M., Zhang, S.A., 2004. Induced systemic resistance and promotion of plant growth by *Bacillus* spp. Phytopathology 94, 1259–1266.

Kogez, A., Pavko, A., 2001. Laboratory experiments of lead biosorption by self immobilized *Rhizopus nigricans* pellets in the stirred tank reactor and packed bed column. Chem. Biochem. Eng. 15, 75–79.

Koul, B., Taak, P., 2018. Soil Pollution: Causes and Consequences. Biotechnological Strategies for Effective Remediation of Polluted Soils. Springer, Singapore, pp. 1–37.

Kumar, B.M., Rao, V.G., 2011. Removal of Cu^{2+} and Pb^{2+} ions from aqueous solutions by free, immobilized and co-immobilized cells of *Saccharomyces cerevisiae* and *Lactobacillus sporogenes*. Int. J. Sci. Emerg. Technol. 2 (3), 80–86.

Kumar, A., Kumar, V., Singh, J., 2019. Role of fungi in the removal of heavy metals and dyes from wastewater by biosorption processes. In Recent Advancement. White Biotechnology Through Fungi. Springer, Cham, pp. 397–418.

Kurniati, E., Arfarita, N., Imai, T., Higuchi, T., Kanno, A., Yamamoto, K., et al., 2014. Potential bioremediation of mercury contaminated substrate using filamentous fungi isolated from forest soil. J. Environ. Sci. 26 (6), 1223–1231.

Lamrood, P.Y., Ralegankar, S.D., 2013. Biosorption of Cu, Zn, Fe, Cd, Pb and Ni by non-treated biomass of some edible mushrooms. Asian J. Exp. Biol. Sci. 4, 190–195.

Li, Q., Wu, S., Liu, G., 2004. Simultaneous biosorption of Cd^{2+} and Pb^{2+} ions by pretreated biomass of *Phanerochaete chrysosporium*. Sep. Purif. Technol. 34 (1), 135–142.

Liew, H.H., Tay, C.C., Yong, S.K., Surif, S., Abdul Talib, S., 2010. Biosorption characteristics of lead [Pb(II)] by *Pleurotus ostreatus* biomass. In: Abstracts of the Proceedings of International Conference on Science and Social Research (CSSR), Kuala Lumpur.

Luef, E., Prey, T., Kubicek, C.P., 1991. Biosorption of zinc by fungal mycelial wastes. Appl. Microbiol. Biotechnol. 34 (5), 688–692.

Luo, J.M., Xiao, X.I., 2010. Biosorption of cadmium (II) from aqueous solutions by industrial fungus *Rhizopus cohnii*. Trans. Nonferrous Met. Soc. China 20, 1104–1111.

Luo, D., Yf, X., Tan, Z.L., Li, X.D., 2013. Removal of Cu^{2+} ions from aqueous solution by the abandoned mushroom compost of *Flammulina velutipes*. J. Environ. Biol. 34, 359–365.

Mandal, T.K., Baldrian, P., Gabriel, J., Nerud, F., Zadrazil, F., 1998. Effect of mercury on the growth of wood-rotting basidiomycetes *Pleurotus ostreatus*, *Pycnoporus cinnabarinus* and *Serpula lacrymans*. Chemosphere 36 (3), 435–440.

Mapolelo, M., Torto, N., 2004. Trace enrichment of metal ions in aquatic environments by *Saccharomyces cerevisiae*. Talanta 64, 39–47.

Marandi, R., Ardejani, F.D., Afshar, H.A., 2010. Biosorption of Pb^{2+} and Zn^{2+} ions by pretreated biomass of *Phanerochaete chrysosporium*. Int. J. Min. Environ. Issues 1 (1), 9–16.

Mishra, A., Malik, A., 2014. Novel fungal consortium for bioremediation of metals and dyes from mixed waste stream. Bioresour. Technol. 171, 217–226.

Mohsenzadeh, F., Shahrokhi, F., 2014. Biological removing of Cadmium from contaminated media by fungal biomass of *Trichoderma* species. J. Environ. Health Sci. Eng. 12, 102.

Mueller, K.E., Durrell, L.W., 1957. Sampling tubes for soil fungi. Phytopathology 47 (4), 243.

Muraleedharan, T.R., Venkobachar, C., 1990. Mechanism of biosorption of Cu^{6+} by *Ganoderma lucidum*. Biotechnol. Bioeng. 35, 320–325.

Murugavelh, S., Mohanty, K., 2012. Bioreduction of hexavalent chromium by live and active *Phanerochaete chrysosporium*: kinetics and modelling. Clean. Soil Air Water 40 (7), 746–751.

Nagy, B., Măicăneanu, A., Indolean, C., Mânzatu, C., Silaghi-Dumitrescu, L., Majdik, C., 2013. Comparative study of Cd(II) biosorption on cultivated *Agaricus bisporus* and wild *Lactarius piperatus* based biocomposites. Linear and nonlinear equilibrium modelling and kinetics. J. Taiwan Inst. Chem. E 45 (3), 921–929.

Okparanma, R.N., Ayotamuno, J.M., Davis, D.D., Allagoa, M., 2011. Mycoremediation of polycyclic aromatic hydrocarbons (PAH)-contaminated oil-based drill-cuttings. Afr. J. Biotechnol. 10 (26), 5149–5156.

Olusola, S.A., Anslem, E.E., 2010. Bioremediation of a crude oil polluted soil with *Pleurotus pulmonarius* and *Glomus mosseae* using *Amaranthus hybridus* as a test plant. J. Bioremed. Biodegrad. 1, 111.

Oyetayo, V.O., Adebayo, A.O., Ibileye, A., 2012. Assessment of the biosorption potential of heavy metals by *Pleurotus tuber-regium*. Int. J. Adv. Biol. Res. 2, 293–297.

Pal, T.K., Bhattacharyya, S., Basumajumdar, A., 2010. Improvement of bioaccumulation of cadmium by *Aspergillus niger* as a function of complex nutrient source. J. Indian Chem. Soc. 87, 391–394.

Parkes, R.J., 1982. Methods for enriching, isolating, and analysing microbial communities in laboratory systems. In: Bull, A.T., Slater, J.H. (Eds.), Microbial Interactions and Communities, vol. 1. Academic Press, London, pp. 45–102.

Pimentel, D., Cooperstein, S., Randell, H., Filiberto, D., Sorrentino, S., et al., 2007. Ecology of increasing diseases: population growth and environmental degradation. Human Ecology 35 (6), 653–668.

Prasad, A.S.A., Varatharaju, G., Anushri, C., Dhivyasree, S., 2013. Biosorption of lead by *Pleurotus florida* and *Trichoderma viride*. Br. Biotechnol. J. 3 (1), 66–78.

Preetha, B., Viruthagiri, T., 2005. Biosorption of Zn^{2+} by *Rhizopus arrhizus*: equilibrium and kinetic modeling. Afr. J. Biotechnol. 4 (6), 506–508.

Puentes-Cárdenas, I.J., Pedroza-Rodríguez, A.M., Navarrete-López, M., Villegas-Garrido, T.L., Cristiani-Urbina, E., 2012. Biosorption of trivalent chromium from aqueous solutions by *Pleurotus ostreatus* biomass. Environ. Eng. Manage. J. 11 (10), 1741–1752.

Purkayastha, R.P., Mitra, A.K., 1992. Metal uptake by mycelia during submerged growth and by sporocarps of an edible fungus *Volvariella volvacea*. Indian J. Exp. Biol. 30, 1184–1187.

Rajput, Y., Shit, S., Shukla, A., Shukla, K., 2011. Biodegradation of malachite green by wild mushroom of Chhattisgarh. J. Exp. Sci. 2, 69–72.

Rana, K.L., Kour, D., Sheikh, I., Dhiman, A., Yadav, N., et al., 2019. Endophytic fungi: biodiversity, ecological significance, and potential industrial applications. Recent Advancement In White Biotechnology Through Fungi. Springer, Cham, pp. 1–62.

Rastegari, A.A., Yadav, A.N., Gupta, A., 2019. Prospects of renewable bioprocessing in future energy systems. Springer International Publishing, Cham, pp. 1–50.

Sallau, A.B., Inuwa, H.M., Ibrahim, S., Nok, A.J., 2014. Isolation and properties of chromate reductase from *Aspergillus niger*. Int. J. Mod. Cell Mol. Biol. 3 (1), 10–21.

Sarı, A., Tuzen, M., 2009. Kinetic and equilibrium studies of biosorption of Pb^{2+} and Cd^{2+} from aqueous solution by macrofungus (*Amanita rubescens*) biomass. J. Hazard. Mater. 164 (2), 1004–1011.

Say, R., Denizli, A., Yakup Arıca, M., 2001. Biosorption of Cd^{2+}, Pb^{2+} and Cu^{2+} with the filamentous fungus *Phanerochaete chrysosporium*. Bioresour. Technol. 76 (1), 67–70.

Sepehr, M.N., Nasseri, S., Assadi, M.M., Yaghmaian, K., 2005. Chromium bioremoval from tannery industries by *Aspergillus oryzae*. Iran. J. Environ. Health Sci. Eng. 2 (4), 273–279.

Sharma, S., Adholeya, A., 2011. Detoxification and accumulation of chromium from tannery effluent and spent chrome effluent by *Paecilomyces lilacinus* fungi. Int. Biodeterior. Biodegrad. 65 (2), 309–317.

Shoaib, A., Naureen, A., Tanveer, F., Aslam, N., 2012. Removal of Ni (II) ions from substrate using filamentous fungi. Int. J. Agric. Biol. 14, 151–153.

Srivastava, S., Thakur, I.S., 2006. Isolation and process parameter optimization of *Aspergillus* sp. for removal of chromium from tannery effluent. Bioresour. Technol 97 (10), 1167–1173.

Sutherland, C., Venkobachar, C., 2013. Equilibrium modeling of Cu (II) biosorption onto untreated and treated forest macro-fungus *Fomes fasciatus*. Int. J. Plant. Anim. Environ. Sci. 3, 193–203.

Tam, P.C., 1995. Heavy metal tolerance by ectomycorrhizal fungi and metal amelioration by *Pisolithus tinctorius*. Mycorrhiza 5, 181–187.

Tay, C.C., Liew, H.H., Yong, S.K., Surif, S., Abdul-Talib, S., 2009. Biosorption of lead(II) from aqueous solutions by *Pleurotus* as a toxicity biosorbent. In: Environmental Science and Technology Conference (ESTEC), Kuala Terengganu, Malaysia.

Tay, C.C., Liew, H.H., Yin, C.Y., Abdul-Talib, S., Surif, S., Suhaimi, A.A., et al., 2011. Biosorption of Cadmium ions using *Pleurotus ostreatus*: growth kinetics, isotherm study and biosorption mechanism. Korean J. Chem. Eng. 28, 825–830.

Thakur, I.S., Srivastava, S., 2011. Bioremediation and bioconversion of chromium and pentachlorophenol in tannery effluent by microorganisms. Int. J. Technol. 3, 224–233.

Tsujiyama, S., Muraoka, T., Takada, N., 2013. Biodegradation of 2,4-dichlorophenol by shiitake mushroom (*Lentinula edodes*) using vanillin as an activator. Biotechnol. Lett. 35, 1079–1083.

Verma, A., Singh, A., Bishnoi, N.R., Gupta, A., 2013. Biosorption of Cu(II) using free and immobilized biomass of *Penicillium citrinum*. Ecol. Eng. 61, 486–490.

Warcup, J.H., 1955. On the origin of colonies of fungi developing on soil dilution plates. Trans. Brit. Mycol. Soc 38, 298–301.

Wu, M., Xu, Y., Ding, W., Li, Y., Xu, H., 2016. Mycoremediation of manganese and phenanthrene by *Pleurotus eryngii* mycelium enhanced by tween 80 and saponin. Appl. Microbiol. Biotechnol. 100, 7249–7261.

Xiangliang, P., Jianlong, W., Daoyong, Z., 2005. Biosorption of Pb(II) by *Pleurotus ostreatus* immobilized in calcium alginate gel. Process Biochem. 40, 2799–2803.

Xiangliang, P., Jianlong, W., Daoyong, Z., 2009. Biosorption of Co(II) by immobilised *Pleurotus ostreatus*. Int. J. Environ. Pollut. 37, 289–298.

Xu, P., Zeng, G., Huang, D., Hu, S., Feng, C., Lai, C., et al., 2013. Synthesis of iron oxide nanoparticles and their application in *Phanerochaete chrysosporium* immobilization for Pb^{2+} removal. Colloids Surf. A 419, 147–155.

Xu, X., Yao, W., Xiao, D., Heinz, T.F., 2014. Spin and pseudospins in layered transition metal dichalcogenides. Nat. Phys. 10, 343–350.

Yadav, A.N., Sachan, S.G., Verma, P., Kaushik, R., Saxena, A.K., 2016. Cold active hydrolytic enzymes production by psychrotrophic Bacilli isolated from three sub-glacial lakes of NW Indian Himalayas. J. Basic Microbiol. 56, 294–307.

Yadav, A., Verma, P., Kumar, R., Kumar, V., Kumar, K., 2017a. Current applications and future prospects of eco-friendly microbes. EU Voice 3, 21–22.

Yadav, A.N., Kumar, R., Kumar, S., Kumar, V., Sugitha, T., et al., 2017b. Beneficial microbiomes: biodiversity and potential biotechnological applications for sustainable agriculture and human health. J. Appl. Biol. Biotechnol. 5, 45–57.

Yadav, A.N., Mishra, S., Singh, S., Gupta, A., 2019a. Recent advancement in white biotechnology through fungi: volume 1: diversity and enzymes perspectives. Springer International Publishing, Cham, pp. 1–62.

Yadav, A.N., 2019b. Fungal white biotechnology: conclusion and future prospects. In: Yadav, A., Singh, S., Mishra, S., Gupta, A. (Eds.), Recent Advancement in White Biotechnology Through Fungi. Fungal Biology. Springer, Cham.

Yagub, M.T., Sen, T.K., Afroze, S., Ang, H.M., 2014. Dye and its removal from aqueous solution by adsorption: a review. Adv. Colloid Interface Sci. 209, 172–184.

Yan, G., Viraraghavan, T., 2003. Heavy-metal removal from aqueous solution by fungus *Mucor rouxii*. Water Res. 37, 4486–4496.

Yayçinkaya, Y., Arica, M.Y., Soysal, L., Bektaş, S., 2002. Cadmium and mercury uptake by immobilized *Pleurotus sapidus*. Turk. J. Chem. 26 (3), 441–452.

Yin, K., Wang, Q., Lv, M., Chen, L., 2018. Microorganism remediation strategies towards heavy metals. Chem. Eng. J 360, 1153–1563.

Zeng, X.X., Chai, L.Y., Tang, J.X., Liu, X.D., Yang, Z.H., 2013. Taxonomy characterization and cadmium biosorption of fungus strain. Trans. Nonferrous Met. Soc. China 23 (9), 2759–2765.

Zhang, D., Jipengwang Zeng, X., Falandysz, J., 2011. Competitive sorption efficiency studies of Cd^{2+}, Cu^{2+} and Pb^{2+} by powdered mycelium of cloud ear fungus. J. Environ. Sci. Health 46, 1776–1782.

Zhang, X., Yang, H., Cui, Z., 2017. *Mucor circinelloides*: efficiency of bioremediation response to heavy metal pollution. Toxicol. Res. 6, 442–447.

Chapter 31

Heavy metals toxicity to food crops and application of microorganisms in bioremediation

Mohammad Tarique Zeyad[1], Waquar Akhter Ansari[2], Mohd Aamir[2], Ram Krishna[3], Sushil Kumar Singh[4], Neelam Atri[2] and Abdul Malik[1]

[1]*Department of Agricultural Microbiology, Faculty of Agricultural Science, Aligarh Muslim University, Aligarh, India,* [2]*Department of Botany, M.M. V, Banaras Hindu University, Varanasi, India,* [3]*Directorate of Onion and Garlic Research, Pune, India,* [4]*Department of Agricultural Biotechnology, Assam Agricultural University, Jorhat, India*

31.1 Introduction

Pollution from HMs is one of the critical problems now a day's affecting environmental sustainability. The HMs of toxic nature originating from industries, mining, tannery, drainage, and many different sources affect the diversity and physicochemical performance of soil microflora (Xie et al., 2016). Due to the very rapid development of industries, gradual decline in source of water for irrigation in agricultural practices is documented. Metal contamination has spread widely all over the world, disturbing the environment and causing serious health hazards to humans. HM contamination in vegetables, serials, and other crops of the agricultural significance greatly affects quality and assurance of the food security (Ali and Al-Qahtani, 2012). Rapid industrial development in developing countries such as India (Sharma et al., 2008; Yadav et al., 2013; Kumar and Kumar, 2017), Iran (Maleki and Zarasvand, 2008), Egypt, and China (Radwan and Salama, 2006; Zhuo et al., 2019) are unorganized.

Several dangerous HMs and metalloids are known to be nonessential for biological functions and other metabolic activities; however, these are hazardous in different prospects (Gall et al., 2015) and therefore listed in 20 hazardous substances by the United States Environmental Protection Agency (USEPA) and the Agency for Toxic Substances and Disease Registry (ATSDR) (2007) (Rai, 2018; Michalek et al., 2019). Copper (Cu), iron (Fe), zinc (Zn), and chromium [Cr(III)] are those HMs that are essential for various metabolic activities, including cytochrome and function of enzymes (Marschner, 2011). Nickle (Ni) is essential constituents of enzyme urease; however, it is dangerous to human health at high dose (Zhuang et al., 2009; Barman et al., 2020). Thus soil provides a classical example of biotic and abiotic interaction for various food crops/vegetables.

Soil, a necessary prerequisite for any crop, might be polluted by HMs coming from different sources (e.g., thermal power plants, coal mines, electroplating, gold mines, textiles, and leather industries waste) and other different sources, which includes (erosion of soil sediments, agricultural overflow, and open storage). Apart from deleterious effects of HM on human health, it negatively affects the soil microbial diversity, soil fertility, and microbial processes, including nitrogen fixation, phosphate solubilization, mineralization, etc. (Gadd, 2010; Zeyad et al., 2019). In addition to soil insects (friendly for soil fertility) as well as invertebrates, HM contamination also affects different small and large mammals (Rai et al., 2018). Many medicinal plants are reported to bioaccumulate various HMs (e.g., Cd, As, Cu, Cr, Pb, and Fe) when grown near different industries (EI Hamiani et al., 2015; Parveen et al., 2020).

Detailed information of the processes for the transmission of HMs from soil to food crops is important for the creation of successful remediation systems. This chapter first describes various sources of soil system HM pollution, their relevance in public health hazards, and fundamental procedures regulating such contamination and the related food interaction of contaminated crops. Present chapter focused to provide a general overview of the global geographical

Microbe Mediated Remediation of Environmental Contaminants. DOI: https://doi.org/10.1016/B978-0-12-821199-1.00035-3

trend of agroecosystem HM points in association with anthropogenic factors and methods. In addition, chapter includes bioremediation approaches such as biosorption of metal ions using microorganisms.

31.2 Heavy metals: a brief account

HMs are those metals that have relatively high densities, atomic weight, and atomic number. Moreover, HMs are defined differently according to their chemical properties, atomic weight, and atomic number (Wijayawardena et al., 2016). For instance, any chemical elements or HMs known are toxic or dangerous in least concentration (Ali et al., 2019). However, for example, Cd, Cu, Ni, Hg, and Pb are closely linked with pollution and significant toxicity. Although, on the basis of their value as a nutrient, HMs were categorized as (1) essential (Cu, Zn, Mn, Fe, and Se), (2) probably essential (V and Co), and (3) possibly toxic (As, Hg, Ni, Pb, and Cd) (Ebdon et al., 2001). HMs are employed in several alternative areas for thousands of years. Pb has been utilized for at least 5000 years, in essential applications just as building materials, colors for coating earthenware production, and funnels for shipping water. Lead acetic acid derivation was utilized to gloss over in old wine in antiquated Rome and numerous Romans expended even in grams of Pb every day. Mercury was purportedly utilized by the Romans as a moisturizer to fix the agony of teeth in newborn children and later (1300s to the late 1800s) it was utilized as medication for syphilis (Jarup, 2003).

31.3 Sources of heavy metals contamination in soil

The main source of HMs pollution in soil are atmospheric deposition, wastewater irrigation, livestock manure, metal-based pesticides or herbicides, phosphate fertilizers, and sewage sludge near agricultural lands (Elgallal et al., 2016; Liu et al., 2019, 2020) (Fig. 31.1). Besides the natural sources of HMs contamination, conventional/emergent anthropogenic pollutants pose significant public health risk by nutritional consumption of root or other plant tissues contaminated food crops from soil. Direct atmospheric deposition of metals on the plant surfaces are caused by industries and vehicles which at the end are deposited in the soil (França et al., 2017). Coal power plants are main sources of soil mercury (Hg) pollution. The long-term effect on consumption of various food crops contaminated with HMs are stated dangerous to human health (Li et al., 2017a,b), such as water spinach, cowpea, amaranth, lettuce, and rice were cultivated in Hg polluted soils (nearby thermal power plant). Many powerful source systems (likewise wastewater as a drainage outlet, sludge from the sewage, and particulate matter pollution/deposition on soil/plant) are a bleak circumstance for HMs contamination and, therefore, world food security (Table 31.1).

HM exposure, especially for plants and vegetables, is now becoming a great challenge due to soil contamination. HMs are discharged from city industrial run-off, mining and smelting, pesticides and fertilizers (Wang et al., 2013; Zhao et al., 2012; Nacke et al., 2013), and electronic waste and automobile waste (Liu et al., 2013) (Fig. 31.1). Moreover, sewage water is one of the key sources of HMs contamination in the soil in which wastewater is utilized for

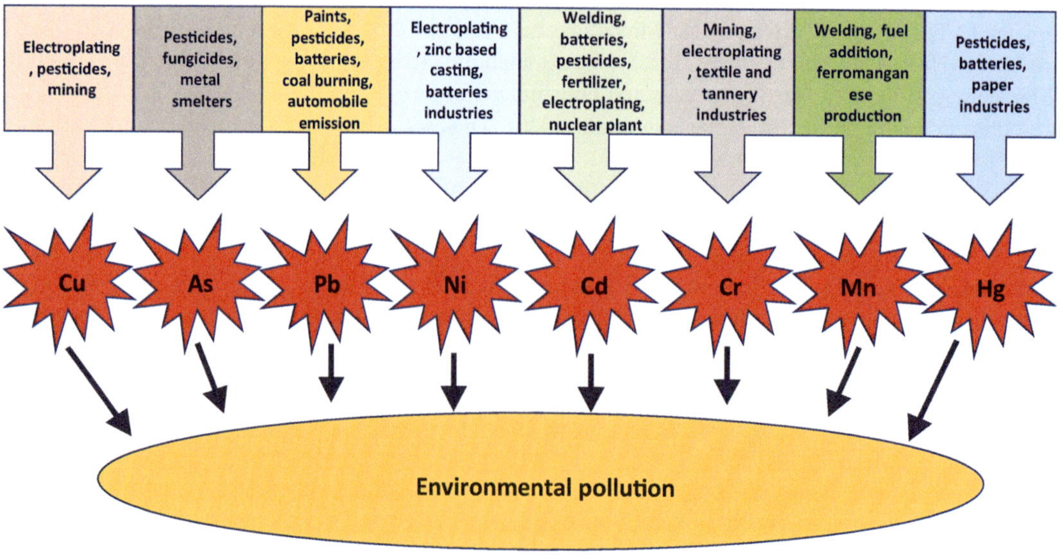

FIGURE 31.1 Sources of HMs contamination in the environment.*HMs*, Heavy metals.

TABLE 31.1 Heavy metals guidelines of their safe limitations.

Samples/source	Standard	Cd	Zn	Mn	Ni	Cu	Pb	Cr
Soil	Indian (Awashthi, 2000)	3–6	300–600	–	75–150	135–270	250–500	–
	European Union (EU) (2002)	3.0	300	–	75	140	300	150
Water	Indian (Awashthi, 2000)	0.01	5.0	0.1	–	0.05	0.1	0.05
	FAO (1985)	0.01	2.0	0.2	0.2	0.2	5.0	0.1
Plant	Indian (Awashthi, 2000)	1.5	50	–	1.5	30	2.5	20
	World Health Organization/Food and Agriculture Organization (WHO/FAO) (2007)	0.2	60	–	–	40	5	–
	European Union (EU) (2006)	0.2	–	–	–	–	0.3	–

Units: soil—μg/g; water—μg/mL; plant—μg/mL.

the water system of the crops (Khan and Malik, 2018). Plant developing territories that are frequently situated in or close wellsprings of climatic reserves have an expanded danger of potential sullying. Ingestion of food that have been developed in metal polluted soil or irrigated with wastewater causes serious health hazards, including numerous long-lasting illnesses, particularly after long period of such food consumption (Sharma et al., 2009; Ugya et al., 2019). Various vegetable species are reported to accumulate HM-based ecological condition (Lokeshwari and Chandrappa, 2006). Several factors are responsible for the accumulation of HMs in crops, for example, soil pH, stages of plant growth, content of metal pollutants, fertilizers, as well as type of crops (Naser et al., 2011; DalCorso et al., 2019). Cu, Cd, As, Cr, Zn, Co, Pb, and Ni are commonly reported in the food crops (Mehmood et al., 2019). By comparing the accumulative HM concentrations with the acceptable limit of Indian Standard (Awashthi, 2000) and safe limits specified by World Health Organization/Food and Agriculture Organization (WHO/FAO) (2007); many endeavors found that HM concentration was higher in food crops cultivated in metal polluted soil or irrigated with industrial wastewater (Ene et al., 2019; Cooper et al., 2020) (Table 31.2). Apart from safety issue, excess metal concentration unfavorably impacts the fertility of soil as well as the crop growth and quality (Rizwan et al., 2017). Many reports showed that vegetables, more specifically leafy vegetables cultivated in soil contaminated with HMs, have enhanced the level of HMs comparatively those cultivated in noncontaminated soil (Islam and Hoque, 2014; Zhou et al., 2016). Therefore it is necessary to have the information related to pollution levels and to formulate strategies for mitigating contamination to reduce the risk to human health through regular testing of HM concentrations on soil as well as crops.

31.4 Importance of food crops

The yield and efficiency of agricultural production decrease for food crop due to pathogens and pests. They cause significant economic losses and lower the domestic, national, and global food security (Savary et al., 2019). Pathogens and pests are generally known as critical barrier to normal and stable food systems. On the other hand, farmers face a challenge of growing enough crops to satisfy rising consumer demand, while preserving the quality and quantity of resources for generations to come. Food crops consist of cereals, legumes (including dried beans), vegetables, seeds and nuts, fruits, herbs and spices, tea and coffee beverage plants, etc. Agriculture plays a crucial role in Indian economy so it is considered as backbone of Indian economy. Agriculture accounts for around 16% of overall GDP. India exports large quantities of agricultural products, including fruits, pulses, spices, vegetables, tea, etc. More than 100 years ago, breakthroughs in botany, linguistics, plant geography, and genetics allowed the geographical origins of food crops to be identified (Khoury et al., 2016).

31.5 Vegetables toxicity through metals

Interactions between various HMs and plants in soil could lead to greater intake of one metal or another metal by crops causing toxicity to humans as well as animals through a food chain. HMs cause severe toxicity to vegetable crops. To confirm the toxicity of HMs, several meticulous researchers reported the impact of HMs on many vegetable crops using

TABLE 31.2 Heavy metals (HMs) contamination in different food crops.

Food crops	Location	Presence of metals in food	Techniques used in the detection of HMs	Source of metal pollution	Reference
Chenopodium sp., *Brassica* sp., grains, leaf and vegetables	India	Cu, Pb, Zn, Cr	ICP–AES	Sewage effluent (inadequately treated)	Rattan et al. (2005)
Spinach	India	Cu, Cr, Zn, Ni, Pb	ICP–MS	Sewage wastewater	Chary et al. (2008)
Rice, wheat, corn, potato, soybean	Brazil	As, Cd, Pd	AAS	Industrial/modern intensive urban agriculture	Corguinha et al. (2015)
Lettuce, leafy vegetable	Spain	Ni, Hg, As, Cd	ICP–MS	Air (PM) from industries and vehicles	Ercilla-Montserrat et al. (2018)
Brassica sp., leafy vegetables and food grains	China	Cr, Pb, Cu, Zn	ICP–OES	Both sewage and industrial waste (from smelter) drained into river water used for irrigation	Liu et al. (2005)
Soybean	Argentina	Pb, Zn, Cu	AAS	Industrial (battery) waste in soil	Rodriguez et al. (2014), Blanco et al. (2017)
Maize (*Zea mays*)	India	Cu, Pb and Cd	AAS	Pot experiment with exogenous supply of HM	Rizvi and Khan (2018)
Grain, maize, spinach, radish, *Brassica napus*, green cabbage, *Brassica juncea* L., turnip, cauliflower, and lettuce	China	Cr, Pb, Cu, Zn	GF–AAS	Sewage effluent (inadequately treated using a biological approach)	Khan et al. (2008)
Cicer arietinum (chickpea)	India	Cr and Ni	AAS	Pot experiment with exogenous supply of HM	Saif and Khan (2018)
Wheat (*Triticum aestivum*)	India	Ni and Pb	AAS and EDX	Pot experiment with exogenous supply of HM	Rizvi et al. (2019)
Sorghum (*Sorghum bicolor*)	China	Cu, Pb, Cd, Cr, As	HPLC	Field experiment with exogenous application of HMs	Xiang et al. (2018)
Luffa acutangula, *Ipomoea batatas*, *Oryza sativa*, and *Raphanus sativus*	Bangladesh	Fe, Zn, Cu, Mn, Pb, Cd	AAS	Landfill open area	Alam et al. (2020)
Vegetables	China	Cd	AAS	Metal contaminated area	Yang et al. (2018)
French beans, beetroot, and kale	Australia	Cr, Pb, Cu, Zn	ICP–MS	Urban stormwater	Tom et al. (2014)
Processed foodstuffs and products	America, Spain, Portugal, Belgium, England, and Chile	Cr, Ni, Cu, Zn, Pb	ICP–MS	Food processing industries/pesticides-based agriculture	González-Martín et al. (2018)
Potato/other foodstuffs	Egypt	Cu, Cr, Pb, Cd, Zn, Ni	AAS	Wastewater partially treated	Radwan and Salama (2006)
Lettuce	United States (Florida)	As	ICP–MS	Contaminated soil	El-Kady and Abdel-Wahhab (2018)

(*Continued*)

TABLE 31.2 (Continued)

Food crops	Location	Presence of metals in food	Techniques used in the detection of HMs	Source of metal pollution	Reference
Cereals	Australia	Pb and Cd	ICP–MS	Metal mining site	Tinggi and Schoendorfer (2018)
Cabbage	China	Cd	ICP–MS	Pot experiment with supply of Cd metal	Junhe et al. (2017)
Vegetables, cereals and fruits	Saudi Arabia	Cu, Fe, Mn, Cd, Pb, Zn, Hg	AAS	Collected from market	Ali and Al-Qahtani (2012)
Mango	Pakistan	Fe, Zn, Mn, Cr, Pb, Cu, Ag, As, Ni, Co, Cd	ICP–MS	Calcium carbide	Hassan et al. (2019)
Tomatoes, pumpkin leaves, cabbage and okra	Zimbabwe	Cr, Pb, Zn, Cu, Ni, As, Co, Sb, Cd	ICP–MS	Greenstone and granitic substrate farmland	Meck et al. (2020)

AAS, Atomic absorption spectrometer; *AES*, atomic emission spectrophotometer; *EDX*, energy dispersive X-ray; *GF*, graphite furnace; *HPLC*, high-performance liquid chromatograph; *ICP*, inductively coupled plasma; *MS*, mass spectrometer; *OES*, optical emission spectrometry.

different experimental tools. In this regard an in vitro experiment was reported by Sabeen et al. (2019) to determine the effect of varying concentrations (125–1000 mg/L) of HMs (Cr, Cd, Fe, Pb, and Ni) on two vegetable crops (*Capsicum annuum* and *Lycopersicum esculentum*). At the end of the experiments, they observed the abnormalities in the root tissues of vegetables in the form of binucleated cells, vagrant chromosomes, laggards, stickiness, giant cells and nuclear lesions, etc. In a similar manner, Musilova et al. (2016) reported that Pb, Zn, and Cd accumulated in potato tubers in a concentration-dependent manner. Likewise, Cooper et al. (2020) described accumulation of HMs and metalloids in some edible portion of fruits as well as in vegetables. The phytotoxicity and accumulation of most severely hazardous HMs such as Cr, Cu, Zn, Pb, Ni, As, Cd, Co, and Sb in edible parts of most commonly grown vegetable crops has been detected by inductively coupled plasma–mass spectrometer technique in Zimbabwe (Meck et al., 2020).

31.6 Heavy metal toxicity to legume crops

To assess the toxic and deleterious impacts of HMs on different biological parameters, dry biomass accumulation and physiology of cowpea (*Vigna unguiculata*) were determined. To achieve this, different concentrations (0.0, 0.6, 1.2, 1.8 and 2.4 mM) of Ni was exogenously employed in pot house conditions under soil environment. At the harvest, it was found that the lower doses of Ni showed less toxicity to plant parameters (root and shoot length), photosynthetic pigments (chlorophyll and carotenoid), and phenol content as compared to the higher levels of HMs (Mujeeb et al., 2019). Similarly, the biochemicals features and physiological activities of chickpea plant was inhibited when raised in soils supplemented with varied level of Ni and Cr. Apart from these the HMs significantly enhanced the stress biomarkers (proline and malondialdehyde) and antioxidant defense enzymes in chickpea foliage (Saif and Khan, 2018). In soil system, after their uptake, HMs accumulated in different organs of plants and caused noxious effects. As an example, root systems are the primary organs of HMs accumulation. Several workers have reported the harmful effect of HMs on primary organs of plants, that is, root system. The distortion in root tip morphology of various crops has been observed under scanning electron microscope (Saif and Khan, 2018; Rizvi and Khan, 2019).

31.7 Heavy metals toxicity to cereals

Cereals are staple food crops and are an essential nutritional source, which are grown as well as consumed all over the world. Cereals and its product are valuable source of energy, carbohydrates, protein, and fibers and contain wide range of micronutrients such as zinc, magnesium, and several vitamins. For domestic organic cereals of Greek local market and imported one, the accumulation level of HMs such as Zn, Cr, Ni, Co, Cu, Fe, Mn, and Pb was found to be similar (Skendi et al., 2019). While the findings from this study are relevant for human health risk, they neither support nor deny the hypothesis of a lower contamination of organic cereals than conventional cereals. Rahman and Islam (2019) reported trace amount availability of five HMs—Cr, Ni, As, Cd, and Pb—elements from cereals, fruits, and vegetables in the range of 0.090−2.5, 0.03−2.6, 0.13−1.7, 0.010−0.74, and 0.37−2.2 mg/g, respectively. The concentrations were more than WHO/FAO permissible limits. Similarly, the concentration of Fe, Ni, Mn, Cr, Cd, Pb, and Zn were observed by using atomic absorption spectrophotometry. By comparing the accumulation level of these metals to WHO/FAO and USEPA, it was found hazardous to public health (Yaradua et al., 2019).

31.8 Heavy metal transport in crops

HMs are transported to crops from the soil through small pores. HM biospeciation can also differ according to food crops. Cu, Cd, Zn, Pb, and As get accumulated in 22 different vegetables in following order: leaves > stalk/root/solanaceous > legumes/melon vegetables. The target hazard quotient suggested high potential of all vegetables to cause severe health risks when eaten (Zhou et al., 2016). The roots of plants perform the main activity in HM absorption and transfer through the entire plants. HMs entry into roots depends on their anatomy and environmental adaptability (in particular, the cell wall). Zn in mangrove plants, for instance, negatively impacts their environmental sustainability through radial loss of oxygen (Cheng et al., 2010). HMs enter the soil by admitting water blended in with minerals and supplements and afterward tie to low-methyl-esterified gelatins, whose levels increased under metal severity (Krzesłowska, 2011).

31.9 Bioremediation strategies adopted for heavy metals removal

Bioremediation process is primarily used to detoxify contamination in soil as well as other environments through microbes, plants, and enzymes of microbial or plant origin. The idea incorporates biodegradation, which alludes to the partial and in some cases aggregate, change, or detoxification of contaminants by microorganisms and plants. Bioremediation can be a sustainable process for the treatment of environmental contaminants. In so many cases, bioremediation is cheaper and much more feasible than other mitigation possibilities.

The restoration and adaptation of contaminated/deteriorated lands has been one of the biggest challenges. Numerous physicochemical techniques have been adapted to this purpose, but most of them are ineffectual because they are costly and destructive for the soil and cannot be implemented in a wider area. Nevertheless, in recent times bioremediation is an exciting and cost-effective alternative to physical−chemical approaches for restoring contaminated and deteriorated land suitable for agriculture production to avoid those problems. With an aim to accomplish such targets, in situ bioremediation approaches has been utilized, which includes the treatment of pollutants at the site of inception (Wang et al., 2020) and ex situ remediation which includes the treatment of HMs polluted soil (Wadgaonkar et al., 2019) off site. As ex situ procedures are expensive, eco-accommodating, and labor intensive, in situ strategies are generally more accepted. In short, plant bacterial systems have been found in general to be increasingly successful in reducing the HMs toxicity to contaminated soils. Related to this, various steps are (1) biostimulation of indigenous microbial communities, (2) bioaugmentation, (3) reduction of metals, (4) bioaccumulations, and (5) biosorption. Every method can remove metals in versatile operating conditions. Absorption, if properly applied, may decrease capital, operating, and aggregate cost of treatment, respectively, by 20%, 36%, and 28%, compared to traditional methods, among these specific bioremediation strategies (Loukidou et al., 2004). Many examples of metal removal/reduction capabilities of some promoting rhizobacteria are presented in Table 31.3. So far, substantial accomplishment has been obtained by employing bioremediation for the removal of HMs. Like all other methods, however, bioremediation also has some drawbacks, such as uncertain performance, that may be due to complex and diverse biological sources applied in biological recovery. For example, after the soil treatment, the residual pollutants form complexes with the soil and, therefore, exert toxicity to microbes more than the parent compounds. Further contamination can be caused by the contaminants deposited in soils over the years. Given all these limitations, bioremediation is regarded as a feasible way to alleviate the metal toxicity of polluted environments. The better performance and lower cost are possible factors making biological approaches a very exciting choice for HM removal of polluted sites as compared to traditional physicochemical methods (He and Chen, 2014).

Bioremediation technique involves both plants (phytoremediation) as well as microorganisms (Prabhu et al., 2017) for HMs elimination from polluted soils. In phytoremediation, plants are used to eliminate or make contaminants harmless to the environment (Manoj et al., 2020). Either these methods "decontaminate" soil or they "stabilize" pollutant in

TABLE 31.3 Microbes involved in bioremediation of heavy metals.

Microorganism	Heavy metal involved	Bioremediation types	Condition	Crop	Reference
Pseudomonas aeruginosa	Cr, Ni, Pb	Biosorption	Field trials	*Cicer arietinum*	Saif and Khan (2018)
Gibberella intermedia, Aspergillus quadrilineatus, Bacillus marisflavi, Bacillus anthracis, Bacillus altitudinis, and *Penicillium citrinum*	Zn, Cu, Mn	Bioaccumulation	Field trials	*Avicennia marina*	Kayalvizhi and Kathiresan (2019)
Aspergillus sydowii	Cd	Cadmium-trichlorfon	Pot trials	*Brassica juncea*	Zhang et al. (2020)
P. aeruginosa	Zn, Cu, Pb,	Bioaugmentation	Pot trials in growth chamber	*Medicago sativa*	Agnello et al. (2016)
Pseudomonas taiwanensis	As and Cd	–	–	–	Satapute et al. (2019)
Cellulosimicrobium cellulans	Cr	Reduction of Cr^{6+} to Cr^{3+}	Field condition	*Capsicum annuum*	Chatterjee et al. (2009)
Mucor sp.	Cr, Mn, Co, Zn, Cu,	Biotransformation and bioaccumulation	Field condition	*Brassica campestris*	Zahoor et al. (2017)
Serratia spp.	Ni and Cd	Bioaccumulation	Pot experiments	*Agrocybe aegerita*	Li et al. (2016)
Providencia vermicola	Cu	Bioinoculation	Pot experiments	*Lens culinaris*	Islam et al. (2016)
Kocuria flava and *Bacillus vietnamensis*	As	Adsorption and accumulation	Field condition	*Rhizophora mangle*	Mallick et al. (2018)
Curtobacterium sp., *Microbacterium* sp.	Zn, Pb, Cu, As	Bioinoculation	Lab condition	*Brassica nigra*	Roman-Ponce et al. (2017)
Raoultella sp.	Cd	Bioinoculation	Pot experiments	*Brassica chinensis*	Xu et al. (2019)
Rhizopus sp., *Fusarium proliferatum, Aspergillus fumigatus*, and *Penicillium radicum*	Cr	Bioaccumulation	Pot trials	*Lactuca sativa*	Hussain et al. (2018)
Enterobacter, Klebsiella, Leifsonia, Bacillus	Cd	Bioaccumulation	Axenic condition	*Zea mays*	Ahmad et al. (2016)
Kocuria sp.	Cu	Bioaccumulation	Lab condition	*B. nigra*	Hansda and Kumar (2017)

situ. Decontamination declines the amounts of contaminants in the soil by eliminating them; however, stabilization does not decrease the amount of contaminant in a site but allows use of soil adjustments to modify the soil composition and sequester or absorb a pollutant into it so that environmental risks are reduced or eliminated (Cunningham et al., 1995; Ayabe et al., 2017).

31.10 Phytoremediation

Bioremediation of HMs from soil is a major issue for researcher throughout the world. The traditional methods used for the removal of HMs often include specific physicochemicals agents (Mir et al., 2017), which can destroy the soil fertility and adversely impact the agroecosystem. Considering these, various approaches have been suggested to eliminate metals from polluted areas. Nevertheless, these methodologies are generally considered inefficient due to operational challenges, massive cost, and poor efficiency of metal leaching. Because of these problems, some viable alternative needs have been found straight away. Bioremediation, which is the method of removing hazardous materials requiring the use of microbes or plants, is considered to be the best way to clear contaminated soil (Dixit et al., 2015; Ramírez-García et al., 2019). Plant-based HMs remediation, also known as green remediation and botanoremediation, is accepted to be an effective and much more feasible process for reducing/cleaning metals contamination from water and soil (Lasat, 1999). Also, no hazardous substance is produced during the phytoremediation practices. This remediation process is usually plant based and is a renewable energy–based biological system (Mahimairaja and Bolan, 2011). Crops associated with phytoremediation were classified as:

Excluders: Plants which are susceptible to metal via a vast range of soil through restriction mechanisms. This group includes grass family members, such as bromegrass, sudan, and fescue.

Indicators: Plants which exhibit limited range of HMs absorption and transport processes and accordingly react to the concentration of HMs in soil. The cereals and grains, such as maize, corn, soybean, wheat, and oats, are included in this category.

Accumulators: Plants that do not prohibit the entrance of metals into the roots but have established unique detoxification mechanisms that are accumulating in the cell. This group includes tobacco, mustard, and member of composite family (e.g., lettuce and spinach).

31.11 Importance of plant and microbe interaction for sustainable agriculture

In the advancement of agriculture and the rehabilitation of the environment the microbes and plant relation have an extremely critical role. Microbial interaction with plant products can be negative or positive, leading to disease or plant growth stimulation and tolerance against various stresses through the support of microbiota having beneficial impact (Abhilash et al., 2012). In relation to this the interaction system between crop and microbe additionally assists in distribution resources throughout the plant body below or above ground and provides resistance to its competitors. The secondary metabolites, released through plant root (root exudates), are responsible for developing a system for the interaction between a plant and a microorganism, though it is more or less difficult and complex to get the perfect mechanisms of relation because they take place in various plant-system dimensions like endosphere, phyllosphere, and rhizosphere. It is, therefore, essential to understand the exact mechanisms of plant–microbe interaction to evaluate the contributions of beneficial plant biota to sustainable agriculture, environmental clean-up, and ecosystem restoration (Dubey et al., 2015). Today agriculture sustainability without pollution is a major global concern (Dubey et al., 2020). That is why a good alternative of this problem can be the beneficial impact of plant–microbe associations. In this respect, number of bioagents have drawn attention from researchers to their utilization in biofertilizers and to other positive impacts, for example, healthy plant growth and sustainable agroecosystem development through alterations in physical, chemical, and biological factors that promote better plant/soil interaction (Kumari et al., 2019). In this context, studies have been undertaken and some bioagent species such as *Pseudomonas*, *Bacillus*, *Azotobacter*, *Azospirillum*, *Variovorax*, *Klebsiella*, *Serratia* sp., and *Enterobacter* have been commercialized (Glick, 2012).

31.12 Microbe-based bioremediation

One interesting and emerging feature of microorganism is the remarkable capacity of microbial populations to change some natural and synthetic chemicals into products that might be subsequently utilized as a source of energy and raw material for their survival. Because of this, microbial communities, in particular, were considered a potential replacement for various effective remediation processes. Even though most microflorae live within diverse ecological niches, those have not been investigated in environmental biotechnology. Therefore need of time is to identify which microorganism can be used effectively and efficiently in metal removal from the polluted environment. In this direction, radiation tolerant *Deinococcus radiodurans* was established to remove uranium via biofilm formation (Manobala et al., 2019). The rate of removal of uranium was also found to be directly related to the age of biofilm. In other study, Chiboub et al. (2020) observed that *Rhizobium sullae* and *Pseudomonas fluorescens* promote plant growth in different concentration (0, 100, and 200 μM) of Cd. Several other specific microbial communities have developed certain systems to withstand and detoxify metals from the defiled condition (Mosa et al., 2016). Strikingly, the high surface–to-volume proportion of microorganisms and their ability to stay away from the metal harmfulness of such life forms make them a suitable and modest option in contrast to compound strategies for metal-remediation (Magyarosy et al.,

2002). Natural procedures related in microbial solidness in a metal-focused on condition are complexation, cell divider and color biosorption, extracellular precipitation and crystallization, metal change, diminished vehicle or impermeability, efflux, intracellular compartmentation, and sequestration (Kang et al., 2016).

31.13 Biosorption

The HMs removal from the environment is extremely important. Processes for the removal of HMs through aqueous solution are mostly based on physical, chemical, and biological techniques. Traditional approaches have been proposed for extracting metal ions from aqueous solution, including evaporation, filtration, chemical precipitation, electrochemical treatment, membrane technology, ion exchange, activated carbon adsorption, etc. (Soni et al., 2019). Nevertheless, electrochemical treatment and chemical precipitation become ineffective, particularly when concentration of metal ion in aqueous solutions ranges from 1 to 100 mg/L (Crini and Lichtfouse, 2019). Membrane, and activated carbon adsorption, ion exchange are outrageously costly because big quantities of heavy-metal-containing wastewater cannot be treated at low concentration, thus cannot be used at larger scale (Pan and An, 2019). In recent times, more concentration has been focused on biotechnology in control and elimination of metal contamination. Substitute arrangement is biosorption, which utilizes various normal materials of organic cause, including microscopic organisms, fungi, yeast, algae, etc. All such biosorbents have metal-sequestering properties and can be utilized to diminish the level of HM particles in solution from ppm to ppb. In this way, it can securely and quickly sequester disintegrated metal particles from weaken complex arrangements and is subsequently appropriate for treating enormous volume and low-fixation complex wastewaters (Chigondo et al., 2013).

Biosorption can be characterized as the disposal of metal or metalloid species, substances, and particulate issue from solution by organic material (Gadd, 2010). A lot of metals can be gathered through a wide scope of techniques reliant and free of digestion. In HM biosorption, living and dead biomass alongside auxiliary metabolites, for example, exopolysaccharides can likewise be utilized. The capacity of some living microorganisms to gather metallic components was at first seen from a toxicological viewpoint (Volesky, 1990). Nevertheless, further studies have shown about inactive or dead microbial biomass passive binding with metal ions through different physical and chemical mechanisms. Consequently, microbe-assisted biosorption is one of the dynamic research areas. The biological activity of metallic ions depends on the chemical composition of its microbial cells (Ayangbenro and Babalola, 2017).

Bacillus sp. was extracted from the Cd hyperaccumulator *Solanum nigrum* a multimetal tolerant bacterial endophytic strain (Guo et al., 2010). The isolated strain has been found at a relatively small amount (10 mg/L) in presence of divalent HMs (Cu, Cd, and Pb). This was the effect of abnormal ATPase activity that provided energy to help *Bacillus* sp. reduce the HM toxicity through the export of cation. It was observed that the strain could specifically consume around 76%, 80%, and 21% Cd, Pb, and Cu, respectively in a 24-h incubation period from a pregiven solution. Subcellular fractionation experiments showed that about 81% and 76% of the cells absorbed by the Cd and Pb are contained in the membrane fraction, while approximately 5% and 7% in the cytoplasm, while the cell wall with 14% and 16% of the overall absorption, respectively.

31.14 Conclusion

HMs are one of the most significant environmental contaminants and so the key source of pollution for food crops as a result of improper irrigation with industrial effluents or sludges. HMs are accumulated by food crops after their entry into soil systems. These metals after their uptake displayed various symptoms in plants such as necrosis, chlorosis, retarded growth, oxidative stress, cytotoxicity, and DNA damage that ultimately lead to the death of plants. These contaminated and bioaccumulated foods, when consumed by humans, cause several disorders such as sarcomas, lymphomas, skin, and lip cancer and respiratory diseases. Since cereals, vegetables, and pulses are the key foods for human beings worldwide, there is an urgent and desperate need of remediation of these contaminated areas for the betterment of food crops. Various physicochemical techniques are applied to clean the HMs from the environments, but due to several limitations, these approaches are unsuccessful. In this regard, soil microbes having the potential of biosorption, bioreduction, and bioaccumulation of HMs are used, and this is an environment friendly and cost-effective strategy for bioremediation of HMs contaminated sites. Conclusively, the problem of toxicity in food crops and also HM contaminated sites can be circumvented by using the soil microbial entities, and this approach might be helpful for the better crop productivity.

References

Abhilash, P.C., Powell, J.R., Singh, H.B., Singh, B.K., 2012. Plant–microbe interactions: novel applications for exploitation in multipurpose remediation technologies. Trends Biotechnol. 30 (8), 416–420.

Agnello, A.C., Bagard, M., van Hullebusch, E.D., Esposito, G., Huguenot, D., 2016. Comparative bioremediation of heavy metals and petroleum hydrocarbons co-contaminated soil by natural attenuation, phytoremediation, bioaugmentation and bioaugmentation-assisted phytoremediation. Sci. Total Environ. 563, 693–703.

Ahmad, I., Akhtar, M.J., Asghar, H.N., Ghafoor, U., Shahid, M., 2016. Differential effects of plant growth-promoting rhizobacteria on maize growth and cadmium uptake. J. Plant Growth Regul. 35 (2), 303–315.

Alam, R., Ahmed, Z., Howladar, M.F., 2020. Evaluation of heavy metal contamination in water, soil and plant around the open landfill site Mogla Bazar in Sylhet, Bangladesh. Groundw. Sustain. Dev. 10, 100311.

Ali, H., Khan, E., Ilahi, I., 2019. Environmental chemistry and ecotoxicology of hazardous heavy metals: environmental persistence, toxicity, and bioaccumulation. J. Chem. 2019.

Ali, M.H., Al-Qahtani, K.M., 2012. Assessment of some heavy metals in vegetables, cereals and fruits in Saudi Arabian markets. Egypt. J. Aquat. Res. 38 (1), 31–37.

ATSDR, U., 2007. Toxicological profile for arsenic. In: Agency for Toxic Substances and Disease Registry, Division of Toxicology, Atlanta, GA.

Awashthi S.K., Central and State Rules as Amended for 1999: Prevention of Food Adulteration Act no 37 of 1954, 2000.

Ayabe, Y., Hijii, N., Takenaka, C., 2017. Effects of local-scale decontamination in a secondary forest contaminated after the Fukushima nuclear power plant accident. Environ. Pollut. 228, 344–353.

Ayangbenro, A.S., Babalola, O.O., 2017. A new strategy for heavy metal polluted environments: a review of microbial biosorbents. Int. J. Environ. Res. Public Health 14 (1), 94.

Barman, M., Datta, S.P., Rattan, R.K., Meena, M.C., 2020. Critical limits of deficiency of nickel in intensively cultivated alluvial soils. J Soil Sci. Plant Nutr. 20, 284–292.

Blanco, A., Salazar, M.J., Cid, C.V., Pignata, M.L., Rodriguez, J.H., 2017. Accumulation of lead and associated metals (Cu and Zn) at different growth stages of soybean crops in lead-contaminated soils: food security and crop quality implications. Environ. Earth Sci. 76 (4), 182.

Chary, N.S., Kamala, C.T., Raj, D.S., 2008. Assessing risk of heavy metals from consuming food grown on sewage irrigated soils and food chain transfer. Ecotoxicol. Environ. Saf. 69 (3), 513–524.

Chatterjee, S., Sau, G.B., Mukherjee, S.K., 2009. Plant growth promotion by a hexavalent chromium reducing bacterial strain, *Cellulosimicrobium cellulans* KUCr3. World J. Microbiol. Biotechnol. 25 (10), 1829–1836.

Cheng, H., Liu, Y., Tam, N.F., Wang, X., Li, S.Y., Chen, G.Z., et al., 2010. The role of radial oxygen loss and root anatomy on zinc uptake and tolerance in mangrove seedlings. Environ. Pollut. 158 (5), 1189–1196.

Chiboub, M., Jebara, S.H., Abid, G., Jebara, M., 2020. Co-inoculation effects of *Rhizobium sullae* and *Pseudomonas* sp. on growth, antioxidant status, and expression pattern of genes associated with heavy metal tolerance and accumulation of cadmium in *Sulla coronaria*. J. Plant Growth Regul. 39, 216–228.

Chigondo, F., Nyamunda, B.C., Sithole, S.C., Gwatidzo, L., 2013. Removal of lead (II) and copper (II) ions from aqueous solution by baobab (*Adononsia digitata*) fruit shells biomass. IOSR J. Appl. Chem. 5 (1), 43–50.

Cooper, A.M., Felix, D., Alcantara, F., Zaslavsky, I., Work, A., Watson, P.L., et al., 2020. Monitoring and mitigation of toxic heavy metals and arsenic accumulation in food crops: a case study of an urban community garden. Plant Direct 4 (1), e00198.

Corguinha, A.P., de Souza, G.A., Gonçalves, V.C., de Andrade Carvalho, C., de Lima, W.E., Martins, F.A., et al., 2015. Assessing arsenic, cadmium, and lead contents in major crops in Brazil for food safety purposes. J. Food Compos. Anal. 37, 143–150.

Crini, G., Lichtfouse, E., 2019. Advantages and disadvantages of techniques used for wastewater treatment. Environ. Chem. Lett. 17 (1), 145–155.

Cunningham, S.D., Berti, W.R., Huang, J.W., 1995. Phytoremediation of contaminated soils. Trends Biotechnol. 13 (9), 393–397.

DalCorso, G., Fasani, E., Manara, A., Visioli, G., Furini, A., 2019. Heavy metal pollutions: state of the art and innovation in phytoremediation. Int. J. Mol. Sci. 20 (14), 3412.

Dixit, R., Malaviya, D., Pandiyan, K., Singh, U.B., Sahu, A., Shukla, R., et al., 2015. Bioremediation of heavy metals from soil and aquatic environment: an overview of principles and criteria of fundamental processes. Sustainability 7 (2), 2189–2212.

Dubey, P.K., Singh, G.S., Abhilash, P.C., 2020. Agriculture in a changing climate. Adaptive Agricultural Practices. Springer, Cham, pp. 1–10.

Dubey, R.K., Tripathi, V., Abhilash, P.C., 2015. Book review: principles of plant-microbe interactions: microbes for sustainable agriculture. Front. Plant Sci. 6, 986.

Ebdon, L., Pitts, L., Cornelis, R., Crews, H., Quevauviller, P., Donard, O.F. (Eds.), 2001. Trace Element Speciation for Environment, Food and Health. Royal Society of Chemistry.

Elgallal, M., Fletcher, L., Evans, B., 2016. Assessment of potential risks associated with chemicals in wastewater used for irrigation in arid and semiarid zones: a review. Agric. Water Manage. 177, 419–431.

El-Kady, A.A., Abdel-Wahhab, M.A., 2018. Occurrence of trace metals in foodstuffs and their health impact. Trends Food Sci. Technol. 75, 36–45.

El Hamiani, O., El Khalil, H., Sirguey, C., Ouhammou, A., Bitton, G., Schwartz, C., et al., 2015. Metal concentrations in plants from mining areas in South Morocco: health risks assessment of consumption of edible and aromatic plants. CLEAN–Soil Air Water 43 (3), 399–407.

Ene, A., Moraru, S.S., Moraru, D.I., 2019. Assessment of bioaccumulation of heavy metals in sunflower cultivated in the agricultural area next to steel industry. Int. Multidiscip. Sci. GeoConf. 19 (3.2), 25–32.

Ercilla-Montserrat, M., Muñoz, P., Montero, J.I., Gabarrell, X., Rieradevall, J., 2018. A study on air quality and heavy metals content of urban food produced in a Mediterranean city (Barcelona). J. Cleaner Prod. 195, 385–395.

European Union (EU), 2002. Heavy Metals in Wastes. European Commission on Environment.

European Commission, 2006. Commission Regulation (EC) No 1881/2006 of 19 December 2006 setting maximum levels for certain contaminants in foodstuffs. Off. J. Eur. Union 364, 5–24.

Food and Agriculture Organization (FAO). 1985. Water Quality for Agriculture. Paper No. 29 (Rev. 1) UNESCO, Publication, Rome.

França, F.C., Albuuerque, A.M., Almeida, A.C., Silveira, P.B., Crescêncio Filho, A., Hazin, C.A., et al., 2017. Heavy metals deposited in the culture of lettuce (*Lactuca sativa* L.) by the influence of vehicular traffic in Pernambuco, Brazil. Food Chem. 215, 171–176.

Gadd, G.M., 2010. Metals, minerals and microbes: geomicrobiology and bioremediation. Microbiology 156 (3), 609–643.

Gall, J.E., Boyd, R.S., Rajakaruna, N., 2015. Transfer of heavy metals through terrestrial food webs: a review. Environ. Monit. Assess. 187 (4), 201.

Glick, B.R., 2012. Plant growth-promoting bacteria: mechanisms and applications. Scientifica 2012.

González-Martín, M.I., Revilla, I., Betances-Salcedo, E.V., Vivar-Quintana, A.M., 2018. Pesticide residues and heavy metals in commercially processed propolis. Microch. J. 143.

Guo, H., Luo, S., Chen, L., Xiao, X., Xi, Q., Wei, W., et al., 2010. Bioremediation of heavy metals by growing hyperaccumulaor endophytic bacterium *Bacillus* sp. L14. Bioresour. Technol. 101 (22), 8599–8605.

Hansda, A., Kumar, V., 2017. Cu-resistant *Kocuria* sp. CRB15: a potential PGPR isolated from the dry tailing of Rakha copper mine. 3 Biotech 7 (2), 132.

Hassan, S., Mazhar, W., Farooq, S., Ali, A., Musharraf, S.G., 2019. Assessment of heavy metals in calcium carbide treated mangoes by inductively coupled plasma-mass spectrometry (ICP-MS). Food Addit. Contam., A 36 (12), 1769–1776.

He, J., Chen, J.P., 2014. A comprehensive review on biosorption of heavy metals by algal biomass: materials, performances, chemistry, and modeling simulation tools. Bioresour. Technol. 160, 67–78. <http://ec.europa.eu/environment/waste/studies/elv/heavymetals_report.pdf>.

Hussain, A., Hamayun, M., Rahman, H., Iqbal, A., Shah, M., Irshad, M., et al., 2018. Bioremediation of hexavalent chromium by endophytic fungi; safe and improved production of *Lactuca sativa* L. Chemosphere 211, 653–663.

Islam, F., Yasmeen, T., Ali, Q., Mubin, M., Ali, S., Arif, M.S., et al., 2016. Copper-resistant bacteria reduces oxidative stress and uptake of copper in lentil plants: potential for bacterial bioremediation. Environ. Sci. Pollut. Res. 23 (1), 220–233.

Islam, M.S., Hoque, M.F., 2014. Concentrations of heavy metals in vegetables around the industrial area of Dhaka city, Bangladesh and health risk assessment. Int. Food Res. J. 21 (6).

Jarup, L., 2003. Hazards of heavy metal contamination. Br. Med. Bull. 68 (1), 167–182.

Junhe, L.U., Xinping, Y.A., Xuchao, M.E., Guoqing, W.A., Yusuo, L.I., Yujun, W.A., et al., 2017. Predicting cadmium safety thresholds in soils based on cadmium uptake by Chinese cabbage. Pedosphere 27 (3), 475–481.

Kang, C.H., Kwon, Y.J., So, J.S., 2016. Bioremediation of heavy metals by using bacterial mixtures. Ecol. Eng. 89, 64–69.

Kayalvizhi, K., Kathiresan, K., 2019. Microbes from wastewater treated mangrove soil and their heavy metal accumulation and Zn solubilization. Biocatal. Agric. Biotechnol. 22, 101379.

Khan, S., Malik, A., 2018. Toxicity evaluation of textile effluents and role of native soil bacterium in biodegradation of a textile dye. Environ. Sci. Pollut. Res. 25 (5), 4446–4458.

Khan, S., Cao, Q., Zheng, Y.M., Huang, Y.Z., Zhu, Y.G., 2008. Health risks of heavy metals in contaminated soils and food crops irrigated with wastewater in Beijing, China. Environ. Pollut. 152 (3), 686–692.

Khoury, C.K., Achicanoy, H.A., Bjorkman, A.D., Navarro-Racines, C., Guarino, L., Flores-Palacios, X., et al., 2016. Origins of food crops connect countries worldwide. Proc. R. Soc. B: Biol. Sci. 283 (1832), 20160792.

Krzesłowska, M., 2011. The cell wall in plant cell response to trace metals: polysaccharide remodeling and its role in defense strategy. Acta Physiol. Plant 33 (1), 35–51.

Kumar, A., Kumar, V., 2017. Risk assessment of heavy metals via consumption of contaminated vegetables collected from different agricultural fields and market sites. Adv. Biochem. 5 (3), 47.

Kumari, A., Prasad, B., Chaurasiya, A., Kumar, P., Ranjan, T., 2019. Soil heavy metal toxicity reduction by bioagents/living organisms. Biofertil. Biopesticides Sustain. Agric. 359.

Lasat, M.M., 1999. Phytoextraction of metals from contaminated soil: a review of plant/soil/metal interaction and assessment of pertinent agronomic issues. J. Hazard. Subst. Res. 2 (1), 5.

Li, F.L., Shi, W., Jin, Z.F., Wu, H.M., Sheng, G.D., 2017a. Excessive uptake of heavy metals by greenhouse vegetables. J. Geochem. Explor. 173, 76–84.

Li, R., Wu, H., Ding, J., Fu, W., Gan, L., Li, Y., 2017b. Mercury pollution in vegetables, grains and soils from areas surrounding coal-fired power plants. Sci. Rep. 7, 46545.

Li, X., Dong, S., Yao, Y., Shi, W., Wu, M., Xu, H., 2016. Inoculation of bacteria for the bioremediation of heavy metals contaminated soil by *Agrocybe aegerita*. RSC Adv. 6 (70), 65816–65824.

Liu, M., Huang, B., Bi, X., Ren, Z., Sheng, G., Fu, J., 2013. Heavy metals and organic compounds contamination in soil from an e-waste region in South China. Environ. Sci.: Process Impacts 15 (5), 919–929.

Liu, Q., Jia, Z., Li, S., Hu, J., 2019. Assessment of heavy metal pollution, distribution and quantitative source apportionment in surface sediments along a partially mixed estuary (Modaomen, China). Chemosphere 225, 829–838.

Liu, W.H., Zhao, J.Z., Ouyang, Z.Y., Söderlund, L., Liu, G.H., 2005. Impacts of sewage irrigation on heavy metal distribution and contamination in Beijing, China. Environ. Int. 31 (6), 805–812.

Lokeshwari, H., Chandrappa, G.T., 2006. Impact of heavy metal contamination of Bellandur Lake on soil and cultivated vegetation. Curr. Sci. 622–627.

Loukidou, M.X., Zouboulis, A.I., Karapantsios, T.D., Matis, K.A., 2004. Equilibrium and kinetic modeling of chromium (VI) biosorption by *Aeromonas caviae*. Colloids Surf. A: Physicochem. Eng. Aspects 242 (1–3), 93–104.

Lu, Q., Wang, S., Bai, X., Liu, F., Li, C., Deng, Y., et al., 2020. Quantitative assessment of human health risks under different land uses based on soil heavy metal pollution sources. Hum. Ecol. Risk Assess. 1–7.

Magyarosy, A., Laidlaw, R., Kilaas, R., Echer, C., Clark, D., Keasling, J., 2002. Nickel accumulation and nickel oxalate precipitation by *Aspergillus niger*. Appl. Microbiol. Biotechnol. 59 (2–3), 382–388.

Mahimairaja, S., Bolan, N.S., 2011. Phytoremediation of arsenic-contaminated soil and water (Symposium 3.5. 1 Heavy Metal Contaminated Soils, <Special Issue> International Symposium: Soil Degradation Control, Remediation, and Reclamation, Tokyo Metropolitan University Symposium Series No. 2, 2010). Pedologist 54 (3),241–248.

Maleki A., Zarasvand, M.A., 2008. Heavy metals in selected edible vegetables and estimation of their daily intake in Sanandaj, Iran. Southeast Asian J Trop Med Public Health 39 (2), 335.

Mallick, I., Bhattacharyya, C., Mukherji, S., Dey, D., Sarkar, S.C., Mukhopadhyay, U.K., et al., 2018. Effective rhizoinoculation and biofilm formation by arsenic immobilizing halophilic plant growth promoting bacteria (PGPB) isolated from mangrove rhizosphere: a step towards arsenic rhizoremediation. Sci. Total Environ. 610, 1239–1250.

Manobala, T., Shukla, S.K., Rao, T.S., Kumar, M.D., 2019. A new uranium bioremediation approach using radio-tolerant *Deinococcus radiodurans* biofilm. J. Biosci. 44 (5), 122.

Manoj, S.R., Karthik, C., Kadirvelu, K., Arulselvi, P.I., Shanmugasundaram, T., Bruno, B., et al., 2020. Understanding the molecular mechanisms for the enhanced phytoremediation of heavy metals through plant growth promoting rhizobacteria: a review. J. Environ. Manage. 254, 109779.

Marschner, H., 2011. Marschner's Mineral Nutrition of Higher Plants. Academic Press.

Meck, M.L., Mudimbu, D., Davies, T.C., 2020. Accumulation of potentially harmful elements in edible parts of vegetables grown on two different geological substrates in Zimbabwe. J. Geochem. Explor. 208, 106392.

Mehmood, A., Mirza, M.A., Choudhary, M.A., Kim, K.H., Raza, W., Raza, N., et al., 2019. Spatial distribution of heavy metals in crops in a wastewater irrigated zone and health risk assessment. Environ. Res. 168, 382–388.

Michalek, I.M., Benn, E.K., dos Santos, F.L., Gordon, S., Wen, C., Liu, B., 2019. A systematic review of global legal regulations on the permissible level of heavy metals in cosmetics with particular emphasis on skin lightening products. Environ. Res. 170, 187–193.

Mir, Z.A., Bharose, R., Lone, A.H., Malik, Z.A., 2017. Review on phytoremediation: an ecofriendly and green technology for removal of heavy metals. Crop Res. 52 (1–3), 74–82.

Mosa, K.A., Saadoun, I., Kumar, K., Helmy, M., Dhankher, O.P., 2016. Potential biotechnological strategies for the cleanup of heavy metals and metalloids. Front. Plant Sci. 7, 303.

Mujeeb, A., Iqbal, M.Z., Shafiq, M., Kabir, M., 2019. Effects of nickel toxicity on seedling growth, photosynthetic pigments, carotenoids and phenols contents of cowpea *Vigna unguiculata* (L.). Int. J. Environ. Agric. Biotechnol. 4 (2).

Musilova, J., Bystricka, J., Lachman, J., Harangozo, L., Trebichalsky, P., Volnova, B., 2016. Potatoes–a crop resistant against input of heavy metals from the metallicaly contaminated soil. Int J Phytoremediat. 18 (6), 547–552.

Nacke, H., Goncalves, A.C., Schwantes, D., Nava, I.A., Strey, L., Coelho, G.F., 2013. Availability of heavy metals (Cd, Pb, and Cr) in agriculture from commercial fertilizers. Arch. Environ. Contam. Toxicol. 64 (4), 537–544.

Naser, H.M., Sultana, S., Mahmud, N.U., Gomes, R., Noor, S., 2011. Heavy metal levels in vegetables with growth stage and plant species variations. Bangladesh J. Agric. Res. 36 (4), 563–574.

Pan, Z.F., An, L., 2019. Removal of heavy metal from wastewater using ion exchange membranes. Applications of Ion Exchange Materials in the Environment. Springer, Cham, pp. 25–46.

Parveen, R., Abbasi, A.M., Shaheen, N., Shah, M.H., 2020. Accumulation of selected metals in the fruits of medicinal plants grown in urban environment of Islamabad, Pakistan. Arabian J. Chem. 13 (1), 308–317.

Prabhu, A.A., Chityala, S., Jayachandran, D., Naik, N., Dasu, V.V., 2017. Rhizoremediation of environmental contaminants using microbial communities. Plant-Microbe Interactions in Agro-Ecological Perspectives. Springer, Singapore, pp. 433–453.

Radwan, M.A., Salama, A.K., 2006. Market basket survey for some heavy metals in Egyptian fruits and vegetables. Food Chem. Toxicol. 44 (8), 1273–1278.

Rahman, M., Islam, M.A., 2019. Concentrations and health risk assessment of trace elements in cereals, fruits, and vegetables of Bangladesh. Biol. Trace Elem. Res. 191 (1), 243–253.

Rai, P.K., Lee, J., Kailasa, S.K., Kwon, E.E., Tsang, Y.F., Ok, Y.S., et al., 2018. A critical review of ferrate (VI)-based remediation of soil and groundwater. Environ. Res. 160, 420–448.

Rai, P.K., 2018. Phytoremediation of Emerging Contaminants in Wetlands. CRC Press.

Ramírez-García, R., Gohil, N., Singh, V., 2019. Recent advances, challenges, and opportunities in bioremediation of hazardous materials. Phytomanagement of Polluted Sites. Elsevier, pp. 517–568.

Rattan, R.K., Datta, S.P., Chhonkar, P.K., Suribabu, K., Singh, A.K., 2005. Long-term impact of irrigation with sewage effluents on heavy metal content in soils, crops and groundwater—a case study. Agric. Ecosyst. Environ. 109 (3–4), 310–322.

Rizvi, A., Ahmed, B., Zaidi, A., Khan, M.S., 2019. Heavy metal mediated phytotoxic impact on winter wheat: oxidative stress and microbial management of toxicity by Bacillus subtilis BM2. RSC Adv. 9 (11), 6125–6142.

Rizvi, A., Khan, M.S., 2018. Heavy metal induced oxidative damage and root morphology alterations of maize (*Zea mays* L.) plants and stress mitigation by metal tolerant nitrogen fixing *Azotobacter chroococcum*. Ecotoxicol. Environ. Saf. 157, 9–20.

Rizvi, A., Khan, M.S., 2019. Heavy metal-mediated toxicity to maize: oxidative damage, antioxidant defence response and metal distribution in plant organs. Int. J. Environ. Sci. Technol. 16 (8), 4873–4886.

Rizwan, M., Ali, S., Qayyum, M.F., Ok, Y.S., Adrees, M., Ibrahim, M., et al., 2017. Effect of metal and metal oxide nanoparticles on growth and physiology of globally important food crops: a critical review. J. Hazard. Mater. 322, 2–16.

Rodriguez, J.H., Salazar, M.J., Steffan, L., Pignata, M.L., Franzaring, J., Klumpp, A., et al., 2014. Assessment of Pb and Zn contents in agricultural soils and soybean crops near to a former battery recycling plant in Córdoba, Argentina. J. Geochem. Explor. 145, 129–134.

Roman-Ponce, B., Reza-Vázquez, D.M., Gutierrez-Paredes, S., María de Jesús, D.E., Maldonado-Hernandez, J., Bahena-Osorio, Y., et al., 2017. Plant growth-promoting traits in rhizobacteria of heavy metal-resistant plants and their effects on *Brassica nigra* seed germination. Pedosphere 27 (3), 511–526.

Sabeen, M., Mahmood, Q., Bhatti, Z.A., Irshad, M., Bilal, M., Hayat, M.T., et al., 2019. Allium cepa assay based comparative study of selected vegetables and the chromosomal aberrations due to heavy metal accumulation. Saudi J. Biol. Sci. 27.

Saif, S., Khan, M.S., 2018. Assessment of toxic impact of metals on proline, antioxidant enzymes, and biological characteristics of *Pseudomonas aeruginosa* inoculated *Cicer arietinum* grown in chromium and nickel-stressed sandy clay loam soils. Environ. Monit. Assess. 190 (5), 290.

Satapute, P., Paidi, M.K., Kurjogi, M., Jogaiah, S., 2019. Physiological adaptation and spectral annotation of Arsenic and Cadmium heavy metal-resistant and susceptible strain *Pseudomonas taiwanensis*. Environ. Pollut. 251, 555–563.

Savary, S., Willocquet, L., Pethybridge, S.J., Esker, P., McRoberts, N., Nelson, A., 2019. The global burden of pathogens and pests on major food crops. Nat. Ecol. Evol. 3 (3), 430–439.

Sharma, R.K., Agrawal, M., Marshall, F.M., 2008. Heavy metal (Cu, Zn, Cd and Pb) contamination of vegetables in urban India: a case study in Varanasi. Environ. Pollut. 154 (2), 254–263.

Sharma, R.K., Agrawal, M., Marshall, F.M., 2009. Heavy metals in vegetables collected from production and market sites of a tropical urban area of India. Food Chem. Toxicol. 47 (3), 583–591.

Skendi, A., Papageorgiou, M., Irakli, M., Katsantonis, D., 2019. Presence of mycotoxins, heavy metals and nitrate residues in organic commercial cereal-based foods sold in the Greek market. J. Consum. Prot. Food Saf. 15.

Soni, R., Dash, B., Kumar, P., Mishra, U.N., Goel, R., 2019. Microbes for bioremediation of heavy metals. Microbial Interventions in Agriculture and Environment. Springer, Singapore, pp. 129–141.

Tinggi, U., Schoendorfer, N., 2018. Analysis of lead and cadmium in cereal products and duplicate diets of a small group of selected Brisbane children for estimation of daily metal exposure. J. Trace Elem. Med. Biol. 50, 671–675.

Tom, M., Fletcher, T.D., McCarthy, D.T., 2014. Heavy metal contamination of vegetables irrigated by urban stormwater: a matter of time? PLoS One 9 (11), e112441.

Ugya, A.Y., Ahmad, A.M., Adamu, H.I., Giwa, S.M., Imam, T.S., 2019. Phytoextraction of heavy metals and risk associated with vegetables grown from soil irrigated with refinery wastewater. J. Appl. Biol. Biotechnol. 7 (2), 14–19.

Volesky B., 1990. Removal and recovery of heavy metals by biosorption. In: Biosorption of Heavy Metals. CRC Press, Florida. pp. 7–43.

Wadgaonkar, S.L., Ferraro, A., Nancharaiah, Y.V., Dhillon, K.S., Fabbricino, M., Esposito, G., et al., 2019. In situ and ex situ bioremediation of seleniferous soils from northwestern India. J. Soils Sediment. 19 (2), 762–773.

Wang, C., He, S., Zou, Y., Liu, J., Zhao, R., Yin, X., et al., 2020. Quantitative evaluation of in-situ bioremediation of compound pollution of oil and heavy metal in sediments from the Bohai Sea, China. Mar. Pollut. Bull. 150, 110787.

Wang, Y., Qiu, Q., Xin, G., Yang, Z., Zheng, J., Ye, Z., et al., 2013. Heavy metal contamination in a vulnerable mangrove swamp in South China. Environ. Monit. Assess. 185 (7), 5775–5787.

World Health Organization/Food and Agriculture Organization (WHO/FAO), 2007. Joint FAO/WHO Food Standard Programme Codex Alimentarius Commission 13th session. In: Report of the Thirty-Eight Session of the Codex Committee on Food Hygiene, Houston, TX, ALINORM 07/30/1.

Wijayawardena, M.A., Megharaj, M., Naidu, R., 2016. Exposure, toxicity, health impacts, and bioavailability of heavy metal mixtures, Advances in Agronomy, vol. 138. Academic Press, pp. 175–234.

Xiang, C., Tian, D., Wang, W., Shen, F., Zhao, G., Ni, X., et al., 2018. Fates of heavy metals in anaerobically digesting the stover of grain sorghum harvested from heavy metal-contaminated farmland. Waste Biomass Valorization 11, 1–2.

Xie, Y., Fan, J., Zhu, W., Amombo, E., Lou, Y., Chen, L., et al., 2016. Effect of heavy metals pollution on soil microbial diversity and Bermuda grass genetic variation. Front. Plant Sci. 7, 755.

Xu, S., Xing, Y., Liu, S., Huang, Q., Chen, W., 2019. Role of novel bacterial *Raoultella* sp. strain X13 in plant growth promotion and cadmium bioremediation in soil. Appl. Microbiol. Biotechnol. 103 (9), 3887–3897.

Yadav, A., Yadav, P.K., Shukla, D.N., 2013. Investigation of heavy metal status in soil and vegetables grown in urban area of Allahabad, Uttar Pradesh, India. Int. J. Sci. Res. Publ. 3 (9), 1–7.

Yang, Y., Chang, A.C., Wang, M., Chen, W., Peng, C., 2018. Assessing cadmium exposure risks of vegetables with plant uptake factor and soil property. Environ. Pollut. 238, 263–269.

Yaradua, A.I., Alhassan, A.J., Saulawa, L.A., Nasir, A., Matazu, K.I., Usman, A., et al., 2019. Evaluation of health effect of some selected heavy metals in maize cultivated in Katsina State, North West Nigeria. Asian Plant Res. J. 1–2.

Zahoor, M., Irshad, M., Rahman, H., Qasim, M., Afridi, S.G., Qadir, M., et al., 2017. Alleviation of heavy metal toxicity and phytostimulation of *Brassica campestris* L. by endophytic *Mucor* sp. MHR-7. Ecotoxicol. Environ. Saf. 142, 139–149.

Zeyad, M.T., Kumar, M., Malik, A., 2019. Mutagenicity, genotoxicity and oxidative stress induced by pesticide industry wastewater using bacterial and plant bioassays. Biotechnol. Rep. 24, e00389.

Zhang, C., Tao, Y., Li, S., Ke, T., Wang, P., Wei, S., et al., 2020. Bioremediation of cadmium-trichlorfon co-contaminated soil by Indian mustard (*Brassica juncea*) associated with the trichlorfon-degrading microbe *Aspergillus sydowii*: related physiological responses and soil enzyme activities. Ecotoxicol. Environ. Saf. 188, 109756.

Zhao, H., Xia, B., Fan, C., Zhao, P., Shen, S., 2012. Human health risk from soil heavy metal contamination under different land uses near Dabaoshan Mine, Southern China. Sci. Total Environ. 417, 45–54.

Zhuang, P., McBride, M.B., Xia, H., Li, N., Li, Z., 2009. Health risk from heavy metals via consumption of food crops in the vicinity of Dabaoshan mine, South China. Sci Total Environ. 407 (5), 1551–1561.

Zhou, H., Yang, W.T., Zhou, X., Liu, L., Gu, J.F., Wang, W.L., et al., 2016. Accumulation of heavy metals in vegetable species planted in contaminated soils and the health risk assessment. Int. J. Environ. Res. Public Health 13 (3), 289.

Zhuo, H., Wang, X., Liu, H., Fu, S., Song, H., Ren, L., 2019. Source analysis and risk assessment of heavy metals in development zones: a case study in Rizhao, China. Environ. Geochem. Health 1–2.

Chapter 32

Role of omics approaches in microbial bioremediation

Manisha Mishra[1], Sandeep Kumar Singh[2] and Ajay Kumar[3]

[1]University Department of Botany, T. M. Bhagalpur University, Bhagalpur, India, [2]Department of Botany, Centre of Advance Study in Science, Institute of Science, Banaras Hindu University, Varanasi, India, [3]Agriculture Research Organization, Ministry of Agriculture and Rural Development, Volcani Center, Rishon LeZion, Israel

32.1 Introduction

The rapid industrialization and increasing population adversely affect the environment by increasing a range of contaminants, including heavy metals, hydrocarbons, industrial effluents xenobiotics, radionuclides, pharmaceutical substances, polyaromatic hydrocarbons (PAHs), and polychlorinated biphenyls in the soil, water, or surrounding environment. The accumulation of contaminants or pollutants also leads to the deterioration of global climate, natural texture, and also productivity of plants soil. Most importantly, their release poses as a major health concern of human beings (Singh et al., 2020a). To bring about remediation of the environmental contaminants, several biological methods such as bioremediation and phytoremediation have been practiced, which are not only cost-effective but eco-friendly as well. Bioremediation is a broad term and it is used to connote the remediation of environmental contaminants via transforming or degrading the toxic or complex contaminants to less toxic or simpler form or even complete degradation utilizing biological means such as microbes or microbial products (Bharagava et al., 2017; Singh et al., 2020b; Gupta et al., 2020; Datta et al., 2020). Intrinsic bioremediation involves the removal of contaminants from the environment without human intervention (Lovley, 2003). However, microbial activity can be enhanced to achieve higher clean up rates by manipulation of the environment in which the microorganisms are growing at the contaminated site. In such cases the bioremediation process is be said to be engineered (Fig. 32.1).

During bioremediation, polluted site can be got rid of contaminants by their microbial breakdown. Bacteria, fungi, and yeasts have been frequently used to remediate the contaminants from the pollution site (Goutam et al., 2018; Saxena and Bharagava, 2016). However, microbes have been found to vary in their metabolic potential to detoxify/transform/breakdown the pollutants. The detoxification or breakdown of contaminants as such is also affected by bioavailability and accessibility of pollutants to the microbes (Antizar-Ladislao, 2010; Singh et al., 2018). In an effort to make pure cultures of microorganisms that are capable of bringing about bioremediation, microbes with such degradation capability are isolated and characterized (Fulekar, 2005).

In most cases, pollutants are highly recalcitrant and do not undergo complete degradation. This results in environmental accumulation. It might, therefore, seem essential to explore novel catabolic pathways and mechanisms responsible for regulating the production of primary and secondary metabolites in the microorganisms. However, before moving toward metabolites and metabolic pathways, gene sequences must be worked out. Advances in genomics are paving way for the isolation and establishment of pure culture of microorganisms for the bioremediation of organic and inorganic pollutants (Nierman and Nelson, 2002). For several microorganisms that are involved in the bioremediation process, their genome sequences are presently available in the public domain (Lovley, 2003). The availability of whole genome sequences is proving greatly helpful in studying gene expression analysis under various environmental conditions utilizing DNA microarray technology (Nierman and Nelson, 2002). High throughput DNA sequencing by next-generation sequencing (NGS) technology allows sequencing of those microbes which cannot be cultured but are viable. Information about gene sequences is further proving helpful in elucidating the physiological and genomic features of microorganisms involved in bioremediation process (Bihari, 2013) (Figs. 32.2 and 32.3).

Microbe Mediated Remediation of Environmental Contaminants. DOI: https://doi.org/10.1016/B978-0-12-821199-1.00036-5

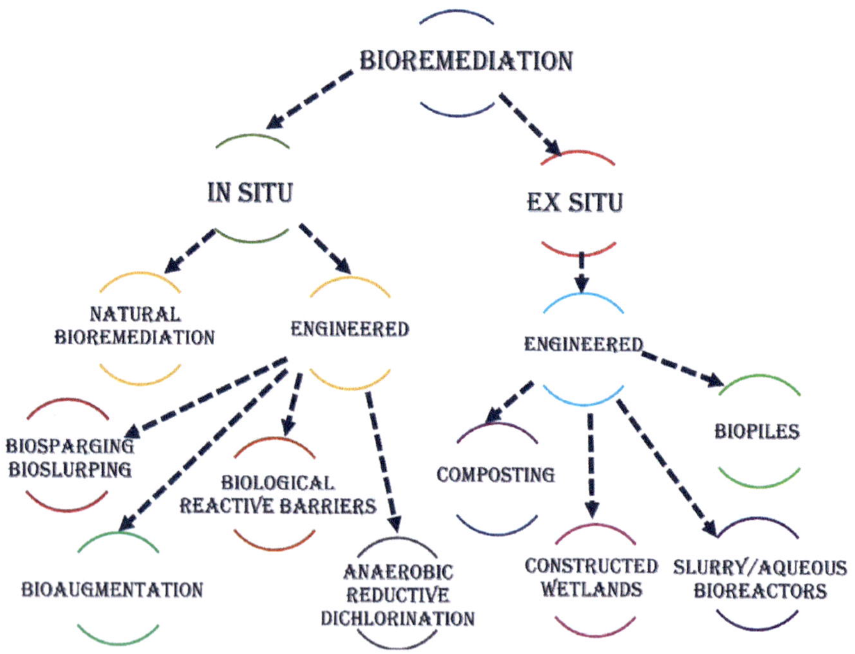

FIGURE 32.1 Bioremediation engineering process.

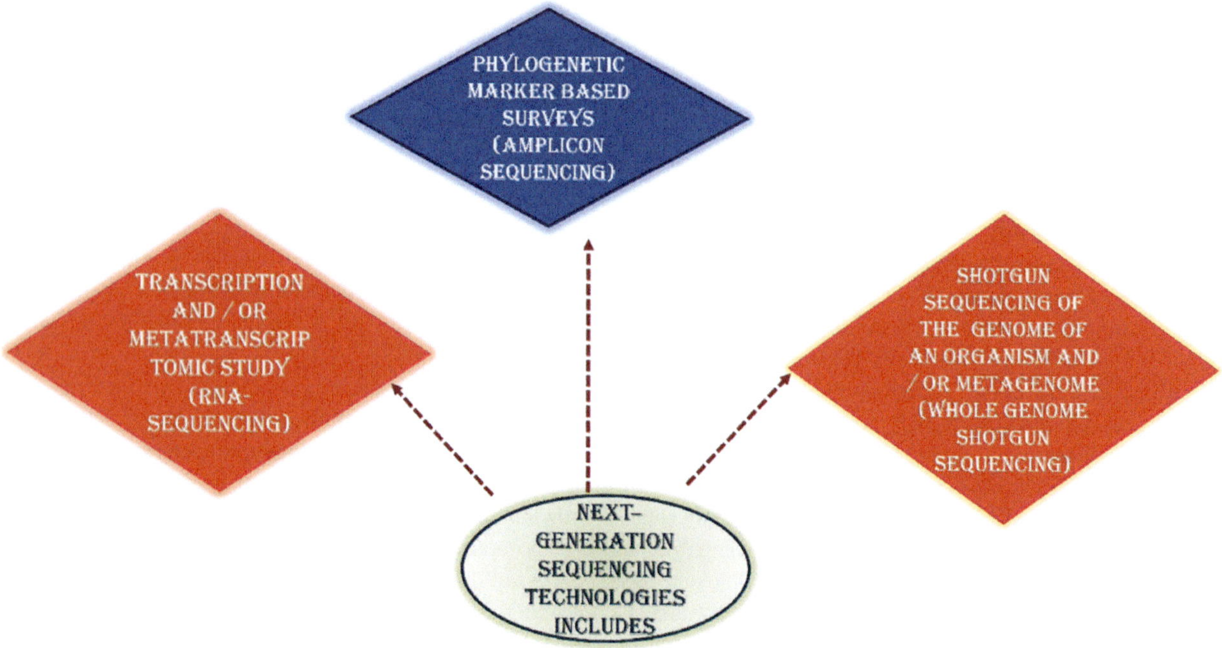

FIGURE 32.2 Concept of next-generation sequencing.

For microorganisms involved in in situ and ex situ bioremediation, it is important to study the profile of their global proteome. Analysis of the entire set of proteins inside a cell is carried out in a proteomic study. Present scientific thought is greatly influenced by proteomics which provides detailed insight into the physiological state of the microorganisms (Singh and Nagaraj, 2006.). Microorganisms that have adapted to the contaminated site possess complex physiology and this might be due to presence of certain unique genes. As microorganisms grow at the site of bioremediation, the in situ mineralization dynamically evolves. This impacts the heterogeneous microbial communities growing in a particular environment. Changes in the bioremediation environment can produce an entirely different pack of metabolites by way of cellular proteins. Metabolomics approaches help in analyzing these metabolites by their dynamic

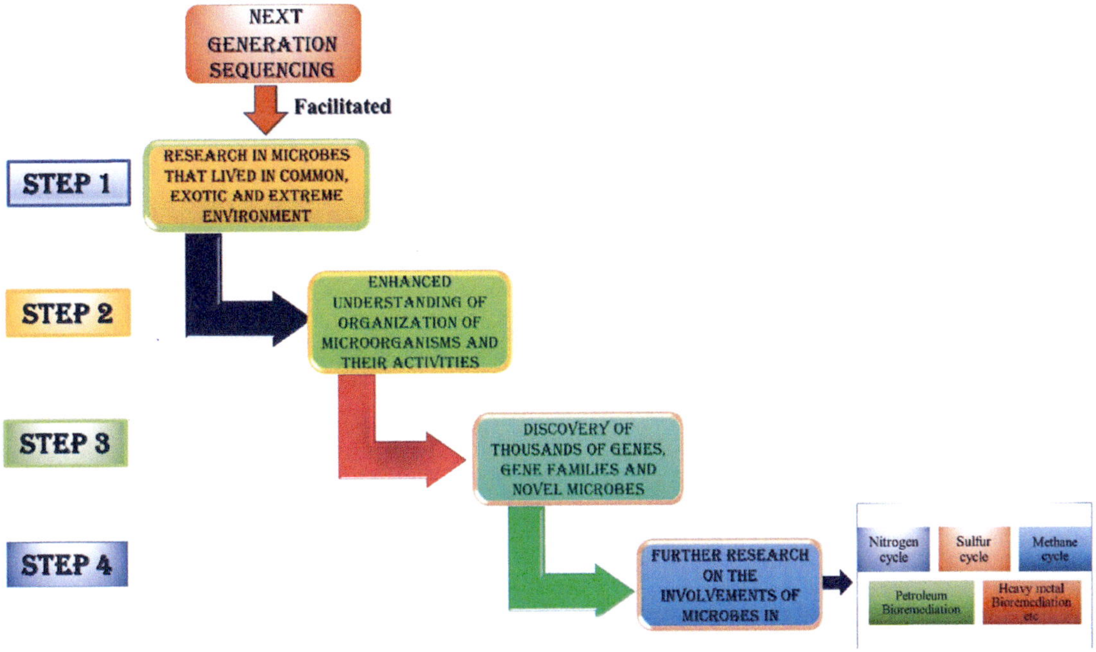

FIGURE 32.3 Application of next-generation sequencing in bioremediation.

monitoring when the microbe comes in direct contact with the contaminant (Singh, 2006). However, global expression and regulation of mRNA transcripts is studied in transcriptomics.

Earlier researches in bioremediation were principally based on chromatographic and spectroscopic studies that allowed for the characterization of metabolites produced during or after biodegradation process. But very little is known about the mechanism underlying the catalytic and respiratory genes involved in bioremediation (Lovley, 2003). Decoding the genes whose function is unknown is expected to help in a keen understanding of the bioremediation process. This is anticipated as detailed information about genetic system of environmentally significant organisms is available (Fulekar, 2005). Identification of regulatory circuits in microorganisms under various environmental stresses provides a global picture of genetic interplay during expression analysis (Baldi and Hatfield, 2002).

32.2 Conventional gene-sequencing technologies

The following techniques have been used for conventional gene sequencing:

1. polymerase chain reaction
2. fluorescence in situ hybridization
3. microbial community profiling
4. amplified ribosomal DNA restriction analysis
5. ribosomal intergenic spacer analysis
6. denaturing gradient gel electrophoresis/temperature gradient gel electrophoresis
7. terminal restriction fragment length polymorphism

32.3 Pregenomics approach to bioremediation

Much before the advent of genomics approaches in bioremediation, treatability study was conducted to study the degradation of the contaminant by the microbes. It involved the incubation of the contaminated samples from the environment in the lab and documentation of the immobilization or breakdown rate of the contaminant (Rogers and McClure, 2003). Treatability study is advantageous as it roughly highlighted the catabolic and anabolic activities of the microbial community. However, these studies had the following limitations:

1. It did not provide complete information about the microbes that were helpful in the bioremediation process.

2. It was not able identify why certain amendments were not able to enhance the efficiency of the bioremediation process by the microbes.
3. Uncertainty also persisted as to whether the isolated microorganisms were involved in the bioremediation process or that they were unwanted (weeds) microbes that rapidly proliferated in in vitro conditions (Lovley, 2003).
4. In their natural state, 99% of the microbes are noncultivable, that is, posed difficulty in growing under culture conditions (Vartoukian et al., 2010; Bursle and Robson, 2016).

It was, therefore, desirable that advanced studies in bioremediation be conducted to study the physiological response of microorganisms that had been isolated from the contamination site. However, for any further molecular analysis, an important prerequisite would be the isolation and characterization of pure cultures of microorganisms from the contaminated site.

32.4 16S rRNA approach

The 16S rRNA gene sequence analyses studies enhance the understanding as well as evolutionary lineage of any microorganism due to the presence of unique conserved sequence of each species (Ju and Zhang, 2015; Pace et al., 1986; Amann et al., 1995). However, this approach is not completely capable of showing the important physiological aspects of the microorganisms (Pace, 1997; Achenbach and Coates, 2000). The absence of closely related organisms in the culture further creates problem in physiology prediction from its phylogeny (Lovley, 2003). Microbial diversity is also underestimated because microbial populations that are dominant show discrete bands and a single band may obscure a number of microbial members (Satokari et al., 2001; Green et al., 2010). So, instead of just analyzing the 16S rRNA genes, identification of the expression of bioremediation associated gene/genes is expected be more informative (Rogers and McClure, 2003). However, bioremediation-associated genes may be present in the microorganism but are not expressed. Quantification of mRNA of such genes is supposed to be helpful. For example, one of the genes involved in the oxygenic breakdown of naphthalene is nah A. Rate of naphthalene degradation increased with the increase in concentration of nah A mRNA (Fleming et al., 1993). However, sometimes this is not always true. This led to the understanding that the bioremediation rates could be affected by factors other than transcription (Nazaret et al., 1994; Jeffrey et al., 1996).

32.5 Omics approach in bioremediation

The microbial-based remediation technology utilizes indigenous microbial communities to clean up a range of environmental contaminants, including heavy metals, dyes, industrial effluents, petrochemicals, and xenobiotics. However, the extent of contamination management depends upon several factors, such as microbial composition, nature or extent of pollutants, as well as the surrounding environmental conditions (Chakraborty et al., 2012; Singh et al., 2019a,b). Therefore to understand the efficient and exact mechanism involved in microbial-based contamination management, latest omics and molecular approaches such as proteomics, genomics, metagenomics, and metabolomics have been broadly studied (Maphosa et al., 2010).

Genomic studies are finding great applicability in studying pure cultures of microorganisms involved in bioremediation. Whole genome sequencing allows sequencing of the genomes of those microbes, whose physiological functioning is not known. In bioremediation associated microbes, plasmids, transposable elements, and environmental phages together constitute the mobilome. It has been found that the transfer of these plasmids is responsible for the spread of catabolic pathways among the microorganisms. pGRT 1 is a recently sequenced plasmid and it has been found to enhance the resistance to solvents, UV light, and environmental contaminants in the microorganisms harboring this plasmid (Molina et al., 2011). Jumping, hopping, or switching off of the mobilome confers unique degradation properties to the members of microbial communities. Advancement in several complex biological interactions is made possible due to elaboration in proteomic studies. Proteomic studies are greatly facilitated due to MS. For organisms that have not been cultured yet, metagenomics allows study of microbial diversity and interaction among its members. In a metagenomic study, DNA is isolated from the metagenome (sample obtained from the environment) and sequenced (Daniel, 2005). The metagenomic study of diverse microbial communities in contaminated samples is now getting practical and economically viable. In a metabolomics study, all the metabolites produced by an organism are quantified after their characterization. The data generated by different omics approaches can now be put together to have greater understanding of the metabolic processes of the microbes during the bioremediation process. This can lead to the development of strains that are capable of degrading a range of xenobiotics (Desai et al., 2010). For the prediction of microbial activity during different strategies of bioremediation models can be made for pure cultures as well as environmental samples using the omics techniques.

The rhizosphere has been thoroughly investigated. However, only 1% of its microbes can be grown in vitro conditions (Bakken, 1997). In contaminant-loaded environments the microbial black box is being opened up due to NGS technologies and in silico analysis (Maphosa et al., 2010). Culture-based methods leads to the loss of information during culturing and this loss can be reduced by metagenomic study. This is because metagenomics involves direct sequencing of environmental DNA. The application of NGS technologies is proving vital in elucidating the organization and activities of microorganisms growing in extreme and stress environmental conditions. It has led to the discovery of several genes, gene families, and several novel microorganisms. Development of in silico models can help in predicting metabolic processes taking place inside cell and its functions in contaminated environment.

32.5.1 The genomics approach

The genomics approach is helpful in understanding the physiology of microorganisms involved in bioremediation. For most of the microorganisms which are involved in bioremediation, their genome sequences (completed or nearly complete) are available. For instance, it has been found that *Geobacter* sp. (important in the bioremediation of organic and metal contaminants) which occurs in subsurface environments was earlier thought to be nonmotile. However, it has now been established to be motile as it harbors flagella coding genes. Also, in *Geobacter sp.*, c type cytochromes (in numbers greater than 100) have been elucidated to be involved in the transfer of electrons to metals during their degradation (Lloyd et al., 2003; Leang et al., 2003). Whole genome DNA microarrays can be spotted as complete genome sequence is available. This can further help in studying expression of genes in the presence of contaminants. Expression of proteins can be studied by proteomic techniques (Baldi and Hatfield, 2002). At least 15 dehalogenase genes have been identified in the genome of *Dehalococcoides*, increasing the range of contaminants it can degrade (Bunge et al., 2003).

32.5.2 The proteomics approach

Phenotypes of microorganisms, which may not be possible to know from the genomic sequences, can be known by proteomics (Fulekar and Sharma, 2008). Proteomics helps in deciphering the physiological responses of microorganisms during contaminant degradation as all the proteins present inside a cell can be analyzed. It also allows to study changes in protein abundance during bioremediation process. Protein separation is done using techniques such as two-dimensional gel electrophoresis and mass spectrometry (MS). Two-dimensional electrophoresis can help in finding out the differential protein expression in microorganisms growing in contaminant free and contaminated site as evident from various studies (Wang et al., 2000; Kim et al., 2004; Santos et al., 2004; Kühner et al., 2005). It is called two-dimensional electrophoresis because in the first dimension proteins are separated by isoelectric focusing, and in the second dimension, they are separated by SDS–PAGE. Use of IPG (immobilized pH gradient) strips have tremendously enhanced the usefulness of 2-DE gels by making it more comparable and reproducible (Singh et al., 2006). The mineralization of pollutants is brought about by a chain of enzymatic reactions. Those enzymes and proteins that are involved in various physiological stages can be isolated and characterized after their extraction by various proteomic studies.

Analysis of small peptides in bioremediation is carried out using MS. Utilizing peptide mass fingerprinting, protein of interest can be identified using MALDI–TOF–MS (matrix associated laser desorption/ionization time of flight MS) (Aebersold and Mann, 2003). SELDI–TOF–MS (surface enhanced laser desorption ionization–time of flight mass spectroscopy) utilizes a chip on which sample fractions are combined with MALDI–TOF–MS analysis (Merchant and Weinberger, 2000; Seibert et al., 2005>; Hivrale et al., 2016; Singh and Nagaraj, 2006). Utilizing SELDI–TOF–MS some biomarker proteins have been identified in *Mytilus edulis* (blue mussels) that were exposed to heavy metals and PAH. However, the reproducibility and specificity of SELDI–TOF–MS in identifying potential biomarkers (Jacobs and Menon, 2004; Diamandis, 2004; Singh and Nagaraj, 2006) sometimes creates problem.

32.5.3 Membrane proteomics

During biodegradation of pollutants that have benzene ring (aromatic ring) including polycyclic aromatic compounds, organophosphates, the membrane proteins present on the cell surface get modified (Pessione et al., 2003, Wang et al., 1995; Richins et al., 1997). Hence it becomes necessary to study these membrane proteins as they get modified during the bioremediation process. However, membrane proteins are hydrophobic in nature; therefore their solubility in lysis buffers used in 2DE is poor. Further, they precipitate at their isoelectric point during electrophoresis in the first dimension which leads to the loss of membrane proteins during 2DE. With the objective of omitting the electroseparation of proteins in 2DE due to its limitations as outlined before, MudPIT (multidimensional protein identification technology) platform is proving to be a beneficial technique that allows the analysis of complex protein mixtures. It is based on the separation of peptide mixtures and complex proteins utilizing two dimensional LC system (Paoletti et al., 2004; Somiari et al., 2005) on a microcapillary column made out of fused silica with detection in a tandem mass spectrometer.

In this, sequence specific digestion of proteins is carried out. The mixture of peptide fragments that results is then separated in a column of a strong cation exchanger which is coupled to a column containing reversed phase HPLC materials. Separated peptide fragments that result are analyzed by MS that generates a spectrum of the fragmented ions. Further, information about the protein is obtained using DTASelect algorithm. The identification of protein fragments is done by SEQUEST algorithm (Singh, 2006; Somiari et al., 2005).

32.5.4 Metagenomics

The modern concept of metagenomics was conceptualized by Handelsman et al. in 1998, and sometimes terms such as ecogenomics, community genomics, or environmental genomics have been used alternatively for metagenomics. Metagenomics encompasses the entire genetic material existing in the environmental sample, which has been recovered from all the samples obtained from the environment. It is also referred to as function-based screening as it screens environmental communities. After metagenome isolation a metagenomic library is constructed and screened for the desired clones. Metagenomics sequence-based screening approaches got popularity due to the advances in throughput NGS technologies including sequencing by oligonucleotide ligation and detection, Solexa (Illumina) sequencing and 454 pyrosequencing. Microbial community of Sargasso Sea revealed huge phylogenetic and metabolic using diversity using genome shot gun sequencing (Yun and Ryu, 2005; Venter et al., 2004) (Fig. 32.4).

The following account gives a brief description of the metagenomics approaches used in bioremediation applications.

32.5.5 Metatranscriptomics (environmental transcriptomics)

Metatranscriptomics approach captures the total mRNA of a sample of microorganisms at specific time under specific conditions (Aguiar-Pulido et al., 2016), which provides expression profiles and regulatory mechanisms operational in microbial communities of samples isolated from the environment. However, limitations do exist in this approach as

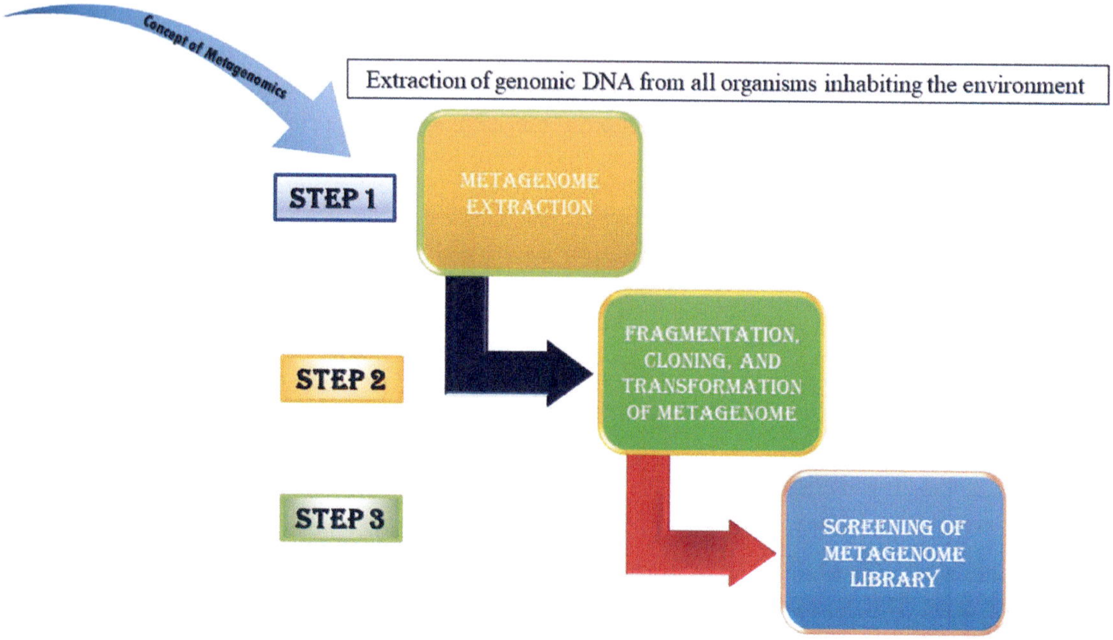

FIGURE 32.4 Concept of metagenomics.

mRNA has very short half-life. Also, problems aggravate due to the availability of less efficient methods of mRNA recovery and its isolation from microbial species (Simon and Daniel, 2011).

The diverse microorganisms in a microbial community can be found out by pyrosequencing (Roesch et al., 2007). For an entire microbial community, direct sequencing of cDNA can quantify the transcripts and allow expression profiling of the whole genome (Simon and Daniel, 2011). By sequencing of random mRNA transcripts of microbial communities, at specific time and place, universal transcription of genes can be studied (Gonzalez et al., 2015). Studies related to transcription of mRNA during PAH degradation, however, has been very limited in members of Ascomycota (Aydin et al., 2017).

At a specific point and time, metaproteomics involves the study of microbial communities (Wilmes and Bond, 2006; Wilmes et al., 2008; Keller and Hettich, 2009). Cytochrome 450 was found to be overexpressed in *Fusarium solani*, when benzo pyrene was present as the contaminant (Koukkou and Drainas, 2008). However, limitations do exist in metatranscriptomics since microbial populations' exhibit enormous genetic diversity, broad protein expression levels, and the distribution of microbial species is not uniform. These factors impede advances in metaproteomic studies. However, in a complex microbial community, it is difficult to identify and detect all the proteins (Simon and Daniel, 2011; Bhargava et al., 2019).

32.5.6 Metabolomics

Microorganisms, during their physiological and biochemical functioning, release certain metabolites in their proximal environment. The identification and quantification of these metabolites is done in metabolomics. The detection of metabolites is carried out by MS and NMR spectroscopy, which produce characteristic peaks. The peak pattern helps in identification as well as quantification of the metabolite. Techniques such as liquid chromatography and gas chromatography have also been used for the detection and quantification of metabolites. According to Aguiar-Pulido et al. (2016), metabolomics can help in the following aspects:

1. Understanding the homeostasis of the host environment, as environment gets affected due to the transformation of pollutants.
2. Understanding signaling during quorum sensing in bacteria as gene expression level changes with the change in microbial density.
3. It is important in pharmacogenomics, drug discovery, and pathway analyses.
4. It is also an important indicator of environmental health.

32.5.7 Fluxomics

Fluxomics studies capture metabolic fluxes in an organism. In a given time period the real time flux of cellular metabolites is called fluxomics (Wiechert et al., 2007). The fluxome or metabolic fluxes gives information about how the genome and environment affect metabolic fluxes. The metabolites that are released are always lesser in number as compared to genes or proteins. So, fluxomics is considered advantageous in comparison to genomics or proteomics. It has been found that *Shewanella* sp. can bring about bioremediation of radionuclides and halogenated organic compounds as their degradation share a metabolic pathway in common. By making use of various methods such as GC–MS and applying various statistical, biochemical, and genetic algorithms, it has been found that *Shewanella* species is capable of growing on alternative sources of carbon, because it exhibits flexibility in metabolic flux (Tang et al., 2009).

32.6 Role of bioinformatics in bioremediation

The prediction of responses and biological system organization can be understood by the network theory (Feist and Palsson, 2008). Although network of metabolism and regulatory interactions within an organism is complex (Pazos et al., 2003; Trigo et al., 2008). According to Ellis et al. (2008), UM-BBD (The University of Minnesota Biocatalysis or Biodegradation Database) is available free of cost and provides vital resources. "Reaction rules" for a particular functional group can predict the most probable metabolic pathway for a given component (Ellis et al., 2008; Megharaj et al., 2011). On the basis of the information contained in the UM-BBD, Pazos et al. (2005) developed a database called "MetaRouter." Utilizing MetaRouter, a correlation was shown to exist between the frequency of 149 chemotypes (triads) common to organic contaminants and the capacity of microbes to metabolize these (Gomez et al., 2007). A predictive tool was also developed which on the basis of predefined environmental factors could classify the pollutants as biodegradable or recalcitrant (http://www.pdg.cnb.uam.es/BDPSERVER) (Megharaj et al., 2011). Another example of a predictive tool is "the environmental contaminant biotransformation pathway" (The enviPath tool, which is an updated version of EAWAGBBD/PPS) (Wicker et al., 2016). Other predictive databases include BioCyc and MetaCyc that were made by SRI International (Caspi et al., 2016; Malla et al., 2018).

32.7 Stable isotope probing

Microbial populations that utilize xenobiotic compounds have been separated by stable isotope probing (SIP). In SIP, microbial communities are grown in C^{13} labeled substrates and then the separation of their entire nucleic acid pool is done by isopycnic density gradient ultracentrifugation (Radajewski et al., 2000). After ultracentrifugation, two zones are produced, each corresponding to C^{12} and C^{13} labeled DNA fragments (corresponding to their low and high buoyant densities). DNA fragments containing C^{12} labeling serves as the negative control. The purpose of the negative control is to separate active C^{13} labeled and inactive C^{12}-labeled population of microbes (molecular techniques could be used to identify C^{13} labeled fragments). In one DNA-SIP experiment, C^{13} naphthalene, C^{13} salicylate, and C^{13} phenanthrene were made available to soils that were contaminated by PAH in a bioreactor. The aim was to identify those bacterial species that were capable of degrading PAH (poly aromatic hydrocarbons). It was found that *Ralstonia* species were degraders of naphthalene and salicylate, and *Acidovorax* species were found to degrade phenanthrene by analyzing 16S rRNA in the taxa (Singleton et al., 2005). Although the SIP approach may be complemented with sequencing and/or molecular fingerprinting techniques to identify microorganisms that can degrade xenobiotics, but the limitation in using this approach lies in the fact that it cannot identify the actual biodegradation pathway (Desai et al., 2010).

32.8 Conclusion

Since xenobiotics and many harmful/radioactive contaminants pose immense threat to the environment, it is imperative that measures may be taken to decontaminate or remediate the contaminated environment in sustainable ways. Bioremediation uses microbial communities to bring about bioremediation in an eco-friendly and economically feasible manner. It is difficult to achieve complete bioremediation of all the environmental contaminants, however, utilizing intrinsic or even genetically engineered microbes. Understanding of microbial physiology and the metabolites involved in the bioremediation pathway may prove helpful in designing bioremediation strategies that are free from the cell. The fast growing novel omics approaches are providing insights into the mechanism of bioremediation process. Enormous data that is generated by the omics techniques need to be assembled in databases. The prediction of the metabolic pathway of a microorganism in a contaminated environment can be done by utilizing omics tools. This would be helpful in the effective elimination of the pollutants from the environment by tracing the microorganisms responsible for the bioremediation pathway. Systems biology along with high throughput techniques will be further required to unfold the mystery behind breakdown potential of other microbes present in the microbial community. To sum up, cell free bioremediation could be achieved by a better understanding of microbial physiology by combining various omics approaches.

References

Achenbach, L.A., Coates, J.D., 2000. Disparity between bacterial phylogeny and physiology. ASM News 66, 714–715.

Aebersold, R., Mann, M., 2003. Mass spectrometry-based proteomics. Nature 422, 198–207.

Aguiar-Pulido, V., Huang, W., Suarez-Ulloa, V., Cickovski, T., Mathee, K., Narasimhan, G., 2016. Metagenomics, metatranscriptomics, and metabolomics approaches for microbiome analysis: supplementary issue: bioinformatics methods and applications for big metagenomics data. Evol. Bioinform. 12, EBO-S36436.

Amann, R.I., Ludwig, W., Schleifer, K.H., 1995. Phylogenetic identification and in situ detection of individual microbial cells without cultivation. Microbiol. Rev. 59, 143–169.

Antizar-Ladislao, B., 2010. Bioremediation: working with bacteria. Elements 6, 389–394.

Aydin, S., Karaçay, H.A., Shahi, A., Gökçe, S., Ince, B., Ince, O., 2017. Aerobic and anaerobic fungal metabolism and Omics insights for increasing polycyclic aromatic hydrocarbons biodegradation. Fungal Biol. Rev. 31, 61–72.

Bakken, L.R., 1997. Culturable and nonculturable bacteria in soil. In: van Elsas, J.D., Trevors, J.T., Wellington, E.M.H. (Eds.), Modern Soil Microbiology. Marcel Dekker, Inc., New York. USA, pp. 47–61.

Baldi, P., Hatfield, G.W., 2002. Systems Biology. Cambridge University Press, Cambridge, pp. 135–176.

Bharagava, R.N., Chowdhary, P., Saxena, G., 2017. Bioremediation, an ecosustainable green technology: its applications and limitations. In: Bharagava, R.N. (Ed.), Environmental Pollutants and Their Bioremediation Approaches. CRC Press, Taylor & Francis Group, Boca Raton, FL, pp. 1–22.

Bhargava, P., Khan, M., Verma, A., Singh, A., Singh, S., Vats, S., et al., 2019. Metagenomics as a tool to explore new insights from plant-microbe interface. Plant Microbe Interface. Springer, Cham, pp. 271–289.

Bihari, Z., 2013. Current trends in bioremediation and biodegradation: next-generation sequencing. J. Bioremed. Biodeg 4, e138.

Bunge, M., Adrian, L., Kraus, A., Opel, M., Lorenz, W.G., Andreesen, J.R., et al., 2003. Reductive dehalogenation of chlorinated dioxins by an anaerobic bacterium. Nature 421, 357–360.

Caspi, R., Billington, R., Ferrer, L., Foerster, H., Fulcher, C.A., Keseler, I.M., et al., 2016. The MetaCyc database of metabolic pathways and enzymes and the BioCyc collection of pathway/genome databases. Nucleic Acids Res. 44, D471–D480.

Chakraborty, R., Wu, C.H., Hazen, T.C., 2012. Systems biology approach to bioremediation. Curr. Opin. Biotech. 23, 483–490.

Datta, S., Singh, S., Kumar, V., Dhanjal, D.S., Sidhu, G.K., Amin, D.S., et al., 2020. Endophytic bacteria in Xenobiotic degradation. In: Kumar, A., Kumar Singh, V. (Eds.), Microbial Endophytes: Prospects of Sustainable Agriculture. Elsevier, pp. 125–156.

Desai, C., Pathak, H., Madamwar, D., 2010. Advances in molecular and "omics" technologies to gauge microbial communities and bioremediation at xenobiotic/anthropogen contaminated sites. Bioresour. Technol. 101, 1558–1569.

Diamandis, E.P., 2004. Mass spectrometry as a diagnostic and a cancer biomarker discovery tool: opportunities and potential limitations. Mol. Cell Proteom. 3, 367–378.

Ellis, L.B., Gao, J., Fenner, K., Wackett, L.P., 2008. The University of Minnesota pathway prediction system: predicting metabolic logic. Nucleic Acids Res. 36, W427–W432.

Feist, A.M., Palsson, B.O., 2008. The growing scope of applications of genome-scale metabolic reconstructions using *Escherichia coli*. Nat. Biotechnol. 26, 659–667.

Fleming, J.T., Sanseverino, J., Sayler, G.S., 1993. Quantitative relationship between naphthalene catabolic gene frequency and expression in predicting PAH degradation in soils at town gas manufacturing sites. Environ. Sci. Technol. 27, 1068–1074.

Fulekar, M.H., 2005. Bioremediation technologies for environment. Indian J. Environ. Prot. 25, 358–364.

Fulekar, M.H., Sharma, J., 2008. Bioinformatics applied in bioremediation. Innov. Rom. Food Biotechnol. 3, 28.

Gomez, M.J., Pazos, F., Guijarro, F.J., de Lorenzo, V., Valencia, A., 2007. The environmental fate of organic pollutants through the global microbial metabolism. Mol. Syst. Biol. 3, 114.

Gonzalez, E., Brereton, N.J., Marleau, J., Nissim, W.G., Labrecque, M., Pitre, F.E., et al., 2015. Meta-transcriptomics indicates biotic cross-tolerance in willow trees cultivated on petroleum hydrocarbon contaminated soil. BMC Plant. Biol. 15, 246.

Goutam, S.P., Saxena, G., Singh, V., Yadav, A.K., Bharagava, R.N., 2018. Green synthesis of TiO_2 nanoparticles using leaf extract of *Jatropha curcas* L. for photocatalytic degradation of tannery wastewater. Chem. Eng. J. 336, 386–396.

Green, S.J., Leigh, M.B., Neufeld, J.D., 2010. Denaturing gradient gel electrophoresis (DGGE) for microbial community analysis. In: Timmis, K.N. (Ed.), Handbook of Hydrocarbon and Lipid Microbiology. Springer, Berlin, Heidelberg, pp. 4137–4158.

Gupta, A., Singh, S.K., Singh, V.K., Singh, M.K., Modi, A., Zhimo, V.Y., et al., 2020. Endophytic microbe approaches in bioremediation of organic pollutants. In: Kumar, A., Kumar Singh, V. (Eds.), Microbial Endophytes: Prospects of Sustainable Agriculture. Elsevier, pp. 157–174.

Hivrale, A.U., Pawar, P.K., Rane, N.R., Govindwar, S.P., 2016. Application of genomics and proteomics in bioremediation. In: Toxicity and Waste Management Using Bioremediation. IGI Global, pp. 97–112.

Jacobs, I.J., Menon, U., 2004. Progress and challenges in screening for early detection of ovarian cancer. Mol. Cell Prot. 3, 55–366.

Jeffrey, W.H., Nazaret, S., Barkay, T., 1996. Detection of the merA gene and its expression in the environment. Microb. Ecol. 32, 293–303.

Ju, F., Zhang, T., 2015. 16S rRNA gene high-throughput sequencing data mining of microbial diversity and interactions. Appl. Microbiol. Biotechnol. 99, 4119–4129.

Keller, M., Hettich, R., 2009. Environmental proteomics: a paradigm shift in characterizing microbial activities at the molecular level. Microbiol. Mol. Biol. Rev. 73, 62e70.

Kim, S.J., Jones, R.C., Cha, C.J., Kweon, O., Edmondson, R.D., Cerniglia, C.E., 2004. Identification of proteins induced by polycyclic aromatic hydrocarbon in *Mycobacterium vanbaalenii* PYR-1 using two-dimensional polyacrylamide gel electrophoresis and de novo sequencing methods. Proteomics 4, 3899–3908.

Koukkou, A.I., Drainas, C., 2008. Addressing PAH biodegradation in Greece: biochemical and molecular approaches. IUBMB Life 60, 275e280.

Kühner, S., Wöhlbrand, L., Fritz, I., Wruck, W., Hultschig, C., Hufnagel, P., et al., 2005. Substrate-dependent regulation of anaerobic degradation pathways for toluene and ethylbenzene in a denitrifying bacterium, strain EbN1. J. Bacteriol. 187, 1493–1503.

Leang, C., Coppi, M.V., Lovley, D.R., 2003. OmcB, a c-type polyheme cytochrome, involved in Fe (III) reduction in Geobacter sulfurreducens. Journal of Bacteriology 185 (7), 2096–2103.

Lloyd, V.K., Dyment, D., Sinclair, D.A.R., Grigliatti, T.A., 2003. Different patterns of gene silencing in position-effect variegation. Genome 46 (6), 1104–1117.

Lovley, D.R., 2003. Cleaning up with genomics: applying molecular biology to bioremediation. Nat. Rev. Microbiol. 1, 35–44.

Malla, M.A., Dubey, A., Yadav, S., Kumar, A., Hashem, A., Abd_Allah, E.F., 2018. Understanding and designing the strategies for the microbe-mediated remediation of environmental contaminants using omics approaches. Front. Microbiol. 9, 1132.

Maphosa, F., de Vos, W.M., Smidt, H., 2010. Exploiting the ecogenomics toolbox for environmental diagnostics of organohalide-respiring bacteria. Trends Biotechnol. 28, 308–316.

Megharaj, M., Ramakrishnan, B., Venkateswarlu, K., Sethunathan, N., Naidu, R., 2011. Bioremediation approaches for organic pollutants: a critical perspective. Environ. Int. 37, 1362–1375.
Merchant, M., Weinberger, S.R., 2000. Recent advancements in surface-enhanced laser desorption/ionization-time of flight-mass spectrometry. Electrophoresis 21, 1164–1177.
Molina, L., Duque, E., Gómez, M.J., Krell, T., Lacal, J., García-Puente, A., et al., 2011. The pGRT1 plasmid of *Pseudomonas putida* DOT-T1E encodes functions relevant for survival under harsh conditions in the environment. Environ. Microbiol. 13, 2315–2327.
Nazaret, S., Jeffrey, W.H., Saouter, E., von Haven, R., Barkay, T., 1994. merA gene expression in aquatic environments measured by mRNA production and Hg(II) volatilization. Appl. Environ. Microbiol. 60, 4059–4065.
Nierman, W.C., Nelson, K.E., 2002. Genomics for applied microbiology. Adv. Appl. Microbiol. 51, 201–248.
Pace, N.R., 1997. A molecular view of microbial diversity and the biosphere. Science 276, 734–740.
Pace, N.R., Stahl, D.A., Lane, D.J., Olsen, G.J., 1986. The analysis of natural populations by ribosomal RNA sequence. Adv. Gen. Microbiol. Ecol. 9, 1–55.
Paoletti, A.C., Zybailov, B., Washburn, M.P., 2004. Principles and applications of multidimensional protein identification technology. Expert. Rev. Prot. 1, 275–282.
Pazos, F., Valencia, A., De Lorenzo, V., 2003. The organization of the microbial biodegradation network from a systems-biology perspective. EMBO Rep. 4, 994–999.
Pazos, F., Guijas, D., Valencia, A., De Lorenzo, V., 2005. MetaRouter: bioinformatics for bioremediation. Nucleic Acids Res. 33 (Suppl. 1), D588–D592.
Pessione, E., Giuffrida, M.G., Prunotto, L., Barello, C., Mazzoli, R., Fortunato, D., et al., 2003. Membrane proteome of *Acinetobacter radioresistens* S13 during aromatic exposure. Proteomics 3, 1070–1076.
Radajewski, S., Ineson, P., Parekh, N.R., Murrell, J.C., 2000. Stable-isotope probing as a tool in microbial ecology. Nature 403, 646–649.
Richins, R.D., Kaneva, I., Mulchandani, A., Chen, W., 1997. Biodegradation of organophosphorus pesticides by surface-expressed organophosphorus hydrolase. Nat. Biotechnol. 15, 984–987.
Roesch, L.F., Fulthorpe, R.R., Riva, A., Casella, G., Hadwin, A.K., Kent, A.D., et al., 2007. Pyrosequencing enumerates and contrasts soil microbial diversity. ISME J. 1, 283–290.
Rogers, S.L., McClure, N., 2003. Bioremediation: A Critical Review. Horizon Scientific Press, Wymondham, pp. 27–59.
Santos, P.M., Benndorf, D., Sá-Correia, I., 2004. Insights into *Pseudomonas putida* KT2440 response to phenol-induced stress by quantitative proteomics. Proteomics 4, 2640–2652.
Satokari, R.M., Vaughan, E.E., Akkermans, A.D.L., Saarela, M., De Vos, W.M., 2001. Polymerase chain reaction and denaturing gradient gel electrophoresis monitoring of fecal *bifidobacterium* populations in a prebiotic and probiotic feeding trial. VTT Publ. 231, 85–89.
Saxena, G., Bharagava, R.N., 2016. Ram Chandra: advances in biodegradation and bioremediation of industrial waste. Clean. Technol. Environ. Policy 18, 979–980.
Seibert, V., Ebert, M.P., Buschmann, T., 2005. Advances in clinical cancer proteomics: SELDI-ToF-mass spectrometry and biomarker discovery. Brief. Funct. Genomics 4, 16–26.
Simon, C., Daniel, R., 2011. Metagenomic analyses: past and future trends. Appl. Environ. Microbiol. 77, 41153–41161.
Singh, O.V., 2006. Proteomics and metabolomics: the molecular make-up of toxic aromatic pollutant bioremediation. Proteomics 6, 5481–5492.
Singh, O.V., Nagaraj, N.S., 2006. Transcriptomics, proteomics and interactomics: unique approaches to track the insights of bioremediation. Brief. Funct. Genomics 4, 355–362.
Singh, V.K., Singh, A.L., Singh, R., Kumar, A., 2018. Iron oxidizing bacteria: insights on diversity, mechanism of iron oxidation and role in management of metal pollution. Environ. Sustain. 1 (3), 221–231.
Singh, S.K., Singh, P.P., Gupta, A., Singh, A.K., Keshri, J., 2019a. Tolerance of heavy metal toxicity using PGPR strains of pseudomonas species. PGPR Amelioration in Sustainable Agriculture. Woodhead Publishing, pp. 239–252.
Singh, V.K., Singh, M., Singh, S.K., Kumar, C., Kumar, A., 2019b. Sustainable agricultural practices using beneficial fungi under changing climate scenario. Climate Change and Agricultural Ecosystems. Woodhead Publishing, pp. 25–42.
Singh, P., Singh, V.K., Singh, R., Borthakur, A., Madhav, S., Ahamad, A., et al., 2020a. Bioremediation: a sustainable approach for management of environmental contaminants. In: Singh, P., Kumar, A., Borthakur, A. (Eds.), Abatement of Environmental Pollutants. Elsevier, pp. 1–23.
Singh, D., Singh, S.K., Singh, V.K., Gupta, A., Aamir, M., Kumar, A., 2020b. Plant growth–promoting bacteria and their role in environmental management. In: Singh, P., Kumar, A., Borthakur, A. (Eds.), Abatement of Environmental Pollutants. Elsevier, pp. 161–175.
Singleton, D.R., Powell, S.N., Sangaiah, R., Gold, A., Ball, L.M., Aitken, M.D., 2005. Stable-isotope probing of bacteria capable of degrading salicylate, naphthalene, or phenanthrene in a bioreactor treating contaminated soil. Appl. Environ. Microbiol. 71, 1202–1209.
Somiari, R.I., Somiari, S., Russell, S., Shriver, C.D., 2005. Proteomics of breast carcinoma. J. Chromatogr. B 815, 215–225.
Tang, Y.J., Martin, H.G., Dehal, P.S., Deutschbauer, A., Llora, X., Meadows, A., et al., 2009. Metabolic flux analysis of *Shewanella* spp. reveals evolutionary robustness in central carbon metabolism. Biotechnol. Bioeng. 102, 1161–1169.
Trigo, A., Valencia, A., Cases, I., 2008. Systemic approaches to biodegradation. FEMS Microbiol. Rev. 33, 98–108.
Venter, J.C., Remington, K., Heidelberg, J.F., Halpern, A.L., Rusch, D., Eisen, J.A., et al., 2004. Environmental genome shotgun sequencing of the Sargasso Sea. Science 304, 66–74.
Wang, Y., Rawlings, M., Gibson, D.T., Labbé, D., Bergeron, H., Brousseau, R., et al., 1995. Identification of a membrane protein and a truncated LysR-type regulator associated with the toluene degradation pathway in *Pseudomonas putida* F1. MGG 246, 570–579.

Wang, R.F., Wennerstrom, D., Cao, W.W., Khan, A.A., Cerniglia, C.E., 2000. Cloning, expression, and characterization of the katGGene, encoding catalase-peroxidase, from the polycyclic aromatic hydrocarbon-degrading bacterium *Mycobacterium* sp. strain PYR-1. Appl. Environ. Microbiol. 66, 4300–4304.
Wicker, J., Lorsbach, T., Gütlein, M., Schmid, E., Latino, D., Kramer, S., et al., 2016. enviPath—the environmental contaminant biotransformation pathway resource. Nucleic Acids Res. 44, D502–D508.
Wiechert, W., Schweissgut, O., Takanaga, H., Frommer, W.B., 2007. Fluxomics: mass spectrometry versus quantitative imaging. Curr. Opin. Plant. Biol. 10, 323–330.
Wilmes, P., Bond, P.L., 2006. Metaproteomics: studying functional gene expression in microbial ecosystems. Trends Microbiol. 14, 92e97.
Wilmes, P., Wexler, M., Bond, P.L., 2008. Metaproteomics provides functional insight into activated sludge wastewater treatment. PLoS One 3, 3.
Yun, J., Ryu, S., 2005. Screening for novel enzymes from metagenome and SIGEX, as a way to improve it. Microb. Cell Fact. 4, 8.
Bursle, E., Robson, J., 2016. Non-culture methods for detecting infection. Austr. Prescriber 39 (5), 171.
Vartoukian, S.R., Palmer, R.M., Wade, W.G., 2010. Strategies for culture of 'unculturable' bacteria. FEMS Microbiol. Lett. 309 (1), 1–7.

Further reading

Chen, W., Mulchandani, A., Deshusses, M.A., 2005. Environmental biotechnology: challenges and opportunities for chemical engineers. AIChE J. 51, 690–695.
Handelsman, J., Rondon, M.R., Brady, S.F., Clardy, J., Goodman, R.M., 1998. Molecular biological access to the chemistry of unknown soil microbes: a new frontier for natural products. Chem. Biol. 5, R245–R249.
Watanabe, K., Baker, P.W., 2000. Environmentally relevant microorganisms. J. Biosci. Bioeng. 89, 1–11.
Watanabe, K., Hamamura, N., 2003. Molecular and physiological approaches to understanding the ecology of pollutant degradation. Curr. Opin. 14, 289–295.

Index

Note: Page numbers followed by "*f*" and "*t*" refer to figures and tables, respectively.

C

D

E

F

G

M

N

O

Q

R

S

CPI Antony Rowe
Eastbourne, UK
May 28, 2021